BIG IDEAS MATH®

Pre-Algebra

A Common Core Curriculum

FLORIDA EDITION

Ron Larson
Laurie Boswell

BIG IDEAS LEARNING®
Erie, Pennsylvania
BigIdeasLearning.com

Big Ideas Learning, LLC
1762 Norcross Road
Erie, PA 16510-3838
USA

For product information and customer support, contact Big Ideas Learning
at **1-877-552-7766** or visit us at ***BigIdeasLearning.com***.

About the Cover
The cover images on the *Big Ideas Math* series illustrate the advancements in aviation from the hot-air balloon to spacecraft. This progression symbolizes the launch of a student's successful journey in mathematics. The sunrise in the background is representative of the dawn of the Common Core era in math education, while the cradle signifies the balanced instruction that is a pillar of the *Big Ideas Math* series.

Printed in the U.S.A.

ISBN 13: 978-1-60840-548-0
ISBN 10: 1-60840-548-6

2 3 4 5 6 7 8 9 10 WEB 17 16 15 14 13

AUTHORS

Ron Larson is a professor of mathematics at Penn State Erie, The Behrend College, where he has taught since receiving his Ph.D. in mathematics from the University of Colorado. Dr. Larson is well known as the lead author of a comprehensive program for mathematics that spans middle school, high school, and college courses. His high school and Advanced Placement books are published by Holt McDougal. Ron's numerous professional activities keep him in constant touch with the needs of students, teachers, and supervisors. Ron and Laurie Boswell began writing together in 1992. Since that time, they have authored over two dozen textbooks. In their collaboration, Ron is primarily responsible for the pupil edition and Laurie is primarily responsible for the teaching edition of the text.

Laurie Boswell is the Head of School and a mathematics teacher at the Riverside School in Lyndonville, Vermont. Dr. Boswell received her Ed.D. from the University of Vermont in 2010. She is a recipient of the Presidential Award for Excellence in Mathematics Teaching. Laurie has taught math to students at all levels, elementary through college. In addition, Laurie was a Tandy Technology Scholar, and served on the NCTM Board of Directors from 2002 to 2005. She currently serves on the board of NCSM, and is a popular national speaker. Along with Ron, Laurie has co-authored numerous math programs.

ABOUT THE BOOK

The *Florida Big Ideas Math* series uses the same research-based strategy of a balanced approach to instruction that made the first *Florida Big Ideas Math* series so successful. This approach opens doors to abstract thought, reasoning, and inquiry as students persevere to answer the Essential Questions that introduce each section. The foundation of the program is the Florida Common Core Standards for Mathematical Content and Standards for Mathematical Practice. Students are subtly introduced to "Habits of Mind" that help them internalize concepts for a greater depth of understanding. These habits serve students well not only in mathematics, but across all curricula throughout their academic careers.

The *Florida Big Ideas Math* series exposes students to highly motivating and relevant problems. Woven throughout the series are the depth and rigor students need to prepare for career-readiness and other college-level courses. In addition, the *Florida Big Ideas Math* series prepares students to meet the challenge of PARCC testing.

We are excited to bring the *Florida Big Ideas Math* series to the students in Florida as they prepare for college and career readiness.

Ron Larson   Laurie Boswell

TEACHER REVIEWERS

- Lisa Amspacher
  Milton Hershey School
  Hershey, PA
- Mary Ballerina
  Orange County Public Schools
  Orlando, FL
- Lisa Bubello
  School District of Palm Beach County
  Lake Worth, FL
- Sam Coffman
  North East School District
  North East, PA
- Kristen Karbon
  Troy School District
  Rochester Hills, MI
- Laurie Mallis
  Westglades Middle School
  Coral Springs, FL
- Dave Morris
  Union City Area School District
  Union City, PA
- Bonnie Pendergast
  Tolleson Union High School District
  Tolleson, AZ
- Valerie Sullivan
  Lamoille South Supervisory Union
  Morrisville, VT
- Becky Walker
  Appleton Area School District
  Appleton, WI
- Zena Wiltshire
  Dade County Public Schools
  Miami, FL

STUDENT REVIEWERS

- Mike Carter
- Matthew Cauley
- Amelia Davis
- Wisdom Dowds
- John Flatley
- Nick Ganger
- Hannah Iadeluca
- Paige Lavine
- Emma Louie
- David Nichols
- Mikala Parnell
- Jordan Pashupathi
- Stephen Piglowski
- Robby Quinn
- Michael Rawlings
- Garrett Sample
- Andrew Samuels
- Addie Sedelmyer
- Tyler Steffy
- Erin Taylor
- Reid Wilson

CONSULTANTS

- Patsy Davis
  Educational Consultant
  Knoxville, Tennessee
- Bob Fulenwider
  Mathematics Consultant
  Bakersfield, California
- Linda Hall
  Mathematics Assessment Consultant
  Norman, Oklahoma
- Ryan Keating
  Special Education Advisor
  Gilbert, Arizona
- Michael McDowell
  Project-Based Instruction Specialist
  Fairfax, California
- Sean McKeighan
  Interdisciplinary Advisor
  Norman, Oklahoma
- Bonnie Spence
  Differentiated Instruction Consultant
  Missoula, Montana

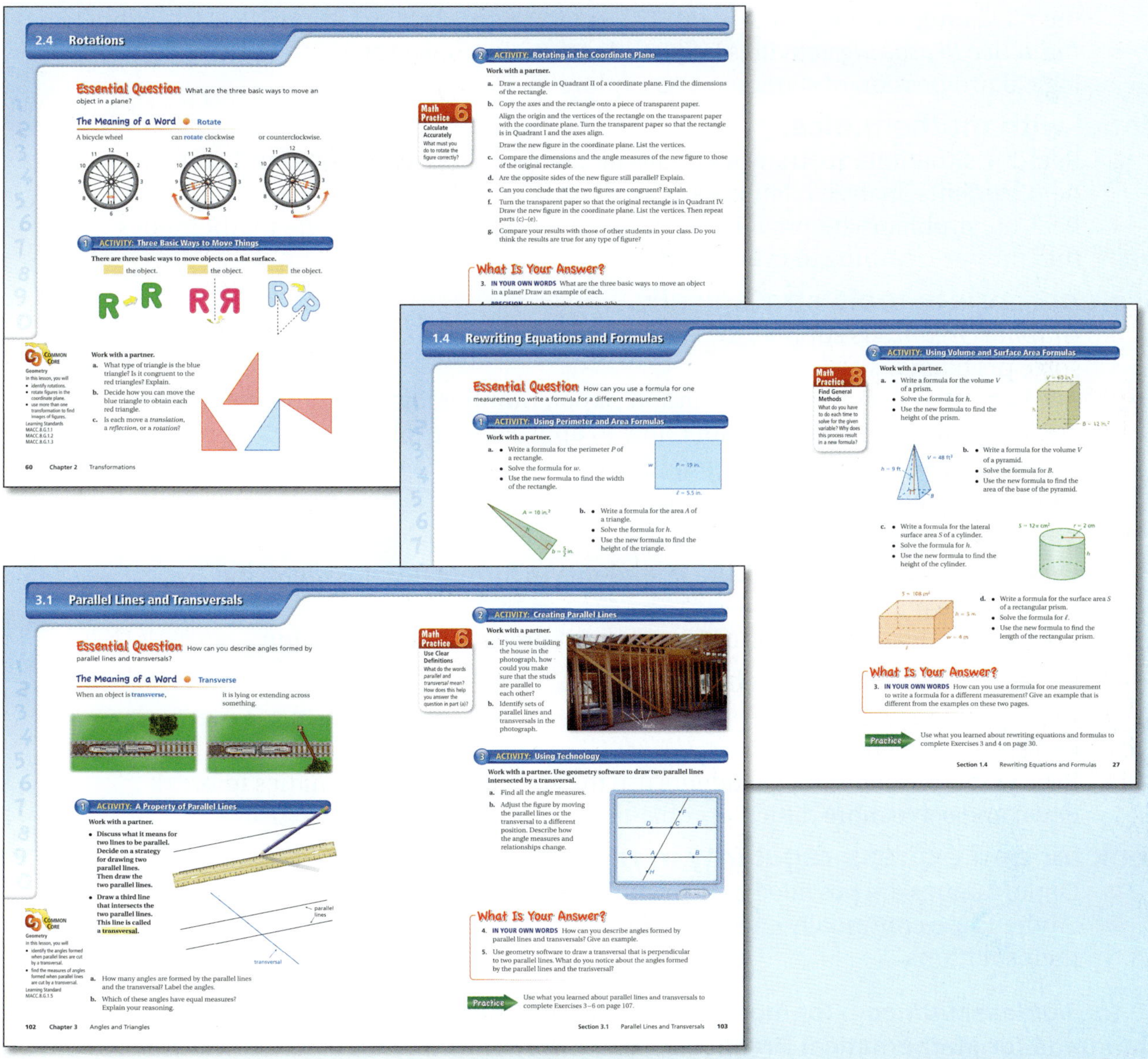

Florida Common Core Standards for Mathematical Practice

Make sense of problems and persevere in solving them.

- Multiple representations are presented to help students move from concrete to representative and into abstract thinking
- *Essential Questions* help students focus and analyze
- *In Your Own Words* provide opportunities for students to look for meaning and entry points to a problem

Reason abstractly and quantitatively.

- Visual problem solving models help students create a coherent representation of the problem
- Opportunities for students to decontextualize and contextualize problems are presented in every lesson

Construct viable arguments and critique the reasoning of others.

- *Error Analysis*; *Different Words, Same Question*; and *Which One Doesn't Belong* features provide students the opportunity to construct arguments and critique the reasoning of others
- *Inductive Reasoning* activities help students make conjectures and build a logical progression of statements to explore their conjecture

Model with mathematics.

- Real-life situations are translated into diagrams, tables, equations, and graphs to help students analyze relations and to draw conclusions
- Real-life problems are provided to help students learn to apply the mathematics that they are learning to everyday life

Use appropriate tools strategically.

- *Graphic Organizers* support the thought process of what, when, and how to solve problems
- A variety of tool papers, such as graph paper, number lines, and manipulatives, are available as students consider how to approach a problem
- Opportunities to use the web, graphing calculators, and spreadsheets support student learning

Attend to precision.

- *On Your Own* questions encourage students to formulate consistent and appropriate reasoning
- Cooperative learning opportunities support precise communication

Look for and make use of structure.

- *Inductive Reasoning* activities provide students the opportunity to see patterns and structure in mathematics
- Real-world problems help students use the structure of mathematics to break down and solve more difficult problems

Look for and express regularity in repeated reasoning.

- Opportunities are provided to help students make generalizations
- Students are continually encouraged to check for reasonableness in their solutions

Go to *BigIdeasMath.com* for more information on the Florida Common Core Standards for Mathematical Practice.

Florida Common Core Standards for Mathematical Content for Grade 8

Chapter Coverage for Standards

1 – 2 – 3 – 4 – 5 – 6 – **7** – 8 – 9 – 10

Domain The Number System

- Know that there are numbers that are not rational, and approximate them by rational numbers.

1 – 2 – 3 – **4** – **5** – 6 – **7** – 8 – 9 – **10**

Domain Expressions and Equations

- Work with radicals and integer exponents.
- Understand the connections between proportional relationships, lines, and linear equations.
- Analyze and solve linear equations and pairs of simultaneous equations.

1 – 2 – 3 – 4 – 5 – **6** – 7 – 8 – 9 – 10

Domain Functions

- Define, evaluate, and compare functions.
- Use functions to model relationships between quantities.

1 – **2** – **3** – 4 – 5 – 6 – **7** – **8** – 9 – 10

Domain Geometry

- Understand congruence and similarity using physical models, transparencies, or geometry software.
- Understand and apply the Pythagorean Theorem.
- Solve real-world and mathematical problems involving volume of cylinders, cones, and spheres.

1 – 2 – 3 – 4 – 5 – 6 – 7 – 8 – **9** – 10

Domain Statistics and Probability

- Investigate patterns of association in bivariate data.

Go to *BigIdeasMath.com* for more information on the Florida Common Core Standards for Mathematical Content.

1

> *"I like talking about math, and working with a partner allows me to do that."*

Equations

Transformations

3

Angles and Triangles

"I like that the Essential Question helps me begin thinking about the lesson."

4 Graphing and Writing Linear Equations

“I really enjoy the projects at the end of the book because they help connect the math to other subjects, like science or art.”

5 Systems of Linear Equations

“I like Newton and Descartes! The cartoons are funny and I like that they model the math that we are learning.”

Functions

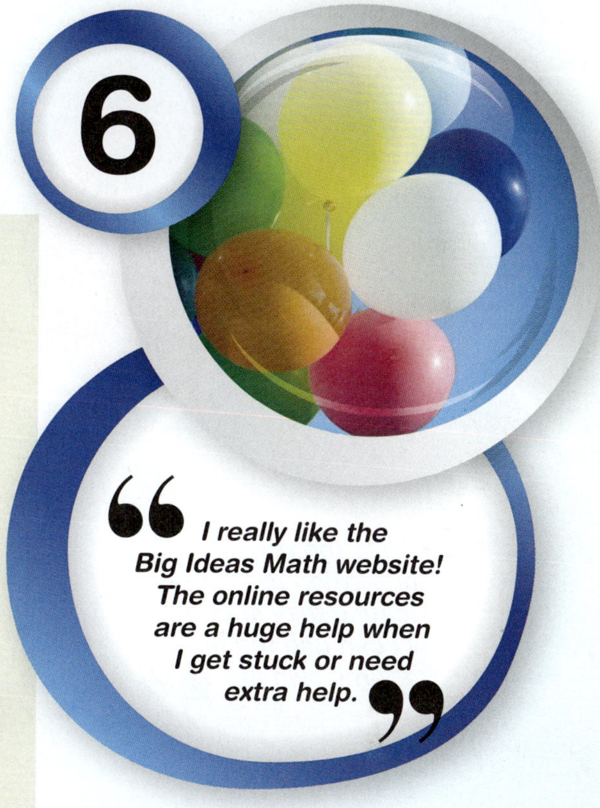
6
I really like the Big Ideas Math website! The online resources are a huge help when I get stuck or need extra help.

7

Real Numbers and the Pythagorean Theorem

"I like the real-life application exercises because they show me how I can use the math in my own life."

Volume and Similar Solids

8

"I like playing the games in the Game Closet! They are a fun way to practice concepts we are learning in class."

9

Data Analysis and Displays

"With the BigIdeasMath.com website I don't have to worry if I forget my book or my workbook at school."

Exponents and Scientific Notation

10

"*I like that I can print the Glossary Flashcards to review before a test.*"

Appendix A: My Big Ideas Projects

How to Use Your Math Book

- Read the **Essential Question** in the activity.

  Discuss the **Math Practice** question with your partner.

  Work with a partner to decide **What Is Your Answer?**

  Now you are ready to do the **Practice** problems.

- Find the **Key Vocabulary** words, **highlighted in yellow**.

  Read their definitions. Study the concepts in each **Key Idea**.

  If you forget a definition, you can look it up online in the

  Multi-Language Glossary at BigIdeasMath.com.

- After you study each **EXAMPLE**, do the exercises in the **On Your Own**.

  Now You're Ready to do the exercises that correspond to the example.

  As you study, look for a **Study Tip** or a **Common Error**.

- The exercises are divided into 3 parts.

  Vocabulary and Concept Check

  Practice and Problem Solving

  Fair Game Review

  If an exercise has a 1 next to it, look back at Example 1 for help with that exercise.

  More help is available at **Check It Out Lesson Tutorials BigIdeasMath.com**.

- To help study for your test, use the following.

  Quiz **Study Help**

  Chapter Review **Chapter Test**

SCAVENGER HUNT

Use this *Scavenger Hunt* to find where things are in **Chapter 1**.

Newton

Find an **Essential Question**.

Find a **Math Practice**.

Find a **Practice**.

Find the definition of a **literal equation**.

Find the **Key Idea** Division Property of Equality.

Use the index in the back of the book.

Find an **EXAMPLE** of solving a two-step equation.

Find an **On Your Own** about $-3x = 2x + 19$.

Find **Now You're Ready** Exercises 18–29.

Find a **Remember** about formulas.

Find a **Vocabulary** question about inverse operations.

Find a **Problem Solving** question about **ANTS**.

Find the answer in the back of the book.

Which examples **?** help with Exercise 8 on page 14?

Find 2 **Quiz**zes.

How many questions are on the **Chapter Test**?

Descartes

1 Equations

"Dear Sir: Here is my suggestion for a good math problem."

"A box contains a total of 30 dog and cat treats. There are 5 times more dog treats than cat treats."

"How many of each type of treat are there?"

"Push faster, Descartes! According to the formula $R = D \div T$, the time needs to be 10 minutes or less to break our all-time speed record!"

What You Learned Before

"Once upon a time, there lived the most handsome dog who just happened to be a genius at math. He..."

Simplifying Algebraic Expressions (MACC.6.EE.1.3)

Example 1 **Simplify $10b + 13 - 6b + 4$.**

$10b + 13 - 6b + 4 = 10b - 6b + 13 + 4$ Commutative Property of Addition

$= (10 - 6)b + 13 + 4$ Distributive Property

$= 4b + 17$ Simplify.

Example 2 **Simplify $5(x + 4) + 2x$.**

$5(x + 4) + 2x = 5(x) + 5(4) + 2x$ Distributive Property

$= 5x + 20 + 2x$ Multiply.

$= 5x + 2x + 20$ Commutative Property of Addition

$= 7x + 20$ Combine like terms.

Try It Yourself

Simplify the expression.

1. $9m - 7m + 2m$

2. $3g - 9 + 11g - 21$

3. $6(3 - y)$

4. $12(a - 4)$

5. $22.5 + 7(n - 3.4)$

6. $15k + 8(11 - k)$

Adding and Subtracting Integers (MACC.7.NS.1.1d)

Example 3 **Find $4 + (-12)$.**

$4 + (-12) = -8$

$|-12| > |4|$. So, subtract $|4|$ from $|-12|$.

Use the sign of -12.

Example 4 **Find $-7 - (-16)$.**

$-7 - (-16) = -7 + 16$ Add the opposite of -16.

$= 9$ Add.

Try It Yourself

Add or subtract.

7. $-5 + (-2)$

8. $0 + (-13)$

9. $-6 + 14$

10. $19 - (-13)$

1.1 Solving Simple Equations

Essential Question How can you use inductive reasoning to discover rules in mathematics? How can you test a rule?

1 ACTIVITY: Sum of the Angles of a Triangle

Work with a partner. Use a protractor to measure the angles of each triangle. Copy and complete the table to organize your results.

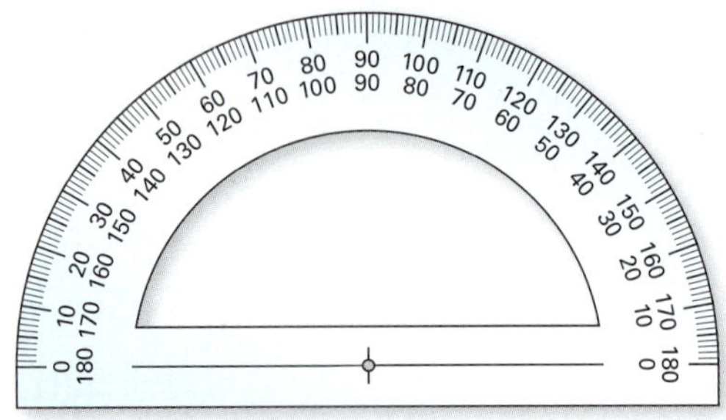

a.

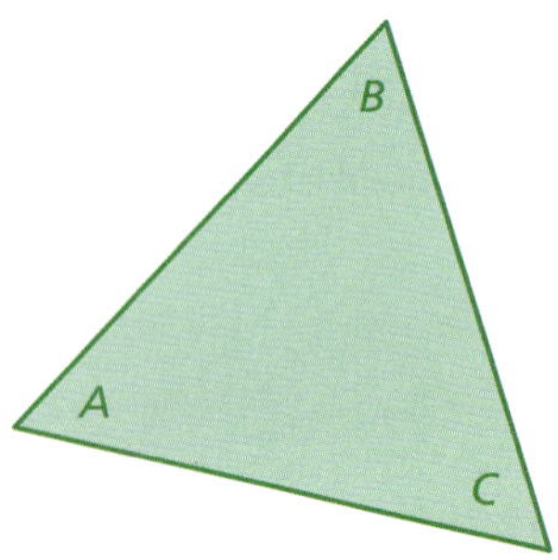

b.

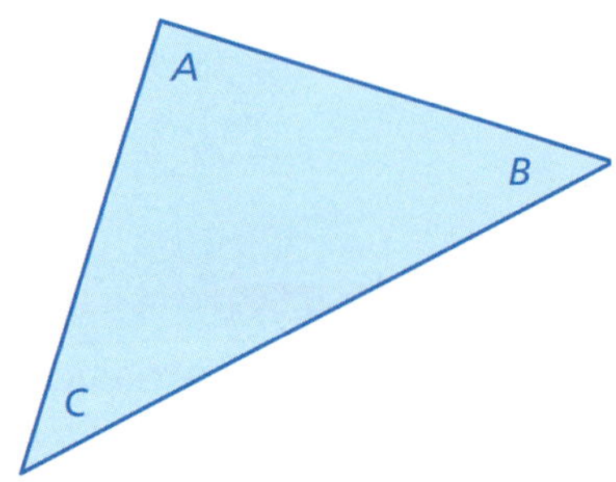

c.

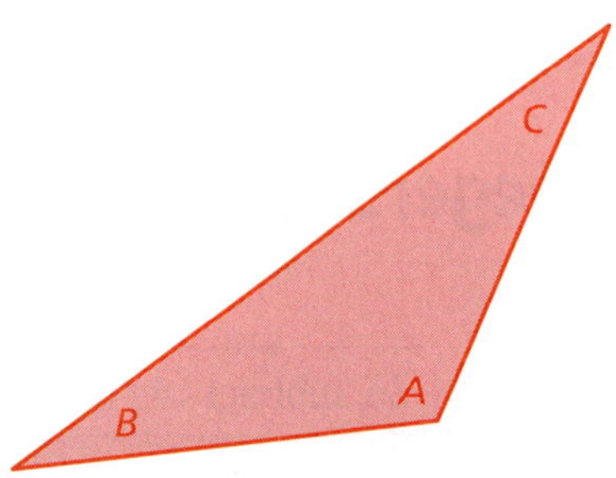

d.

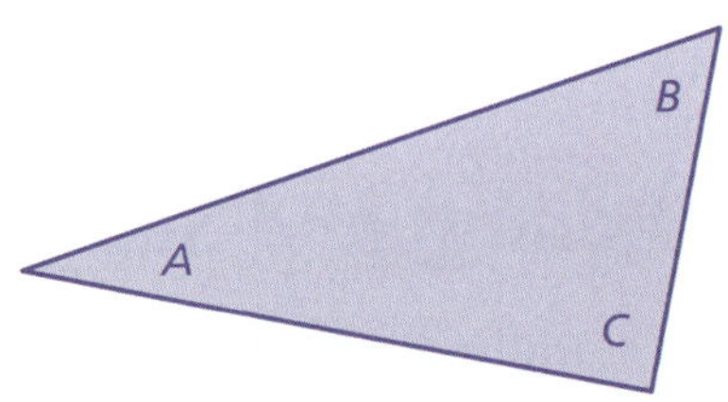

COMMON CORE

Solving Equations

In this lesson, you will

- solve simple equations using addition, subtraction, multiplication, or division.

Learning Standards
MACC.8.EE.3.7a
MACC.8.EE.3.7b

Triangle	Angle A (degrees)	Angle B (degrees)	Angle C (degrees)	$A + B + C$
a.				
b.				
c.				
d.				

2 ACTIVITY: Writing a Rule

Work with a partner. Use inductive reasoning to write and test a rule.

a. **STRUCTURE** Use the completed table in Activity 1 to write a rule about the sum of the angle measures of a triangle.

b. **TEST YOUR RULE** Draw four triangles that are different from those in Activity 1. Measure the angles of each triangle. Organize your results in a table. Find the sum of the angle measures of each triangle.

3 ACTIVITY: Applying Your Rule

> **Math Practice 3**
> **Analyze Conjectures**
> Do your results support the rule you wrote in Activity 2? Explain.

Work with a partner. Use the rule you wrote in Activity 2 to write an equation for each triangle. Then solve the equation to find the value of *x*. Use a protractor to check the reasonableness of your answer.

a.

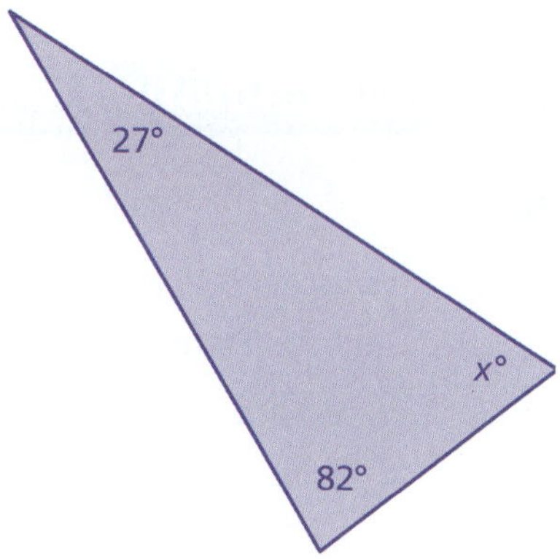

b.

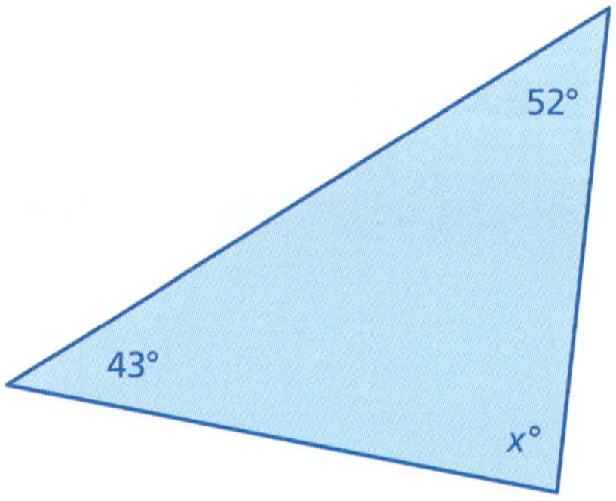

c.

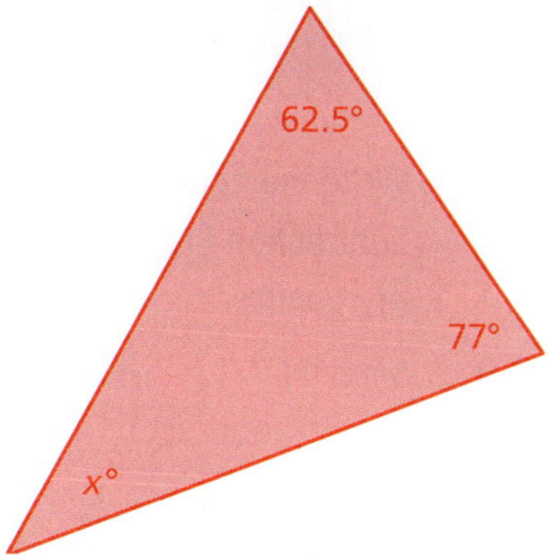

d.

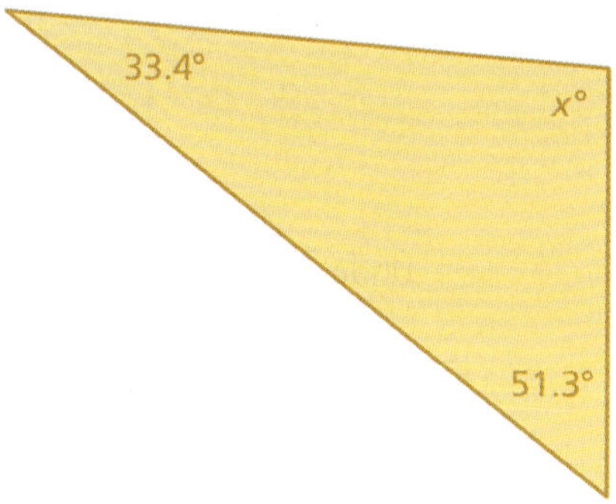

What Is Your Answer?

4. **IN YOUR OWN WORDS** How can you use inductive reasoning to discover rules in mathematics? How can you test a rule? How can you use a rule to solve problems in mathematics?

Practice Use what you learned about solving simple equations to complete Exercises 4–6 on page 7.

1.1 Lesson

Check It Out
Lesson Tutorials
BigIdeasMath.com

Remember

Addition and subtraction are inverse operations.

Key Ideas

Addition Property of Equality

Words Adding the same number to each side of an equation produces an equivalent equation.

Algebra If $a = b$, then $a + c = b + c$.

Subtraction Property of Equality

Words Subtracting the same number from each side of an equation produces an equivalent equation.

Algebra If $a = b$, then $a - c = b - c$.

EXAMPLE 1 Solving Equations Using Addition or Subtraction

a. Solve $x - 7 = -6$.

$x - 7 = -6$ — Write the equation.

Undo the subtraction. → $+7 \quad +7$ — Addition Property of Equality

$x = 1$ — Simplify.

∴ The solution is $x = 1$.

Check

$x - 7 = -6$

$1 - 7 \stackrel{?}{=} -6$

$-6 = -6$ ✓

b. Solve $y + 3.4 = 0.5$.

$y + 3.4 = 0.5$ — Write the equation.

Undo the addition. → $-3.4 \quad -3.4$ — Subtraction Property of Equality

$y = -2.9$ — Simplify.

∴ The solution is $y = -2.9$.

Check

$y + 3.4 = 0.5$

$-2.9 + 3.4 \stackrel{?}{=} 0.5$

$0.5 = 0.5$ ✓

c. Solve $h + 2\pi = 3\pi$.

$h + 2\pi = 3\pi$ — Write the equation.

Undo the addition. → $-2\pi \quad -2\pi$ — Subtraction Property of Equality

$h = \pi$ — Simplify.

∴ The solution is $h = \pi$.

On Your Own

Now You're Ready
Exercises 7–15

Solve the equation. Check your solution.

1. $b + 2 = -5$
2. $g - 1.7 = -0.9$
3. $-3 = k + 3$
4. $r - \pi = \pi$
5. $t - \frac{1}{4} = -\frac{3}{4}$
6. $5.6 + z = -8$

Remember

Multiplication and division are inverse operations.

Key Ideas

Multiplication Property of Equality

Words Multiplying each side of an equation by the same number produces an equivalent equation.

Algebra If $a = b$, then $a \cdot c = b \cdot c$.

Division Property of Equality

Words Dividing each side of an equation by the same number produces an equivalent equation.

Algebra If $a = b$, then $a \div c = b \div c$, $c \neq 0$.

EXAMPLE 2 Solving Equations Using Multiplication or Division

a. Solve $-\frac{3}{4}n = -2$.

$-\frac{3}{4}n = -2$ Write the equation.

Use the reciprocal. → $-\frac{4}{3} \cdot \left(-\frac{3}{4}n\right) = -\frac{4}{3} \cdot (-2)$ Multiplication Property of Equality

$n = \frac{8}{3}$ Simplify.

The solution is $n = \frac{8}{3}$.

b. Solve $\pi x = 3\pi$.

$\pi x = 3\pi$ Write the equation.

Undo the multiplication. → $\frac{\pi x}{\pi} = \frac{3\pi}{\pi}$ Division Property of Equality

$x = 3$ Simplify.

The solution is $x = 3$.

Check

$\pi x = 3\pi$

$\pi(3) \stackrel{?}{=} 3\pi$

$3\pi = 3\pi$ ✓

On Your Own

Now You're Ready
Exercises 18–26

Solve the equation. Check your solution.

7. $\frac{y}{4} = -7$
8. $6\pi = \pi x$
9. $0.09w = 1.8$

EXAMPLE 3 Identifying the Solution of an Equation

What value of k makes the equation $k + 4 \div 0.2 = 5$ true?

Ⓐ -15 Ⓑ -5 Ⓒ -3 Ⓓ 1.5

$k + 4 \div 0.2 = 5$	Write the equation.
$k + 20 = 5$	Divide 4 by 0.2.
$\underline{-20} \quad \underline{-20}$	Subtraction Property of Equality
$k = -15$	Simplify.

∴ The correct answer is Ⓐ.

EXAMPLE 4 Real-Life Application

The melting point of bromine is −7°C.

The *melting point* of a solid is the temperature at which the solid becomes a liquid. The melting point of bromine is $\frac{1}{30}$ of the melting point of nitrogen. Write and solve an equation to find the melting point of nitrogen.

Words	The melting point of bromine	is	$\frac{1}{30}$	of	the melting point of nitrogen.
Variable	Let n be the melting point of nitrogen.				
Equation	-7	$=$	$\frac{1}{30}$	$\cdot$	n

$-7 = \frac{1}{30}n$	Write the equation.
$30 \cdot (-7) = 30 \cdot \left(\frac{1}{30}n\right)$	Multiplication Property of Equality
$-210 = n$	Simplify.

∴ So, the melting point of nitrogen is −210°C.

On Your Own

Now You're Ready
Exercises 33–38

10. Solve $p - 8 \div \frac{1}{2} = -3$.

11. Solve $q + |-10| = 2$.

12. The melting point of mercury is about $\frac{1}{4}$ of the melting point of krypton. The melting point of mercury is −39°C. Write and solve an equation to find the melting point of krypton.

1.1 Exercises

Check It Out
Help with Homework
BigIdeasMath.com

Vocabulary and Concept Check

1. **VOCABULARY** Which of the operations $+$, $-$, $\times$, and $\div$ are inverses of each other?
2. **VOCABULARY** Are the equations $3x = -9$ and $4x = -12$ equivalent? Explain.
3. **WHICH ONE DOESN'T BELONG?** Which equation does *not* belong with the other three? Explain your reasoning.

$x - 2 = 4$ $\qquad$ $x - 3 = 6$ $\qquad$ $x - 5 = 1$ $\qquad$ $x - 6 = 0$

Practice and Problem Solving

CHOOSE TOOLS **Find the value of x. Check the reasonableness of your answer.**

4. 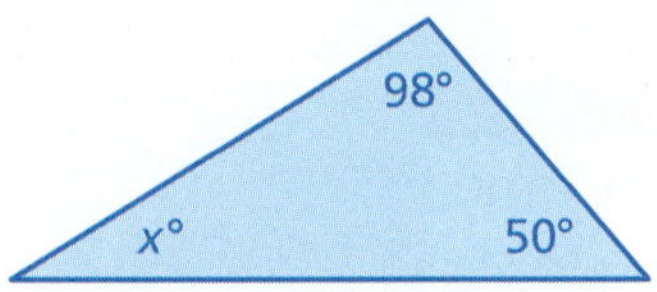

5. 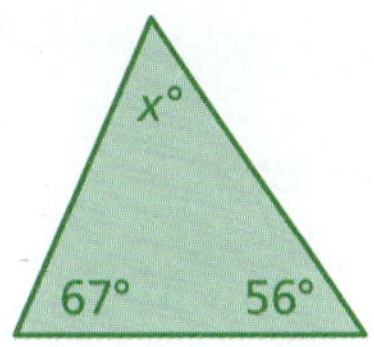

6. 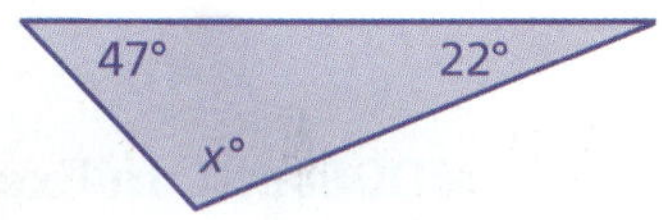

Solve the equation. Check your solution.

1 7. $x + 12 = 7$ $\qquad$ 8. $g - 16 = 8$ $\qquad$ 9. $-9 + p = 12$

10. $0.7 + y = -1.34$ $\qquad$ 11. $x - 8\pi = \pi$ $\qquad$ 12. $4\pi = w - 6\pi$

13. $\frac{5}{6} = \frac{1}{3} + d$ $\qquad$ 14. $\frac{3}{8} = r + \frac{2}{3}$ $\qquad$ 15. $n - 1.4 = -6.3$

16. **CONCERT** A discounted concert ticket costs \$14.50 less than the original price p. You pay \$53 for a discounted ticket. Write and solve an equation to find the original price.

17. **BOWLING** Your friend's final bowling score is 105. Your final bowling score is 14 pins less than your friend's final score.

    a. Write and solve an equation to find your final score.

    b. Your friend made a spare in the 10th frame. Did you? Explain.

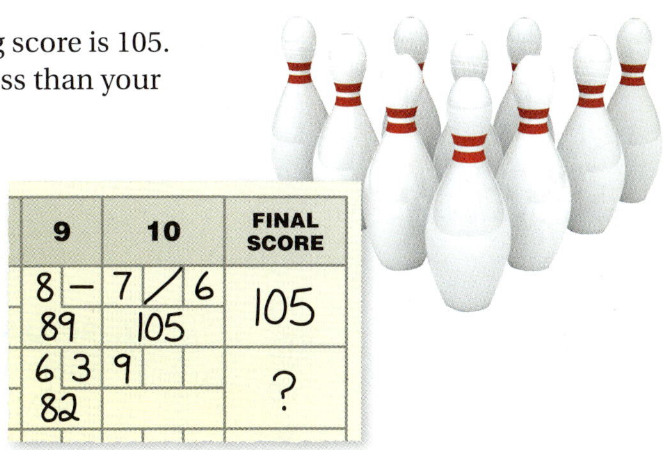

Solve the equation. Check your solution.

2 **18.** $7x = 35$

19. $4 = -0.8n$

20. $6 = -\frac{w}{8}$

21. $\frac{m}{\pi} = 7.3$

22. $-4.3g = 25.8$

23. $\frac{3}{2} = \frac{9}{10}k$

24. $-7.8x = -1.56$

25. $-2 = \frac{6}{7}p$

26. $3\pi d = 12\pi$

27. ERROR ANALYSIS Describe and correct the error in solving the equation.

✗
$$-1.5 + k = 8.2$$
$$k = 8.2 + (-1.5)$$
$$k = 6.7$$

28. TENNIS A gym teacher orders 42 tennis balls. Each package contains 3 tennis balls. Which of the following equations represents the number x of packages?

$x + 3 = 42$ $\quad$ $3x = 42$ $\quad$ $\frac{x}{3} = 42$ $\quad$ $x = \frac{3}{42}$

MODELING In Exercises 29–32, write and solve an equation to answer the question.

29. PARK You clean a community park for 6.5 hours. You earn \$42.25. How much do you earn per hour?

30. ROCKET LAUNCH A rocket is scheduled to launch from a command center in 3.75 hours. What time is it now?

31. BANKING After earning interest, the balance of an account is \$420. The new balance is $\frac{7}{6}$ of the original balance. How much interest did it earn?

Roller Coasters at Cedar Point	
Coaster	**Height (feet)**
Top Thrill Dragster	420
Millennium Force	310
Magnum XL-200	205
Mantis	?

32. ROLLER COASTER Cedar Point amusement park has some of the tallest roller coasters in the United States. The Mantis is 165 feet shorter than the Millennium Force. What is the height of the Mantis?

Solve the equation. Check your solution.

3 **33.** $-3 = h + 8 \div 2$

34. $12 = w - |-7|$

35. $q + |6.4| = 9.6$

36. $d - 2.8 \div 0.2 = -14$

37. $\frac{8}{9} = x + \frac{1}{3}(7)$

38. $p - \frac{1}{4} \cdot 3 = -\frac{5}{6}$

39. **LOGIC** Without solving, determine whether the solution of $-2x = -15$ is *greater than* or *less than* -15. Explain.

40. **OPEN-ENDED** Write a subtraction equation and a division equation so that each has a solution of -2.

41. **ANTS** Some ant species can carry 50 times their body weight. It takes 32 ants to carry the cherry. About how much does each ant weigh?

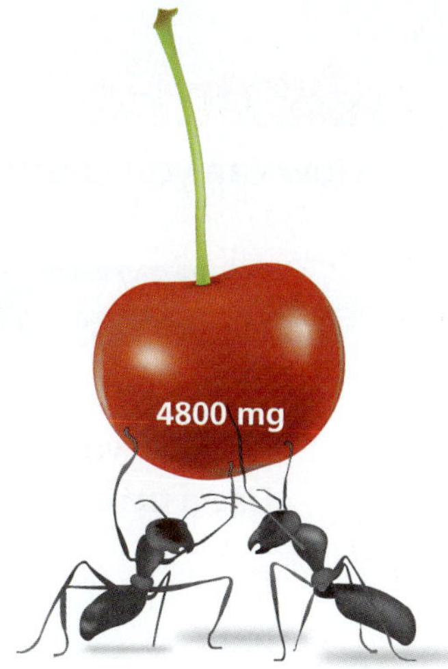

42. **REASONING** One-fourth of the girls and one-eighth of the boys in a class retake their school pictures. The photographer retakes pictures for 16 girls and 7 boys. How many students are in the class?

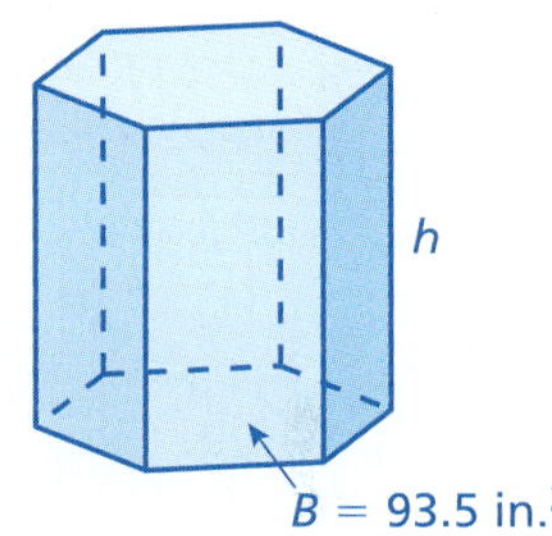

43. **VOLUME** The volume V of the prism is 1122 cubic inches. Use the formula $V = Bh$ to find the height h of the prism.

44. **Critical Thinking** A neighbor pays you and two friends \$90 to paint her garage. You divide the money three ways in the ratio 2 : 3 : 5.

a. How much does each person receive?

b. What is one possible reason the money is not divided evenly?

Fair Game Review What you learned in previous grades & lessons

Simplify the expression. *(Skills Review Handbook)*

45. $2(x - 2) + 5x$

46. $0.4b - 3.2 + 1.2b$

47. $\frac{1}{4}g + 6g - \frac{2}{3}$

48. **MULTIPLE CHOICE** The temperature at 4:00 P.M. was -12 °C. By 11:00 P.M., the temperature had dropped 14 °C. What was the temperature at 11:00 P.M.? *(Skills Review Handbook)*

(A) -26 °C (B) -2 °C (C) 2 °C (D) 26 °C

1.2 Solving Multi-Step Equations

Essential Question How can you solve a multi-step equation? How can you check the reasonableness of your solution?

1 ACTIVITY: Solving for the Angles of a Triangle

Work with a partner. Write an equation for each triangle. Solve the equation to find the value of the variable. Then find the angle measures of each triangle. Use a protractor to check the reasonableness of your answer.

a.

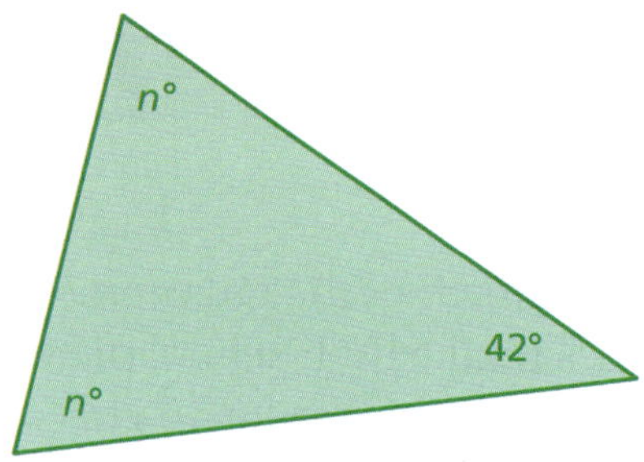

b.

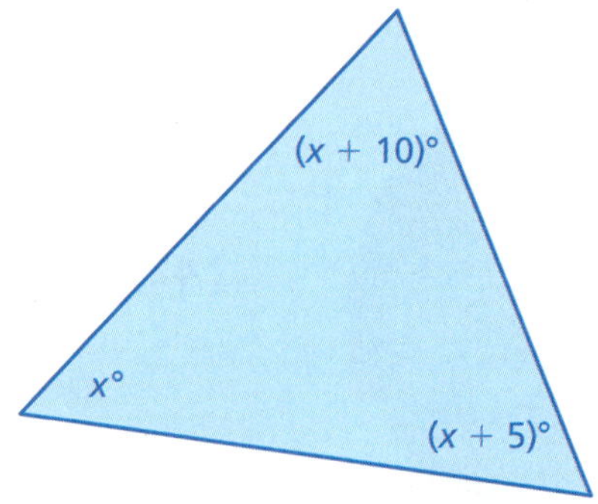

c.

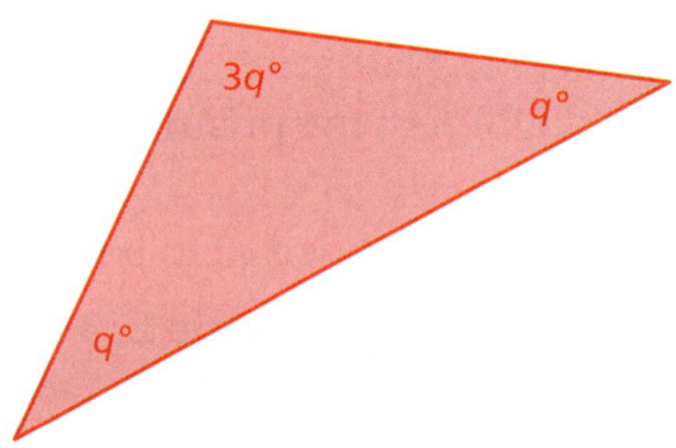

d.

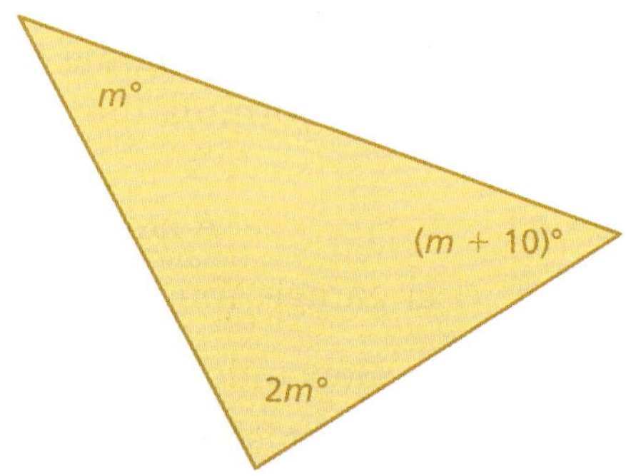

e.

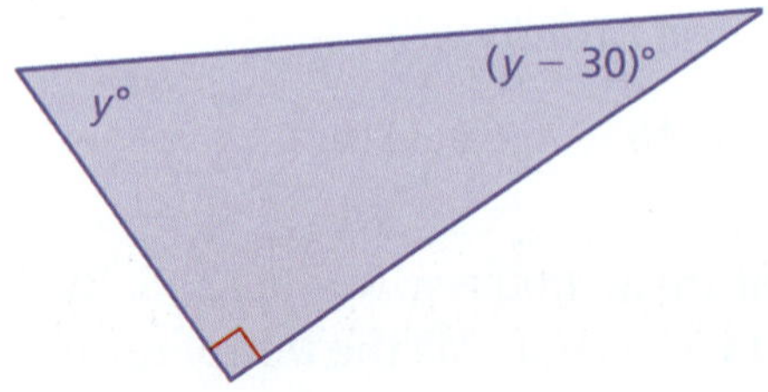

f.

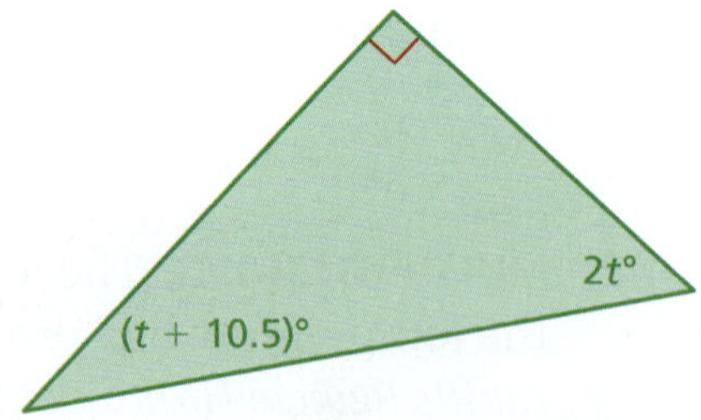

COMMON CORE

Solving Equations

In this lesson, you will

- use inverse operations to solve multi-step equations.
- use the Distributive Property to solve multi-step equations.

Learning Standards
MACC.8.EE.3.7a
MACC.8.EE.3.7b

2 ACTIVITY: Problem-Solving Strategy

Math Practice 1

Find Entry Points

How do you decide which triangle to solve first? Explain.

Work with a partner.

The six triangles form a rectangle.

Find the angle measures of each triangle. Use a protractor to check the reasonableness of your answers.

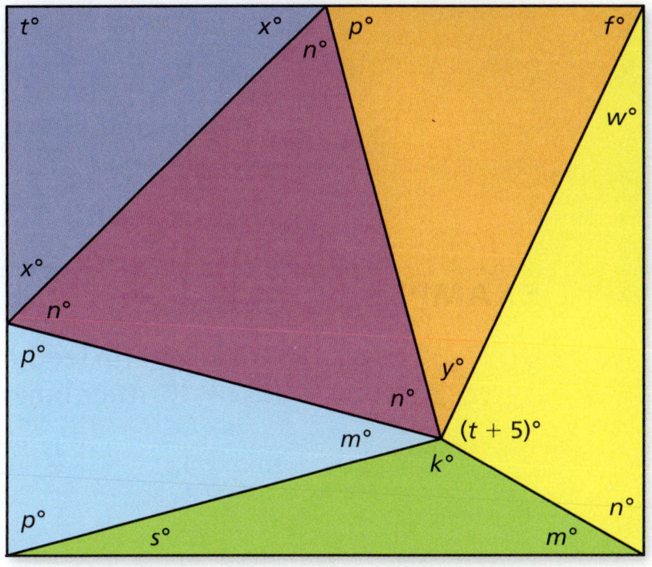

3 ACTIVITY: Puzzle

Work with a partner. A survey asked 200 people to name their favorite weekday. The results are shown in the circle graph.

a. How many degrees are in each part of the circle graph?

b. What percent of the people chose each day?

c. How many people chose each day?

d. Organize your results in a table.

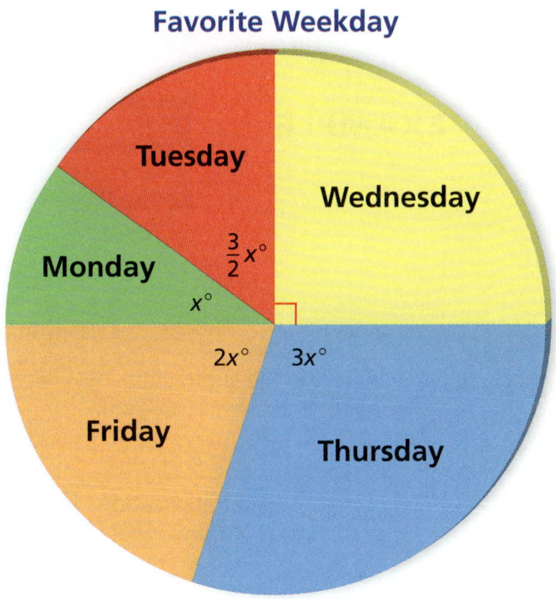

What Is Your Answer?

4. **IN YOUR OWN WORDS** How can you solve a multi-step equation? How can you check the reasonableness of your solution?

Practice Use what you learned about solving multi-step equations to complete Exercises 3–5 on page 14.

1.2 Lesson

Check It Out
Lesson Tutorials
BigIdeasMath.com

Key Idea

Solving Multi-Step Equations

To solve multi-step equations, use inverse operations to isolate the variable.

EXAMPLE 1 Solving a Two-Step Equation

The height (in feet) of a tree after x years is $1.5x + 15$. After how many years is the tree 24 feet tall?

	$1.5x + 15 = 24$	Write an equation.
Undo the addition. →	$\underline{-15} \quad \underline{-15}$	Subtraction Property of Equality
	$1.5x = 9$	Simplify.
Undo the multiplication. →	$\dfrac{1.5x}{1.5} = \dfrac{9}{1.5}$	Division Property of Equality
	$x = 6$	Simplify.

So, the tree is 24 feet tall after 6 years.

EXAMPLE 2 Combining Like Terms to Solve an Equation

Solve $8x - 6x - 25 = -35$.

	$8x - 6x - 25 = -35$	Write the equation.
	$2x - 25 = -35$	Combine like terms.
Undo the subtraction. →	$\underline{+25} \quad \underline{+25}$	Addition Property of Equality
	$2x = -10$	Simplify.
Undo the multiplication. →	$\dfrac{2x}{2} = \dfrac{-10}{2}$	Division Property of Equality
	$x = -5$	Simplify.

The solution is $x = -5$.

On Your Own

Now You're Ready
Exercises 6–9

Solve the equation. Check your solution.

1. $-3z + 1 = 7$
2. $\frac{1}{2}x - 9 = -25$
3. $-4n - 8n + 17 = 23$

EXAMPLE 3 Using the Distributive Property to Solve an Equation

Solve $2(1 - 5x) + 4 = -8$.

$2(1 - 5x) + 4 = -8$	Write the equation.
$2(1) - 2(5x) + 4 = -8$	Distributive Property
$2 - 10x + 4 = -8$	Multiply.
$-10x + 6 = -8$	Combine like terms.
$-6 \quad -6$	Subtraction Property of Equality
$-10x = -14$	Simplify.
$\frac{-10x}{-10} = \frac{-14}{-10}$	Division Property of Equality
$x = 1.4$	Simplify.

Study Tip

Here is another way to solve the equation in Example 3.

$$2(1 - 5x) + 4 = -8$$
$$2(1 - 5x) = -12$$
$$1 - 5x = -6$$
$$-5x = -7$$
$$x = 1.4$$

EXAMPLE 4 Real-Life Application

Use the table to find the number of miles x you need to run on Friday so that the mean number of miles run per day is 1.5.

Day	Miles
Monday	2
Tuesday	0
Wednesday	1.5
Thursday	0
Friday	x

Write an equation using the definition of *mean*.

sum of the data / number of values	$\frac{2 + 0 + 1.5 + 0 + x}{5} = 1.5$	Write the equation.
	$\frac{3.5 + x}{5} = 1.5$	Combine like terms.
Undo the division.	$5 \cdot \frac{3.5 + x}{5} = 5 \cdot 1.5$	Multiplication Property of Equality
	$3.5 + x = 7.5$	Simplify.
Undo the addition.	$-3.5 \quad -3.5$	Subtraction Property of Equality
	$x = 4$	Simplify.

So, you need to run 4 miles on Friday.

On Your Own

Now You're Ready
Exercises 10 and 11

Solve the equation. Check your solution.

4. $-3(x + 2) + 5x = -9$

5. $5 + 1.5(2d - 1) = 0.5$

6. You scored 88, 92, and 87 on three tests. Write and solve an equation to find the score you need on the fourth test so that your mean test score is 90.

1.2 Exercises

Check It Out
Help with Homework
BigIdeasMath.com

Vocabulary and Concept Check

1. **WRITING** Write the verbal statement as an equation. Then solve.

   2 more than 3 times a number is 17.

2. **OPEN-ENDED** Explain how to solve the equation $2(4x - 11) + 9 = 19$.

Practice and Problem Solving

CHOOSE TOOLS **Find the value of the variable. Then find the angle measures of the polygon. Use a protractor to check the reasonableness of your answer.**

3. 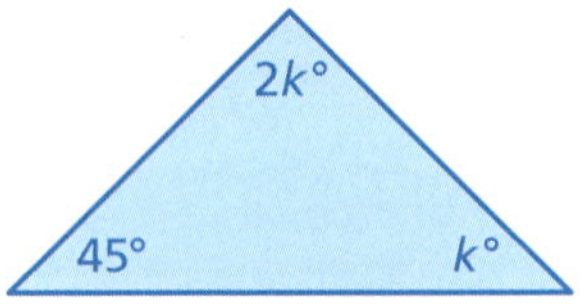

Sum of angle measures: 180°

4. 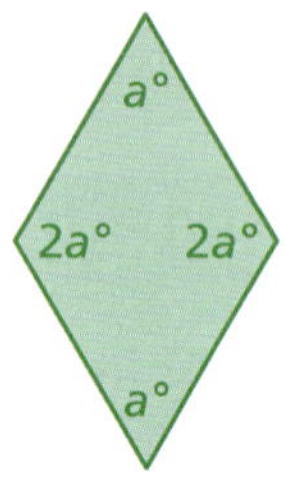

Sum of angle measures: 360°

5. 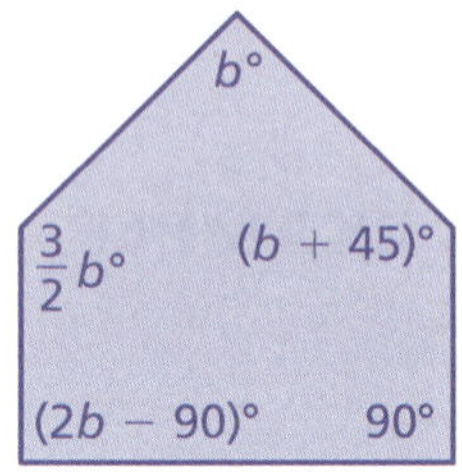

Sum of angle measures: 540°

Solve the equation. Check your solution.

① ② 6. $10x + 2 = 32$

7. $19 - 4c = 17$

8. $1.1x + 1.2x - 5.4 = -10$

9. $\frac{2}{3}h - \frac{1}{3}h + 11 = 8$

③ 10. $6(5 - 8v) + 12 = -54$

11. $21(2 - x) + 12x = 44$

12. **ERROR ANALYSIS** Describe and correct the error in solving the equation.

    ✗
    $$-2(7 - y) + 4 = -4$$
    $$-14 - 2y + 4 = -4$$
    $$-10 - 2y = -4$$
    $$-2y = 6$$
    $$y = -3$$

13. **WATCHES** The cost C (in dollars) of making n watches is represented by $C = 15n + 85$. How many watches are made when the cost is \$385?

14. **HOUSE** The height of the house is 26 feet. What is the height x of each story?

In Exercises 15–17, write and solve an equation to answer the question.

15. POSTCARD The area of the postcard is 24 square inches. What is the width b of the message (in inches)?

16. BREAKFAST You order two servings of pancakes and a fruit cup. The cost of the fruit cup is \$1.50. You leave a 15% tip. Your total bill is \$11.50. How much does one serving of pancakes cost?

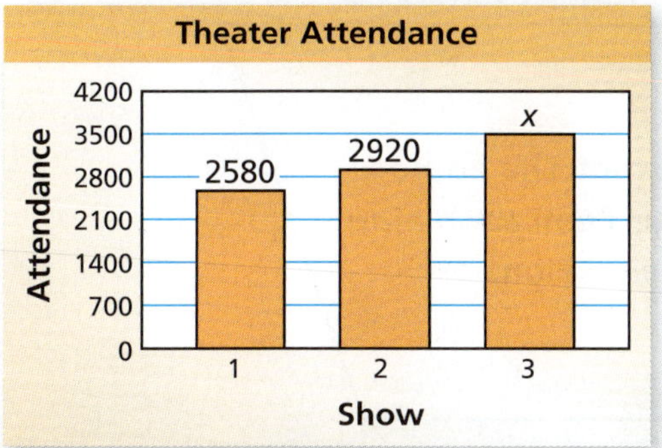

17. THEATER How many people must attend the third show so that the average attendance per show is 3000?

18. DIVING Divers in a competition are scored by an international panel of judges. The highest and the lowest scores are dropped. The total of the remaining scores is multiplied by the degree of difficulty of the dive. This product is multiplied by 0.6 to determine the final score.

a. A diver's final score is 77.7. What is the degree of difficulty of the dive?

Judge	Russia	China	Mexico	Germany	Italy	Japan	Brazil
Score	7.5	8.0	6.5	8.5	7.0	7.5	7.0

b. Critical Thinking The degree of difficulty of a dive is 4.0. The diver's final score is 97.2. Judges award half or whole points from 0 to 10. What scores could the judges have given the diver?

Fair Game Review What you learned in previous grades & lessons

Let $a = 3$ and $b = -2$. Copy and complete the statement using <, >, or =.
(Skills Review Handbook)

19. $-5a$ ▢ 4 **20.** 5 ▢ $b + 7$ **21.** $a - 4$ ▢ $10b + 8$

22. MULTIPLE CHOICE What value of x makes the equation $x + 5 = 2x$ true?
(Skills Review Handbook)

Ⓐ -1 Ⓑ 0 Ⓒ 3 Ⓓ 5

Study Help

Check It Out
Graphic Organizer
BigIdeasMath.com

You can use a **Y chart** to compare two topics. List differences in the branches and similarities in the base of the Y. Here is an example of a Y chart that compares solving simple equations using addition to solving simple equations using subtraction.

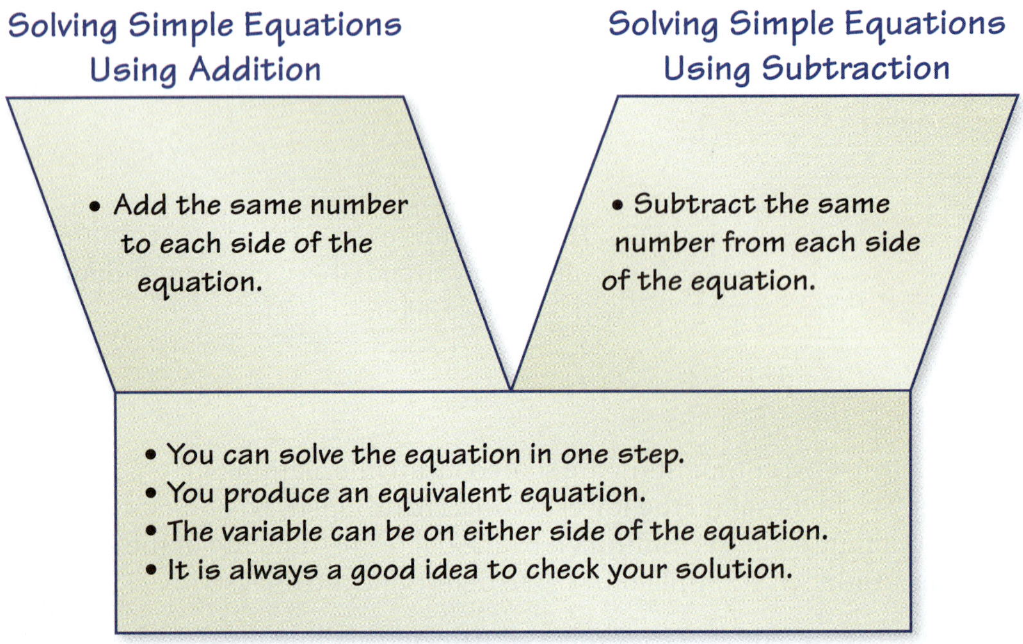

On Your Own

Make Y charts to help you study and compare these topics.

1. solving simple equations using multiplication and solving simple equations using division
2. solving simple equations and solving multi-step equations

After you complete this chapter, make Y charts for the following topics.

3. solving equations with the variable on one side and solving equations with variables on both sides
4. solving multi-step equations and solving equations with variables on both sides
5. solving multi-step equations and rewriting literal equations

"I made a Y chart to compare and contrast Fluffy's characteristics with yours."

1.1–1.2 Quiz

Check It Out
Progress Check
BigIdeasMath.com

Solve the equation. Check your solution. *(Section 1.1)*

1. $-\frac{1}{2} = y - 1$

2. $-3\pi + w = 2\pi$

3. $1.2m = 0.6$

4. $q + 2.7 = -0.9$

Solve the equation. Check your solution. *(Section 1.2)*

5. $-4k + 17 = 1$

6. $\frac{1}{4}z + 8 = 12$

7. $-3(2n + 1) + 7 = -5$

8. $2.5(t - 2) - 6 = 9$

Find the value of x. Then find the angle measures of the polygon. *(Section 1.2)*

9.

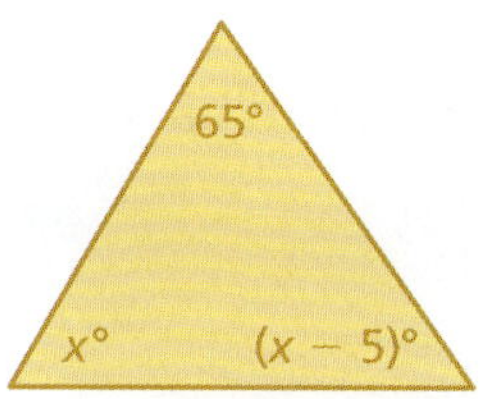

Sum of angle measures: 180°

10.

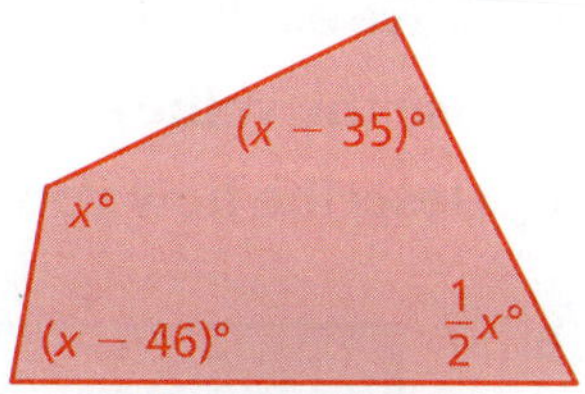

Sum of angle measures: 360°

11. JEWELER The equation $P = 2.5m + 35$ represents the price P (in dollars) of a bracelet, where m is the cost of the materials (in dollars). The price of a bracelet is \$115. What is the cost of the materials? *(Section 1.2)*

12. PASTURE A 455-foot fence encloses a pasture. What is the length of each side of the pasture? *(Section 1.2)*

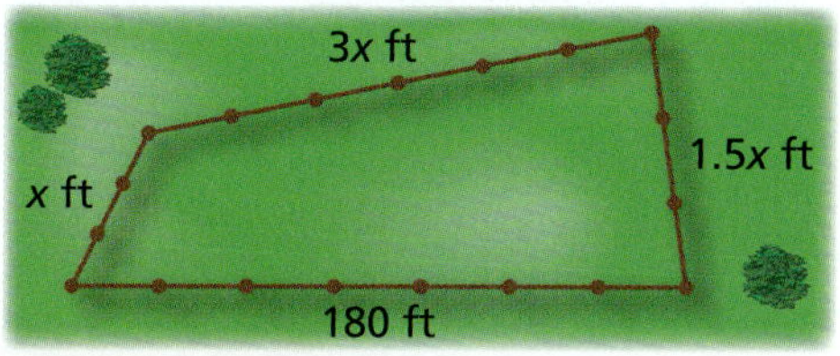

13. POSTERS A machine prints 230 movie posters each hour. Write and solve an equation to find the number of hours it takes the machine to print 1265 posters. *(Section 1.1)*

14. BASKETBALL Use the table to write and solve an equation to find the number of points p you need to score in the fourth game so that the mean number of points is 20. *(Section 1.2)*

Game	Points
1	25
2	15
3	18
4	p

1.3 Solving Equations with Variables on Both Sides

Essential Question How can you solve an equation that has variables on both sides?

1 ACTIVITY: Perimeter and Area

Work with a partner.

- **Each figure has the unusual property that the value of its perimeter (in feet) is equal to the value of its area (in square feet). Write an equation for each figure.**
- **Solve each equation for *x*.**
- **Use the value of *x* to find the perimeter and the area of each figure.**
- **Describe how you can check your solution.**

a.

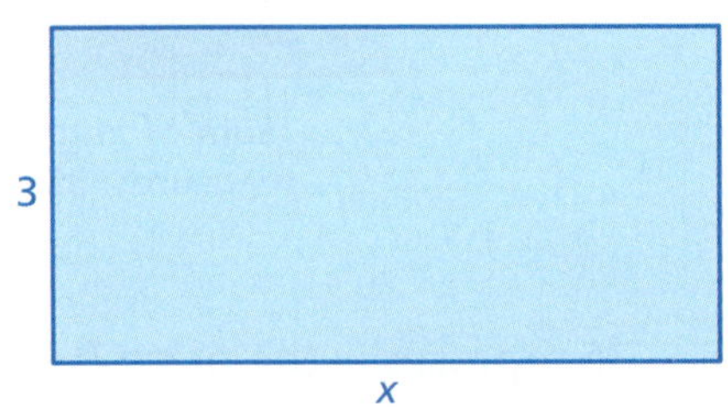

b.

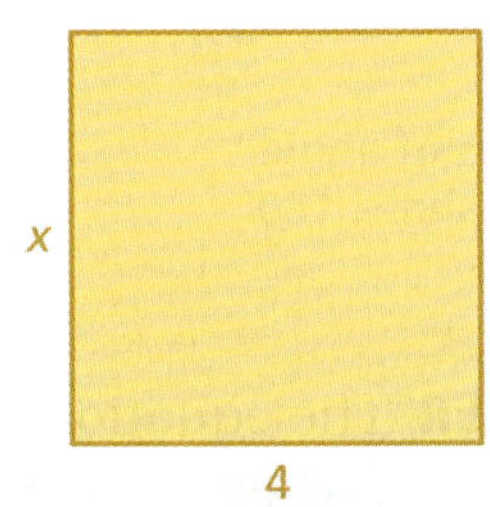

c.

d.

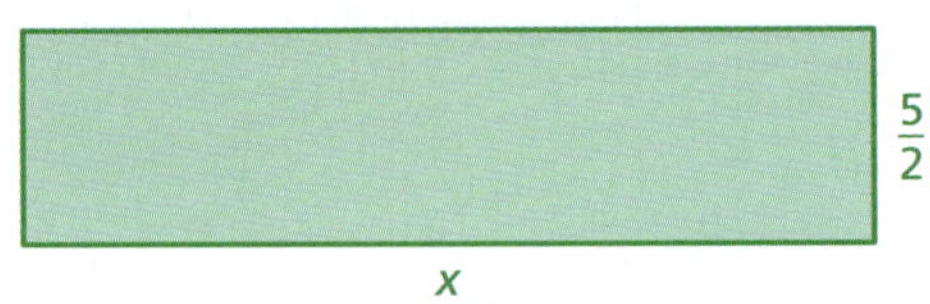

e.

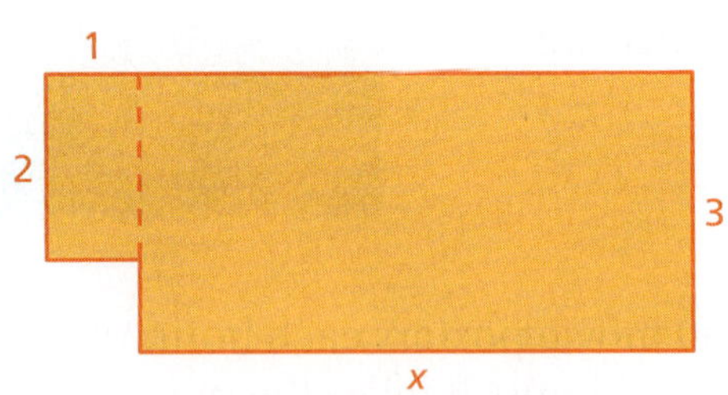

f.

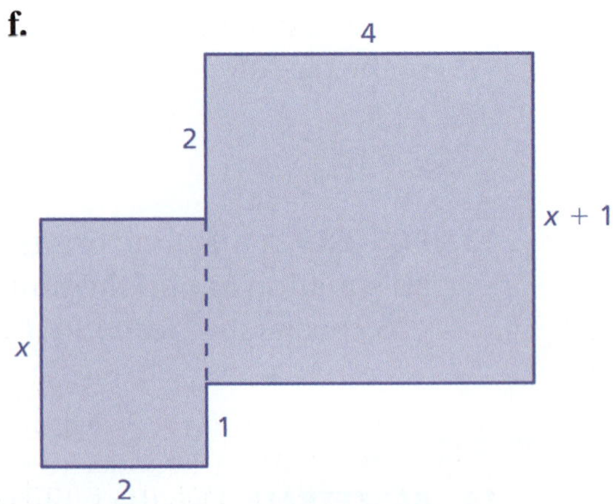

g.

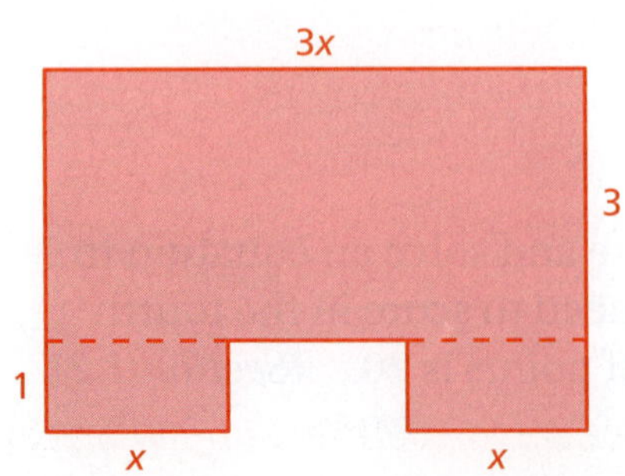

COMMON CORE

Solving Equations

In this lesson, you will

- solve equations with variables on both sides.
- determine whether equations have no solution or infinitely many solutions.

Learning Standards
MACC.8.EE.3.7a
MACC.8.EE.3.7b

2 ACTIVITY: Surface Area and Volume

Math Practice 2

Use Operations

What properties of operations do you need to use in order to find the value of x?

Work with a partner.

- **Each solid has the unusual property that the value of its surface area (in square inches) is equal to the value of its volume (in cubic inches). Write an equation for each solid.**
- **Solve each equation for x.**
- **Use the value of x to find the surface area and the volume of each solid.**
- **Describe how you can check your solution.**

a.

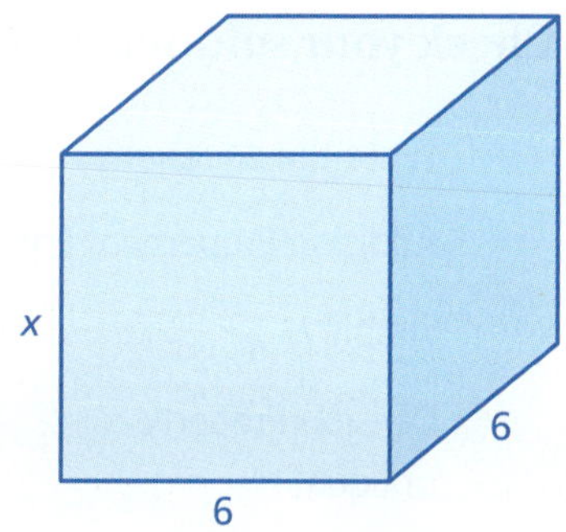

b.

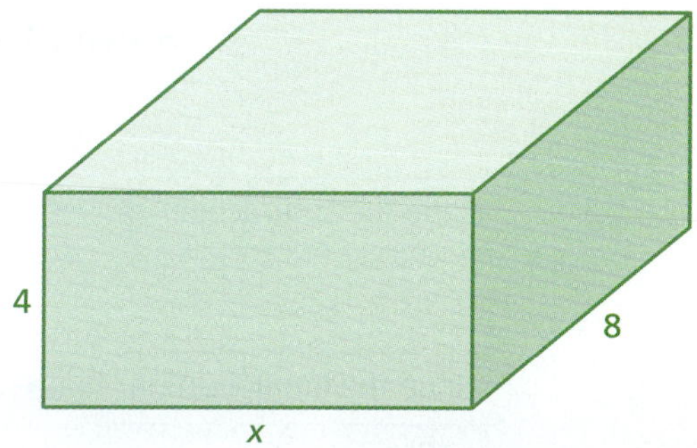

3 ACTIVITY: Puzzle

Work with a partner. The perimeter of the larger triangle is 150% of the perimeter of the smaller triangle. Find the dimensions of each triangle.

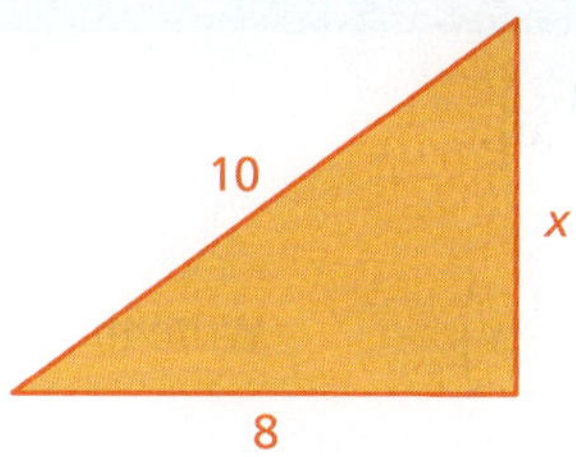

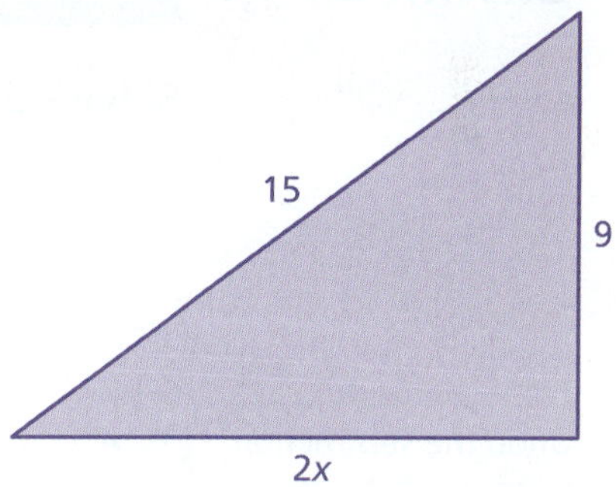

What Is Your Answer?

4. **IN YOUR OWN WORDS** How can you solve an equation that has variables on both sides? How do you move a variable term from one side of the equation to the other?

5. Write an equation that has variables on both sides. Solve the equation.

Practice Use what you learned about solving equations with variables on both sides to complete Exercises 3–5 on page 23.

1.3 Lesson

Check It Out
Lesson Tutorials
BigIdeasMath.com

Key Idea

Solving Equations with Variables on Both Sides

To solve equations with variables on both sides, collect the variable terms on one side and the constant terms on the other side.

EXAMPLE 1 Solving an Equation with Variables on Both Sides

Solve $15 - 2x = -7x$. Check your solution.

	$15 - 2x = -7x$	Write the equation.
Undo the subtraction. →	$+\,2x \qquad +\,2x$	Addition Property of Equality
	$15 = -5x$	Simplify.
Undo the multiplication. →	$\dfrac{15}{-5} = \dfrac{-5x}{-5}$	Division Property of Equality
	$-3 = x$	Simplify.

∴ The solution is $x = -3$.

Check

$15 - 2x = -7x$

$15 - 2(-3) \stackrel{?}{=} -7(-3)$

$21 = 21$ ✓

EXAMPLE 2 Using the Distributive Property to Solve an Equation

Solve $-2(x - 5) = 6\left(2 - \frac{1}{2}x\right)$.

	$-2(x - 5) = 6\left(2 - \frac{1}{2}x\right)$	Write the equation.
	$-2x + 10 = 12 - 3x$	Distributive Property
Undo the subtraction. →	$+\,3x \qquad\qquad +\,3x$	Addition Property of Equality
	$x + 10 = 12$	Simplify.
Undo the addition. →	$-\,10 \qquad -\,10$	Subtraction Property of Equality
	$x = 2$	Simplify.

∴ The solution is $x = 2$.

On Your Own

Now You're Ready
Exercises 6–14

Solve the equation. Check your solution.

1. $-3x = 2x + 19$
2. $2.5y + 6 = 4.5y - 1$
3. $6(4 - z) = 2z$

Some equations do not have one solution. Equations can also have no solution or infinitely many solutions.

When solving an equation that has no solution, you will obtain an equivalent equation that is not true for any value of the variable, such as $0 = 2$.

EXAMPLE 3 Solving Equations with No Solution

Solve $3 - 4x = -7 - 4x$.

$3 - 4x = -7 - 4x$ Write the equation.

Undo the subtraction. → $+4x \quad +4x$ Addition Property of Equality

$3 = -7$ ✗ Simplify.

The equation $3 = -7$ is never true. So, the equation has no solution.

When solving an equation that has infinitely many solutions, you will obtain an equivalent equation that is true for all values of the variable, such as $-5 = -5$.

EXAMPLE 4 Solving Equations with Infinitely Many Solutions

Solve $6x + 4 = 4\left(\frac{3}{2}x + 1\right)$.

$6x + 4 = 4\left(\frac{3}{2}x + 1\right)$ Write the equation.

$6x + 4 = 6x + 4$ Distributive Property

Undo the addition. → $-6x \quad -6x$ Subtraction Property of Equality

$4 = 4$ Simplify.

The equation $4 = 4$ is always true. So, the equation has infinitely many solutions.

On Your Own

Now You're Ready
Exercises 18–29

Solve the equation.

4. $2x + 1 = 2x - 1$

5. $\frac{1}{2}(6t - 4) = 3t - 2$

6. $\frac{1}{3}(2b + 9) = \frac{2}{3}\left(b + \frac{9}{2}\right)$

7. $6(5 - 2v) = -4(3v + 1)$

EXAMPLE 5 Writing and Solving an Equation

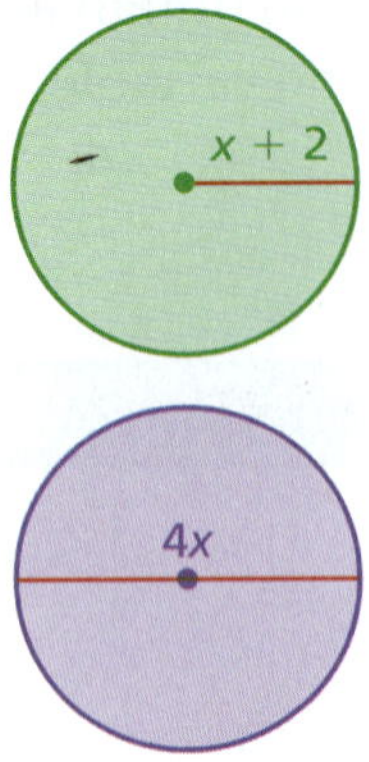

The circles are identical. What is the area of each circle?

Ⓐ 2　　Ⓑ 4　　Ⓒ 16π　　Ⓓ 64π

The circles are identical, so the radius of each circle is the same.

$x + 2 = 2x$	Write an equation. The radius of the purple circle is $\frac{4x}{2} = 2x$.
$-x \quad\quad -x$	Subtraction Property of Equality
$2 = x$	Simplify.

Because the radius of each circle is 4, the area of each circle is $\pi r^2 = \pi(4)^2 = 16\pi$.

∴ So, the correct answer is Ⓒ.

EXAMPLE 6 Real-Life Application

A boat travels *x* miles per hour upstream on the Mississippi River. On the return trip, the boat travels 2 miles per hour faster. How far does the boat travel upstream?

The speed of the boat on the return trip is $(x + 2)$ miles per hour.

Distance upstream = Distance of return trip

$3x = 2.5(x + 2)$	Write an equation.
$3x = 2.5x + 5$	Distributive Property
$-2.5x \quad\quad -2.5x$	Subtraction Property of Equality
$0.5x = 5$	Simplify.
$\frac{0.5x}{0.5} = \frac{5}{0.5}$	Division Property of Equality
$x = 10$	Simplify.

∴ The boat travels 10 miles per hour for 3 hours upstream. So, it travels 30 miles upstream.

On Your Own

8. **WHAT IF?** In Example 5, the diameter of the purple circle is $3x$. What is the area of each circle?
9. A boat travels x miles per hour from one island to another island in 2.5 hours. The boat travels 5 miles per hour faster on the return trip of 2 hours. What is the distance between the islands?

1.3 Exercises

Check It Out
Help with Homework
BigIdeasMath.com

Vocabulary and Concept Check

1. **WRITING** Is $x = 3$ a solution of the equation $3x - 5 = 4x - 9$? Explain.

2. **OPEN-ENDED** Write an equation that has variables on both sides and has a solution of -3.

Practice and Problem Solving

The value of the solid's surface area is equal to the value of the solid's volume. Find the value of x.

3. 

4. 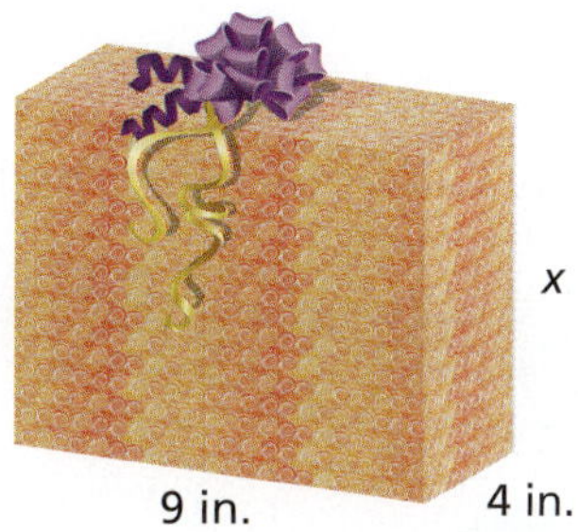

5. 

Solve the equation. Check your solution.

6. $m - 4 = 2m$

7. $3k - 1 = 7k + 2$

8. $6.7x = 5.2x + 12.3$

9. $-24 - \frac{1}{8}p = \frac{3}{8}p$

10. $12(2w - 3) = 6w$

11. $2(n - 3) = 4n + 1$

12. $2(4z - 1) = 3(z + 2)$

13. $0.1x = 0.2(x + 2)$

14. $\frac{1}{6}d + \frac{2}{3} = \frac{1}{4}(d - 2)$

15. **ERROR ANALYSIS** Describe and correct the error in solving the equation.

✗
$$3x - 4 = 2x + 1$$
$$3x - 4 - 2x = 2x + 1 - 2x$$
$$x - 4 = 1$$
$$x - 4 + 4 = 1 - 4$$
$$x = -3$$

16. **TRAIL MIX** The equation $4.05p + 14.40 = 4.50(p + 3)$ represents the number p of pounds of peanuts you need to make trail mix. How many pounds of peanuts do you need for the trail mix?

17. **CARS** Write and solve an equation to find the number of miles you must drive to have the same cost for each of the car rentals.

$15 plus $0.50 per mile

$25 plus $0.25 per mile

Solve the equation. Check your solution, if possible.

18. $x + 6 = x$

19. $3x - 1 = 1 - 3x$

20. $4x - 9 = 3.5x - 9$

21. $\frac{1}{2}x + \frac{1}{2}x = x + 1$

22. $3x + 15 = 3(x + 5)$

23. $\frac{1}{3}(9x + 3) = 3x + 1$

24. $5x - 7 = 4x - 1$

25. $2x + 4 = -(-7x + 6)$

26. $5.5 - x = -4.5 - x$

27. $10x - \frac{8}{3} - 4x = 6x$

28. $-3(2x - 3) = -6x + 9$

29. $6(7x + 7) = 7(6x + 6)$

30. ERROR ANALYSIS Describe and correct the error in solving the equation.

✗
$-4(2n - 3) = 12 - 8n$
$-8n + 12 = 12 - 8n$
$-8n = -8n$
$0 = 0$
The solution is $n = 0$.

31. OPEN-ENDED Write an equation with variables on both sides that has no solution. Explain why it has no solution.

32. GEOMETRY Are there any values of x for which the areas of the figures are the same? Explain.

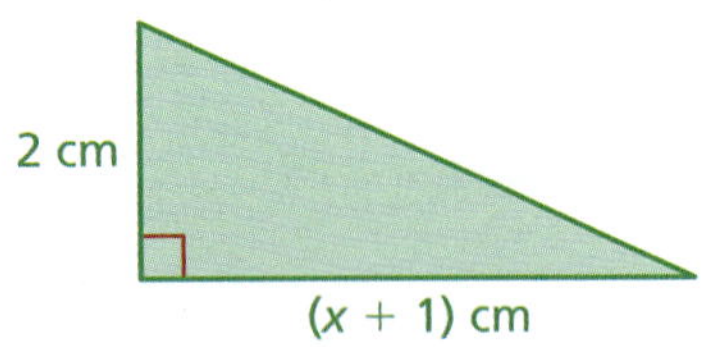

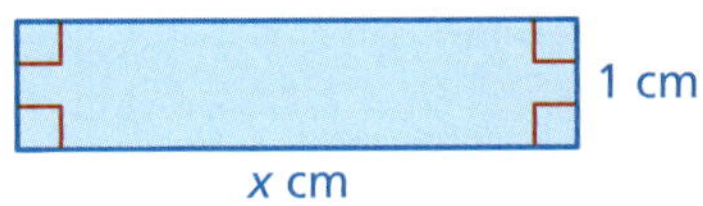

33. SATELLITE TV Provider A charges \$75 for installation and charges \$39.95 per month for the basic package. Provider B offers free installation and charges \$39.95 per month for the basic package. Your neighbor subscribes to Provider A the same month you subscribe to Provider B. After how many months is your neighbor's total cost the same as your total cost for satellite TV?

34. PIZZA CRUST Pepe's Pizza makes 52 pizza crusts the first week and 180 pizza crusts each subsequent week. Dianne's Delicatessen makes 26 pizza crusts the first week and 90 pizza crusts each subsequent week. In how many weeks will the total number of pizza crusts made by Pepe's Pizza equal twice the total number of pizza crusts made by Dianne's Delicatessen?

35. PRECISION Is the triangle an equilateral triangle? Explain.

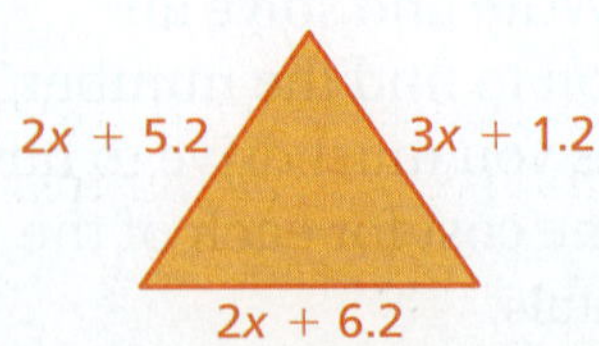

A polygon is *regular* if each of its sides has the same length. Find the perimeter of the regular polygon.

36.

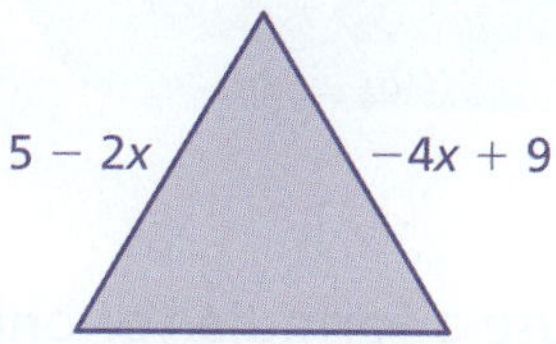

37.

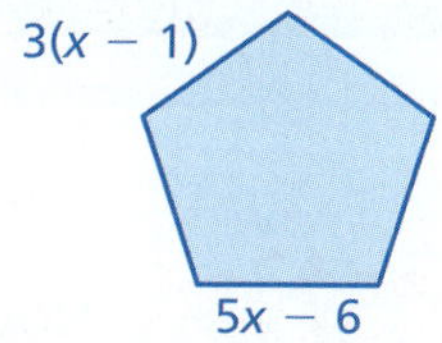

38.

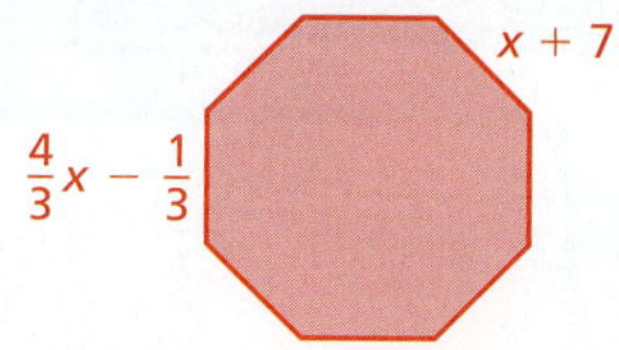

39. **PRECISION** The cost of mailing a DVD in an envelope by Express Mail® is equal to the cost of mailing a DVD in a box by Priority Mail®. What is the weight of the DVD with its packing material? Round your answer to the nearest hundredth.

	Packing Material	Priority Mail®	Express Mail®
Box	$2.25	$2.50 per lb	$8.50 per lb
Envelope	$1.10	$2.50 per lb	$8.50 per lb

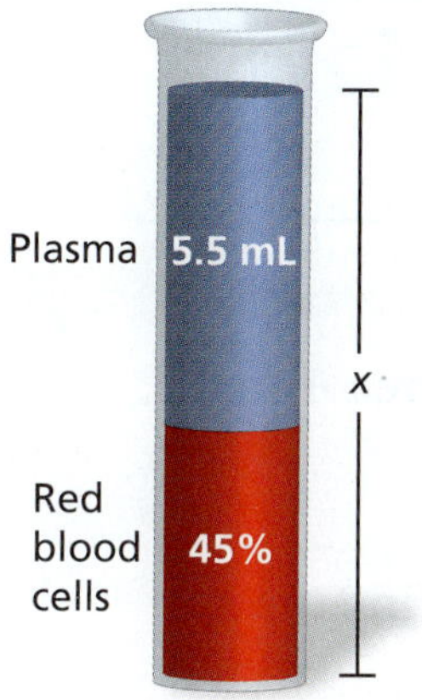

40. **PROBLEM SOLVING** Would you solve the equation $0.25x + 7 = \frac{1}{3}x - 8$ using fractions or decimals? Explain.

41. **BLOOD SAMPLE** The amount of red blood cells in a blood sample is equal to the total amount in the sample minus the amount of plasma. What is the total amount x of blood drawn?

42. **NUTRITION** One serving of oatmeal provides 16% of the fiber you need daily. You must get the remaining 21 grams of fiber from other sources. How many grams of fiber should you consume daily?

43. **Geometry** A 6-foot-wide hallway is painted as shown, using equal amounts of white and black paint.

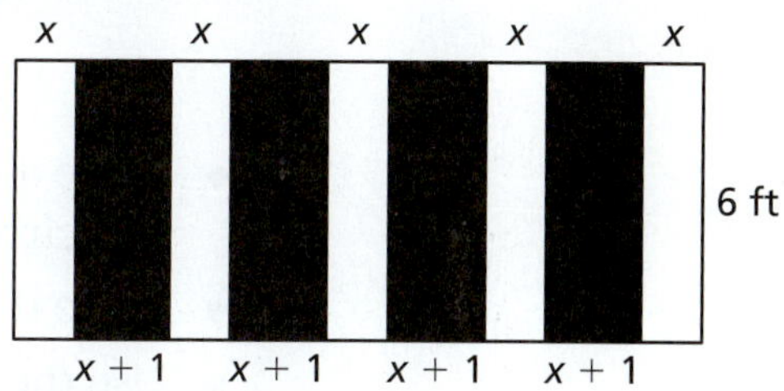

a. How long is the hallway?

b. Can this same hallway be painted with the same pattern, but using twice as much black paint as white paint? Explain.

Fair Game Review What you learned in previous grades & lessons

Find the volume of the solid. *(Skills Review Handbook)*

44.

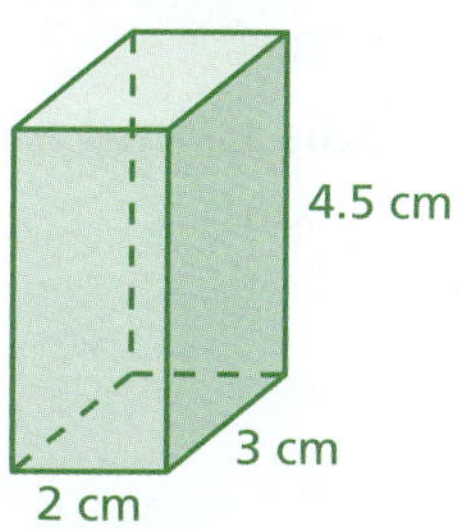

45.

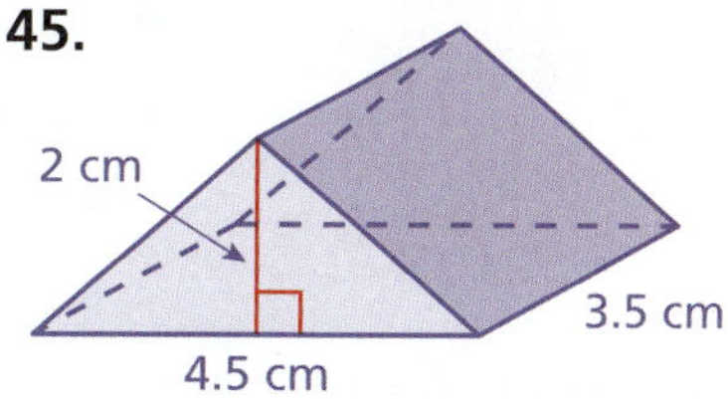

46.

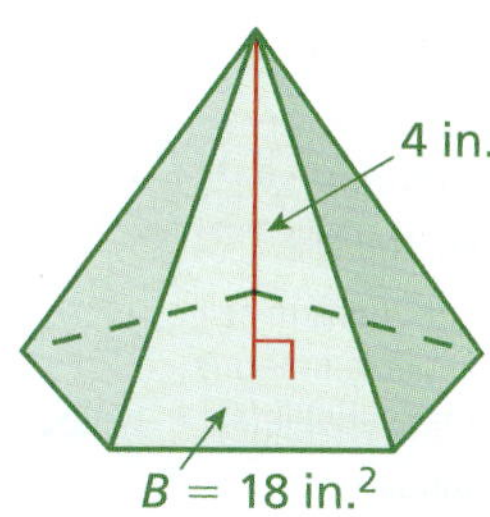

47. **MULTIPLE CHOICE** A car travels 480 miles on 15 gallons of gasoline. How many miles does the car travel per gallon? *(Skills Review Handbook)*

Ⓐ 28 mi/gal Ⓑ 30 mi/gal Ⓒ 32 mi/gal Ⓓ 35 mi/gal

1.4 Rewriting Equations and Formulas

Essential Question How can you use a formula for one measurement to write a formula for a different measurement?

1 ACTIVITY: Using Perimeter and Area Formulas

Work with a partner.

a.
- Write a formula for the perimeter P of a rectangle.
- Solve the formula for w.
- Use the new formula to find the width of the rectangle.

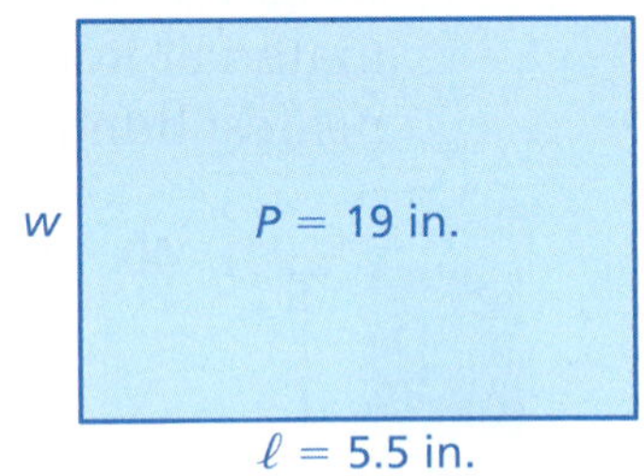

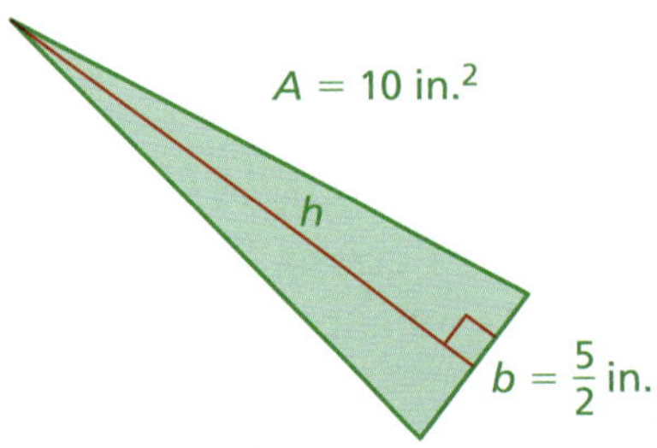

b.
- Write a formula for the area A of a triangle.
- Solve the formula for h.
- Use the new formula to find the height of the triangle.

c.
- Write a formula for the circumference C of a circle.
- Solve the formula for r.
- Use the new formula to find the radius of the circle.

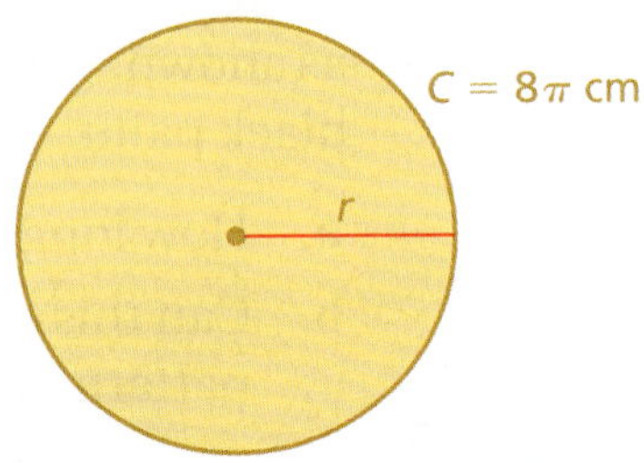

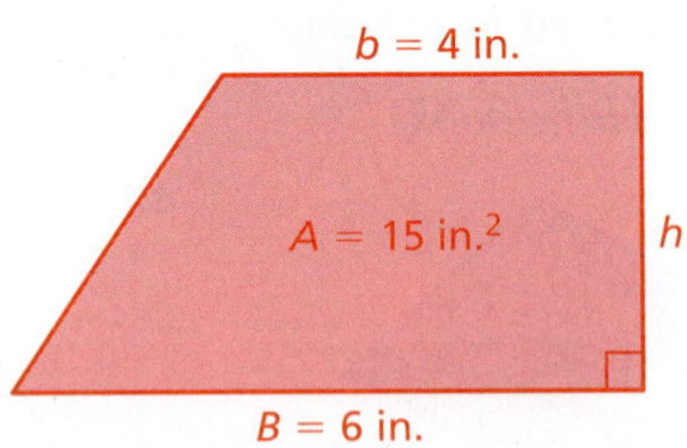

d.
- Write a formula for the area A of a trapezoid.
- Solve the formula for h.
- Use the new formula to find the height of the trapezoid.

e.
- Write a formula for the area A of a parallelogram.
- Solve the formula for h.
- Use the new formula to find the height of the parallelogram.

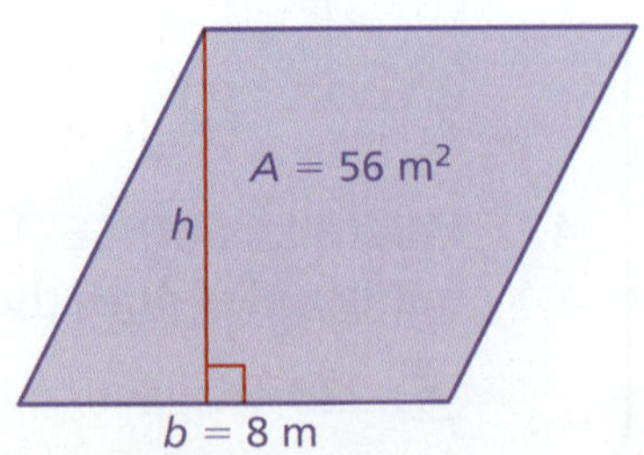

COMMON CORE

Solving Equations

In this lesson, you will
- rewrite equations to solve for one variable in terms of the other variable(s).

Applying Standard MACC.8.EE.3.7

2 ACTIVITY: Using Volume and Surface Area Formulas

Math Practice 8

Find General Methods

What do you have to do each time to solve for the given variable? Why does this process result in a new formula?

Work with a partner.

a.
- Write a formula for the volume V of a prism.
- Solve the formula for h.
- Use the new formula to find the height of the prism.

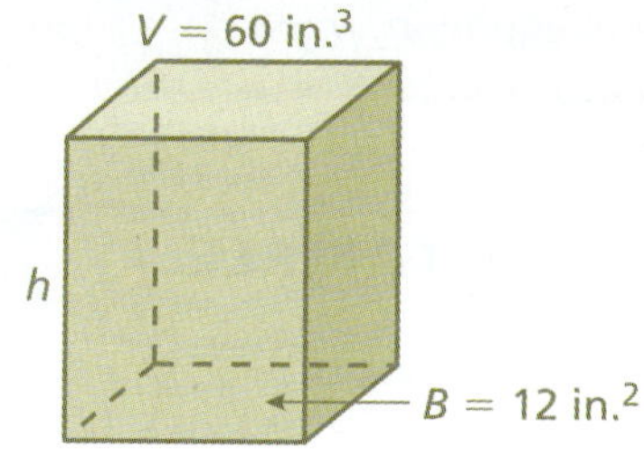

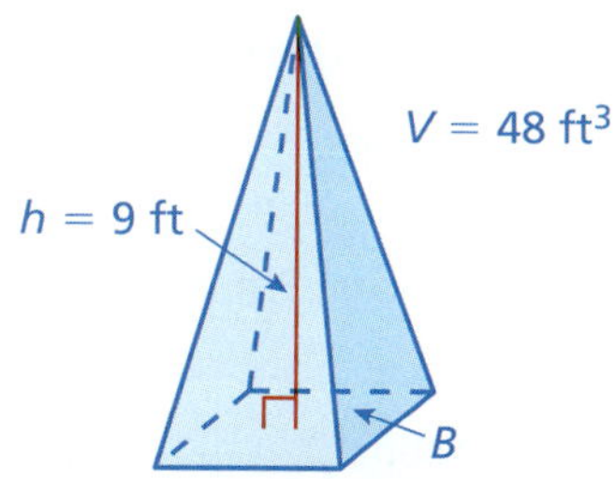

b.
- Write a formula for the volume V of a pyramid.
- Solve the formula for B.
- Use the new formula to find the area of the base of the pyramid.

c.
- Write a formula for the lateral surface area S of a cylinder.
- Solve the formula for h.
- Use the new formula to find the height of the cylinder.

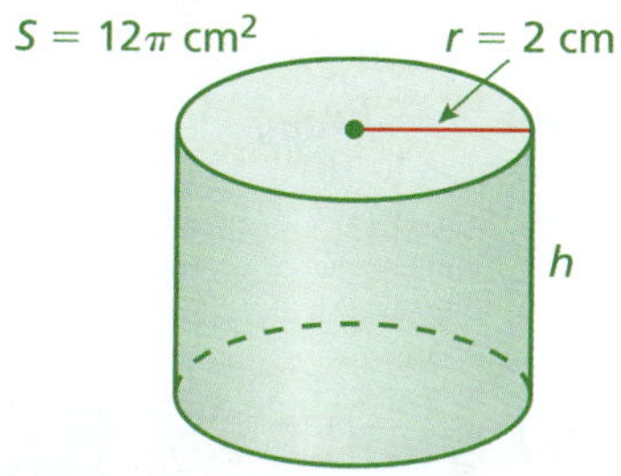

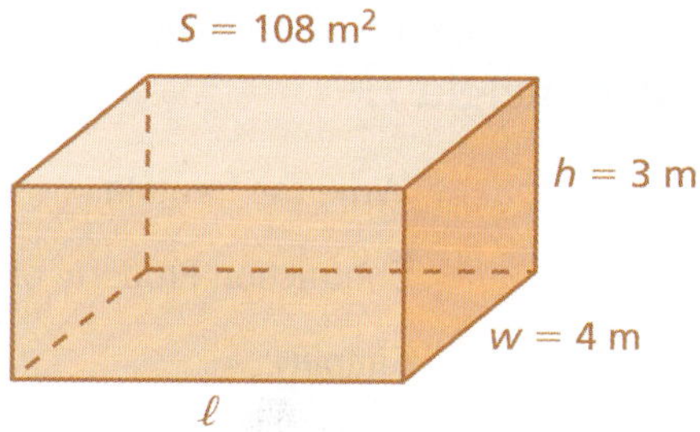

d.
- Write a formula for the surface area S of a rectangular prism.
- Solve the formula for ℓ.
- Use the new formula to find the length of the rectangular prism.

What Is Your Answer?

3. IN YOUR OWN WORDS How can you use a formula for one measurement to write a formula for a different measurement? Give an example that is different from the examples on these two pages.

Practice Use what you learned about rewriting equations and formulas to complete Exercises 3 and 4 on page 30.

1.4 Lesson

Check It Out
Lesson Tutorials
BigIdeasMath.com

Key Vocabulary
literal equation, *p. 28*

An equation that has two or more variables is called a **literal equation**. To rewrite a literal equation, solve for one variable in terms of the other variable(s).

EXAMPLE 1 Rewriting an Equation

Solve the equation $2y + 5x = 6$ for y.

	$2y + 5x = 6$	Write the equation.
Undo the addition.	$2y + 5x - 5x = 6 - 5x$	Subtraction Property of Equality
	$2y = 6 - 5x$	Simplify.
Undo the multiplication.	$\frac{2y}{2} = \frac{6 - 5x}{2}$	Division Property of Equality
	$y = 3 - \frac{5}{2}x$	Simplify.

On Your Own

Now You're Ready
Exercises 5–10

Solve the equation for y.

1. $5y - x = 10$
2. $4x - 4y = 1$
3. $12 = 6x + 3y$

EXAMPLE 2 Rewriting a Formula

Remember
A *formula* shows how one variable is related to one or more other variables. A formula is a type of literal equation.

The formula for the surface area S of a cone is $S = \pi r^2 + \pi r \ell$. Solve the formula for the slant height ℓ.

$S = \pi r^2 + \pi r \ell$	Write the formula.
$S - \pi r^2 = \pi r^2 - \pi r^2 + \pi r \ell$	Subtraction Property of Equality
$S - \pi r^2 = \pi r \ell$	Simplify.
$\frac{S - \pi r^2}{\pi r} = \frac{\pi r \ell}{\pi r}$	Division Property of Equality
$\frac{S - \pi r^2}{\pi r} = \ell$	Simplify.

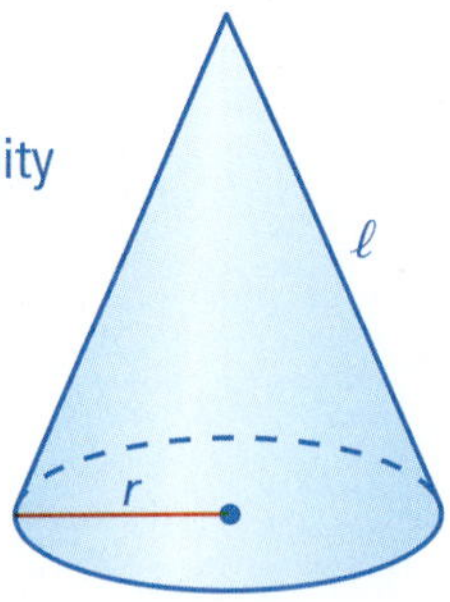

On Your Own

Now You're Ready
Exercises 14–19

Solve the formula for the red variable.

4. Area of rectangle: $A = bh$
5. Simple interest: $I = Prt$
6. Surface area of cylinder: $S = 2\pi r^2 + 2\pi rh$

Multi-Language Glossary at BigIdeasMath.com

Key Idea

Temperature Conversion

A formula for converting from degrees Fahrenheit F to degrees Celsius C is

$$C = \frac{5}{9}(F - 32).$$

EXAMPLE 3 Rewriting the Temperature Formula

Solve the temperature formula for F.

$C = \frac{5}{9}(F - 32)$ — Write the temperature formula.

Use the reciprocal. → $\frac{9}{5} \cdot C = \frac{9}{5} \cdot \frac{5}{9}(F - 32)$ — Multiplication Property of Equality

$\frac{9}{5}C = F - 32$ — Simplify.

Undo the subtraction. → $\frac{9}{5}C + 32 = F - 32 + 32$ — Addition Property of Equality

$\frac{9}{5}C + 32 = F$ — Simplify.

The rewritten formula is $F = \frac{9}{5}C + 32$.

EXAMPLE 4 Real-Life Application

Which has the greater temperature?

Convert the Celsius temperature of lightning to Fahrenheit.

$F = \frac{9}{5}C + 32$ — Write the rewritten formula from Example 3.

$= \frac{9}{5}(30{,}000) + 32$ — Substitute 30,000 for C.

$= 54{,}032$ — Simplify.

Because 54,032 °F is greater than 11,000 °F, lightning has the greater temperature.

On Your Own

7. Room temperature is considered to be 70 °F. Suppose the temperature is 23 °C. Is this greater than or less than room temperature?

1.4 Exercises

Check It Out
Help with Homework
BigIdeasMath.com

Vocabulary and Concept Check

1. **VOCABULARY** Is $-2x = \frac{3}{8}$ a literal equation? Explain.

2. **DIFFERENT WORDS, SAME QUESTION** Which is different? Find "both" answers.

Solve $4x - 2y = 6$ for y.	Solve $6 = 4x - 2y$ for y.
Solve $4x - 2y = 6$ for y in terms of x.	Solve $4x - 2y = 6$ for x in terms of y.

Practice and Problem Solving

3. **a.** Write a formula for the area A of a triangle.
   b. Solve the formula for b.
   c. Use the new formula to find the base of the triangle.

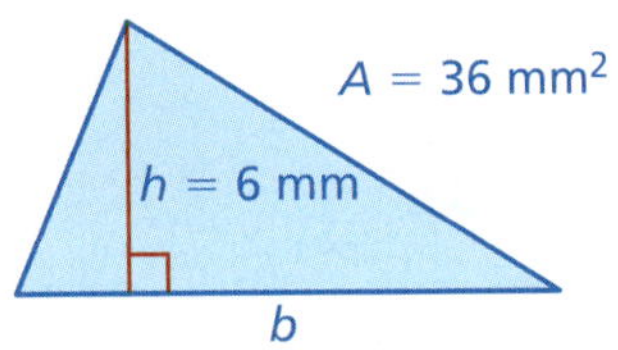

4. **a.** Write a formula for the volume V of a prism.
   b. Solve the formula for B.
   c. Use the new formula to find the area of the base of the prism.

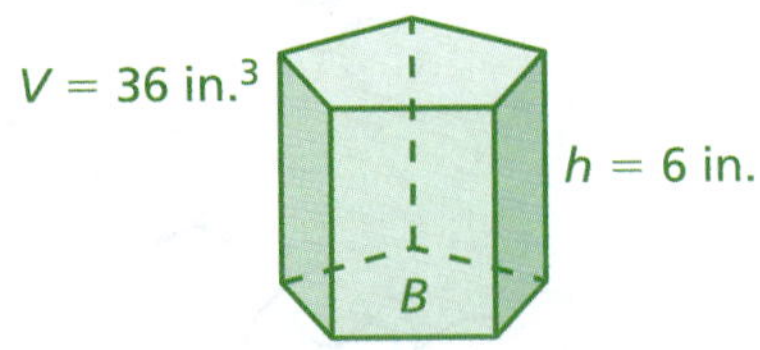

Solve the equation for *y*.

1 5. $\frac{1}{3}x + y = 4$

6. $3x + \frac{1}{5}y = 7$

7. $6 = 4x + 9y$

8. $\pi = 7x - 2y$

9. $4.2x - 1.4y = 2.1$

10. $6y - 1.5x = 8$

11. **ERROR ANALYSIS** Describe and correct the error in rewriting the equation.

✗
$2x - y = 5$
$y = -2x + 5$

12. **TEMPERATURE** The formula $K = C + 273.15$ converts temperatures from Celsius C to Kelvin K.
   a. Solve the formula for C.
   b. Convert 300 Kelvin to Celsius.

13. **INTEREST** The formula for simple interest is $I = Prt$.
   a. Solve the formula for t.
   b. Use the new formula to find the value of t in the table.

I	\$75
P	\$500
r	5%
t	

Solve the equation for the red variable.

2 **14.** $d = rt$

15. $e = mc^2$

16. $R - C = P$

17. $A = \frac{1}{2}\pi w^2 + 2\ell w$

18. $B = 3\frac{V}{h}$

19. $g = \frac{1}{6}(w + 40)$

20. LOGIC Why is it useful to rewrite a formula in terms of another variable?

21. REASONING The formula $K = \frac{5}{9}(F - 32) + 273.15$ converts temperatures from Fahrenheit F to Kelvin K.

a. Solve the formula for F.

b. The freezing point of water is 273.15 Kelvin. What is this temperature in Fahrenheit?

c. The temperature of dry ice is −78.5 °C. Which is colder, dry ice or liquid nitrogen?

Navy Pier Ferris Wheel

22. FERRIS WHEEL The Navy Pier Ferris Wheel in Chicago has a circumference that is 56% of the circumference of the first Ferris wheel built in 1893.

a. What is the radius of the Navy Pier Ferris Wheel?

b. What was the radius of the first Ferris wheel?

c. The first Ferris wheel took 9 minutes to make a complete revolution. How fast was the wheel moving?

23. Repeated Reasoning The formula for the volume of a sphere is $V = \frac{4}{3}\pi r^3$. Solve the formula for r^3. Use Guess, Check, and Revise to find the radius of the sphere.

$V = 381.51$ in.^{3} r

Fair Game Review What you learned in previous grades & lessons

Multiply. *(Skills Review Handbook)*

24. $5 \times \frac{3}{4}$

25. $-2 \times \frac{8}{3}$

26. $\frac{1}{4} \times \frac{3}{2} \times \frac{8}{9}$

27. $25 \times \frac{3}{5} \times \frac{1}{12}$

28. MULTIPLE CHOICE Which of the following is not equivalent to $\frac{3}{4}$? *(Skills Review Handbook)*

Ⓐ 0.75 Ⓑ 3:4 Ⓒ 75% Ⓓ 4:3

1.3–1.4 Quiz

Check It Out
Progress Check
BigIdeasMath.com

Solve the equation. Check your solution, if possible. *(Section 1.3)*

1. $2(x + 4) = -5x + 1$

2. $\frac{1}{2}s = 4s - 21$

3. $8.3z = 4.1z + 10.5$

4. $3(b + 5) = 4(2b - 5)$

5. $n + 7 - n = 4$

6. $\frac{1}{4}(4r - 8) = r - 2$

Solve the equation for *y*. *(Section 1.4)*

7. $6x - 3y = 9$

8. $8 = 2y - 10x$

Solve the formula for the red variable. *(Section 1.4)*

9. Volume of a cylinder: $V = \pi r^2 h$

10. Area of a trapezoid: $A = \frac{1}{2}h(b + B)$

11. TEMPERATURE In which city is the water temperature higher? *(Section 1.4)*

12. SAVINGS ACCOUNT You begin with \$25 in a savings account and \$50 in a checking account. Each week you deposit \$5 into savings and \$10 into checking. After how many weeks is the amount in checking twice the amount in savings? *(Section 1.3)*

13. INTEREST The formula for simple interest I is $I = Prt$. Solve the formula for the interest rate r. What is the interest rate r if the principal P is \$1500, the time t is 2 years, and the interest earned I is \$90? *(Section 1.4)*

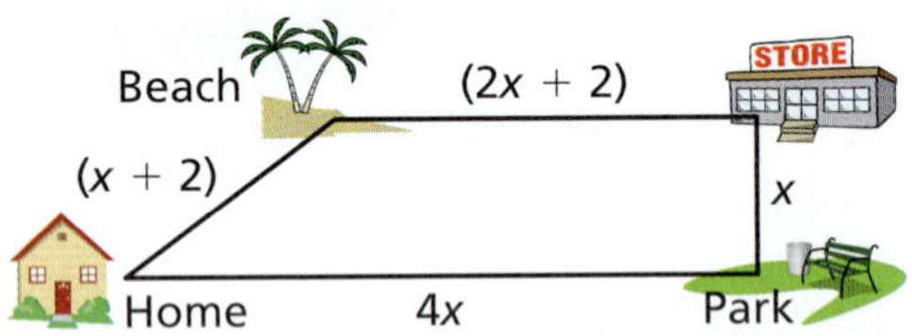

14. ROUTES From your home, the route to the store that passes the beach is 2 miles shorter than the route to the store that passes the park. What is the length of each route? *(Section 1.3)*

15. PERIMETER Use the triangle shown. *(Section 1.4)*

a. Write a formula for the perimeter P of the triangle.

b. Solve the formula for b.

c. Use the new formula to find b when a is 10 feet and c is 17 feet.

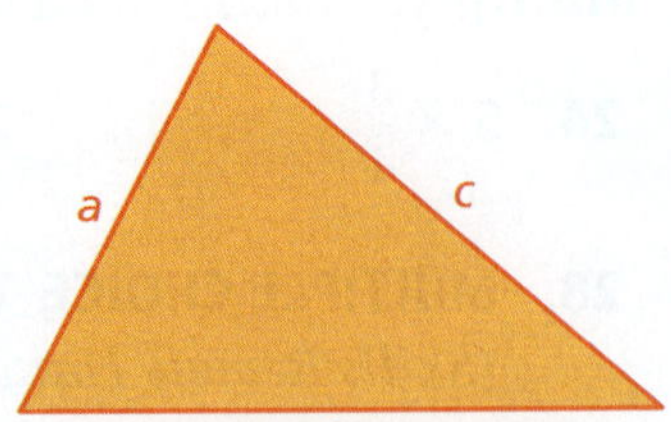

Perimeter = 42 feet

1 Chapter Review

Check It Out
Vocabulary Help
BigIdeasMath.com

Review Key Vocabulary

literal equation, *p. 28*

Review Examples and Exercises

1.1 Solving Simple Equations *(pp. 2–9)*

The *boiling point* of a liquid is the temperature at which the liquid becomes a gas. The boiling point of mercury is about $\frac{41}{200}$ of the boiling point of lead. Write and solve an equation to find the boiling point of lead.

Let x be the boiling point of lead.

$$\frac{41}{200}x = 357 \quad \text{Write the equation.}$$

$$\frac{200}{41} \cdot \left(\frac{41}{200}x\right) = \frac{200}{41} \cdot 357 \quad \text{Multiplication Property of Equality}$$

$$x \approx 1741 \quad \text{Simplify.}$$

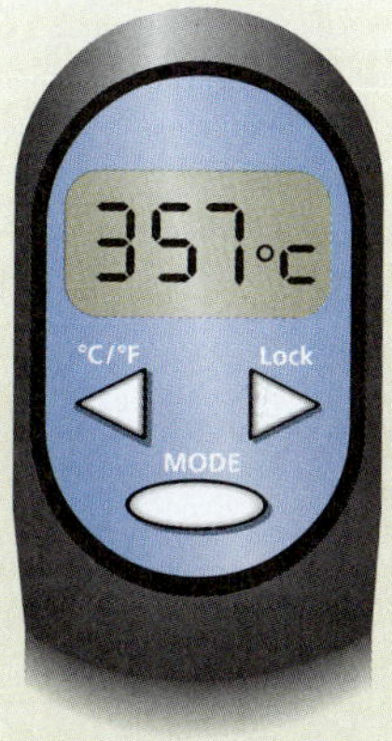

Mercury 357°C

The boiling point of lead is about 1741°C.

Exercises

Solve the equation. Check your solution.

1. $y + 8 = -11$
2. $3.2 = -0.4n$
3. $-\frac{t}{4} = -3\pi$

1.2 Solving Multi-Step Equations *(pp. 10–15)*

Solve $-14x + 28 + 6x = -44$.

$$-14x + 28 + 6x = -44 \quad \text{Write the equation.}$$

$$-8x + 28 = -44 \quad \text{Combine like terms.}$$

$$\underline{-28} \quad \underline{-28} \quad \text{Subtraction Property of Equality}$$

$$-8x = -72 \quad \text{Simplify.}$$

$$\frac{-8x}{-8} = \frac{-72}{-8} \quad \text{Division Property of Equality}$$

$$x = 9 \quad \text{Simplify.}$$

The solution is $x = 9$.

Exercises

Find the value of x. Then find the angle measures of the polygon.

4.

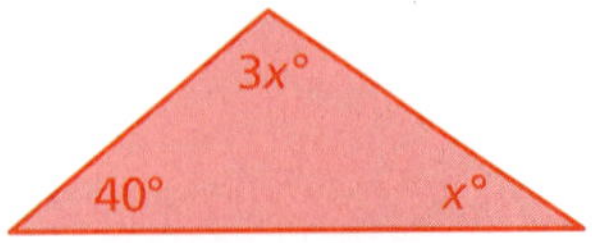

Sum of angle measures: 180°

5.

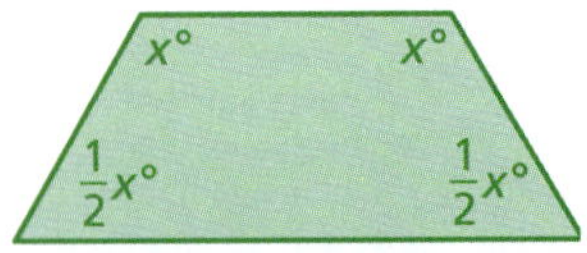

Sum of angle measures: 360°

6.

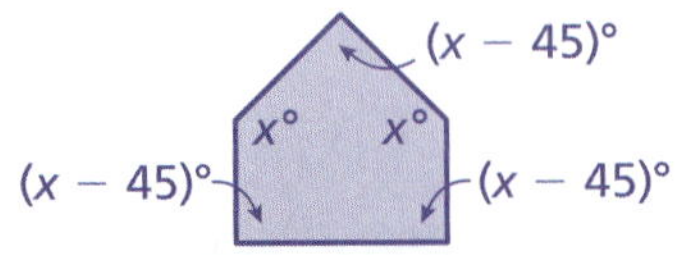

Sum of angle measures: 540°

1.3 Solving Equations with Variables on Both Sides *(pp. 18–25)*

a. Solve $3(x - 4) = -2(4 - x)$.

$3(x - 4) = -2(4 - x)$	Write the equation.
$3x - 12 = -8 + 2x$	Distributive Property
$-2x \qquad -2x$	Subtraction Property of Equality
$x - 12 = -8$	Simplify.
$+12 \qquad +12$	Addition Property of Equality
$x = 4$	Simplify.

The solution is $x = 4$.

b. Solve $4 - 5k = -8 - 5k$.

$4 - 5k = -8 - 5k$	Write the equation.
$+5k \qquad +5k$	Addition Property of Equality
$4 = -8$ ✗	Simplify.

The equation $4 = -8$ is never true. So, the equation has no solution.

c. Solve $2\left(7g + \frac{2}{3}\right) = 14g + \frac{4}{3}$.

$2\left(7g + \frac{2}{3}\right) = 14g + \frac{4}{3}$	Write the equation.
$14g + \frac{4}{3} = 14g + \frac{4}{3}$	Distributive Property
$-14g \qquad -14g$	Subtraction Property of Equality
$\frac{4}{3} = \frac{4}{3}$	Simplify.

The equation $\frac{4}{3} = \frac{4}{3}$ is always true. So, the equation has infinitely many solutions.

Exercises

Solve the equation. Check your solution, if possible.

7. $5m - 1 = 4m + 5$

8. $3(5p - 3) = 5(p - 1)$

9. $\frac{2}{5}n + \frac{1}{10} = \frac{1}{2}(n + 4)$

10. $7t + 3 = 8 + 7t$

11. $\frac{1}{5}(15b - 7) = 3b - 9$

12. $\frac{1}{6}(12z - 18) = 2z - 3$

1.4 Rewriting Equations and Formulas *(pp. 26–31)*

a. Solve $7y + 6x = 4$ for y.

$7y + 6x = 4$	Write the equation.
$7y + 6x - 6x = 4 - 6x$	Subtraction Property of Equality
$7y = 4 - 6x$	Simplify.
$\frac{7y}{7} = \frac{4 - 6x}{7}$	Division Property of Equality
$y = \frac{4}{7} - \frac{6}{7}x$	Simplify.

b. The equation for a line in slope-intercept form is $y = mx + b$. Solve the equation for x.

$y = mx + b$	Write the equation.
$y - b = mx + b - b$	Subtraction Property of Equality
$y - b = mx$	Simplify.
$\frac{y - b}{m} = \frac{mx}{m}$	Division Property of Equality
$\frac{y - b}{m} = x$	Simplify.

Exercises

Solve the equation for y.

13. $6y + x = 8$

14. $10x - 5y = 15$

15. $20 = 5x + 10y$

16. **a.** The formula $F = \frac{9}{5}(K - 273.15) + 32$ converts a temperature from Kelvin K to Fahrenheit F. Solve the formula for K.

b. Convert 240 °F to Kelvin K. Round your answer to the nearest hundredth.

17. **a.** Write the formula for the area A of a trapezoid.

b. Solve the formula for h.

c. Use the new formula to find the height h of the trapezoid.

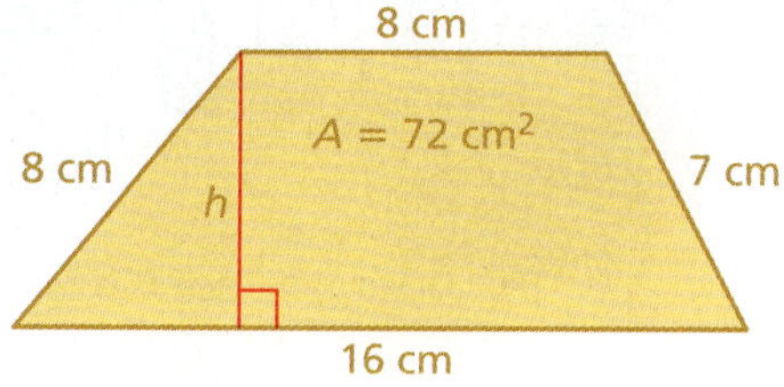

1 Chapter Test

Check It Out
Test Practice
BigIdeasMath.com

Solve the equation. Check your solution, if possible.

1. $4 + y = 9.5$

2. $-\frac{x}{9} = -8$

3. $z - \frac{2}{3} = \frac{1}{8}$

4. $3.8n - 13 = 1.4n + 5$

5. $9(8d - 5) + 13 = 12d - 2$

6. $9j - 8 = 8 + 9j$

7. $2.5(2p + 5) = 5p + 12.5$

8. $\frac{3}{4}t + \frac{1}{8} = \frac{3}{4}(t + 8)$

9. $\frac{1}{7}(14r + 28) = 2(r + 2)$

Find the value of *x*. Then find the angle measures of the polygon.

10.

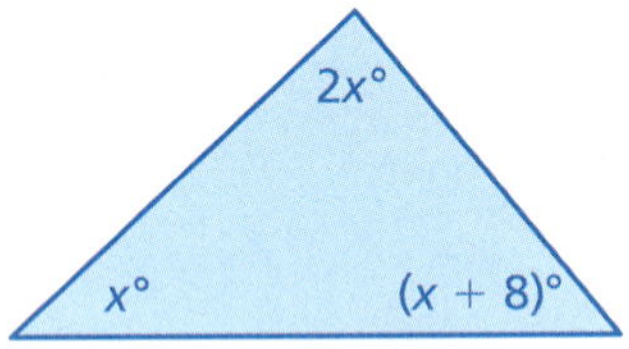

Sum of angle measures: 180°

11.

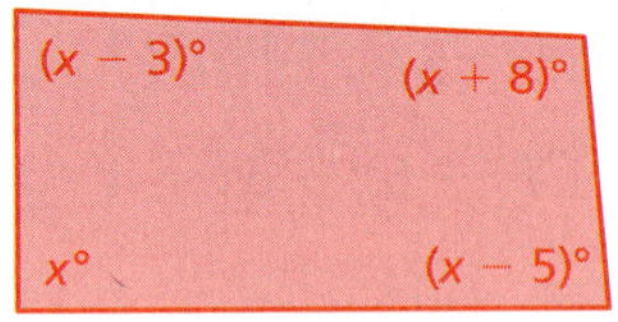

Sum of angle measures: 360°

Solve the equation for *y*.

12. $1.2x - 4y = 28$

13. $0.5 = 0.4y - 0.25x$

Solve the formula for the red variable.

14. Perimeter of a rectangle: $P = 2\ell + 2w$

15. Distance formula: $d = rt$

16. BASKETBALL Your basketball team wins a game by 13 points. The opposing team scores 72 points. Explain how to find your team's score.

17. CYCLING You are biking at a speed of 18 miles per hour. You are 3 miles behind your friend, who is biking at a speed of 12 miles per hour. Write and solve an equation to find the amount of time it takes for you to catch up to your friend.

18. VOLCANOES Two scientists are measuring lava temperatures. One scientist records a temperature of 1725°F. The other scientist records a temperature of 950°C. Which is the greater temperature? $\left(\text{Use } C = \frac{5}{9}(F - 32).\right)$

19. JOBS Your profit for mowing lawns this week is \$24. You are paid \$8 per hour and you paid \$40 for gas for the lawn mower. How many hours did you work this week?

1 Standards Assessment

Test-Taking Strategy
Solve Directly or Eliminate Choices

"You can eliminate A and D. Then, solve directly to determine that the correct answer is B."

1. Which value of x makes the equation true? *(MACC.8.EE.3.7b)*

$$4x = 32$$

A. 8 **C.** 36

B. 28 **D.** 128

2. A taxi ride costs \$3 plus \$2 for each mile driven. When you rode in a taxi, the total cost was \$39. This can be modeled by the equation below, where m represents the number of miles driven.

$$2m + 3 = 39$$

How long was your taxi ride? *(MACC.8.EE.3.7b)*

F. 72 mi **H.** 21 mi

G. 34 mi **I.** 18 mi

3. Which of the following equations has exactly one solution? *(MACC.8.EE.3.7a)*

A. $\frac{2}{3}(x + 6) = \frac{2}{3}x + 4$ **C.** $\frac{4}{5}\left(n + \frac{1}{3}\right) = \frac{4}{5}n + \frac{1}{3}$

B. $\frac{3}{7}y + 13 = 13 - \frac{3}{7}y$ **D.** $\frac{7}{8}\left(2t + \frac{1}{8}\right) = \frac{7}{4}t$

4. The perimeter of the square is equal to the perimeter of the triangle. What are the side lengths of the square? *(MACC.8.EE.3.7a)*

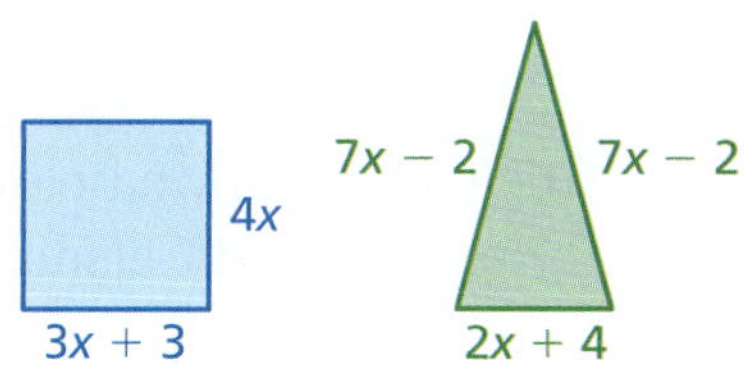

5. The formula below relates distance, rate, and time.

$$d = rt$$

Solve this formula for t. *(MACC.8.EE.3.7b)*

F. $t = dr$ **H.** $t = d - r$

G. $t = \frac{d}{r}$ **I.** $t = \frac{r}{d}$

6. What could be the first step to solve the equation shown below? *(MACC.8.EE.3.7b)*

$$3x + 5 = 2(x + 7)$$

A. Combine $3x$ and 5.

B. Multiply x by 2 and 7 by 2.

C. Subtract x from $3x$.

D. Subtract 5 from 7.

7. You work as a sales representative. You earn \$400 per week plus 5% of your total sales for the week. *(MACC.8.EE.3.7b)*

Think
Solve
Explain

Part A Last week, you had total sales of \$5000. Find your total earnings. Show your work.

Part B One week, you earned \$1350. Let s represent your total sales that week. Write an equation that you could use to find s.

Part C Using your equation from Part B, find s. Show all steps clearly.

8. In 10 years, Maria will be 39 years old. Let m represent Maria's age today. Which equation can you use to find m? *(MACC.8.EE.3.7b)*

F. $m = 39 + 10$

G. $m - 10 = 39$

H. $m + 10 = 39$

I. $10m = 39$

9. Which value of y makes the equation below true? *(MACC.8.EE.3.7b)*

$$3y + 8 = 7y + 11$$

A. -4.75

B. -0.75

C. 0.75

D. 4.75

10. The equation below is used to convert a Fahrenheit temperature F to its equivalent Celsius temperature C.

$$C = \frac{5}{9}(F - 32)$$

Which formula can be used to convert a Celsius temperature to its equivalent Fahrenheit temperature? *(MACC.8.EE.3.7b)*

F. $F = \frac{5}{9}(C - 32)$

G. $F = \frac{9}{5}(C + 32)$

H. $F = \frac{9}{5}C + \frac{32}{5}$

I. $F = \frac{9}{5}C + 32$

11. You have already saved \$35 for a new cell phone. You need \$175 in all. You think you can save \$10 per week. At this rate, how many more weeks will you need to save money before you can buy the new cell phone? *(MACC.8.EE.3.7b)*

12. What is the greatest angle measure in the triangle below? *(MACC.8.EE.3.7b)*

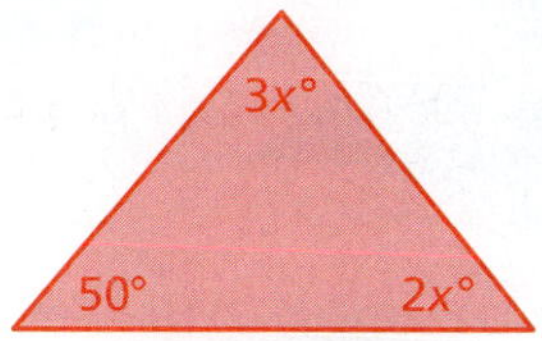

Sum of angle measures: 180°

A. 26°

B. 78°

C. 108°

D. 138°

13. Which value of x makes the equation below true? *(MACC.8.EE.3.7b)*

$$6(x - 3) = 4x - 7$$

F. -5.5

G. -2

H. 1.1

I. 5.5

14. The drawing below shows equal weights on two sides of a balance scale.

What can you conclude from the drawing? *(MACC.8.EE.3.7b)*

A. A mug weighs one-third as much as a trophy.

B. A mug weighs one-half as much as a trophy.

C. A mug weighs twice as much as a trophy.

D. A mug weighs three times as much as a trophy.

2 Transformations

"Just 2 more minutes. I'm almost done with my 'cat tessellation' painting."

"If you hold perfectly still..." "...each frame becomes a horizontal..." "...translation of the previous frame..."

What You Learned Before

Reflecting Points
(MACC.6.NS.3.6b)

Example 1 **Reflect (3, −4) in the *x*-axis.**

Plot (3, −4).

To reflect (3, −4) in the *x*-axis, use the same *x*-coordinate, 3, and take the opposite of the *y*-coordinate. The opposite of −4 is 4.

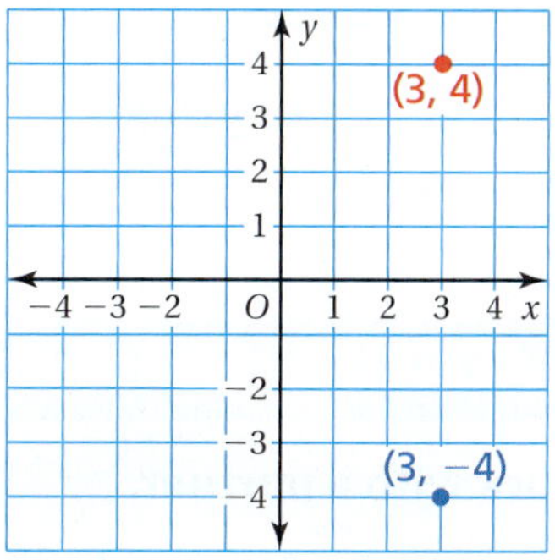

∴ So, the reflection of (3, −4) in the *x*-axis is (3, 4).

Try It Yourself

Reflect the point in (a) the *x*-axis and (b) the *y*-axis.

1. (7, 3) **2.** (−4, 6) **3.** (5, −5) **4.** (−8, −3)

5. (0, 1) **6.** (−5, 0) **7.** (4, −6.5) **8.** $\left(-3\frac{1}{2}, -4\right)$

Drawing a Polygon in a Coordinate Plane (MACC.6.G.1.3)

Example 2 **The vertices of a quadrilateral are *A*(1, 5), *B*(2, 9), *C*(6, 8), and *D*(8, 1). Draw the quadrilateral in a coordinate plane.**

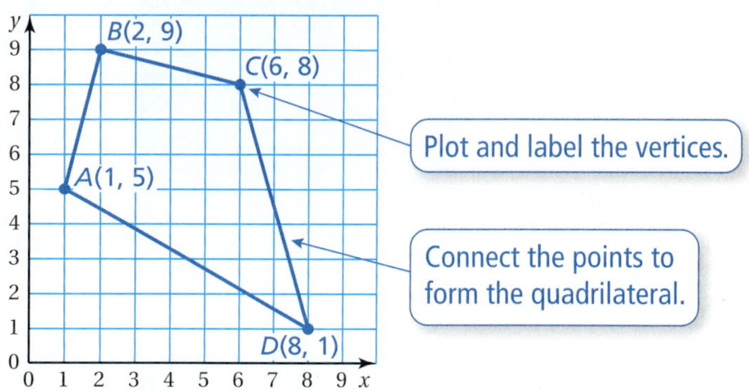

Try It Yourself

Draw the polygon with the given vertices in a coordinate plane.

9. *J*(1, 1), *K*(5, 6), *M*(9, 3) **10.** *Q*(2, 3), *R*(2, 8), *S*(7, 8), *T*(7, 3)

2.1 Congruent Figures

Essential Question How can you identify congruent triangles?

Two figures are congruent when they have the same size and the same shape.

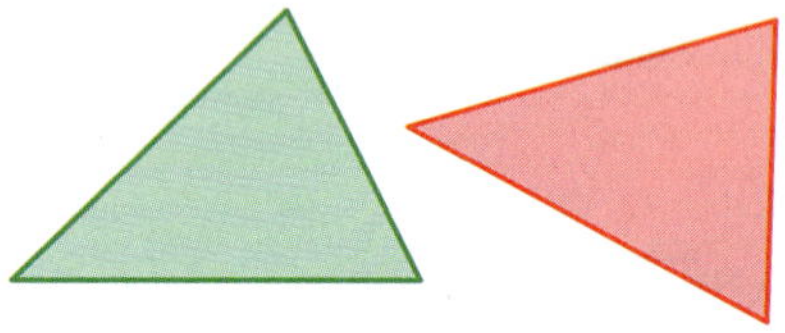

Congruent
Same size *and* shape

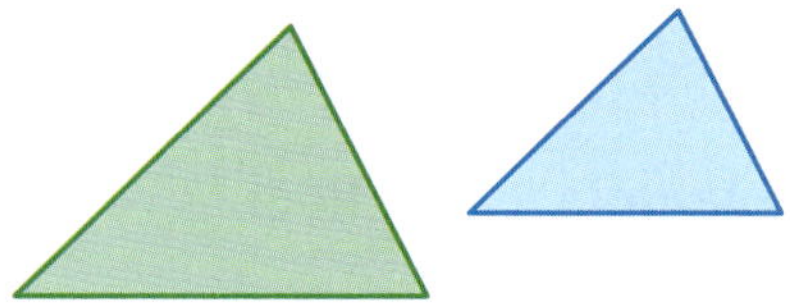

Not Congruent
Same shape, but not same size

1 ACTIVITY: Identifying Congruent Triangles

Work with a partner.

- **Which of the geoboard triangles below are congruent to the geoboard triangle at the right?**
- **Form each triangle on a geoboard.**
- **Measure each side with a ruler. Record your results in a table.**
- **Write a conclusion about the side lengths of triangles that are congruent.**

a.

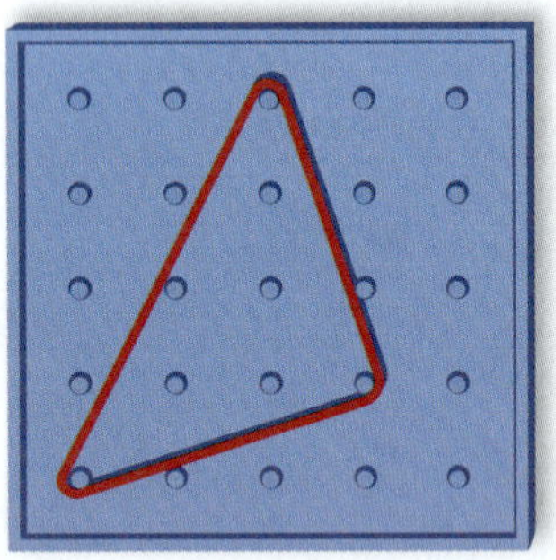

b.

c.

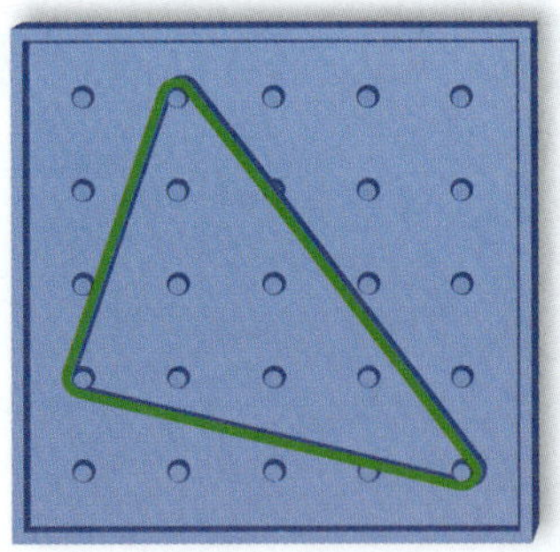

d.

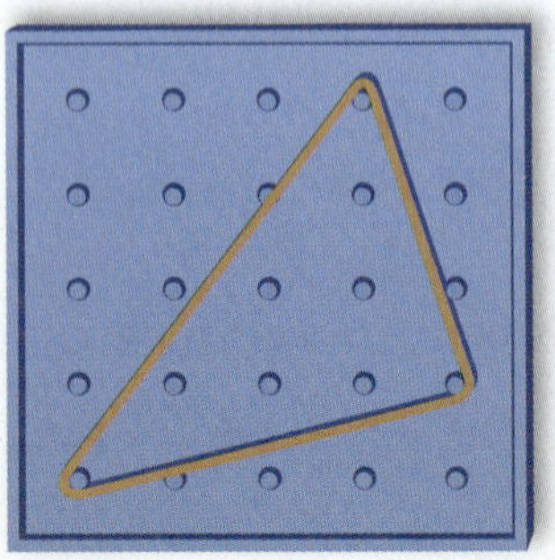

e.

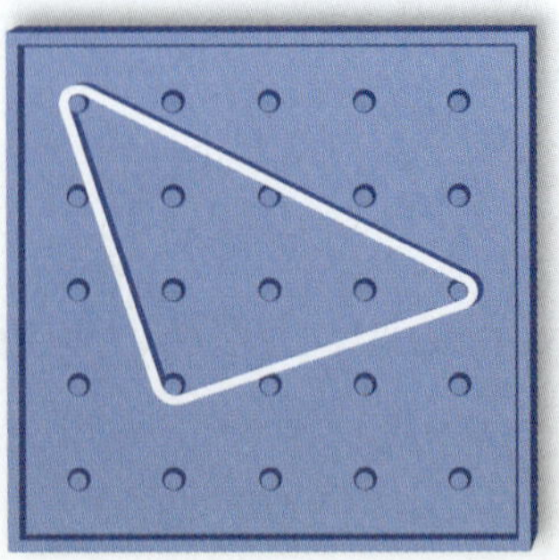

f.

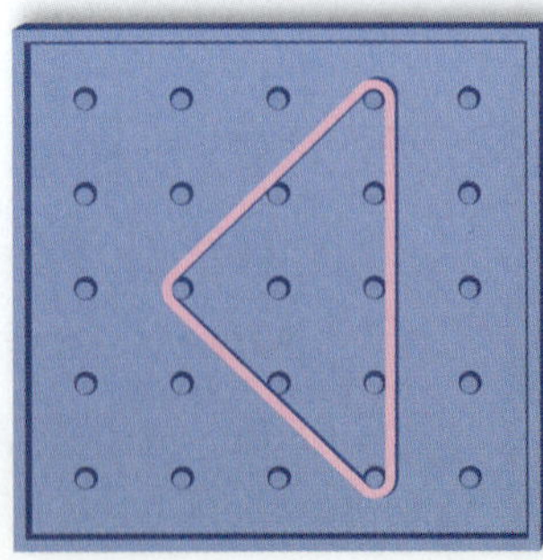

COMMON CORE

Geometry

In this lesson, you will

- name corresponding angles and corresponding sides of congruent figures.
- identify congruent figures.

Preparing for Standard MACC.8.G.1.2

Math Practice 5

Recognize Usefulness of Tools

What are some advantages and disadvantages of using a geoboard to construct congruent triangles?

The geoboard at the right shows three congruent triangles.

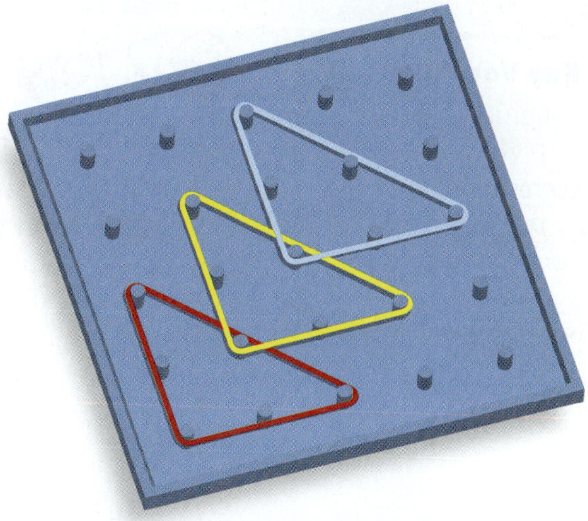

2 ACTIVITY: Forming Congruent Triangles

Work with a partner.

a. Form the yellow triangle in Activity 1 on your geoboard. Record the triangle on geoboard dot paper.

b. Move each vertex of the triangle one peg to the right. Is the new triangle congruent to the original triangle? How can you tell?

c. On a 5-by-5 geoboard, make as many different triangles as possible, each of which is congruent to the yellow triangle in Activity 1. Record each triangle on geoboard dot paper.

What Is Your Answer?

3. **IN YOUR OWN WORDS** How can you identify congruent triangles? Use the conclusion you wrote in Activity 1 as part of your answer.

4. Can you form a triangle on your geoboard whose side lengths are 3, 4, and 5 units? If so, draw such a triangle on geoboard dot paper.

Practice Use what you learned about congruent triangles to complete Exercises 4 and 5 on page 46.

Check It Out
Lesson Tutorials
BigIdeasMath.com

Key Vocabulary
congruent figures, *p. 44*
corresponding angles, *p. 44*
corresponding sides, *p. 44*

Key Idea

Congruent Figures

Figures that have the same size and the same shape are called **congruent figures**. The triangles below are congruent.

EXAMPLE 1 Naming Corresponding Parts

The figures are congruent. Name the corresponding angles and the corresponding sides.

Corresponding Angles	*Corresponding Sides*
$\angle A$ and $\angle W$	Side AB and Side WX
$\angle B$ and $\angle X$	Side BC and Side XY
$\angle C$ and $\angle Y$	Side CD and Side YZ
$\angle D$ and $\angle Z$	Side AD and Side WZ

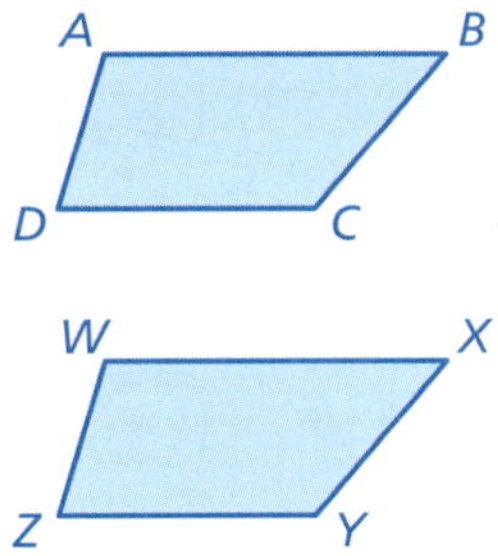

On Your Own

Now You're Ready
Exercises 6 and 7

1. The figures are congruent. Name the corresponding angles and the corresponding sides.

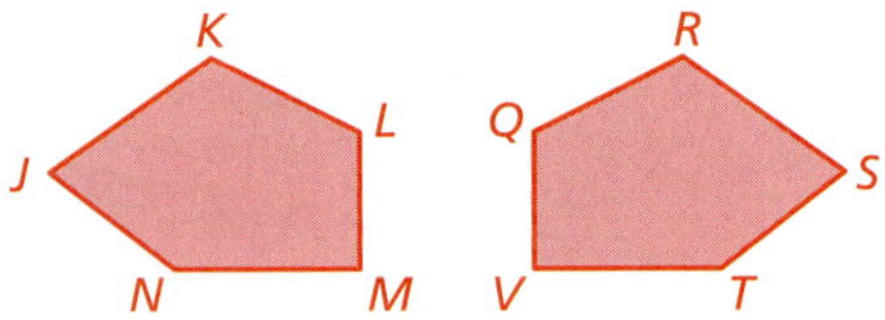

Key Idea

Identifying Congruent Figures

Two figures are congruent when corresponding angles and corresponding sides are congruent.

Triangle ABC is congruent to Triangle DEF.

$\triangle ABC \cong \triangle DEF$

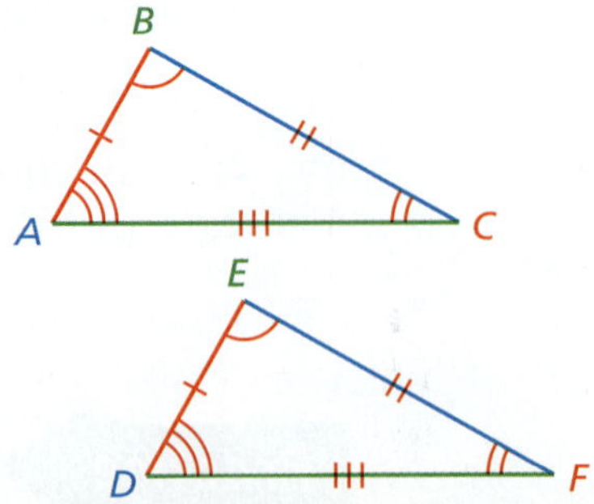

Reading

The symbol $\cong$ means *is congruent to*.

Multi-Language Glossary at BigIdeasMath.com

EXAMPLE 2 Identifying Congruent Figures

Which square is congruent to Square A?

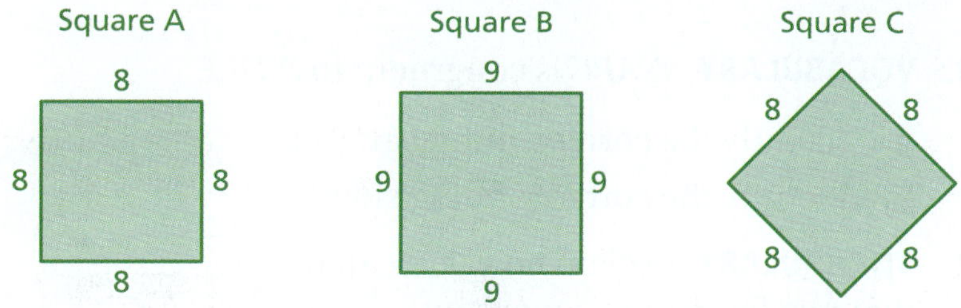

Each square has four right angles. So, corresponding angles are congruent. Check to see if corresponding sides are congruent.

Square A and Square B

Each side length of Square A is 8, and each side length of Square B is 9. So, corresponding sides are not congruent.

Square A and Square C

Each side length of Square A and Square C is 8. So, corresponding sides are congruent.

So, Square C is congruent to Square A.

EXAMPLE 3 Using Congruent Figures

Trapezoids *ABCD* and *JKLM* are congruent.

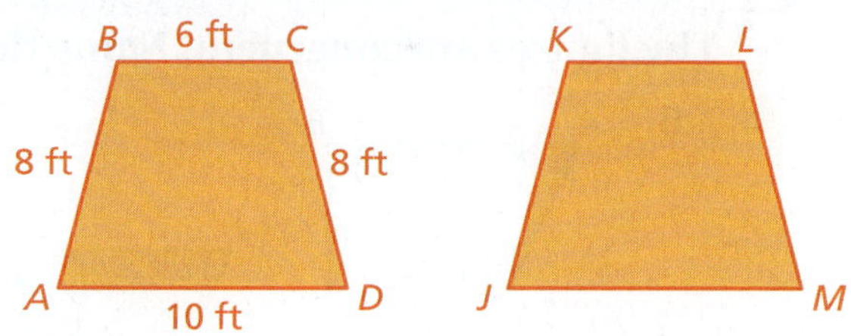

a. What is the length of side *JM*?

Side *JM* corresponds to side *AD*.

So, the length of side *JM* is 10 feet.

b. What is the perimeter of *JKLM*?

The perimeter of *ABCD* is 10 + 8 + 6 + 8 = 32 feet. Because the trapezoids are congruent, their corresponding sides are congruent.

So, the perimeter of *JKLM* is also 32 feet.

On Your Own

Now You're Ready
Exercises 8, 9, and 12

2. Which square in Example 2 is congruent to Square D?

3. In Example 3, which angle of *JKLM* corresponds to $\angle C$? What is the length of side *KJ*?

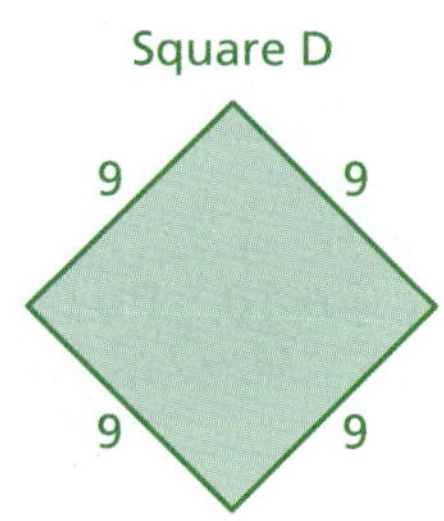

2.1 Exercises

Check It Out
Help with Homework
BigIdeasMath.com

Vocabulary and Concept Check

1. **VOCABULARY** $\triangle ABC$ is congruent to $\triangle DEF$.

   a. Identify the corresponding angles.

   b. Identify the corresponding sides.

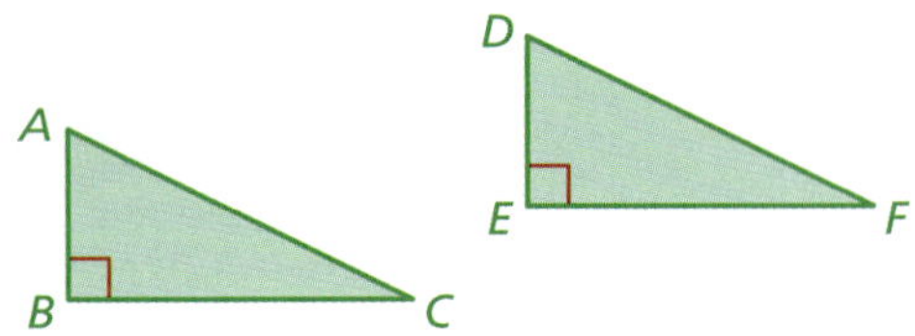

2. **VOCABULARY** Explain how you can tell that two figures are congruent.

3. **WHICH ONE DOESN'T BELONG?** Which one does *not* belong with the other three? Explain your reasoning.

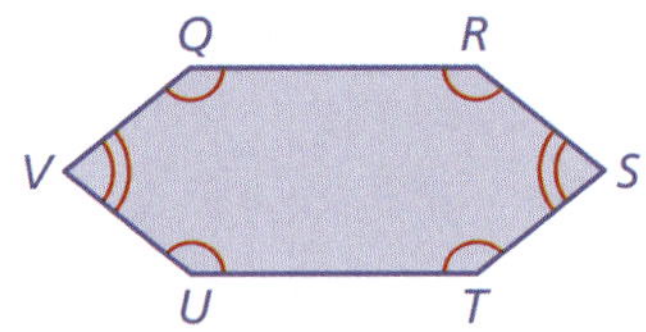

$\angle R$ $\angle U$ $\angle V$ $\angle Q$

Practice and Problem Solving

Tell whether the triangles are *congruent* or *not congruent*.

4.

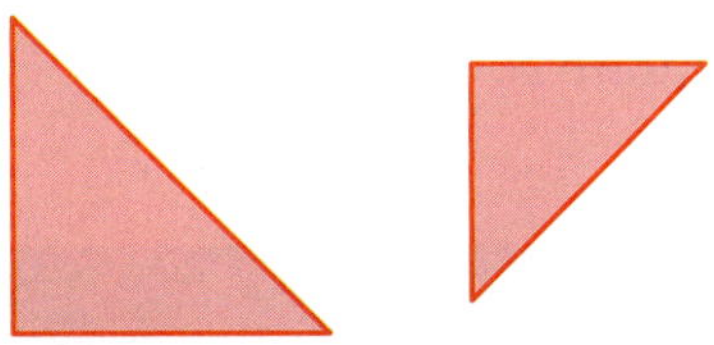

5.

The figures are congruent. Name the corresponding angles and the corresponding sides.

① 6.

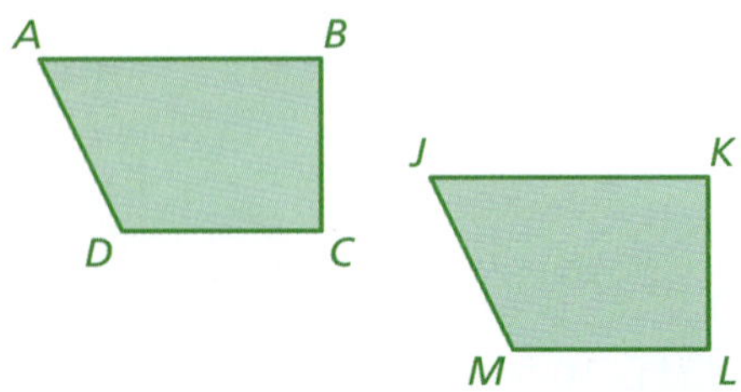

7.

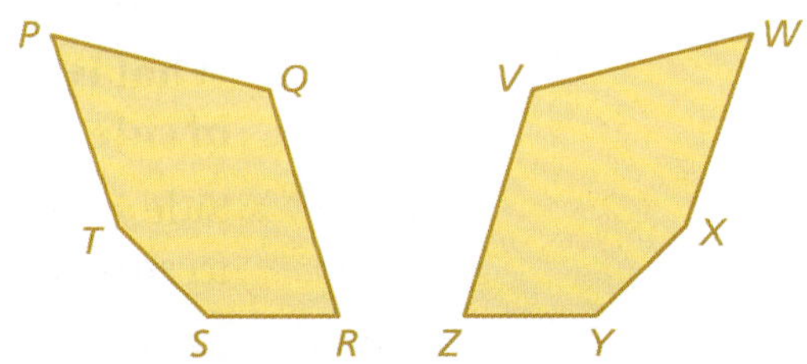

Tell whether the two figures are congruent. Explain your reasoning.

② 8.

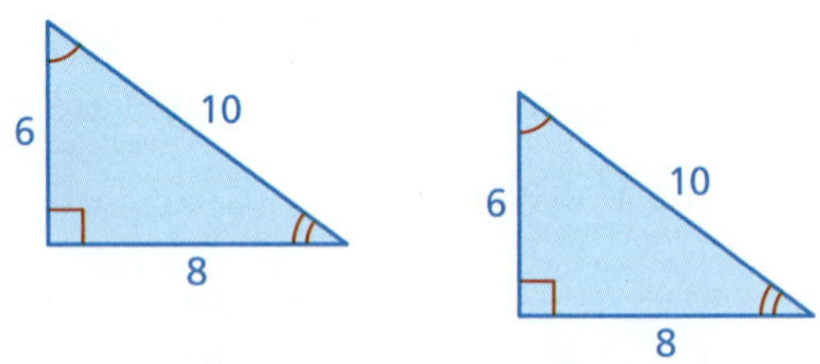

9.

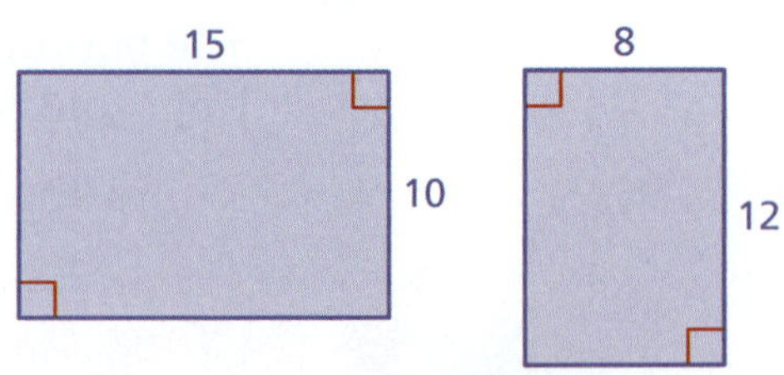

10. **PUZZLE** Describe the relationship between the unfinished puzzle and the missing piece.

11. ERROR ANALYSIS Describe and correct the error in telling whether the two figures are congruent.

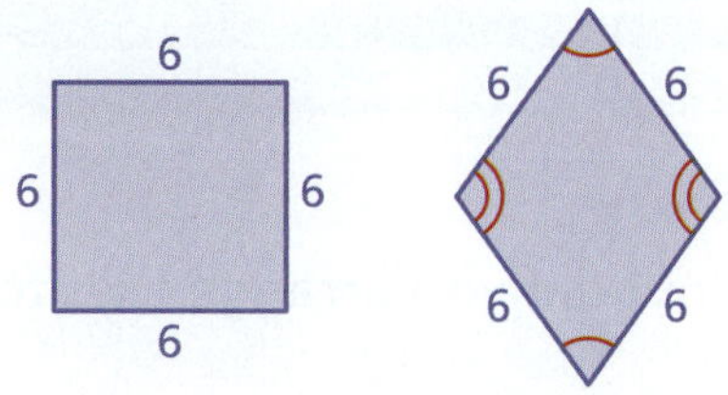

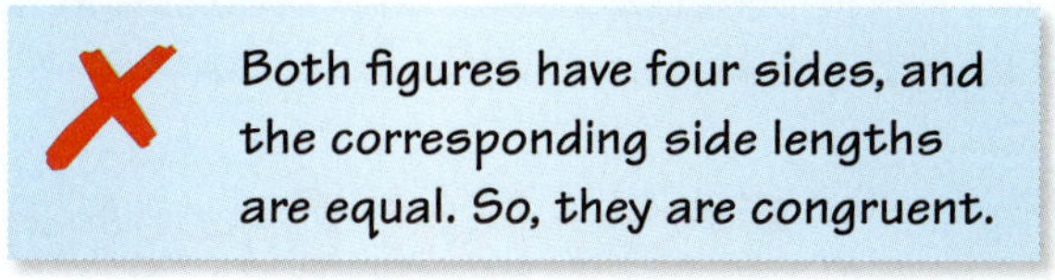

12. HOUSES The fronts of the houses are identical.

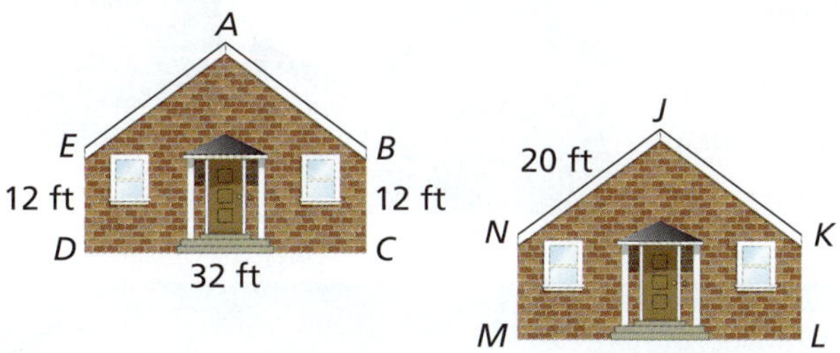

a. What is the length of side LM?

b. Which angle of $JKLMN$ corresponds to $\angle D$?

c. Side AB is congruent to side AE. What is the length of side AB?

d. What is the perimeter of $ABCDE$?

13. REASONING Here are two ways to draw *one* line to divide a rectangle into two congruent figures. Draw three other ways.

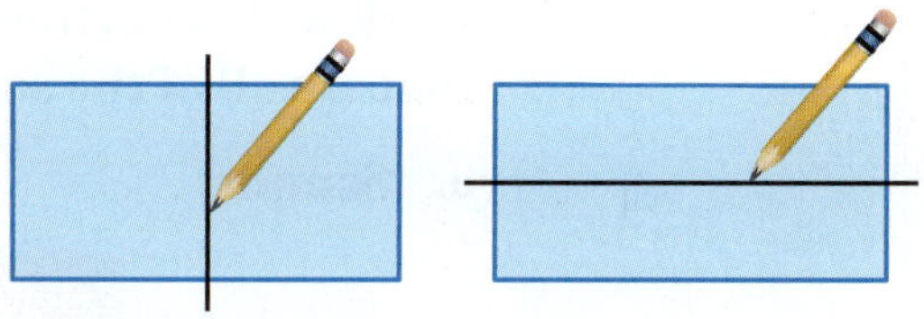

14. CRITICAL THINKING Are the areas of two congruent figures equal? Explain. Draw a diagram to support your answer.

15. True or False? The trapezoids are congruent. Determine whether the statement is *true* or *false*. Explain your reasoning.

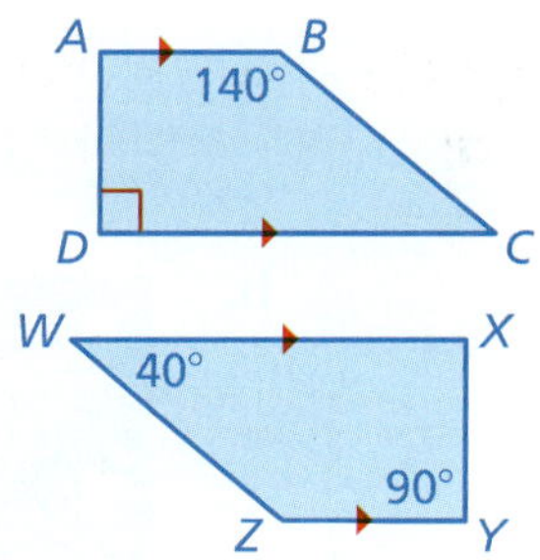

a. Side AB is congruent to side YZ.

b. $\angle A$ is congruent to $\angle X$.

c. $\angle A$ corresponds to $\angle X$.

d. The sum of the angle measures of $ABCD$ is 360°.

Fair Game Review What you learned in previous grades & lessons

Plot and label the ordered pair in a coordinate plane. *(Skills Review Handbook)*

16. $A(5, 3)$ **17.** $B(4, -1)$ **18.** $C(-2, 6)$ **19.** $D(-4, -2)$

20. MULTIPLE CHOICE You have 2 quarters and 5 dimes in your pocket. Write the ratio of quarters to the total number of coins. *(Skills Review Handbook)*

Ⓐ $\frac{2}{5}$ Ⓑ 2 : 7 Ⓒ 5 to 7 Ⓓ $\frac{7}{2}$

2.2 Translations

Essential Question How can you arrange tiles to make a tessellation?

The Meaning of a Word ● Translate

When you **translate** a tile, you slide it from one place to another.

When tiles cover a floor with no empty spaces, the collection of tiles is called a *tessellation*.

1 ACTIVITY: Describing Tessellations

Work with a partner. Can you make the tessellation by translating single tiles that are all of the same shape and design? If so, show how.

a. Sample: Tile Pattern Single Tiles

b.

c.

Common Core

Geometry

In this lesson, you will

- identify translations.
- translate figures in the coordinate plane.

Learning Standards
MACC.8.G.1.1
MACC.8.G.1.2
MACC.8.G.1.3

2 ACTIVITY: Tessellations and Basic Shapes

Work with a partner.

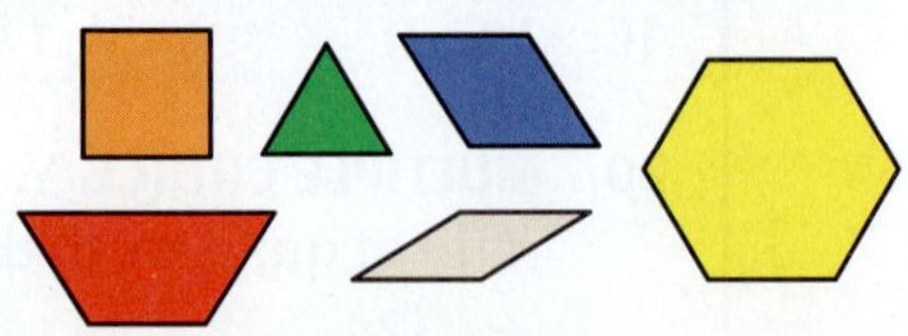

a. Which pattern blocks can you use to make a tessellation? For each one that works, draw the tessellation.

b. Can you make the tessellation by translating? Or do you have to rotate or flip the pattern blocks?

3 ACTIVITY: Designing Tessellations

Work with a partner. Design your own tessellation. Use one of the basic shapes from Activity 2.

Sample:

Step 1: Start with a square.

Step 2: Cut a design out of one side.

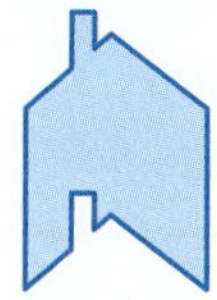

Step 3: Tape it to the other side to make your pattern.

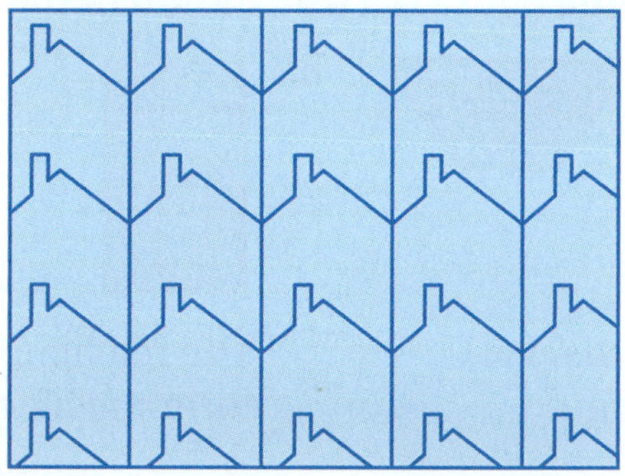

Step 4: Translate the pattern to make your tessellation.

Step 5: Color the tessellation.

4 ACTIVITY: Translating in the Coordinate Plane

Work with a partner.

a. Draw a rectangle in a coordinate plane. Find the dimensions of the rectangle.

b. Move each vertex 3 units right and 4 units up. Draw the new figure. List the vertices.

c. Compare the dimensions and the angle measures of the new figure to those of the original rectangle.

d. Are the opposite sides of the new figure still parallel? Explain.

e. Can you conclude that the two figures are congruent? Explain.

f. Compare your results with those of other students in your class. Do you think the results are true for any type of figure?

> **Math Practice 3**
>
> **Justify Conclusions**
>
> What information do you need to conclude that two figures are congruent?

What Is Your Answer?

5. **IN YOUR OWN WORDS** How can you arrange tiles to make a tessellation? Give an example.

6. **PRECISION** Explain why any parallelogram can be translated to make a tessellation.

Practice

Use what you learned about translations to complete Exercises 4–6 on page 52.

2.2 Lesson

Check It Out
Lesson Tutorials
BigIdeasMath.com

Key Vocabulary
transformation, *p. 50*
image, *p. 50*
translation, *p. 50*

A **transformation** changes a figure into another figure. The new figure is called the **image**.

A **translation** is a transformation in which a figure *slides* but does not turn. Every point of the figure moves the same distance and in the same direction.

EXAMPLE 1 Identifying a Translation

Tell whether the blue figure is a translation of the red figure.

a.

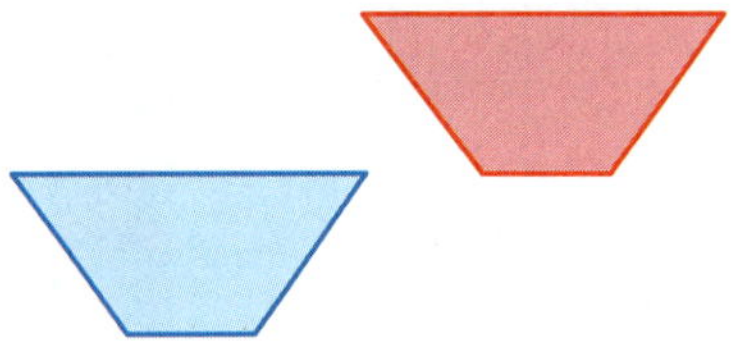

The red figure *slides* to form the blue figure.

So, the blue figure is a translation of the red figure.

b.

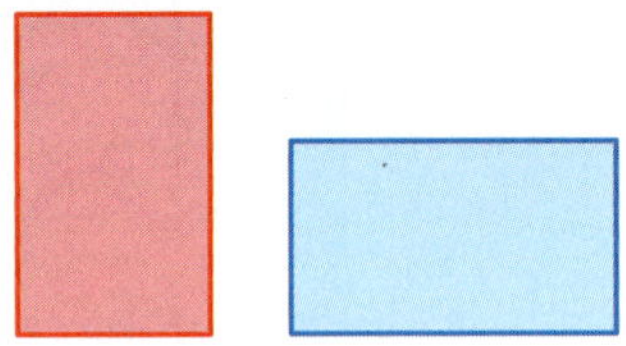

The red figure *turns* to form the blue figure.

So, the blue figure is *not* a translation of the red figure.

On Your Own

Now You're Ready
Exercises 4–9

Tell whether the blue figure is a translation of the red figure. Explain.

1.

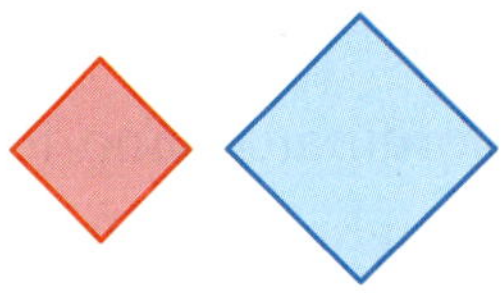

2.

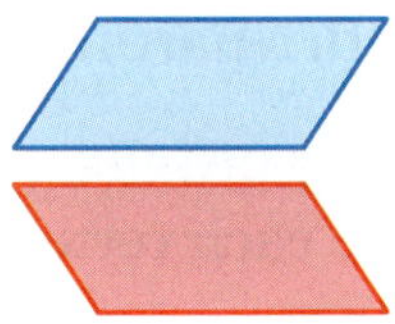

3.

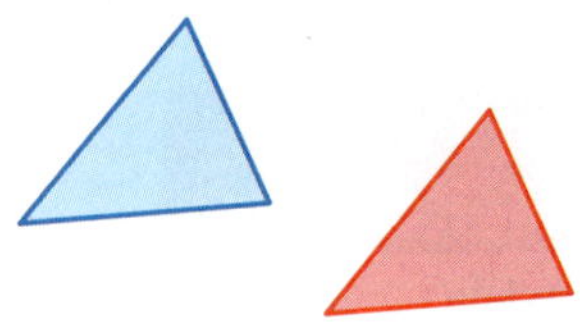

Key Idea

Translations in the Coordinate Plane

Words To translate a figure a units horizontally and b units vertically in a coordinate plane, add a to the x-coordinates and b to the y-coordinates of the vertices.

Positive values of a and b represent translations up and right. Negative values of a and b represent translations down and left.

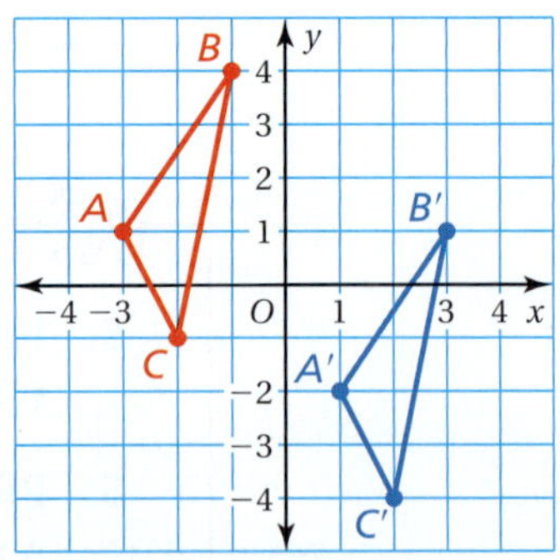

Algebra $(x, y) \rightarrow (x + a, y + b)$

In a translation, the original figure and its image are congruent.

Reading

A' is read "A prime." Use *prime* symbols when naming an image.

$A \rightarrow A'$
$B \rightarrow B'$
$C \rightarrow C'$

Multi-Language Glossary at BigIdeasMath.com

EXAMPLE 2 Translating a Figure in the Coordinate Plane

Translate the red triangle 3 units right and 3 units down. What are the coordinates of the image?

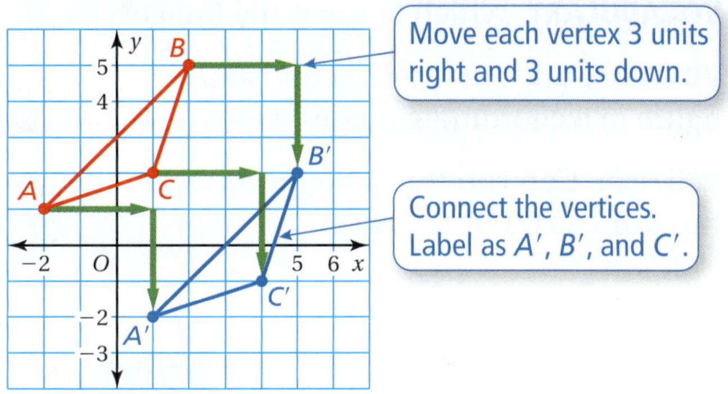

∴ The coordinates of the image are $A'(1, -2)$, $B'(5, 2)$, and $C'(4, -1)$.

On Your Own

Now You're Ready
Exercises 10 and 11

4. **WHAT IF?** The red triangle is translated 4 units left and 2 units up. What are the coordinates of the image?

EXAMPLE 3 Translating a Figure Using Coordinates

The vertices of a square are $A(1, -2)$, $B(3, -2)$, $C(3, -4)$, and $D(1, -4)$. Draw the figure and its image after a translation 4 units left and 6 units up.

Add −4 to each x-coordinate. So, subtract 4 from each x-coordinate.

Add 6 to each y-coordinate.

Vertices of $ABCD$	$(x - 4, y + 6)$	Vertices of $A'B'C'D'$
$A(1, -2)$	$(1 - 4, -2 + 6)$	$A'(-3, 4)$
$B(3, -2)$	$(3 - 4, -2 + 6)$	$B'(-1, 4)$
$C(3, -4)$	$(3 - 4, -4 + 6)$	$C'(-1, 2)$
$D(1, -4)$	$(1 - 4, -4 + 6)$	$D'(-3, 2)$

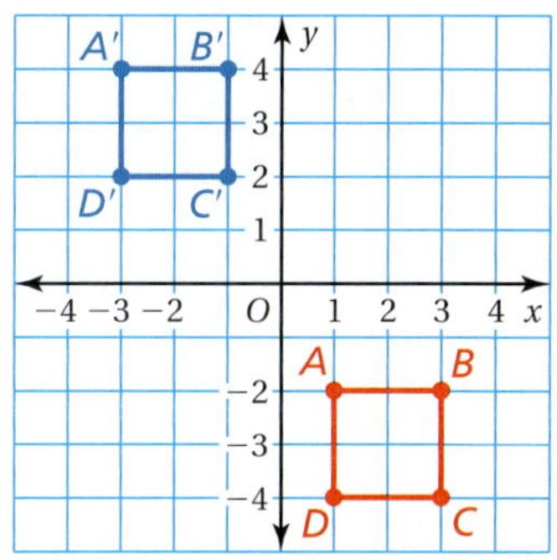

∴ The figure and its image are shown at the above right.

On Your Own

Now You're Ready
Exercises 12–15

5. The vertices of a triangle are $A(-2, -2)$, $B(0, 2)$, and $C(3, 0)$. Draw the figure and its image after a translation 1 unit left and 2 units up.

2.2 Exercises

Check It Out
Help with Homework
BigIdeasMath.com

Vocabulary and Concept Check

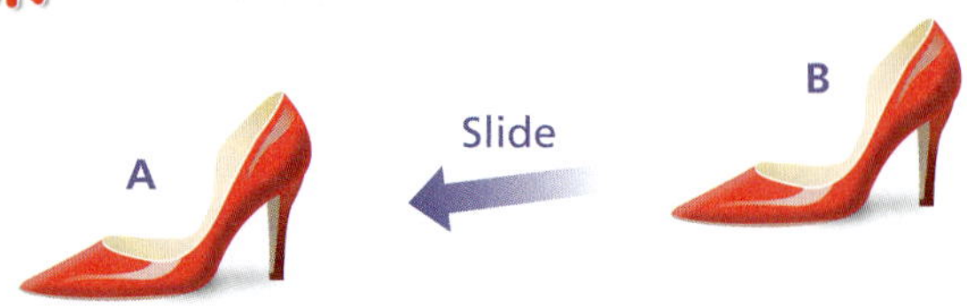

1. **VOCABULARY** Which figure is the image?
2. **VOCABULARY** How do you translate a figure in a coordinate plane?
3. **WRITING** Can you translate the letters in the word TOKYO to form the word KYOTO? Explain.

Practice and Problem Solving

Tell whether the blue figure is a translation of the red figure.

(1) 4. 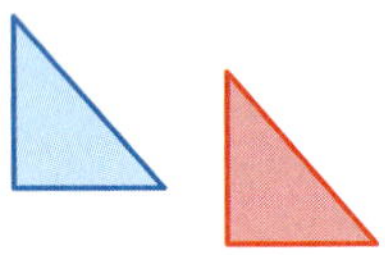

5. 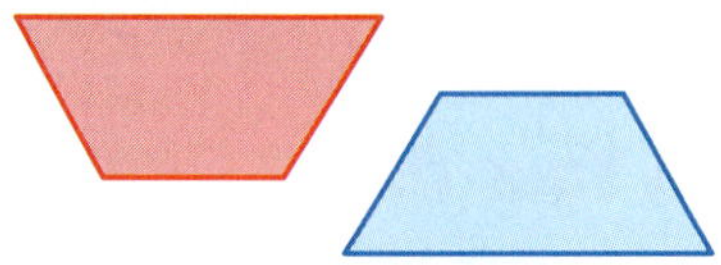

6. 

7. 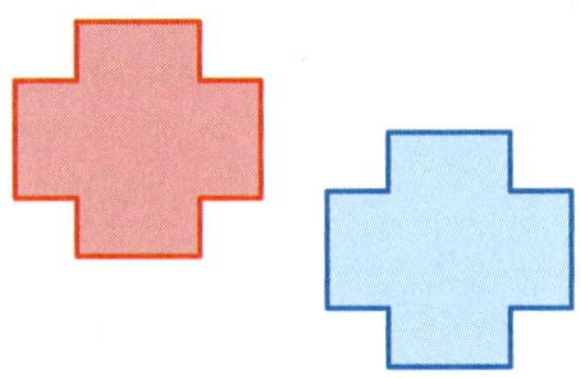

8. 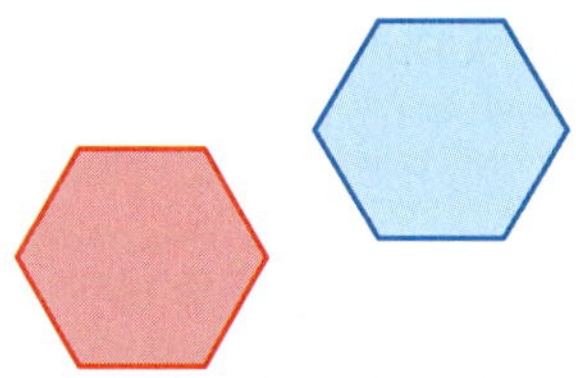

9. 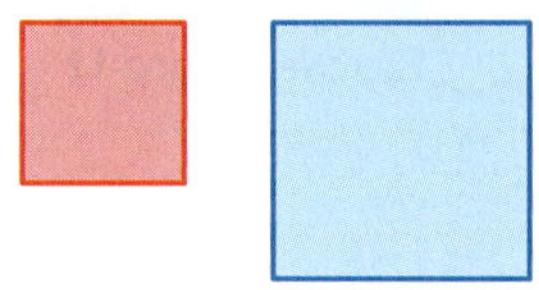

(2) 10. Translate the triangle 4 units right and 3 units down. What are the coordinates of the image?

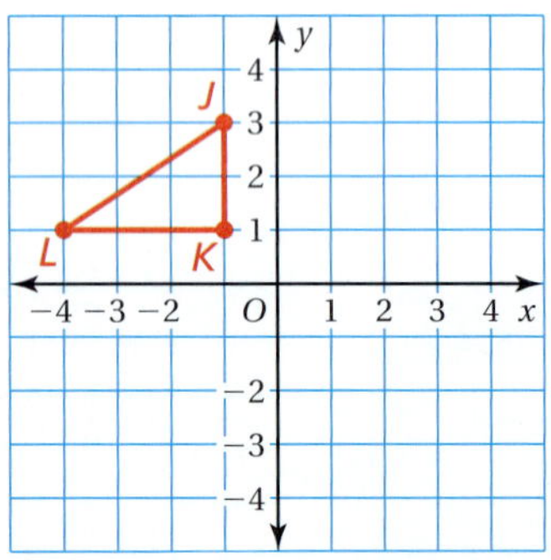

11. Translate the figure 2 units left and 4 units down. What are the coordinates of the image?

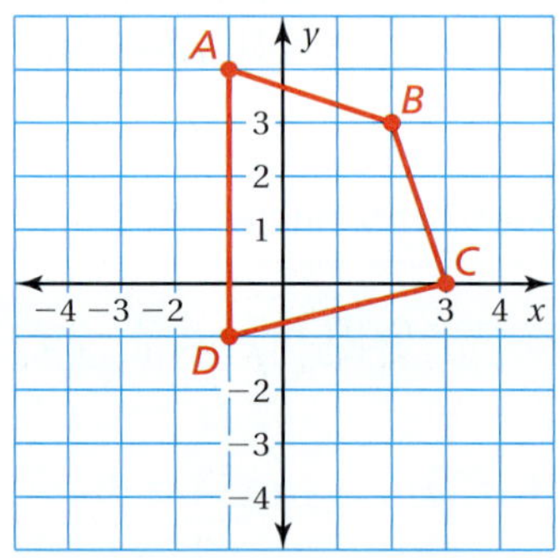

The vertices of a triangle are $L(0, 1)$, $M(1, -2)$, and $N(-2, 1)$. Draw the figure and its image after the translation.

(3) 12. 1 unit left and 6 units up

13. 5 units right

14. $(x + 2, y + 3)$

15. $(x - 3, y - 4)$

16. **ICONS** You can click and drag an icon on a computer screen. Is this an example of a translation? Explain.

Describe the translation of the point to its image.

17. (3, −2) → (1, 0)

18. (−8, −4) → (−3, 5)

Describe the translation from the red figure to the blue figure.

19.

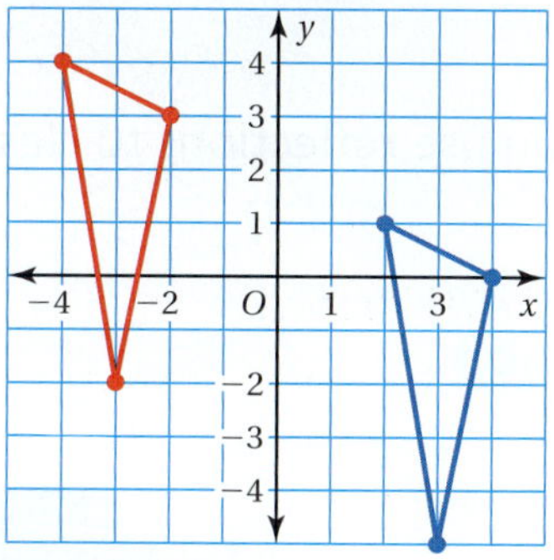

20.

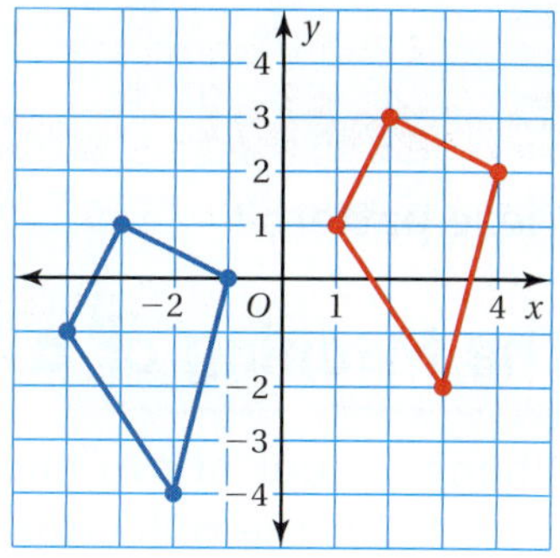

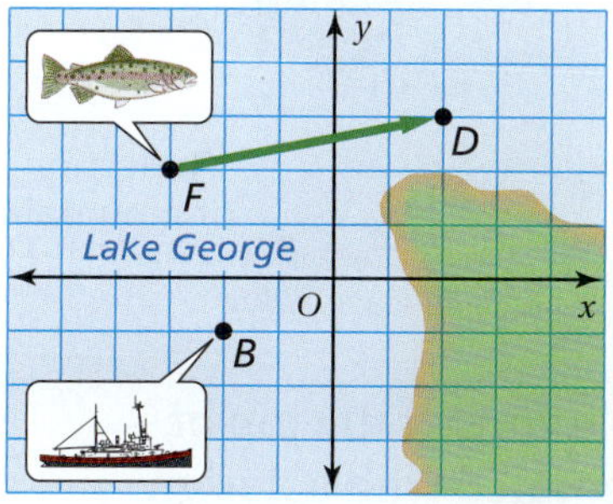

21. **FISHING** A school of fish translates from point F to point D.

a. Describe the translation of the school of fish.

b. Can the fishing boat make the same translation? Explain.

c. Describe a translation the fishing boat could make to get to point D.

22. **REASONING** The vertices of a triangle are $A(0, -3)$, $B(2, -1)$, and $C(3, -3)$. You translate the triangle 5 units right and 2 units down. Then you translate the image 3 units left and 8 units down. Is the original triangle congruent to the final image? If so, give two ways to show that they are congruent.

23. Problem Solving In chess, a knight can move only in an L-shaped pattern:

- *two* vertical squares, then *one* horizontal square;
- *two* horizontal squares, then *one* vertical square;
- *one* vertical square, then *two* horizontal squares; or
- *one* horizontal square, then *two* vertical squares.

Write a series of translations to move the knight from g8 to g5.

Fair Game Review What you learned in previous grades & lessons

Tell whether you can fold the figure in half so that one side matches the other. *(Skills Review Handbook)*

24. 

25. 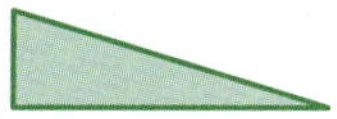

26. 

27. 

28. **MULTIPLE CHOICE** You put \$550 in an account that earns 4.4% simple interest per year. How much interest do you earn in 6 months? *(Skills Review Handbook)*

Ⓐ \$1.21 Ⓑ \$12.10 Ⓒ \$121.00 Ⓓ \$145.20

2.3 Reflections

Essential Question How can you use reflections to classify a frieze pattern?

The Meaning of a Word ● Reflection

When you look at a mountain by a lake, you can see the **reflection**, or mirror image, of the mountain in the lake.

If you fold the photo on its axis, the mountain and its reflection will align.

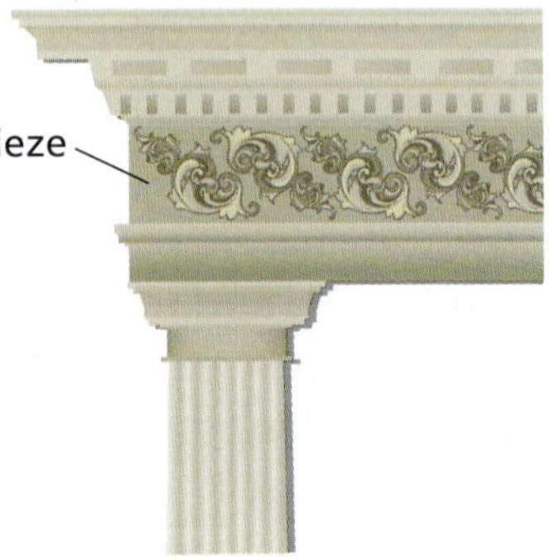

A *frieze* is a horizontal band that runs at the top of a building. A frieze is often decorated with a design that repeats.

- All frieze patterns are translations of themselves.
- Some frieze patterns are reflections of themselves.

1 ACTIVITY: Frieze Patterns and Reflections

Work with a partner. Consider the frieze pattern shown.

a. Is the frieze pattern a reflection of itself when folded horizontally? Explain.

b. Is the frieze pattern a reflection of itself when folded vertically? Explain.

COMMON CORE

Geometry

In this lesson, you will

- identify reflections.
- reflect figures in the x-axis or the y-axis of the coordinate plane.

Learning Standards
MACC.8.G.1.1
MACC.8.G.1.2
MACC.8.G.1.3

2 ACTIVITY: Frieze Patterns and Reflections

Work with a partner. Is the frieze pattern a reflection of itself when folded *horizontally*, *vertically*, or *neither*?

a.

b.

3 ACTIVITY: Reflecting in the Coordinate Plane

Work with a partner.

a. Draw a rectangle in Quadrant I of a coordinate plane. Find the dimensions of the rectangle.

b. Copy the axes and the rectangle onto a piece of transparent paper.

Flip the transparent paper once so that the rectangle is in Quadrant IV. Then align the origin and the axes with the coordinate plane.

Draw the new figure in the coordinate plane. List the vertices.

c. Compare the dimensions and the angle measures of the new figure to those of the original rectangle.

d. Are the opposite sides of the new figure still parallel? Explain.

e. Can you conclude that the two figures are congruent? Explain.

f. Flip the transparent paper so that the original rectangle is in Quadrant II. Draw the new figure in the coordinate plane. List the vertices. Then repeat parts (c)–(e).

g. Compare your results with those of other students in your class. Do you think the results are true for any type of figure?

Math Practice 7

Look for Patterns

What do you notice about the vertices of the original figure and the image? How does this help you determine whether the figures are congruent?

What Is Your Answer?

4. **IN YOUR OWN WORDS** How can you use reflections to classify a frieze pattern?

Practice Use what you learned about reflections to complete Exercises 4–6 on page 58.

2.3 Lesson

Check It Out
Lesson Tutorials
BigIdeasMath.com

Key Vocabulary
reflection, *p. 56*
line of reflection, *p. 56*

A **reflection**, or *flip*, is a transformation in which a figure is reflected in a line called the **line of reflection**. A reflection creates a mirror image of the original figure.

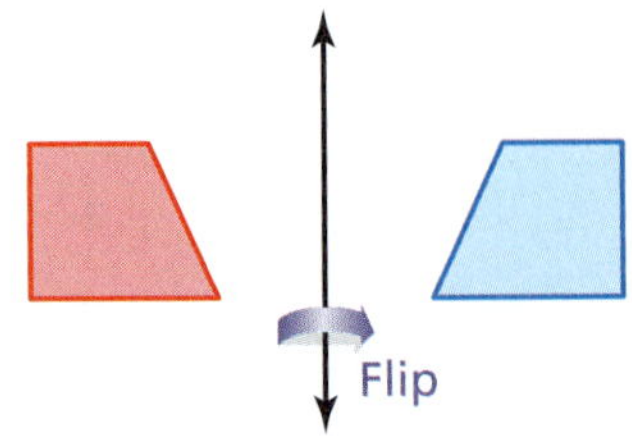

EXAMPLE 1 Identifying a Reflection

Tell whether the blue figure is a reflection of the red figure.

a.

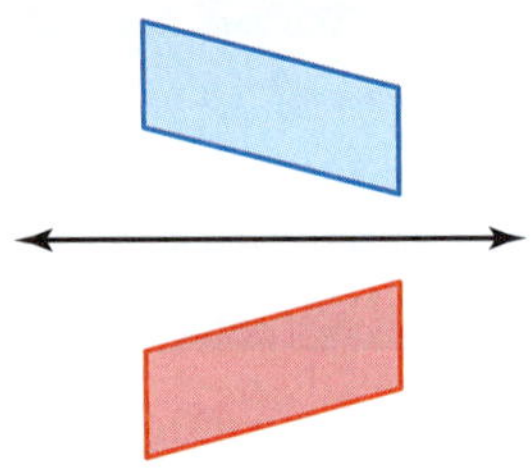

The red figure can be *flipped* to form the blue figure.

So, the blue figure is a reflection of the red figure.

b.

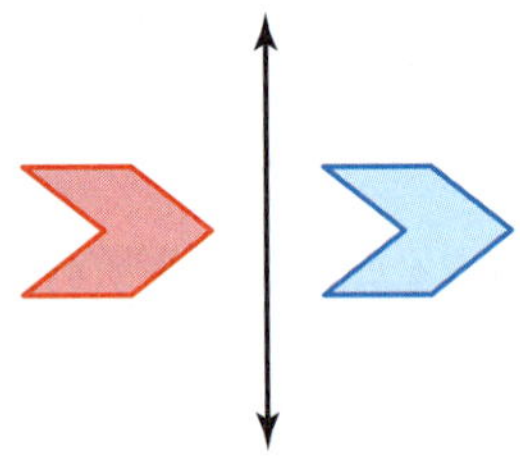

If the red figure were *flipped*, it would point to the left.

So, the blue figure is *not* a reflection of the red figure.

On Your Own

Now You're Ready
Exercises 4–9

Tell whether the blue figure is a reflection of the red figure. Explain.

1.

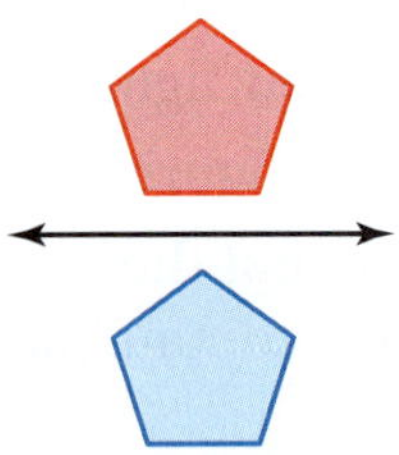

2.

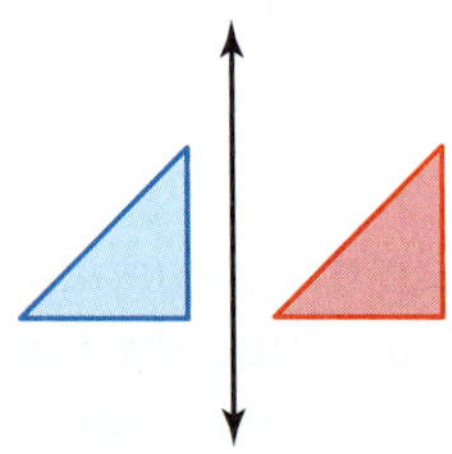

3.

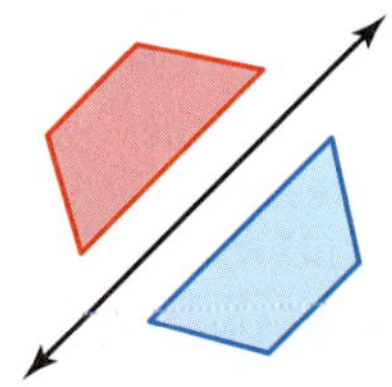

Key Idea

Reflections in the Coordinate Plane

Words To reflect a figure in the x-axis, take the opposite of the y-coordinate.

To reflect a figure in the y-axis, take the opposite of the x-coordinate.

Algebra Reflection in x-axis: $(x, y) \rightarrow (x, -y)$

Reflection in y-axis: $(x, y) \rightarrow (-x, y)$

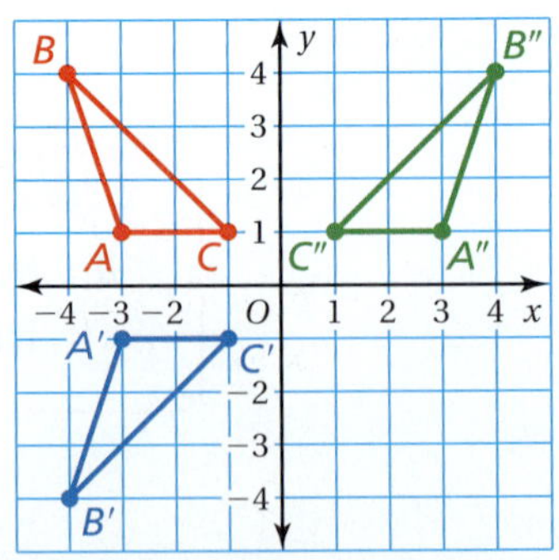

In a reflection, the original figure and its image are congruent.

Multi-Language Glossary at BigIdeasMath.com

EXAMPLE 2 Reflecting a Figure in the *x*-axis

The vertices of a triangle are $A(-1, 1)$, $B(-1, 3)$, and $C(6, 3)$. Draw the figure and its reflection in the x-axis. What are the coordinates of the image?

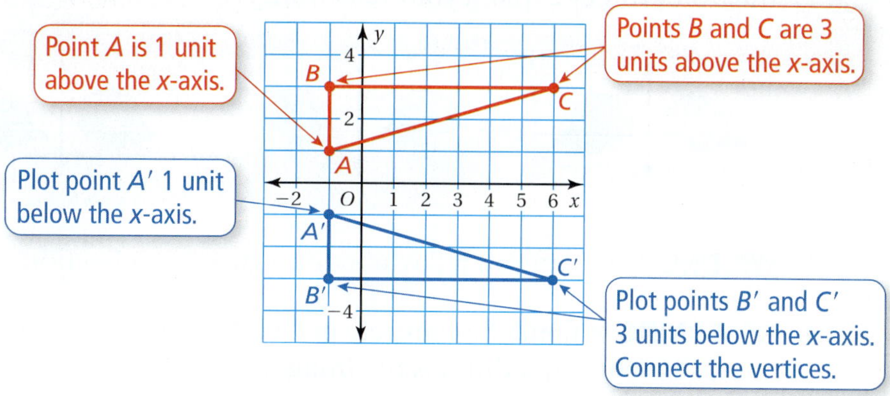

∴ The coordinates of the image are $A'(-1, -1)$, $B'(-1, -3)$, and $C'(6, -3)$.

EXAMPLE 3 Reflecting a Figure in the *y*-axis

The vertices of a quadrilateral are $P(-2, 5)$, $Q(-1, -1)$, $R(-4, 2)$, and $S(-4, 4)$. Draw the figure and its reflection in the y-axis.

Take the opposite of the x-coordinate. → $(-x, y)$ ← The y-coordinate does not change.

Vertices of *PQRS*	$(-x, y)$	Vertices of *P′Q′R′S′*
$P(-2, 5)$	$(-(-2), 5)$	$P'(2, 5)$
$Q(-1, -1)$	$(-(-1), -1)$	$Q'(1, -1)$
$R(-4, 2)$	$(-(-4), 2)$	$R'(4, 2)$
$S(-4, 4)$	$(-(-4), 4)$	$S'(4, 4)$

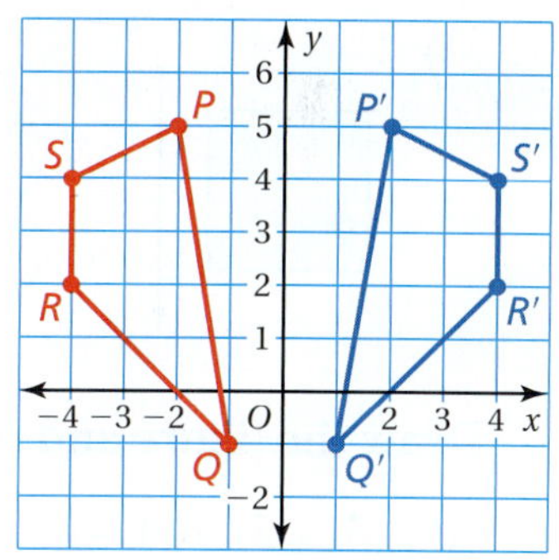

∴ The figure and its image are shown at the above right.

On Your Own

Now You're Ready
Exercises 10–17

4. The vertices of a rectangle are $A(-4, -3)$, $B(-4, -1)$, $C(-1, -1)$, and $D(-1, -3)$.
   a. Draw the figure and its reflection in the x-axis.
   b. Draw the figure and its reflection in the y-axis.
   c. Are the images in parts (a) and (b) congruent? Explain.

2.3 Exercises

Check It Out
Help with Homework
BigIdeasMath.com

Vocabulary and Concept Check

1. **WHICH ONE DOESN'T BELONG?** Which transformation does *not* belong with the other three? Explain your reasoning.

2. **WRITING** How can you tell when one figure is a reflection of another figure?

3. **REASONING** A figure lies entirely in Quadrant I. The figure is reflected in the x-axis. In which quadrant is the image?

Practice and Problem Solving

Tell whether the blue figure is a reflection of the red figure.

① 4.

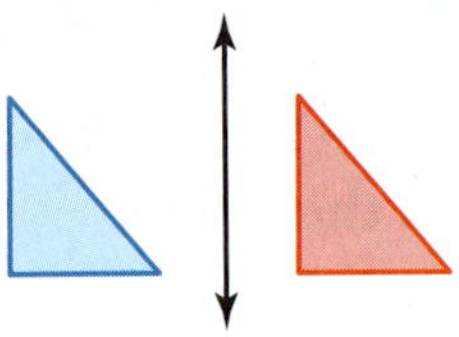

5.

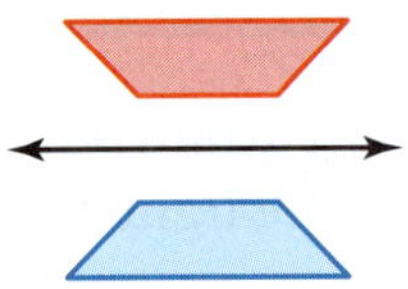

6.

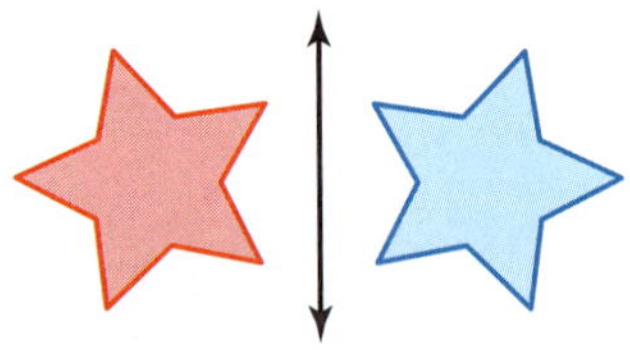

7.

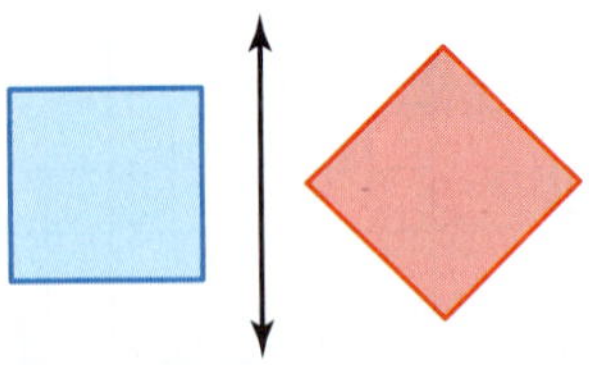

8.

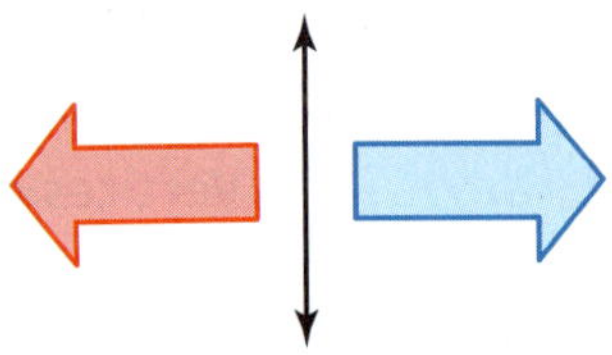

9.

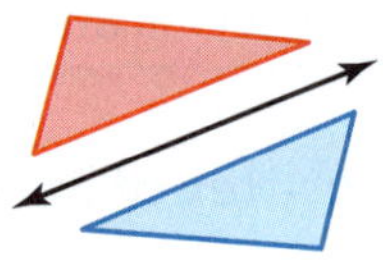

Draw the figure and its reflection in the x-axis. Identify the coordinates of the image.

② 10. $A(3, 2)$, $B(4, 4)$, $C(1, 3)$

11. $M(-2, 1)$, $N(0, 3)$, $P(2, 2)$

12. $H(2, -2)$, $J(4, -1)$, $K(6, -3)$, $L(5, -4)$

13. $D(-2, -1)$, $E(0, -1)$, $F(0, -5)$, $G(-2, -5)$

Draw the figure and its reflection in the y-axis. Identify the coordinates of the image.

③ 14. $Q(-4, 2)$, $R(-2, 4)$, $S(-1, 1)$

15. $T(4, -2)$, $U(4, 2)$, $V(6, -2)$

16. $W(2, -1)$, $X(5, -2)$, $Y(5, -5)$, $Z(2, -4)$

17. $J(2, 2)$, $K(7, 4)$, $L(9, -2)$, $M(3, -1)$

18. **ALPHABET** Which letters look the same when reflected in the line?

A B C D E F G H I J K L M N O P Q R S T U V W X Y Z

The coordinates of a point and its image are given. Is the reflection in the *x-axis* or *y-axis*?

19. $(2, -2) \rightarrow (2, 2)$

20. $(-4, 1) \rightarrow (4, 1)$

21. $(-2, -5) \rightarrow (2, -5)$

22. $(-3, -4) \rightarrow (-3, 4)$

Find the coordinates of the figure after the transformations.

23. Translate the triangle 1 unit right and 5 units down. Then reflect the image in the y-axis.

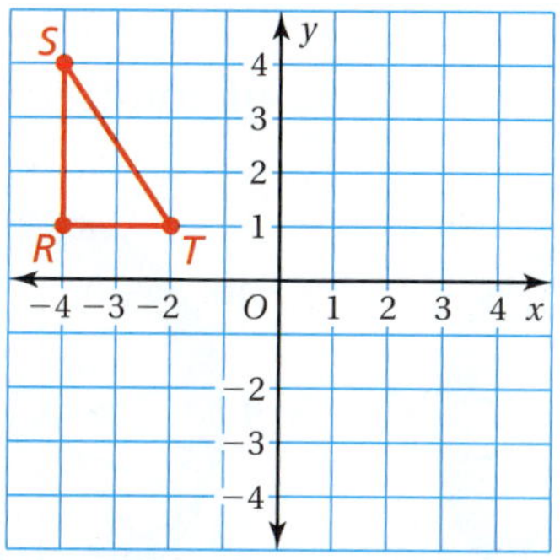

24. Reflect the trapezoid in the x-axis. Then translate the trapezoid 2 units left and 3 units up.

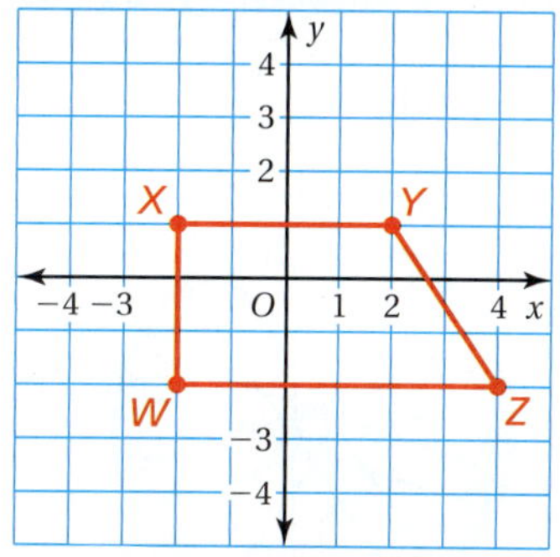

25. REASONING In Exercises 23 and 24, is the original figure congruent to the final image? Explain.

26. NUMBER SENSE You reflect a point (x, y) in the x-axis, and then in the y-axis. What are the coordinates of the final image?

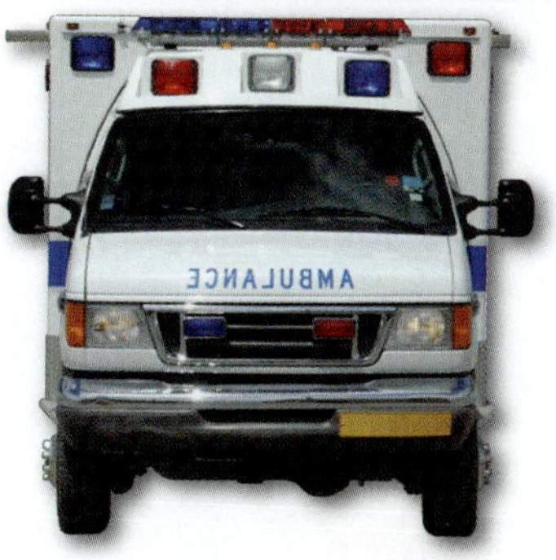

27. EMERGENCY VEHICLE Hold a mirror to the left side of the photo of the vehicle.

a. What word do you see in the mirror?

b. Why do you think it is written that way on the front of the vehicle?

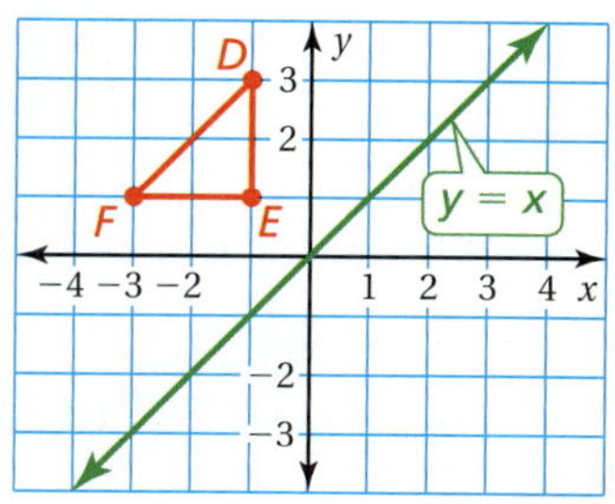

28. Critical Thinking Reflect the triangle in the line $y = x$. How are the x- and y-coordinates of the image related to the x- and y-coordinates of the original triangle?

Fair Game Review What you learned in previous grades & lessons

Classify the angle as *acute, right, obtuse,* or *straight*. *(Skills Review Handbook)*

29. 

30. 

31. 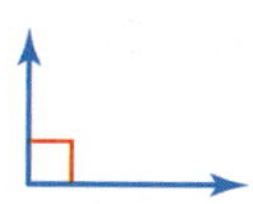

32. 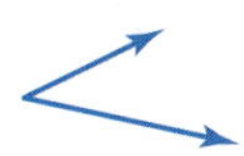

33. MULTIPLE CHOICE 36 is 75% of what number? *(Skills Review Handbook)*

Ⓐ 27 Ⓑ 48 Ⓒ 54 Ⓓ 63

2.4 Rotations

Essential Question What are the three basic ways to move an object in a plane?

The Meaning of a Word ● Rotate

A bicycle wheel can **rotate** clockwise or counterclockwise.

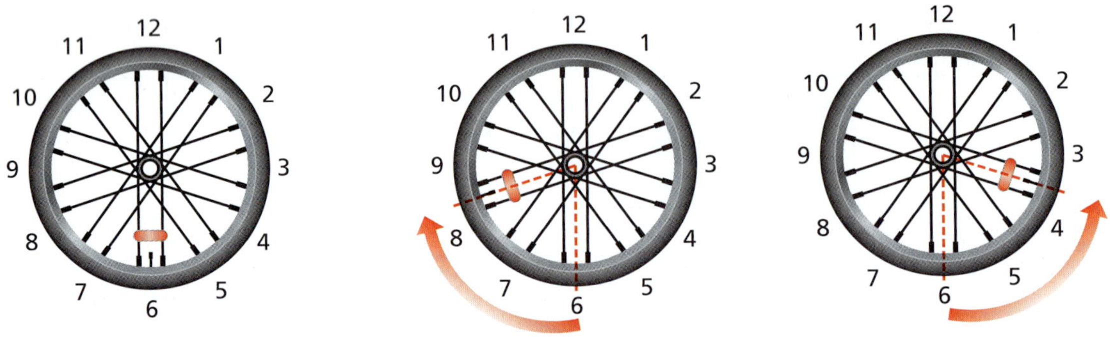

1 ACTIVITY: Three Basic Ways to Move Things

There are three basic ways to move objects on a flat surface.

_____ the object. _____ the object. _____ the object.

Work with a partner.

a. What type of triangle is the blue triangle? Is it congruent to the red triangles? Explain.

b. Decide how you can move the blue triangle to obtain each red triangle.

c. Is each move a *translation*, a *reflection*, or a *rotation*?

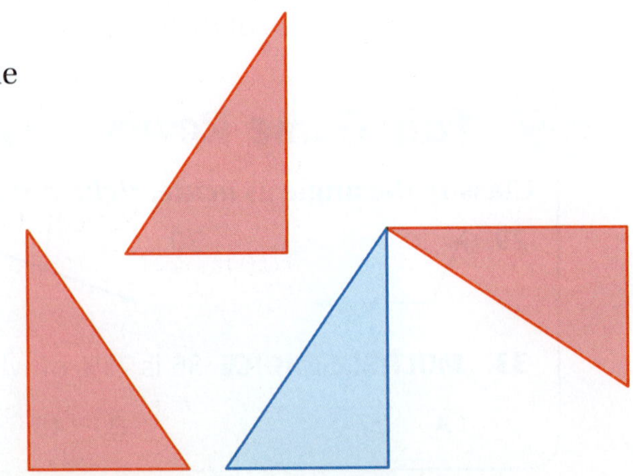

COMMON CORE

Geometry

In this lesson, you will

- identify rotations.
- rotate figures in the coordinate plane.
- use more than one transformation to find images of figures.

Learning Standards
MACC.8.G.1.1
MACC.8.G.1.2
MACC.8.G.1.3

2 ACTIVITY: Rotating in the Coordinate Plane

Work with a partner.

a. Draw a rectangle in Quadrant II of a coordinate plane. Find the dimensions of the rectangle.

b. Copy the axes and the rectangle onto a piece of transparent paper.

Align the origin and the vertices of the rectangle on the transparent paper with the coordinate plane. Turn the transparent paper so that the rectangle is in Quadrant I and the axes align.

Draw the new figure in the coordinate plane. List the vertices.

c. Compare the dimensions and the angle measures of the new figure to those of the original rectangle.

d. Are the opposite sides of the new figure still parallel? Explain.

e. Can you conclude that the two figures are congruent? Explain.

f. Turn the transparent paper so that the original rectangle is in Quadrant IV. Draw the new figure in the coordinate plane. List the vertices. Then repeat parts (c)–(e).

g. Compare your results with those of other students in your class. Do you think the results are true for any type of figure?

Math Practice 6

Calculate Accurately

What must you do to rotate the figure correctly?

What Is Your Answer?

3. **IN YOUR OWN WORDS** What are the three basic ways to move an object in a plane? Draw an example of each.

4. **PRECISION** Use the results of Activity 2(b).

   a. Draw four angles using the conditions below.

   - The origin is the vertex of each angle.
   - One side of each angle passes through a vertex of the original rectangle.
   - The other side of each angle passes through the corresponding vertex of the rotated rectangle.

   b. Measure each angle in part (a). For each angle, measure the distances between the origin and the vertices of the rectangles. What do you notice?

   c. How can the results of part (b) help you rotate a figure?

5. **PRECISION** Repeat the procedure in Question 4 using the results of Activity 2(f).

Practice Use what you learned about transformations to complete Exercises 7–9 on page 65.

2.4 Lesson

Check It Out
Lesson Tutorials
BigIdeasMath.com

Key Vocabulary
rotation, *p. 62*
center of rotation, *p. 62*
angle of rotation, *p. 62*

Key Idea

Rotations

A **rotation**, or *turn*, is a transformation in which a figure is rotated about a point called the **center of rotation**. The number of degrees a figure rotates is the **angle of rotation**.

In a rotation, the original figure and its image are congruent.

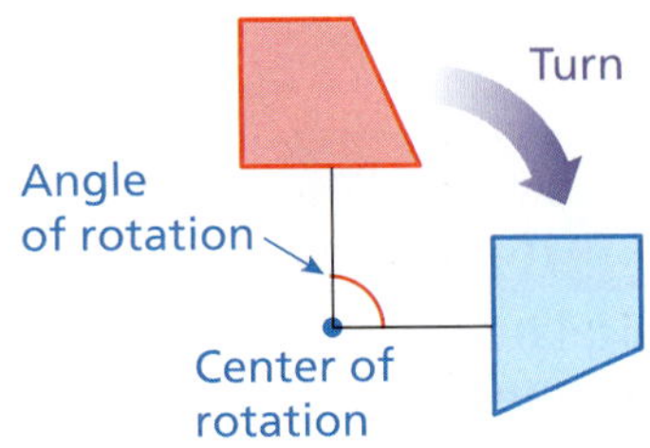

EXAMPLE 1 Identifying a Rotation

You must rotate the puzzle piece 270° clockwise about point *P* to fit it into a puzzle. Which piece fits in the puzzle as shown?

Rotate the puzzle piece 270° clockwise about point *P*.

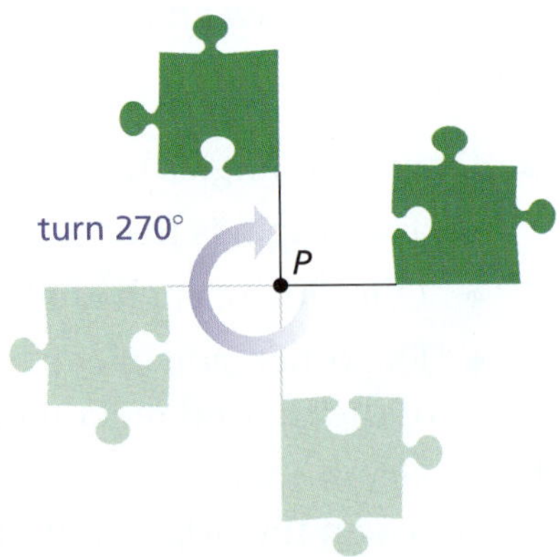

Study Tip
When rotating figures, it may help to sketch the rotation in several steps, as shown in Example 1.

So, the correct answer is Ⓒ.

On Your Own

Now You're Ready
Exercises 10–12

1. Which piece is a 90° counterclockwise rotation about point *P*?
2. Is Choice D a rotation of the original puzzle piece? If not, what kind of transformation does the image show?

Multi-Language Glossary at BigIdeasMath.com

EXAMPLE 2 Rotating a Figure

The vertices of a trapezoid are $W(-4, 2)$, $X(-3, 4)$, $Y(-1, 4)$, and $Z(-1, 2)$. Rotate the trapezoid 180° about the origin. What are the coordinates of the image?

Study Tip

A 180° clockwise rotation and a 180° counterclockwise rotation have the same image. So, you do not need to specify direction when rotating a figure 180°.

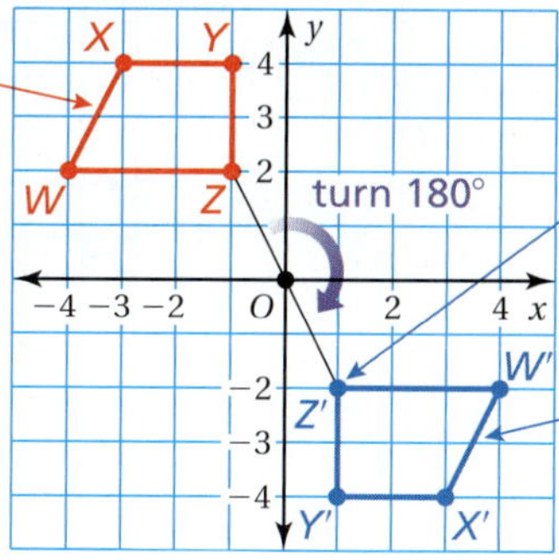

∴ The coordinates of the image are $W'(4, -2)$, $X'(3, -4)$, $Y'(1, -4)$, and $Z'(1, -2)$.

EXAMPLE 3 Rotating a Figure

The vertices of a triangle are $J(1, 2)$, $K(4, 2)$, and $L(1, -3)$. Rotate the triangle 90° counterclockwise about vertex L. What are the coordinates of the image?

Common Error

Be sure to pay attention to whether a rotation is clockwise or counterclockwise.

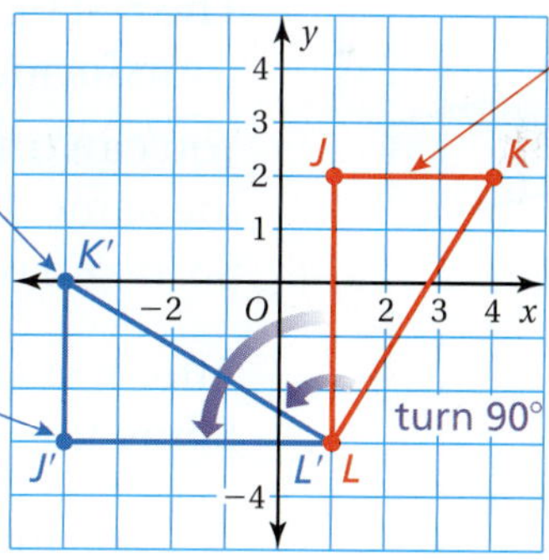

∴ The coordinates of the image are $J'(-4, -3)$, $K'(-4, 0)$, and $L'(1, -3)$.

On Your Own

Now You're Ready
Exercises 13–18

3. A triangle has vertices $Q(4, 5)$, $R(4, 0)$, and $S(1, 0)$.
   - **a.** Rotate the triangle 90° counterclockwise about the origin.
   - **b.** Rotate the triangle 180° about vertex S.
   - **c.** Are the images in parts (a) and (b) congruent? Explain.

EXAMPLE 4 Using More than One Transformation

The vertices of a rectangle are $A(-3, -3)$, $B(1, -3)$, $C(1, -5)$, and $D(-3, -5)$. Rotate the rectangle 90° clockwise about the origin, and then reflect it in the y-axis. What are the coordinates of the image?

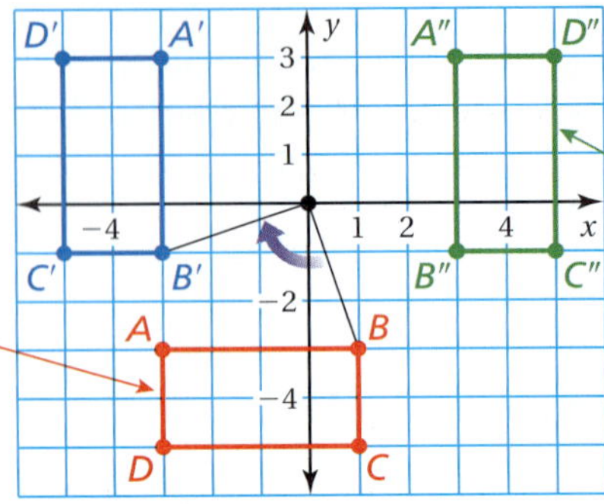

The coordinates of the image are $A''(3, 3)$, $B''(3, -1)$, $C''(5, -1)$ and $D''(5, 3)$.

The image of a translation, reflection, or rotation is congruent to the original figure. So, two figures are congruent when one can be obtained from the other by a sequence of translations, reflections, and rotations.

EXAMPLE 5 Describing a Sequence of Transformations

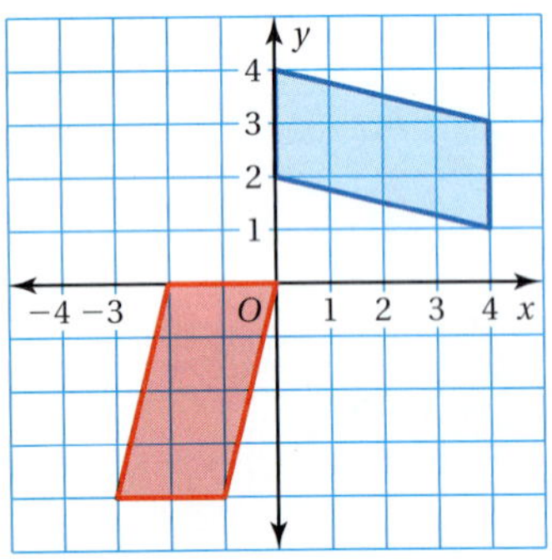

The red figure is congruent to the blue figure. Describe a sequence of transformations in which the blue figure is the image of the red figure.

You can turn the red figure 90° so that it has the same orientation as the blue figure. So, begin with a rotation.

After rotating, you need to slide the figure up.

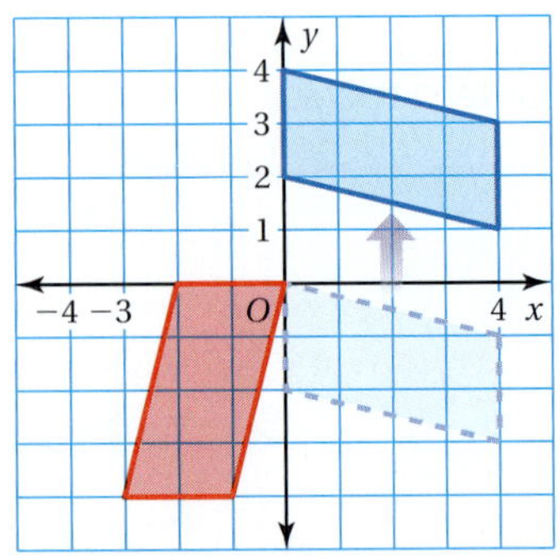

So, one possible sequence of transformations is a 90° counterclockwise rotation about the origin followed by a translation 4 units up.

On Your Own

Now You're Ready
Exercises 22–25

4. The vertices of a triangle are $P(-1, 2)$, $Q(-1, 0)$, and $R(2, 0)$. Rotate the triangle 180° about vertex R, and then reflect it in the x-axis. What are the coordinates of the image?

5. In Example 5, describe a different sequence of transformations in which the blue figure is the image of the red figure.

2.4 Exercises

Check It Out
Help with Homework
BigIdeasMath.com

Vocabulary and Concept Check

1. **VOCABULARY** What are the coordinates of the center of rotation in Example 2? Example 3?

MENTAL MATH **A figure lies entirely in Quadrant II. In which quadrant will the figure lie after the given clockwise rotation about the origin?**

2. 90°
3. 180°
4. 270°
5. 360°

6. **DIFFERENT WORDS, SAME QUESTION** Which is different? Find "both" answers.

What are the coordinates of the figure after a 90° clockwise rotation about the origin?

What are the coordinates of the figure after a 270° clockwise rotation about the origin?

What are the coordinates of the figure after turning the figure 90° to the right about the origin?

What are the coordinates of the figure after a 270° counterclockwise rotation about the origin?

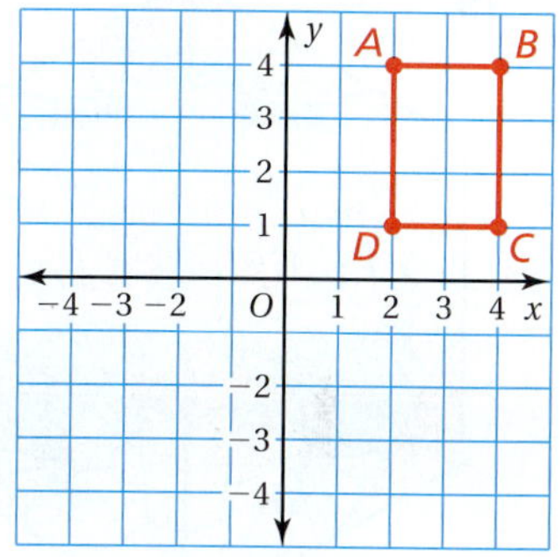

Practice and Problem Solving

Identify the transformation.

7.
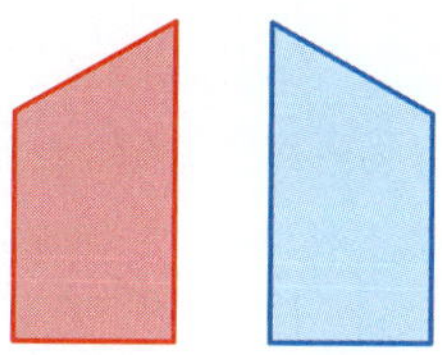

8.
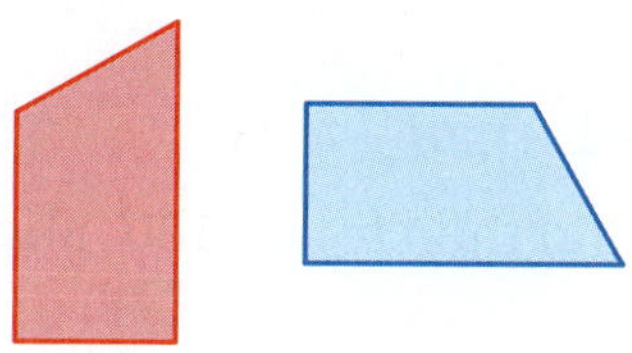

9.
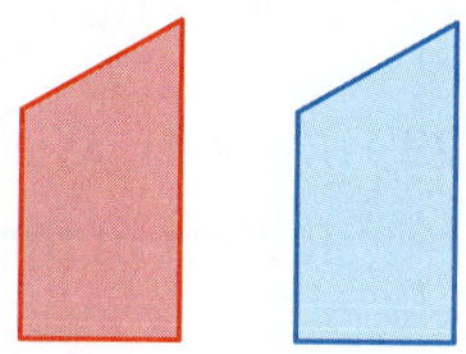

Tell whether the blue figure is a rotation of the red figure about the origin. If so, give the angle and direction of rotation.

10.
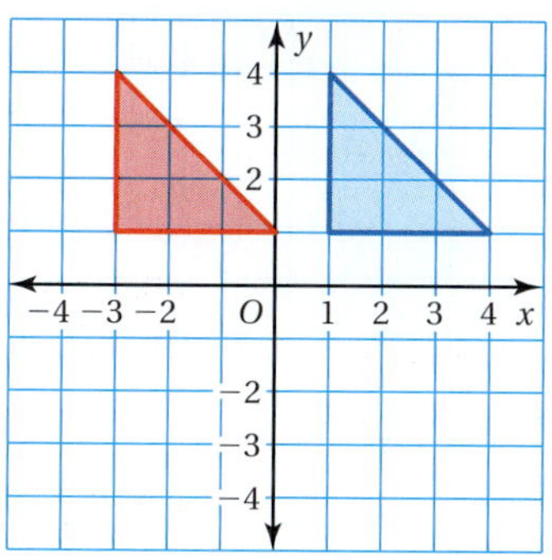

11.
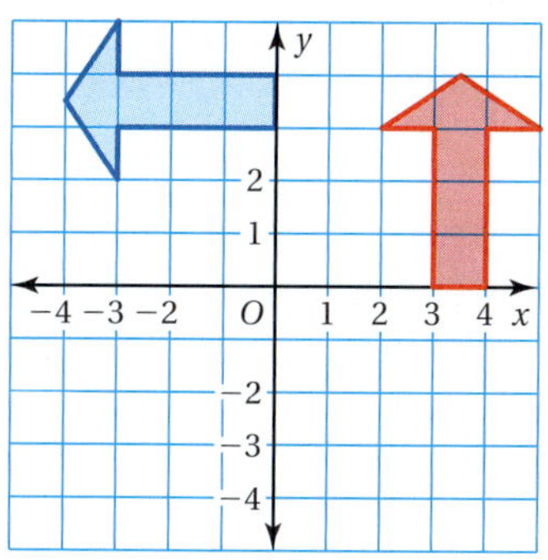

12.
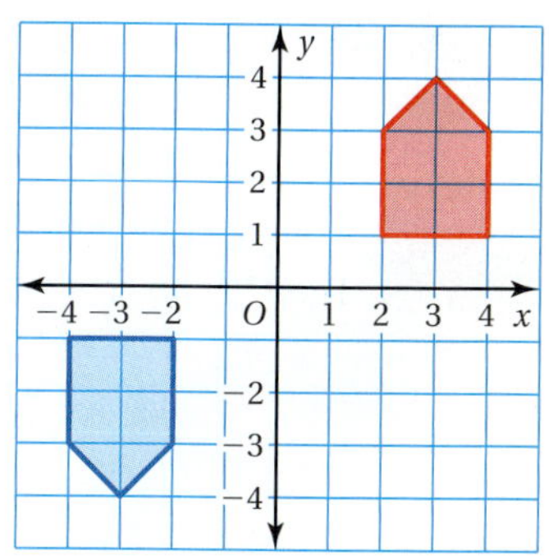

The vertices of a figure are given. Rotate the figure as described. Find the coordinates of the image.

2 3 **13.** $A(2, -2)$, $B(4, -1)$, $C(4, -3)$, $D(2, -4)$
90° counterclockwise about the origin

14. $F(1, 2)$, $G(3, 5)$, $H(3, 2)$
180° about the origin

15. $J(-4, 1)$, $K(-2, 1)$, $L(-4, -3)$
90° clockwise about vertex L

16. $P(-3, 4)$, $Q(-1, 4)$, $R(-2, 1)$, $S(-4, 1)$
180° about vertex R

17. $W(-6, -2)$, $X(-2, -2)$, $Y(-2, -6)$, $Z(-5, -6)$
270° counterclockwise about the origin

18. $A(1, -1)$, $B(5, -6)$, $C(1, -6)$
90° counterclockwise about vertex A

A figure has *rotational symmetry* if a rotation of 180° or less produces an image that fits exactly on the original figure. Explain why the figure has rotational symmetry.

19.

20.

21.

The vertices of a figure are given. Find the coordinates of the figure after the transformations given.

4 **22.** $R(-7, -5)$, $S(-1, -2)$, $T(-1, -5)$

Rotate 90° counterclockwise about the origin. Then translate 3 units left and 8 units up.

23. $J(-4, 4)$, $K(-3, 4)$, $L(-1, 1)$, $M(-4, 1)$

Reflect in the x-axis, and then rotate 180° about the origin.

The red figure is congruent to the blue figure. Describe two different sequences of transformations in which the blue figure is the image of the red figure.

5 **24.**

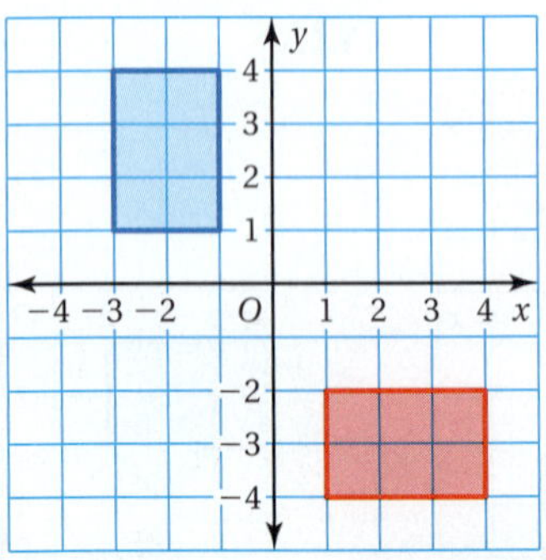

25.

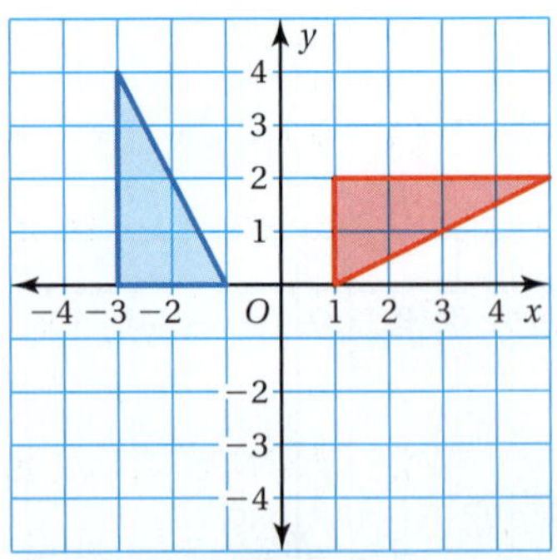

26. **REASONING** A trapezoid has vertices $A(-6, -2)$, $B(-3, -2)$, $C(-1, -4)$, and $D(-6, -4)$.

a. Rotate the trapezoid 180° about the origin. What are the coordinates of the image?

b. Describe a way to obtain the same image without using rotations.

27. **TREASURE MAP** You want to find the treasure located on the map at ✕. You are located at ●. The following transformations will lead you to the treasure, but they are not in the correct order. Find the correct order. Use each transformation exactly once.

- Rotate 180° about the origin.
- Reflect in the y-axis.
- Rotate 90° counterclockwise about the origin.
- Translate 1 unit right and 1 unit up.

28. **CRITICAL THINKING** Consider $\triangle JKL$.

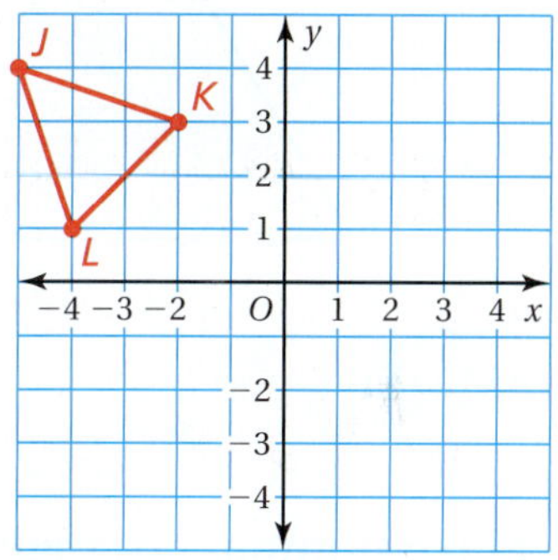

a. Rotate $\triangle JKL$ 90° clockwise about the origin. How are the x- and y-coordinates of $\triangle J'K'L'$ related to the x- and y-coordinates of $\triangle JKL$?

b. Rotate $\triangle JKL$ 180° about the origin. How are the x- and y-coordinates of $\triangle J'K'L'$ related to the x- and y-coordinates of $\triangle JKL$?

c. Do you think your answers to parts (a) and (b) hold true for any figure? Explain.

29. **Reasoning** You rotate a triangle 90° counterclockwise about the origin. Then you translate its image 1 unit left and 2 units down. The vertices of the final image are $(-5, 0)$, $(-2, 2)$, and $(-2, -1)$. What are the vertices of the original triangle?

Fair Game Review What you learned in previous grades & lessons

Tell whether the ratios form a proportion. *(Skills Review Handbook)*

30. $\frac{3}{5}, \frac{15}{20}$ 31. $\frac{2}{3}, \frac{12}{18}$ 32. $\frac{7}{28}, \frac{12}{48}$ 33. $\frac{54}{72}, \frac{36}{45}$

34. **MULTIPLE CHOICE** What is the solution of the equation $x + 6 \div 2 = 5$? *(Section 1.1)*

Ⓐ $x = -16$ Ⓑ $x = 2$ Ⓒ $x = 4$ Ⓓ $x = 16$

2 Study Help

Check It Out
Graphic Organizer
BigIdeasMath.com

You can use a **summary triangle** to explain a concept. Here is an example of a summary triangle for translating a figure.

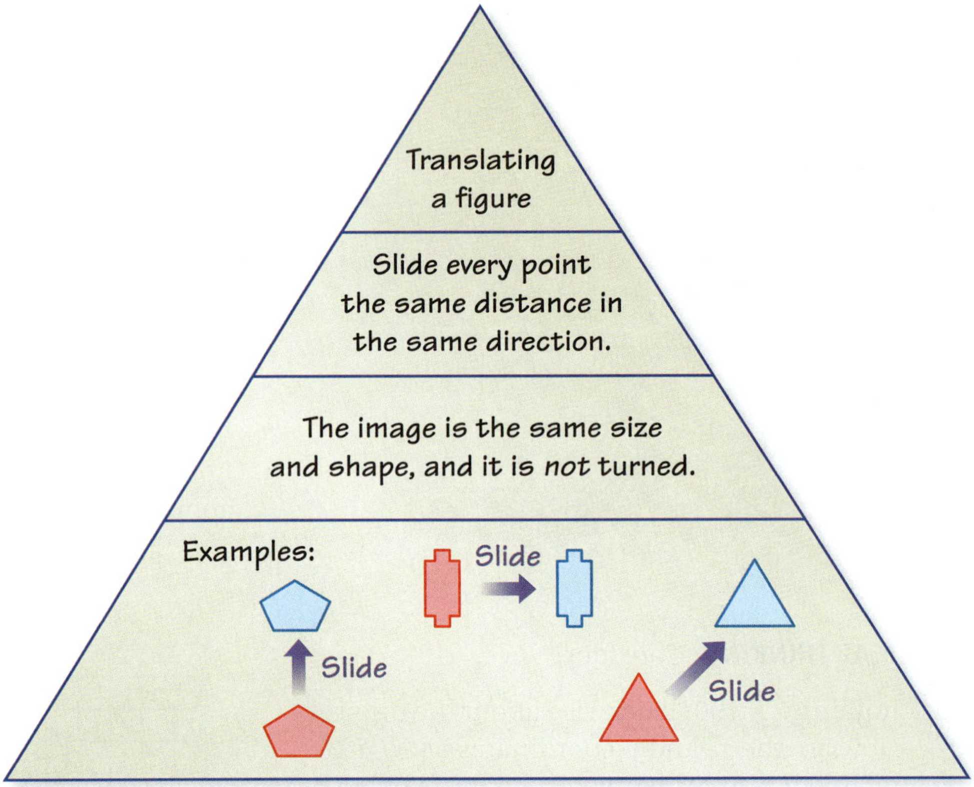

On Your Own

Make summary triangles to help you study these topics.

1. congruent figures
2. reflecting a figure
3. rotating a figure

After you complete this chapter, make summary triangles for the following topics.

4. similar figures
5. perimeters of similar figures
6. areas of similar figures
7. dilating a figure
8. transforming a figure

"I hope my owner sees my summary triangle. I just can't seem to learn 'roll over.' "

2.1–2.4 Quiz

Tell whether the two figures are congruent. Explain your reasoning. *(Section 2.1)*

1.

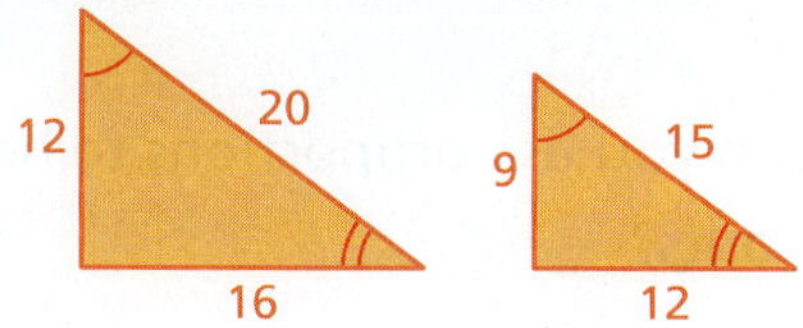

2.

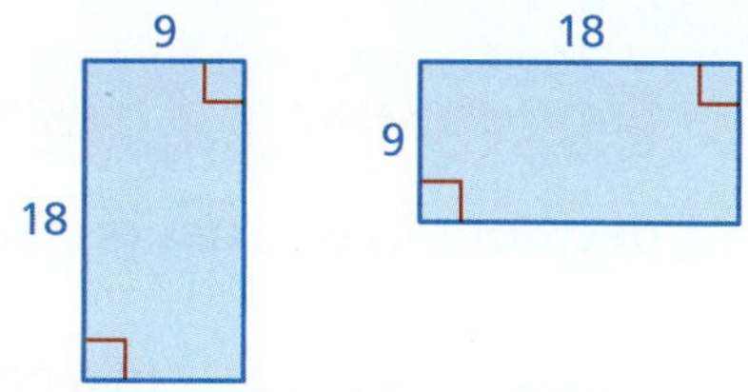

Tell whether the blue figure is a translation of the red figure. *(Section 2.2)*

3.

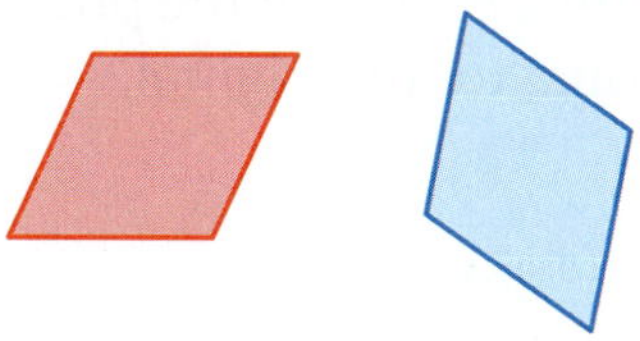

4.

Tell whether the blue figure is a reflection of the red figure. *(Section 2.3)*

5.

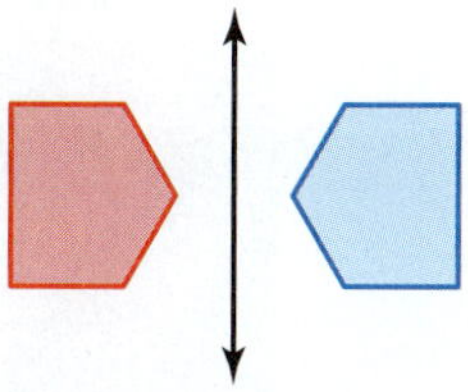

6.

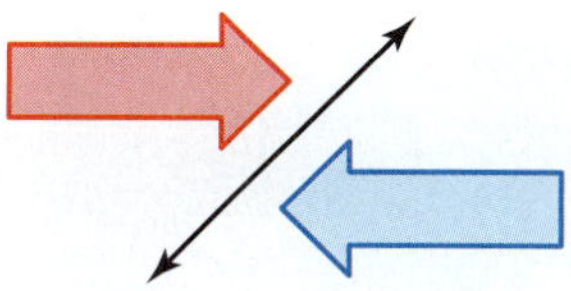

The red figure is congruent to the blue figure. Describe two different sequences of transformations in which the blue figure is the image of the red figure. *(Section 2.4)*

7.

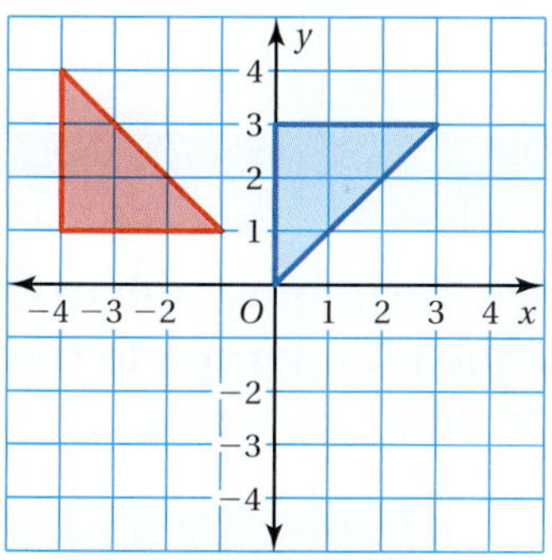

8.

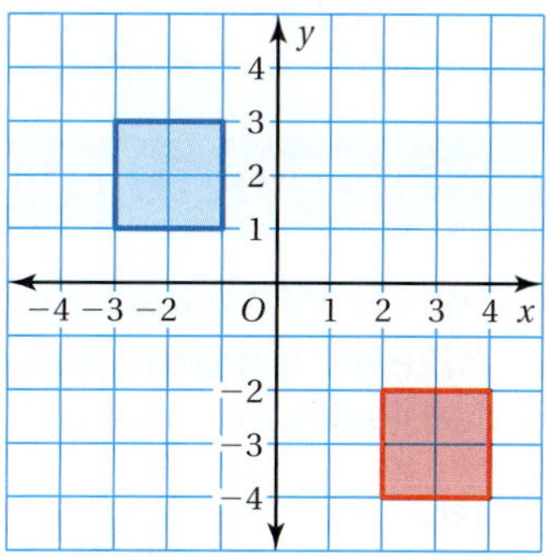

9. **AIRPLANE** Describe a translation of the airplane from point *A* to point *B*. *(Section 2.2)*

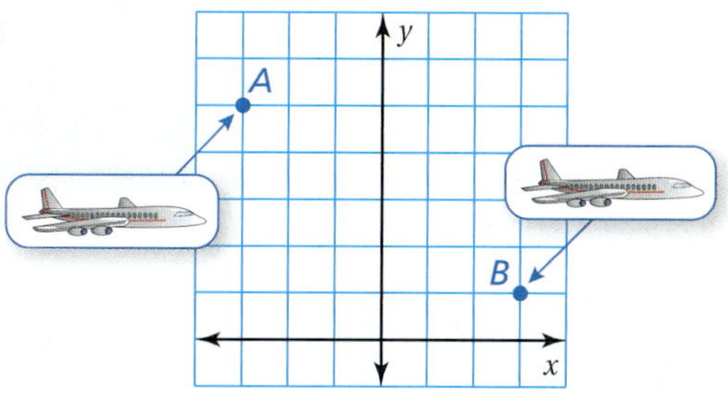

10. **MINIGOLF** You hit the golf ball along the red path so that its image will be a reflection in the *y*-axis. Does the golf ball land in the hole? Explain. *(Section 2.3)*

2.5 Similar Figures

Essential Question How can you use proportions to help make decisions in art, design, and magazine layouts?

Original photograph

In a computer art program, when you click and drag on a side of a photograph, you distort it.

But when you click and drag on a corner of the photograph, the dimensions remain proportional to the original.

Distorted

Distorted

Proportional

1 ACTIVITY: Reducing Photographs

Work with a partner. You are trying to reduce the photograph to the indicated size for a nature magazine. Can you reduce the photograph to the indicated size without distorting or cropping? Explain your reasoning.

a.

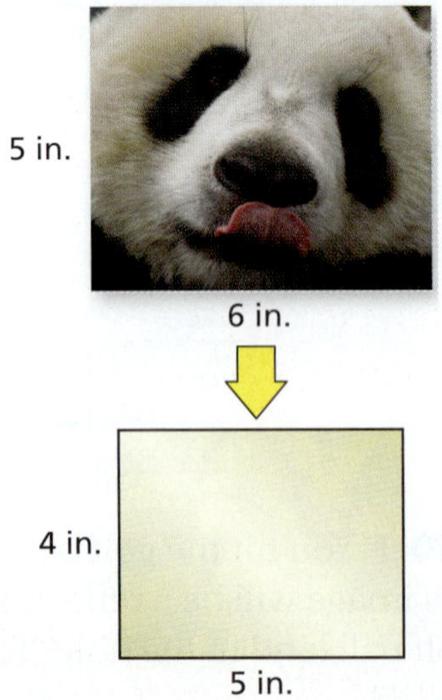

b.

COMMON CORE

Geometry

In this lesson, you will

- name corresponding angles and corresponding sides of similar figures.
- identify similar figures.
- find unknown measures of similar figures.

Preparing for Standard MACC.8.G.1.4

2 ACTIVITY: Creating Designs

Work with a partner.

> **Math Practice 4**
>
> **Analyze Relationships**
>
> How can you use mathematics to determine whether the dimensions are proportional?

a. Tell whether the dimensions of the new designs are proportional to the dimensions of the original design. Explain your reasoning.

Original **Design 1** **Design 2**

b. Draw two designs whose dimensions are proportional to the given design. Make one bigger and one smaller. Label the sides of the designs with their lengths.

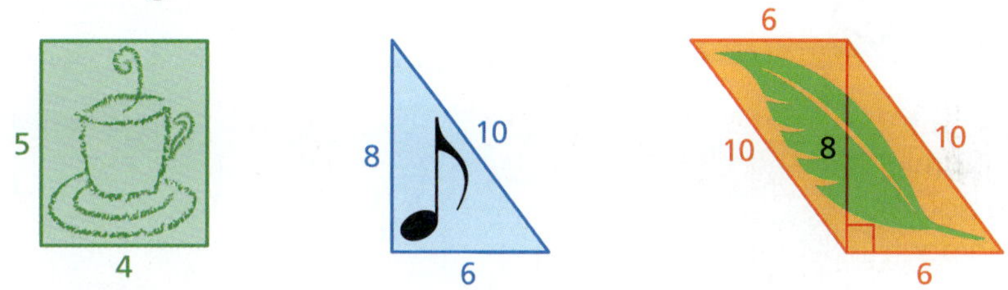

What Is Your Answer?

3. **IN YOUR OWN WORDS** How can you use proportions to help make decisions in art, design, and magazine layouts? Give two examples.

4. **a.** Use a computer art program to draw two rectangles whose dimensions are proportional to each other.

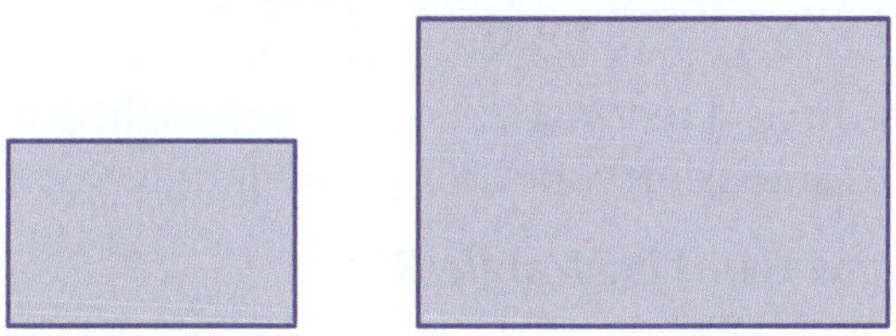

b. Print the two rectangles on the same piece of paper.

c. Use a centimeter ruler to measure the length and the width of each rectangle.

d. Find the following ratios. What can you conclude?

$$\frac{\text{Length of larger}}{\text{Length of smaller}} \qquad \frac{\text{Width of larger}}{\text{Width of smaller}}$$

"I love this statue. It seems similar to a big statue I saw in New York."

Practice Use what you learned about similar figures to complete Exercises 4 and 5 on page 74.

2.5 Lesson

Check It Out
Lesson Tutorials
BigIdeasMath.com

Key Vocabulary
similar figures, *p. 72*

Key Idea

Similar Figures

Figures that have the same shape but not necessarily the same size are called **similar figures**.

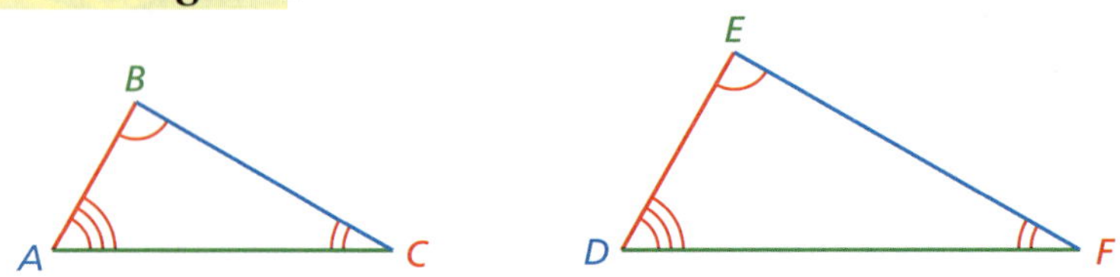

Triangle ABC is similar to Triangle DEF.

Words Two figures are similar when

- corresponding side lengths are proportional and
- corresponding angles are congruent.

Symbols

Side Lengths	*Angles*	*Figures*
$\frac{AB}{DE} = \frac{BC}{EF} = \frac{AC}{DF}$	$\angle A \cong \angle D$	$\triangle ABC \sim \triangle DEF$
	$\angle B \cong \angle E$	
	$\angle C \cong \angle F$	

Reading

The symbol ~ means *is similar to*.

Common Error

When writing a similarity statement, make sure to list the vertices of the figures in the correct order.

EXAMPLE 1 Identifying Similar Figures

Which rectangle is similar to Rectangle A?

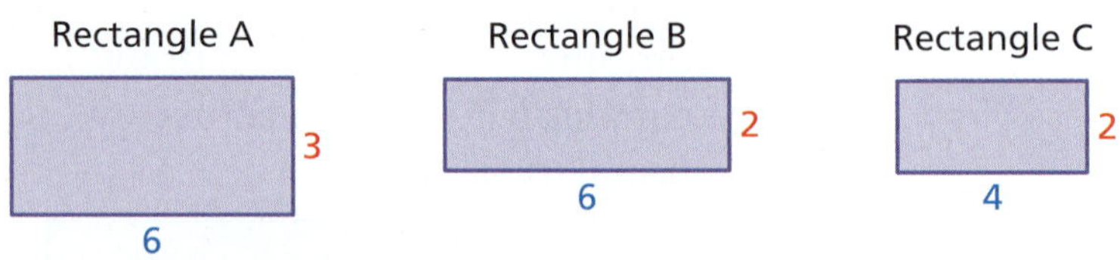

Each figure is a rectangle. So, corresponding angles are congruent. Check to see if corresponding side lengths are proportional.

Rectangle A and Rectangle B

$$\frac{\text{Length of A}}{\text{Length of B}} = \frac{6}{6} = 1 \qquad \frac{\text{Width of A}}{\text{Width of B}} = \frac{3}{2} \qquad \text{Not proportional}$$

Rectangle A and Rectangle C

$$\frac{\text{Length of A}}{\text{Length of C}} = \frac{6}{4} = \frac{3}{2} \qquad \frac{\text{Width of A}}{\text{Width of C}} = \frac{3}{2} \qquad \text{Proportional}$$

So, Rectangle C is similar to Rectangle A.

On Your Own

Now You're Ready
Exercises 4–7

1. Rectangle D is 3 units long and 1 unit wide. Which rectangle is similar to Rectangle D?

Multi-Language Glossary at BigIdeasMath.com

EXAMPLE 2 Finding an Unknown Measure in Similar Figures

The triangles are similar. Find x.

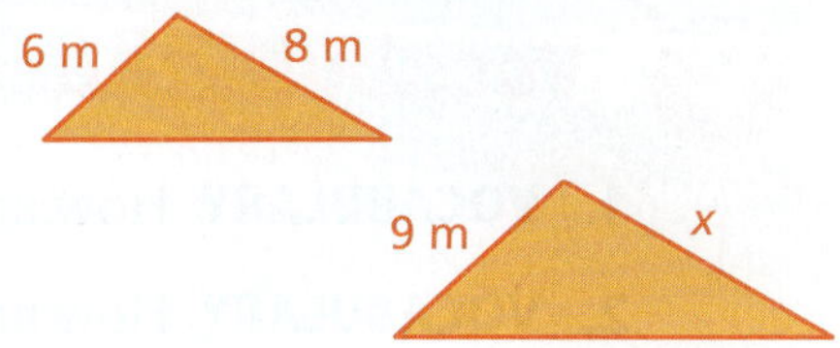

Because the triangles are similar, corresponding side lengths are proportional. So, write and solve a proportion to find x.

$$\frac{6}{9} = \frac{8}{x}$$ Write a proportion.

$$6x = 72$$ Cross Products Property

$$x = 12$$ Divide each side by 6.

So, x is 12 meters.

On Your Own

Now You're Ready
Exercises 8–11

The figures are similar. Find x.

2. 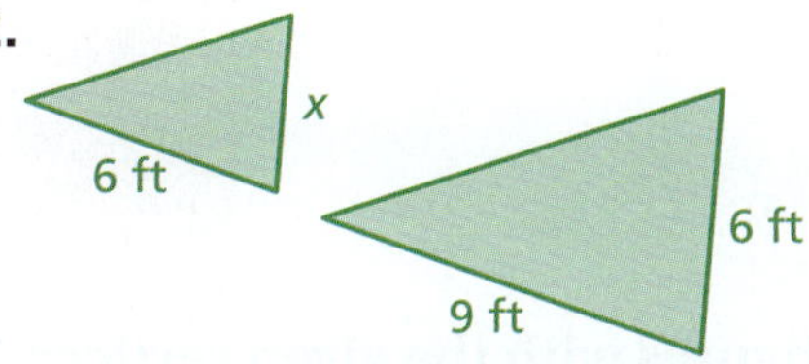

3. 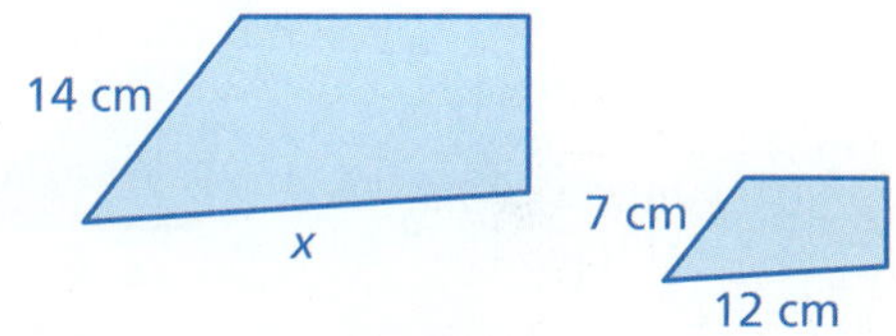

EXAMPLE 3 Real-Life Application

An artist draws a replica of a painting that is on the Berlin Wall. The painting includes a red trapezoid. The shorter base of the similar trapezoid in the replica is 3.75 inches. What is the height h of the trapezoid in the replica?

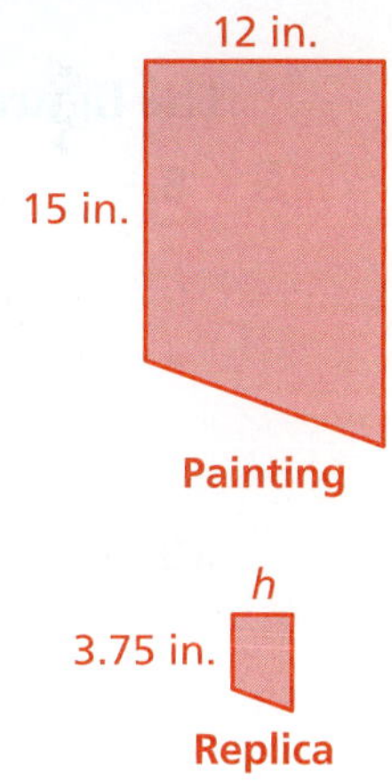

Because the trapezoids are similar, corresponding side lengths are proportional. So, write and solve a proportion to find h.

$$\frac{3.75}{15} = \frac{h}{12}$$ Write a proportion.

$$12 \cdot \frac{3.75}{15} = 12 \cdot \frac{h}{12}$$ Multiplication Property of Equality

$$3 = h$$ Simplify.

So, the height of the trapezoid in the replica is 3 inches.

On Your Own

4. **WHAT IF?** The longer base in the replica is 4.5 inches. What is the length of the longer base in the painting?

2.5 Exercises

Check It Out
Help with Homework
BigIdeasMath.com

Vocabulary and Concept Check

1. **VOCABULARY** How are corresponding angles of two similar figures related?
2. **VOCABULARY** How are corresponding side lengths of two similar figures related?
3. **CRITICAL THINKING** Are two figures that have the same size and shape similar? Explain.

Practice and Problem Solving

Tell whether the two figures are similar. Explain your reasoning.

1 **4.**

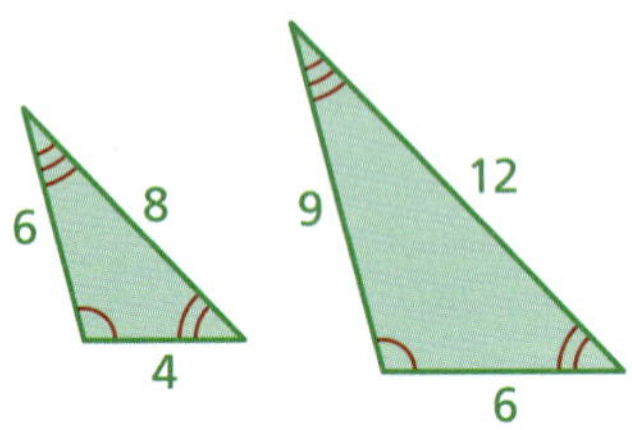

5.

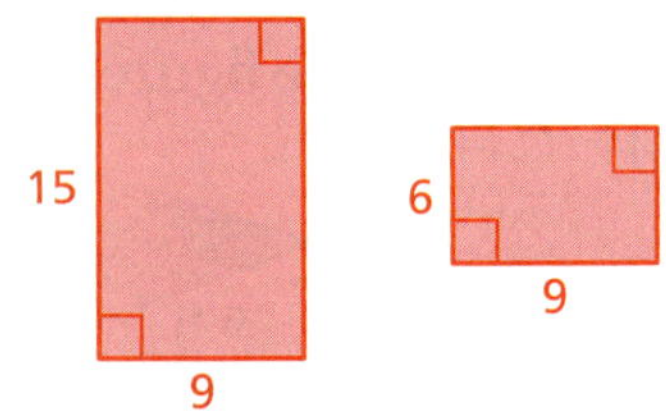

In a coordinate plane, draw the figures with the given vertices. Which figures are similar? Explain your reasoning.

6. Rectangle A: (0, 0), (4, 0), (4, 2), (0, 2)
Rectangle B: (0, 0), $(-6, 0)$, $(-6, 3)$, (0, 3)
Rectangle C: (0, 0), (4, 0), (4, 2), (0, 2)

7. Figure A: $(-4, 2)$, $(-2, 2)$, $(-2, 0)$, $(-4, 0)$
Figure B: (1, 4), (4, 4), (4, 1), (1, 1)
Figure C: $(2, -1)$, $(5, -1)$, $(5, -3)$, $(2, -3)$

The figures are similar. Find *x*.

2 **8.**

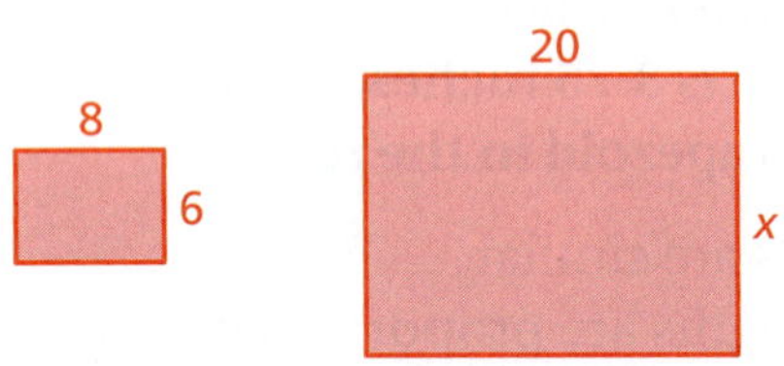

9.

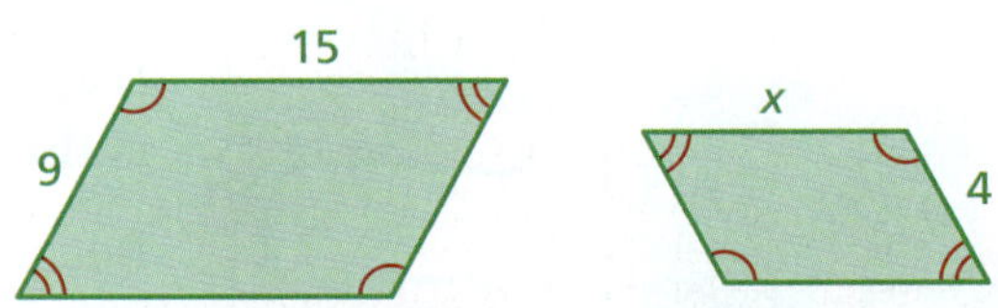

10.

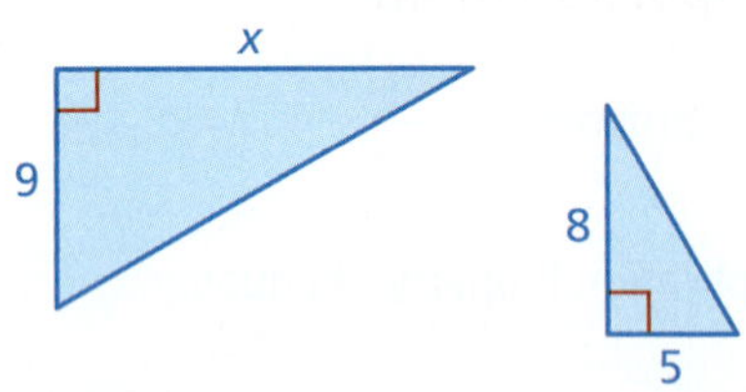

11.

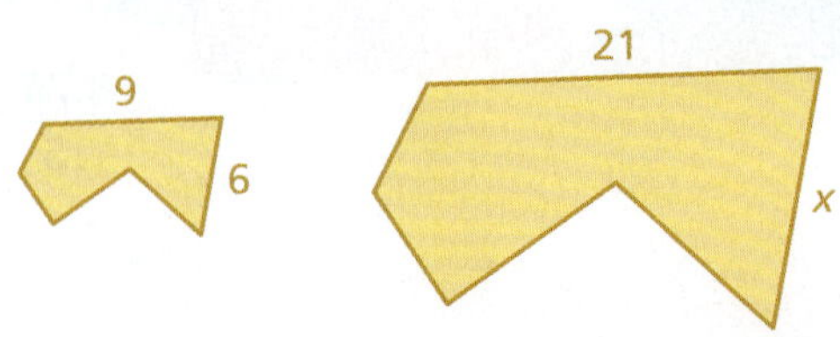

12. **MEXICO** A Mexican flag is 63 inches long and 36 inches wide. Is the drawing at the right similar to the Mexican flag?

13. **DESKS** A student's rectangular desk is 30 inches long and 18 inches wide. The teacher's desk is similar to the student's desk and has a length of 50 inches. What is the width of the teacher's desk?

14. **LOGIC** Are the following figures *always, sometimes,* or *never* similar? Explain.

- **a.** two triangles
- **b.** two squares
- **c.** two rectangles
- **d.** a square and a triangle

15. **CRITICAL THINKING** Can you draw two quadrilaterals each having two 130° angles and two 50° angles that are *not* similar? Justify your answer.

16. **SIGN** All the angle measures in the sign are 90°.

- **a.** You increase each side length by 20%. Is the new sign similar to the original?
- **b.** You increase each side length by 6 inches. Is the new sign similar to the original?

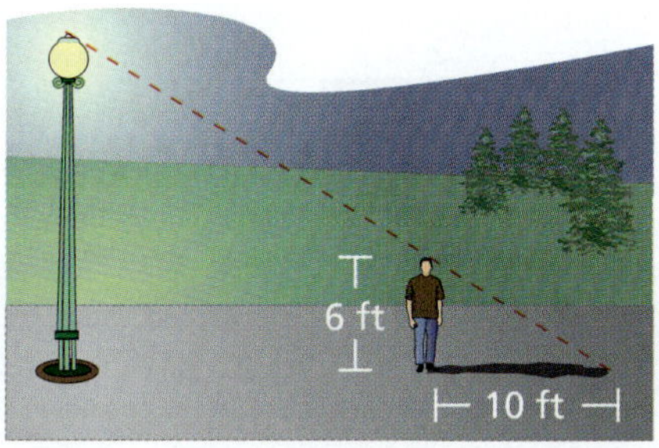

17. **STREETLIGHT** A person standing 20 feet from a streetlight casts a shadow as shown. How many times taller is the streetlight than the person? Assume the triangles are similar.

18. **REASONING** Is an object similar to a scale drawing of the object? Explain.

19. **GEOMETRY** Use a ruler to draw two different isosceles triangles similar to the one shown. Measure the heights of each triangle to the nearest centimeter.

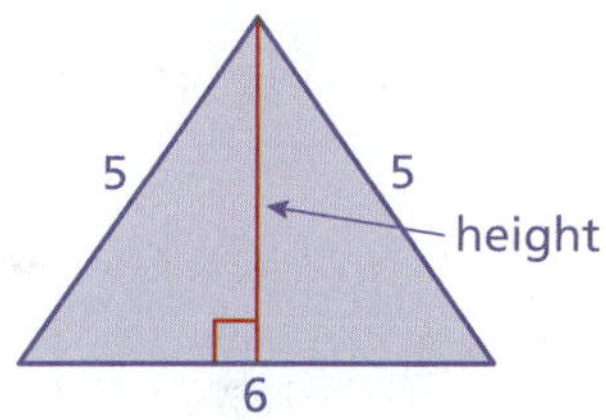

- **a.** Is the ratio of the corresponding heights proportional to the ratio of the corresponding side lengths?
- **b.** Do you think this is true for all similar triangles? Explain.

20. **Critical Thinking** Given $\triangle ABC \sim \triangle DEF$ and $\triangle DEF \sim \triangle JKL$, is $\triangle ABC \sim \triangle JKL$? Give an example or a non-example.

Fair Game Review What you learned in previous grades & lessons

Simplify. *(Skills Review Handbook)*

21. $\left(\frac{4}{9}\right)^2$ 22. $\left(\frac{3}{8}\right)^2$ 23. $\left(\frac{7}{4}\right)^2$ 24. $\left(\frac{6.5}{2}\right)^2$

25. **MULTIPLE CHOICE** You solve the equation $S = \ell w + 2wh$ for w. Which equation is correct? *(Section 1.4)*

Ⓐ $w = \frac{S - \ell}{2h}$ Ⓑ $w = \frac{S - 2h}{\ell}$ Ⓒ $w = \frac{S}{\ell + 2h}$ Ⓓ $w = S - \ell - 2h$

2.6 Perimeters and Areas of Similar Figures

Essential Question How do changes in dimensions of similar geometric figures affect the perimeters and the areas of the figures?

1 ACTIVITY: Creating Similar Figures

Work with a partner. Use pattern blocks to make a figure whose dimensions are 2, 3, and 4 times greater than those of the original figure.

a. Square

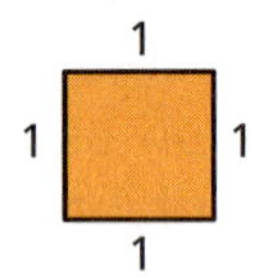

b. Rectangle

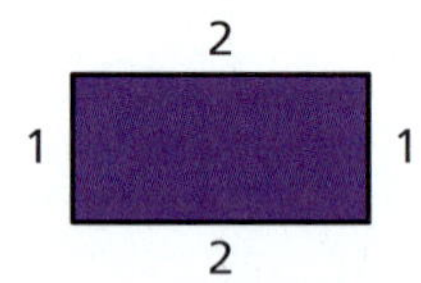

2 ACTIVITY: Finding Patterns for Perimeters

Work with a partner. Copy and complete the table for the perimeter P of each figure in Activity 1. Describe the pattern.

Figure	Original Side Lengths	Double Side Lengths	Triple Side Lengths	Quadruple Side Lengths
	$P =$			
	$P =$			

3 ACTIVITY: Finding Patterns for Areas

Work with a partner. Copy and complete the table for the area A of each figure in Activity 1. Describe the pattern.

Figure	Original Side Lengths	Double Side Lengths	Triple Side Lengths	Quadruple Side Lengths
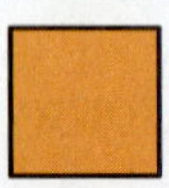	$A =$			
	$A =$			

COMMON CORE

Geometry

In this lesson, you will

- understand the relationship between perimeters of similar figures.
- understand the relationship between areas of similar figures.
- find ratios of perimeters and areas for similar figures.

Preparing for Standard MACC.8.G.1.4

4 ACTIVITY: Drawing and Labeling Similar Figures

Work with a partner.

a. Find a blue rectangle that is similar to the red rectangle and has one side from $(-1, -6)$ to $(5, -6)$. Label the vertices.

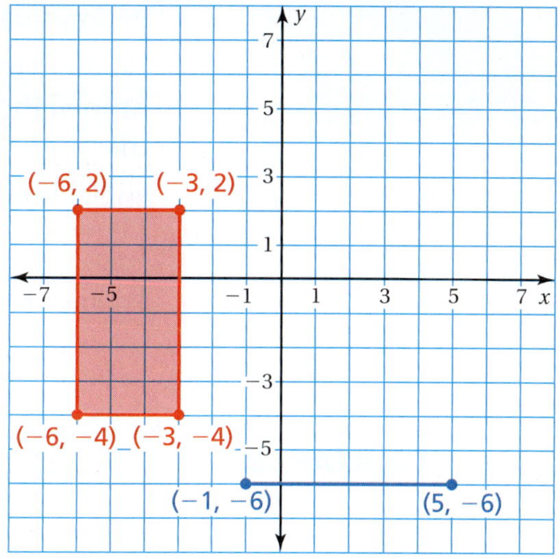

Check that the two rectangles are similar by showing that the ratios of corresponding sides are equal.

$$\frac{\text{Red Length}}{\text{Blue Length}} \stackrel{?}{=} \frac{\text{Red Width}}{\text{Blue Width}}$$

$$\frac{\text{change in } y}{\text{change in } y} \stackrel{?}{=} \frac{\text{change in } x}{\text{change in } x}$$

$$\frac{\square}{\square} \stackrel{?}{=} \frac{\square}{\square}$$

$$\frac{\square}{\square} \stackrel{?}{=} \frac{\square}{\square}$$

∴ The ratios are equal. So, the rectangles are similar.

Math Practice 1

Analyze Givens

What values should you use to fill in the proportion? Does it matter where each value goes? Explain.

b. Compare the perimeters and the areas of the figures. Are the results the same as your results from Activities 2 and 3? Explain.

c. There are three other blue rectangles that are similar to the red rectangle and have the given side.

- Draw each one. Label the vertices of each.
- Show that each is similar to the original red rectangle.

What Is Your Answer?

5. **IN YOUR OWN WORDS** How do changes in dimensions of similar geometric figures affect the perimeters and the areas of the figures?

6. What information do you need to know to find the dimensions of a figure that is similar to another figure? Give examples to support your explanation.

Practice Use what you learned about perimeters and areas of similar figures to complete Exercises 8 and 9 on page 80.

2.6 Lesson

Check It Out
Lesson Tutorials
BigIdeasMath.com

Key Idea

Perimeters of Similar Figures

When two figures are similar, the ratio of their perimeters is equal to the ratio of their corresponding side lengths.

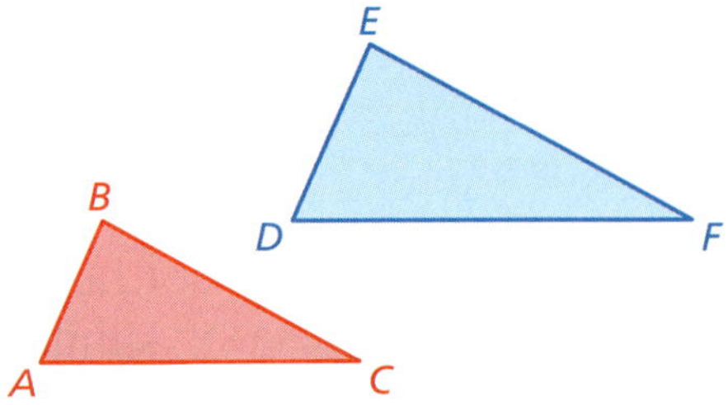

$$\frac{\text{Perimeter of } \triangle ABC}{\text{Perimeter of } \triangle DEF} = \frac{AB}{DE} = \frac{BC}{EF} = \frac{AC}{DF}$$

EXAMPLE 1 Finding Ratios of Perimeters

Find the ratio (red to blue) of the perimeters of the similar rectangles.

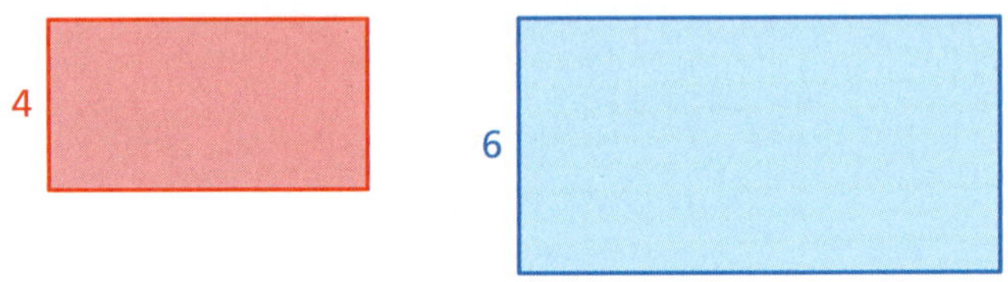

$$\frac{\text{Perimeter of red rectangle}}{\text{Perimeter of blue rectangle}} = \frac{4}{6} = \frac{2}{3}$$

∴ The ratio of the perimeters is $\frac{2}{3}$.

On Your Own

1. The height of Figure A is 9 feet. The height of a similar Figure B is 15 feet. What is the ratio of the perimeter of A to the perimeter of B?

Key Idea

Areas of Similar Figures

When two figures are similar, the ratio of their areas is equal to the *square* of the ratio of their corresponding side lengths.

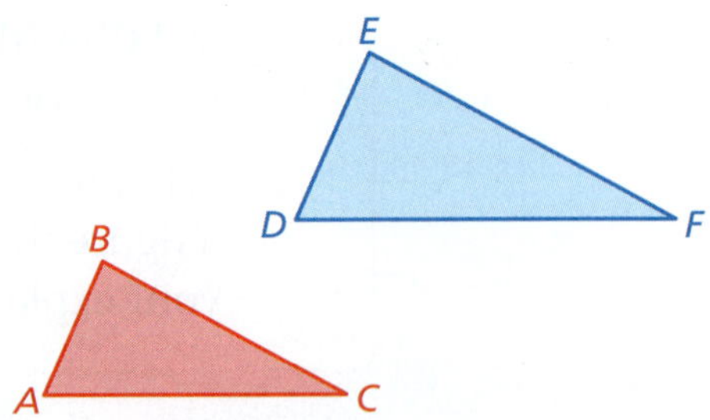

$$\frac{\text{Area of } \triangle ABC}{\text{Area of } \triangle DEF} = \left(\frac{AB}{DE}\right)^2 = \left(\frac{BC}{EF}\right)^2 = \left(\frac{AC}{DF}\right)^2$$

EXAMPLE 2 Finding Ratios of Areas

Find the ratio (red to blue) of the areas of the similar triangles.

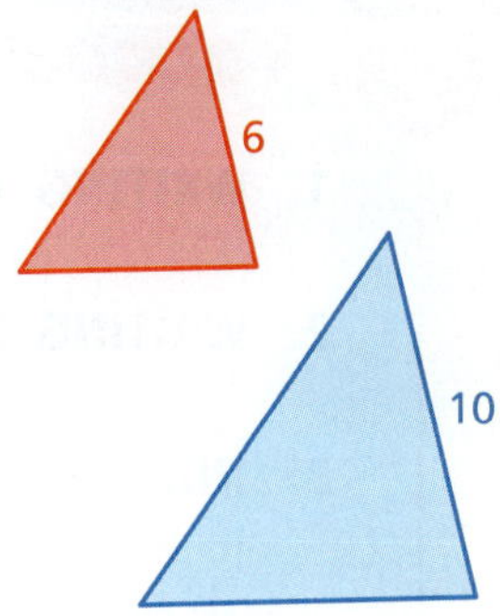

$$\frac{\text{Area of red triangle}}{\text{Area of blue triangle}} = \left(\frac{6}{10}\right)^2$$

$$= \left(\frac{3}{5}\right)^2 = \frac{9}{25}$$

∴ The ratio of the areas is $\frac{9}{25}$.

On Your Own

Now You're Ready
Exercises 4–7

2. The base of Triangle P is 8 meters. The base of a similar Triangle Q is 7 meters. What is the ratio of the area of P to the area of Q?

EXAMPLE 3 Using Proportions to Find Perimeters and Areas

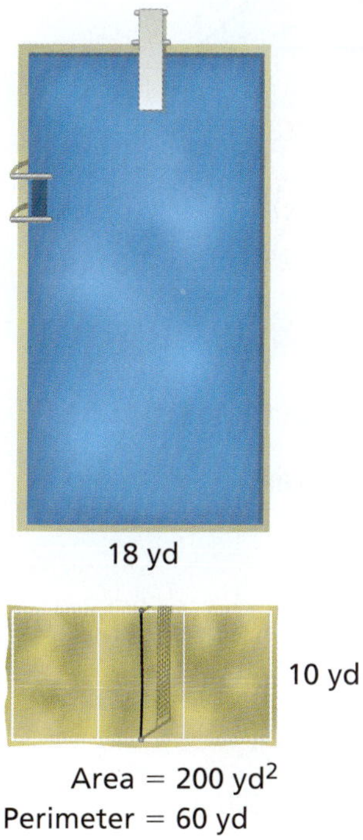

A swimming pool is similar in shape to a volleyball court. Find the perimeter *P* and the area *A* of the pool.

The rectangular pool and the court are similar. So, use the ratio of corresponding side lengths to write and solve proportions to find the perimeter and the area of the pool.

Perimeter

$$\frac{\text{Perimeter of court}}{\text{Perimeter of pool}} = \frac{\text{Width of court}}{\text{Width of pool}}$$

$$\frac{60}{P} = \frac{10}{18}$$

$$1080 = 10P$$

$$108 = P$$

Area

$$\frac{\text{Area of court}}{\text{Area of pool}} = \left(\frac{\text{Width of court}}{\text{Width of pool}}\right)^2$$

$$\frac{200}{A} = \left(\frac{10}{18}\right)^2$$

$$\frac{200}{A} = \frac{100}{324}$$

$$64{,}800 = 100A$$

$$648 = A$$

∴ So, the perimeter of the pool is 108 yards, and the area is 648 square yards.

On Your Own

3. **WHAT IF?** The width of the pool is 16 yards. Find the perimeter *P* and the area *A* of the pool.

2.6 Exercises

Check It Out
Help with Homework
BigIdeasMath.com

Vocabulary and Concept Check

1. **WRITING** How are the perimeters of two similar figures related?

2. **WRITING** How are the areas of two similar figures related?

3. **NUMBER SENSE** Rectangle *ABCD* is similar to Rectangle *WXYZ*. The area of *ABCD* is 30 square inches. Explain how to find the area of *WXYZ*.

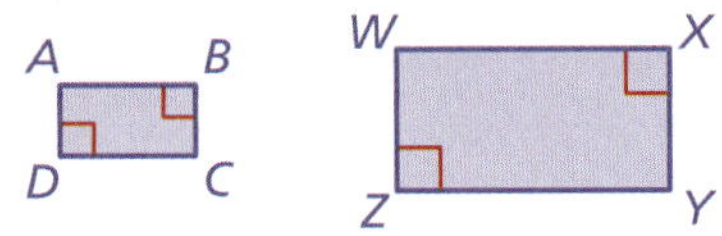

$\frac{AD}{WZ} = \frac{1}{2}$ $\frac{AB}{WX} = \frac{1}{2}$

Practice and Problem Solving

The two figures are similar. Find the ratios (red to blue) of the perimeters and of the areas.

1 2 4.

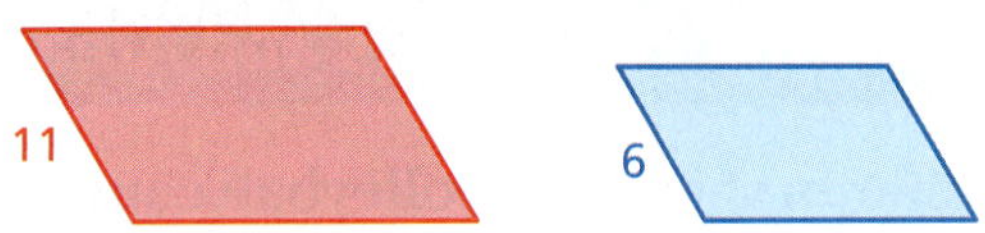

5.

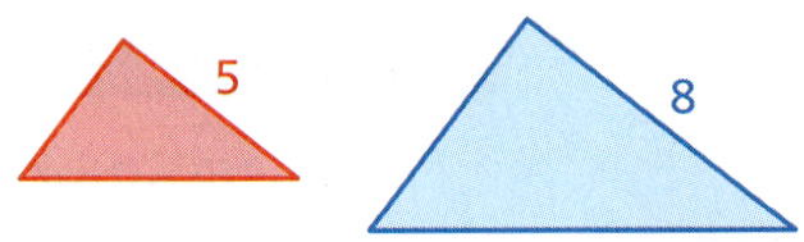

6.

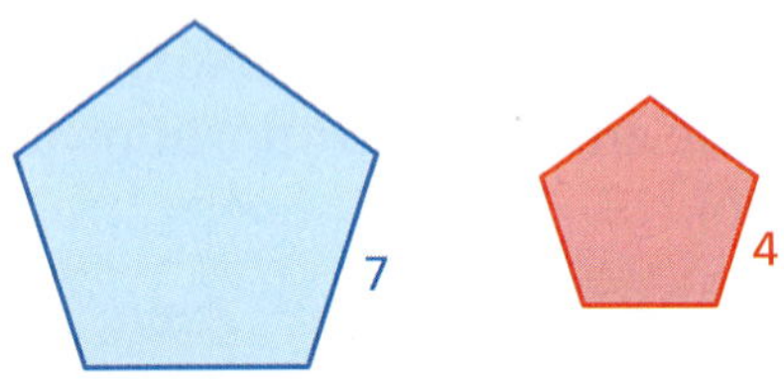

7.

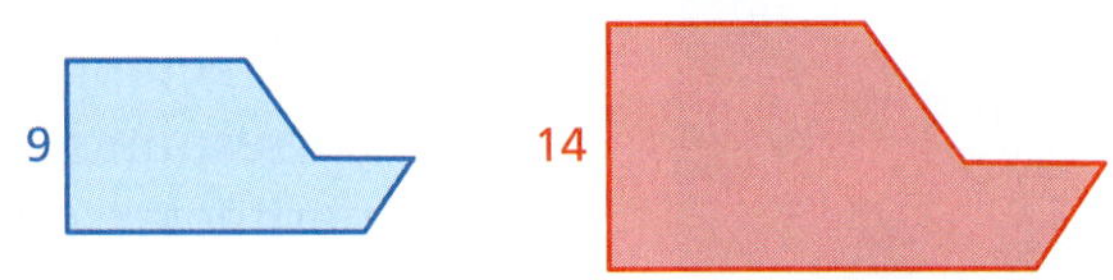

8. **PERIMETER** How does doubling the side lengths of a right triangle affect its perimeter?

9. **AREA** How does tripling the side lengths of a right triangle affect its area?

The figures are similar. Find *x*.

10. The ratio of the perimeters is 7 : 10.

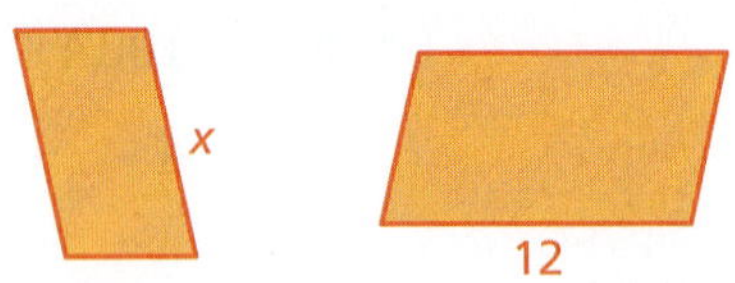

11. The ratio of the perimeters is 8 : 5.

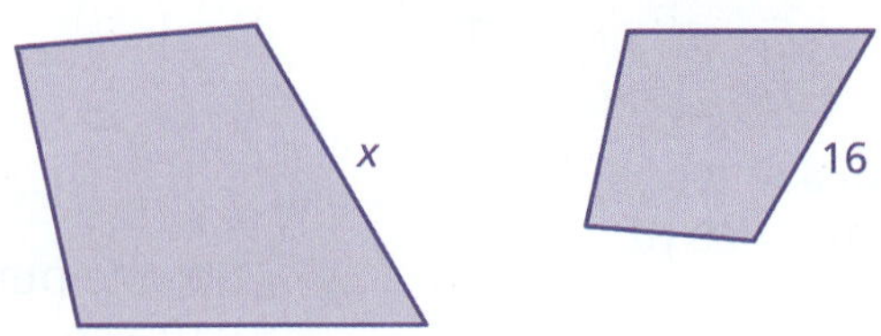

12. **FOOSBALL** The playing surfaces of two foosball tables are similar. The ratio of the corresponding side lengths is 10 : 7. What is the ratio of the areas?

13. **CHEERLEADING** A rectangular school banner has a length of 44 inches, a perimeter of 156 inches, and an area of 1496 square inches. The cheerleaders make signs similar to the banner. The length of a sign is 11 inches. What is its perimeter and its area?

14. REASONING The vertices of two rectangles are $A(-5, -1)$, $B(-1, -1)$, $C(-1, -4)$, $D(-5, -4)$ and $W(1, 6)$, $X(7, 6)$, $Y(7, -2)$, $Z(1, -2)$. Compare the perimeters and the areas of the rectangles. Are the rectangles similar? Explain.

15. SQUARE The ratio of the side length of Square A to the side length of Square B is 4 : 9. The side length of Square A is 12 yards. What is the perimeter of Square B?

16. FABRIC The cost of the fabric is \$1.31. What would you expect to pay for a similar piece of fabric that is 18 inches by 42 inches?

17. AMUSEMENT PARK A scale model of a merry-go-round and the actual merry-go-round are similar.

a. How many times greater is the base area of the actual merry-go-round than the base area of the scale model? Explain.

b. What is the base area of the actual merry-go-round in square feet?

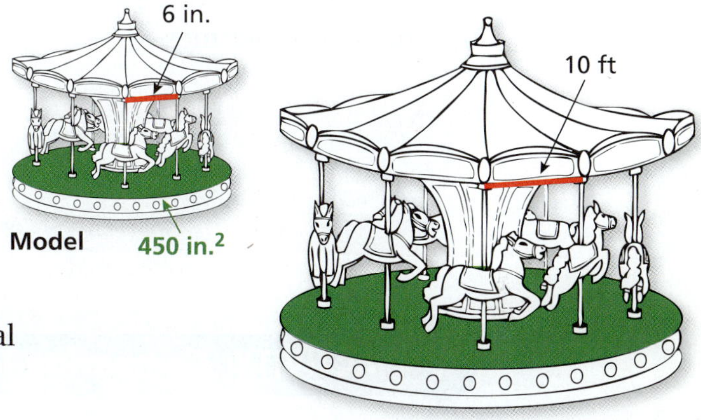

18. STRUCTURE The circumference of Circle K is π. The circumference of Circle L is 4π.

a. What is the ratio of their circumferences? of their radii? of their areas?

b. What do you notice?

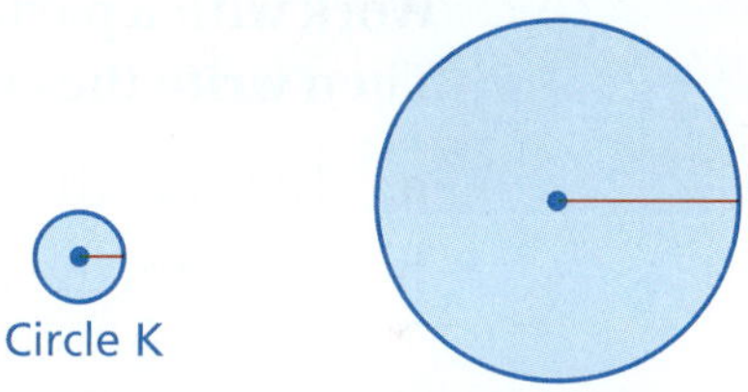

19. GEOMETRY A triangle with an area of 10 square meters has a base of 4 meters. A similar triangle has an area of 90 square meters. What is the *height* of the larger triangle?

20. Problem Solving You need two bottles of fertilizer to treat the flower garden shown. How many bottles do you need to treat a similar garden with a perimeter of 105 feet?

Fair Game Review What you learned in previous grades & lessons

Solve the equation. Check your solution. *(Section 1.3)*

21. $4x + 12 = -2x$ **22.** $2b + 6 = 7b - 2$ **23.** $8(4n + 13) = 6n$

24. MULTIPLE CHOICE Last week, you collected 20 pounds of cans for recycling. This week, you collect 25 pounds of cans for recycling. What is the percent of increase? *(Skills Review Handbook)*

Ⓐ 20% Ⓑ 25% Ⓒ 80% Ⓓ 125%

2.7 Dilations

Essential Question How can you enlarge or reduce a figure in the coordinate plane?

The Meaning of a Word ● Dilate

When you have your eyes checked, the optometrist sometimes **dilates** one or both of the pupils of your eyes.

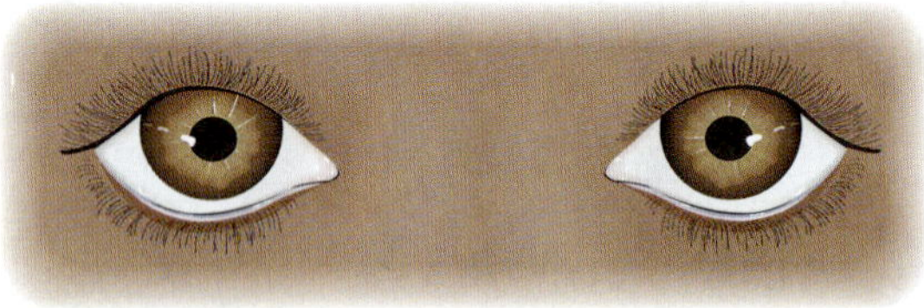

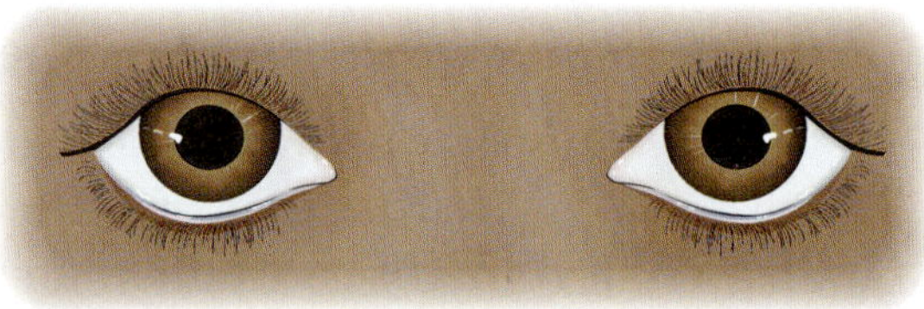

1 ACTIVITY: Comparing Triangles in a Coordinate Plane

Work with a partner. Write the coordinates of the vertices of the blue triangle. Then write the coordinates of the vertices of the red triangle.

a. How are the two sets of coordinates related?

b. How are the two triangles related? Explain your reasoning.

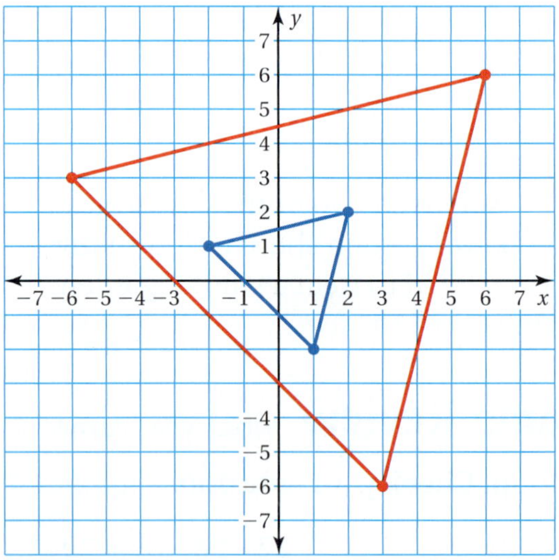

c. Draw a green triangle whose coordinates are twice the values of the corresponding coordinates of the blue triangle. How are the green and blue triangles related? Explain your reasoning.

d. How are the coordinates of the red and green triangles related? How are the two triangles related? Explain your reasoning.

COMMON CORE

Geometry

In this lesson, you will

- identify dilations.
- dilate figures in the coordinate plane.
- use more than one transformation to find images of figures.

Learning Standards
MACC.8.G.1.3
MACC.8.G.1.4

2 ACTIVITY: Drawing Triangles in a Coordinate Plane

Work with a partner.

a. Draw the triangle whose vertices are (0, 2), (−2, 2), and (1, −2).

b. Multiply each coordinate of the vertices by 2 to obtain three new vertices. Draw the triangle given by the three new vertices. How are the two triangles related?

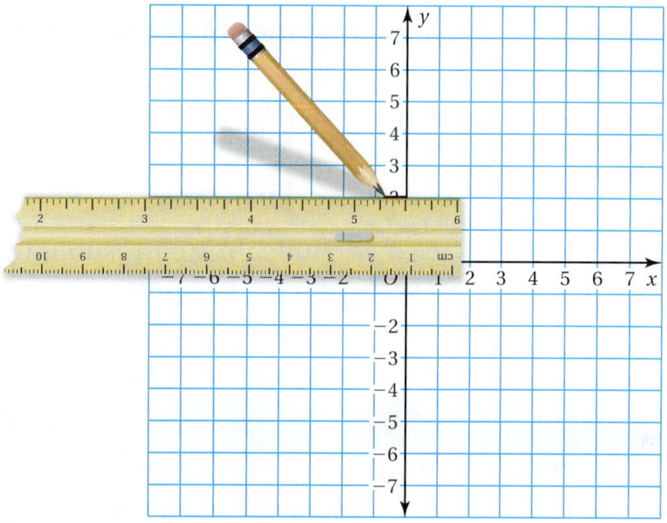

c. Repeat part (b) by multiplying by 3 instead of 2.

Math Practice 3

Use Prior Results

What are the four types of transformations you studied in this chapter? What information can you use to fill in your table?

3 ACTIVITY: Summarizing Transformations

Work with a partner. Make a table that summarizes the relationships between the original figure and its image for the four types of transformations you studied in this chapter.

What Is Your Answer?

4. **IN YOUR OWN WORDS** How can you enlarge or reduce a figure in the coordinate plane?

5. Describe how knowing how to enlarge or reduce figures in a technical drawing is important in a career such as drafting.

Practice Use what you learned about dilations to complete Exercises 4–6 on page 87.

2.7 Lesson

Check It Out
Lesson Tutorials
BigIdeasMath.com

A **dilation** is a transformation in which a figure is made larger or smaller with respect to a point called the **center of dilation**.

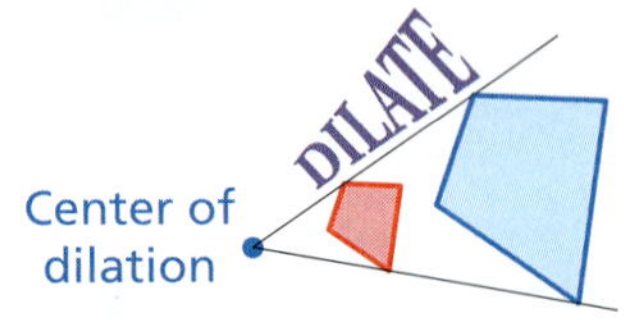

EXAMPLE 1 Identifying a Dilation

Key Vocabulary
dilation, *p. 84*
center of dilation, *p. 84*
scale factor, *p. 84*

Tell whether the blue figure is a dilation of the red figure.

a.

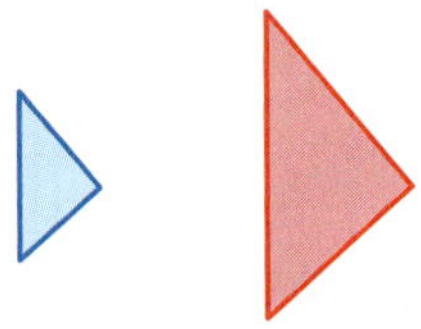

Lines connecting corresponding vertices meet at a point.

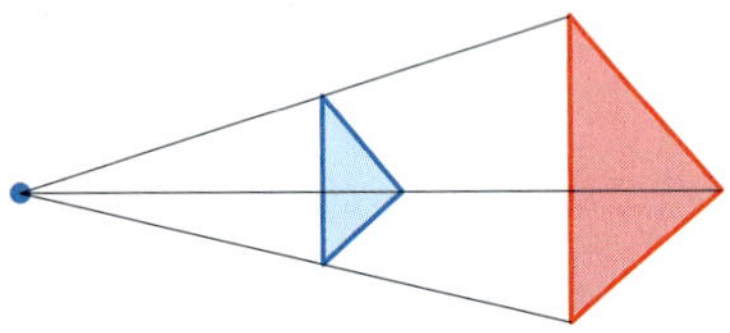

So, the blue figure is a dilation of the red figure.

b.

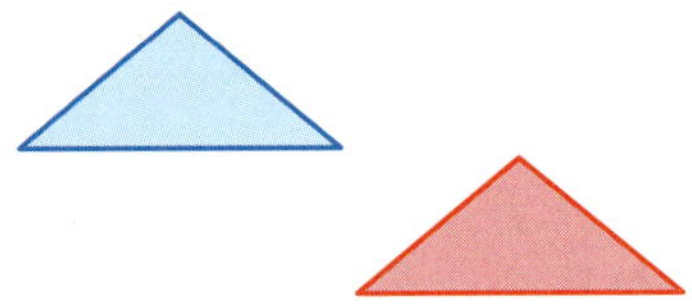

The figures have the same size and shape. The red figure *slides* to form the blue figure.

So, the blue figure is *not* a dilation of the red figure. It is a translation.

On Your Own

Now You're Ready
Exercises 7–12

Tell whether the blue figure is a dilation of the red figure. Explain.

1.

2.

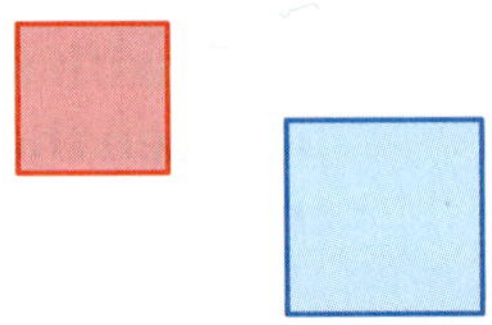

In a dilation, the original figure and its image are similar. The ratio of the side lengths of the image to the corresponding side lengths of the original figure is the **scale factor** of the dilation.

Key Idea

Dilations in the Coordinate Plane

Words To dilate a figure with respect to the origin, multiply the coordinates of each vertex by the scale factor k.

Algebra $(x, y) \rightarrow (kx, ky)$

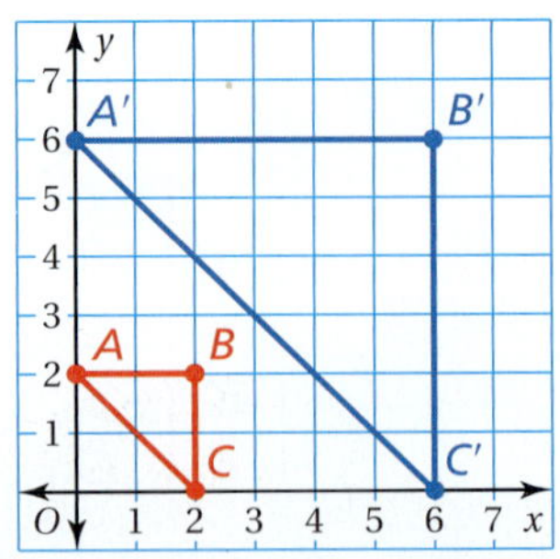

- When $k > 1$, the dilation is an enlargement.
- When $k > 0$ and $k < 1$, the dilation is a reduction.

Multi-Language Glossary at BigIdeasMath.com

EXAMPLE 2 Dilating a Figure

Draw the image of Triangle *ABC* after a dilation with a scale factor of 3. Identify the type of dilation.

Multiply each *x*- and *y*-coordinate by the scale factor 3.

Vertices of *ABC*	(3x, 3y)	Vertices of *A′B′C′*
A(1, 3)	(3 • 1, 3 • 3)	*A′*(3, 9)
B(2, 3)	(3 • 2, 3 • 3)	*B′*(6, 9)
C(2, 1)	(3 • 2, 3 • 1)	*C′*(6, 3)

Study Tip

You can check your answer by drawing a line from the origin through each vertex of the original figure. The vertices of the image should lie on these lines.

The image is shown at the right. The dilation is an *enlargement* because the scale factor is greater than 1.

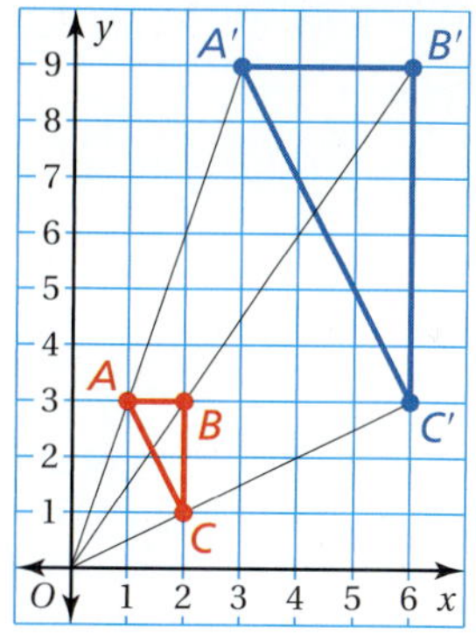

EXAMPLE 3 Dilating a Figure

Draw the image of Rectangle *WXYZ* after a dilation with a scale factor of 0.5. Identify the type of dilation.

Multiply each *x*- and *y*-coordinate by the scale factor 0.5.

Vertices of *WXYZ*	(0.5x, 0.5y)	Vertices of *W′X′Y′Z′*
W(−4, −6)	(0.5 • (−4), 0.5 • (−6))	*W′*(−2, −3)
X(−4, 8)	(0.5 • (−4), 0.5 • 8)	*X′*(−2, 4)
Y(4, 8)	(0.5 • 4, 0.5 • 8)	*Y′*(2, 4)
Z(4, −6)	(0.5 • 4, 0.5 • (−6))	*Z′*(2, −3)

The image is shown at the right. The dilation is a *reduction* because the scale factor is greater than 0 and less than 1.

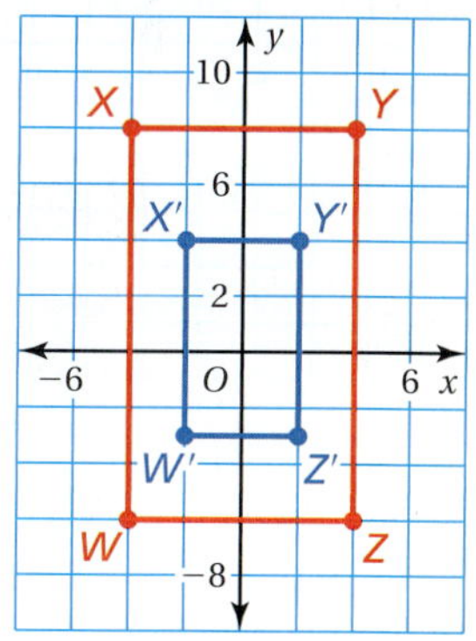

On Your Own

Now You're Ready
Exercises 13–18

3. **WHAT IF?** Triangle *ABC* in Example 2 is dilated by a scale factor of 2. What are the coordinates of the image?

4. **WHAT IF?** Rectangle *WXYZ* in Example 3 is dilated by a scale factor of $\frac{1}{4}$. What are the coordinates of the image?

EXAMPLE 4 Using More than One Transformation

The vertices of a trapezoid are $A(-2, -1)$, $B(-1, 1)$, $C(0, 1)$, and $D(0, -1)$. Dilate the trapezoid with respect to the origin using a scale factor of 2. Then translate it 6 units right and 2 units up. What are the coordinates of the image?

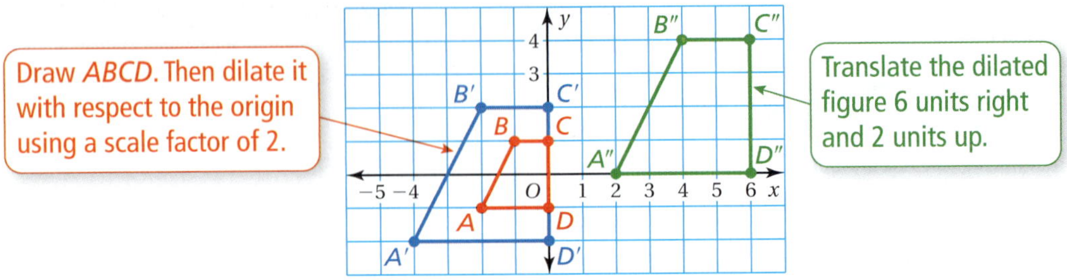

The coordinates of the image are $A''(2, 0)$, $B''(4, 4)$, $C''(6, 4)$, and $D''(6, 0)$.

The image of a translation, reflection, or rotation is congruent to the original figure, and the image of a dilation is similar to the original figure. So, two figures are similar when one can be obtained from the other by a sequence of translations, reflections, rotations, and dilations.

EXAMPLE 5 Describing a Sequence of Transformations

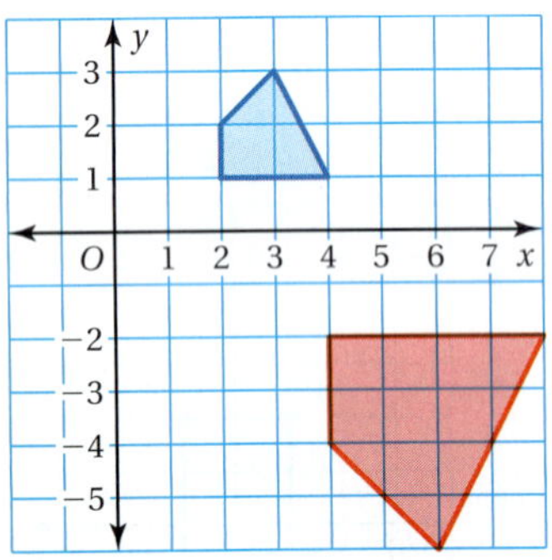

The red figure is similar to the blue figure. Describe a sequence of transformations in which the blue figure is the image of the red figure.

From the graph, you can see that the blue figure is one-half the size of the red figure. So, begin with a dilation with respect to the origin using a scale factor of $\frac{1}{2}$.

After dilating, you need to flip the figure in the x-axis.

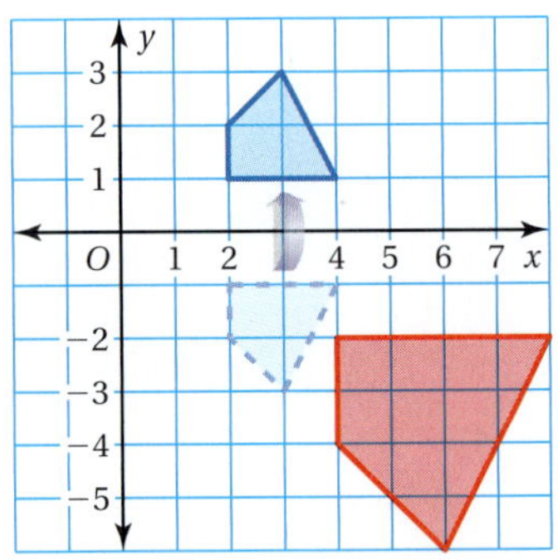

So, one possible sequence of transformations is a dilation with respect to the origin using a scale factor of $\frac{1}{2}$ followed by a reflection in the x-axis.

On Your Own

Now You're Ready
Exercises 23–28

5. In Example 4, use a scale factor of 3 in the dilation. Then rotate the figure 180° about the image of vertex C. What are the coordinates of the image?
6. In Example 5, can you reflect the red figure first, and then perform the dilation to obtain the blue figure? Explain.

2.7 Exercises

Check It Out
Help with Homework
BigIdeasMath.com

Vocabulary and Concept Check

1. **VOCABULARY** How is a dilation different from other transformations?
2. **VOCABULARY** For what values of scale factor k is a dilation called an *enlargement*? a *reduction*?
3. **REASONING** Which figure is *not* a dilation of the blue figure? Explain.

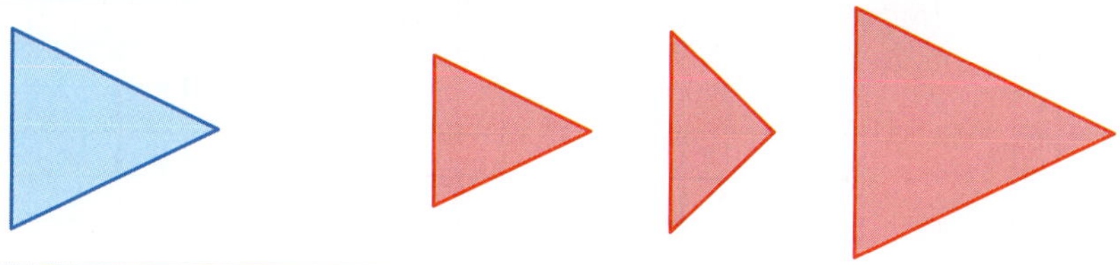

Practice and Problem Solving

Draw the triangle with the given vertices. Multiply each coordinate of the vertices by 3, and then draw the new triangle. How are the two triangles related?

4. (0, 2), (3, 2), (3, 0)
5. (−1, 1), (−1, −2), (2, −2)
6. (−3, 2), (1, 2), (1, −4)

Tell whether the blue figure is a dilation of the red figure.

7.

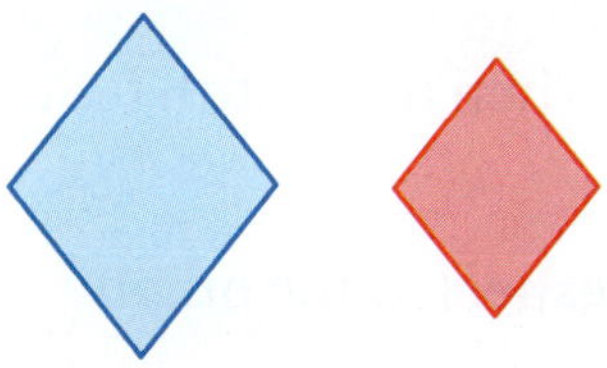

8.

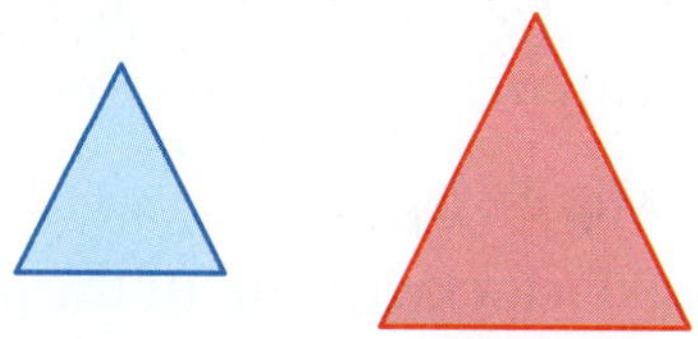

9.

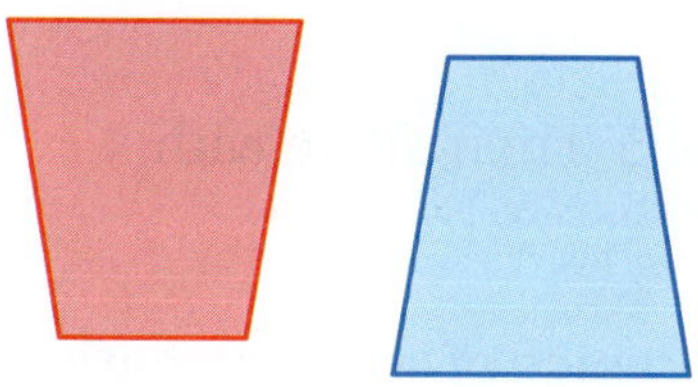

10.

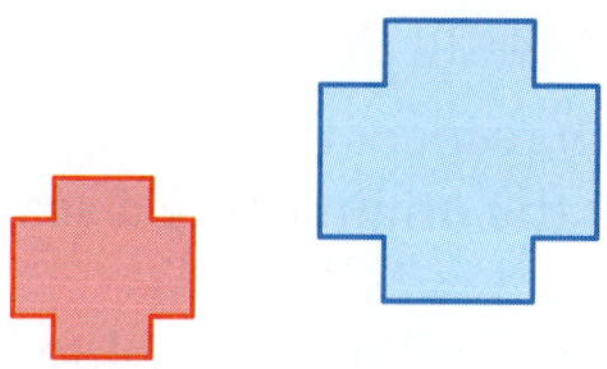

11.

12.

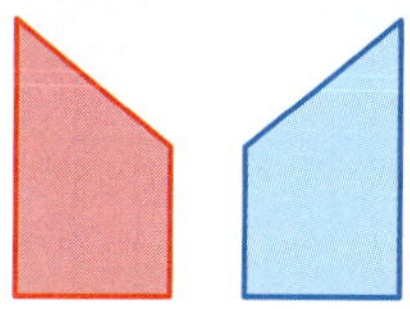

The vertices of a figure are given. Draw the figure and its image after a dilation with the given scale factor. Identify the type of dilation.

13. $A(1, 1)$, $B(1, 4)$, $C(3, 1)$; $k = 4$
14. $D(0, 2)$, $E(6, 2)$, $F(6, 4)$; $k = 0.5$
15. $G(-2, -2)$, $H(-2, 6)$, $J(2, 6)$; $k = 0.25$
16. $M(2, 3)$, $N(5, 3)$, $P(5, 1)$; $k = 3$
17. $Q(-3, 0)$, $R(-3, 6)$, $T(4, 6)$, $U(4, 0)$; $k = \frac{1}{3}$
18. $V(-2, -2)$, $W(-2, 3)$, $X(5, 3)$, $Y(5, -2)$; $k = 5$

19. **ERROR ANALYSIS** Describe and correct the error in listing the coordinates of the image after a dilation with a scale factor of $\frac{1}{2}$.

Vertices of ABC	$(2x, 2y)$	Vertices of $A'B'C'$
$A(2, 5)$	$(2 \cdot 2, 2 \cdot 5)$	$A'(4, 10)$
$B(2, 0)$	$(2 \cdot 2, 2 \cdot 0)$	$B'(4, 0)$
$C(4, 0)$	$(2 \cdot 4, 2 \cdot 0)$	$C'(8, 0)$

The blue figure is a dilation of the red figure. Identify the type of dilation and find the scale factor.

20.
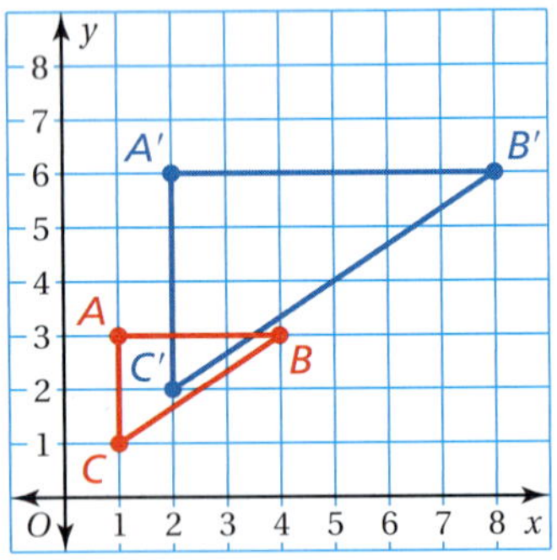

21.
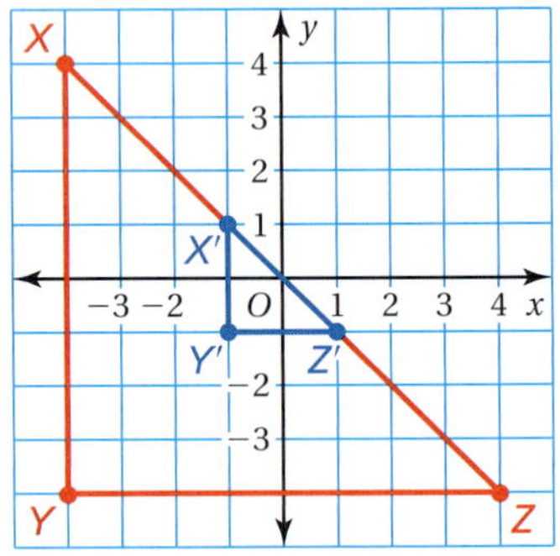

22.
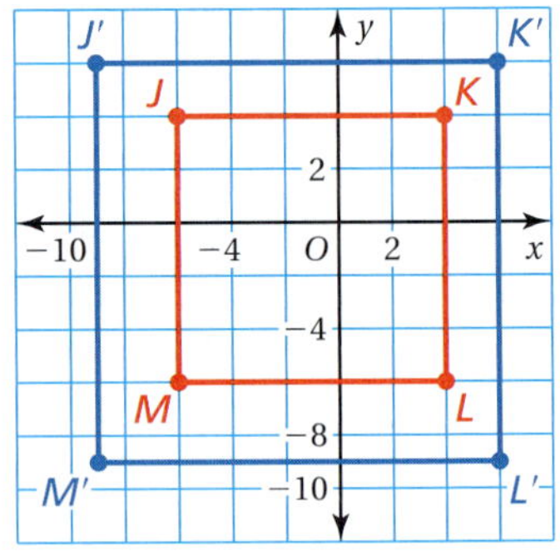

The vertices of a figure are given. Find the coordinates of the figure after the transformations given.

(4) 23. $A(-5, 3)$, $B(-2, 3)$, $C(-2, 1)$, $D(-5, 1)$

Reflect in the y-axis. Then dilate with respect to the origin using a scale factor of 2.

24. $F(-9, -9)$, $G(-3, -6)$, $H(-3, -9)$

Dilate with respect to the origin using a scale factor of $\frac{2}{3}$. Then translate 6 units up.

25. $J(1, 1)$, $K(3, 4)$, $L(5, 1)$

Rotate 90° clockwise about the origin. Then dilate with respect to the origin using a scale factor of 3.

26. $P(-2, 2)$, $Q(4, 2)$, $R(2, -6)$, $S(-4, -6)$

Dilate with respect to the origin using a scale factor of 5. Then dilate with respect to the origin using a scale factor of 0.5.

The red figure is similar to the blue figure. Describe a sequence of transformations in which the blue figure is the image of the red figure.

(5) 27.
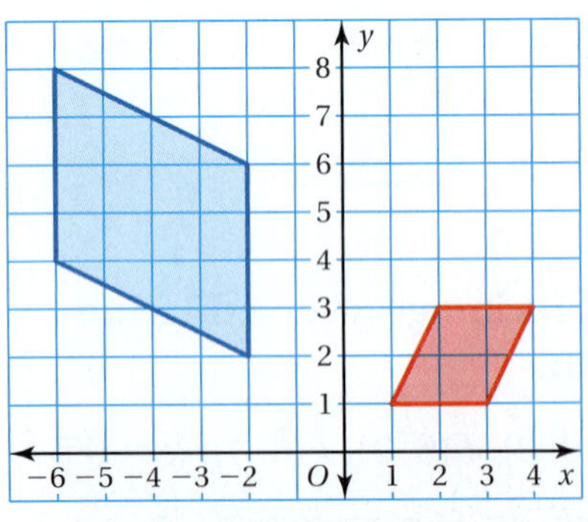

28.
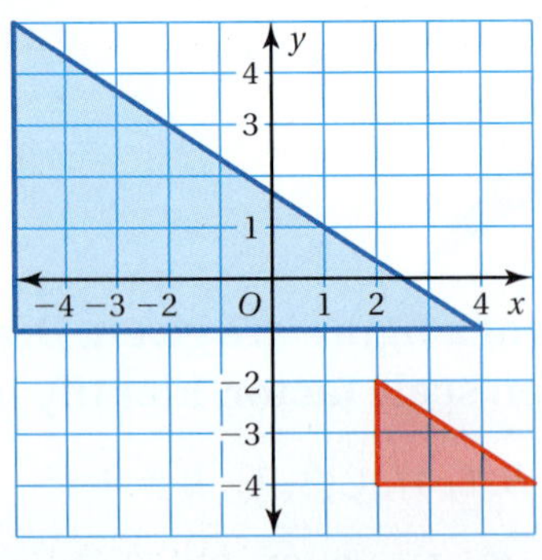

29. **STRUCTURE** In Exercises 27 and 28, is the blue figure still the image of the red figure when you perform the sequence in the opposite order? Explain.

30. **OPEN-ENDED** Draw a rectangle on a coordinate plane. Choose a scale factor of 2, 3, 4, or 5, and then dilate the rectangle. How many times greater is the area of the image than the area of the original rectangle?

31. **SHADOW PUPPET** You can use a flashlight and a shadow puppet (your hands) to project shadows on the wall.

  a. Identify the type of dilation.

  b. What does the flashlight represent?

  c. The length of the ears on the shadow puppet is 3 inches. The length of the ears on the shadow is 4 inches. What is the scale factor?

  d. Describe what happens as the shadow puppet moves closer to the flashlight. How does this affect the scale factor?

32. **REASONING** A triangle is dilated using a scale factor of 3. The image is then dilated using a scale factor of $\frac{1}{2}$. What scale factor could you use to dilate the original triangle to get the final image? Explain.

CRITICAL THINKING The coordinate notation shows how the coordinates of a figure are related to the coordinates of its image after transformations. What are the transformations? Are the figure and its image similar or congruent? Explain.

33. $(x, y) \rightarrow (2x + 4, 2y - 3)$

34. $(x, y) \rightarrow (-x - 1, y - 2)$

35. $(x, y) \rightarrow \left(\frac{1}{3}x, -\frac{1}{3}y\right)$

36. **STRUCTURE** How are the transformations $(2x + 3, 2y - 1)$ and $(2(x + 3), 2(y + 1))$ different?

37. **Problem Solving** The vertices of a trapezoid are $A(-2, 3)$, $B(2, 3)$, $C(5, -2)$, and $D(-2, -2)$. Dilate the trapezoid with respect to vertex A using a scale factor of 2. What are the coordinates of the image? Explain the method you used.

Fair Game Review What you learned in previous grades & lessons

Tell whether the angles are *complementary* or *supplementary*. Then find the value of x. *(Skills Review Handbook)*

38.

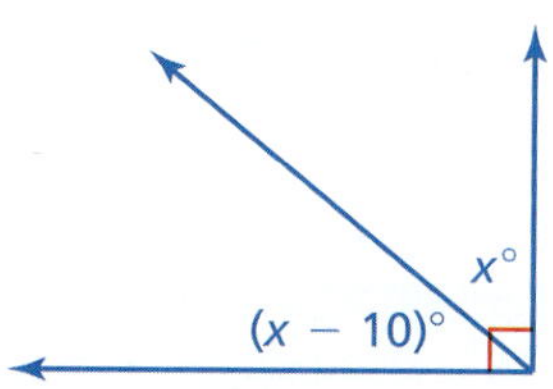

39.

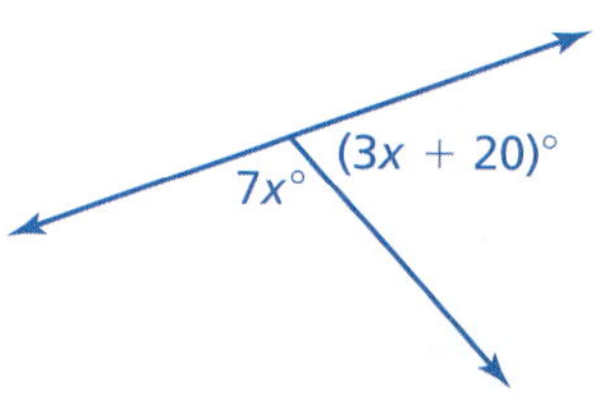

40.

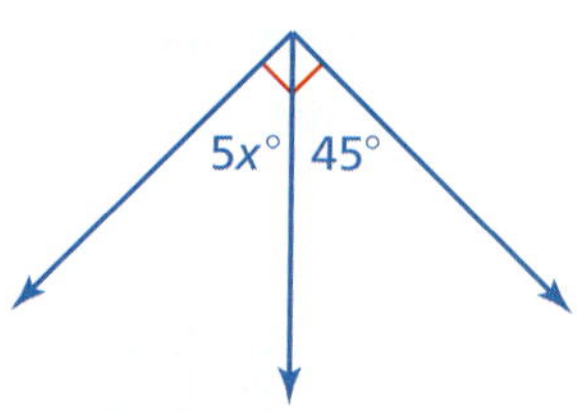

41. **MULTIPLE CHOICE** Which quadrilateral is *not* a parallelogram? *(Skills Review Handbook)*

Ⓐ rhombus Ⓑ trapezoid Ⓒ square Ⓓ rectangle

2.5–2.7 Quiz

Check It Out
Progress Check
BigIdeasMath.com

1. Tell whether the two rectangles are similar. Explain your reasoning. *(Section 2.5)*

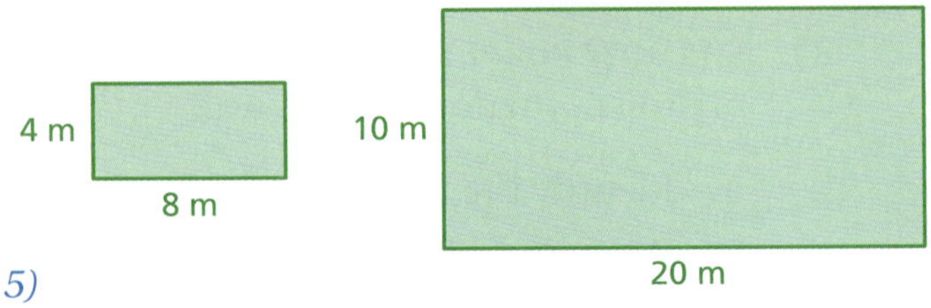

The figures are similar. Find x. *(Section 2.5)*

2.

3.

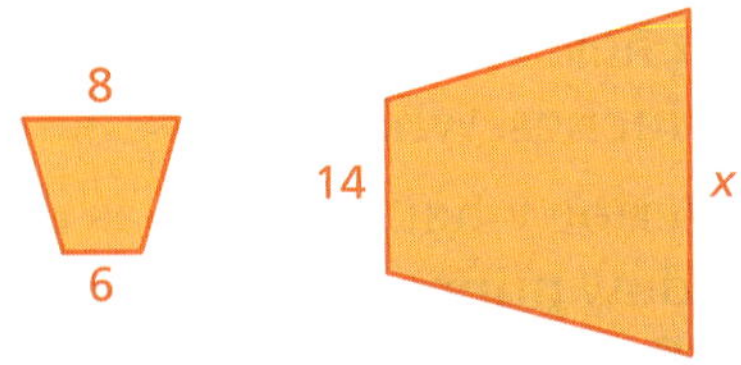

The two figures are similar. Find the ratios (red to blue) of the perimeters and of the areas. *(Section 2.6)*

4.

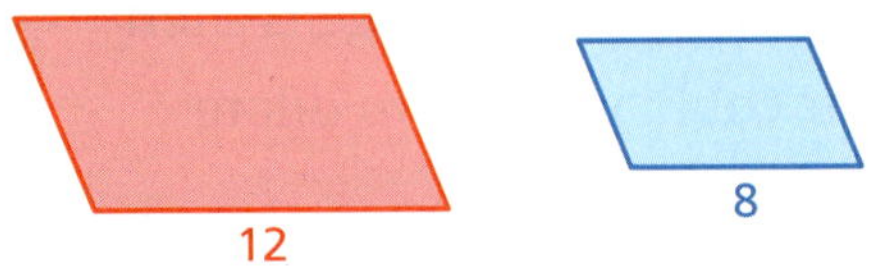

5.

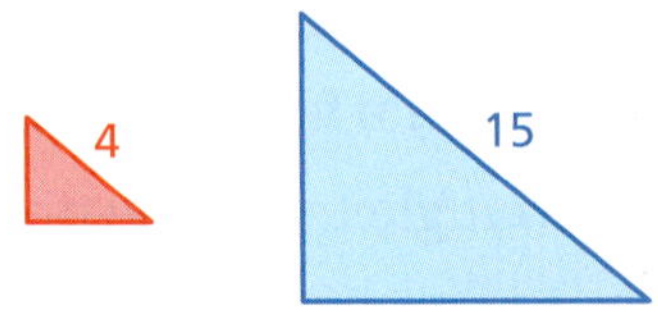

Tell whether the blue figure is a dilation of the red figure. *(Section 2.7)*

6.

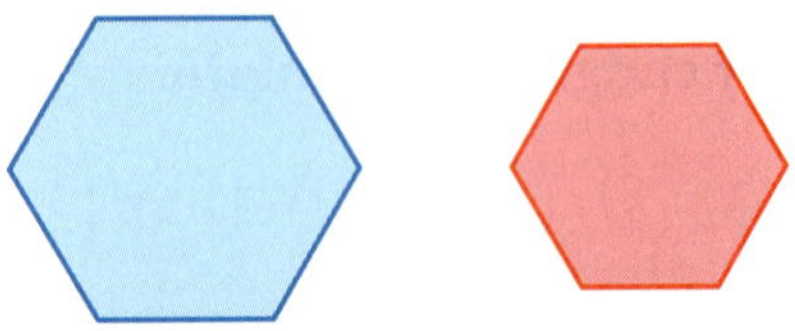

7.

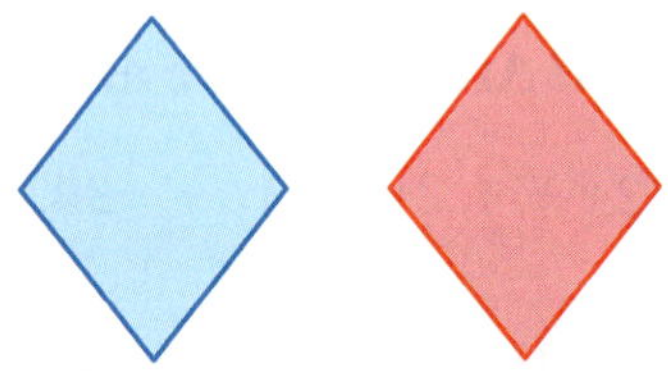

8. **SCREENS** The TV screen is similar to the computer screen. What is the area of the TV screen? *(Section 2.6)*

9. **GEOMETRY** The vertices of a rectangle are $A(2, 4)$, $B(5, 4)$, $C(5, -1)$, and $D(2, -1)$. Dilate the rectangle with respect to the origin using a scale factor of $\frac{1}{2}$. Then translate it 4 units left and 3 units down. What are the coordinates of the image? *(Section 2.7)*

10. **TENNIS COURT** The tennis courts for singles and doubles matches are different sizes. Are the courts similar? Explain. *(Section 2.5)*

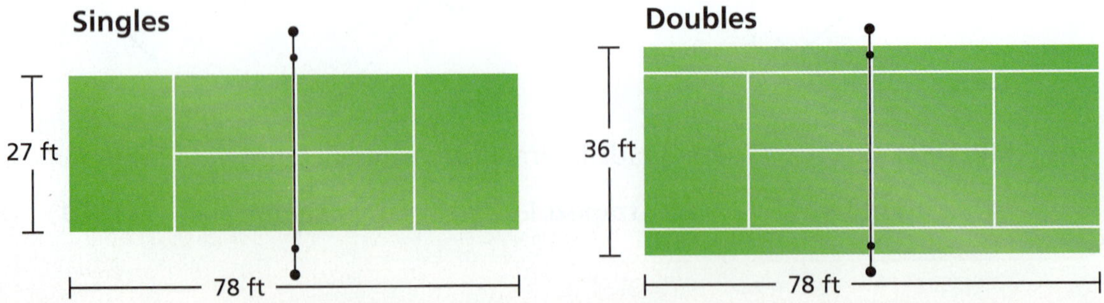

2 Chapter Review

Review Key Vocabulary

congruent figures, *p. 44*
corresponding angles, *p. 44*
corresponding sides, *p. 44*
transformation, *p. 50*
image, *p. 50*
translation, *p. 50*
reflection, *p. 56*
line of reflection, *p. 56*
rotation, *p. 62*
center of rotation, *p. 62*
angle of rotation, *p. 62*
similar figures, *p. 72*
dilation, *p. 84*
center of dilation, *p. 84*
scale factor, *p. 84*

Review Examples and Exercises

2.1 Congruent Figures *(pp. 42–47)*

Trapezoids *EFGH* and *QRST* are congruent.

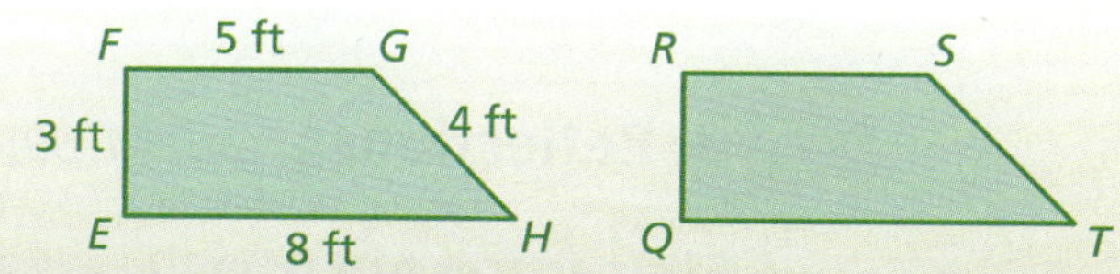

a. What is the length of side *QT*?

Side *QT* corresponds to side *EH*.

So, the length of side *QT* is 8 feet.

b. Which angle of *QRST* corresponds to $\angle H$?

$\angle T$ corresponds to $\angle H$.

Exercises

Use the figures above.

1. What is the length of side *QR*?
2. What is the perimeter of *QRST*?

The figures are congruent. Name the corresponding angles and the corresponding sides.

3.

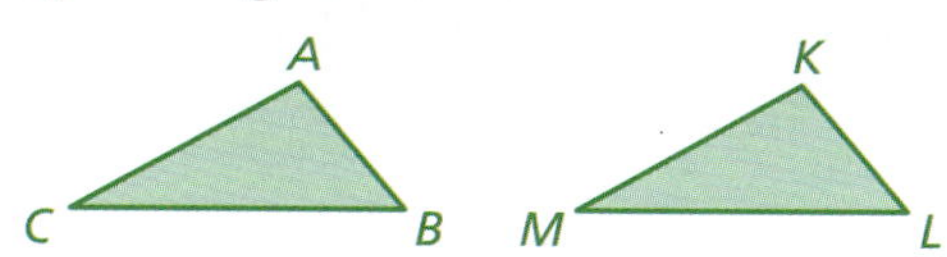

4.

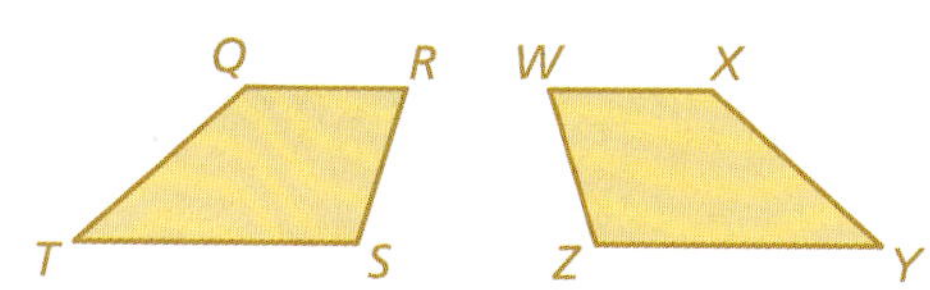

2.2 Translations *(pp. 48–53)*

Translate the red triangle 4 units left and 1 unit down. What are the coordinates of the image?

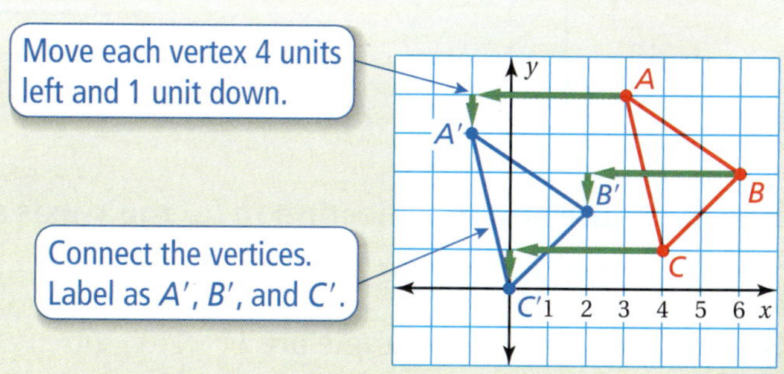

The coordinates of the image are $A'(-1, 4)$, $B'(2, 2)$, and $C'(0, 0)$.

Exercises

Tell whether the blue figure is a translation of the red figure.

5. 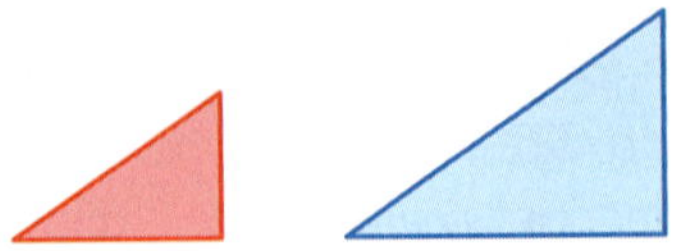

6. 

7. The vertices of a quadrilateral are $W(1, 2)$, $X(1, 4)$, $Y(4, 4)$, and $Z(4, 2)$. Draw the figure and its image after a translation 3 units left and 2 units down.

8. The vertices of a triangle are $A(-1, -2)$, $B(-2, 2)$, and $C(-3, 0)$. Draw the figure and its image after a translation 5 units right and 1 unit up.

2.3 Reflections *(pp. 54–59)*

The vertices of a triangle are $A(-2, 1)$, $B(4, 1)$, and $C(4, 4)$. Draw the figure and its reflection in the x-axis. What are the coordinates of the image?

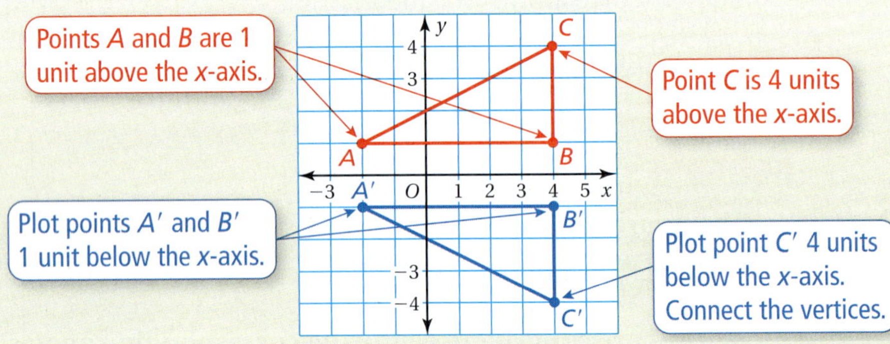

The coordinates of the image are $A'(-2, -1)$, $B'(4, -1)$, and $C'(4, -4)$.

Exercises

Tell whether the blue figure is a reflection of the red figure.

9. 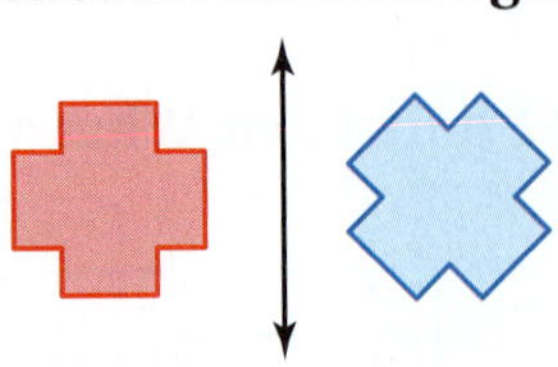

10. 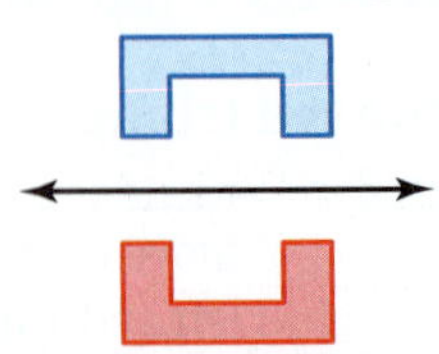

Draw the figure and its reflection in (a) the x-axis and (b) the y-axis.

11. $A(2, 0)$, $B(1, 5)$, $C(4, 3)$

12. $D(-5, -5)$, $E(-5, -1)$, $F(-2, -2)$, $G(-2, -5)$

13. The vertices of a rectangle are $E(-1, 1)$, $F(-1, 3)$, $G(-5, 3)$, and $H(-5, 1)$. Find the coordinates of the figure after reflecting in the x-axis, and then translating 3 units right.

2.4 Rotations (pp. 60–67)

The vertices of a triangle are $A(1, 1)$, $B(3, 2)$, and $C(2, 4)$. Rotate the triangle 90° counterclockwise about the origin. What are the coordinates of the image?

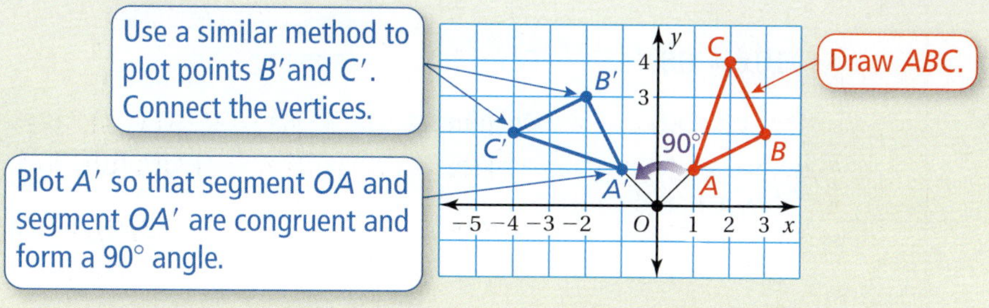

The coordinates of the image are $A'(-1, 1)$, $B'(-2, 3)$, and $C'(-4, 2)$.

Exercises

Tell whether the blue figure is a rotation of the red figure about the origin. If so, give the angle and the direction of rotation.

14.

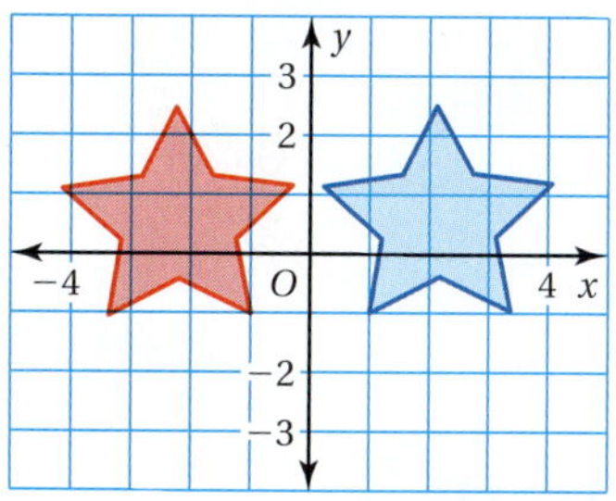

15.

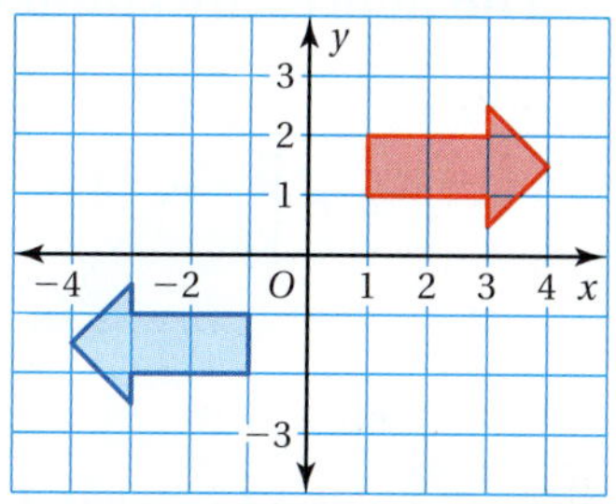

The vertices of a triangle are $A(-4, 2)$, $B(-2, 2)$, and $C(-3, 4)$. Rotate the triangle about the origin as described. Find the coordinates of the image.

16. 180°

17. 270° clockwise

2.5 Similar Figures (pp. 70–75)

a. Is Rectangle A similar to Rectangle B?

Each figure is a rectangle. So, corresponding angles are congruent. Check to see if corresponding side lengths are proportional.

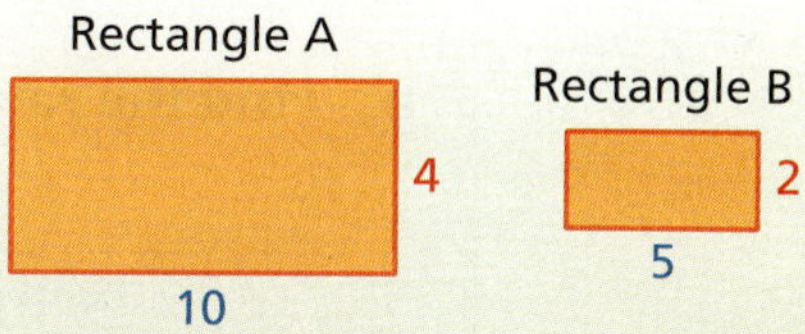

$\dfrac{\text{Length of A}}{\text{Length of B}} = \dfrac{10}{5} = 2$ $\quad\dfrac{\text{Width of A}}{\text{Width of B}} = \dfrac{4}{2} = 2$ Proportional

So, Rectangle A is similar to Rectangle B.

b. The two rectangles are similar. Find x.

Because the rectangles are similar, corresponding side lengths are proportional. So, write and solve a proportion to find x.

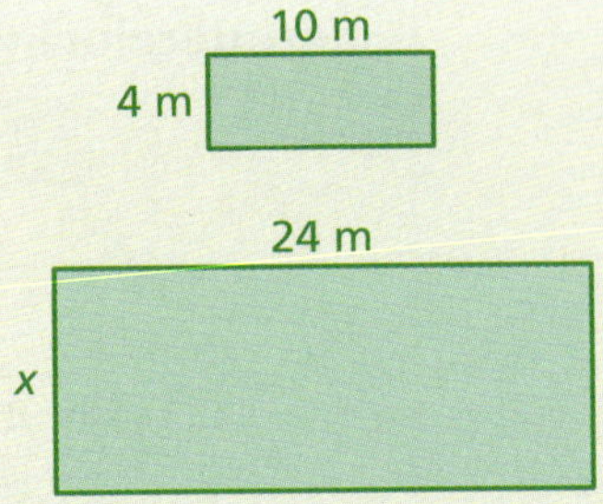

$\frac{10}{24} = \frac{4}{x}$	Write a proportion.
$10x = 96$	Cross Products Property
$x = 9.6$	Divide each side by 10.

∴ So, x is 9.6 meters.

Exercises

Tell whether the two figures are similar. Explain your reasoning.

18.

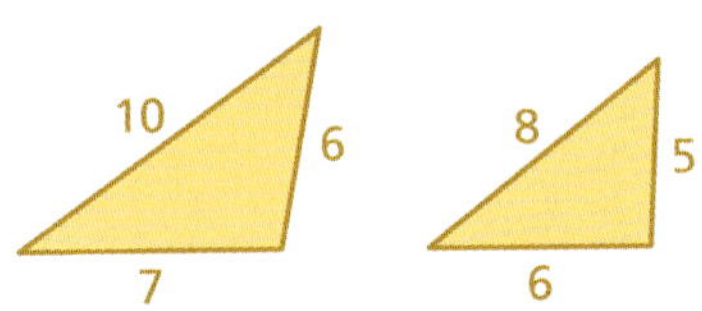

19.

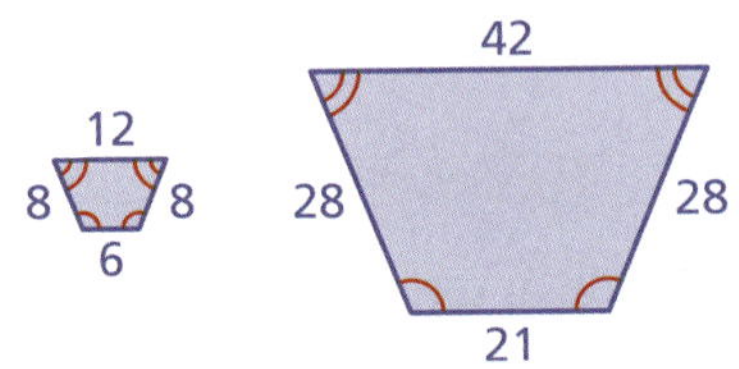

The figures are similar. Find x.

20.

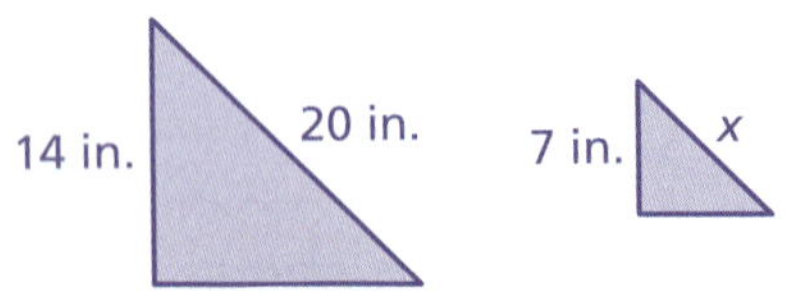

21.

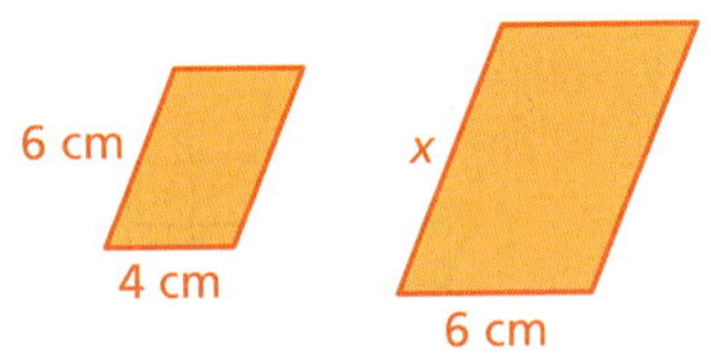

2.6 Perimeters and Areas of Similar Figures (pp. 76–81)

a. Find the ratio (red to blue) of the perimeters of the similar parallelograms.

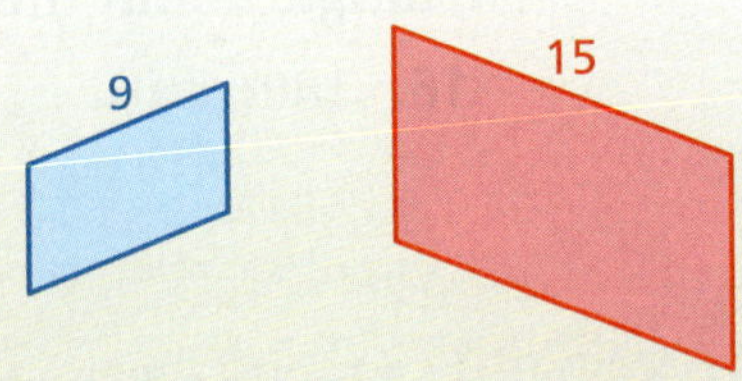

$$\frac{\text{Perimeter of red parallelogram}}{\text{Perimeter of blue parallelogram}} = \frac{15}{9}$$

$$= \frac{5}{3}$$

∴ The ratio of the perimeters is $\frac{5}{3}$.

b. Find the ratio (red to blue) of the areas of the similar figures.

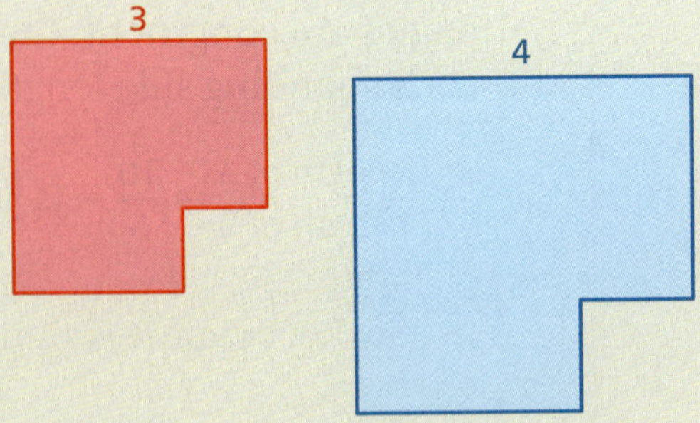

$$\frac{\text{Area of red figure}}{\text{Area of blue figure}} = \left(\frac{3}{4}\right)^2$$

$$= \frac{9}{16}$$

∴ The ratio of the areas is $\frac{9}{16}$.

Exercises

The two figures are similar. Find the ratios (red to blue) of the perimeters and of the areas.

22.

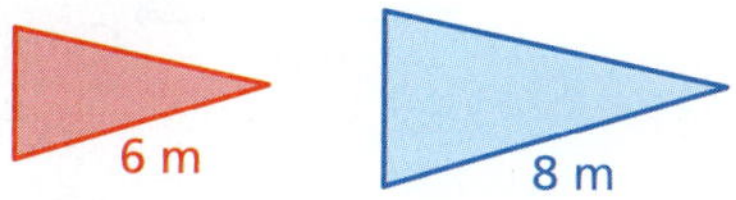

23.

24. **PHOTOS** Two photos are similar. The ratio of the corresponding side lengths is 3 : 4. What is the ratio of the areas?

2.7 Dilations *(pp. 82–89)*

Draw the image of Triangle *ABC* after a dilation with a scale factor of 2. Identify the type of dilation.

Multiply each x- and y-coordinate by the scale factor 2.

Vertices of ABC	$(2x, 2y)$	Vertices of $A'B'C'$
$A(1, 1)$	$(2 \cdot 1, 2 \cdot 1)$	$A'(2, 2)$
$B(1, 2)$	$(2 \cdot 1, 2 \cdot 2)$	$B'(2, 4)$
$C(3, 2)$	$(2 \cdot 3, 2 \cdot 2)$	$C'(6, 4)$

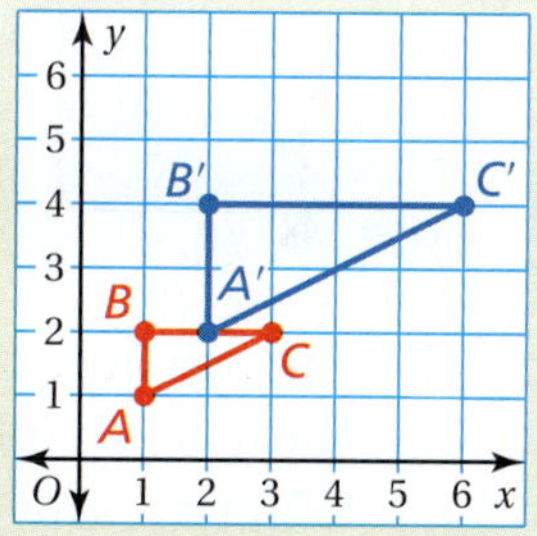

∴ The image is shown at the above right. The dilation is an *enlargement* because the scale factor is greater than 1.

Exercises

Tell whether the blue figure is a dilation of the red figure.

25.

26.

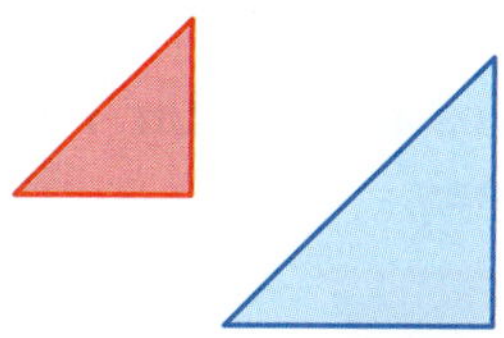

The vertices of a figure are given. Draw the figure and its image after a dilation with the given scale factor. Identify the type of dilation.

27. $P(-3, -2)$, $Q(-3, 0)$, $R(0, 0)$; $k = 4$

28. $B(3, 3)$, $C(3, 6)$, $D(6, 6)$, $E(6, 3)$; $k = \frac{1}{3}$

29. The vertices of a rectangle are $Q(-6, 2)$, $R(6, 2)$, $S(6, -4)$, and $T(-6, -4)$. Dilate the rectangle with respect to the origin using a scale factor of $\frac{3}{2}$. Then translate it 5 units right and 1 unit down. What are the coordinates of the image?

2 Chapter Test

Check It Out
Test Practice
BigIdeasMath.com

Triangles *ABC* and *DEF* are congruent.

1. Which angle of *DEF* corresponds to $\angle C$?
2. What is the perimeter of *DEF*?

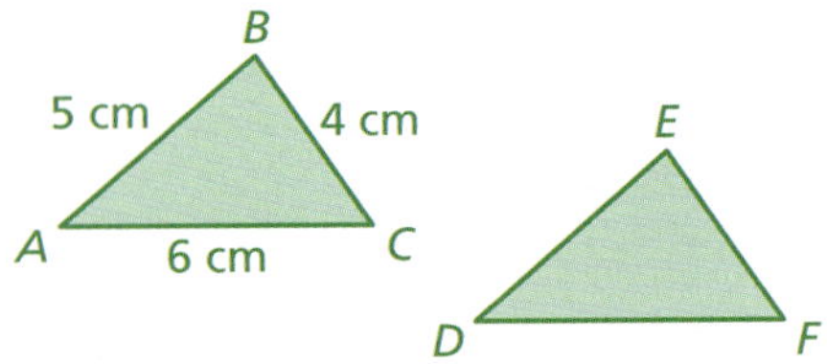

Tell whether the blue figure is a *translation, reflection, rotation,* or *dilation* of the red figure.

3. 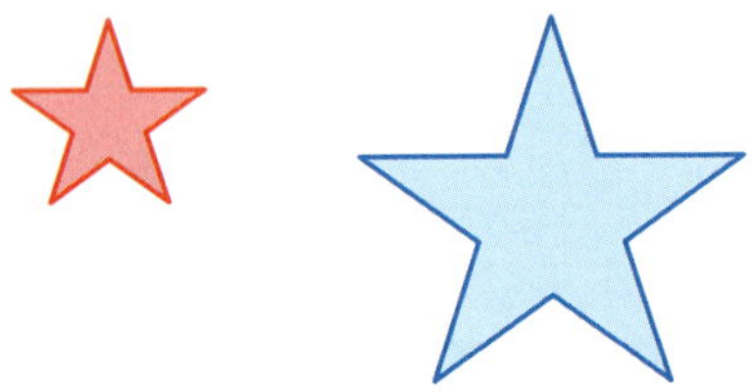

4. 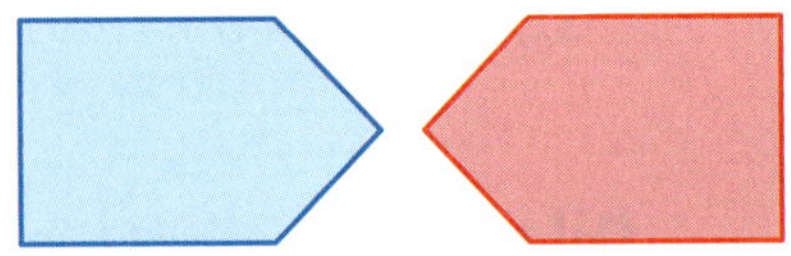

5. 

6. 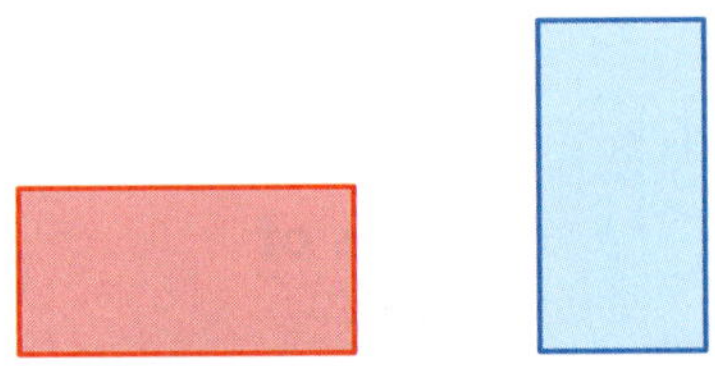

7. The vertices of a triangle are $A(2, 5)$, $B(1, 2)$, and $C(3, 1)$. Reflect the triangle in the x-axis, and then rotate the triangle 90° counterclockwise about the origin. What are the coordinates of the image?

8. The vertices of a triangle are $A(2, 4)$, $B(2, 1)$, and $C(5, 1)$. Dilate the triangle with respect to the origin using a scale factor of 2. Then translate the triangle 2 units left and 1 unit up. What are the coordinates of the image?

9. Tell whether the parallelograms are similar. Explain your reasoning.

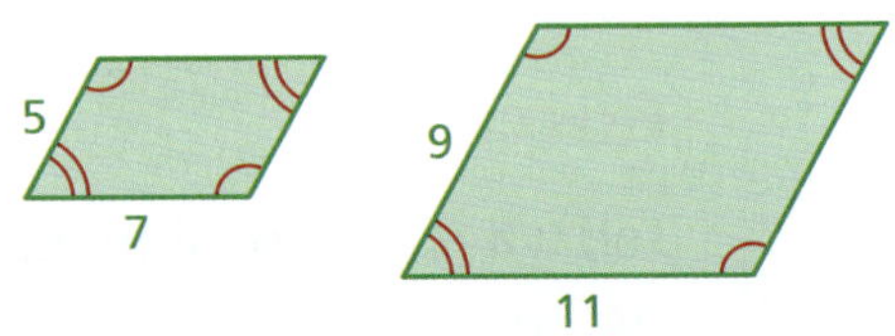

The two figures are similar. Find the ratios (red to blue) of the perimeters and of the areas.

10. 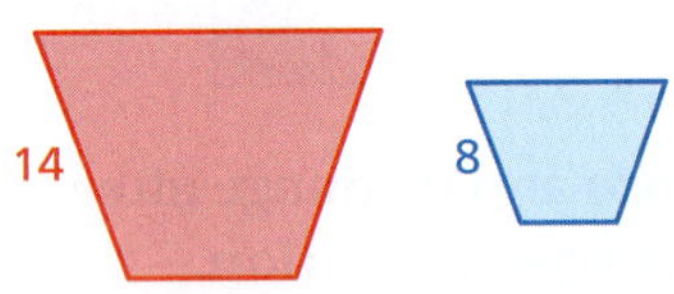

11. 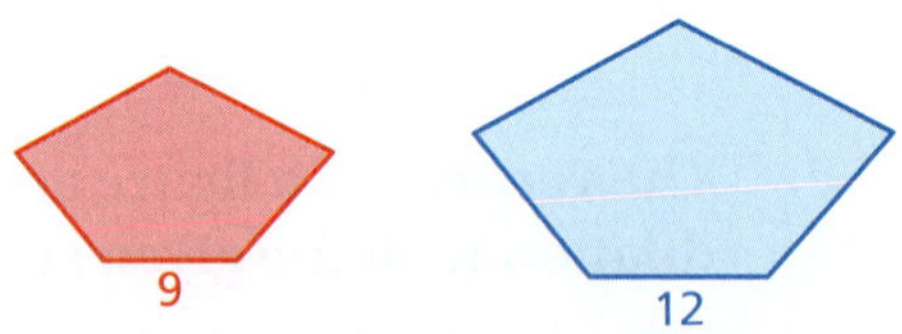

12. **SCREENS** A wide-screen television measures 36 inches by 54 inches. A movie theater screen measures 42 feet by 63 feet. Are the screens similar? Explain.

13. **CURTAINS** You want to use the rectangular piece of fabric shown to make a set of curtains for your window. Name the types of congruent shapes you can make with one straight cut. Draw an example of each type.

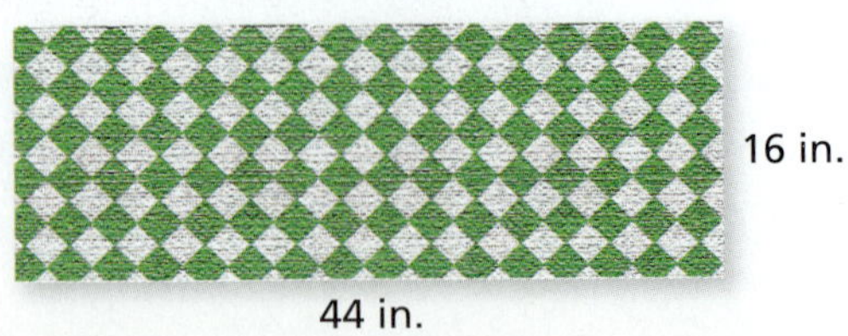

2 Standards Assessment

1. A clockwise rotation of 90° is equivalent to a counterclockwise rotation of how many degrees? *(MACC.8.G.1.2)*

2. The formula $K = C + 273.15$ converts temperatures from Celsius C to Kelvin K. Which of the following formulas is *not* correct? *(MACC.8.EE.3.7a)*

A. $K - C = 273.15$

B. $C = K - 273.15$

C. $C - K = -273.15$

D. $C = K + 273.15$

3. Joe wants to solve the equation $-3(x + 2) = 12x$. What should he do first? *(MACC.8.EE.3.7a)*

F. Subtract 2 from each side.

G. Add 3 to each side.

H. Multiply each side by -3.

I. Divide each side by -3.

4. Which transformation *turns* a figure? *(MACC.8.G.1.1)*

A. translation

B. reflection

C. rotation

D. dilation

5. A triangle is graphed in the coordinate plane below.

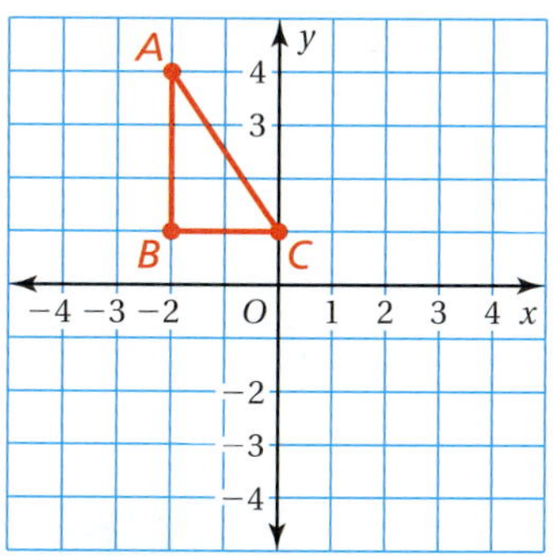

Translate the triangle 3 units right and 2 units down. What are the coordinates of the image? *(MACC.8.G.1.3)*

F. $A'(1, 4)$, $B'(1, 1)$, $C'(3, 1)$

G. $A'(1, 2)$, $B'(1, -1)$, $C'(3, -1)$

H. $A'(-2, 2)$, $B'(-2, -1)$, $C'(0, -1)$

I. $A'(0, 1)$, $B'(0, -2)$, $C'(2, -2)$

6. Dale solved the equation in the box shown. What should Dale do to correct the error that he made? *(MACC.8.EE.3.7b)*

$$-\frac{x}{3} + \frac{2}{5} = -\frac{7}{15}$$
$$-\frac{x}{3} + \frac{2}{5} - \frac{2}{5} = -\frac{7}{15} - \frac{2}{5}$$
$$-\frac{x}{3} = -\frac{13}{15}$$
$$3 \cdot \left(-\frac{x}{3}\right) = 3 \cdot \left(-\frac{13}{15}\right)$$
$$x = -2\frac{3}{5}$$

A. Add $\frac{2}{5}$ to each side to get $-\frac{x}{3} = -\frac{1}{15}$.

B. Multiply each side by -3 to get $x + \frac{2}{5} = \frac{7}{5}$.

C. Multiply each side by -3 to get $x = 2\frac{3}{5}$.

D. Subtract $\frac{2}{5}$ from each side to get $-\frac{x}{3} = -\frac{5}{10}$.

7. Jenny dilates the rectangle below using a scale factor of $\frac{1}{2}$.

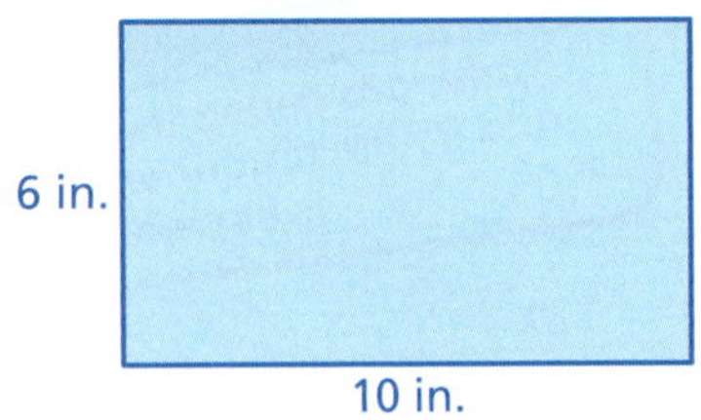

What is the area of the dilated rectangle in square inches? *(MACC.8.G.1.4)*

8. The vertices of a rectangle are $A(-4, 2)$, $B(3, 2)$, $C(3, -5)$, and $D(-4, -5)$. If the rectangle is dilated by a scale factor of 3, what will be the coordinates of vertex C'? *(MACC.8.G.1.3)*

F. $(9, -15)$

G. $(-12, 6)$

H. $(-12, -15)$

I. $(9, 6)$

9. In the figures, Triangle *EFG* is a dilation of Triangle *HIJ*.

Which proportion is *not* necessarily correct for Triangle *EFG* and Triangle *HIJ*? *(MACC.8.G.1.4)*

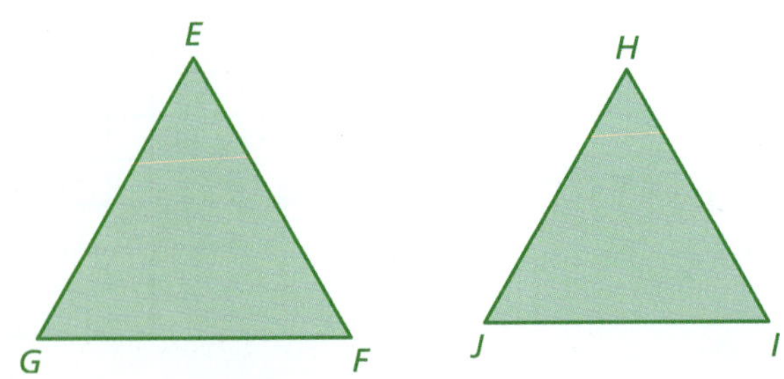

A. $\frac{EF}{FG} = \frac{HI}{IJ}$

B. $\frac{EG}{HI} = \frac{FG}{IJ}$

C. $\frac{GE}{EF} = \frac{JH}{HI}$

D. $\frac{EF}{HI} = \frac{GE}{JH}$

10. In the figures below, Rectangle *EFGH* is a dilation of Rectangle *IJKL*.

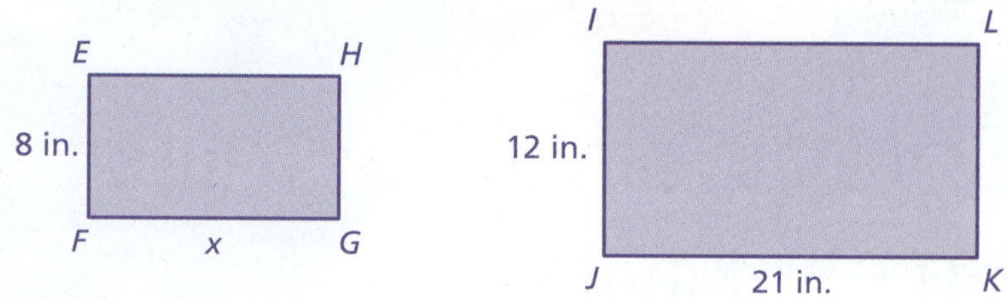

What is x? *(MACC.8.G.1.4)*

F. 14 in.

G. 15 in.

H. 16 in.

I. 17 in.

11. Several transformations are used to create the pattern. *(MACC.8.G.1.2, MACC.8.G.1.4)*

Think Solve Explain

Part A Describe the transformation of Triangle *GLM* to Triangle *DGH*.

Part B Describe the transformation of Triangle *ALQ* to Triangle *GLM*.

Part C Triangle *DFN* is a dilation of Triangle *GHM*. Find the scale factor.

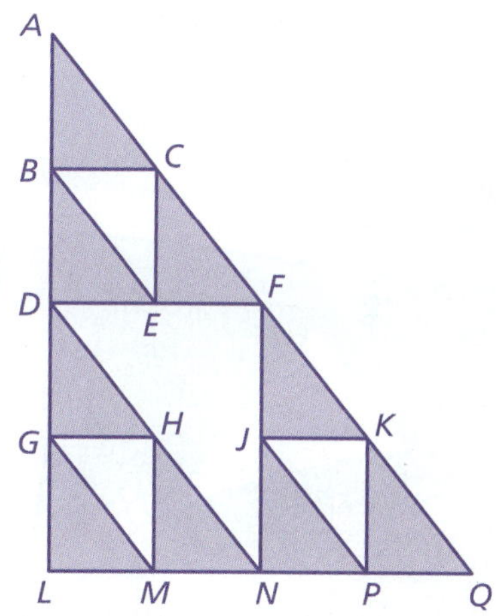

12. A rectangle is graphed in the coordinate plane below.

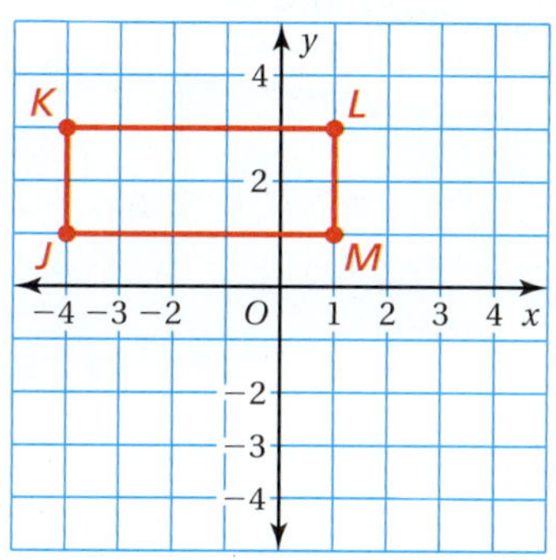

Rotate the triangle 180° about the origin. What are the coordinates of the image? *(MACC.8.G.1.3)*

A. $J'(4, -1), K'(4, -3), L'(-1, -3), M'(-1, -1)$

B. $J'(-4, -1), K'(-4, -3), L'(1, -3), M'(1, -1)$

C. $J'(1, 4), K'(3, 4), L'(3, -1), M'(1, -1)$

D. $J'(-4, 1), K'(-4, 3), L'(1, 3), M'(1, 1)$

3 Angles and Triangles

What You Learned Before

Adjacent and Vertical Angles (MACC.7.G.2.5)

Example 1 **Tell whether the angles are *adjacent* or *vertical*. Then find the value of *x*.**

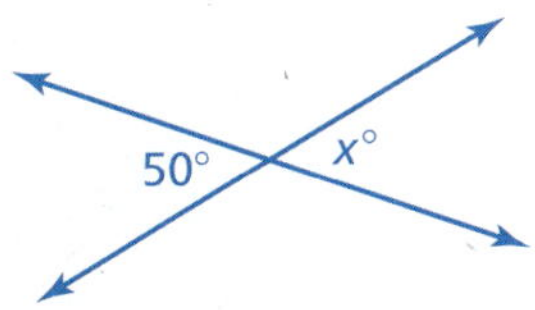

The angles are vertical angles. Because vertical angles are congruent, the angles have the same measure.

∴ So, the value of x is 50.

Try It Yourself

Tell whether the angles are *adjacent* or *vertical*. Then find the value of *x*.

1.

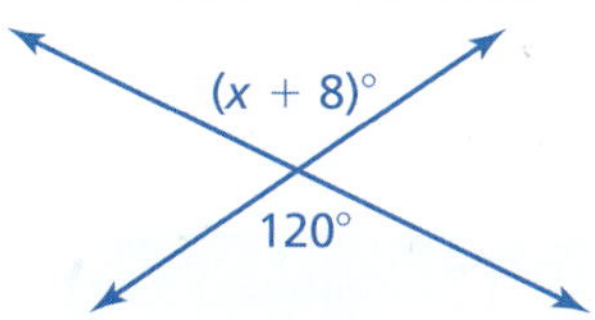

2.

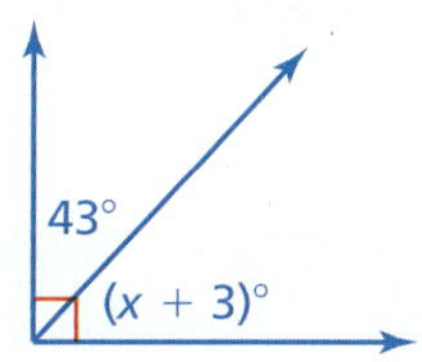

Complementary and Supplementary Angles (MACC.7.G.2.5)

Example 2 **Tell whether the angles are *complementary* or *supplementary*. Then find the value of *x*.**

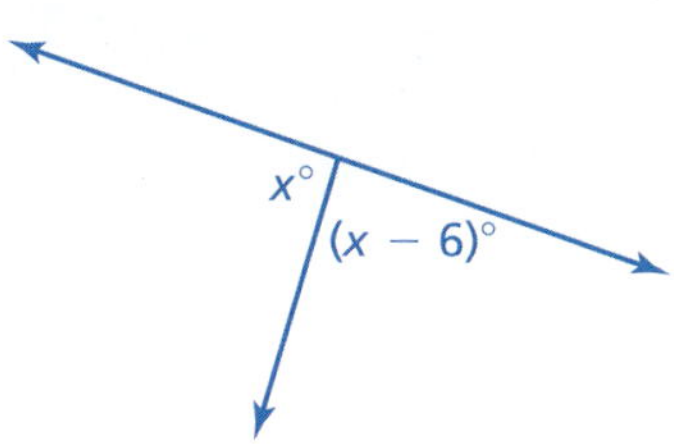

The two angles make up a straight angle. So, the angles are supplementary angles, and the sum of their measures is 180°.

$x + (x - 6) = 180$	Write equation.
$2x - 6 = 180$	Combine like terms.
$2x = 186$	Add 6 to each side.
$x = 93$	Divide each side by 2.

Try It Yourself

Tell whether the angles are *complementary* or *supplementary*. Then find the value of *x*.

3.

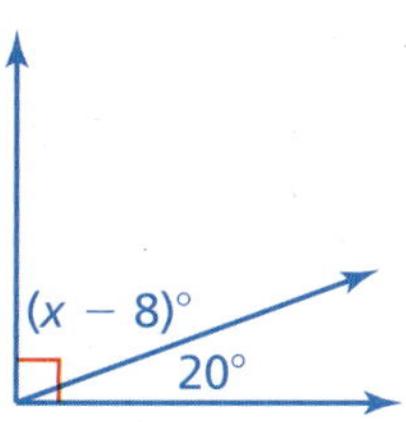

4.

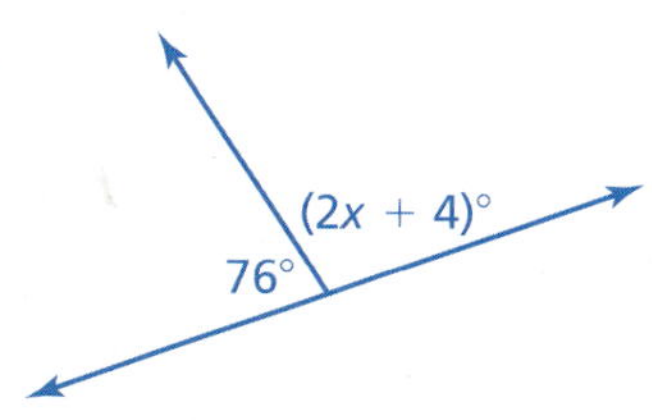

3.1 Parallel Lines and Transversals

Essential Question How can you describe angles formed by parallel lines and transversals?

The Meaning of a Word ● Transverse

When an object is **transverse**, it is lying or extending across something.

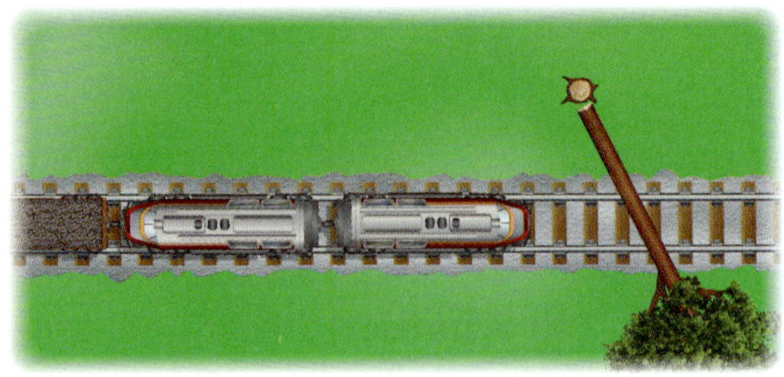

1 ACTIVITY: A Property of Parallel Lines

Work with a partner.

- **Discuss what it means for two lines to be parallel. Decide on a strategy for drawing two parallel lines. Then draw the two parallel lines.**
- **Draw a third line that intersects the two parallel lines. This line is called a transversal.**

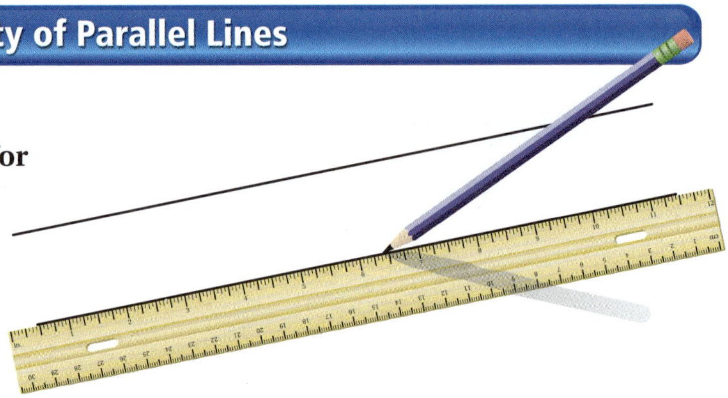

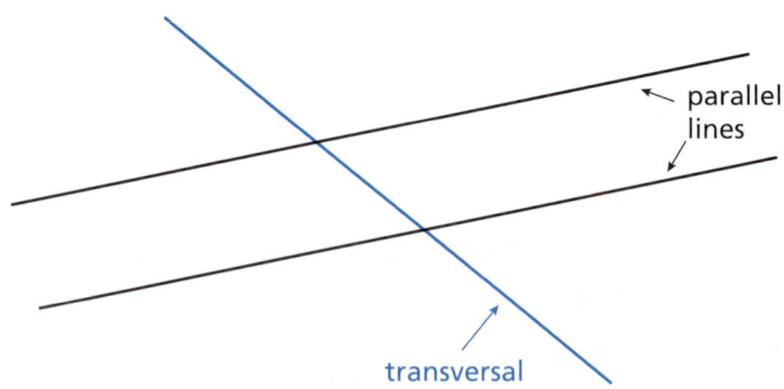

a. How many angles are formed by the parallel lines and the transversal? Label the angles.

b. Which of these angles have equal measures? Explain your reasoning.

COMMON CORE

Geometry

In this lesson, you will

- identify the angles formed when parallel lines are cut by a transversal.
- find the measures of angles formed when parallel lines are cut by a transversal.

Learning Standard
MACC.8.G.1.5

2 ACTIVITY: Creating Parallel Lines

Math Practice 6

Use Clear Definitions

What do the words *parallel* and *transversal* mean? How does this help you answer the question in part (a)?

Work with a partner.

a. If you were building the house in the photograph, how could you make sure that the studs are parallel to each other?

b. Identify sets of parallel lines and transversals in the photograph.

3 ACTIVITY: Using Technology

Work with a partner. Use geometry software to draw two parallel lines intersected by a transversal.

a. Find all the angle measures.

b. Adjust the figure by moving the parallel lines or the transversal to a different position. Describe how the angle measures and relationships change.

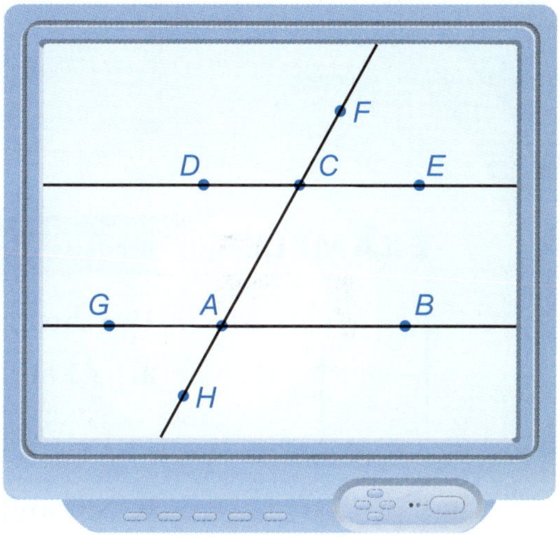

What Is Your Answer?

4. **IN YOUR OWN WORDS** How can you describe angles formed by parallel lines and transversals? Give an example.

5. Use geometry software to draw a transversal that is perpendicular to two parallel lines. What do you notice about the angles formed by the parallel lines and the transversal?

Practice Use what you learned about parallel lines and transversals to complete Exercises 3–6 on page 107.

3.1 Lesson

Check It Out
Lesson Tutorials
BigIdeasMath.com

Key Vocabulary
transversal, *p. 104*
interior angles, *p. 105*
exterior angles, *p. 105*

Lines in the same plane that do not intersect are called *parallel lines*. Lines that intersect at right angles are called *perpendicular lines*.

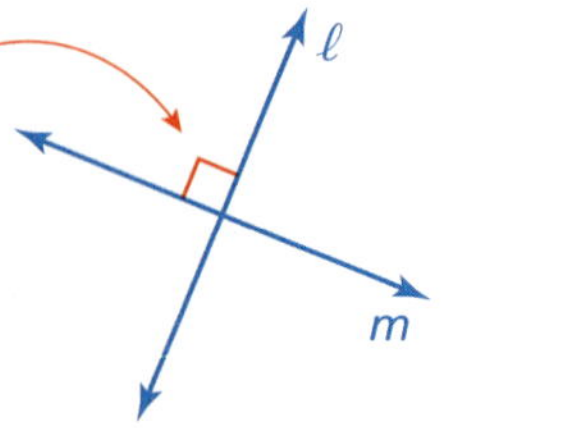

A line that intersects two or more lines is called a **transversal**. When parallel lines are cut by a transversal, several pairs of congruent angles are formed.

Key Idea

Study Tip
Corresponding angles lie on the same side of the transversal in corresponding positions.

Corresponding Angles

When a transversal intersects parallel lines, corresponding angles are congruent.

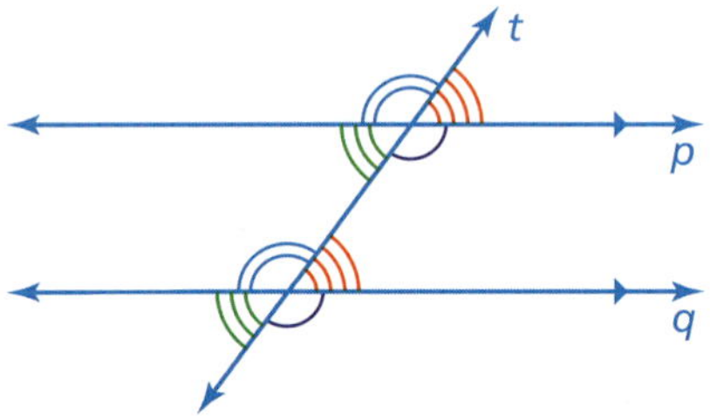

Corresponding angles

EXAMPLE 1 Finding Angle Measures

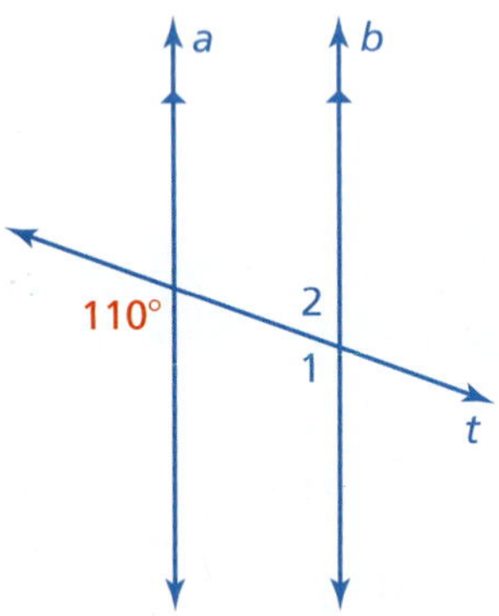

Use the figure to find the measures of (a) ∠1 and (b) ∠2.

a. ∠1 and the 110° angle are corresponding angles. They are congruent.

So, the measure of ∠1 is 110°.

b. ∠1 and ∠2 are supplementary.

$\angle 1 + \angle 2 = 180^\circ$	Definition of supplementary angles
$110^\circ + \angle 2 = 180^\circ$	Substitute 110° for ∠1.
$\angle 2 = 70^\circ$	Subtract 110° from each side.

So, the measure of ∠2 is 70°.

On Your Own

Now You're Ready
Exercises 7–9

Use the figure to find the measure of the angle. Explain your reasoning.

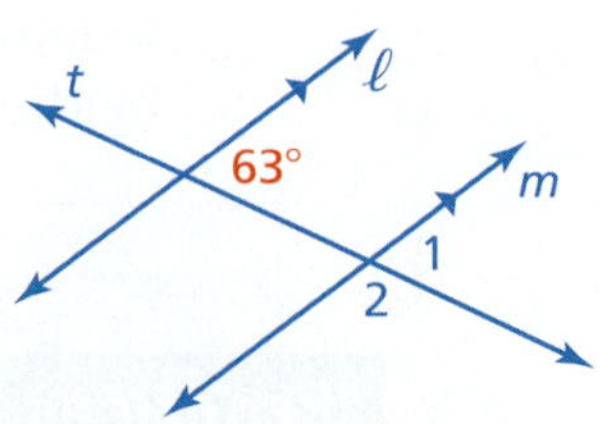

1. ∠1 **2.** ∠2

Multi-Language Glossary at BigIdeasMath.com

EXAMPLE 2 Using Corresponding Angles

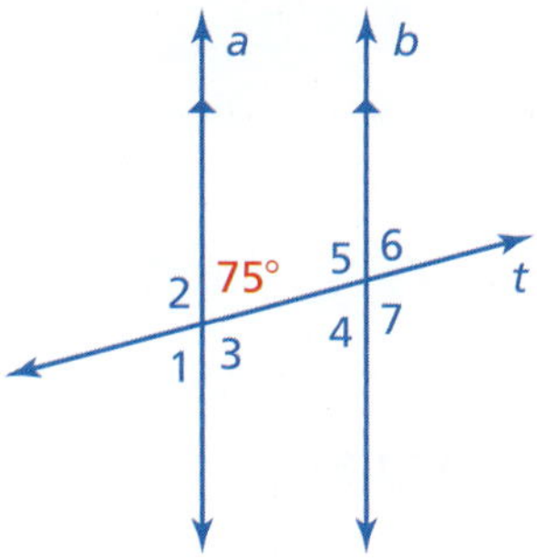

Use the figure to find the measures of the numbered angles.

∠1: ∠1 and the 75° angle are vertical angles. They are congruent.

So, the measure of ∠1 is 75°.

∠2 and ∠3: The 75° angle is supplementary to both ∠2 and ∠3.

$75° + \angle 2 = 180°$	Definition of supplementary angles
$\angle 2 = 105°$	Subtract 75° from each side.

So, the measures of ∠2 and ∠3 are 105°.

∠4, ∠5, ∠6, and ∠7: Using corresponding angles, the measures of ∠4 and ∠6 are 75°, and the measures of ∠5 and ∠7 are 105°.

On Your Own

Now You're Ready
Exercises 15–17

3. Use the figure to find the measures of the numbered angles.

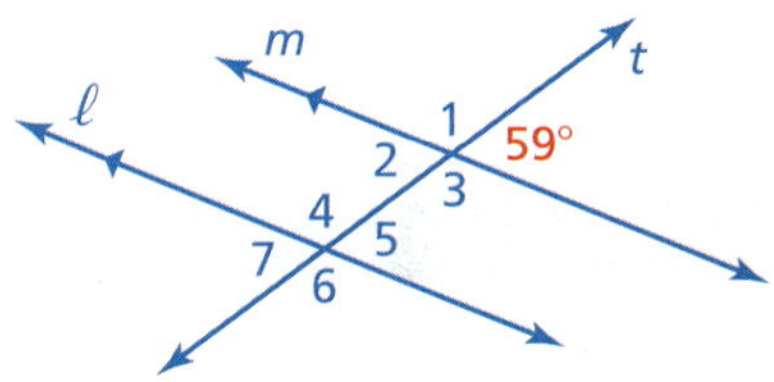

When two parallel lines are cut by a transversal, four **interior angles** are formed on the inside of the parallel lines and four **exterior angles** are formed on the outside of the parallel lines.

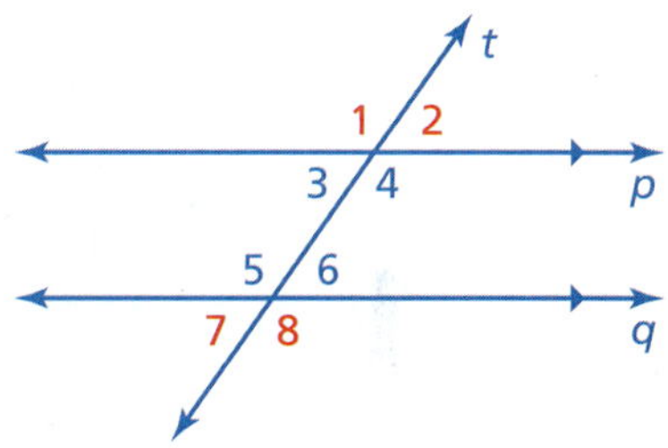

∠3, ∠4, ∠5, and ∠6 are interior angles.
∠1, ∠2, ∠7, and ∠8 are exterior angles.

EXAMPLE 3 Using Corresponding Angles

A store owner uses pieces of tape to paint a window advertisement. The letters are slanted at an 80° angle. What is the measure of ∠1?

(A) 80° (B) 100° (C) 110° (D) 120°

Because all the letters are slanted at an 80° angle, the dashed lines are parallel. The piece of tape is the transversal.

Using corresponding angles, the 80° angle is congruent to the angle that is supplementary to ∠1, as shown.

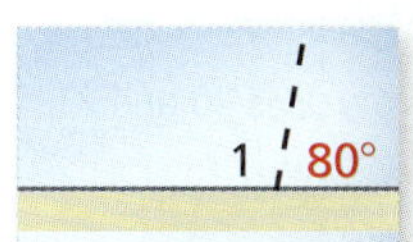

The measure of ∠1 is 180° − 80° = 100°. The correct answer is (B).

On Your Own

Now You're Ready
Exercises 18 and 19

4. **WHAT IF?** In Example 3, the letters are slanted at a 65° angle. What is the measure of ∠1?

Key Idea

Alternate Interior Angles and Alternate Exterior Angles

When a transversal intersects parallel lines, alternate interior angles are congruent and alternate exterior angles are congruent.

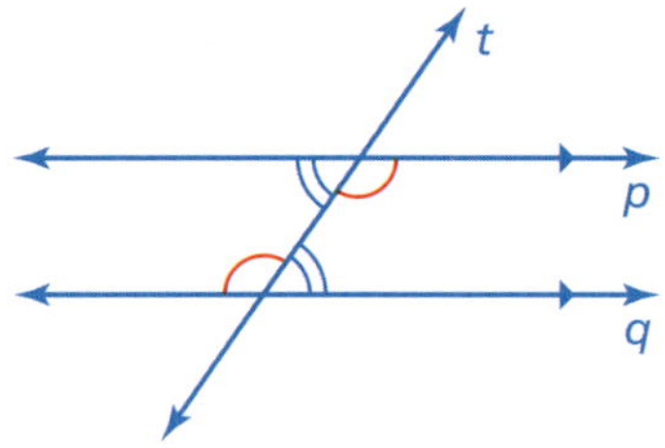

Alternate interior angles

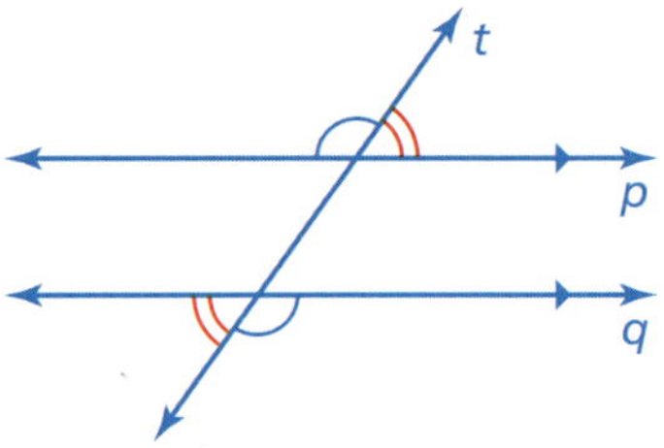

Alternate exterior angles

Study Tip

Alternate interior angles and alternate exterior angles lie on opposite sides of the transversal.

EXAMPLE 4 Identifying Alternate Interior and Alternate Exterior Angles

The photo shows a portion of an airport. Describe the relationship between each pair of angles.

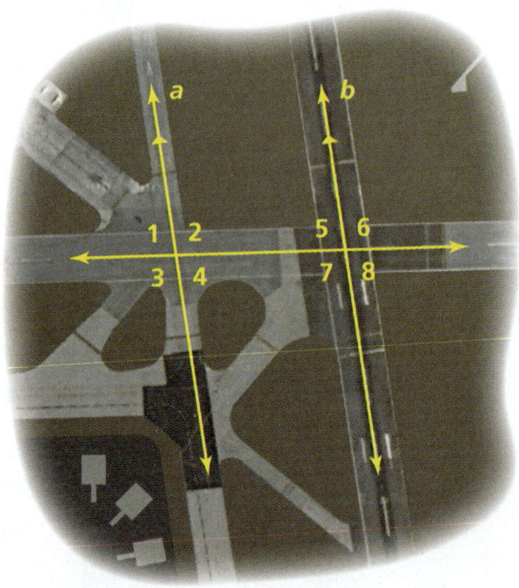

a. ∠3 and ∠6

∠3 and ∠6 are alternate exterior angles.

So, ∠3 is congruent to ∠6.

b. ∠2 and ∠7

∠2 and ∠7 are alternate interior angles.

So, ∠2 is congruent to ∠7.

On Your Own

Now You're Ready
Exercises 20 and 21

In Example 4, the measure of ∠4 is 84°. Find the measure of the angle. Explain your reasoning.

5. ∠3
6. ∠5
7. ∠6

3.1 Exercises

Check It Out
Help with Homework
BigIdeasMath.com

Vocabulary and Concept Check

1. **VOCABULARY** Draw two parallel lines and a transversal. Label a pair of corresponding angles.

2. **WHICH ONE DOESN'T BELONG?** Which statement does *not* belong with the other three? Explain your reasoning. Refer to the figure for Exercises 3–6.

The measure of $\angle 2$	The measure of $\angle 5$
The measure of $\angle 6$	The measure of $\angle 8$

Practice and Problem Solving

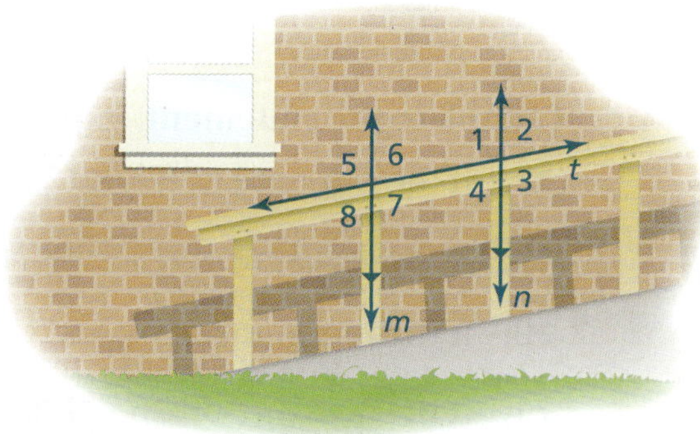

In Exercises 3–6, use the figure.

3. Identify the parallel lines.
4. Identify the transversal.
5. How many angles are formed by the transversal?
6. Which of the angles are congruent?

Use the figure to find the measures of the numbered angles.

1 7.

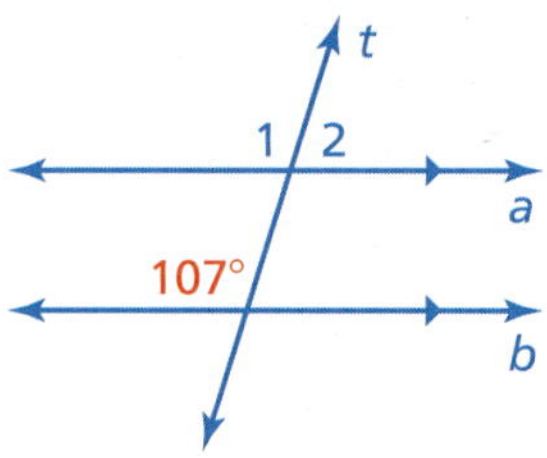

8.

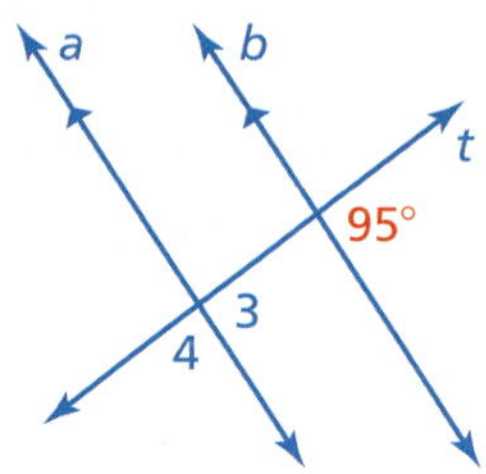

9.

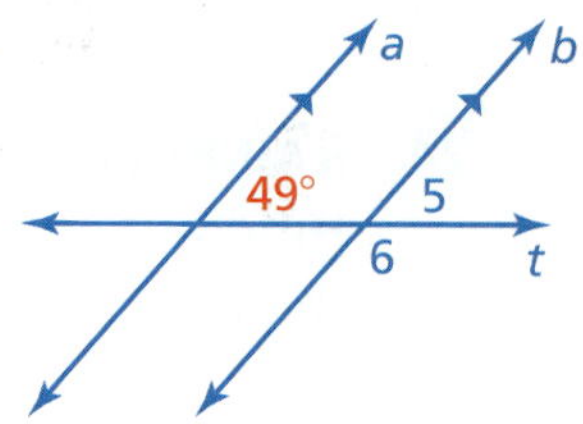

10. **ERROR ANALYSIS** Describe and correct the error in describing the relationship between the angles.

✗ $\angle 5$ is congruent to $\angle 6$.

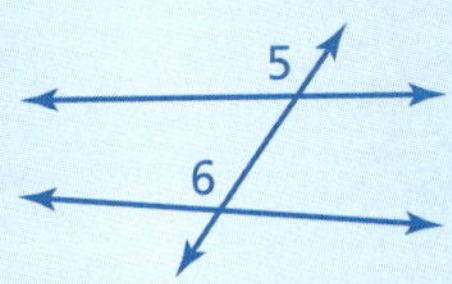

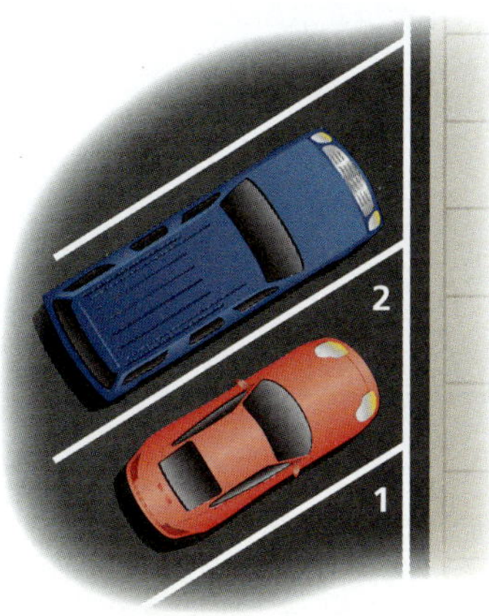

11. **PARKING** The painted lines that separate parking spaces are parallel. The measure of $\angle 1$ is 60°. What is the measure of $\angle 2$? Explain.

12. **OPEN-ENDED** Describe two real-life situations that use parallel lines.

13. **PROJECT** Trace line p and line t on a piece of paper. Label $\angle 1$. Move the paper so that $\angle 1$ aligns with $\angle 8$. Describe the transformations that you used to show that $\angle 1$ is congruent to $\angle 8$.

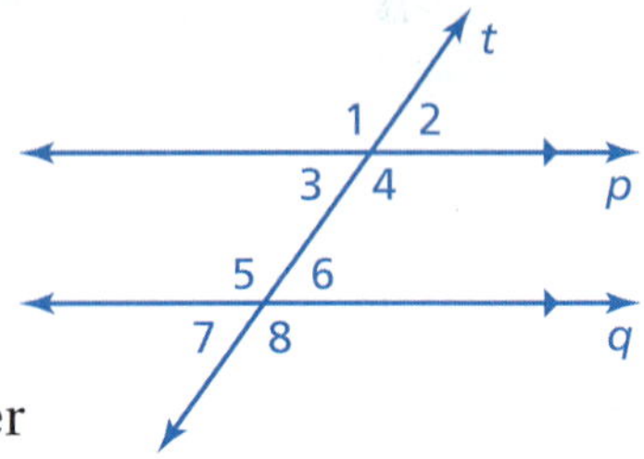

14. **REASONING** Two horizontal lines are cut by a transversal. What is the least number of angle measures you need to know in order to find the measure of every angle? Explain your reasoning.

Use the figure to find the measures of the numbered angles. Explain your reasoning.

2 15.

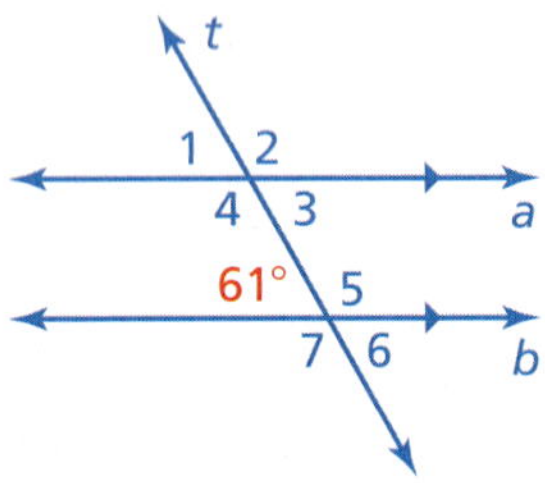

16.

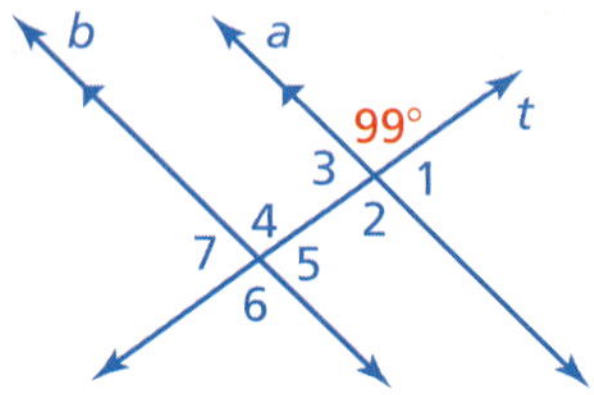

17.

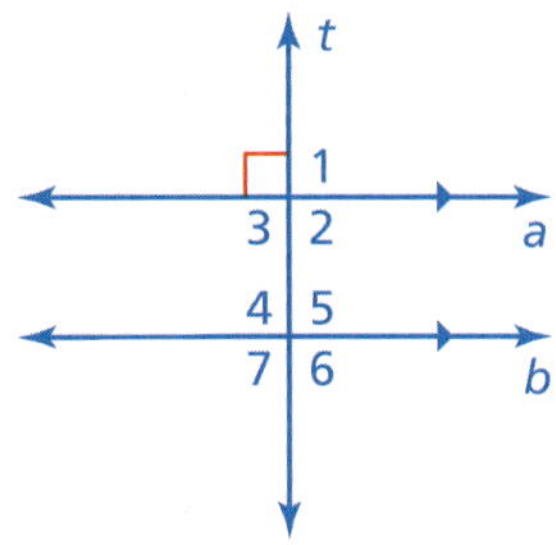

Complete the statement. Explain your reasoning.

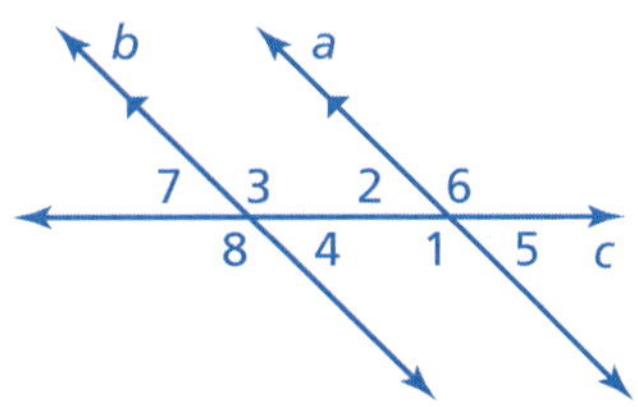

3 18. If the measure of $\angle 1 = 124°$, then the measure of $\angle 4 =$ ______.

19. If the measure of $\angle 2 = 48°$, then the measure of $\angle 3 =$ ______.

4 20. If the measure of $\angle 4 = 55°$, then the measure of $\angle 2 =$ ______.

21. If the measure of $\angle 6 = 120°$, then the measure of $\angle 8 =$ ______.

22. If the measure of $\angle 7 = 50.5°$, then the measure of $\angle 6 =$ ______.

23. If the measure of $\angle 3 = 118.7°$, then the measure of $\angle 2 =$ ______.

24. **RAINBOW** A rainbow forms when sunlight reflects off raindrops at different angles. For blue light, the measure of $\angle 2$ is 40°. What is the measure of $\angle 1$?

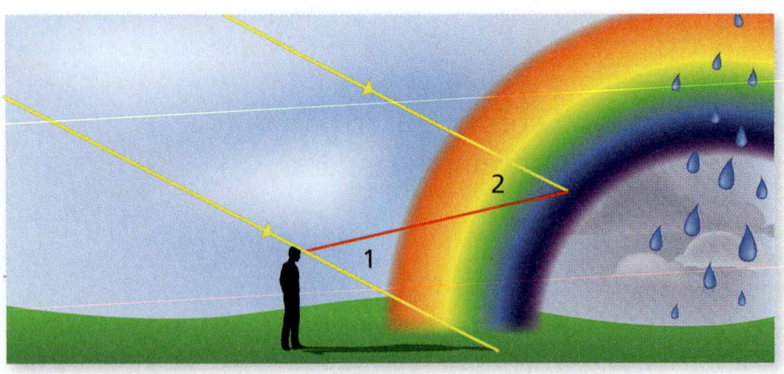

25. **REASONING** When a transversal is perpendicular to two parallel lines, all the angles formed measure 90°. Explain why.

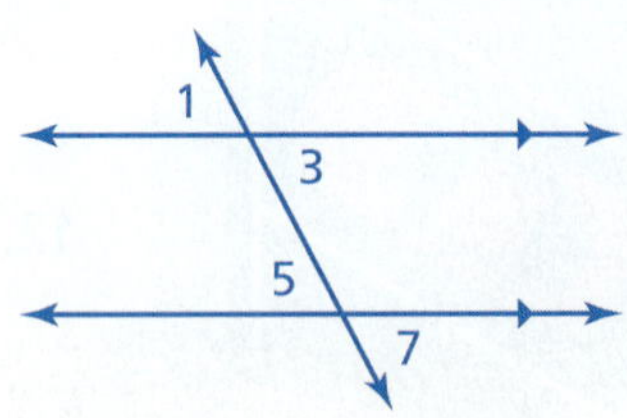

26. **LOGIC** Describe two ways you can show that $\angle 1$ is congruent to $\angle 7$.

CRITICAL THINKING Find the value of x.

27.

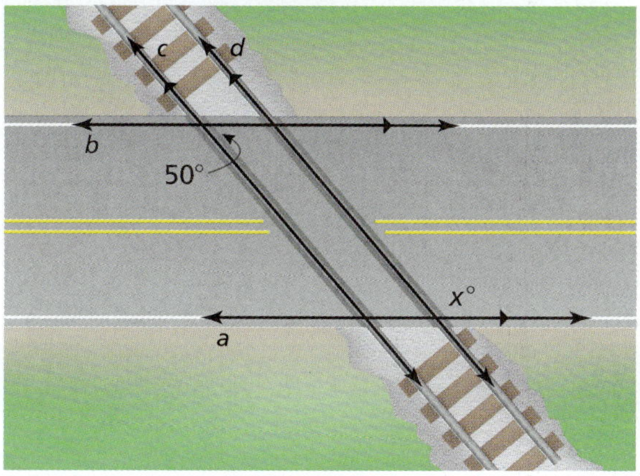

28.

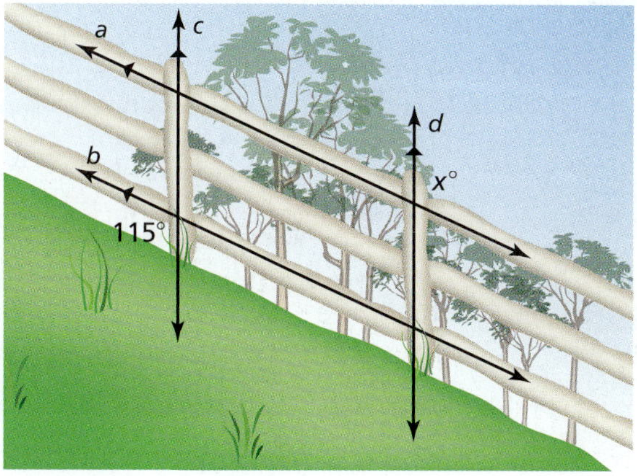

29. OPTICAL ILLUSION Refer to the figure.

a. Do the horizontal lines appear to be parallel? Explain.

b. Draw your own optical illusion using parallel lines.

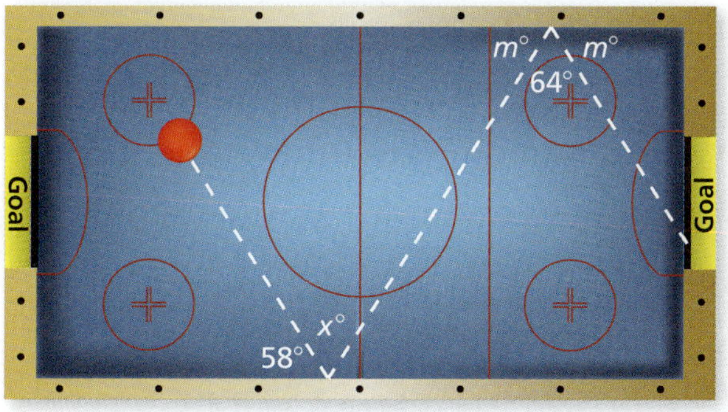

30. Geometry The figure shows the angles used to make a double bank shot in an air hockey game.

a. Find the value of x.

b. Can you still get the red puck in the goal when x is increased by a little? by a lot? Explain.

Fair Game Review What you learned in previous grades & lessons

Evaluate the expression. *(Skills Review Handbook)*

31. $4 + 3^2$ **32.** $5(2)^2 - 6$ **33.** $11 + (-7)^2 - 9$ **34.** $8 \div 2^2 + 1$

35. MULTIPLE CHOICE The triangles are similar. What length does x represent? *(Section 2.5)*

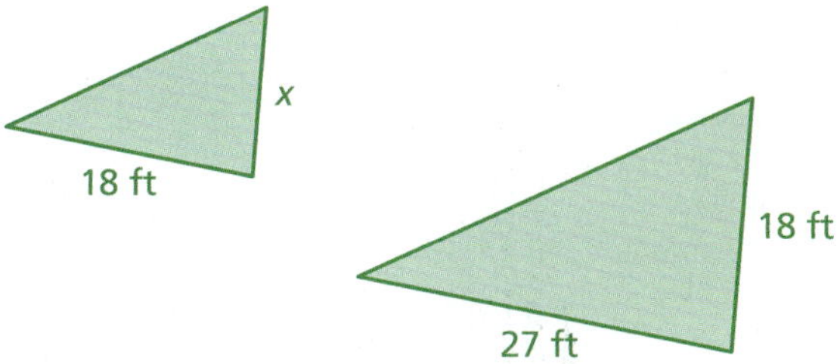

Ⓐ 2 ft Ⓑ 12 ft

Ⓒ 15 ft Ⓓ 27 ft

3.2 Angles of Triangles

Essential Question How can you describe the relationships among the angles of a triangle?

1 ACTIVITY: Exploring the Interior Angles of a Triangle

Work with a partner.

a. Draw a triangle. Label the interior angles *A*, *B*, and *C*.

b. Carefully cut out the triangle. Tear off the three corners of the triangle.

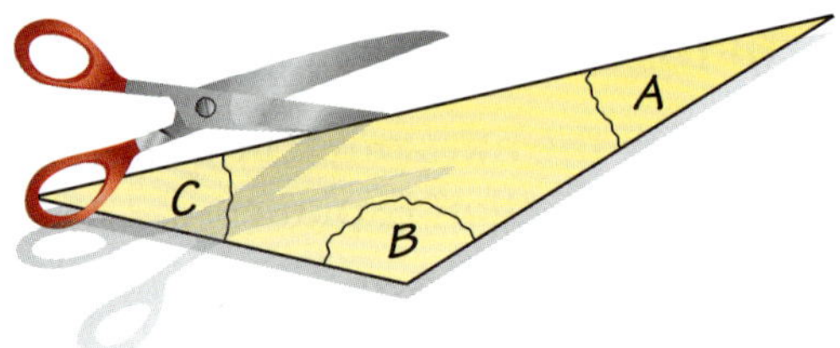

c. Arrange angles *A* and *B* so that they share a vertex and are adjacent.

d. How can you place the third angle to determine the sum of the measures of the interior angles? What is the sum?

e. Compare your results with those of others in your class.

f. **STRUCTURE** How does your result in part (d) compare to the rule you wrote in Lesson 1.1, Activity 2?

2 ACTIVITY: Exploring the Interior Angles of a Triangle

Work with a partner.

a. Describe the figure.

b. **LOGIC** Use what you know about parallel lines and transversals to justify your result in part (d) of Activity 1.

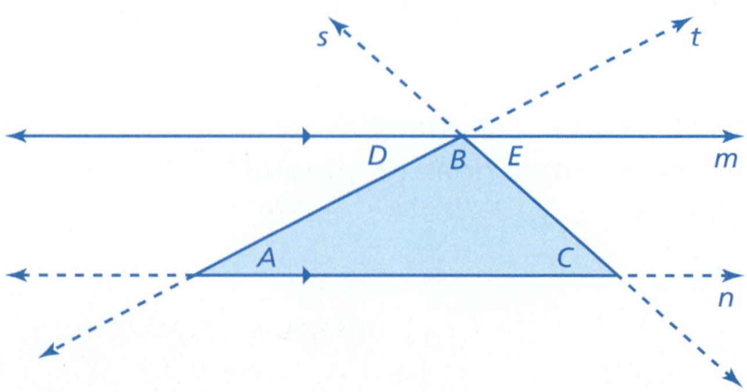

COMMON CORE

Geometry

In this lesson, you will

- understand that the sum of the interior angle measures of a triangle is 180°.
- find the measures of interior and exterior angles of triangles.

Learning Standard
MACC.8.G.1.5

3 ACTIVITY: Exploring an Exterior Angle of a Triangle

Math Practice 8

Maintain Oversight

Do you think your conclusion will be true for the exterior angle of any triangle? Explain.

Work with a partner.

a. Draw a triangle. Label the interior angles *A*, *B*, and *C*.

b. Carefully cut out the triangle.

c. Place the triangle on a piece of paper and extend one side to form *exterior angle D*, as shown.

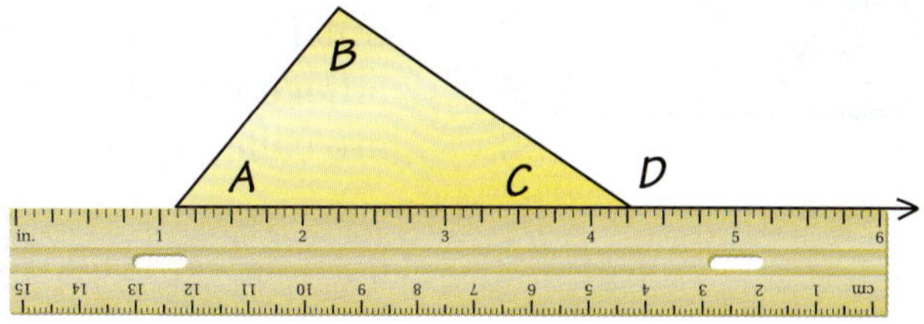

d. Tear off the corners that are not adjacent to the exterior angle. Arrange them to fill the exterior angle, as shown. What does this tell you about the measure of exterior angle *D*?

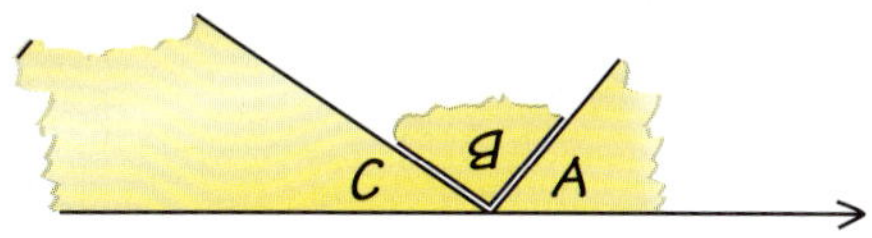

4 ACTIVITY: Measuring the Exterior Angles of a Triangle

Work with a partner.

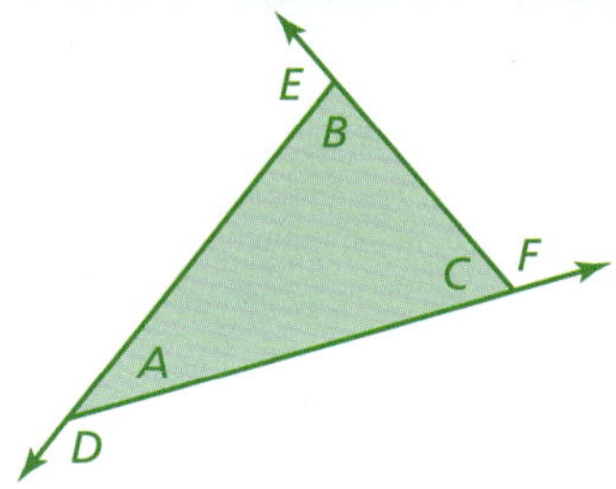

a. Draw a triangle and label the interior and exterior angles, as shown.

b. Use a protractor to measure all six angles. Copy and complete the table to organize your results. What does the table tell you about the measure of an exterior angle of a triangle?

Exterior Angle	$D = \square°$	$E = \square°$	$F = \square°$
Interior Angle	$B = \square°$	$A = \square°$	$A = \square°$
Interior Angle	$C = \square°$	$C = \square°$	$B = \square°$

What Is Your Answer?

5. REPEATED REASONING Draw three triangles that have different shapes. Repeat parts (b)–(d) from Activity 1 for each triangle. Do you get the same results? Explain.

6. IN YOUR OWN WORDS How can you describe the relationships among angles of a triangle?

Practice Use what you learned about angles of a triangle to complete Exercises 4–6 on page 114.

3.2 Lesson

Check It Out
Lesson Tutorials
BigIdeasMath.com

Key Vocabulary
interior angles of a polygon, *p. 112*
exterior angles of a polygon, *p. 112*

The angles inside a polygon are called **interior angles**. When the sides of a polygon are extended, other angles are formed. The angles outside the polygon that are adjacent to the interior angles are called **exterior angles**.

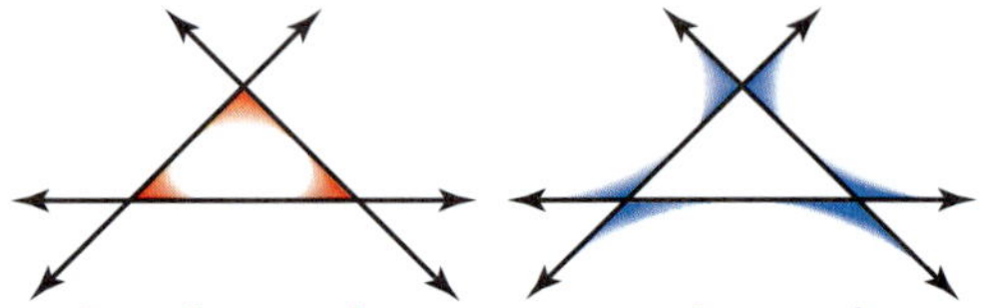

Key Idea

Interior Angle Measures of a Triangle

Words The sum of the interior angle measures of a triangle is 180°.

Algebra $x + y + z = 180$

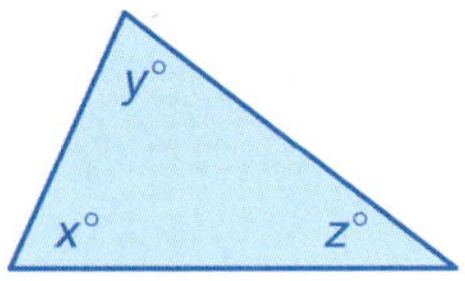

EXAMPLE 1 Using Interior Angle Measures

Find the value of *x*.

a.

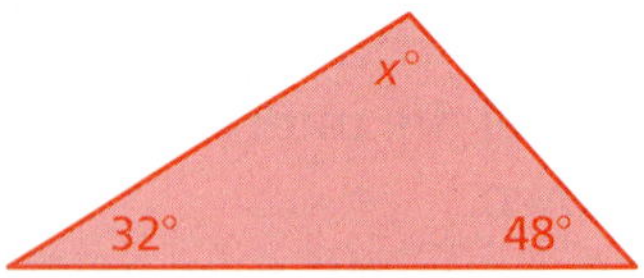

$$x + 32 + 48 = 180$$
$$x + 80 = 180$$
$$x = 100$$

b.

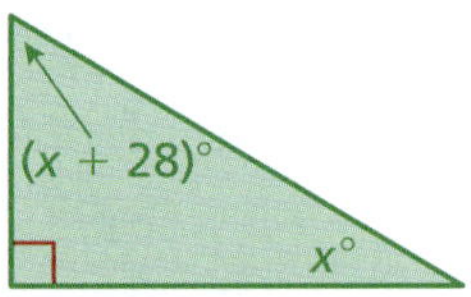

$$x + (x + 28) + 90 = 180$$
$$2x + 118 = 180$$
$$2x = 62$$
$$x = 31$$

On Your Own

Now You're Ready
Exercises 4–9

Find the value of *x*.

1.

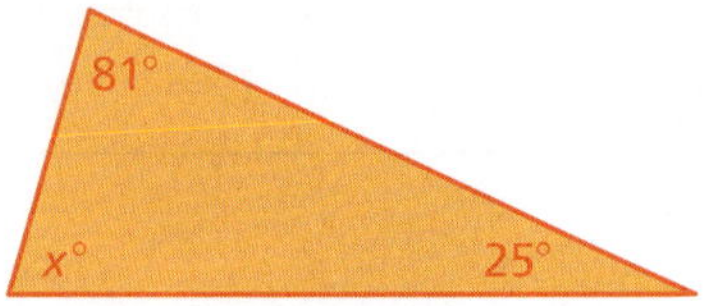

2.

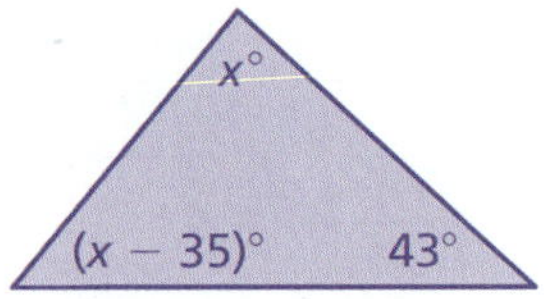

Key Idea

Exterior Angle Measures of a Triangle

Words The measure of an exterior angle of a triangle is equal to the sum of the measures of the two nonadjacent interior angles.

Algebra $z = x + y$

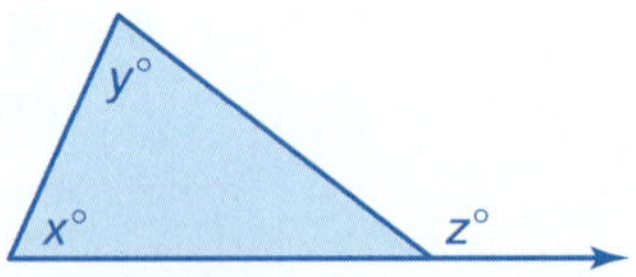

Multi-Language Glossary at BigIdeasMath.com

EXAMPLE 2 Finding Exterior Angle Measures

Study Tip

Each vertex has a pair of congruent exterior angles. However, it is common to show only one exterior angle at each vertex.

Find the measure of the exterior angle.

a.

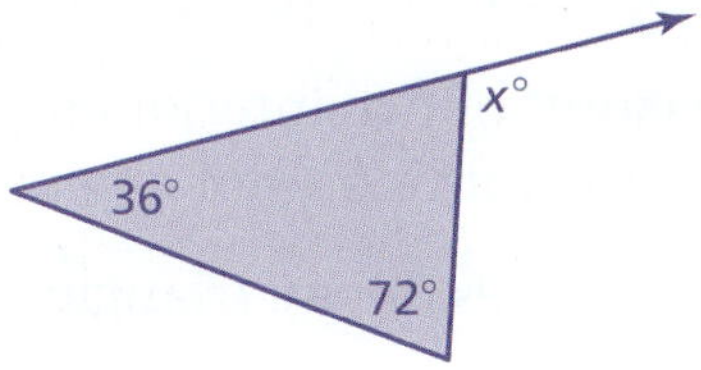

$x = 36 + 72$

$x = 108$

So, the measure of the exterior angle is 108°.

b.

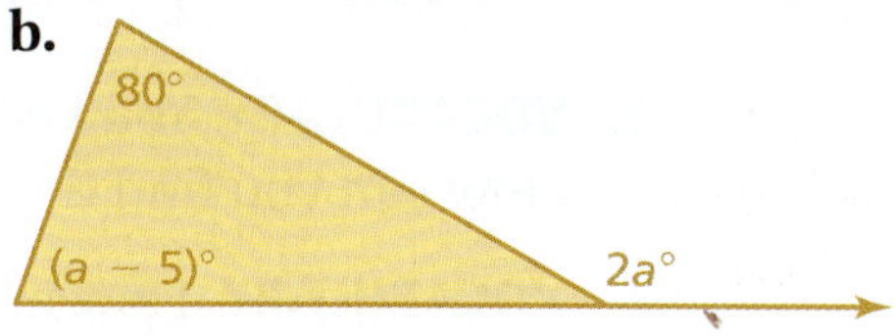

$2a = (a - 5) + 80$

$2a = a + 75$

$a = 75$

So, the measure of the exterior angle is $2(75)° = 150°$.

EXAMPLE 3 Real-Life Application

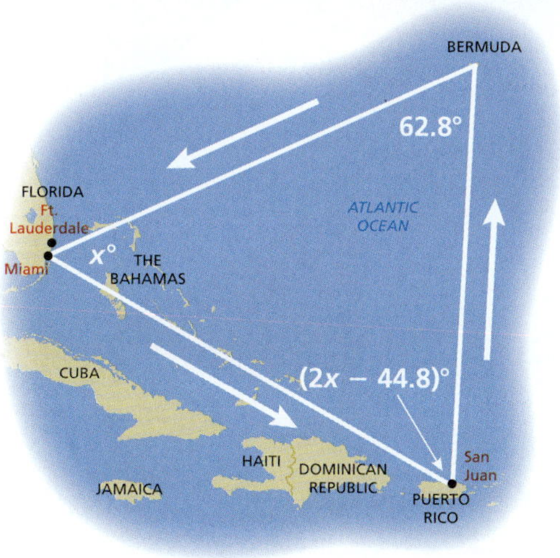

An airplane leaves from Miami and travels around the Bermuda Triangle. What is the value of *x*?

Ⓐ 26.8 Ⓑ 27.2 Ⓒ 54 Ⓓ 64

Use what you know about the interior angle measures of a triangle to write an equation.

$x + (2x - 44.8) + 62.8 = 180$	Write equation.
$3x + 18 = 180$	Combine like terms.
$3x = 162$	Subtract 18 from each side.
$x = 54$	Divide each side by 3.

The value of x is 54. The correct answer is Ⓒ.

On Your Own

Now You're Ready
Exercises 12–14

Find the measure of the exterior angle.

3.

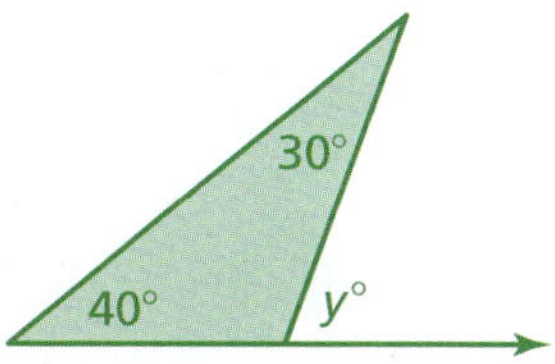

4.

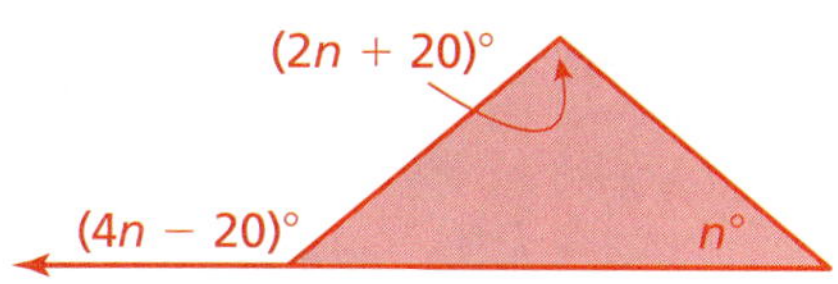

5. In Example 3, the airplane leaves from Fort Lauderdale. The interior angle measure at Bermuda is 63.9°. The interior angle measure at San Juan is $(x + 7.5)°$. Find the value of x.

3.2 Exercises

Check It Out
Help with Homework
BigIdeasMath.com

Vocabulary and Concept Check

1. **VOCABULARY** You know the measures of two interior angles of a triangle. How can you find the measure of the third interior angle?

2. **VOCABULARY** How many exterior angles does a triangle have at each vertex? Explain.

3. **NUMBER SENSE** List the measures of the exterior angles for the triangle shown at the right.

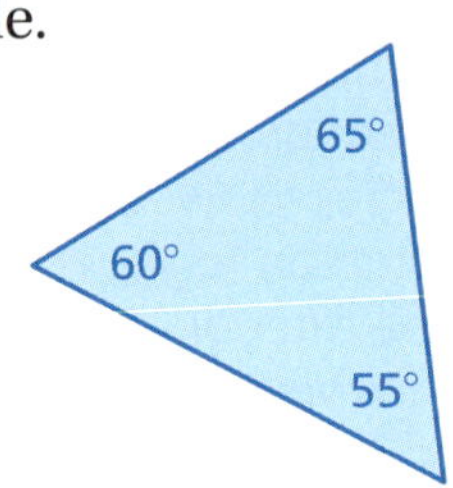

Practice and Problem Solving

Find the measures of the interior angles.

1 4.
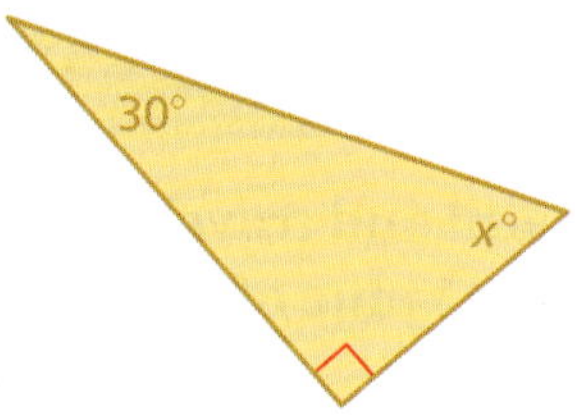

5.
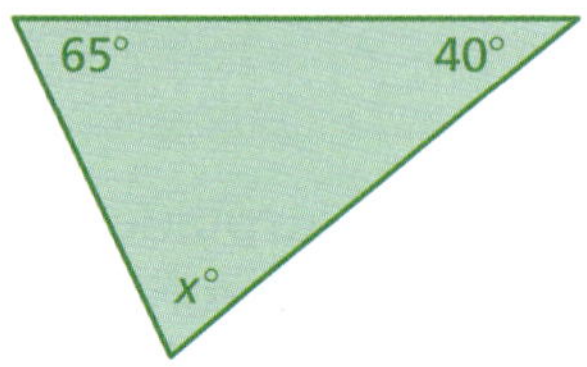

6.
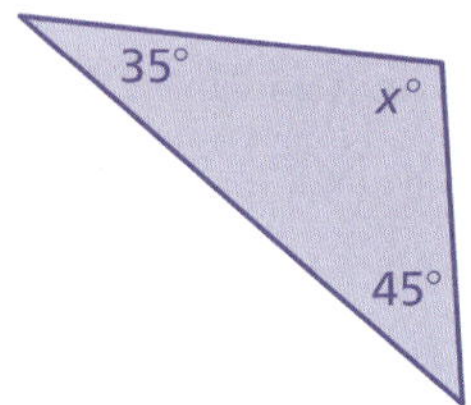

7.
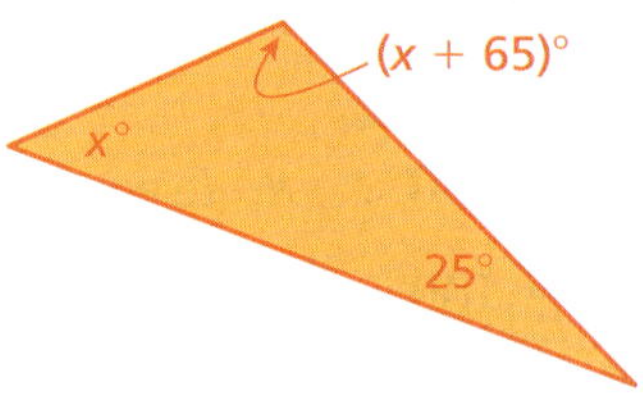

8.
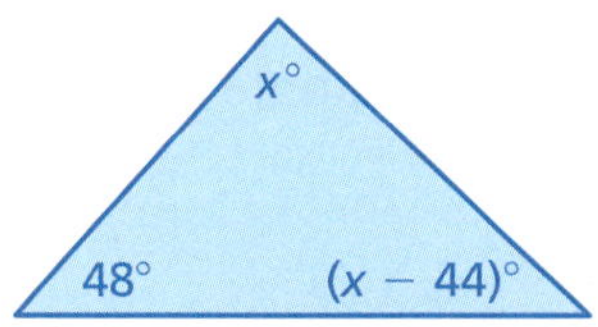

9.
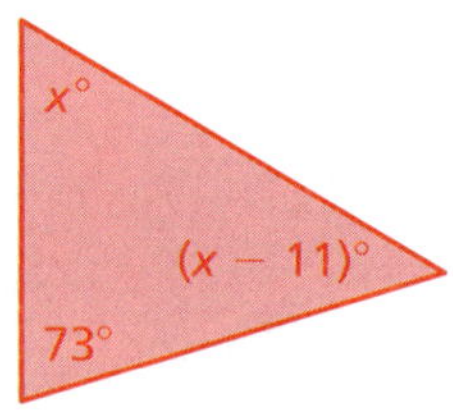

10. **BILLIARD RACK** Find the value of x in the billiard rack.

11. **NO PARKING** The triangle with lines through it designates a no parking zone. What is the value of x?

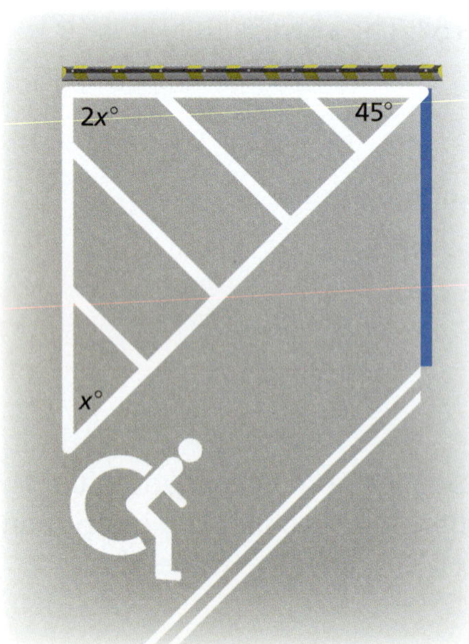

Find the measure of the exterior angle.

2 **12.**

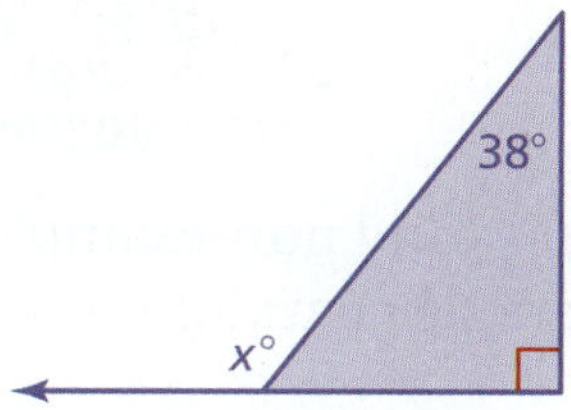

13.

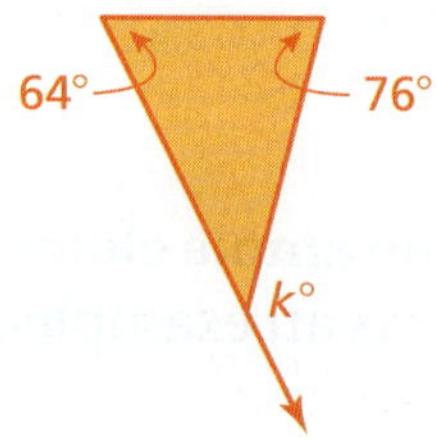

14.

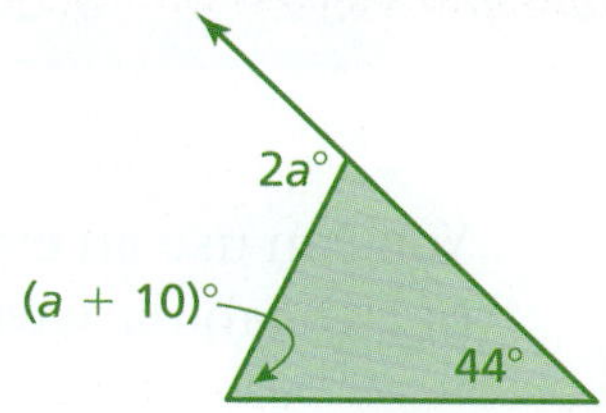

15. ERROR ANALYSIS Describe and correct the error in finding the measure of the exterior angle.

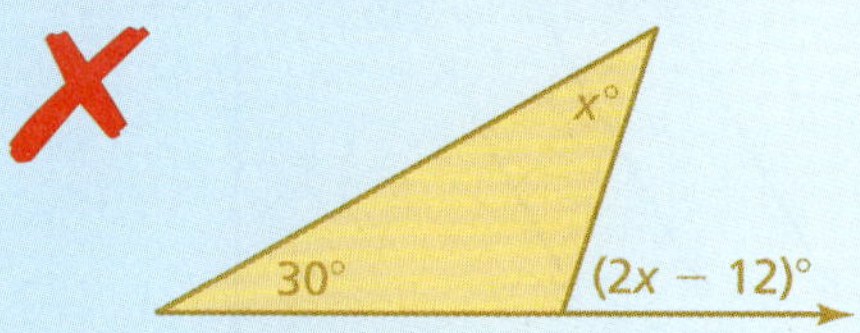

$(2x - 12) + x + 30 = 180$

$3x + 18 = 180$

$x = 54$

The exterior angle is $(2(54) - 12)° = 96°$.

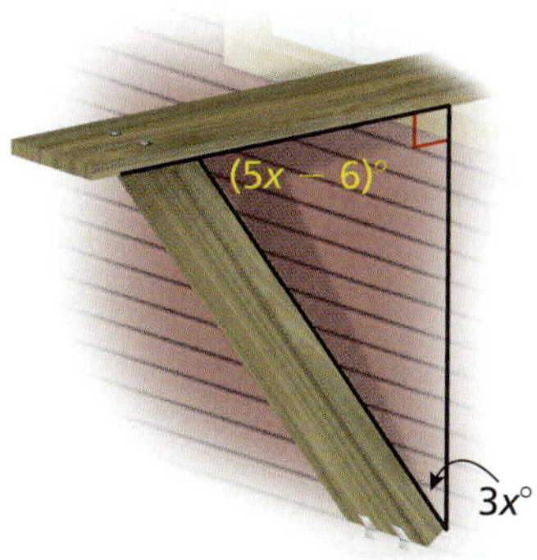

16. RATIO The ratio of the interior angle measures of a triangle is 2 : 3 : 5. What are the angle measures?

17. CONSTRUCTION The support for a window air-conditioning unit forms a triangle and an exterior angle. What is the measure of the exterior angle?

18. REASONING A triangle has an exterior angle with a measure of 120°. Can you determine the measures of the interior angles? Explain.

Determine whether the statement is *always, sometimes,* or *never* true. Explain your reasoning.

19. Given three angle measures, you can construct a triangle.

20. The acute interior angles of a right triangle are complementary.

21. A triangle has more than one vertex with an acute exterior angle.

22. Precision Using the figure at the right, show that $z = x + y$. (*Hint:* Find two equations involving w.)

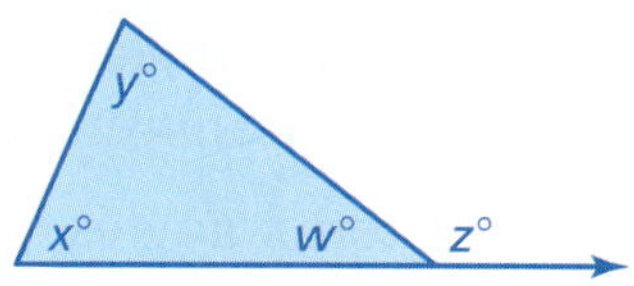

Fair Game Review What you learned in previous grades & lessons

Solve the equation. Check your solution. *(Section 1.2)*

23. $-4x + 3 = 19$ **24.** $2(y - 1) + 6y = -10$ **25.** $5 + 0.5(6n + 14) = 3$

26. MULTIPLE CHOICE Which transformation moves every point of a figure the same distance and in the same direction? *(Section 2.2)*

Ⓐ translation Ⓑ reflection Ⓒ rotation Ⓓ dilation

3 Study Help

Check It Out
Graphic Organizer
BigIdeasMath.com

You can use an **example and non-example chart** to list examples and non-examples of a vocabulary word or item. Here is an example and non-example chart for transversals.

Transversals

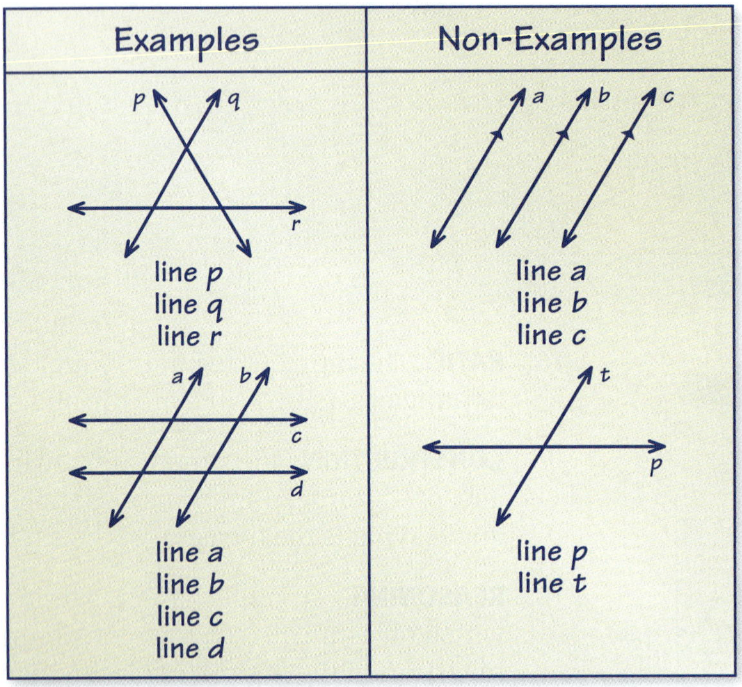

On Your Own

Make example and non-example charts to help you study these topics.

1. interior angles formed by parallel lines and a transversal
2. exterior angles formed by parallel lines and a transversal

After you complete this chapter, make example and non-example charts for the following topics.

3. interior angles of a polygon
4. exterior angles of a polygon
5. regular polygons
6. similar triangles

"What do you think of my example & non-example chart for popular cat toys?"

3.1–3.2 Quiz

Check It Out
Progress Check
BigIdeasMath.com

Use the figure to find the measure of the angle. Explain your reasoning. *(Section 3.1)*

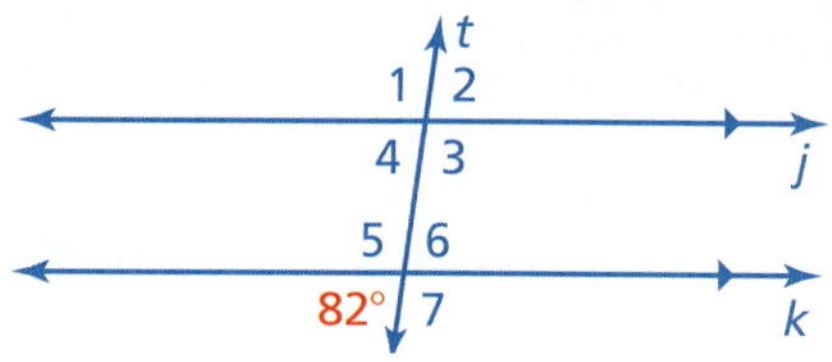

1. $\angle 2$

2. $\angle 6$

3. $\angle 4$

4. $\angle 1$

Complete the statement. Explain your reasoning. *(Section 3.1)*

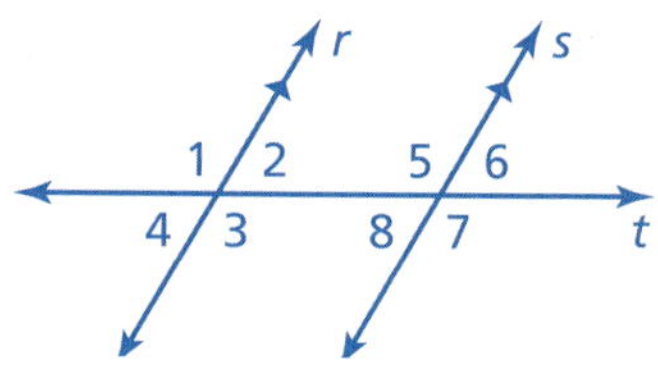

5. If the measure of $\angle 1 = 123°$, then the measure of $\angle 7 =$ ____.

6. If the measure of $\angle 2 = 58°$, then the measure of $\angle 5 =$ ____.

7. If the measure of $\angle 5 = 119°$, then the measure of $\angle 3 =$ ____.

8. If the measure of $\angle 4 = 60°$, then the measure of $\angle 6 =$ ____.

Find the measures of the interior angles. *(Section 3.2)*

9.

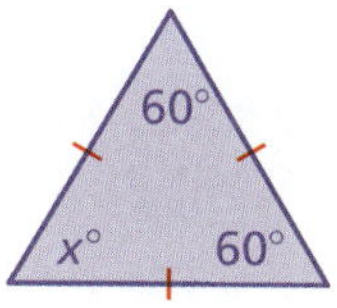

10.

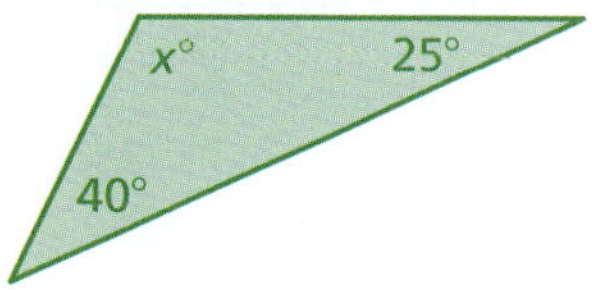

11.

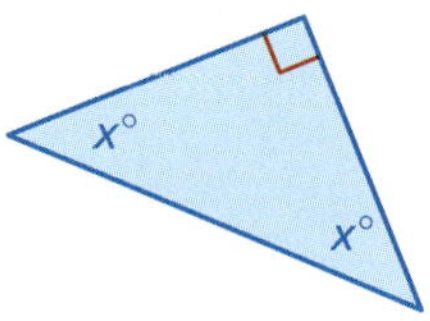

Find the measure of the exterior angle. *(Section 3.2)*

12.

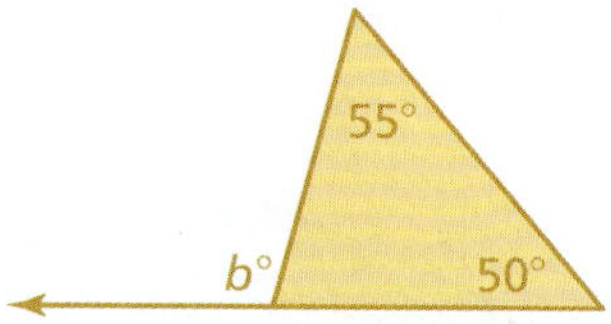

13.

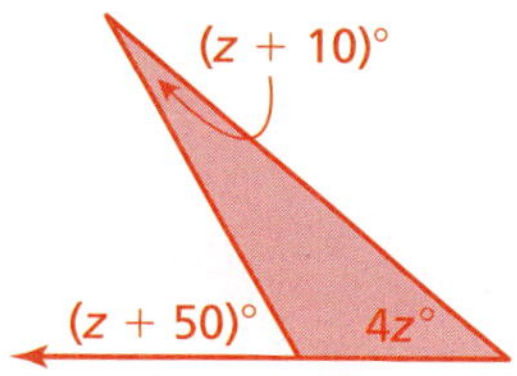

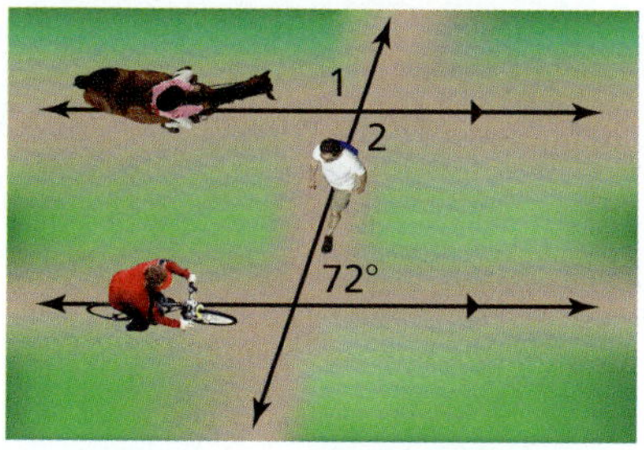

14. PARK In a park, a bike path and a horse riding path are parallel. In one part of the park, a hiking trail intersects the two paths. Find the measures of $\angle 1$ and $\angle 2$. Explain your reasoning. *(Section 3.1)*

15. LADDER A ladder leaning against a wall forms a triangle and exterior angles with the wall and the ground. What are the measures of the exterior angles? Justify your answer. *(Section 3.2)*

3.3 Angles of Polygons

Essential Question How can you find the sum of the interior angle measures and the sum of the exterior angle measures of a polygon?

1 ACTIVITY: Exploring the Interior Angles of a Polygon

Work with a partner. In parts (a) – (e), identify each polygon and the number of sides *n*. Then find the sum of the interior angle measures of the polygon.

a. Polygon: ______ Number of sides: $n =$ ______

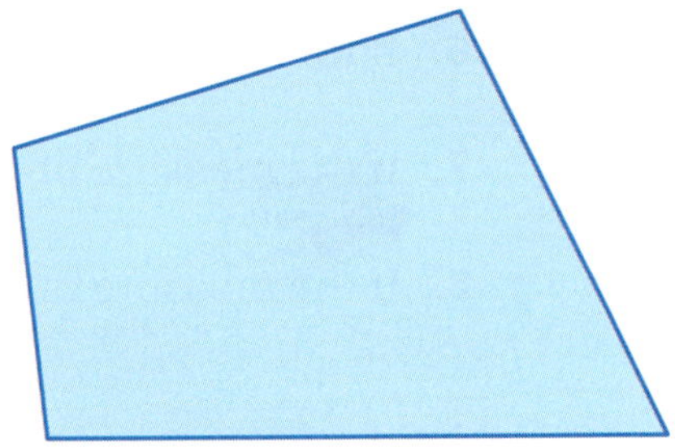

Draw a line segment on the figure that divides it into two triangles. Is there more than one way to do this? Explain.

What is the sum of the interior angle measures of each triangle?

What is the sum of the interior angle measures of the figure?

b.

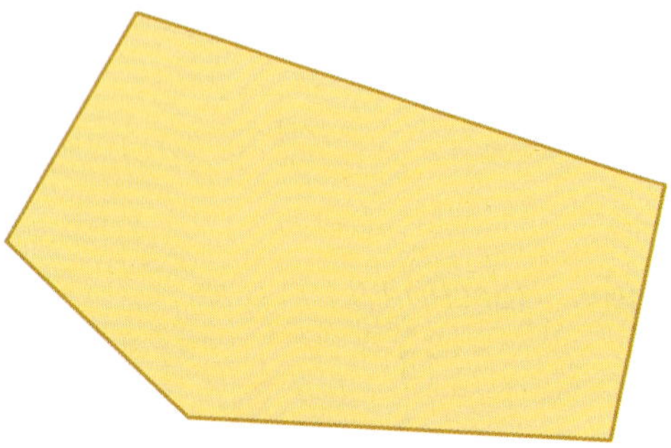

c.

d.

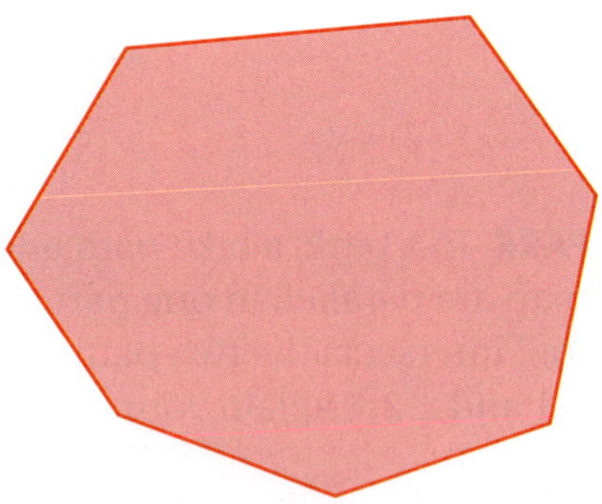

e.

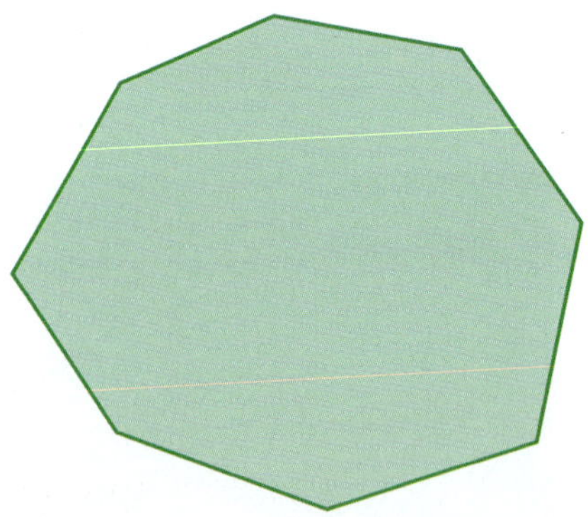

f. **REPEATED REASONING** Use your results to complete the table. Then find the sum of the interior angle measures of a polygon with 12 sides.

Number of Sides, *n*	3	4	5	6	7	8
Number of Triangles						
Angle Sum, *S*						

COMMON CORE

Geometry

In this lesson, you will
- find the sum of the interior angle measures of polygons.
- understand that the sum of the exterior angle measures of a polygon is 360°.
- find the measures of interior and exterior angles of polygons.

Applying Standard
MACC.8.G.1.5

A polygon is **convex** when every line segment connecting any two vertices lies entirely inside the polygon. A polygon is **concave** when at least one line segment connecting any two vertices lies outside the polygon.

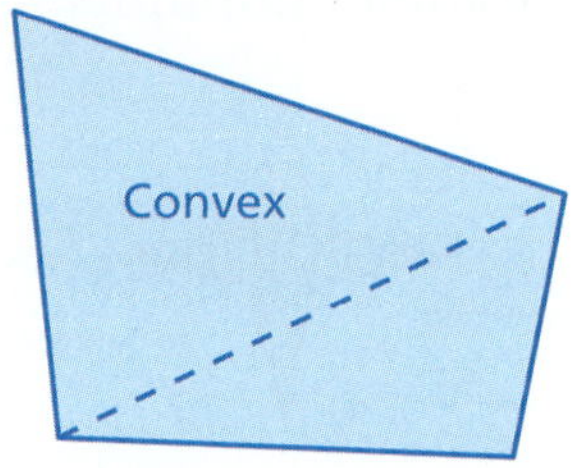

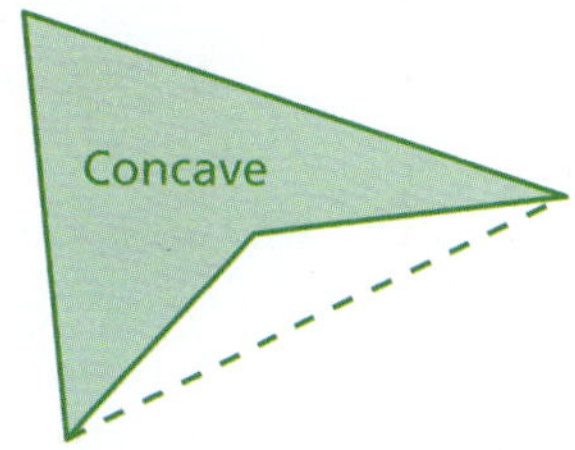

2 ACTIVITY: Exploring the Exterior Angles of a Polygon

Math Practice 3

Analyze Conjectures

Do your observations about the sum of the exterior angles make sense? Do you think they would hold true for any convex polygon? Explain.

Work with a partner.

a. Draw a convex pentagon. Extend the sides to form the exterior angles. Label one exterior angle at each vertex *A*, *B*, *C*, *D*, and *E*, as shown.

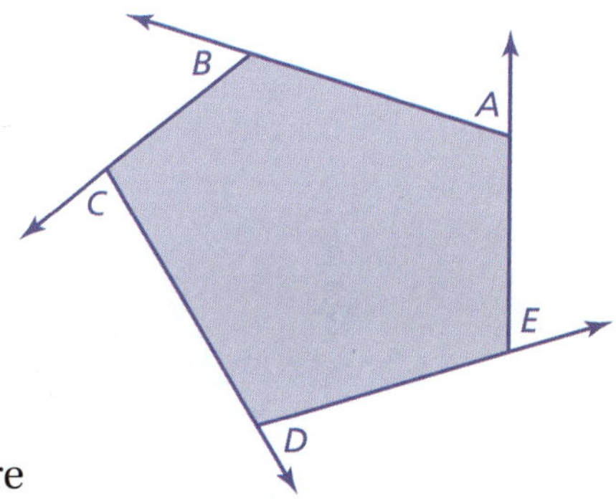

b. Cut out the exterior angles. How can you join the vertices to determine the sum of the angle measures? What do you notice?

c. **REPEATED REASONING** Repeat the procedure in parts (a) and (b) for each figure below.

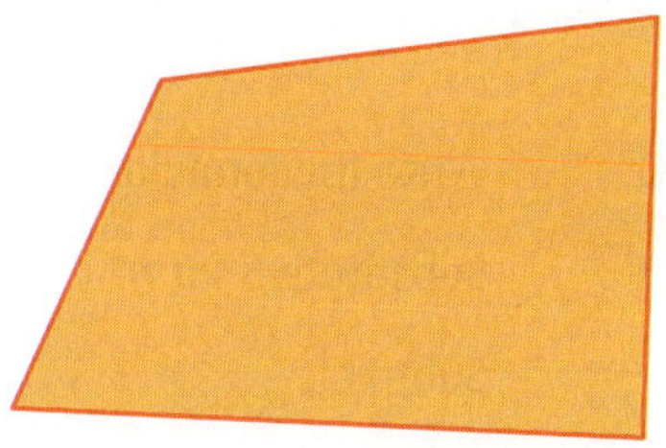

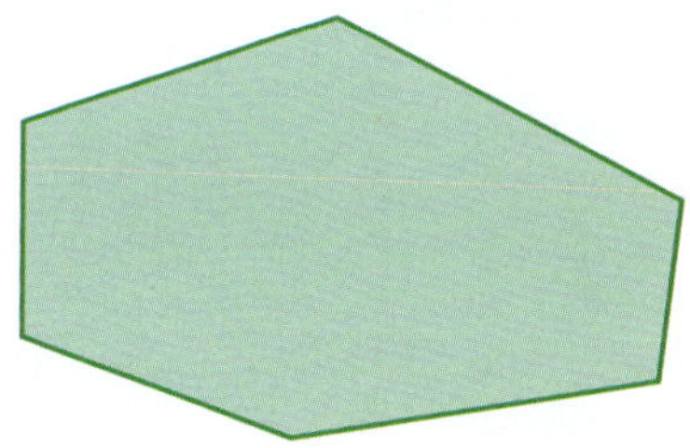

What can you conclude about the sum of the measures of the exterior angles of a convex polygon? Explain.

What Is Your Answer?

3. **STRUCTURE** Use your results from Activity 1 to write an expression that represents the sum of the interior angle measures of a polygon.

4. **IN YOUR OWN WORDS** How can you find the sum of the interior angle measures and the sum of the exterior angle measures of a polygon?

Practice

Use what you learned about angles of polygons to complete Exercises 4–6 on page 123.

3.3 Lesson

Check It Out
Lesson Tutorials
BigIdeasMath.com

Key Vocabulary
convex polygon, *p. 119*
concave polygon, *p. 119*
regular polygon, *p. 121*

A *polygon* is a closed plane figure made up of three or more line segments that intersect only at their endpoints.

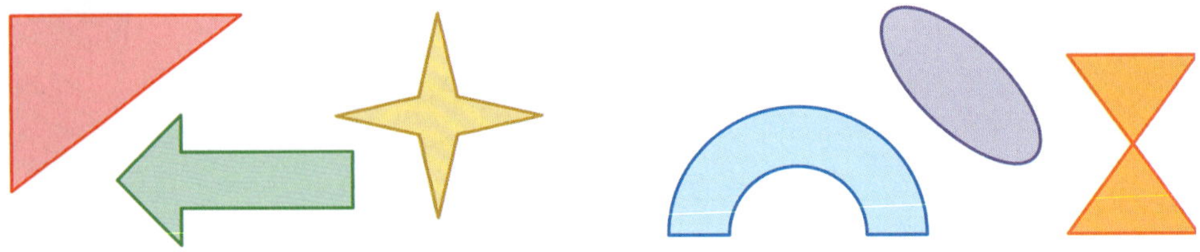

Polygons **Not polygons**

Key Idea

Interior Angle Measures of a Polygon

The sum S of the interior angle measures of a polygon with n sides is

$$S = (n - 2) \cdot 180^\circ.$$

EXAMPLE 1 Finding the Sum of Interior Angle Measures

Reading
For polygons whose names you have not learned, you can use the phrase "*n*-gon," where *n* is the number of sides. For example, a 15-gon is a polygon with 15 sides.

Find the sum of the interior angle measures of the school crossing sign.

The sign is in the shape of a pentagon. It has 5 sides.

$S = (n - 2) \cdot 180^\circ$ Write the formula.

$= (5 - 2) \cdot 180^\circ$ Substitute 5 for n.

$= 3 \cdot 180^\circ$ Subtract.

$= 540^\circ$ Multiply.

∴ The sum of the interior angle measures is 540°.

On Your Own

Now You're Ready
Exercises 7–9

Find the sum of the interior angle measures of the green polygon.

1. 

2. 

EXAMPLE 2 Finding an Interior Angle Measure of a Polygon

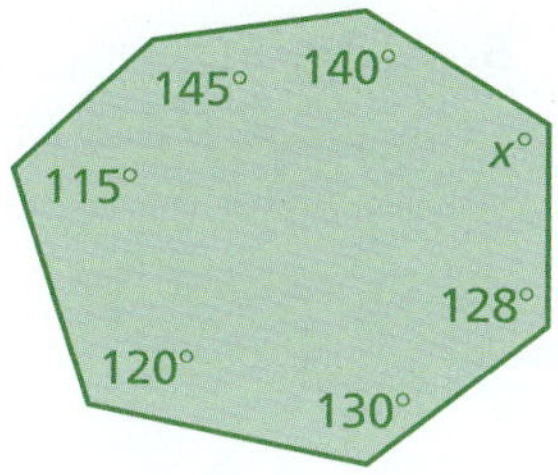

Find the value of x.

Step 1: The polygon has 7 sides. Find the sum of the interior angle measures.

$S = (n - 2) \cdot 180°$ Write the formula.

$= (7 - 2) \cdot 180°$ Substitute 7 for n.

$= 900°$ Simplify. The sum of the interior angle measures is 900°.

Step 2: Write and solve an equation.

$$140 + 145 + 115 + 120 + 130 + 128 + x = 900$$

$$778 + x = 900$$

$$x = 122$$

∴ The value of x is 122.

On Your Own

Now You're Ready
Exercises 12–14

Find the value of x.

3. 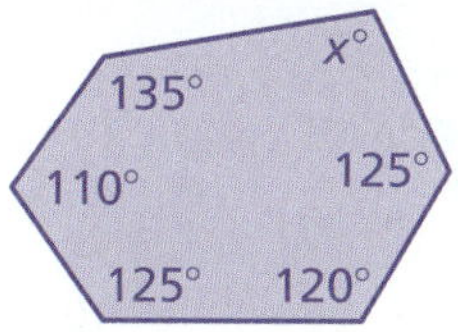

4. 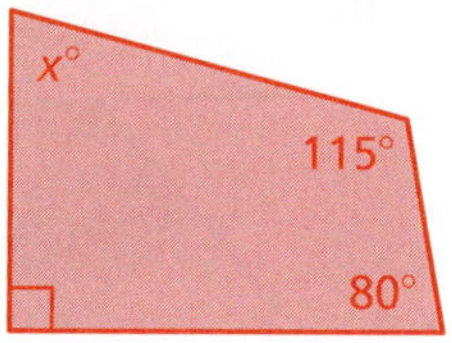

5. 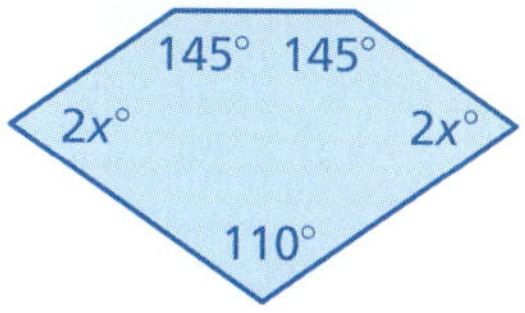

In a **regular polygon**, all the sides are congruent, and all the interior angles are congruent.

EXAMPLE 3 Real-Life Application

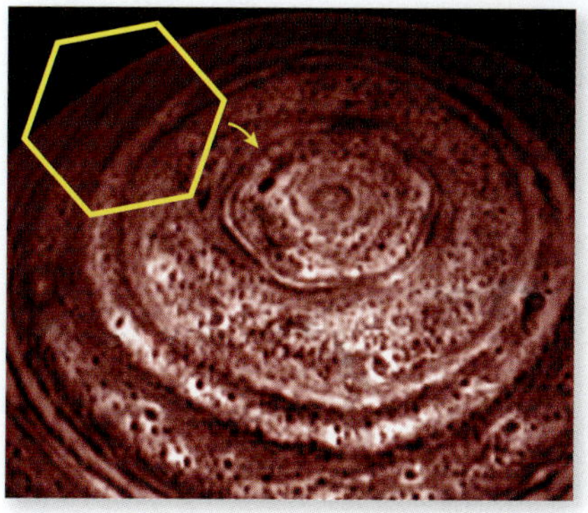

The hexagon is about 15,000 miles across. Approximately four Earths could fit inside it.

A cloud system discovered on Saturn is in the approximate shape of a regular hexagon. Find the measure of each interior angle of the hexagon.

Step 1: A hexagon has 6 sides. Find the sum of the interior angle measures.

$S = (n - 2) \cdot 180°$ Write the formula.

$= (6 - 2) \cdot 180°$ Substitute 6 for n.

$= 720°$ Simplify. The sum of the interior angle measures is 720°.

Step 2: Divide the sum by the number of interior angles, 6.

$720° \div 6 = 120°$

∴ The measure of each interior angle is 120°.

On Your Own

Now You're Ready
Exercises 16–18

Find the measure of each interior angle of the regular polygon.

6. octagon **7.** decagon **8.** 18-gon

Key Idea

Exterior Angle Measures of a Polygon

Words The sum of the measures of the exterior angles of a convex polygon is 360°.

Algebra $w + x + y + z = 360$

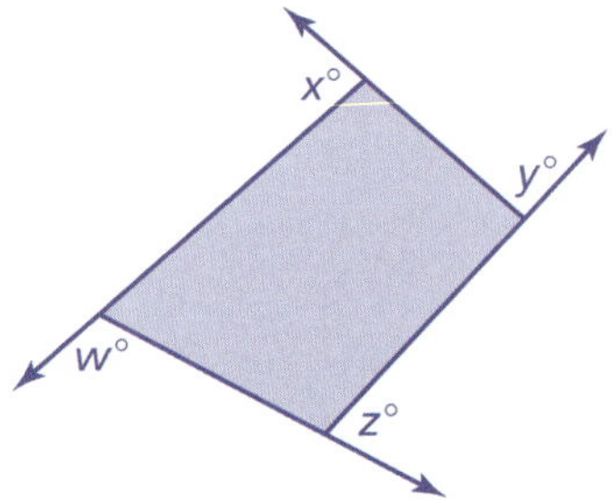

EXAMPLE 4 Finding Exterior Angle Measures

Find the measures of the exterior angles of each polygon.

a.

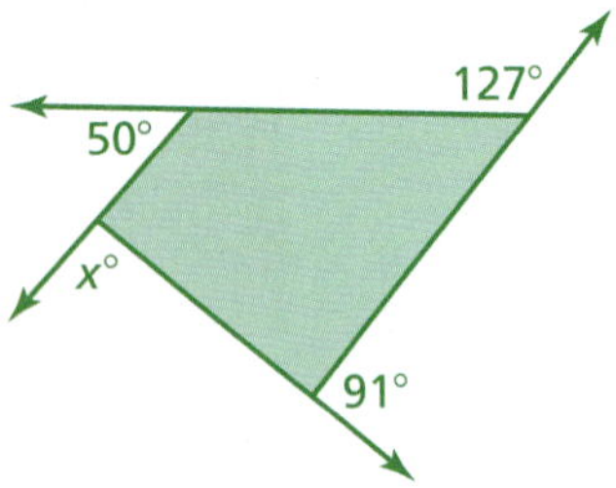

Write and solve an equation for x.

$$x + 50 + 127 + 91 = 360$$
$$x + 268 = 360$$
$$x = 92$$

So, the measures of the exterior angles are 92°, 50°, 127°, and 91°.

b.

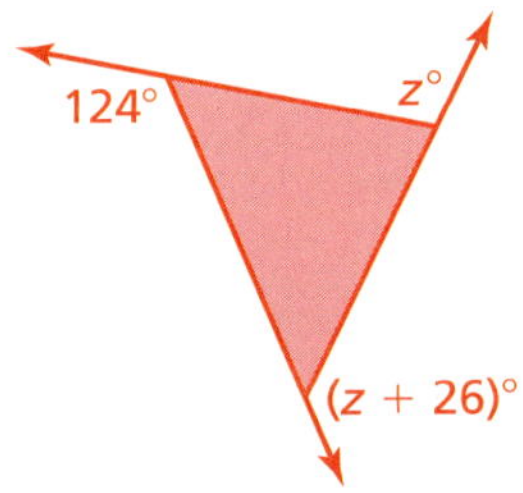

Write and solve an equation for z.

$$124 + z + (z + 26) = 360$$
$$2z + 150 = 360$$
$$z = 105$$

So, the measures of the exterior angles are 124°, 105°, and $(105 + 26)° = 131°$.

On Your Own

Now You're Ready
Exercises 22–28

9. Find the measures of the exterior angles of the polygon.

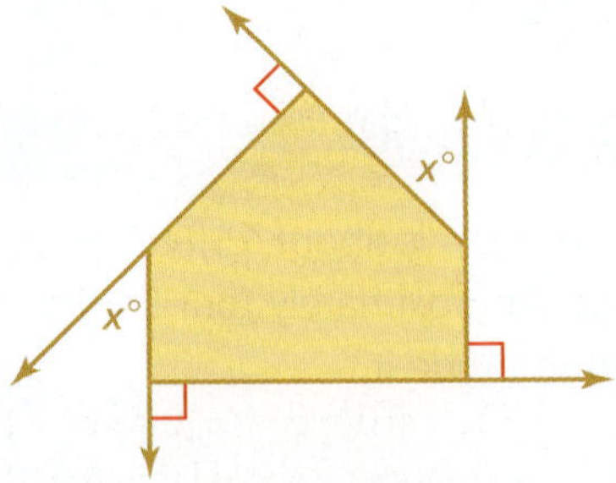

3.3 Exercises

Check It Out
Help with Homework
BigIdeasMath.com

Vocabulary and Concept Check

1. **VOCABULARY** Draw a regular polygon that has three sides.
2. **WHICH ONE DOESN'T BELONG?** Which figure does *not* belong with the other three? Explain your reasoning.

3. **DIFFERENT WORDS, SAME QUESTION** Which is different? Find "both" answers.

What is the measure of an interior angle of a regular pentagon?	What is the sum of the interior angle measures of a convex pentagon?
What is the sum of the interior angle measures of a regular pentagon?	What is the sum of the interior angle measures of a concave pentagon?

Practice and Problem Solving

Use triangles to find the sum of the interior angle measures of the polygon.

4. 
5. 
6. 

Find the sum of the interior angle measures of the polygon.

1 7. 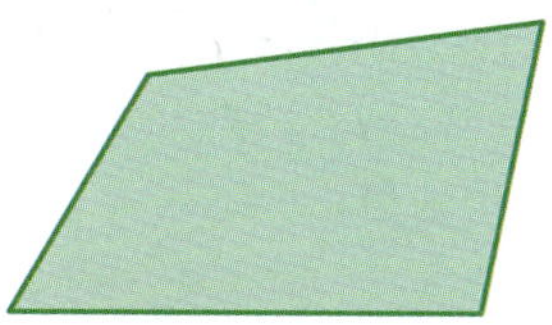
8. 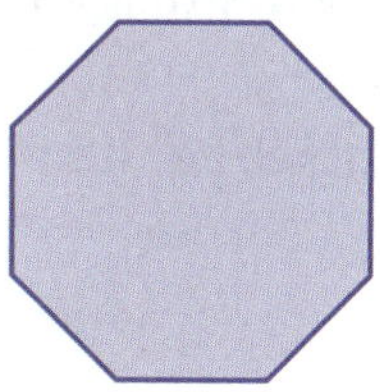
9. 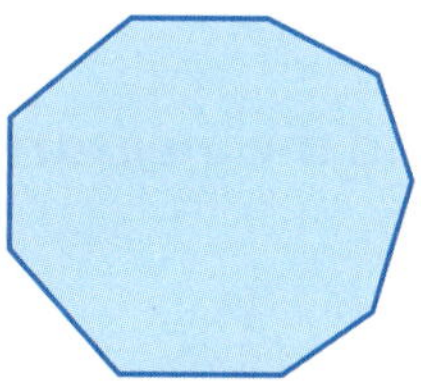

10. **ERROR ANALYSIS** Describe and correct the error in finding the sum of the interior angle measures of a 13-gon.

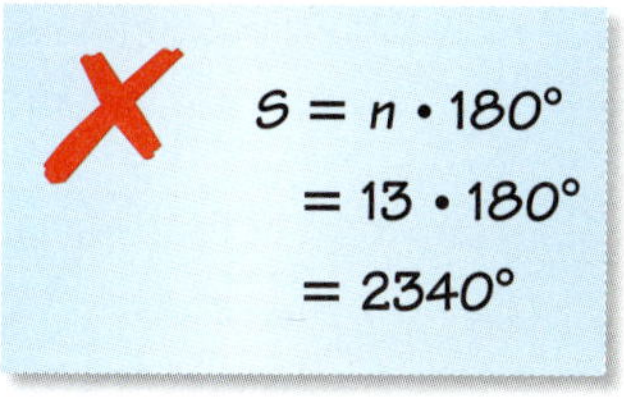

11. **NUMBER SENSE** Can a pentagon have interior angles that measure 120°, 105°, 65°, 150°, and 95°? Explain.

Find the measures of the interior angles.

2 **12.**

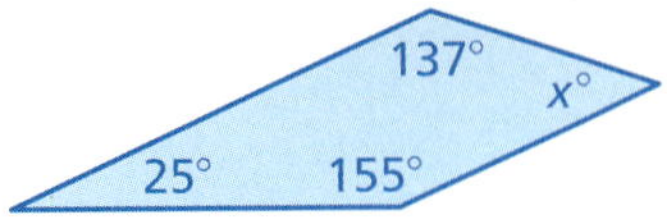

13.

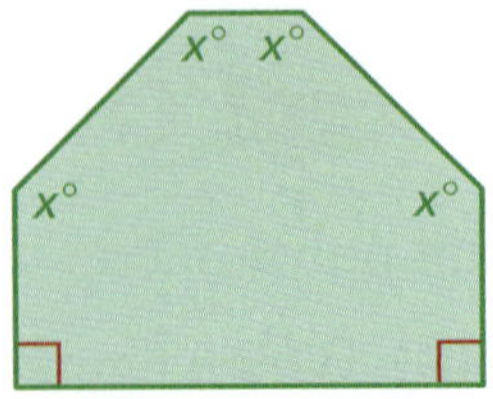

14.

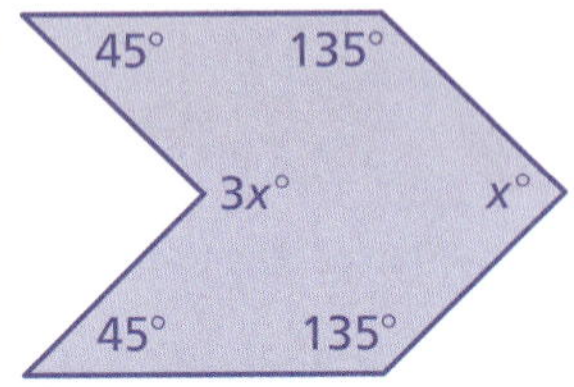

15. REASONING The sum of the interior angle measures in a regular polygon is 1260°. What is the measure of one of the interior angles of the polygon?

Find the measure of each interior angle of the regular polygon.

3 **16.**

17.

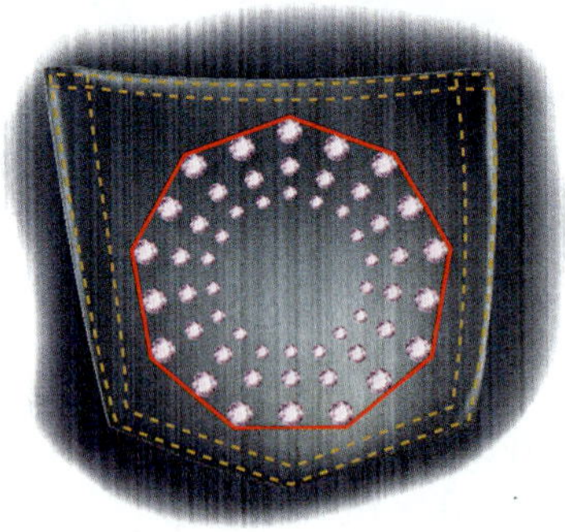

18.

19. ERROR ANALYSIS Describe and correct the error in finding the measure of each interior angle of a regular 20-gon.

✗

$S = (n - 2) \cdot 180°$

$= (20 - 2) \cdot 180°$

$= 18 \cdot 180°$

$= 3240°$

$3240° \div 18 = 180$

The measure of each interior angle is 180°.

20. FIRE HYDRANT A fire hydrant bolt is in the shape of a regular pentagon.

a. What is the measure of each interior angle?

b. Why are fire hydrants made this way?

21. PROBLEM SOLVING The interior angles of a regular polygon each measure 165°. How many sides does the polygon have?

Find the measures of the exterior angles of the polygon.

4 **22.**

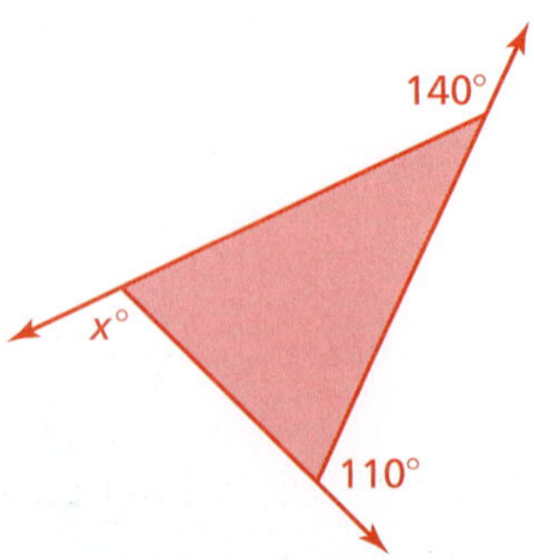

23.

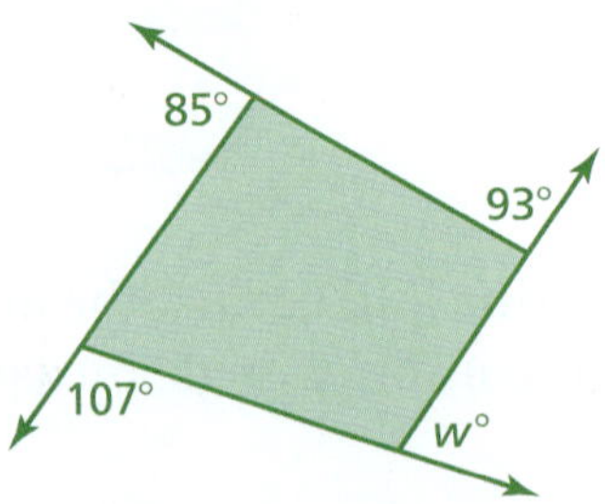

24.

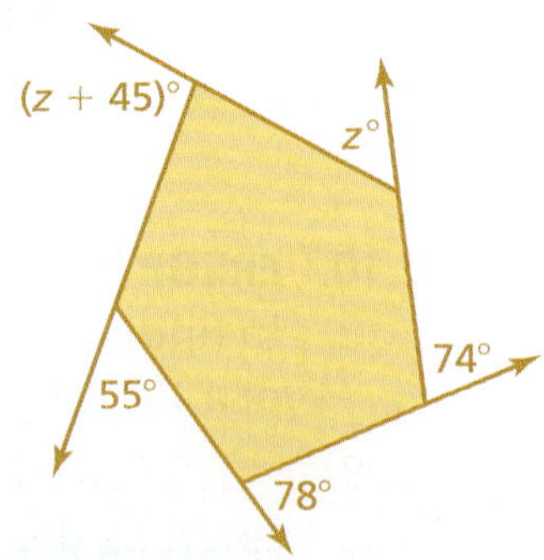

25. REASONING What is the measure of an exterior angle of a regular hexagon? Explain.

Find the measures of the exterior angles of the polygon.

26.

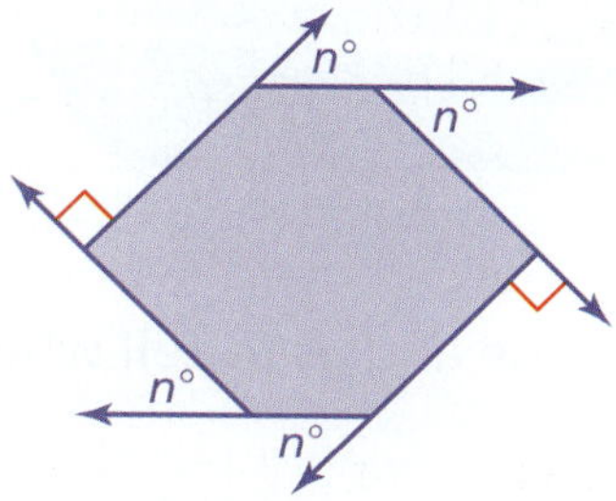

27.

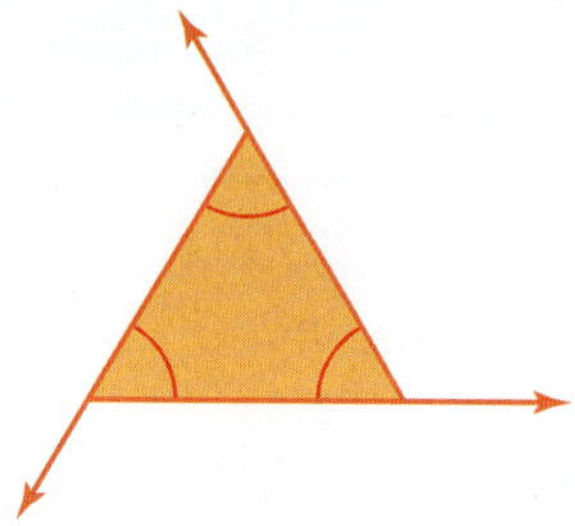

28.

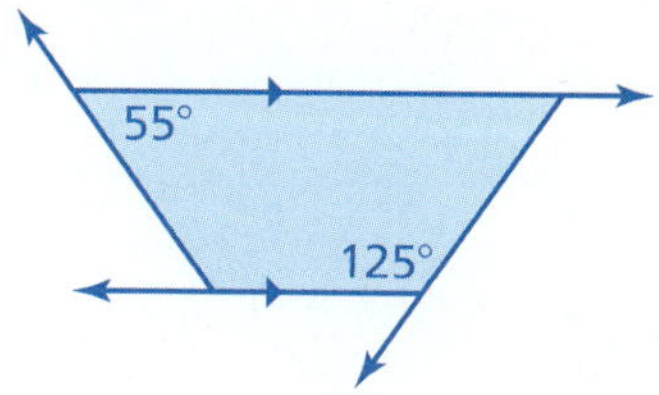

29. **STAINED GLASS** The center of the stained glass window is in the shape of a regular polygon. What is the measure of each interior angle of the polygon? What is the measure of each exterior angle?

30. **PENTAGON** Draw a pentagon that has two right interior angles, two 45° interior angles, and one 270° interior angle.

31. **GAZEBO** The floor of a gazebo is in the shape of a heptagon. Four of the interior angles measure 135°. The other interior angles have equal measures. Find their measures.

32. **MONEY** The border of a Susan B. Anthony dollar is in the shape of a regular polygon.

a. How many sides does the polygon have?

b. What is the measure of each interior angle of the border? Round your answer to the nearest degree.

33. Geometry When tiles can be used to cover a floor with no empty spaces, the collection of tiles is called a *tessellation.*

a. Create a tessellation using equilateral triangles.

b. Find two more regular polygons that form tessellations.

c. Create a tessellation that uses two different regular polygons.

d. Use what you know about interior and exterior angles to explain why the polygons in part (c) form a tessellation.

Fair Game Review What you learned in previous grades & lessons

Solve the proportion. *(Skills Review Handbook)*

34. $\frac{x}{12} = \frac{3}{4}$ **35.** $\frac{14}{21} = \frac{x}{3}$ **36.** $\frac{9}{x} = \frac{6}{2}$ **37.** $\frac{10}{4} = \frac{15}{x}$

38. **MULTIPLE CHOICE** The ratio of tulips to daisies is 3 : 5. Which of the following could be the total number of tulips and daisies? *(Skills Review Handbook)*

Ⓐ 6 Ⓑ 10 Ⓒ 15 Ⓓ 16

3.4 Using Similar Triangles

Essential Question How can you use angles to tell whether triangles are similar?

1 ACTIVITY: Constructing Similar Triangles

Work with a partner.

- **Use a straightedge to draw a line segment that is 4 centimeters long.**
- **Then use the line segment and a protractor to draw a triangle that has a 60° and a 40° angle, as shown. Label the triangle *ABC*.**

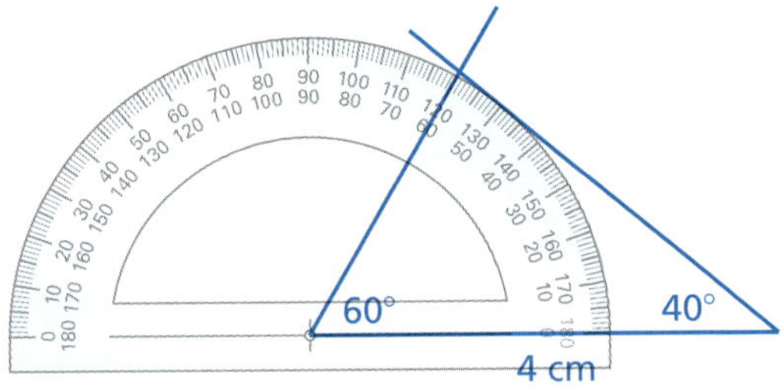

a. Explain how to draw a larger triangle that has the same two angle measures. Label the triangle *JKL*.

b. Explain how to draw a smaller triangle that has the same two angle measures. Label the triangle *PQR*.

c. Are all of the triangles similar? Explain.

2 ACTIVITY: Using Technology to Explore Triangles

Work with a partner. Use geometry software to draw the triangle below.

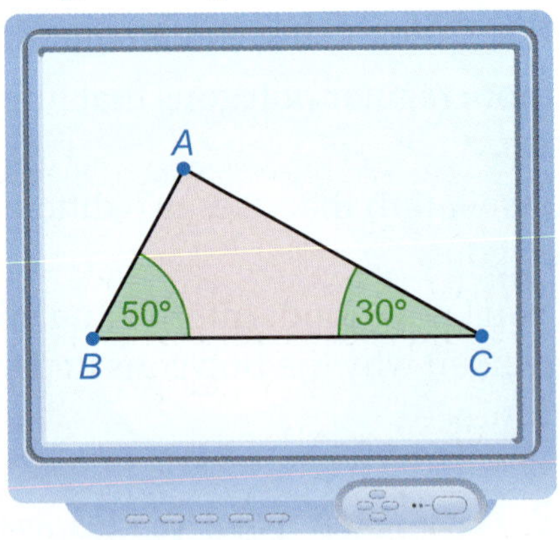

a. Dilate the triangle by the following scale factors.

$2 \qquad \frac{1}{2} \qquad \frac{1}{4} \qquad 2.5$

b. Measure the third angle in each triangle. What do you notice?

c. **REASONING** You have two triangles. Two angles in the first triangle are congruent to two angles in the second triangle. Can you conclude that the triangles are similar? Explain.

COMMON CORE

Geometry

In this lesson, you will

- understand the concept of similar triangles.
- identify similar triangles.
- use indirect measurement to find missing measures.

Learning Standard
MACC.8.G.1.5

3 ACTIVITY: Indirect Measurement

Math Practice 2

Make Sense of Quantities

What do you know about the sides of the triangles when the triangles are similar?

Work with a partner.

a. Use the fact that two rays from the Sun are parallel to explain why $\triangle ABC$ and $\triangle DEF$ are similar.

b. Explain how to use similar triangles to find the height of the flagpole.

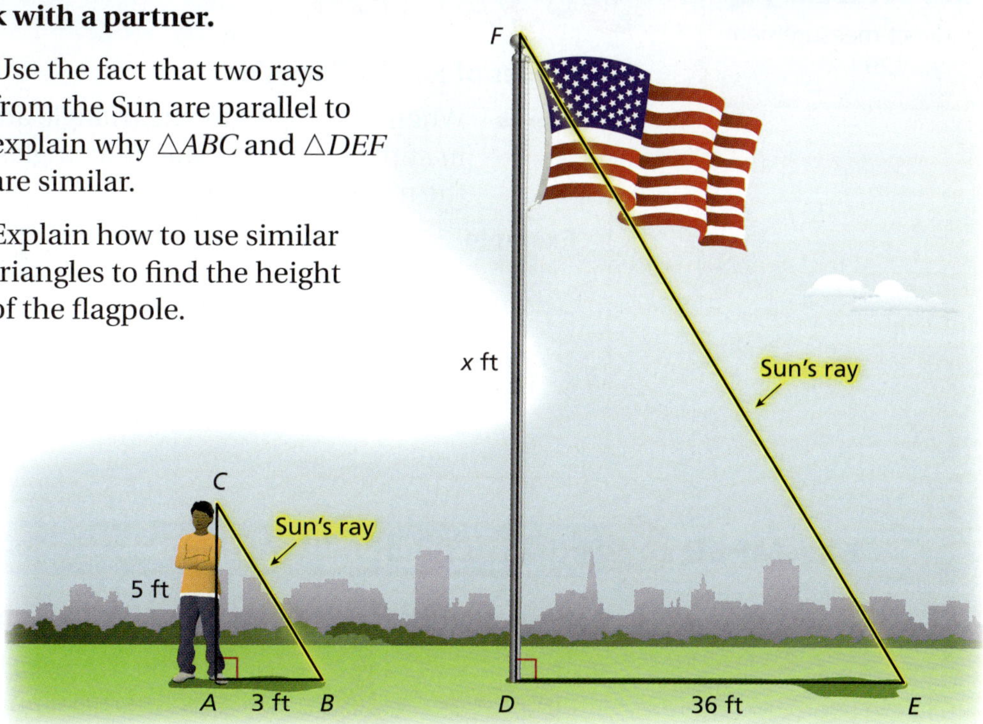

What Is Your Answer?

4. **IN YOUR OWN WORDS** How can you use angles to tell whether triangles are similar?

5. **PROJECT** Work with a partner or in a small group.

   a. Explain why the process in Activity 3 is called "indirect" measurement.

   b. **CHOOSE TOOLS** Use indirect measurement to measure the height of something outside your school (a tree, a building, a flagpole). Before going outside, decide what materials you need to take with you.

   c. **MODELING** Draw a diagram of the indirect measurement process you used. In the diagram, label the lengths that you actually measured and also the lengths that you calculated.

6. **PRECISION** Look back at Exercise 17 in Section 2.5. Explain how you can show that the two triangles are similar.

Practice Use what you learned about similar triangles to complete Exercises 4 and 5 on page 130.

3.4 Lesson

Check It Out
Lesson Tutorials
BigIdeasMath.com

Key Vocabulary
indirect measurement, *p. 129*

Key Idea

Angles of Similar Triangles

Words When two angles in one triangle are congruent to two angles in another triangle, the third angles are also congruent and the triangles are similar.

Example

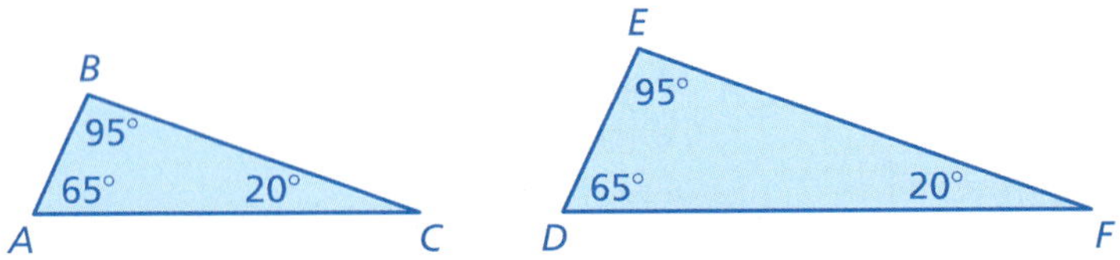

Triangle ABC is similar to Triangle DEF: $\triangle ABC \sim \triangle DEF$.

EXAMPLE 1 Identifying Similar Triangles

Tell whether the triangles are similar. Explain.

a.

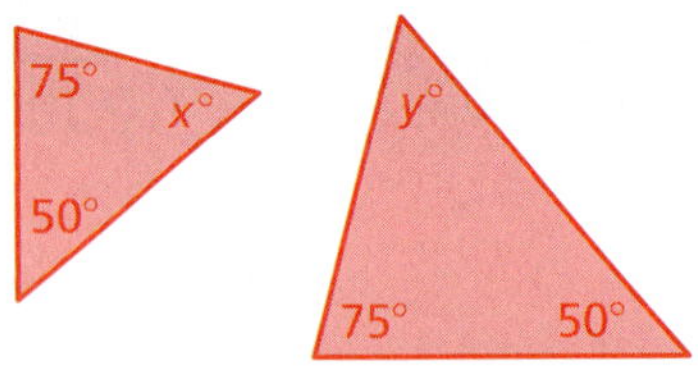

The triangles have two pairs of congruent angles.

So, the third angles are congruent, and the triangles are similar.

b.

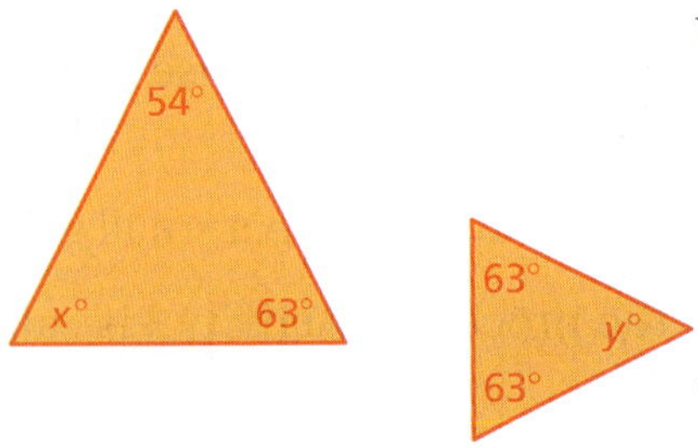

Write and solve an equation to find x.

$$x + 54 + 63 = 180$$
$$x + 117 = 180$$
$$x = 63$$

The triangles have two pairs of congruent angles.

So, the third angles are congruent, and the triangles are similar.

c.

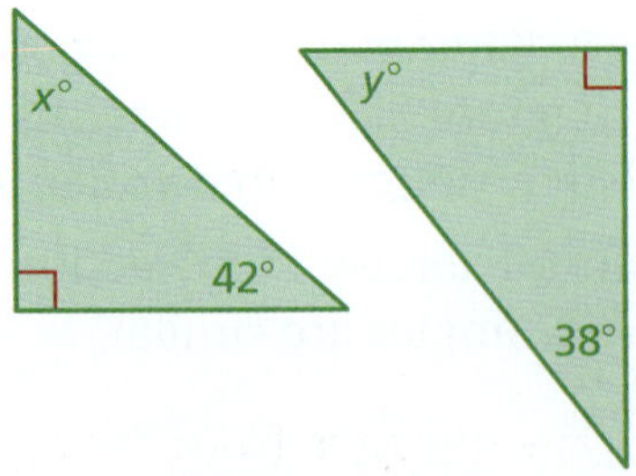

Write and solve an equation to find x.

$$x + 90 + 42 = 180$$
$$x + 132 = 180$$
$$x = 48$$

The triangles do not have two pairs of congruent angles.

So, the triangles are not similar.

Multi-Language Glossary at BigIdeasMath.com

On Your Own

Now You're Ready
Exercises 6–9

Tell whether the triangles are similar. Explain.

1. 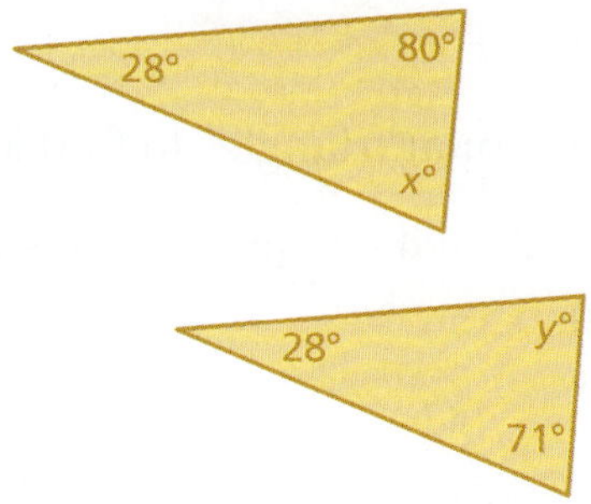

2. 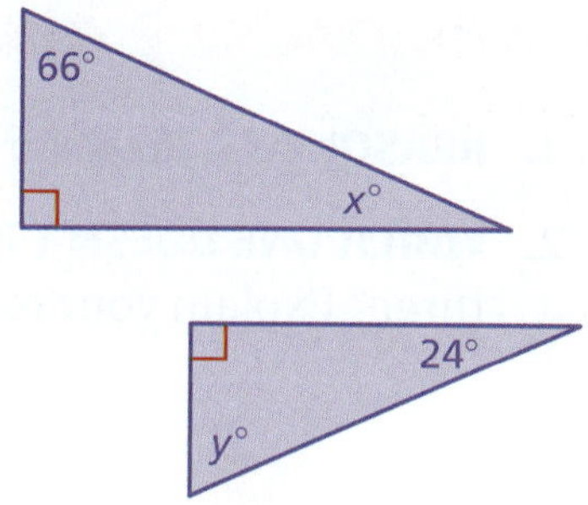

Indirect measurement uses similar figures to find a missing measure when it is difficult to find directly.

EXAMPLE 2 Using Indirect Measurement

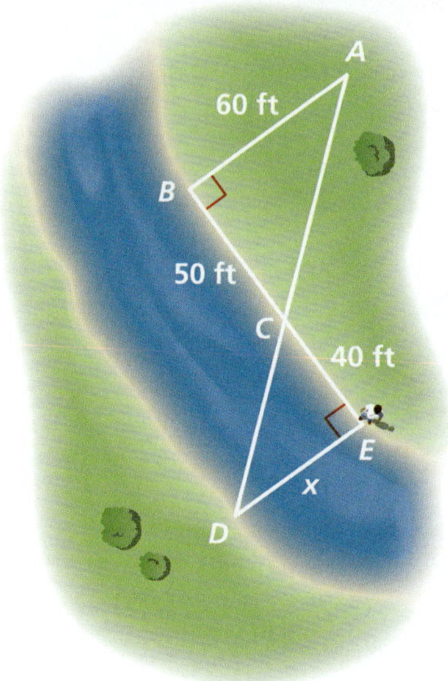

You plan to cross a river and want to know how far it is to the other side. You take measurements on your side of the river and make the drawing shown. (a) Explain why $\triangle ABC$ and $\triangle DEC$ are similar. (b) What is the distance x across the river?

a. $\angle B$ and $\angle E$ are right angles, so they are congruent. $\angle ACB$ and $\angle DCE$ are vertical angles, so they are congruent.

Because two angles in $\triangle ABC$ are congruent to two angles in $\triangle DEC$, the third angles are also congruent and the triangles are similar.

b. The ratios of the corresponding side lengths in similar triangles are equal. Write and solve a proportion to find x.

$$\frac{x}{60} = \frac{40}{50} \qquad \text{Write a proportion.}$$

$$60 \cdot \frac{x}{60} = 60 \cdot \frac{40}{50} \qquad \text{Multiplication Property of Equality}$$

$$x = 48 \qquad \text{Simplify.}$$

So, the distance across the river is 48 feet.

On Your Own

Now You're Ready
Exercise 13

3. **WHAT IF?** The distance from vertex A to vertex B is 55 feet. What is the distance across the river?

3.4 Exercises

Check It Out
Help with Homework
BigIdeasMath.com

Vocabulary and Concept Check

1. **REASONING** How can you use similar triangles to find a missing measurement?
2. **WHICH ONE DOESN'T BELONG?** Which triangle does *not* belong with the other three? Explain your reasoning.

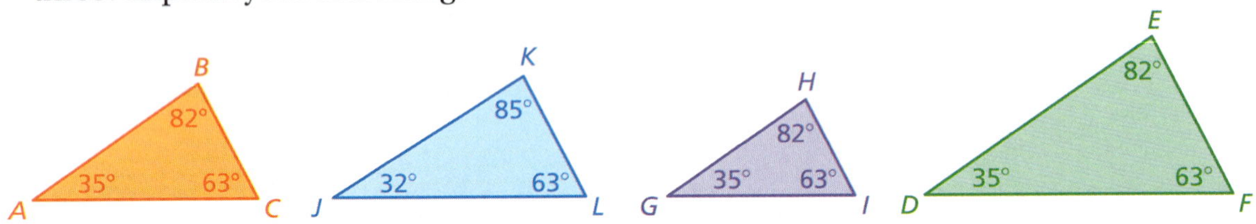

3. **WRITING** Two triangles have two pairs of congruent angles. In your own words, explain why you do not need to find the measures of the third pair of angles to determine that they are congruent.

Practice and Problem Solving

Make a triangle that is larger or smaller than the one given and has the same angle measures. Find the ratios of the corresponding side lengths.

4. 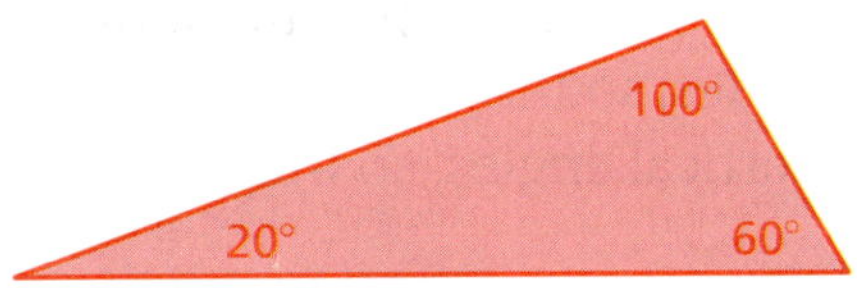

5. 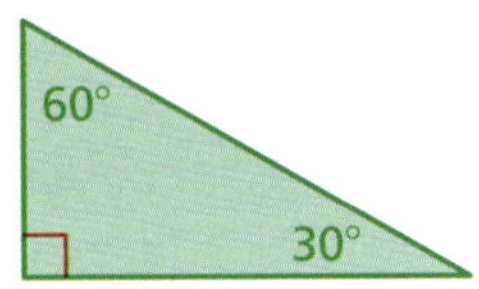

Tell whether the triangles are similar. Explain.

6. 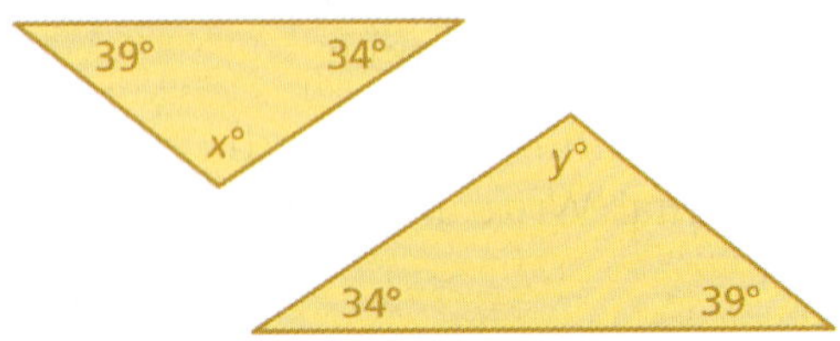

7. 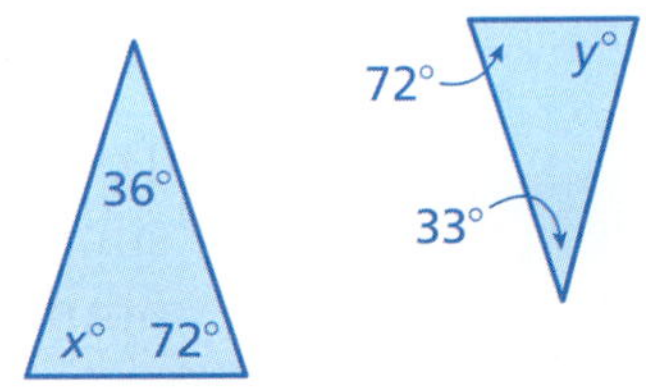

8. 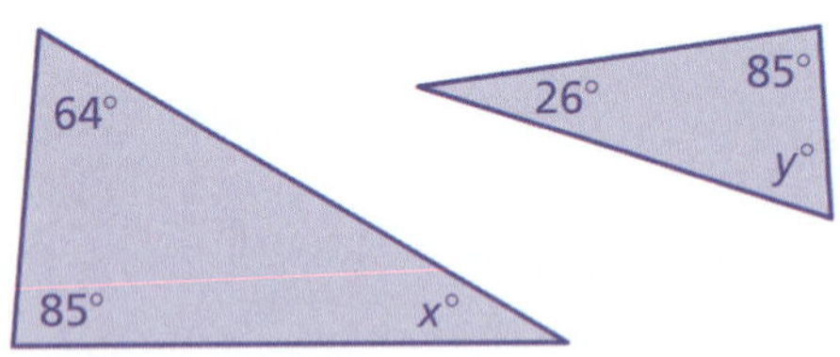

9. 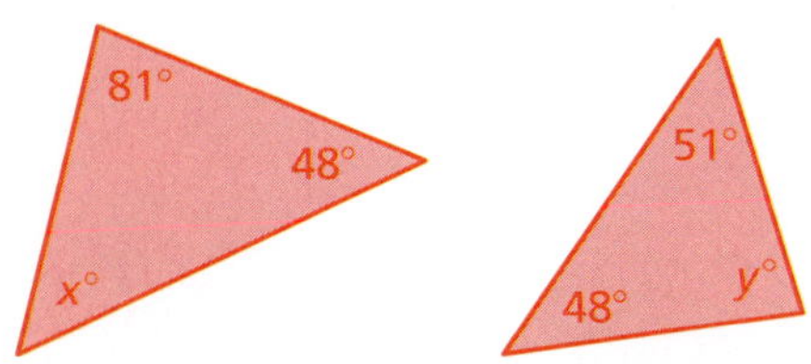

10. **RULERS** Which of the rulers are similar in shape? Explain.

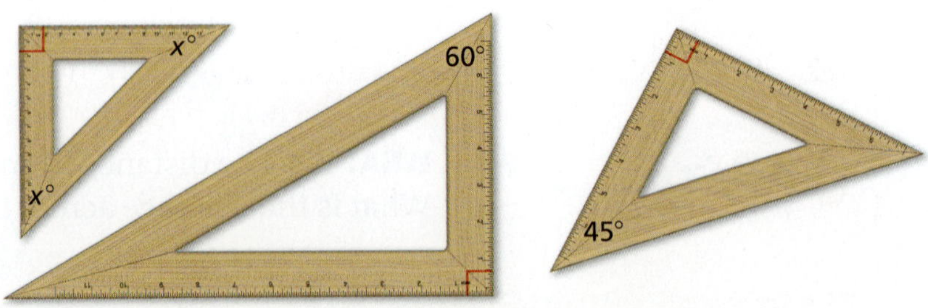

Tell whether the triangles are similar. Explain.

11. 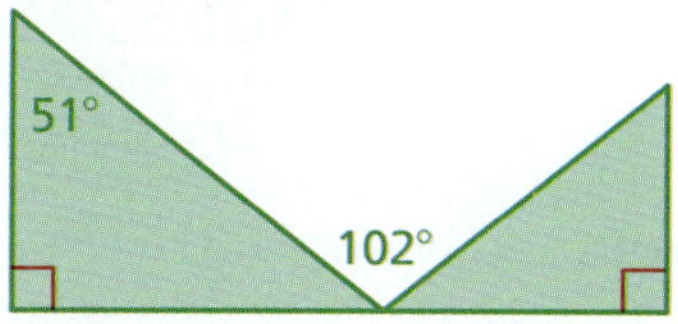

12. 

13. **TREASURE** The map shows the number of steps you must take to get to the treasure. However, the map is old, and the last dimension is unreadable. Explain why the triangles are similar. How many steps do you take from the pyramids to the treasure?

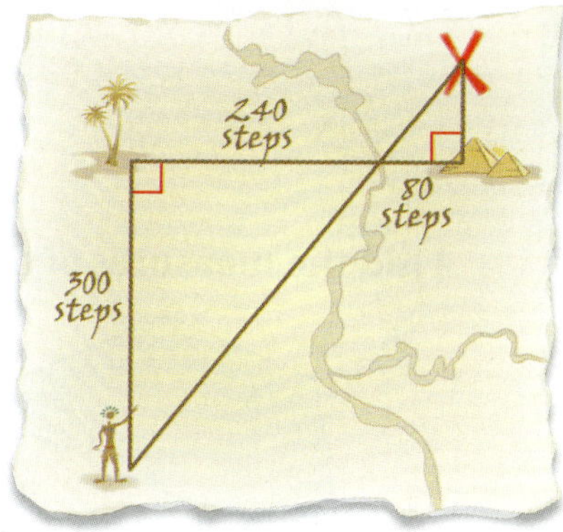

14. **CRITICAL THINKING** The side lengths of a triangle are increased by 50% to make a similar triangle. Does the area increase by 50% as well? Explain.

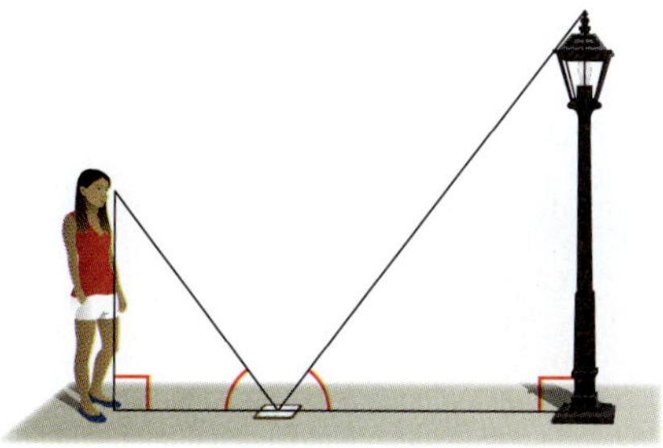

15. **PINE TREE** A person who is 6 feet tall casts a 3-foot-long shadow. A nearby pine tree casts a 15-foot-long shadow. What is the height h of the pine tree?

16. **OPEN-ENDED** You place a mirror on the ground 6 feet from the lamppost. You move back 3 feet and see the top of the lamppost in the mirror. What is the height of the lamppost?

17. **REASONING** In each of two right triangles, one angle measure is two times another angle measure. Are the triangles similar? Explain your reasoning.

18. **Geometry** In the diagram, segments BG, CF, and DE are parallel. The length of segment BD is 6.32 feet, and the length of segment DE is 6 feet. Name all pairs of similar triangles in the diagram. Then find the lengths of segments BG and CF.

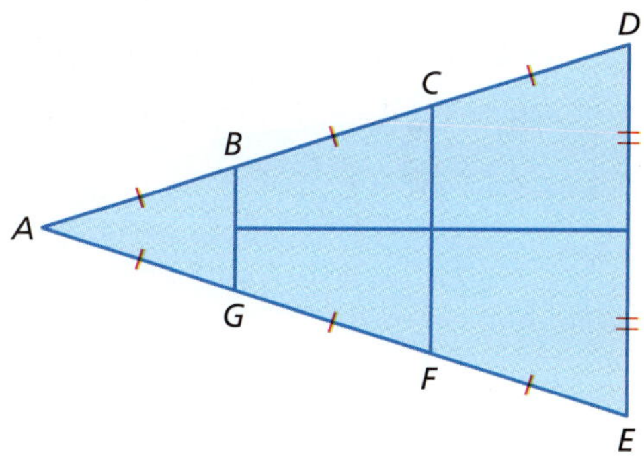

Fair Game Review What you learned in previous grades & lessons

Solete the equation for *y*. *(Section 1.4)*

19. $y - 5x = 3$

20. $4x + 6y = 12$

21. $2x - \frac{1}{4}y = 1$

22. **MULTIPLE CHOICE** What is the value of x? *(Section 3.2)*

Ⓐ 17 Ⓑ 62

Ⓒ 118 Ⓓ 152

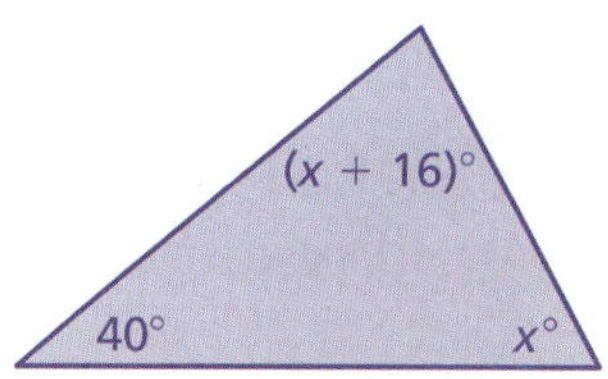

3.3–3.4 Quiz

Check It Out
Progress Check
BigIdeasMath.com

Find the sum of the interior angle measures of the polygon. *(Section 3.3)*

1.

2.

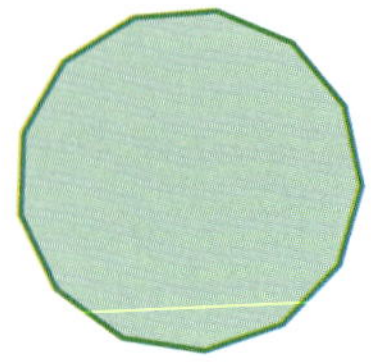

Find the measures of the interior angles of the polygon. *(Section 3.3)*

3.

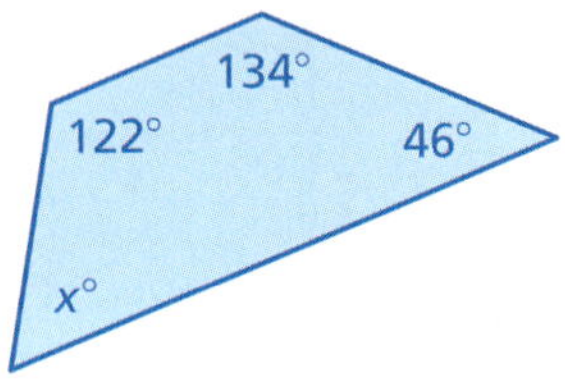

4.

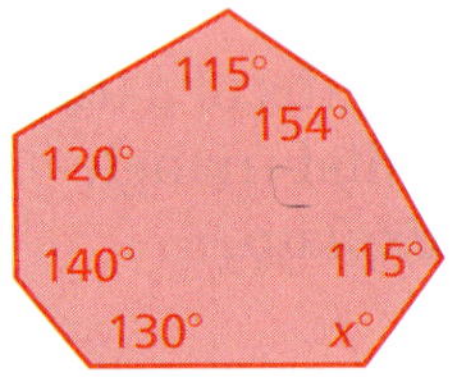

5.

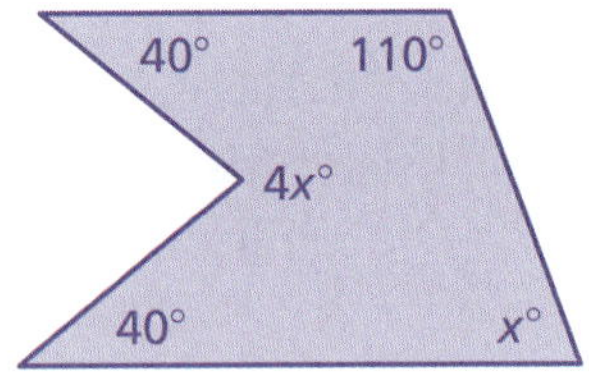

Find the measures of the exterior angles of the polygon. *(Section 3.3)*

6.

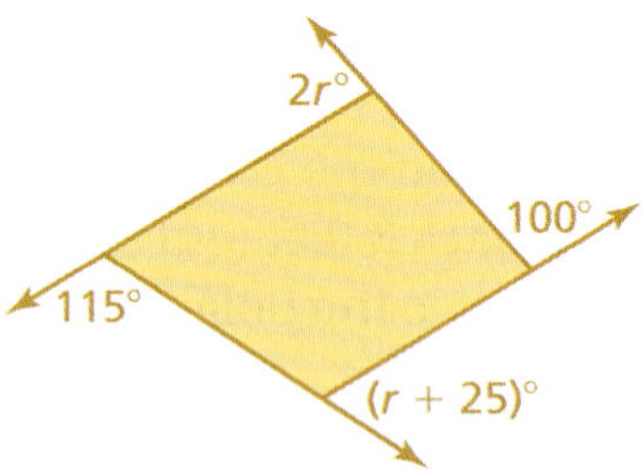

7.

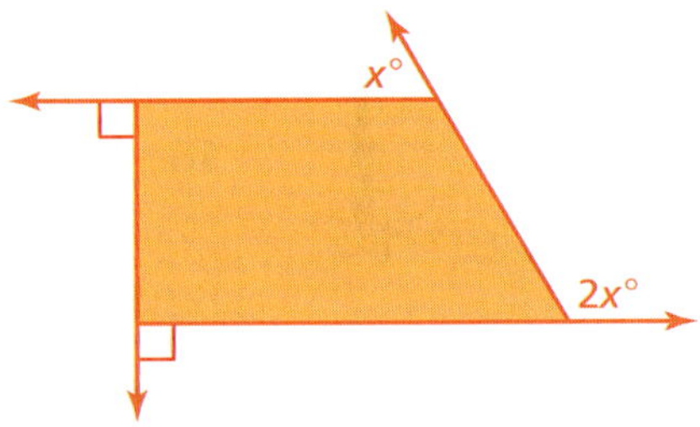

Tell whether the triangles are similar. Explain. *(Section 3.4)*

8.

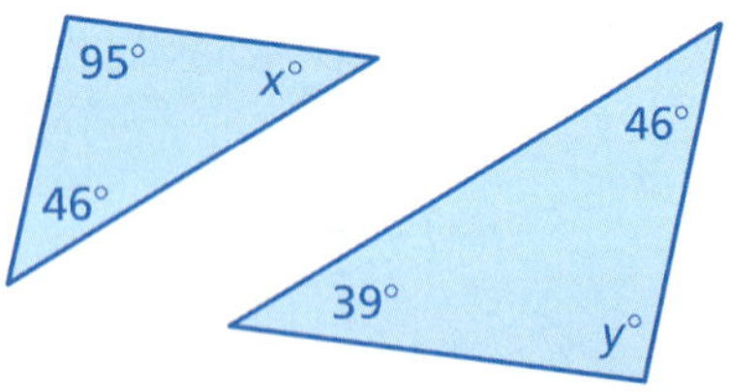

9.

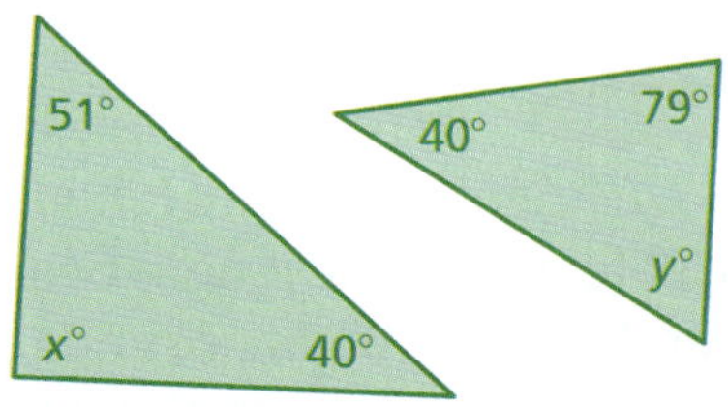

10. REASONING The sum of the interior angle measures of a polygon is 4140°. How many sides does the polygon have? *(Section 3.3)*

11. SWAMP You are trying to find the distance ℓ across a patch of swamp water. *(Section 3.4)*

a. Explain why $\triangle VWX$ and $\triangle YZX$ are similar.

b. What is the distance across the patch of swamp water?

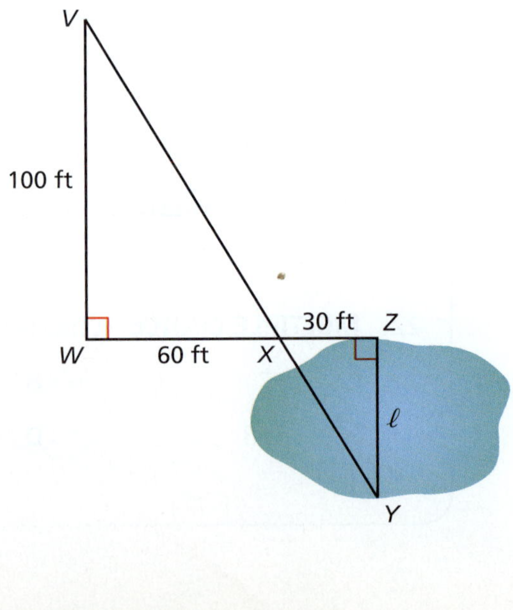

Check It Out
Vocabulary Help
BigIdeasMath.com

Review Key Vocabulary

transversal, *p. 104*
interior angles, *p. 105*
exterior angles, *p. 105*
interior angles of a polygon, *p. 112*
exterior angles of a polygon, *p. 112*
convex polygon, *p. 119*
concave polygon, *p. 119*
regular polygon, *p. 121*
indirect measurement, *p. 129*

Review Examples and Exercises

3.1 Parallel Lines and Transversals *(pp. 102–109)*

Use the figure to find the measure of ∠6.

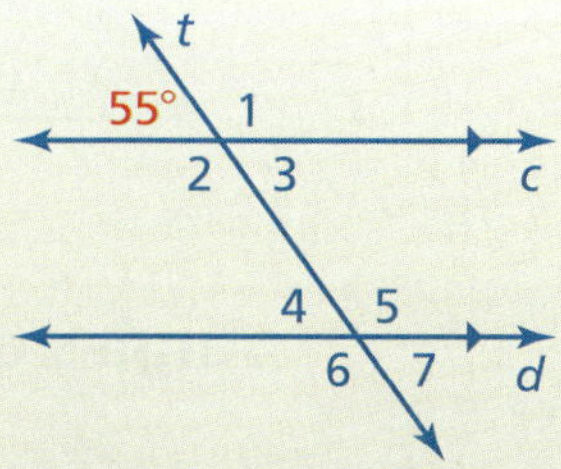

∠2 and the 55° angle are supplementary.
So, the measure of ∠2 is 180° − 55° = 125°.

∠2 and ∠6 are corresponding angles.
They are congruent.

So, the measure of ∠6 is 125°.

Exercises

Use the figure to find the measure of the angle. Explain your reasoning.

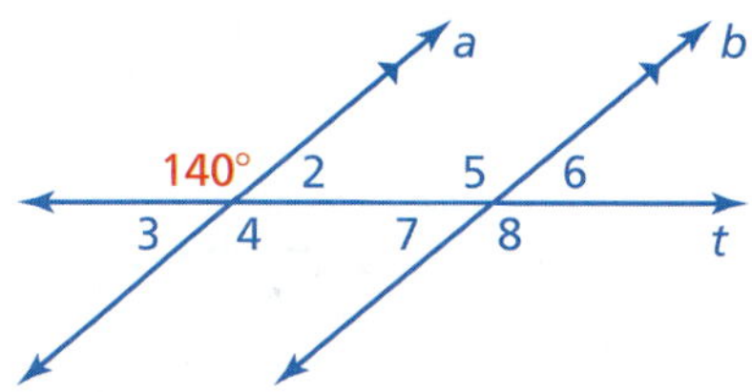

1. ∠8
2. ∠5
3. ∠7
4. ∠2

3.2 Angles of Triangles *(pp. 110–115)*

a. Find the value of x.

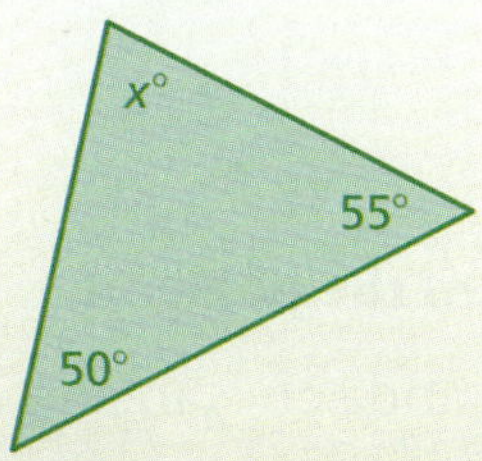

$$x + 50 + 55 = 180$$
$$x + 105 = 180$$
$$x = 75$$

The value of x is 75.

b. Find the measure of the exterior angle.

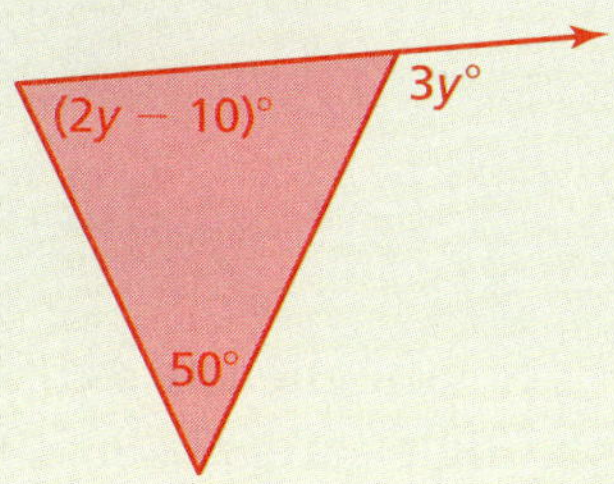

$$3y = (2y - 10) + 50$$
$$3y = 2y + 40$$
$$y = 40$$

So, the measure of the exterior angle is 3(40)° = 120°.

Exercises

Find the measures of the interior angles.

5.

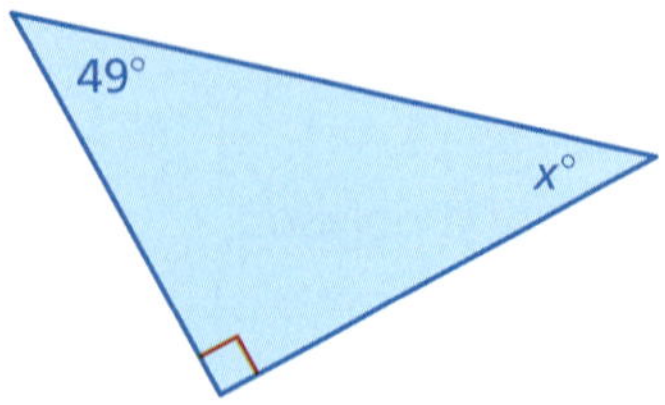

6.

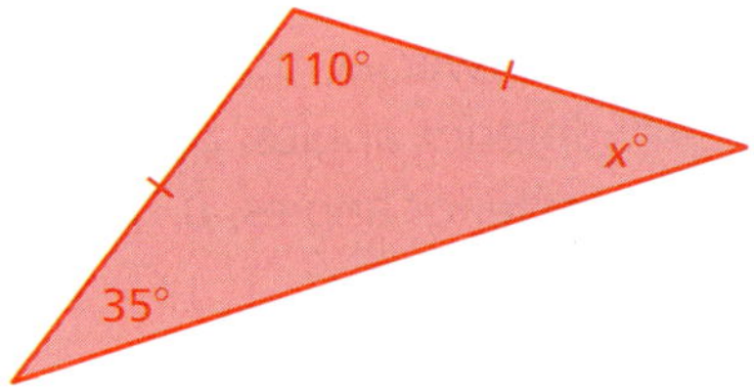

Find the measure of the exterior angle.

7.

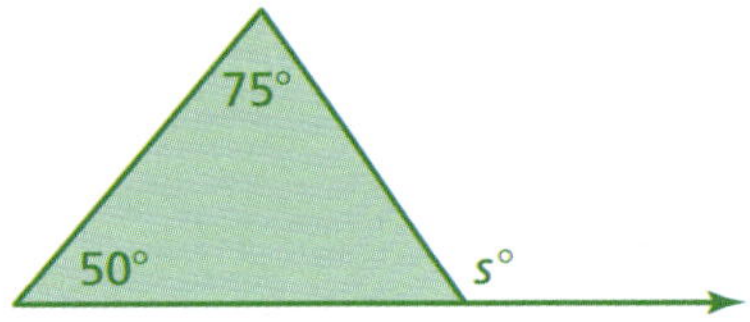

8.

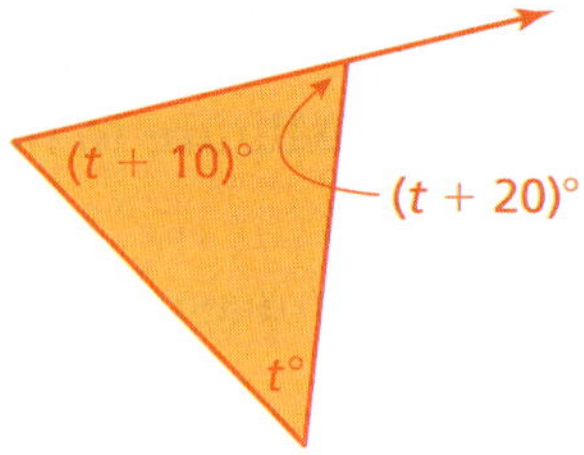

3.3 Angles of Polygons *(pp. 118–125)*

a. Find the value of x.

Step 1: The polygon has 6 sides. Find the sum of the interior angle measures.

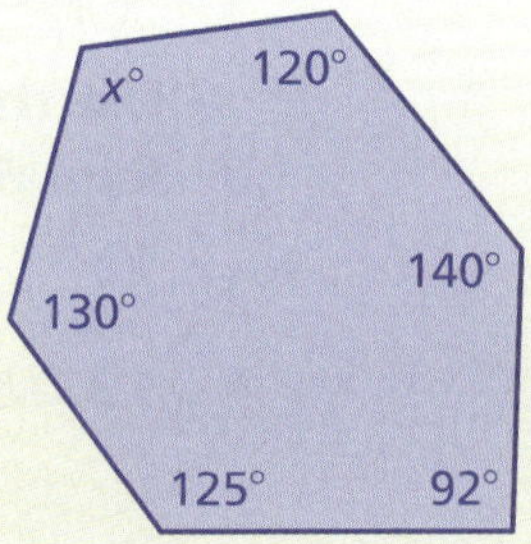

$S = (n - 2) \cdot 180°$ Write the formula.

$= (6 - 2) \cdot 180°$ Substitute 6 for n.

$= 720$ Simplify. The sum of the interior angle measures is 720°.

Step 2: Write and solve an equation.

$$130 + 125 + 92 + 140 + 120 + x = 720$$
$$607 + x = 720$$
$$x = 113$$

The value of x is 113.

b. Find the measures of the exterior angles of the polygon.

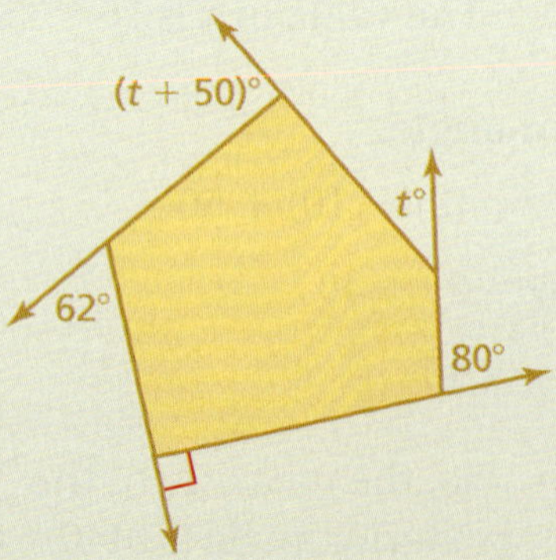

Write and solve an equation for t.

$$t + 80 + 90 + 62 + (t + 50) = 360$$
$$2t + 282 = 360$$
$$2t = 78$$
$$t = 39$$

So, the measures of the exterior angles are 39°, 80°, 90°, 62°, and $(39 + 50)° = 89°$.

Exercises

Find the measures of the interior angles of the polygon.

9. 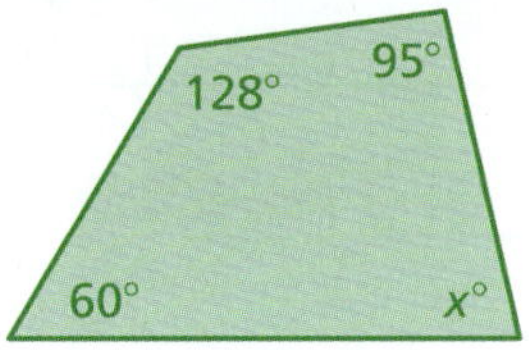

10. 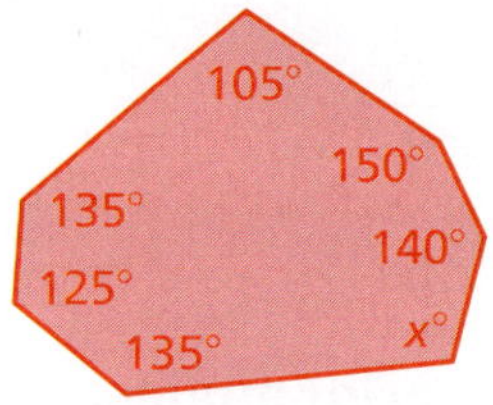

11. 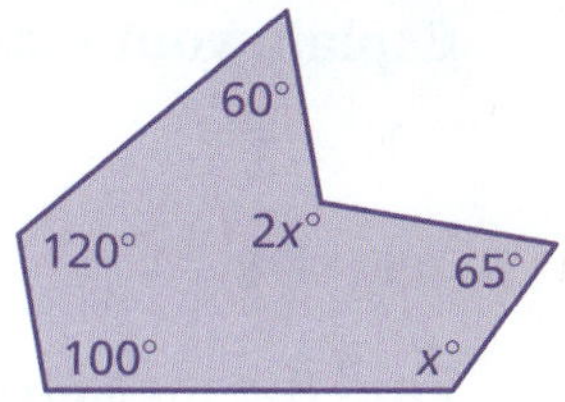

Find the measures of the exterior angles of the polygon.

12. 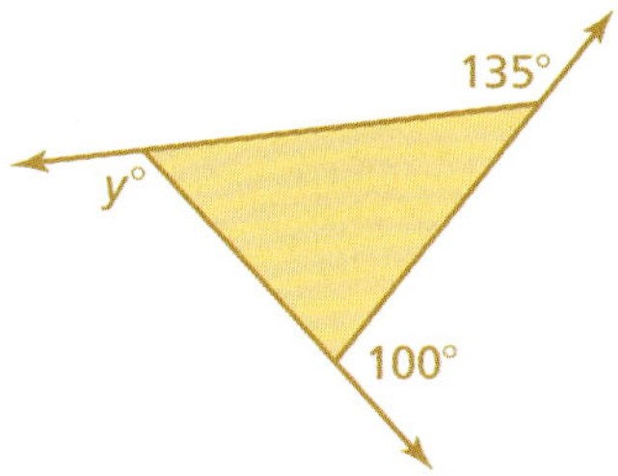

13. 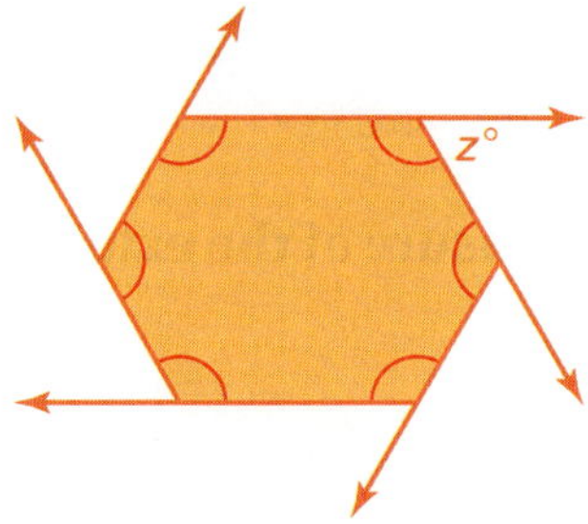

3.4 Using Similar Triangles *(pp. 126–131)*

Tell whether the triangles are similar. Explain.

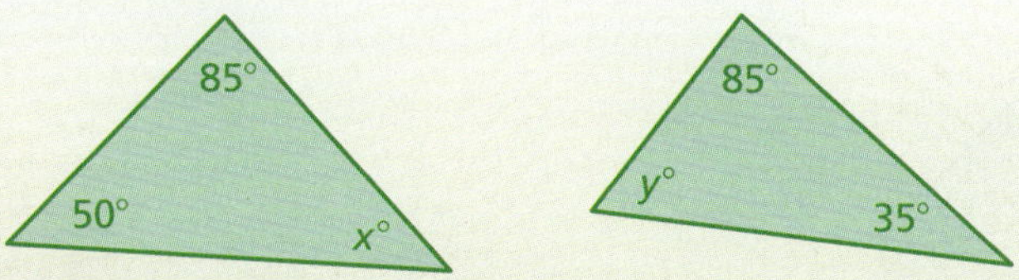

Write and solve an equation to find x.

$$50 + 85 + x = 180$$

$$135 + x = 180$$

$$x = 45$$

The triangles do not have two pairs of congruent angles. So, the triangles are not similar.

Exercises

Tell whether the triangles are similar. Explain.

14. 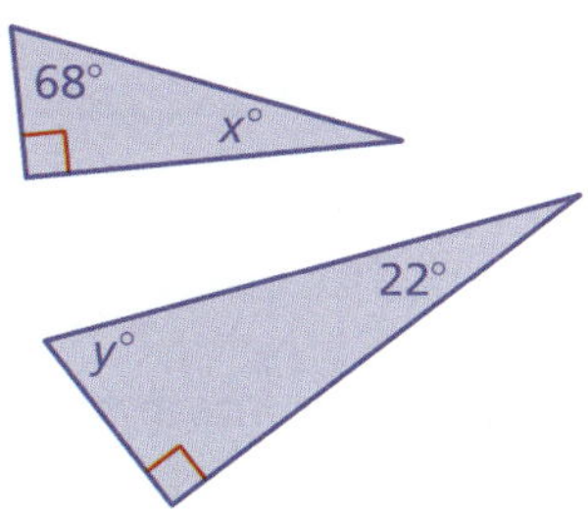

15. 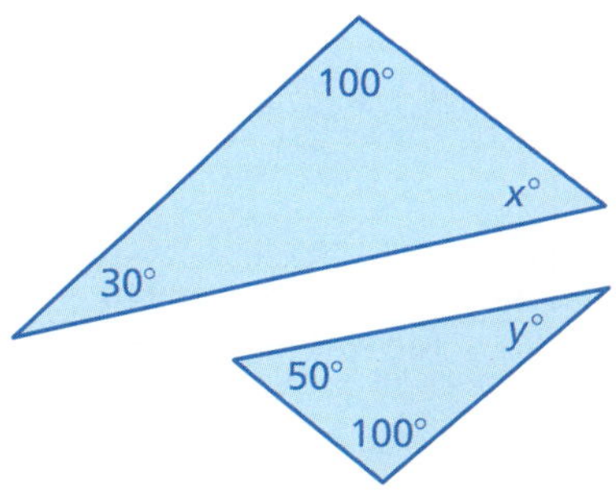

3 Chapter Test

Check It Out
Test Practice
BigIdeasMath.com

Use the figure to find the measure of the angle. Explain your reasoning.

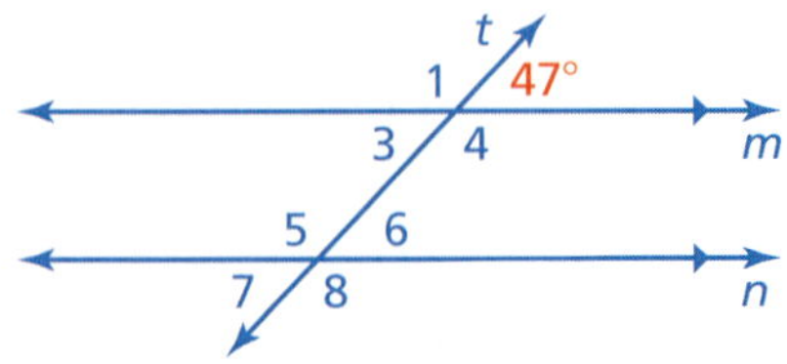

1. ∠1 **2.** ∠8

3. ∠4 **4.** ∠5

Find the measures of the interior angles.

5.

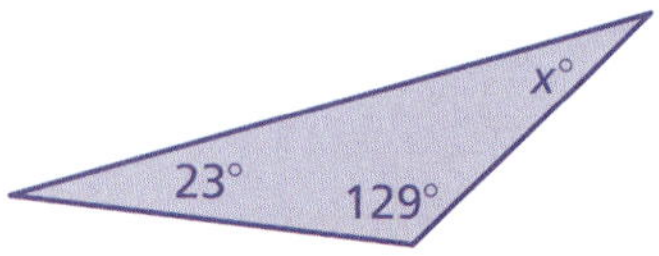

6.

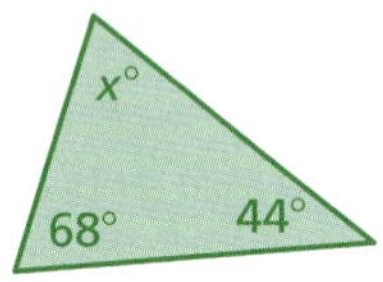

7.

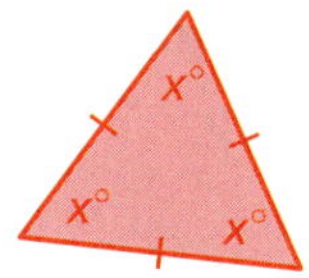

Find the measure of the exterior angle.

8.

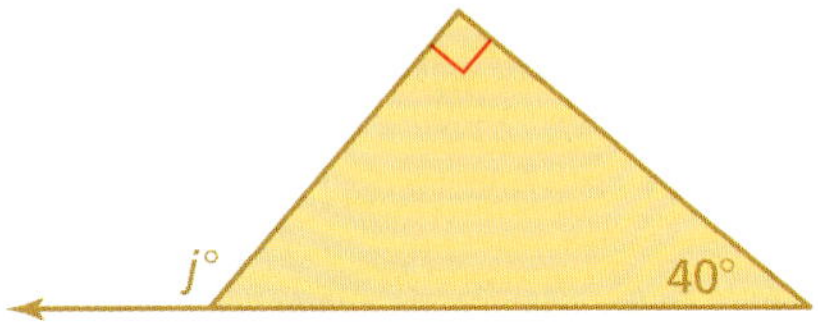

9.

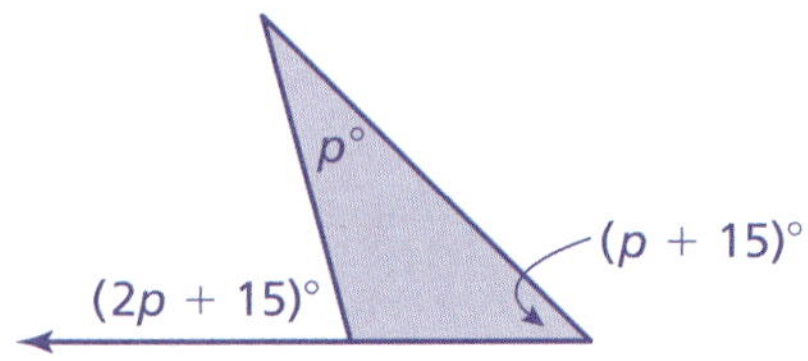

10. Find the measures of the interior angles of the polygon.

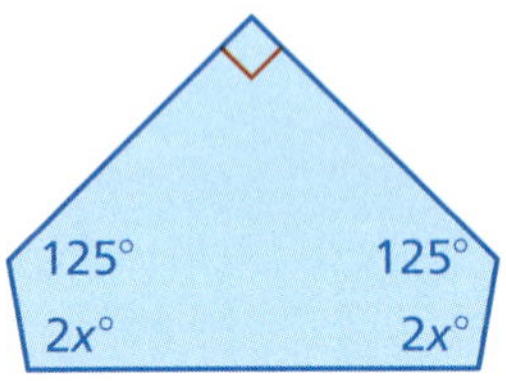

11. Find the measures of the exterior angles of the polygon.

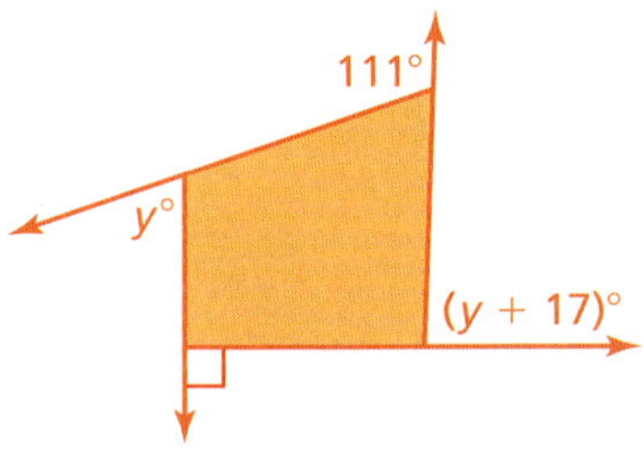

Tell whether the triangles are similar. Explain.

12.

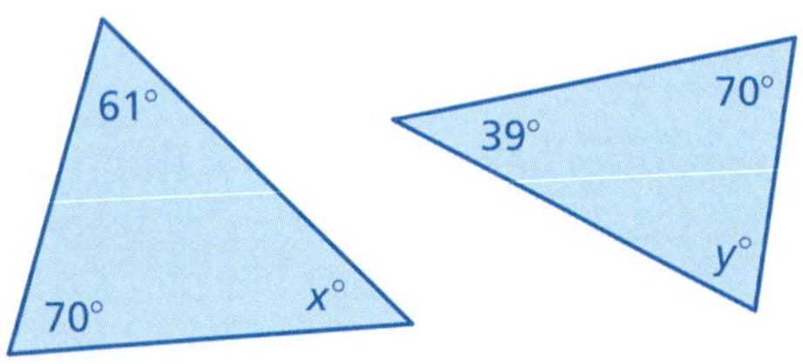

13.

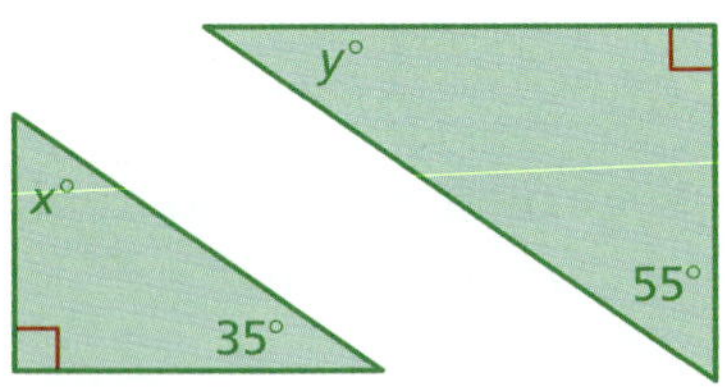

14. **WRITING** Describe two ways you can find the measure of ∠5.

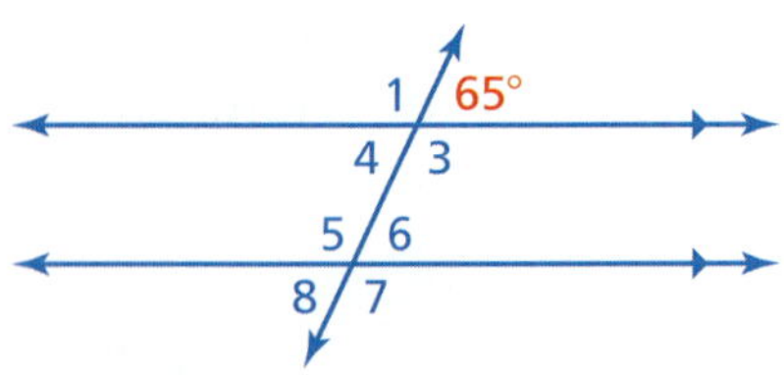

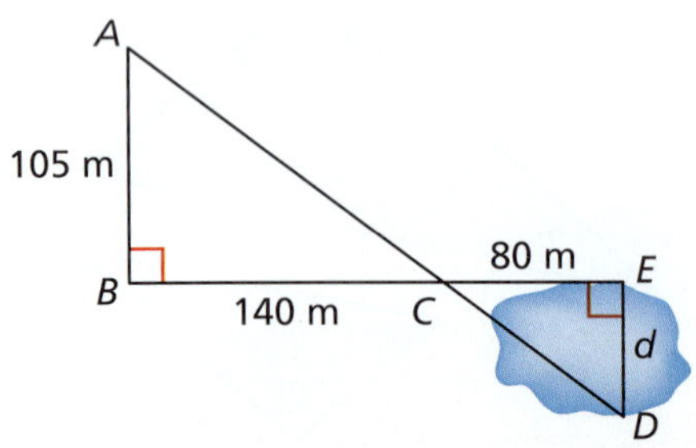

15. **POND** Use the given measurements to find the distance d across the pond.

3 Standards Assessment

1. The border of a Canadian one-dollar coin is shaped like an 11-sided regular polygon. The shape was chosen to help visually impaired people identify the coin. How many degrees are in each angle along the border? Round your answer to the nearest degree. *(MACC.8.G.1.5)*

Test-Taking Strategy
Solve Problem Before Looking at Choices

"Solve the problem before looking at the choices. You know 180 − 2(70) = 40. So, the answer is C."

2. A public utility charges its residential customers for natural gas based on the number of therms used each month. The formula below shows how the monthly cost C in dollars is related to the number t of therms used.

$$C = 11 + 1.6t$$

Solve this formula for t. *(MACC.8.EE.3.7b)*

A. $t = \dfrac{C}{12.6}$

B. $t = \dfrac{C - 11}{1.6}$

C. $t = \dfrac{C}{1.6} - 11$

D. $t = C - 12.6$

3. What is the value of x? *(MACC.8.EE.3.7b)*

$$5(x - 4) = 3x$$

F. -10

G. 2

H. $2\frac{1}{2}$

I. 10

4. In the figures below, $\triangle PQR$ is a dilation of $\triangle STU$.

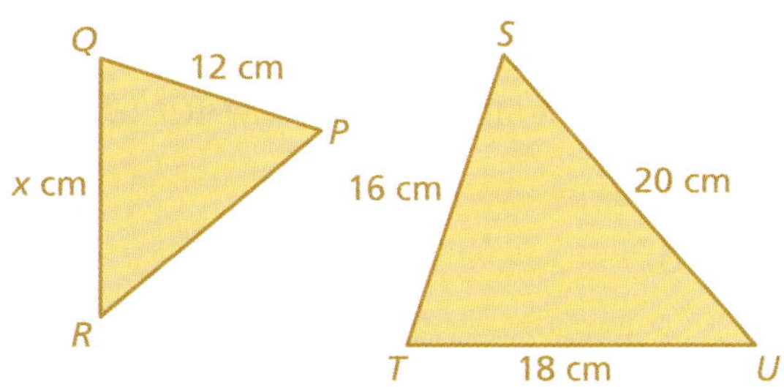

What is the value of x? *(MACC.8.G.1.4)*

A. 9.6

B. $10\frac{2}{3}$

C. 13.5

D. 15

5. What is the value of x? *(MACC.8.G.1.5)*

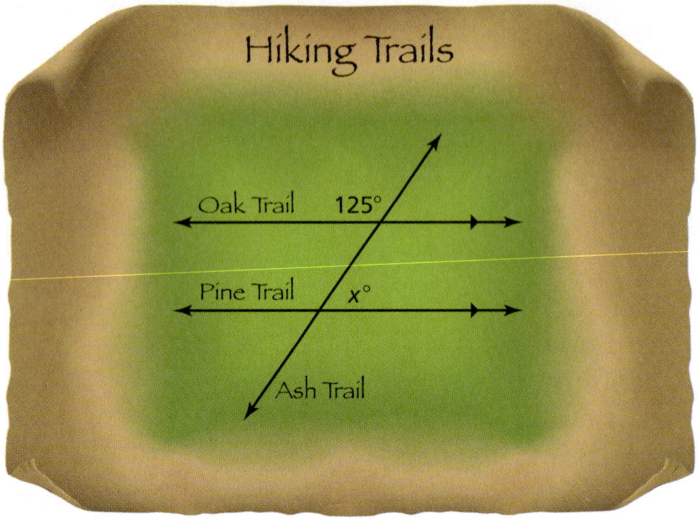

6. Olga was solving an equation in the box shown.

$$-\frac{2}{5}(10x - 15) = -30$$
$$10x - 15 = -30\left(-\frac{2}{5}\right)$$
$$10x - 15 = 12$$
$$10x - 15 + 15 = 12 + 15$$
$$10x = 27$$
$$\frac{10x}{10} = \frac{27}{10}$$
$$x = \frac{27}{10}$$

What should Olga do to correct the error that she made? *(MACC.8.EE.3.7b)*

F. Multiply both sides by $-\frac{5}{2}$ instead of $-\frac{2}{5}$.

G. Multiply both sides by $\frac{2}{5}$ instead of $-\frac{2}{5}$.

H. Distribute $-\frac{2}{5}$ to get $-4x - 6$.

I. Add 15 to -30.

7. In the coordinate plane below, $\triangle XYZ$ is plotted and its vertices are labeled.

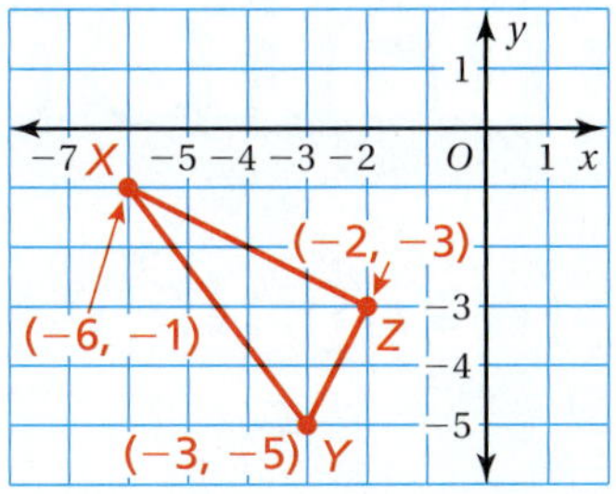

Which of the following shows $\triangle X'Y'Z'$, the image of $\triangle XYZ$ after it is reflected in the y-axis? *(MACC.8.G.1.3)*

A.

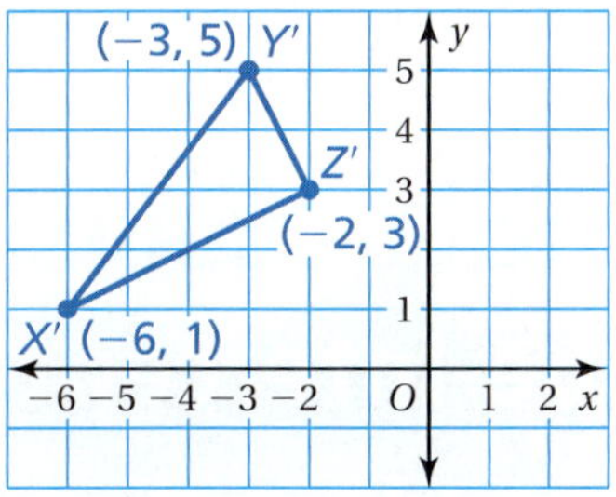

C.

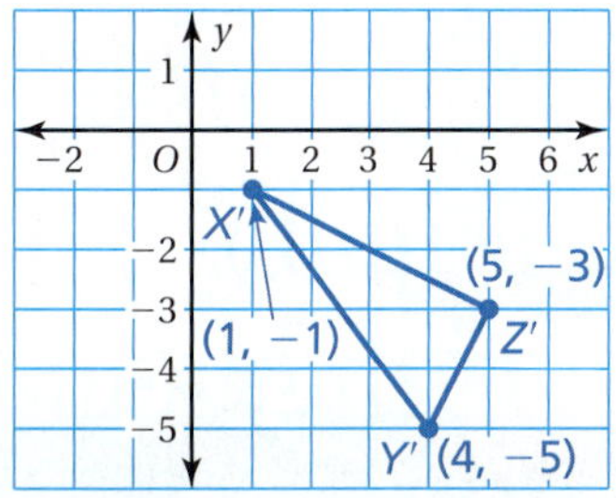

B.

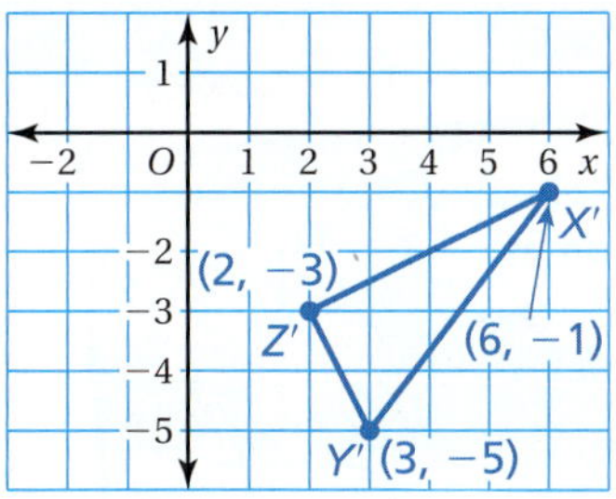

D.

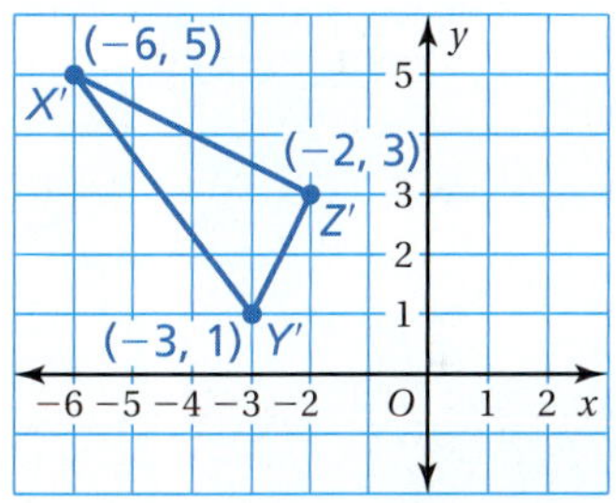

8. The sum S of the interior angle measures of a polygon with n sides can be found by using a formula. *(MACC.8.G.1.5)*

Think Solve Explain

Part A Write the formula.

Part B A quadrilateral has angles measuring 100°, 90°, and 90°. Find the measure of its fourth angle. Show your work and explain your reasoning.

Part C The sum of the measures of the angles of the pentagon shown is 540°. Divide the pentagon into triangles to show why this must be true. Show your work and explain your reasoning.

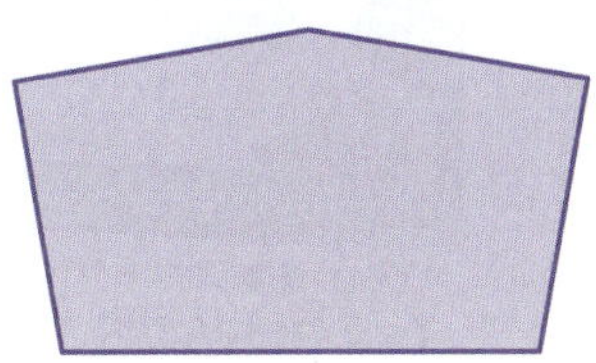

4 Graphing and Writing Linear Equations

4.1 Graphing Linear Equations
4.2 Slope of a Line
4.3 Graphing Proportional Relationships
4.4 Graphing Linear Equations in Slope-Intercept Form
4.5 Graphing Linear Equations in Standard Form
4.6 Writing Equations in Slope-Intercept Form
4.7 Writing Equations in Point-Slope Form

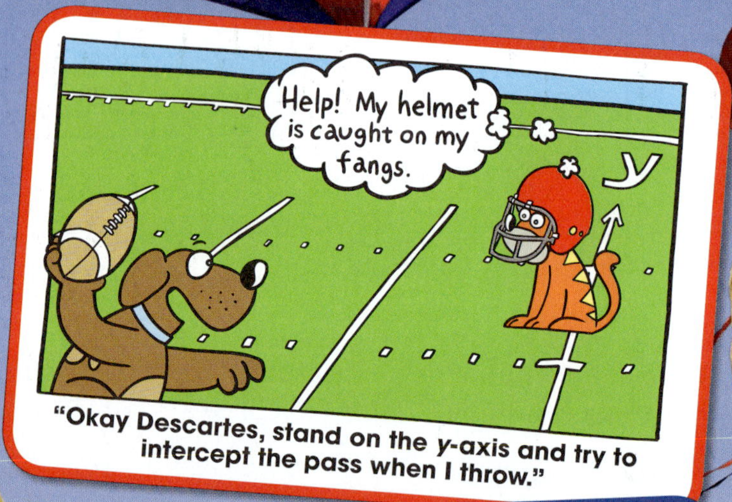

"Okay Descartes, stand on the *y*-axis and try to intercept the pass when I throw."

"Here's an easy example of a line with a slope of 1."

"You eat one mouse treat the first day. Two treats the second day. And so on. Get it?"

What You Learned Before

"I estimate that we are on a slope of about −0.625. What do you think?"

Evaluating Expressions Using Order of Operations (MACC.6.EE.1.2c)

Example 1 **Evaluate $2xy + 3(x + y)$ when $x = 4$ and $y = 7$.**

$2xy + 3(x + y) = 2(4)(7) + 3(4 + 7)$ Substitute 4 for x and 7 for y.

$= 8(7) + 3(4 + 7)$ Use order of operations.

$= 56 + 3(11)$ Simplify.

$= 56 + 33$ Multiply.

$= 89$ Add.

Try It Yourself

Evaluate the expression when $a = \frac{1}{4}$ and $b = 6$.

1. $-8ab$ **2.** $16a^2 - 4b$ **3.** $\frac{5b}{32a^2}$ **4.** $12a + (b - a - 4)$

Plotting Points (MACC.6.NS.3.6c)

Example 2 **Write the ordered pair that corresponds to point U.**

Point U is 3 units to the left of the origin and 4 units down. So, the x-coordinate is -3, and the y-coordinate is -4.

The ordered pair $(-3, -4)$ corresponds to point U.

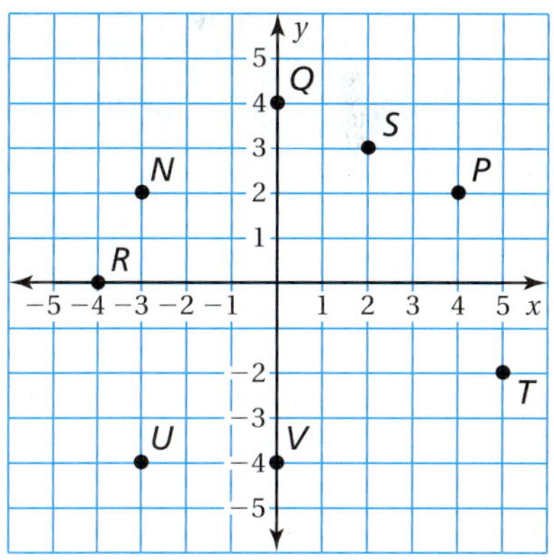

Example 3 **Which point is located at $(5, -2)$?**

Start at the origin. Move 5 units right and 2 units down.

Point T is located at $(5, -2)$.

Try It Yourself

Use the graph to answer the question.

5. Write the ordered pair that corresponds to point Q.

6. Write the ordered pair that corresponds to point P.

7. Which point is located at $(-4, 0)$?

8. Which point is located in Quadrant II?

4.1 Graphing Linear Equations

Essential Question How can you recognize a linear equation? How can you draw its graph?

1 ACTIVITY: Graphing a Linear Equation

Work with a partner.

a. Use the equation $y = \frac{1}{2}x + 1$ to complete the table. (Choose any two x-values and find the y-values.)

	Solution Points	
x		
$y = \frac{1}{2}x + 1$		

b. Write the two ordered pairs given by the table. These are called *solution points* of the equation.

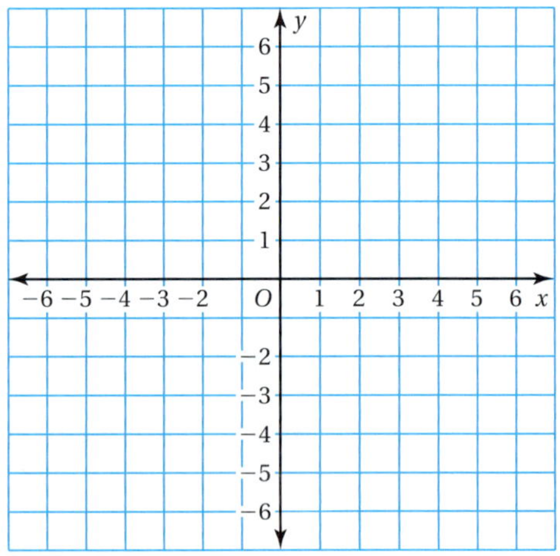

c. **PRECISION** Plot the two solution points. Draw a line *exactly* through the two points.

d. Find a different point on the line. Check that this point is a solution point of the equation $y = \frac{1}{2}x + 1$.

e. **LOGIC** Do you think it is true that *any* point on the line is a solution point of the equation $y = \frac{1}{2}x + 1$? Explain.

f. Choose five additional x-values for the table. (Choose positive and negative x-values.) Plot the five corresponding solution points. Does each point lie on the line?

	Solution Points				
x					
$y = \frac{1}{2}x + 1$					

g. **LOGIC** Do you think it is true that *any* solution point of the equation $y = \frac{1}{2}x + 1$ is a point on the line? Explain.

h. Why do you think $y = ax + b$ is called a *linear equation*?

COMMON CORE

Graphing Equations

In this lesson, you will

- understand that lines represent solutions of linear equations.
- graph linear equations.

Preparing for Standard MACC.8.EE.2.5

2 ACTIVITY: Using a Graphing Calculator

Use a graphing calculator to graph $y = 2x + 5$.

a. Enter the equation $y = 2x + 5$ into your calculator.

b. Check the settings of the *viewing window*. The boundaries of the graph are set by the minimum and the maximum x- and y-values. The numbers of units between the tick marks are set by the x- and y-scales.

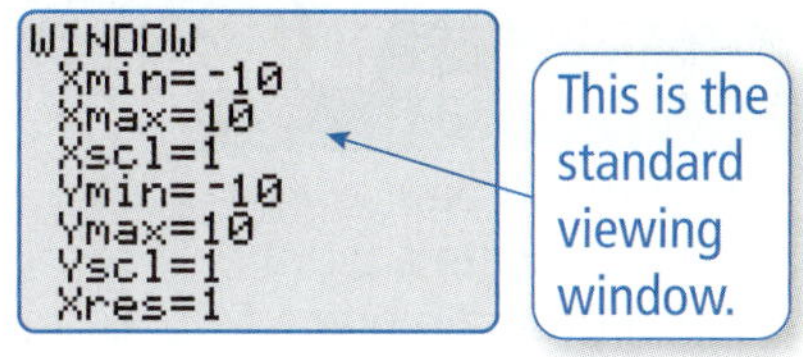

c. Graph $y = 2x + 5$ on your calculator.

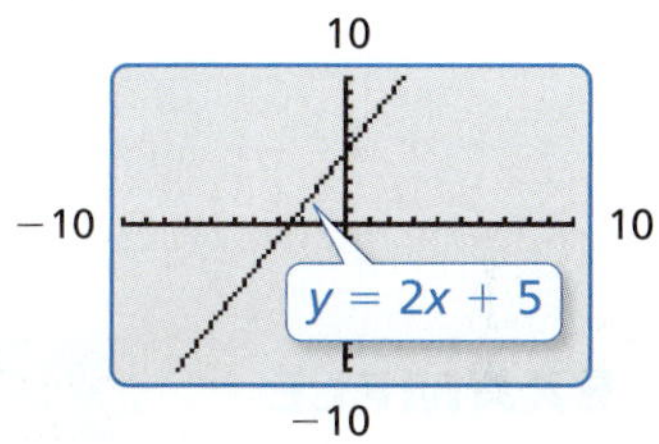

d. Change the settings of the viewing window to match those shown.

Compare the two graphs.

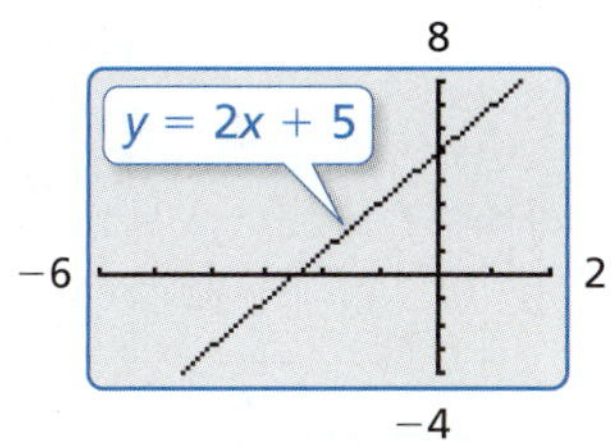

Math Practice 5

Recognize Usefulness of Tools

What are some advantages and disadvantages of using a graphing calculator to graph a linear equation?

What Is Your Answer?

3. **IN YOUR OWN WORDS** How can you recognize a linear equation? How can you draw its graph? Write an equation that is linear. Write an equation that is *not* linear.

4. Use a graphing calculator to graph $y = 5x - 12$ in the standard viewing window.

   a. Can you tell where the line crosses the x-axis? Can you tell where the line crosses the y-axis?

   b. How can you adjust the viewing window so that you can determine where the line crosses the x- and y-axes?

5. **CHOOSE TOOLS** You want to graph $y = 2.5x - 3.8$. Would you graph it by hand or by using a graphing calculator? Why?

Practice Use what you learned about graphing linear equations to complete Exercises 3 and 4 on page 146.

4.1 Lesson

Check It Out
Lesson Tutorials
BigIdeasMath.com

Key Vocabulary
linear equation, *p. 144*
solution of a linear equation, *p. 144*

Remember
An ordered pair (x, y) is used to locate a point in a coordinate plane.

Key Idea

Linear Equations

A **linear equation** is an equation whose graph is a line. The points on the line are **solutions** of the equation.

You can use a graph to show the solutions of a linear equation. The graph below represents the equation $y = x + 1$.

x	y	(x, y)
−1	0	(−1, 0)
0	1	(0, 1)
2	3	(2, 3)

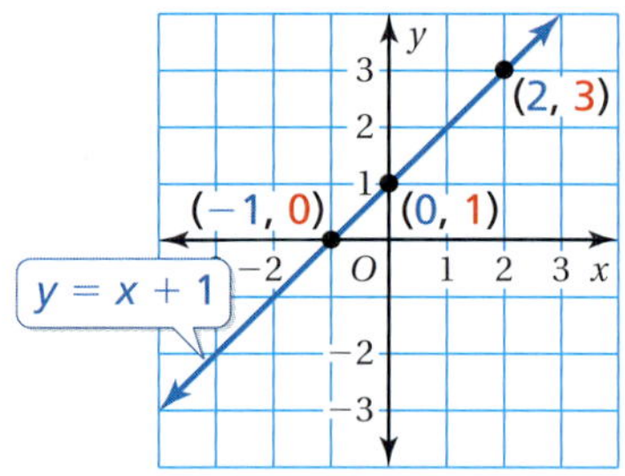

EXAMPLE 1 Graphing a Linear Equation

Graph $y = -2x + 1$.

Step 1: Make a table of values.

x	$y = -2x + 1$	y	(x, y)
−1	$y = -2(-1) + 1$	3	(−1, 3)
0	$y = -2(0) + 1$	1	(0, 1)
2	$y = -2(2) + 1$	−3	(2, −3)

Step 2: Plot the ordered pairs.

Step 3: Draw a line through the points.

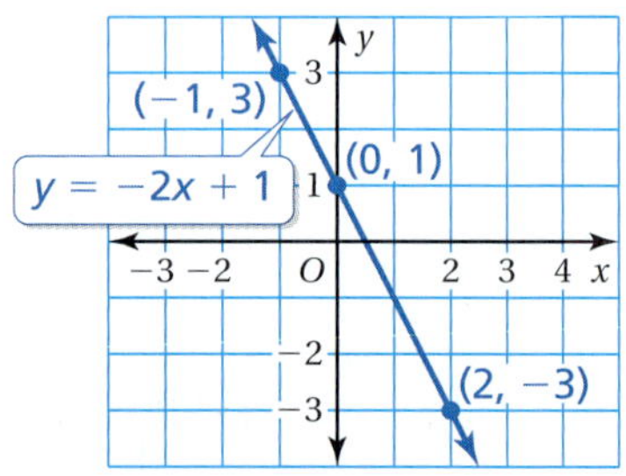

Check

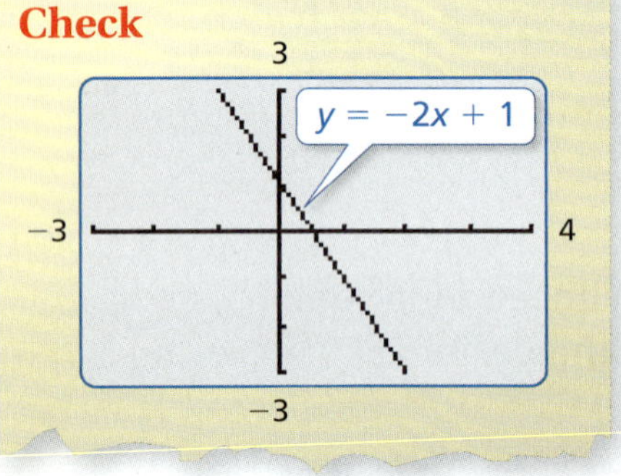

Key Idea

Graphing Horizontal and Vertical Lines

The graph of $y = b$ is a horizontal line passing through $(0, b)$.

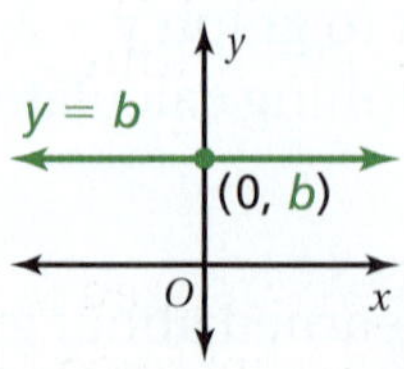

The graph of $x = a$ is a vertical line passing through $(a, 0)$.

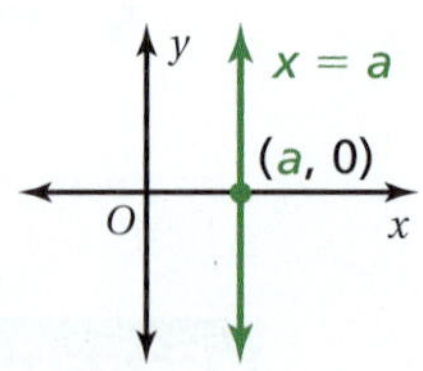

 Multi-Language Glossary at BigIdeasMath.com

EXAMPLE 2 Graphing a Horizontal Line and a Vertical Line

a. Graph $y = -3$.

The graph of $y = -3$ is a horizontal line passing through $(0, -3)$. Draw a horizontal line through this point.

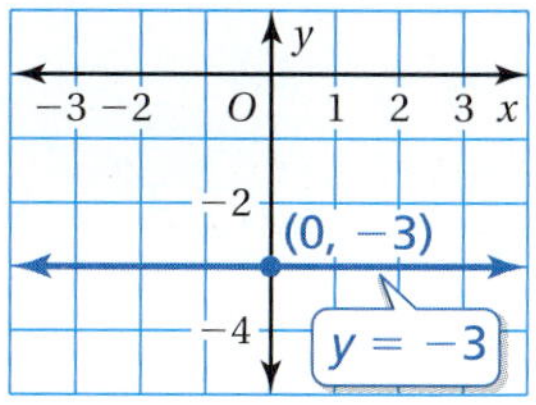

b. Graph $x = 2$.

The graph of $x = 2$ is a vertical line passing through $(2, 0)$. Draw a vertical line through this point.

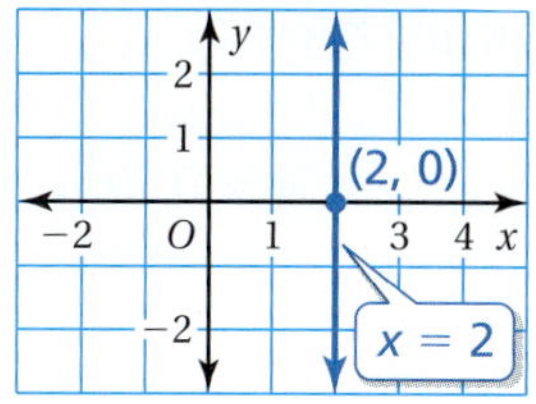

On Your Own

Now You're Ready
Exercises 5–16

Graph the linear equation. Use a graphing calculator to check your graph, if possible.

1. $y = 3x$
2. $y = -\frac{1}{2}x + 2$
3. $x = -4$
4. $y = -1.5$

EXAMPLE 3 Real-Life Application

The wind speed y (in miles per hour) of a tropical storm is $y = 2x + 66$, where x is the number of hours after the storm enters the Gulf of Mexico.

a. Graph the equation.

b. When does the storm become a hurricane?

A tropical storm becomes a hurricane when wind speeds are at least 74 miles per hour.

a. Make a table of values.

x	$y = 2x + 66$	y	(x, y)
0	$y = 2(0) + 66$	66	(0, 66)
1	$y = 2(1) + 66$	68	(1, 68)
2	$y = 2(2) + 66$	70	(2, 70)
3	$y = 2(3) + 66$	72	(3, 72)

Plot the ordered pairs and draw a line through the points.

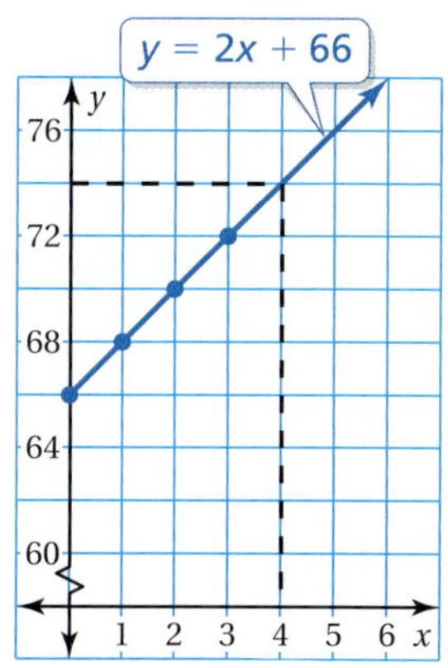

b. From the graph, you can see that $y = 74$ when $x = 4$. So, the storm becomes a hurricane 4 hours after it enters the Gulf of Mexico.

On Your Own

5. **WHAT IF?** The wind speed of the storm is $y = 1.5x + 62$. When does the storm become a hurricane?

4.1 Exercises

Check It Out
Help with Homework
BigIdeasMath.com

Vocabulary and Concept Check

1. **VOCABULARY** What type of graph represents the solutions of the equation $y = 2x + 4$?

2. **WHICH ONE DOESN'T BELONG?** Which equation does *not* belong with the other three? Explain your reasoning.

$y = 0.5x - 0.2$ $\qquad$ $4x + 3 = y$ $\qquad$ $y = x^2 + 6$ $\qquad$ $\frac{3}{4}x + \frac{1}{3} = y$

Practice and Problem Solving

PRECISION **Copy and complete the table. Plot the two solution points and draw a line *exactly* through the two points. Find a different solution point on the line.**

3.

x		
$y = 3x - 1$		

4.

x		
$y = \frac{1}{3}x + 2$		

Graph the linear equation. Use a graphing calculator to check your graph, if possible.

1 2

5. $y = -5x$
6. $y = \frac{1}{4}x$
7. $y = 5$
8. $x = -6$
9. $y = x - 3$
10. $y = -7x - 1$
11. $y = -\frac{x}{3} + 4$
12. $y = \frac{3}{4}x - \frac{1}{2}$
13. $y = -\frac{2}{3}$
14. $y = 6.75$
15. $x = -0.5$
16. $x = \frac{1}{4}$

17. **ERROR ANALYSIS** Describe and correct the error in graphing the equation.

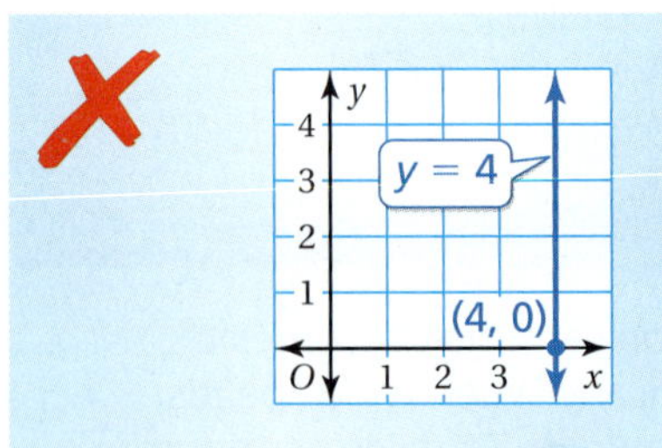

18. **MESSAGING** You sign up for an unlimited text-messaging plan for your cell phone. The equation $y = 20$ represents the cost y (in dollars) for sending x text messages. Graph the equation. What does the graph tell you?

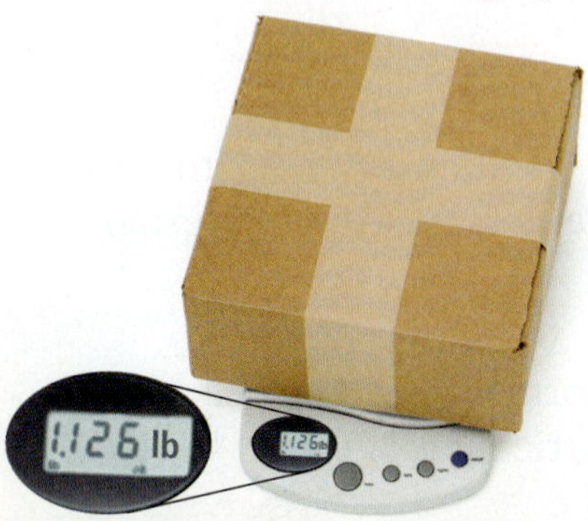

19. **MAIL** The equation $y = 2x + 3$ represents the cost y (in dollars) of mailing a package that weighs x pounds.
    - **a.** Graph the equation.
    - **b.** Use the graph to estimate how much it costs to mail the package.
    - **c.** Use the equation to find exactly how much it costs to mail the package.

Solve for *y*. Then graph the equation. Use a graphing calculator to check your graph.

20. $y - 3x = 1$

21. $5x + 2y = 4$

22. $-\frac{1}{3}y + 4x = 3$

23. $x + 0.5y = 1.5$

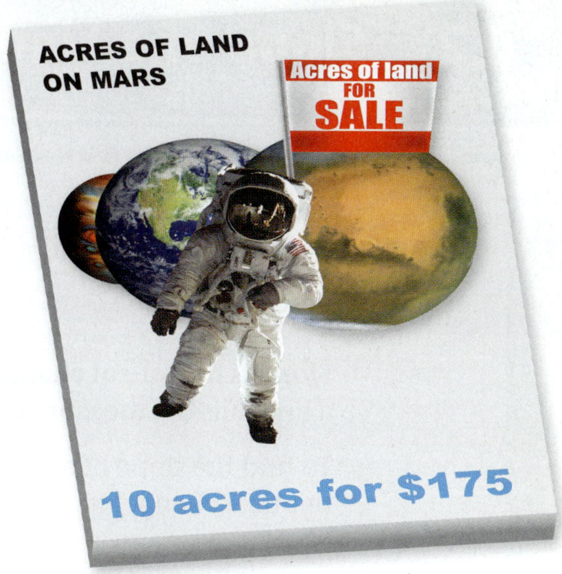

24. **SAVINGS** You have \$100 in your savings account and plan to deposit \$12.50 each month.

a. Graph a linear equation that represents the balance in your account.

b. How many months will it take you to save enough money to buy 10 acres of land on Mars?

25. **GEOMETRY** The sum S of the interior angle measures of a polygon with n sides is $S = (n - 2) \cdot 180°$.

a. Plot four points (n, S) that satisfy the equation. Is the equation a linear equation? Explain your reasoning.

b. Does the value $n = 3.5$ make sense in the context of the problem? Explain your reasoning.

26. **SEA LEVEL** Along the U.S. Atlantic coast, the sea level is rising about 2 millimeters per year. How many millimeters has sea level risen since you were born? How do you know? Use a linear equation and a graph to justify your answer.

27. Problem Solving One second of video on your digital camera uses the same amount of memory as two pictures. Your camera can store 250 pictures.

a. Write and graph a linear equation that represents the number y of pictures your camera can store when you take x seconds of video.

b. How many pictures can your camera store in addition to the video shown?

Fair Game Review What you learned in previous grades & lessons

Write the ordered pair corresponding to the point. *(Skills Review Handbook)*

28. point A

29. point B

30. point C

31. point D

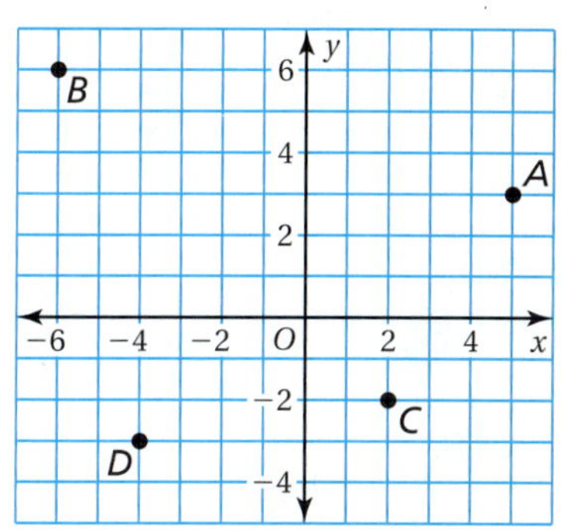

32. **MULTIPLE CHOICE** A debate team has 15 female members. The ratio of females to males is 3 : 2. How many males are on the debate team? *(Skills Review Handbook)*

Ⓐ 6 Ⓑ 10 Ⓒ 22 Ⓓ 25

4.2 Slope of a Line

Essential Question How can you use the slope of a line to describe the line?

Slope is the rate of change between any two points on a line. It is the measure of the *steepness* of the line.

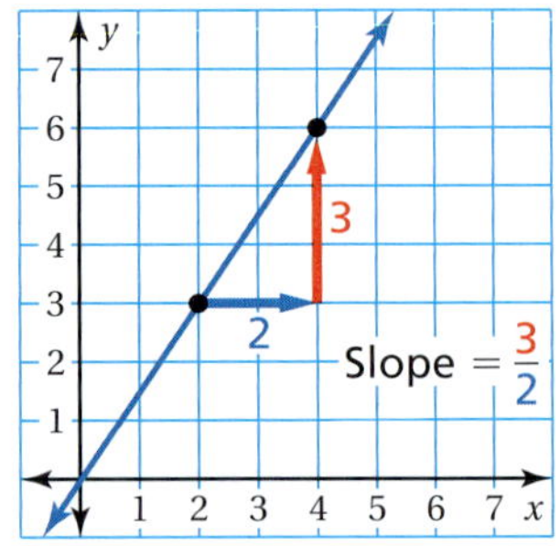

To find the slope of a line, find the ratio of the change in y (vertical change) to the change in x (horizontal change).

$$\text{slope} = \frac{\text{change in } y}{\text{change in } x}$$

1 ACTIVITY: Finding the Slope of a Line

Work with a partner. Find the slope of each line using two methods.

Method 1: Use the two black points. •

Method 2: Use the two pink points. •

Do you get the same slope using each method? Why do you think this happens?

a.

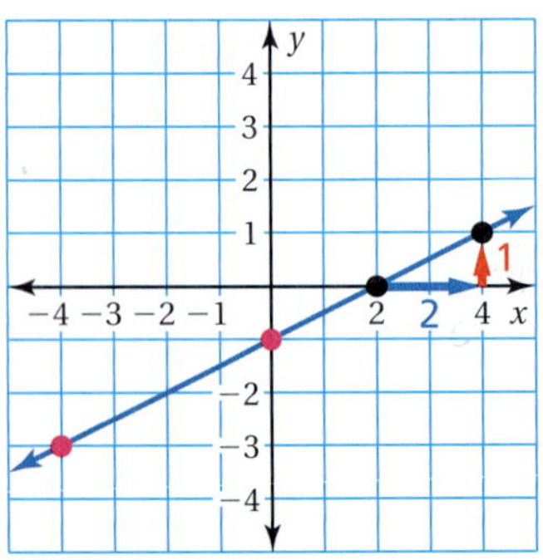

b.

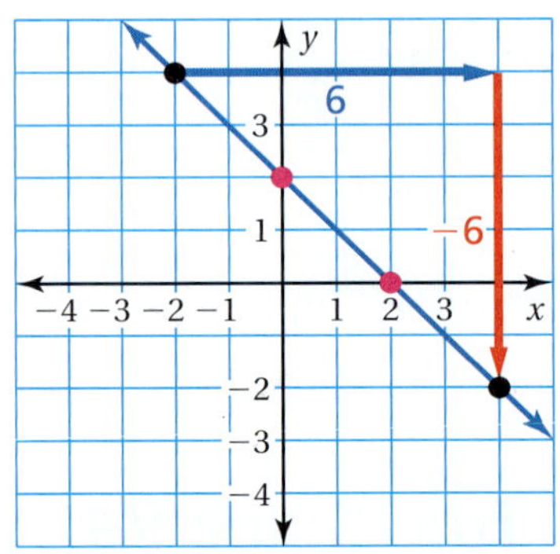

c.

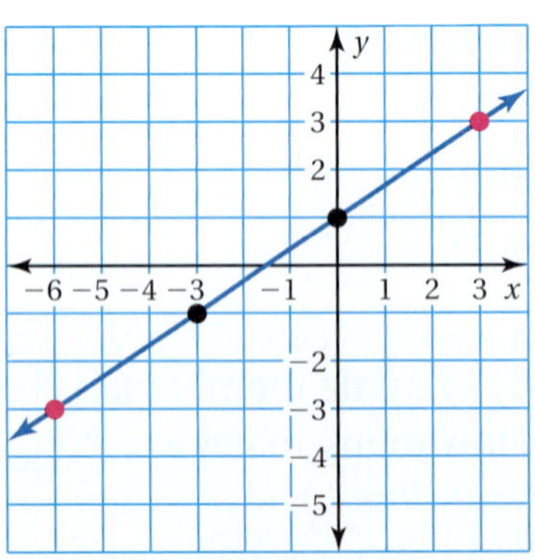

d.

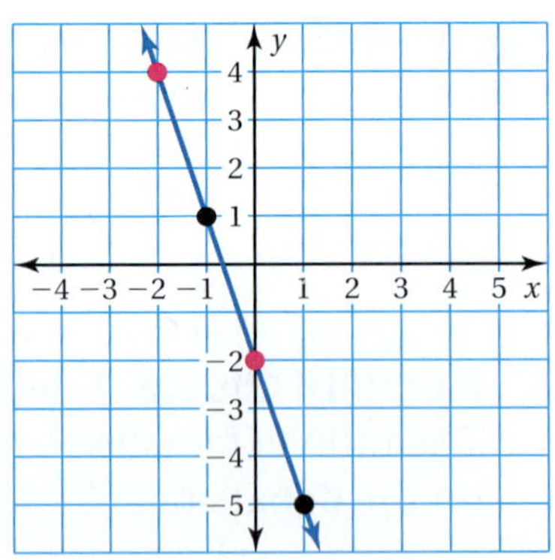

COMMON CORE

Graphing Equations

In this lesson, you will

- find slopes of lines by using two points.
- find slopes of lines from tables.

Learning Standard MACC.8.EE.2.6

2 ACTIVITY: Using Similar Triangles

Work with a partner. Use the figure shown.

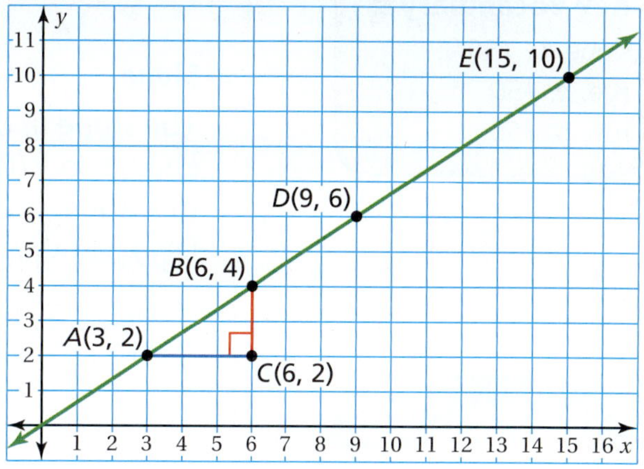

a. $\triangle ABC$ is a right triangle formed by drawing a horizontal line segment from point A and a vertical line segment from point B. Use this method to draw another right triangle, $\triangle DEF$.

b. What can you conclude about $\triangle ABC$ and $\triangle DEF$? Justify your conclusion.

c. For each triangle, find the ratio of the length of the vertical side to the length of the horizontal side. What do these ratios represent?

d. What can you conclude about the slope between any two points on the line?

3 ACTIVITY: Drawing Lines with Given Slopes

Work with a partner.

a. Draw two lines with slope $\frac{3}{4}$. One line passes through $(-4, 1)$, and the other line passes through $(4, 0)$. What do you notice about the two lines?

b. Draw two lines with slope $-\frac{4}{3}$. One line passes through $(2, 1)$, and the other line passes through $(-1, -1)$. What do you notice about the two lines?

c. **CONJECTURE** Make a conjecture about two different nonvertical lines in the same plane that have the same slope.

d. Graph one line from part (a) and one line from part (b) in the same coordinate plane. Describe the angle formed by the two lines. What do you notice about the product of the slopes of the two lines?

e. **REPEATED REASONING** Repeat part (d) for the two lines you did *not* choose. Based on your results, make a conjecture about two lines in the same plane whose slopes have a product of -1.

Math Practice 1

Interpret a Solution

What does the slope tell you about the graph of the line? Explain.

What Is Your Answer?

4. **IN YOUR OWN WORDS** How can you use the slope of a line to describe the line?

Practice Use what you learned about the slope of a line to complete Exercises 4–6 on page 153.

Check It Out
Lesson Tutorials
BigIdeasMath.com

Key Vocabulary
slope, *p. 150*
rise, *p. 150*
run, *p. 150*

Key Idea

Slope

The **slope** m of a line is a ratio of the change in y (the **rise**) to the change in x (the **run**) between any two points, (x_1, y_1) and (x_2, y_2), on the line.

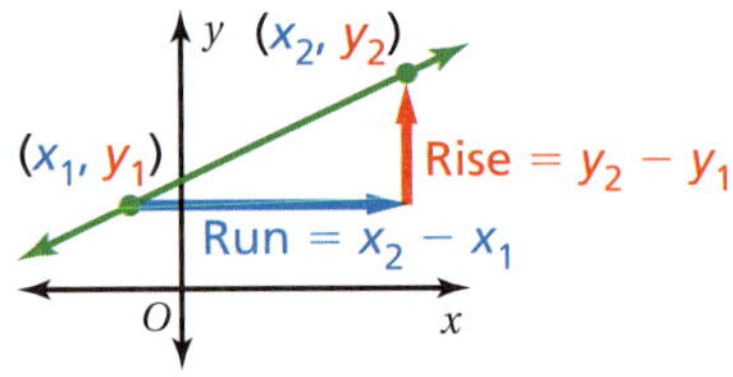

$$m = \frac{\text{rise}}{\text{run}} = \frac{\text{change in } y}{\text{change in } x} = \frac{y_2 - y_1}{x_2 - x_1}$$

Positive Slope

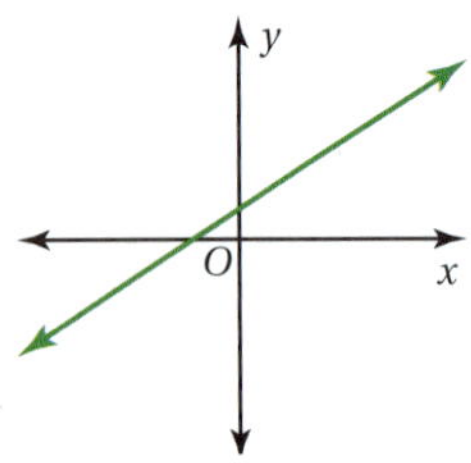

The line rises from left to right.

Negative Slope

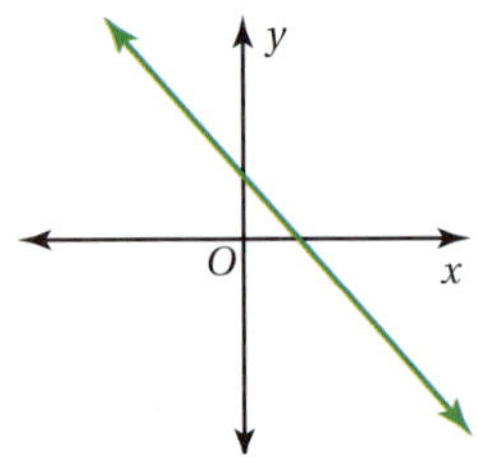

The line falls from left to right.

Reading

In the slope formula, x_1 is read as "x sub one," and y_2 is read as "y sub two." The numbers 1 and 2 in x_1 and y_2 are called *subscripts*.

EXAMPLE 1 Finding the Slope of a Line

Describe the slope of the line. Then find the slope.

a.

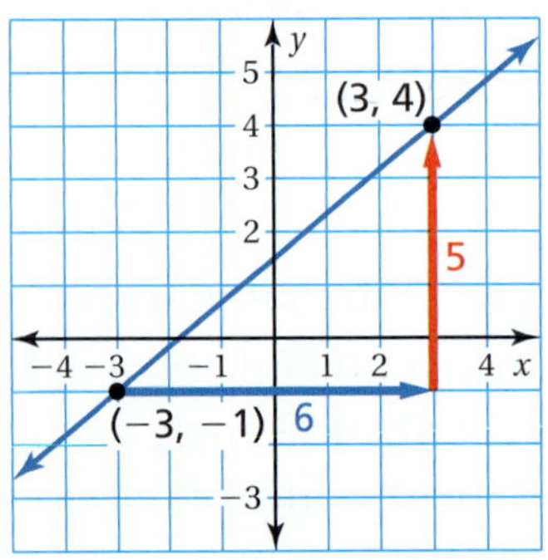

The line rises from left to right. So, the slope is positive. Let $(x_1, y_1) = (-3, -1)$ and $(x_2, y_2) = (3, 4)$.

$$m = \frac{y_2 - y_1}{x_2 - x_1}$$

$$= \frac{4 - (-1)}{3 - (-3)}$$

$$= \frac{5}{6}$$

b.

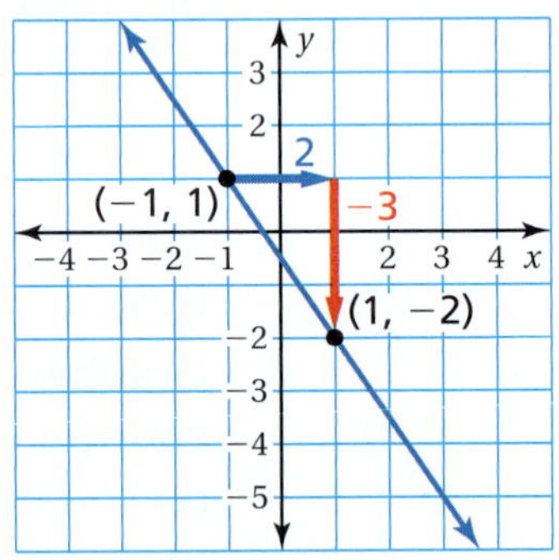

The line falls from left to right. So, the slope is negative. Let $(x_1, y_1) = (-1, 1)$ and $(x_2, y_2) = (1, -2)$.

$$m = \frac{y_2 - y_1}{x_2 - x_1}$$

$$= \frac{-2 - 1}{1 - (-1)}$$

$$= \frac{-3}{2}, \text{ or } -\frac{3}{2}$$

Study Tip

When finding slope, you can label either point as (x_1, y_1) and the other point as (x_2, y_2).

On Your Own

Now You're Ready
Exercises 7–9

Find the slope of the line.

1.

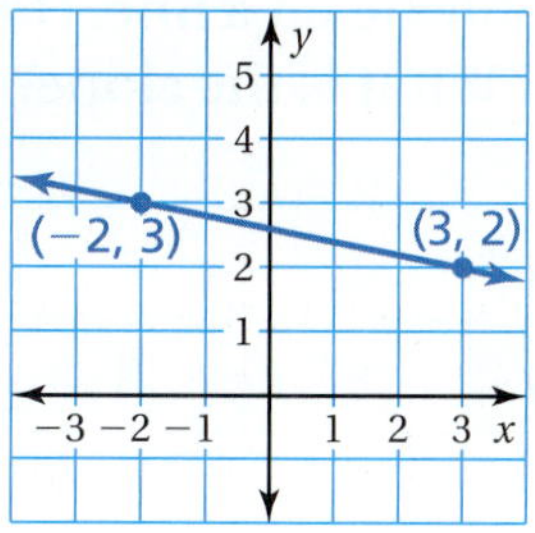

2.

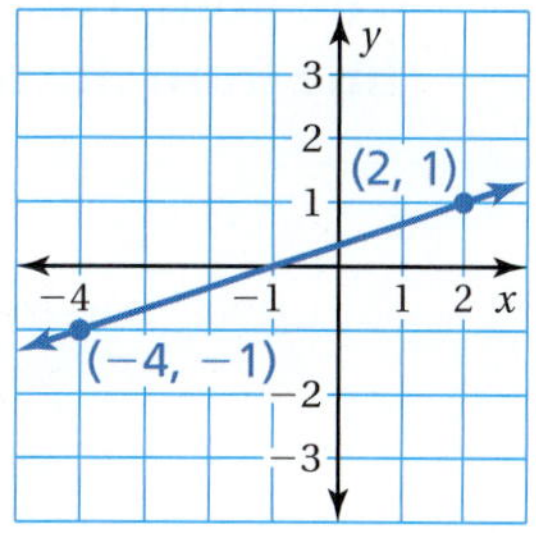

3.

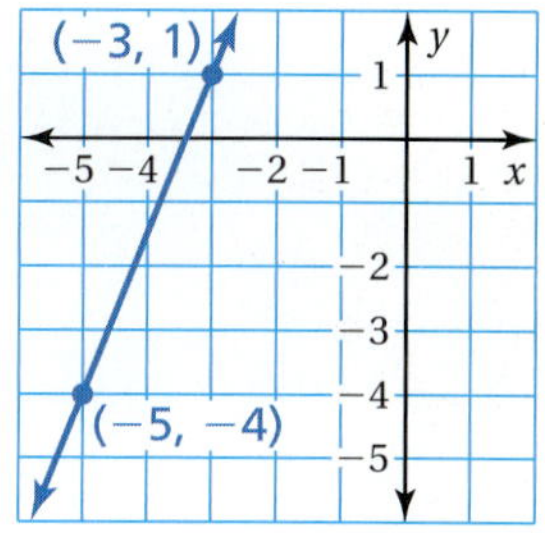

EXAMPLE 2 Finding the Slope of a Horizontal Line

Find the slope of the line.

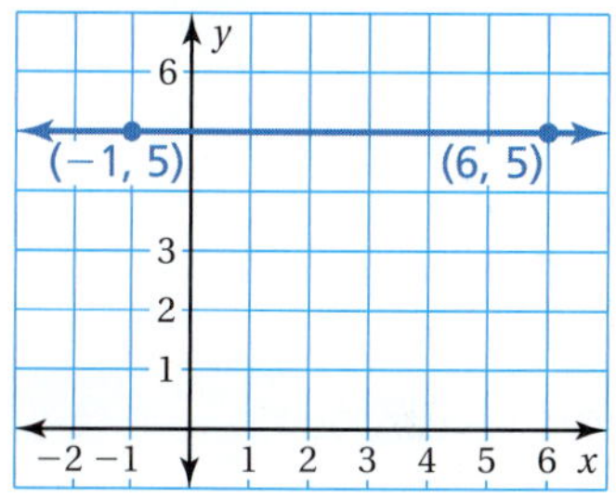

$$m = \frac{y_2 - y_1}{x_2 - x_1}$$

$$= \frac{5 - 5}{6 - (-1)}$$

$$= \frac{0}{7}, \text{ or } 0$$

∴ The slope is 0.

EXAMPLE 3 Finding the Slope of a Vertical Line

Find the slope of the line.

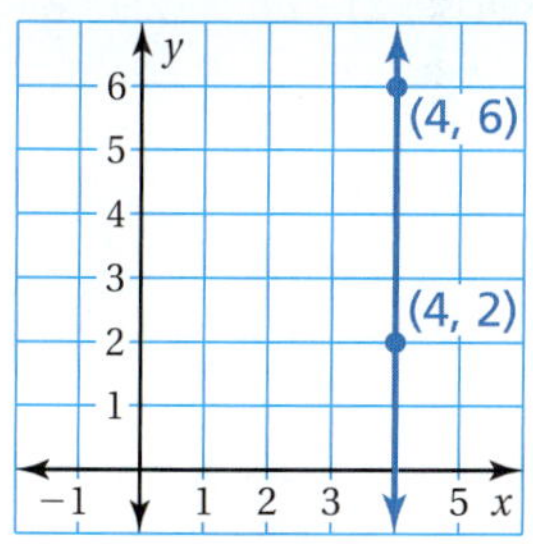

$$m = \frac{y_2 - y_1}{x_2 - x_1}$$

$$= \frac{6 - 2}{4 - 4}$$

$$= \frac{4}{0} \quad ✗$$

∴ Because division by zero is undefined, the slope of the line is undefined.

> **Study Tip**
> The slope of every horizontal line is 0. The slope of every vertical line is undefined.

On Your Own

Now You're Ready
Exercises 13–15

Find the slope of the line through the given points.

4. (1, −2), (7, −2)

5. (−2, 4), (3, 4)

6. (−3, −3), (−3, −5)

7. (0, 8), (0, 0)

8. How do you know that the slope of every horizontal line is 0? How do you know that the slope of every vertical line is undefined?

EXAMPLE 4 Finding Slope from a Table

The points in the table lie on a line. How can you find the slope of the line from the table? What is the slope?

x	1	4	7	10
y	8	6	4	2

Choose any two points from the table and use the slope formula.

Use the points $(x_1, y_1) = (1, 8)$ and $(x_2, y_2) = (4, 6)$.

$$m = \frac{y_2 - y_1}{x_2 - x_1}$$

$$= \frac{6 - 8}{4 - 1}$$

$$= \frac{-2}{3}, \text{ or } -\frac{2}{3}$$

∴ The slope is $-\frac{2}{3}$.

Check

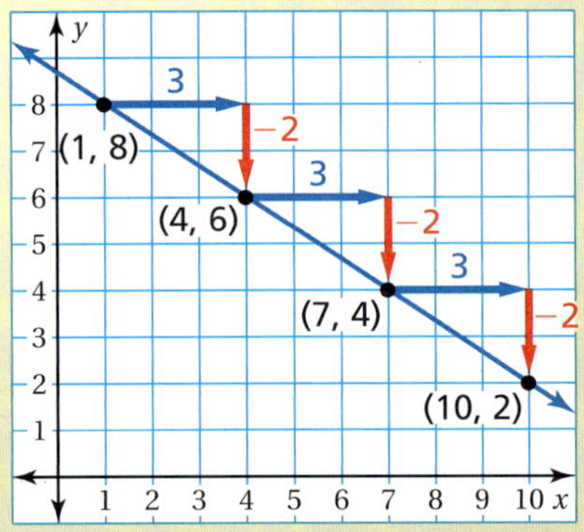

On Your Own

Now You're Ready
Exercises 21–24

The points in the table lie on a line. Find the slope of the line.

9.

x	1	3	5	7
y	2	5	8	11

10.

x	−3	−2	−1	0
y	6	4	2	0

Summary

Slope

Positive Slope	*Negative Slope*	*Slope of 0*	*Undefined Slope*
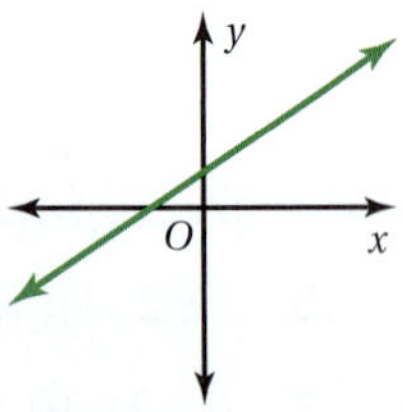	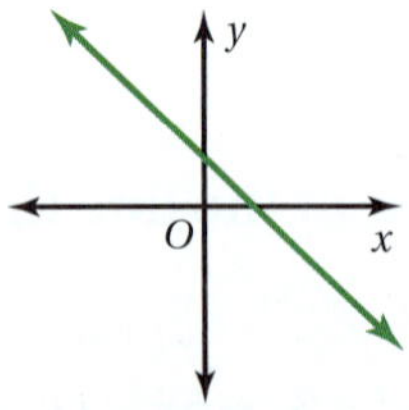	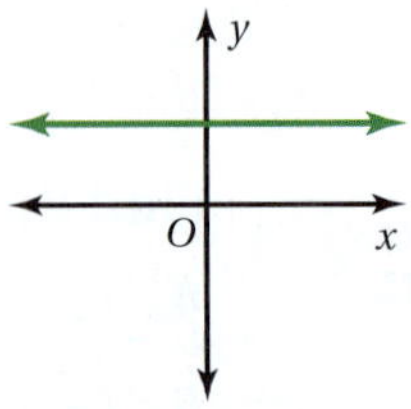	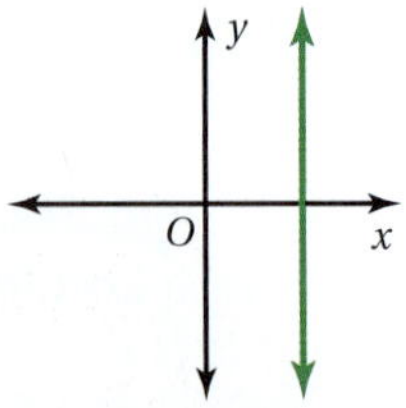
The line rises from left to right.	The line falls from left to right.	The line is horizontal.	The line is vertical.

4.2 Exercises

Check It Out
Help with Homework
BigIdeasMath.com

Vocabulary and Concept Check

1. **CRITICAL THINKING** Refer to the graph.

   a. Which lines have positive slopes?

   b. Which line has the steepest slope?

   c. Do any lines have an undefined slope? Explain.

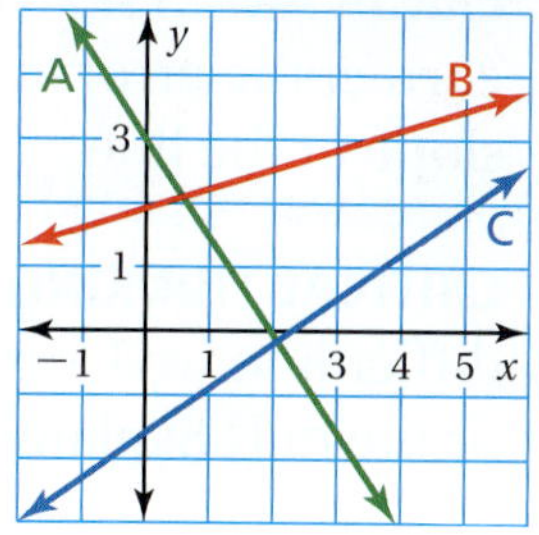

2. **OPEN-ENDED** Describe a real-life situation in which you need to know the slope.

3. **REASONING** The slope of a line is 0. What do you know about the line?

Practice and Problem Solving

Draw a line through each point using the given slope. What do you notice about the two lines?

4. slope $= 1$

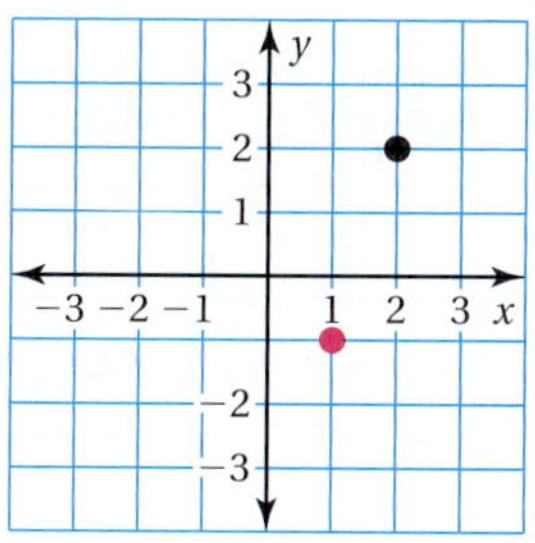

5. slope $= -3$

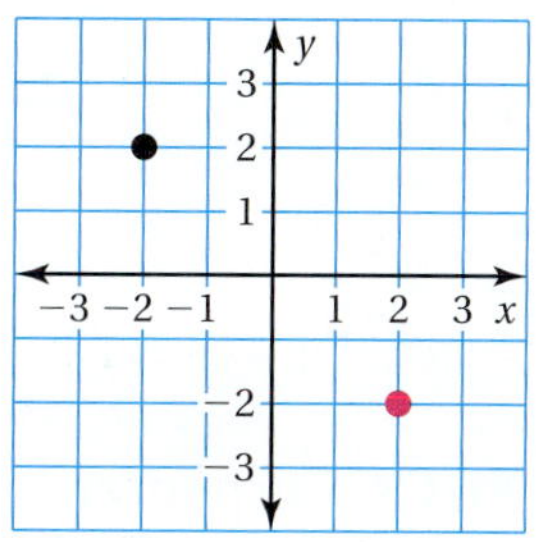

6. slope $= \frac{1}{4}$

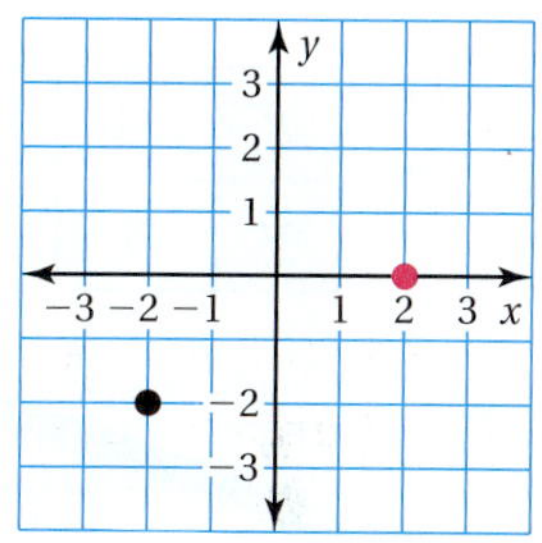

Find the slope of the line.

1 7.

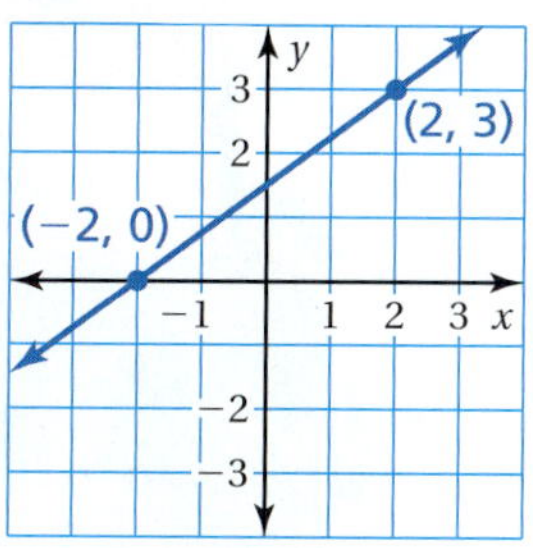

8.

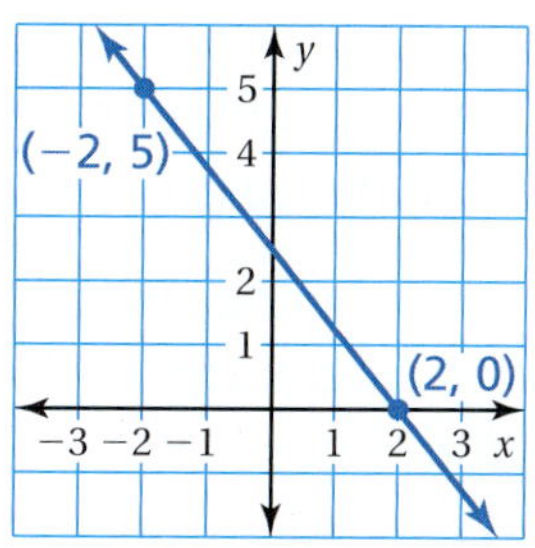

9.

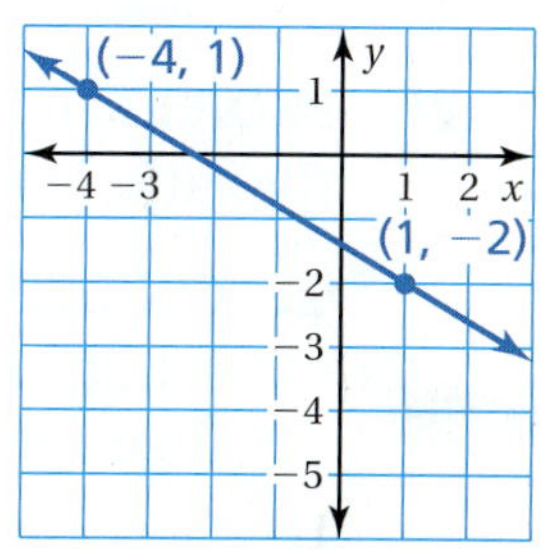

10.

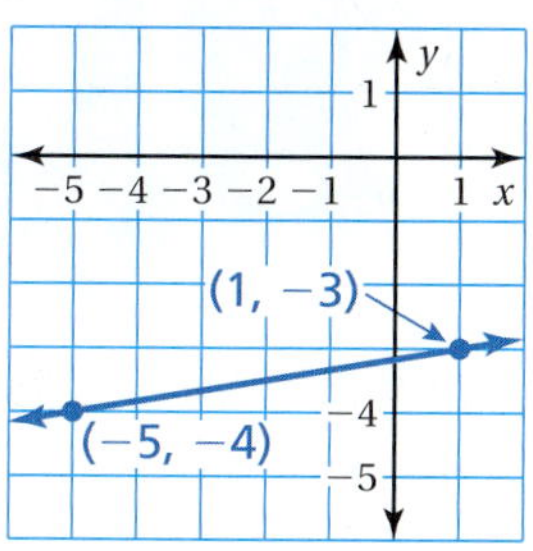

11.

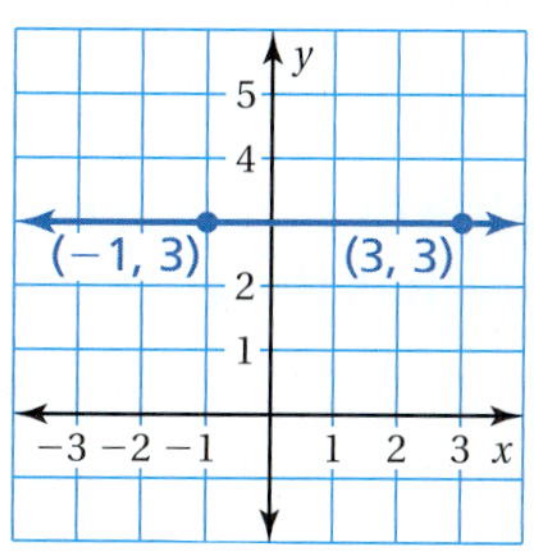

12.

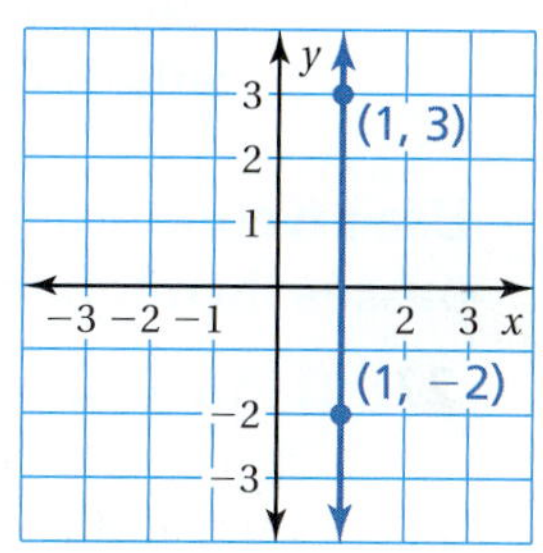

Find the slope of the line through the given points.

2 3 **13.** (4, −1), (−2, −1) **14.** (5, −3), (5, 8) **15.** (−7, 0), (−7, −6)

16. (−3, 1), (−1, 5) **17.** (10, 4), (4, 15) **18.** (−3, 6), (2, 6)

19. ERROR ANALYSIS Describe and correct the error in finding the slope of the line.

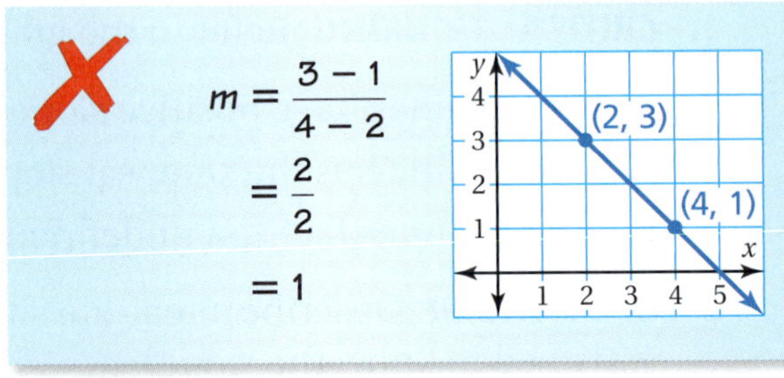

20. CRITICAL THINKING Is it more difficult to walk up the ramp or the hill? Explain.

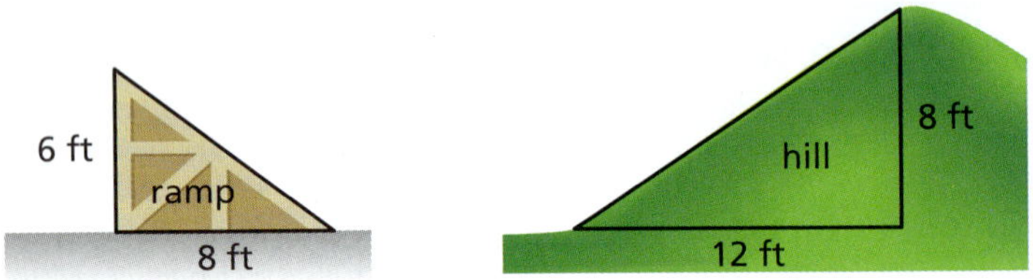

The points in the table lie on a line. Find the slope of the line.

4 **21.**

x	1	3	5	7
y	2	10	18	26

22.

x	−3	2	7	12
y	0	2	4	6

23.

x	−6	−2	2	6
y	8	5	2	−1

24.

x	−8	−2	4	10
y	8	1	−6	−13

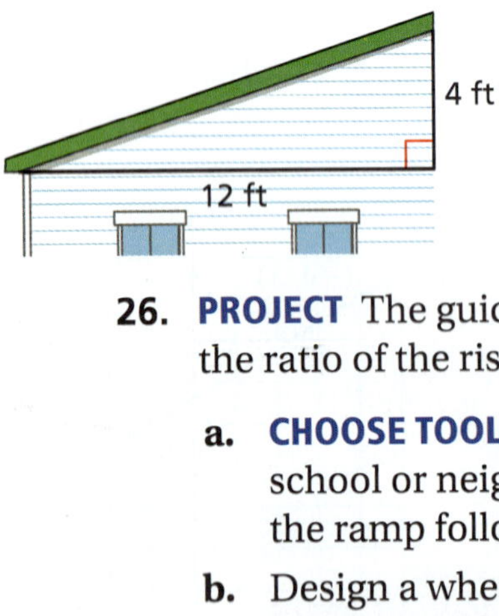

25. PITCH Carpenters refer to the slope of a roof as the *pitch* of the roof. Find the pitch of the roof.

26. PROJECT The guidelines for a wheelchair ramp suggest that the ratio of the rise to the run be no greater than 1 : 12.

a. CHOOSE TOOLS Find a wheelchair ramp in your school or neighborhood. Measure its slope. Does the ramp follow the guidelines?

b. Design a wheelchair ramp that provides access to a building with a front door that is 2.5 feet above the sidewalk. Illustrate your design.

Use an equation to find the value of *k* so that the line that passes through the given points has the given slope.

27. (1, 3), (5, *k*); $m = 2$ **28.** (−2, *k*), (2, 0); $m = -1$

29. (−4, *k*), (6, −7); $m = -\frac{1}{5}$ **30.** (4, −4), (*k*, −1); $m = \frac{3}{4}$

31. **TURNPIKE TRAVEL** The graph shows the cost of traveling by car on a turnpike.

   a. Find the slope of the line.

   b. Explain the meaning of the slope as a rate of change.

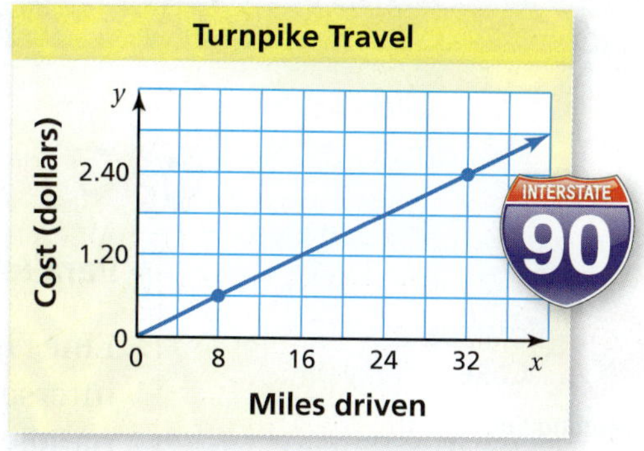

32. **BOAT RAMP** Which is steeper: the boat ramp or a road with a 12% grade? Explain. (*Note:* Road grade is the vertical increase divided by the horizontal distance.)

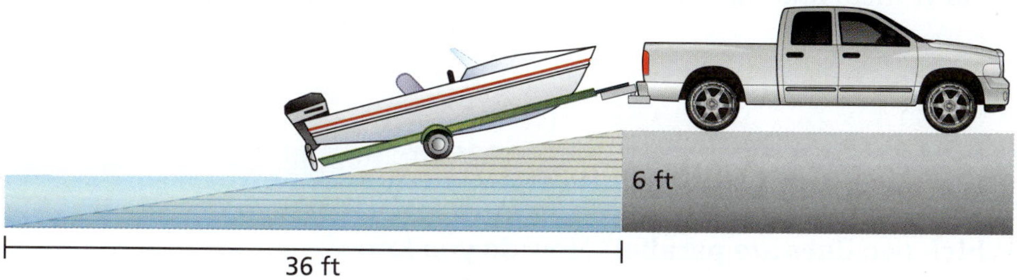

33. **REASONING** Do the points $A(-2, -1)$, $B(1, 5)$, and $C(4, 11)$ lie on the same line? Without using a graph, how do you know?

34. **BUSINESS** A small business earns a profit of $6500 in January and $17,500 in May. What is the rate of change in profit for this time period?

35. **STRUCTURE** Choose two points in the coordinate plane. Use the slope formula to find the slope of the line that passes through the two points. Then find the slope using the formula $\frac{y_1 - y_2}{x_1 - x_2}$. Explain why your results are the same.

36. Critical Thinking The top and the bottom of the slide are level with the ground, which has a slope of 0.

   a. What is the slope of the main portion of the slide?

   b. How does the slope change when the bottom of the slide is only 12 inches above the ground? Is the slide steeper? Explain.

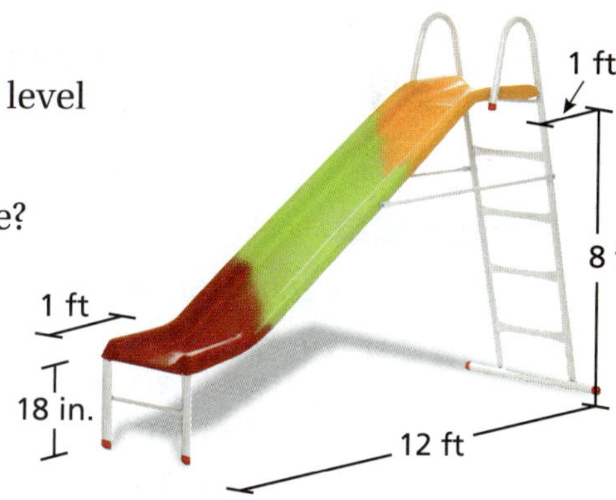

Fair Game Review What you learned in previous grades & lessons

Solve the proportion. *(Skills Review Handbook)*

37. $\frac{b}{30} = \frac{5}{6}$

38. $\frac{7}{4} = \frac{n}{32}$

39. $\frac{3}{8} = \frac{x}{20}$

40. **MULTIPLE CHOICE** What is the prime factorization of 84? *(Skills Review Handbook)*

   (A) $2 \times 3 \times 7$ (B) $2^2 \times 3 \times 7$ (C) $2 \times 3^2 \times 7$ (D) $2^2 \times 21$

Slopes of Parallel and Perpendicular Lines

Check It Out
Lesson Tutorials
BigIdeasMath.com

Key Idea

Common Core

Graphing Equations

In this extension, you will

- identify parallel and perpendicular lines.

Applying Standard MACC.8.EE.2.6

Parallel Lines and Slopes

Lines in the same plane that do not intersect are parallel lines. Nonvertical parallel lines have the same slope.

All vertical lines are parallel.

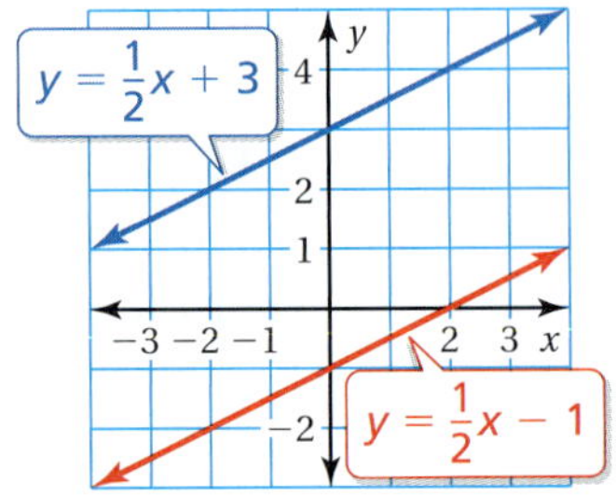

EXAMPLE 1 Identifying Parallel Lines

Which two lines are parallel? How do you know?

Find the slope of each line.

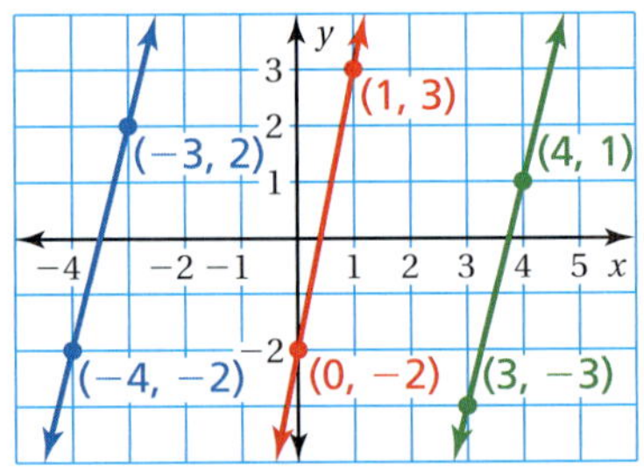

Blue Line

$$m = \frac{y_2 - y_1}{x_2 - x_1} = \frac{-2 - 2}{-4 - (-3)} = \frac{-4}{-1}, \text{ or } 4$$

Red Line

$$m = \frac{y_2 - y_1}{x_2 - x_1} = \frac{-2 - 3}{0 - 1} = \frac{-5}{-1}, \text{ or } 5$$

Green Line

$$m = \frac{y_2 - y_1}{x_2 - x_1} = \frac{-3 - 1}{3 - 4} = \frac{-4}{-1}, \text{ or } 4$$

The slopes of the blue and green lines are 4. The slope of the red line is 5.

∴ The blue and green lines have the same slope, so they are parallel.

Practice

Which lines are parallel? How do you know?

1. 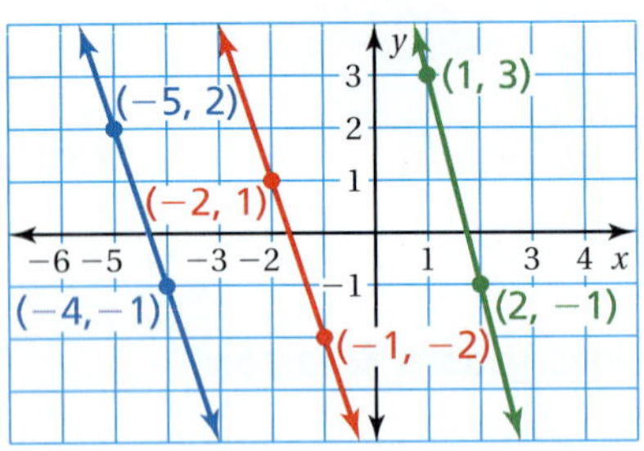

2. 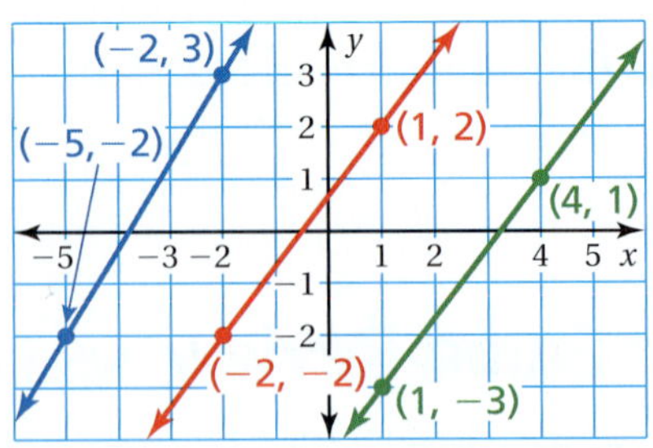

Are the given lines parallel? Explain your reasoning.

3. $y = -5, y = 3$

4. $y = 0, x = 0$

5. $x = -4, x = 1$

6. **GEOMETRY** The vertices of a quadrilateral are $A(-5, 3)$, $B(2, 2)$, $C(4, -3)$, and $D(-2, -2)$. How can you use slope to determine whether the quadrilateral is a parallelogram? Is it a parallelogram? Justify your answer.

Key Idea

Perpendicular Lines and Slope

Lines in the same plane that intersect at right angles are perpendicular lines. Two nonvertical lines are perpendicular when the product of their slopes is -1.

Vertical lines are perpendicular to horizontal lines.

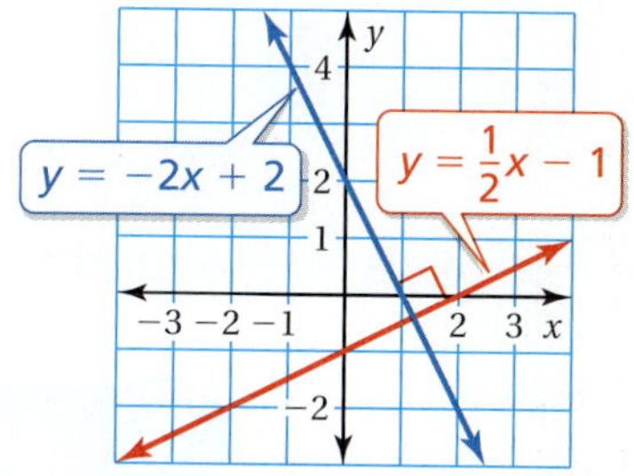

EXAMPLE 2 Identifying Perpendicular Lines

Which two lines are perpendicular? How do you know?

Find the slope of each line.

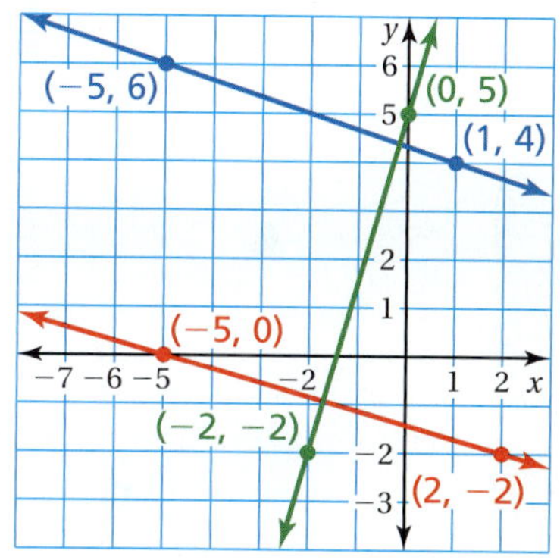

Blue Line

$$m = \frac{y_2 - y_1}{x_2 - x_1} = \frac{4 - 6}{1 - (-5)} = \frac{-2}{6}, \text{ or } -\frac{1}{3}$$

Red Line

$$m = \frac{y_2 - y_1}{x_2 - x_1} = \frac{-2 - 0}{2 - (-5)} = -\frac{2}{7}$$

Green Line

$$m = \frac{y_2 - y_1}{x_2 - x_1} = \frac{5 - (-2)}{0 - (-2)} = \frac{7}{2}$$

The slope of the red line is $-\frac{2}{7}$. The slope of the green line is $\frac{7}{2}$.

Because $-\frac{2}{7} \cdot \frac{7}{2} = -1$, the red and green lines are perpendicular.

Practice

Which lines are perpendicular? How do you know?

7. 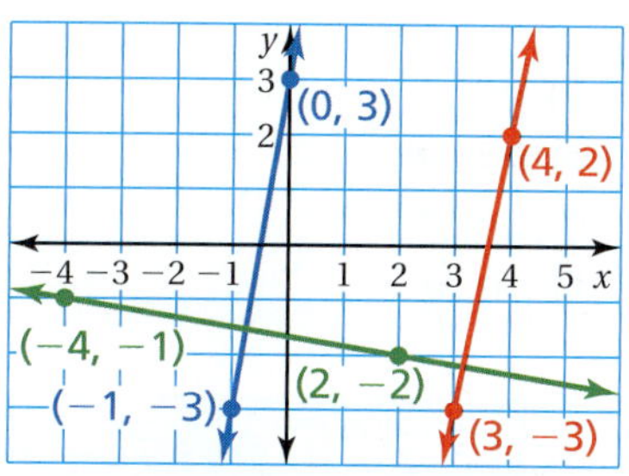

8. 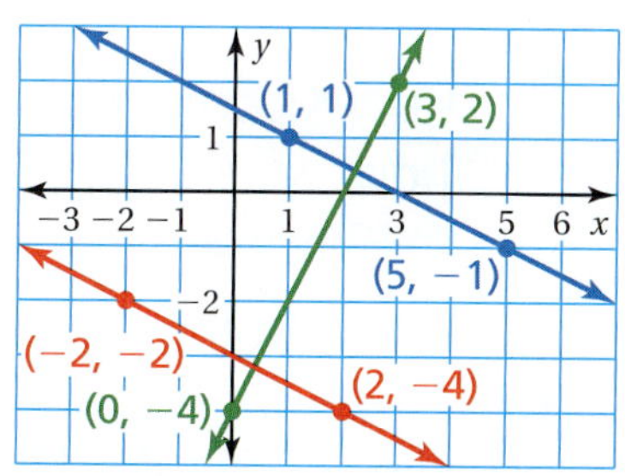

Are the given lines perpendicular? Explain your reasoning.

9. $x = -2, y = 8$

10. $x = -8, x = 7$

11. $y = 0, x = 0$

12. **GEOMETRY** The vertices of a parallelogram are $J(-5, 0)$, $K(1, 4)$, $L(3, 1)$, and $M(-3, -3)$. How can you use slope to determine whether the parallelogram is a rectangle? Is it a rectangle? Justify your answer.

4.3 Graphing Proportional Relationships

Essential Question How can you describe the graph of the equation $y = mx$?

1 ACTIVITY: Identifying Proportional Relationships

Work with a partner. Tell whether x and y are in a proportional relationship. Explain your reasoning.

a. Money

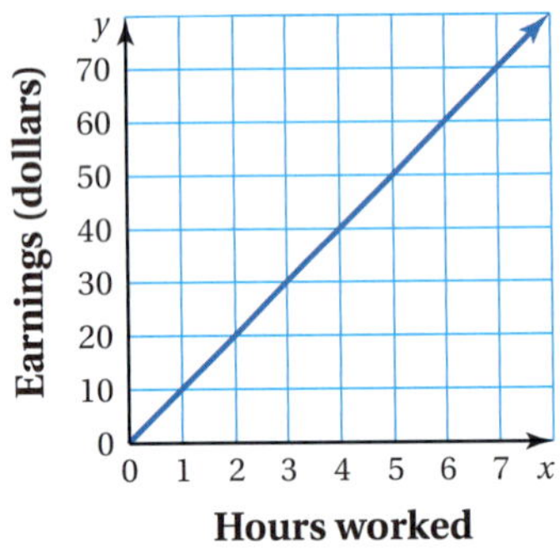

b. Helicopter

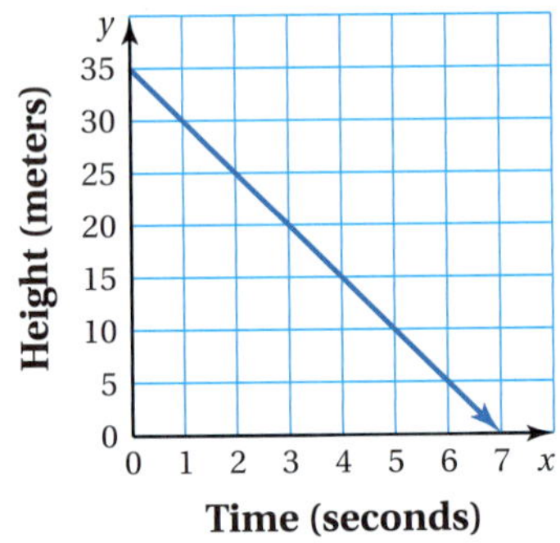

c. Tickets

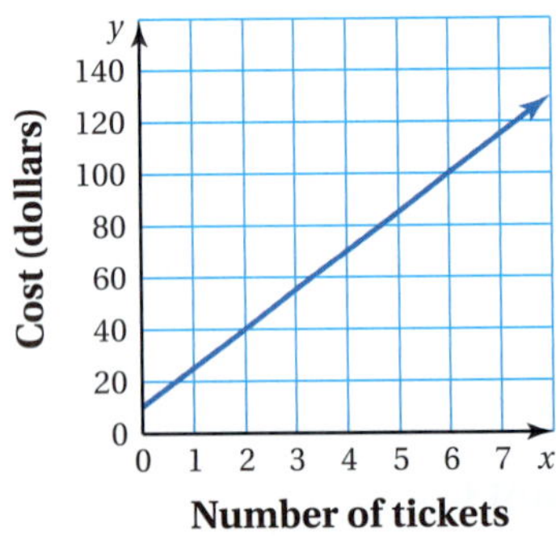

d. Pizzas

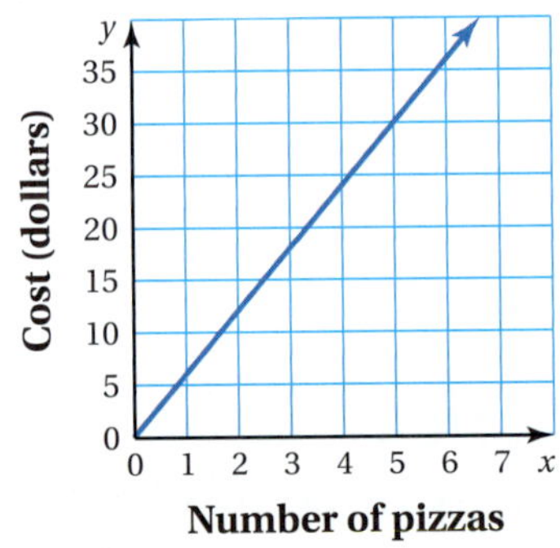

e.

Laps, x	1	2	3	4
Time (seconds), y	90	200	325	480

f.

Cups of Sugar, x	$\frac{1}{2}$	1	$1\frac{1}{2}$	2
Cups of Flour, y	1	2	3	4

COMMON CORE

Graphing Equations

In this lesson, you will

- write and graph proportional relationships.

Learning Standards
MACC.8.EE.2.5
MACC.8.EE.2.6

2 ACTIVITY: Analyzing Proportional Relationships

Work with a partner. Use only the proportional relationships in Activity 1 to do the following.

- **Find the slope of the line.**
- **Find the value of y for the ordered pair $(1, y)$.**

What do you notice? What does the value of y represent?

3 ACTIVITY: Deriving an Equation

Work with a partner. Let (x, y) represent any point on the graph of a proportional relationship.

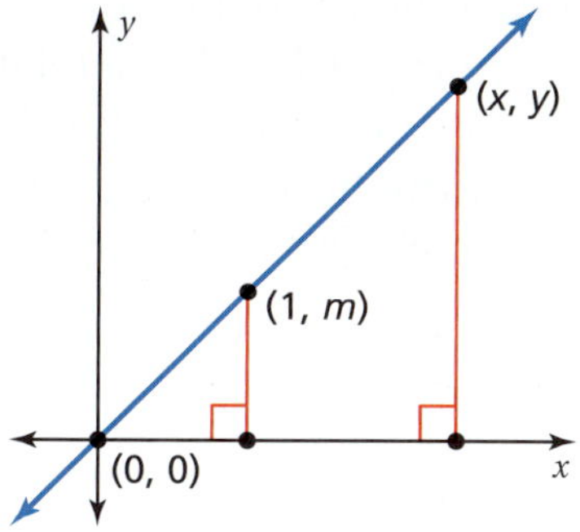

a. Explain why the two triangles are similar.

b. Because the triangles are similar, the corresponding side lengths are proportional. Use the vertical and horizontal side lengths to complete the steps below.

$$\frac{\square}{\square} = \frac{m}{1}$$ Ratios of side lengths

$$\frac{\square}{\square} = m$$ Simplify.

$$\square = m \cdot \square$$ Multiplication Property of Equality

What does the final equation represent?

c. Use your result in part (b) to write an equation that represents each proportional relationship in Activity 1.

Math Practice 7

View as Components

What part of the graph can you use to find the side lengths?

What Is Your Answer?

4. **IN YOUR OWN WORDS** How can you describe the graph of the equation $y = mx$? How does the value of m affect the graph of the equation?

5. Give a real-life example of two quantities that are in a proportional relationship. Write an equation that represents the relationship and sketch its graph.

Practice Use what you learned about proportional relationships to complete Exercises 3–6 on page 162.

4.3 Lesson

Check It Out
Lesson Tutorials
BigIdeasMath.com

Key Idea

Direct Variation

Words When two quantities x and y are proportional, the relationship can be represented by the direct variation equation $y = mx$, where m is the constant of proportionality.

Graph The graph of $y = mx$ is a line with a slope of m that passes through the origin.

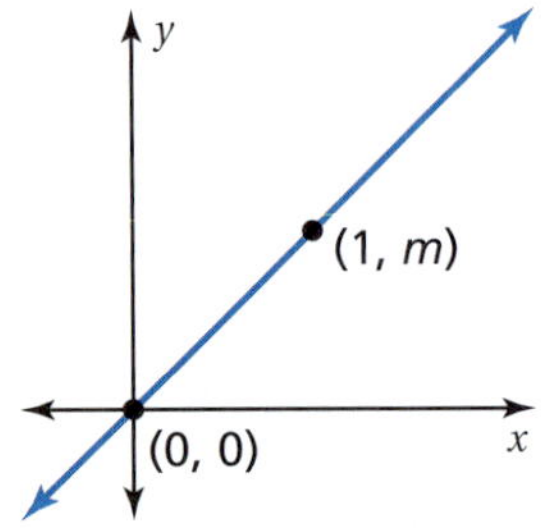

Study Tip

In the direct variation equation $y = mx$, m represents the constant of proportionality, the slope, and the unit rate.

EXAMPLE 1 Graphing a Proportional Relationship

The cost y (in dollars) for x gigabytes of data on an Internet plan is represented by $y = 10x$. Graph the equation and interpret the slope.

The equation shows that the slope m is 10. So, the graph passes through (0, 0) and (1, 10).

Plot the points and draw a line through the points. Because negative values of x do not make sense in this context, graph in the first quadrant only.

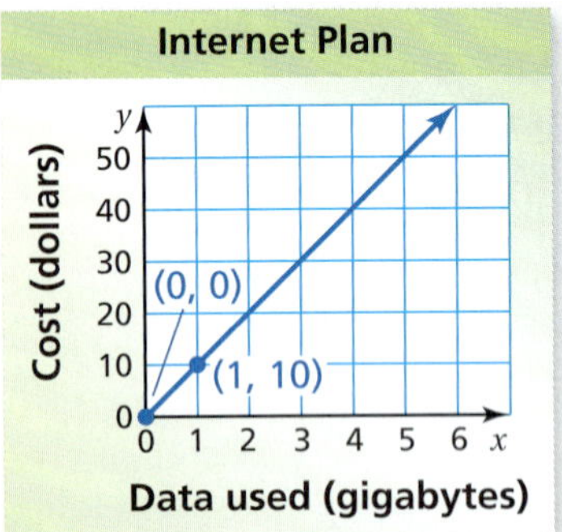

∴ The slope indicates that the unit cost is \$10 per gigabyte.

EXAMPLE 2 Writing and Using a Direct Variation Equation

The weight y of an object on Titan, one of Saturn's moons, is proportional to the weight x of the object on Earth. An object that weighs 105 pounds on Earth would weigh 15 pounds on Titan.

a. Write an equation that represents the situation.

Use the point (105, 15) to find the slope of the line.

$y = mx$	Direct variation equation
$15 = m(105)$	Substitute 15 for y and 105 for x.
$\frac{1}{7} = m$	Simplify.

∴ So, an equation that represents the situation is $y = \frac{1}{7}x$.

Study Tip

In Example 2, the slope indicates that the weight of an object on Titan is one-seventh its weight on Earth.

b. How much would a chunk of ice that weighs 3.5 pounds on Titan weigh on Earth?

$3.5 = \frac{1}{7}x$	Substitute 3.5 for y.
$24.5 = x$	Multiply each side by 7.

∴ So, the chunk of ice would weigh 24.5 pounds on Earth.

On Your Own

Now You're Ready
Exercises 7–8

1. **WHAT IF?** In Example 1, the cost is represented by $y = 12x$. Graph the equation and interpret the slope.

2. In Example 2, how much would a spacecraft that weighs 3500 kilograms on Earth weigh on Titan?

EXAMPLE 3 Comparing Proportional Relationships

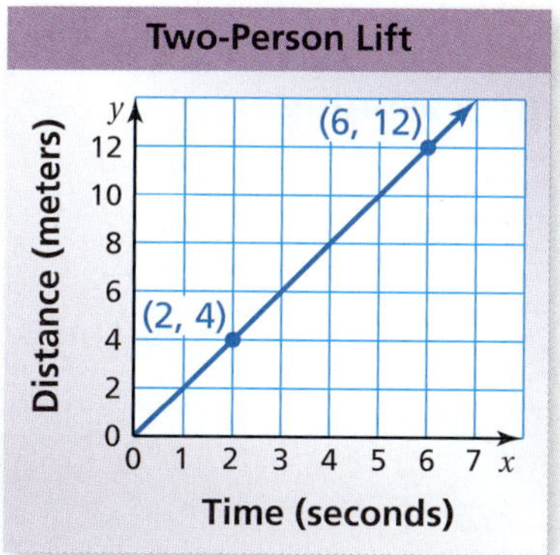

The distance y (in meters) that a four-person ski lift travels in x seconds is represented by the equation $y = 2.5x$. The graph shows the distance that a two-person ski lift travels.

a. Which ski lift is faster?

Interpret each slope as a unit rate.

Four-Person Lift

$y = 2.5x$

The slope is 2.5.

The four-person lift travels 2.5 meters per second.

Two-Person Lift

$$\text{slope} = \frac{\text{change in } y}{\text{change in } x} = \frac{8}{4} = 2$$

The two-person lift travels 2 meters per second.

So, the four-person lift is faster than the two-person lift.

b. Graph the equation that represents the four-person lift in the same coordinate plane as the two-person lift. Compare the steepness of the graphs. What does this mean in the context of the problem?

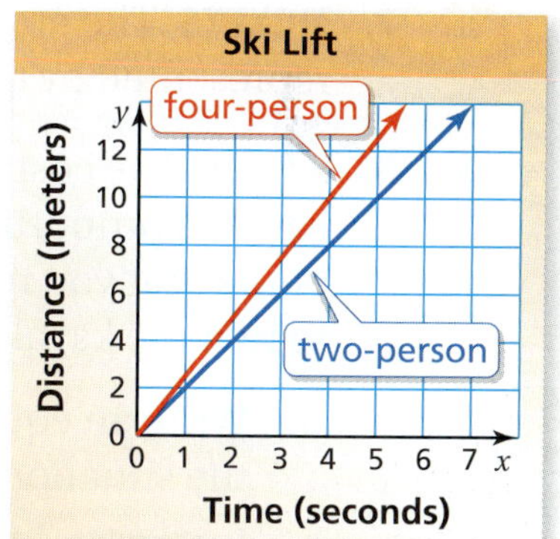

The graph that represents the four-person lift is steeper than the graph that represents the two-person lift. So, the four-person lift is faster.

On Your Own

Now You're Ready
Exercise 9

3. The table shows the distance y (in meters) that a T-bar ski lift travels in x seconds. Compare its speed to the ski lifts in Example 3.

x (seconds)	1	2	3	4
y (meters)	$2\frac{1}{4}$	$4\frac{1}{2}$	$6\frac{3}{4}$	9

4.3 Exercises

Check It Out
Help with Homework
BigIdeasMath.com

Vocabulary and Concept Check

1. **VOCABULARY** What point is on the graph of every direct variation equation?

2. **REASONING** Does the equation $y = 2x + 3$ represent a proportional relationship? Explain.

Practice and Problem Solving

Tell whether x and y are in a proportional relationship. Explain your reasoning. If so, write an equation that represents the relationship.

3. 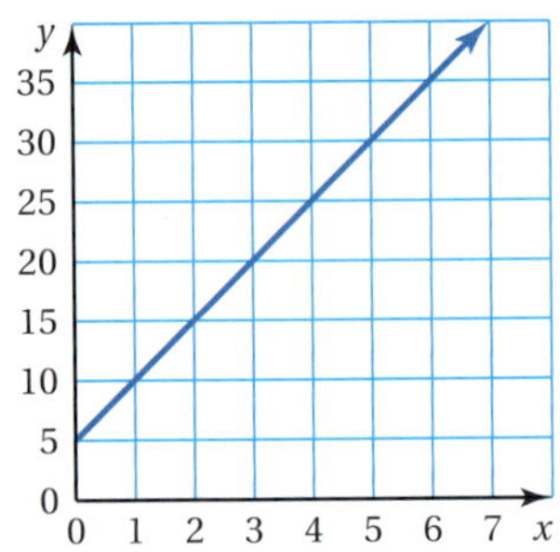

4. 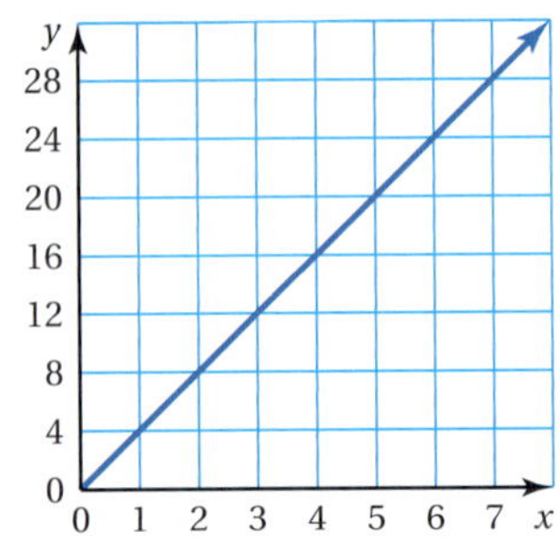

5.

x	3	6	9	12
y	1	2	3	4

6.

x	2	5	8	10
y	4	8	13	23

① 7. **TICKETS** The amount y (in dollars) that you raise by selling x fundraiser tickets is represented by the equation $y = 5x$. Graph the equation and interpret the slope.

② 8. **KAYAK** The cost y (in dollars) to rent a kayak is proportional to the number x of hours that you rent the kayak. It costs \$27 to rent the kayak for 3 hours.

   a. Write an equation that represents the situation.

   b. Interpret the slope.

   c. How much does it cost to rent the kayak for 5 hours?

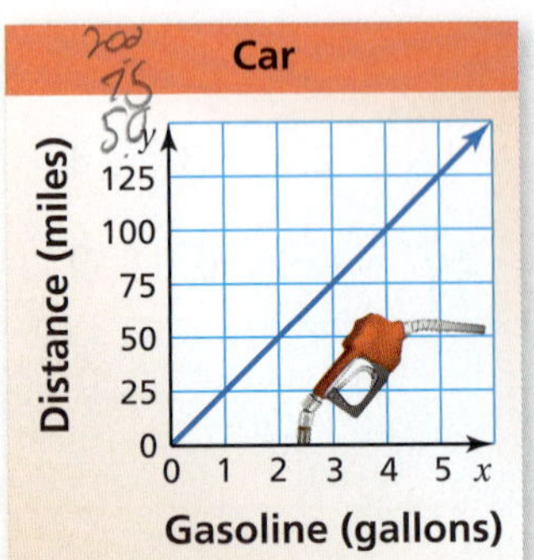

③ 9. **MILEAGE** The distance y (in miles) that a truck travels on x gallons of gasoline is represented by the equation $y = 18x$. The graph shows the distance that a car travels.

   a. Which vehicle gets better gas mileage? Explain how you found your answer.

   b. How much farther can the vehicle you chose in part (a) travel than the other vehicle on 8 gallons of gasoline?

10. **BIOLOGY** Toenails grow about 13 millimeters per year. The table shows fingernail growth.

Weeks	1	2	3	4
Fingernail Growth (millimeters)	0.7	1.4	2.1	2.8

a. Do fingernails or toenails grow faster? Explain.

b. In the same coordinate plane, graph equations that represent the growth rates of toenails and fingernails. Compare the steepness of the graphs. What does this mean in the context of the problem?

11. **REASONING** The quantities x and y are in a proportional relationship. What do you know about the ratio of y to x for any point (x, y) on the line?

12. **PROBLEM SOLVING** The graph relates the temperature change y (in degrees Fahrenheit) to the altitude change x (in thousands of feet).

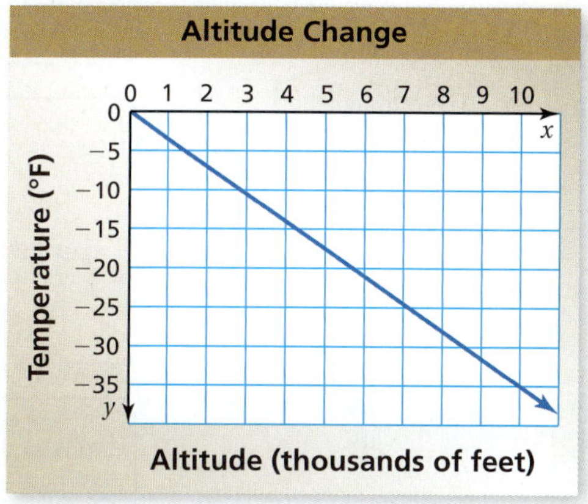

a. Is the relationship proportional? Explain.

b. Write an equation of the line. Interpret the slope.

c. You are at the bottom of a mountain where the temperature is 74°F. The top of the mountain is 5500 feet above you. What is the temperature at the top of the mountain?

13. **Critical Thinking** Consider the distance equation $d = rt$, where d is the distance (in feet), r is the rate (in feet per second), and t is the time (in seconds).

a. You run 6 feet per second. Are distance and time proportional? Explain. Graph the equation.

b. You run for 50 seconds. Are distance and rate proportional? Explain. Graph the equation.

c. You run 300 feet. Are rate and time proportional? Explain. Graph the equation.

d. One of these situations represents *inverse variation*. Which one is it? Why do you think it is called inverse variation?

Fair Game Review What you learned in previous grades & lessons

Graph the linear equation. *(Section 4.1)*

14. $y = -\frac{1}{2}x$

15. $y = 3x - \frac{3}{4}$

16. $y = -\frac{x}{3} - \frac{3}{2}$

17. **MULTIPLE CHOICE** What is the value of x? *(Section 3.3)*

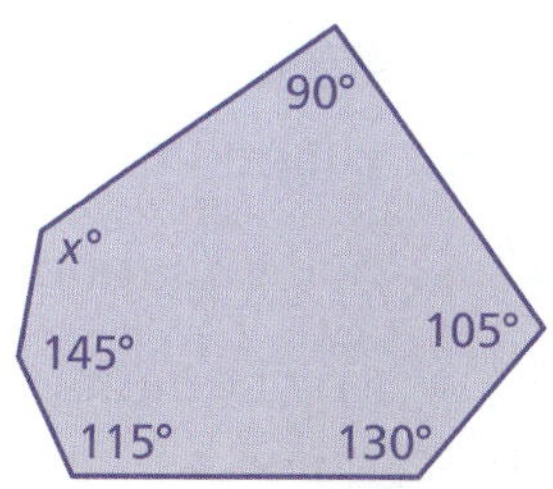

Ⓐ 110 Ⓑ 135

Ⓒ 315 Ⓓ 522

4 Study Help

Check It Out
Graphic Organizer
BigIdeasMath.com

You can use a **process diagram** to show the steps involved in a procedure. Here is an example of a process diagram for graphing a linear equation.

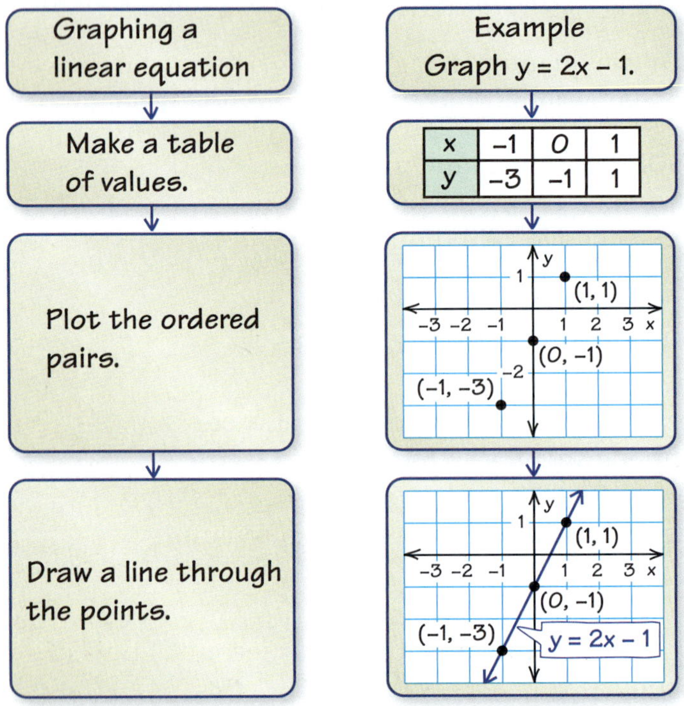

On Your Own

Make process diagrams with examples to help you study these topics.

1. finding the slope of a line
2. graphing a proportional relationship

After you complete this chapter, make process diagrams for the following topics.

3. graphing a linear equation using
   a. slope and y-intercept
   b. x- and y-intercepts
4. writing equations in slope-intercept form
5. writing equations in point-slope form

"Here is a process diagram with suggestions for what to do if a hyena knocks on your door."

4.1–4.3 Quiz

Check It Out
Progress Check
BigIdeasMath.com

Graph the linear equation. *(Section 4.1)*

1. $y = -x + 8$ **2.** $y = \frac{x}{3} - 4$ **3.** $x = -1$ **4.** $y = 3.5$

Find the slope of the line. *(Section 4.2)*

5.

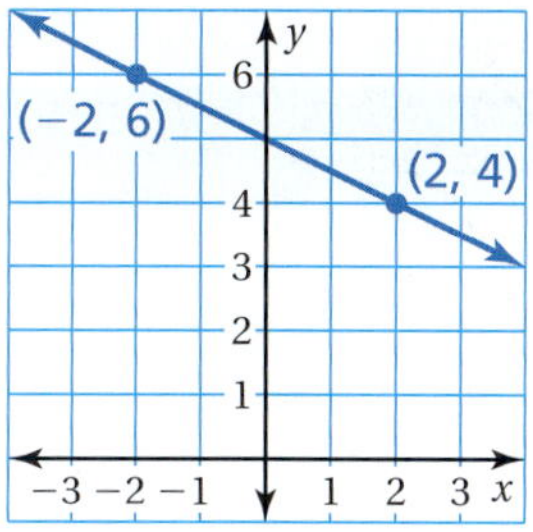

6.

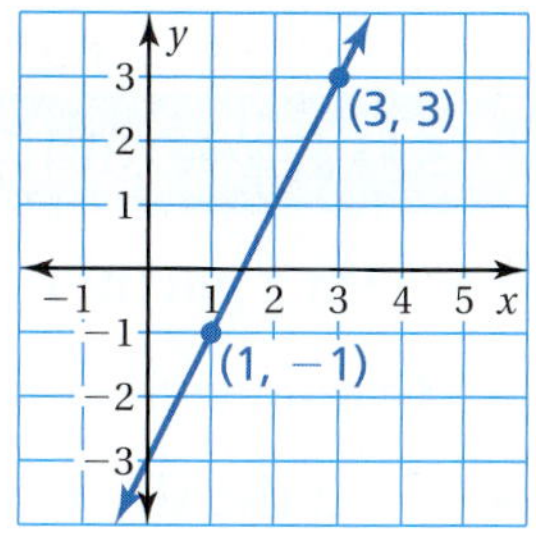

7.

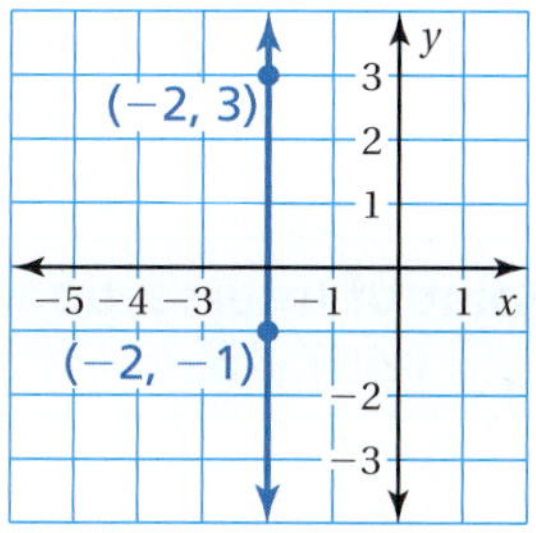

8.

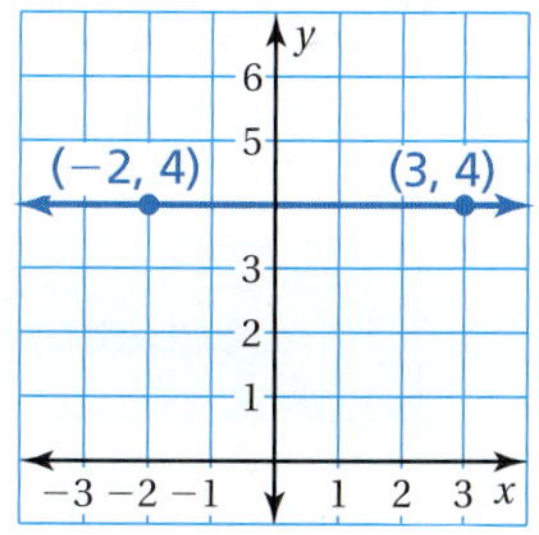

9. What is the slope of a line that is parallel to the line in Exercise 5? What is the slope of a line that is perpendicular to the line in Exercise 5? *(Section 4.2)*

10. Are the lines $y = -1$ and $x = 1$ parallel? Are they perpendicular? Justify your answer. *(Section 4.2)*

11. **BANKING** A bank charges \$3 each time you use an out-of-network ATM. At the beginning of the month, you have \$1500 in your bank account. You withdraw \$60 from your bank account each time you use an out-of-network ATM. Graph a linear equation that represents the balance in your account after you use an out-of-network ATM x times. *(Section 4.1)*

12. **MUSIC** The number y of hours of cello lessons that you take after x weeks is represented by the equation $y = 3x$. Graph the equation and interpret the slope. *(Section 4.3)*

13. **DINNER PARTY** The cost y (in dollars) to provide food for guests at a dinner party is proportional to the number x of guests attending the party. It costs \$30 to provide food for 4 guests. *(Section 4.3)*

a. Write an equation that represents the situation.

b. Interpret the slope.

c. How much does it cost to provide food for 10 guests?

4.4 Graphing Linear Equations in Slope-Intercept Form

Essential Question How can you describe the graph of the equation $y = mx + b$?

1 ACTIVITY: Analyzing Graphs of Lines

Work with a partner.

- **Graph each equation.**
- **Find the slope of each line.**
- **Find the point where each line crosses the *y*-axis.**
- **Complete the table.**

Equation	Slope of Graph	Point of Intersection with *y*-axis
a. $y = -\frac{1}{2}x + 1$		
b. $y = -x + 2$		
c. $y = -x - 2$		
d. $y = \frac{1}{2}x + 1$		
e. $y = x + 2$		
f. $y = x - 2$		
g. $y = \frac{1}{2}x - 1$		
h. $y = -\frac{1}{2}x - 1$		
i. $y = 3x + 2$		
j. $y = 3x - 2$		

k. Do you notice any relationship between the slope of the graph and its equation? between the point of intersection with the *y*-axis and its equation? Compare the results with those of other students in your class.

COMMON CORE

Graphing Equations

In this lesson, you will

- find slopes and *y*-intercepts of graphs of linear equations.
- graph linear equations written in slope-intercept form.

Learning Standard
MACC.8.EE.2.6

2 ACTIVITY: Deriving an Equation

Work with a partner.

a. Look at the graph of each equation in Activity 1. Do any of the graphs represent a proportional relationship? Explain.

b. For a nonproportional linear relationship, the graph crosses the y-axis at some point $(0, b)$, where b does not equal 0. Let (x, y) represent any other point on the graph. You can use the formula for slope to write the equation for a nonproportional linear relationship.

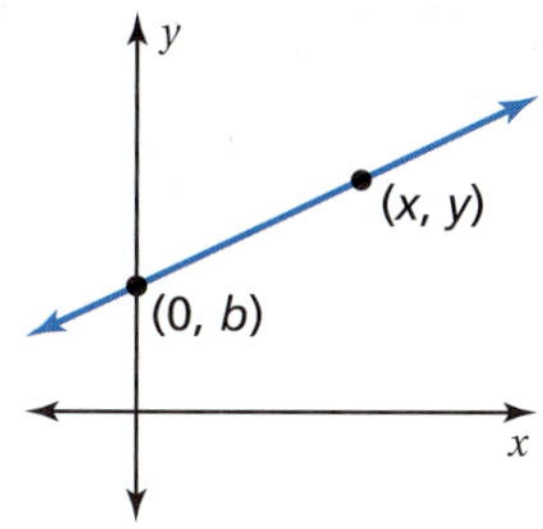

Use the graph to complete the steps.

$\dfrac{y_2 - y_1}{x_2 - x_1} = m$	Slope formula
$\dfrac{y - \square}{x - \square} = m$	Substitute values.
$\dfrac{\square}{\square} = m$	Simplify.
$\dfrac{\square}{\square} \cdot \square = m \cdot \square$	Multiplication Property of Equality
$y - \square = m \cdot \square$	Simplify.
$y = m\square + \square$	Addition Property of Equality

c. What do m and b represent in the equation?

Math Practice 3

Use Prior Results

How can you use the results of Activity 1 to help support your answer?

What Is Your Answer?

3. IN YOUR OWN WORDS How can you describe the graph of the equation $y = mx + b$?

a. How does the value of m affect the graph of the equation?

b. How does the value of b affect the graph of the equation?

c. Check your answers to parts (a) and (b) with three equations that are not in Activity 1.

4. LOGIC Why do you think $y = mx + b$ is called the *slope-intercept form* of the equation of a line? Use drawings or diagrams to support your answer.

Practice Use what you learned about graphing linear equations in slope-intercept form to complete Exercises 4–6 on page 170.

4.4 Lesson

Check It Out
Lesson Tutorials
BigIdeasMath.com

Key Vocabulary
x-intercept, *p. 168*
y-intercept, *p. 168*
slope-intercept form, *p. 168*

Key Ideas

Intercepts

The ***x*-intercept** of a line is the x-coordinate of the point where the line crosses the x-axis. It occurs when $y = 0$.

The ***y*-intercept** of a line is the y-coordinate of the point where the line crosses the y-axis. It occurs when $x = 0$.

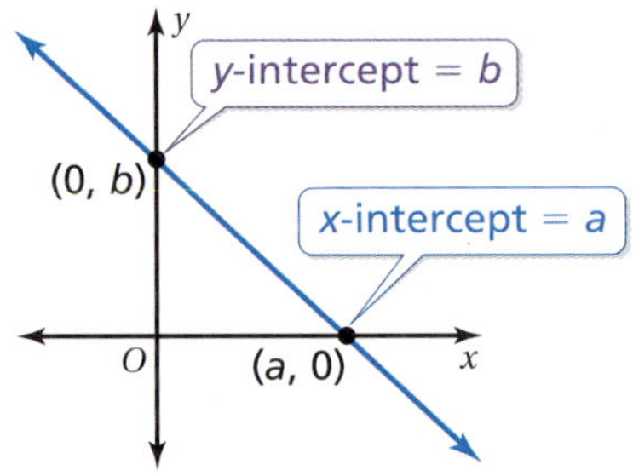

Slope-Intercept Form

Words A linear equation written in the form $y = mx + b$ is in **slope-intercept form**. The slope of the line is m, and the y-intercept of the line is b.

Algebra $y = mx + b$ (slope: m; y-intercept: b)

Study Tip

Linear equations can, but do not always, pass through the origin. So, proportional relationships are a special type of linear equation in which $b = 0$.

EXAMPLE 1 Identifying Slopes and *y*-Intercepts

Find the slope and the *y*-intercept of the graph of each linear equation.

a. $y = -4x - 2$

$y = -4x + (-2)$ Write in slope-intercept form.

The slope is -4, and the y-intercept is -2.

b. $y - 5 = \frac{3}{2}x$

$y = \frac{3}{2}x + 5$ Add 5 to each side.

The slope is $\frac{3}{2}$, and the y-intercept is 5.

On Your Own

Now You're Ready
Exercises 7–15

Find the slope and the *y*-intercept of the graph of the linear equation.

1. $y = 3x - 7$

2. $y - 1 = -\frac{2}{3}x$

Multi-Language Glossary at BigIdeasMath.com

EXAMPLE 2 Graphing a Linear Equation in Slope-Intercept Form

Graph $y = -3x + 3$. Identify the x-intercept.

Step 1: Find the slope and the y-intercept.

$$y = -3x + 3$$

slope: -3; y-intercept: 3

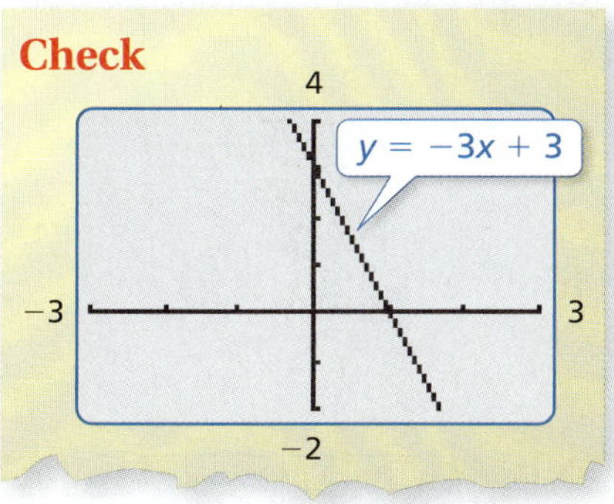

Step 2: The y-intercept is 3. So, plot (0, 3).

Step 3: Use the slope to find another point and draw the line.

$$m = \frac{\text{rise}}{\text{run}} = \frac{-3}{1}$$

Plot the point that is 1 unit right and 3 units down from (0, 3). Draw a line through the two points.

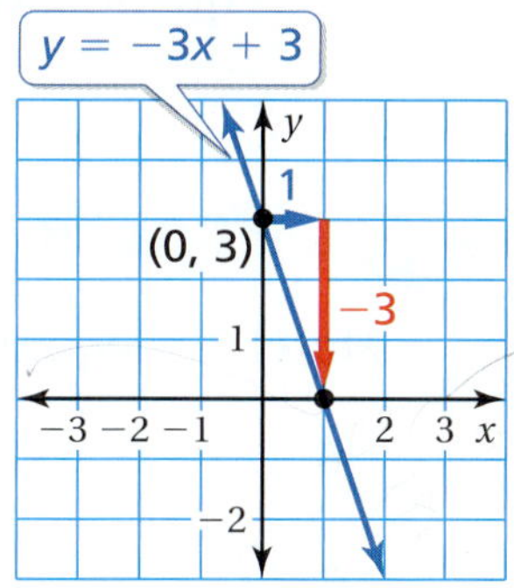

The line crosses the x-axis at (1, 0). So, the x-intercept is 1.

EXAMPLE 3 Real-Life Application

The cost y (in dollars) of taking a taxi x miles is $y = 2.5x + 2$. (a) Graph the equation. (b) Interpret the y-intercept and the slope.

a. The slope of the line is $2.5 = \frac{5}{2}$. Use the slope and the y-intercept to graph the equation.

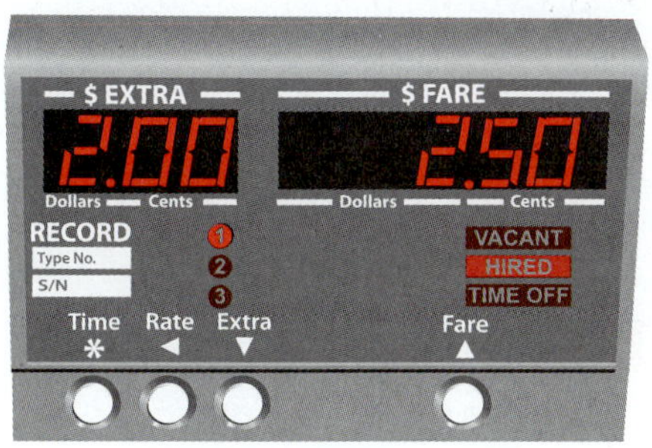

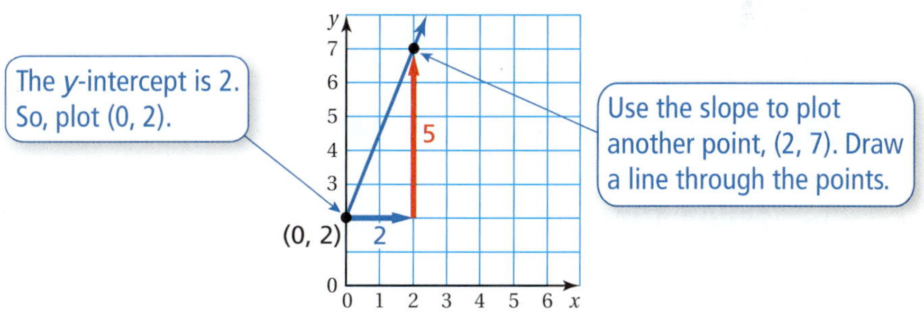

b. The slope is 2.5. So, the cost per mile is \$2.50. The y-intercept is 2. So, there is an initial fee of \$2 to take the taxi.

On Your Own

Now You're Ready
Exercises 18–23

Graph the linear equation. Identify the x-intercept. Use a graphing calculator to check your answer.

3. $y = x - 4$

4. $y = -\frac{1}{2}x + 1$

5. In Example 3, the cost y (in dollars) of taking a different taxi x miles is $y = 2x + 1.5$. Interpret the y-intercept and the slope.

4.4 Exercises

Check It Out
Help with Homework
BigIdeasMath.com

Vocabulary and Concept Check

1. **VOCABULARY** How can you find the x-intercept of the graph of $2x + 3y = 6$?
2. **CRITICAL THINKING** Is the equation $y = 3x$ in slope-intercept form? Explain.
3. **OPEN-ENDED** Describe a real-life situation that you can model with a linear equation. Write the equation. Interpret the y-intercept and the slope.

Practice and Problem Solving

Match the equation with its graph. Identify the slope and the y-intercept.

4. $y = 2x + 1$
5. $y = \frac{1}{3}x - 2$
6. $y = -\frac{2}{3}x + 1$

A.

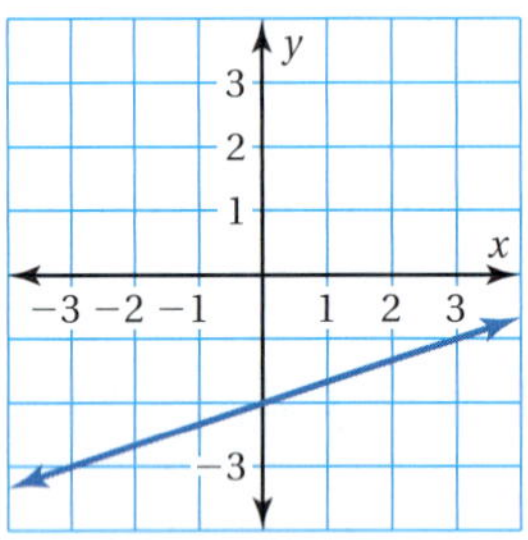

B.

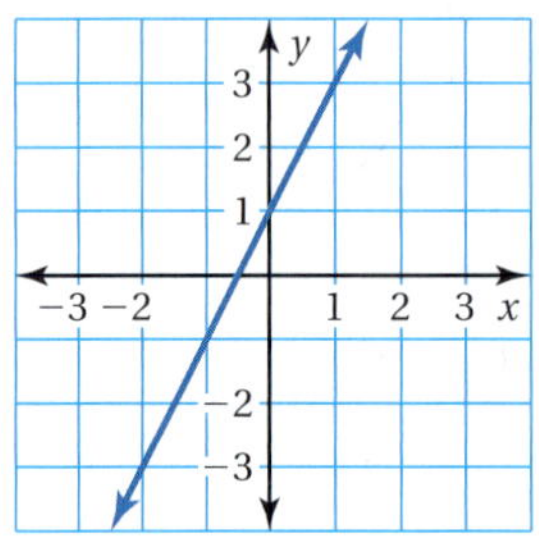

C.

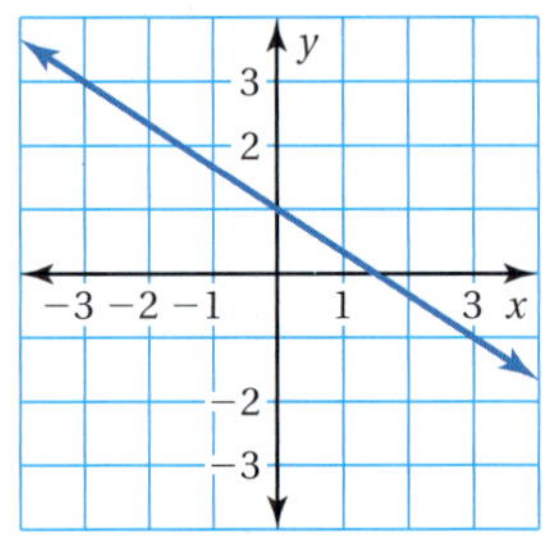

Find the slope and the y-intercept of the graph of the linear equation.

1 7. $y = 4x - 5$
8. $y = -7x + 12$
9. $y = -\frac{4}{5}x - 2$
10. $y = 2.25x + 3$
11. $y + 1 = \frac{4}{3}x$
12. $y - 6 = \frac{3}{8}x$
13. $y - 3.5 = -2x$
14. $y = -5 - \frac{1}{2}x$
15. $y = 11 + 1.5x$

16. **ERROR ANALYSIS** Describe and correct the error in finding the slope and the y-intercept of the graph of the linear equation.

✗ $y = 4x - 3$
The slope is 4, and the y-intercept is 3.

17. **SKYDIVING** A skydiver parachutes to the ground. The height y (in feet) of the skydiver after x seconds is $y = -10x + 3000$.
    a. Graph the equation.
    b. Interpret the x-intercept and the slope.

Graph the linear equation. Identify the x-intercept. Use a graphing calculator to check your answer.

2 **18.** $y = \frac{1}{5}x + 3$ **19.** $y = 6x - 7$ **20.** $y = -\frac{8}{3}x + 9$

21. $y = -1.4x - 1$ **22.** $y + 9 = -3x$ **23.** $y = 4 - \frac{3}{5}x$

24. APPLES You go to a harvest festival and pick apples.

a. Which equation represents the cost (in dollars) of going to the festival and picking x pounds of apples? Explain.

$y = 5x + 0.75$ $y = 0.75x + 5$

b. Graph the equation you chose in part (a).

25. REASONING Without graphing, identify the equations of the lines that are (a) parallel and (b) perpendicular. Explain your reasoning.

$y = 2x + 4$	$y = -\frac{1}{3}x - 1$	$y = -3x - 2$	$y = \frac{1}{2}x + 1$
$y = 3x + 3$	$y = -\frac{1}{2}x + 2$	$y = -3x + 5$	$y = 2x - 3$

26. Critical Thinking Six friends create a website. The website earns money by selling banner ads. The site has 5 banner ads. It costs \$120 a month to operate the website.

a. A banner ad earns \$0.005 per click. Write a linear equation that represents the monthly income y (in dollars) for x clicks.

b. Graph the equation in part (a). On the graph, label the number of clicks needed for the friends to start making a profit.

Fair Game Review What you learned in previous grades & lessons

Solve the equation for y. *(Section 1.4)*

27. $y - 2x = 3$ **28.** $4x + 5y = 13$ **29.** $2x - 3y = 6$ **30.** $7x + 4y = 8$

31. MULTIPLE CHOICE Which point is a solution of the equation $3x - 8y = 11$? *(Section 4.1)*

Ⓐ $(1, 1)$ Ⓑ $(1, -1)$ Ⓒ $(-1, 1)$ Ⓓ $(-1, -1)$

4.5 Graphing Linear Equations in Standard Form

Essential Question How can you describe the graph of the equation $ax + by = c$?

1 ACTIVITY: Using a Table to Plot Points

Work with a partner. You sold a total of $16 worth of tickets to a school concert. You lost track of how many of each type of ticket you sold.

$\dfrac{\square}{\text{adult}}$ • Number of adult tickets + $\dfrac{\square}{\text{student}}$ • Number of student tickets = $\square$

a. Let x represent the number of adult tickets.
Let y represent the number of student tickets.
Write an equation that relates x and y.

b. Copy and complete the table showing the different combinations of tickets you might have sold.

Number of Adult Tickets, x					
Number of Student Tickets, y					

c. Plot the points from the table. Describe the pattern formed by the points.

d. If you remember how many adult tickets you sold, can you determine how many student tickets you sold? Explain your reasoning.

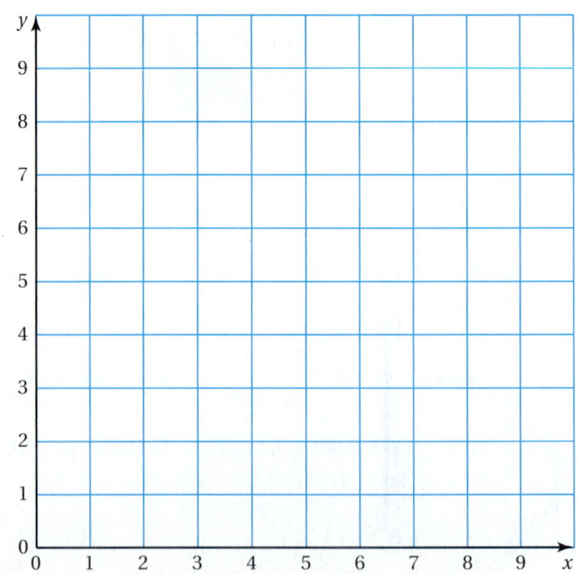

COMMON CORE
Graphing Equations
In this lesson, you will
- graph linear equations written in standard form.

Applying Standard
MACC.8.EE.2.6

2 ACTIVITY: Rewriting an Equation

Work with a partner. You sold a total of $16 worth of cheese. You forgot how many pounds of each type of cheese you sold.

$\frac{\square}{\text{pound}}$ • Pounds of swiss + $\frac{\square}{\text{pound}}$ • Pounds of cheddar = $\square$

Math Practice 2

Understand Quantities

What do the equation and the graph represent? How can you use this information to solve the problem?

a. Let x represent the number of pounds of swiss cheese.

Let y represent the number of pounds of cheddar cheese.

Write an equation that relates x and y.

b. Rewrite the equation in slope-intercept form. Then graph the equation.

c. You sold 2 pounds of cheddar cheese. How many pounds of swiss cheese did you sell?

d. Does the value $x = 2.5$ make sense in the context of the problem? Explain.

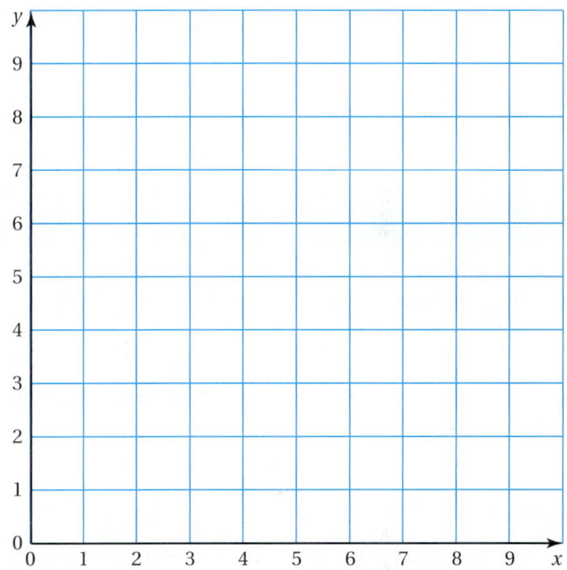

What Is Your Answer?

3. **IN YOUR OWN WORDS** How can you describe the graph of the equation $ax + by = c$?

4. Activities 1 and 2 show two different methods for graphing $ax + by = c$. Describe the two methods. Which method do you prefer? Explain.

5. Write a real-life problem that is similar to those shown in Activities 1 and 2.

6. Why do you think it might be easier to graph $x + y = 10$ without rewriting it in slope-intercept form and then graphing?

Practice Use what you learned about graphing linear equations in standard form to complete Exercises 3 and 4 on page 176.

4.5 Lesson

Check It Out
Lesson Tutorials
BigIdeasMath.com

Key Vocabulary
standard form, *p. 174*

Study Tip
Any linear equation can be written in standard form.

Key Idea

Standard Form of a Linear Equation

The **standard form** of a linear equation is

$$ax + by = c$$

where a and b are not both zero.

EXAMPLE 1 Graphing a Linear Equation in Standard Form

Graph $-2x + 3y = -6$.

Step 1: Write the equation in slope-intercept form.

$-2x + 3y = -6$	Write the equation.
$3y = 2x - 6$	Add $2x$ to each side.
$y = \frac{2}{3}x - 2$	Divide each side by 3.

Step 2: Use the slope and the y-intercept to graph the equation.

$$y = \frac{2}{3}x + (-2)$$

slope: $\frac{2}{3}$; y-intercept: -2

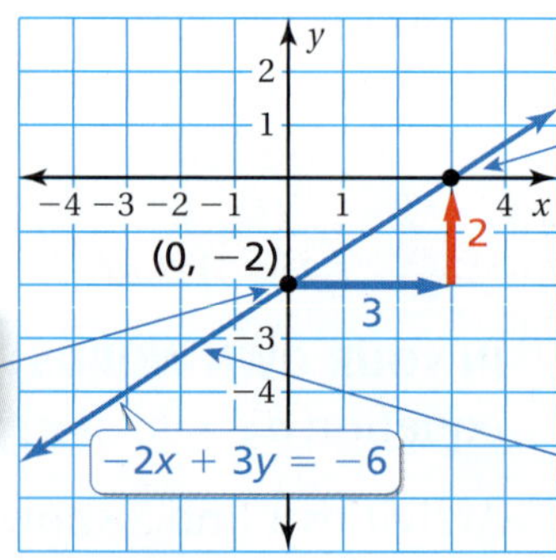

Check

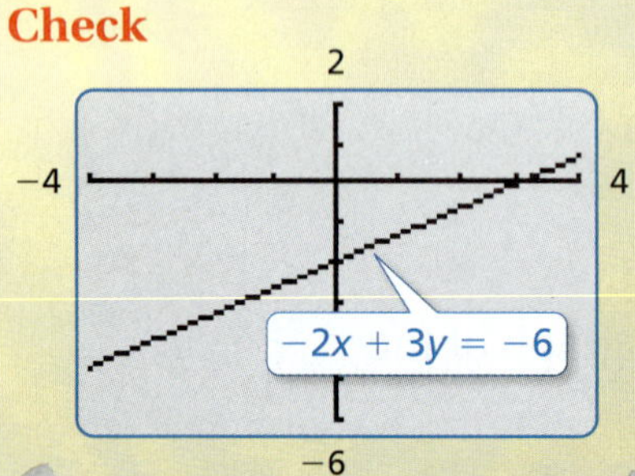

On Your Own

Now You're Ready
Exercises 5–10

Graph the linear equation. Use a graphing calculator to check your graph.

1. $x + y = -2$

2. $-\frac{1}{2}x + 2y = 6$

3. $-\frac{2}{3}x + y = 0$

4. $2x + y = 5$

Multi-Language Glossary at BigIdeasMath.com

EXAMPLE 2 Graphing a Linear Equation in Standard Form

Graph $x + 3y = -3$ using intercepts.

Step 1: To find the x-intercept, substitute 0 for y.

$$x + 3y = -3$$
$$x + 3(0) = -3$$
$$x = -3$$

To find the y-intercept, substitute 0 for x.

$$x + 3y = -3$$
$$0 + 3y = -3$$
$$y = -1$$

Step 2: Graph the equation.

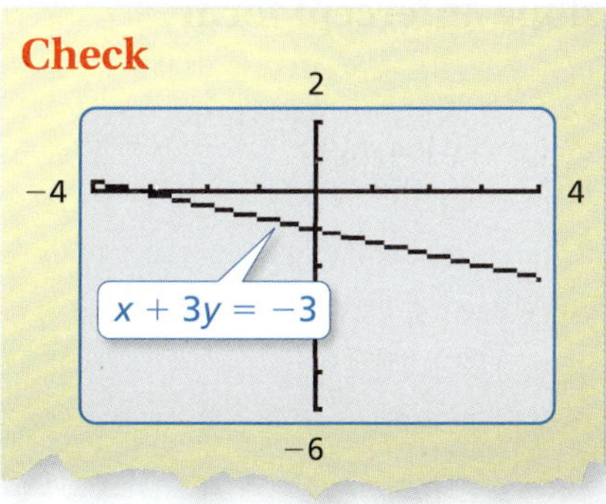

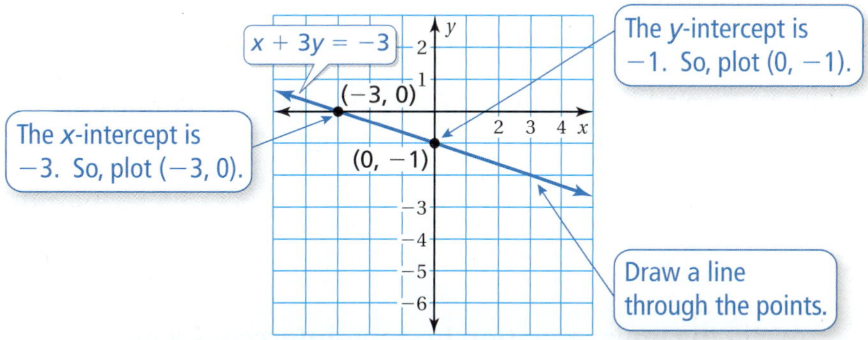

EXAMPLE 3 Real-Life Application

You have \$6 to spend on apples and bananas. (a) Graph the equation $1.5x + 0.6y = 6$, where x is the number of pounds of apples and y is the number of pounds of bananas. (b) Interpret the intercepts.

a. Find the intercepts and graph the equation.

x-intercept

$$1.5x + 0.6y = 6$$
$$1.5x + 0.6(0) = 6$$
$$x = 4$$

y-intercept

$$1.5x + 0.6y = 6$$
$$1.5(0) + 0.6y = 6$$
$$y = 10$$

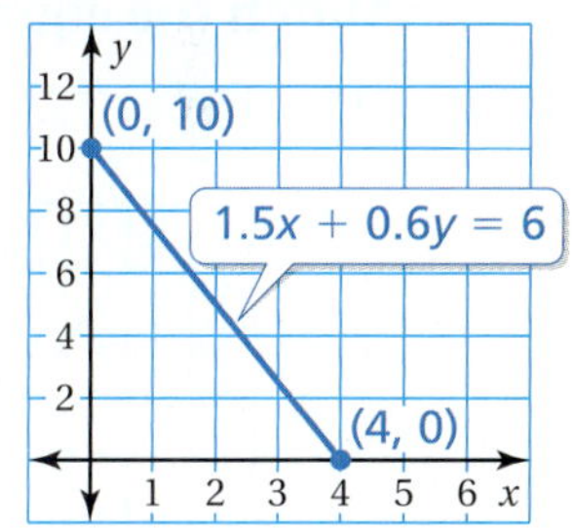

b. The x-intercept shows that you can buy 4 pounds of apples when you do not buy any bananas. The y-intercept shows that you can buy 10 pounds of bananas when you do not buy any apples.

On Your Own

Now You're Ready
Exercises 16–18

Graph the linear equation using intercepts. Use a graphing calculator to check your graph.

5. $2x - y = 8$

6. $x + 3y = 6$

7. **WHAT IF?** In Example 3, you buy y pounds of oranges instead of bananas. Oranges cost \$1.20 per pound. Graph the equation $1.5x + 1.2y = 6$. Interpret the intercepts.

4.5 Exercises

Check It Out
Help with Homework
BigIdeasMath.com

Vocabulary and Concept Check

1. **VOCABULARY** Is the equation $y = -2x + 5$ in standard form? Explain.
2. **WRITING** Describe two ways to graph the equation $4x + 2y = 6$.

Practice and Problem Solving

Define two variables for the verbal model. Write an equation in slope-intercept form that relates the variables. Graph the equation.

3. $\dfrac{\$2.00}{\text{pound}} \cdot \text{Pounds of peaches} + \dfrac{\$1.50}{\text{pound}} \cdot \text{Pounds of apples} = \$15$

4. $\dfrac{16 \text{ miles}}{\text{hour}} \cdot \text{Hours biked} + \dfrac{2 \text{ miles}}{\text{hour}} \cdot \text{Hours walked} = 32 \text{ miles}$

Write the linear equation in slope-intercept form.

(1) 5. $2x + y = 17$ 6. $5x - y = \frac{1}{4}$ 7. $-\frac{1}{2}x + y = 10$

Graph the linear equation. Use a graphing calculator to check your graph.

8. $-18x + 9y = 72$ 9. $16x - 4y = 2$ 10. $\frac{1}{4}x + \frac{3}{4}y = 1$

Match the equation with its graph.

11. $15x - 12y = 60$ 12. $5x + 4y = 20$ 13. $10x + 8y = -40$

A.

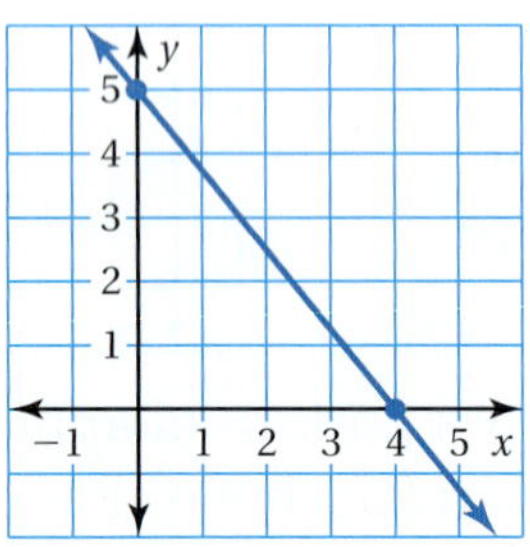

B.

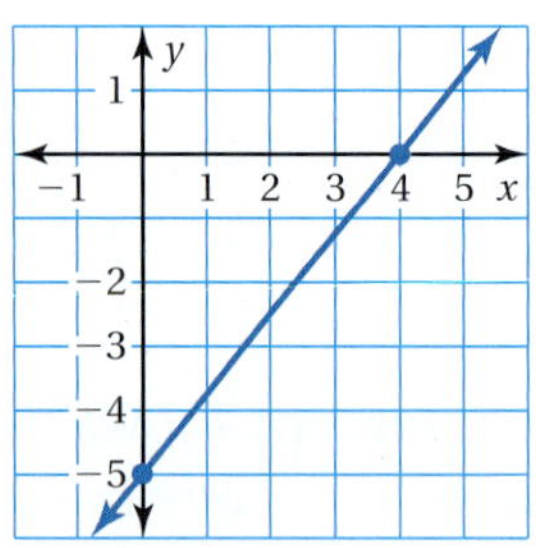

C.

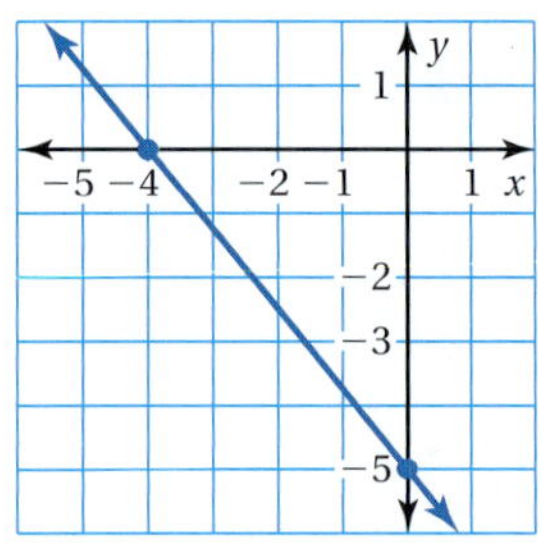

14. **ERROR ANALYSIS** Describe and correct the error in finding the x-intercept.

✗
$$-2x + 3y = 12$$
$$-2(0) + 3y = 12$$
$$3y = 12$$
$$y = 4$$

15. **BRACELET** A charm bracelet costs \$65, plus \$25 for each charm. The equation $-25x + y = 65$ represents the cost y of the bracelet, where x is the number of charms.
    a. Graph the equation.
    b. How much does the bracelet shown cost?

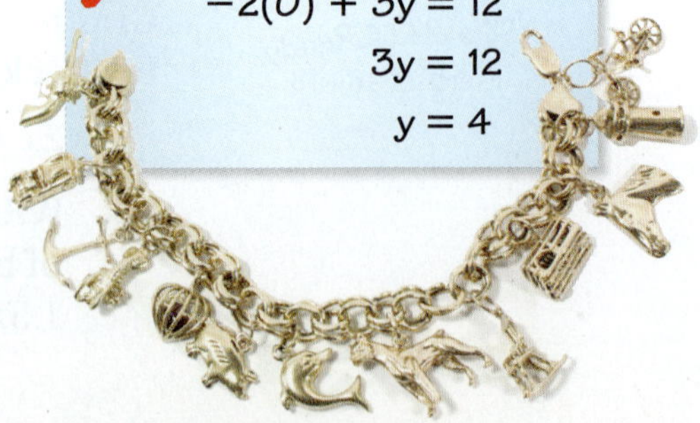

Graph the linear equation using intercepts. Use a graphing calculator to check your graph.

2 **16.** $3x - 4y = -12$ **17.** $2x + y = 8$ **18.** $\frac{1}{3}x - \frac{1}{6}y = -\frac{2}{3}$

19. **SHOPPING** The amount of money you spend on x CDs and y DVDs is given by the equation $14x + 18y = 126$. Find the intercepts and graph the equation.

20. **SCUBA** Five friends go scuba diving. They rent a boat for x days and scuba gear for y days. The total spent is $1000.

a. Write an equation in standard form that represents the situation.

b. Graph the equation and interpret the intercepts.

21. **MODELING** You work at a restaurant as a host and a server. You earn $9.45 for each hour you work as a host and $7.65 for each hour you work as a server.

a. Write an equation in standard form that models your earnings.

b. Graph the equation.

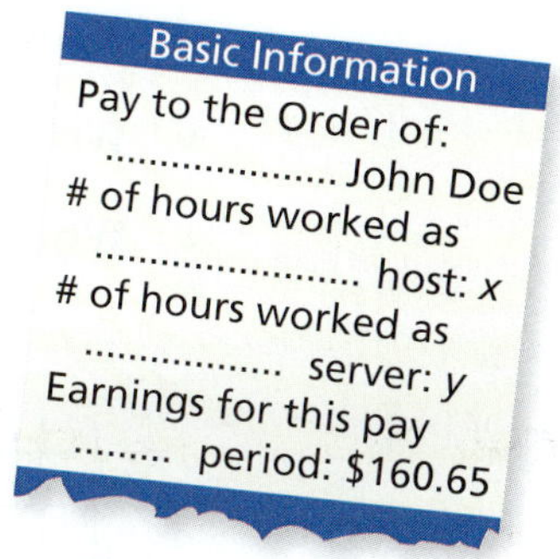

22. **LOGIC** Does the graph of every linear equation have an x-intercept? Explain your reasoning. Include an example.

23. Critical Thinking For a house call, a veterinarian charges $70, plus $40 an hour.

a. Write an equation that represents the total fee y (in dollars) the veterinarian charges for a visit lasting x hours.

b. Find the x-intercept. Does this value make sense in this context? Explain your reasoning.

c. Graph the equation.

Fair Game Review What you learned in previous grades & lessons

The points in the table lie on a line. Find the slope of the line. *(Section 4.2)*

24.

x	-2	-1	0	1
y	-10	-6	-2	2

25.

x	2	4	6	8
y	2	3	4	5

26. **MULTIPLE CHOICE** Which value of x makes the equation $4x - 12 = 3x - 9$ true? *(Section 1.3)*

Ⓐ -1 Ⓑ 0 Ⓒ 1 Ⓓ 3

4.6 Writing Equations in Slope-Intercept Form

Essential Question How can you write an equation of a line when you are given the slope and the *y*-intercept of the line?

1 ACTIVITY: Writing Equations of Lines

Work with a partner.

- **Find the slope of each line.**
- **Find the *y*-intercept of each line.**
- **Write an equation for each line.**
- **What do the three lines have in common?**

a.

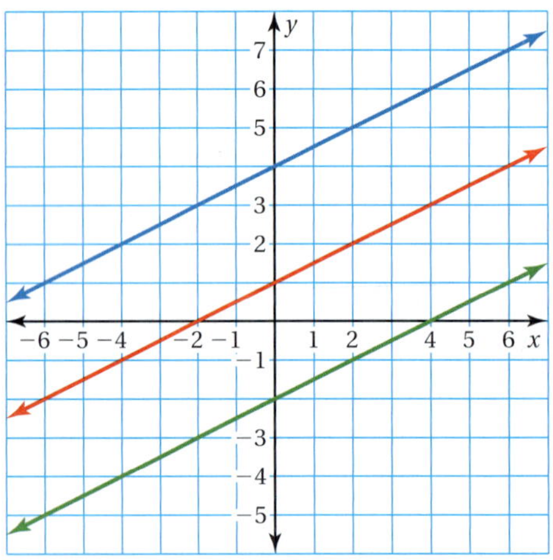

b.

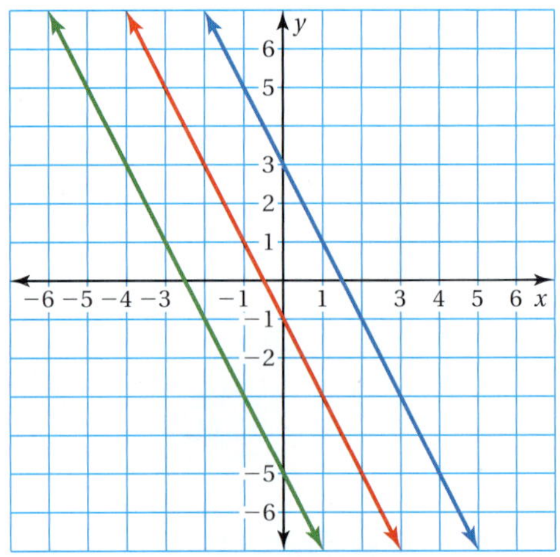

c.

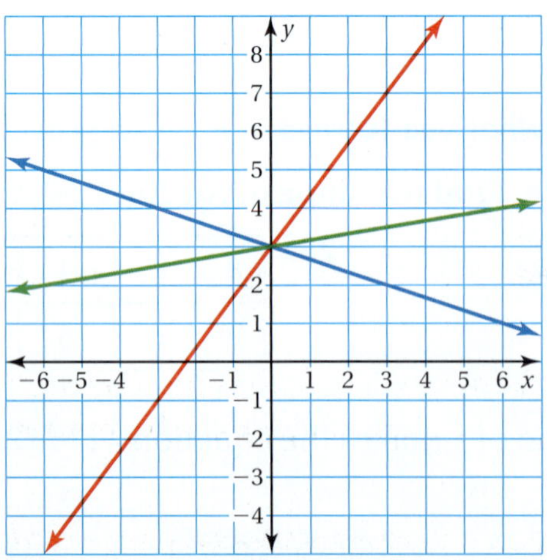

d.

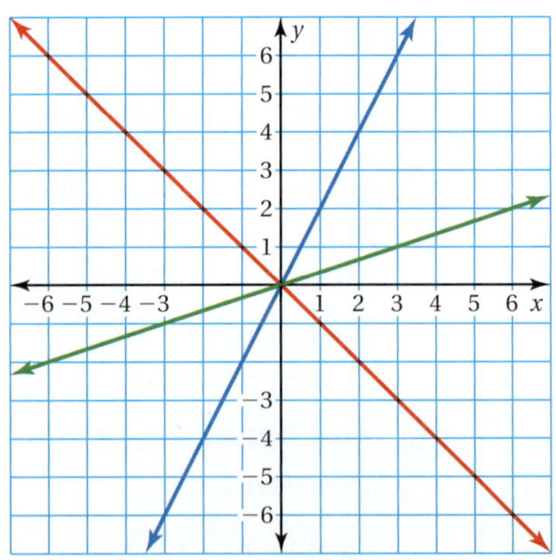

COMMON CORE

Writing Equations

In this lesson, you will

- write equations of lines in slope-intercept form.

Preparing for Standard MACC.8.F.2.4

> **Math Practice 1**
> **Analyze Givens**
> What do you need to know to write an equation?

2 ACTIVITY: Describing a Parallelogram

Work with a partner.

- **Find the area of each parallelogram.**
- **Write an equation that represents each side of each parallelogram.**

a.

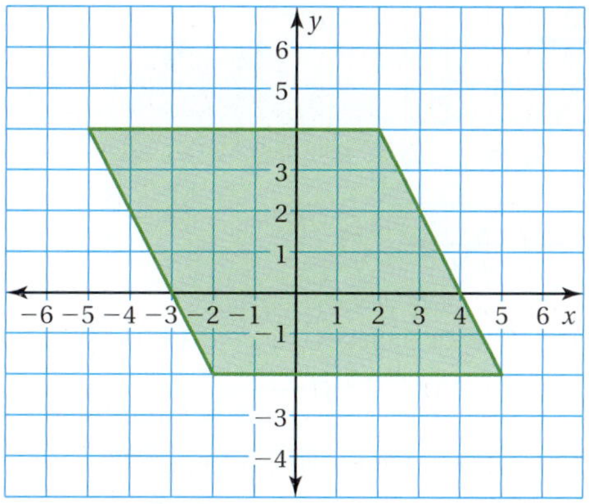

b.

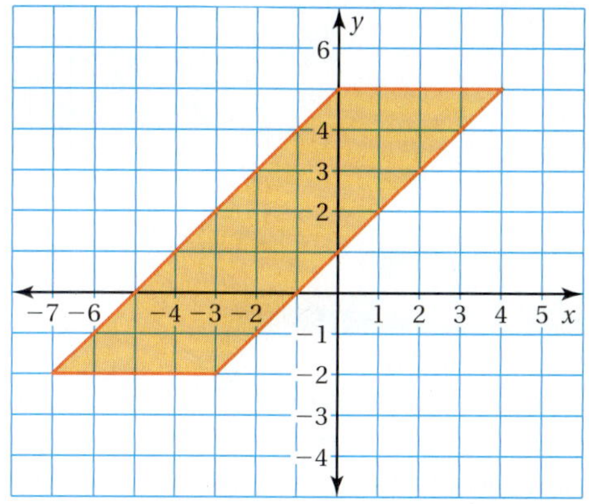

3 ACTIVITY: Interpreting the Slope and the *y*-Intercept

Work with a partner. The graph shows a trip taken by a car, where *t* is the time (in hours) and *y* is the distance (in miles) from Phoenix.

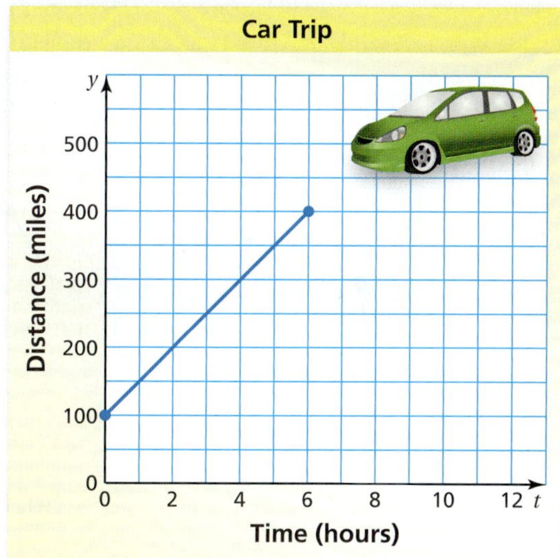

a. Find the *y*-intercept of the graph. What does it represent?

b. Find the slope of the graph. What does it represent?

c. How long did the trip last?

d. How far from Phoenix was the car at the end of the trip?

e. Write an equation that represents the graph.

What Is Your Answer?

4. IN YOUR OWN WORDS How can you write an equation of a line when you are given the slope and the *y*-intercept of the line? Give an example that is different from those in Activities 1, 2, and 3.

5. Two sides of a parallelogram are represented by the equations $y = 2x + 1$ and $y = -x + 3$. Give two equations that can represent the other two sides.

Practice Use what you learned about writing equations in slope-intercept form to complete Exercises 3 and 4 on page 182.

4.6 Lesson

Check It Out
Lesson Tutorials
BigIdeasMath.com

EXAMPLE 1 Writing Equations in Slope-Intercept Form

Write an equation of the line in slope-intercept form.

a.

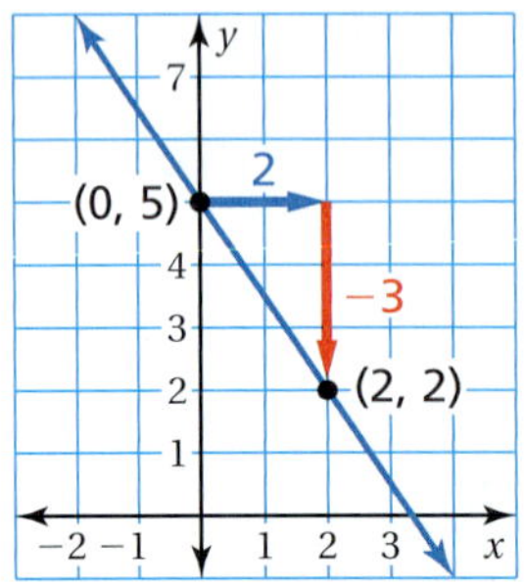

Find the slope and the *y*-intercept.

$$m = \frac{y_2 - y_1}{x_2 - x_1} = \frac{2 - 5}{2 - 0} = \frac{-3}{2}, \text{ or } -\frac{3}{2}$$

Because the line crosses the *y*-axis at (0, 5), the *y*-intercept is 5.

slope y-intercept

So, the equation is $y = -\frac{3}{2}x + 5$.

Study Tip

After writing an equation, check that the given points are solutions of the equation.

b.

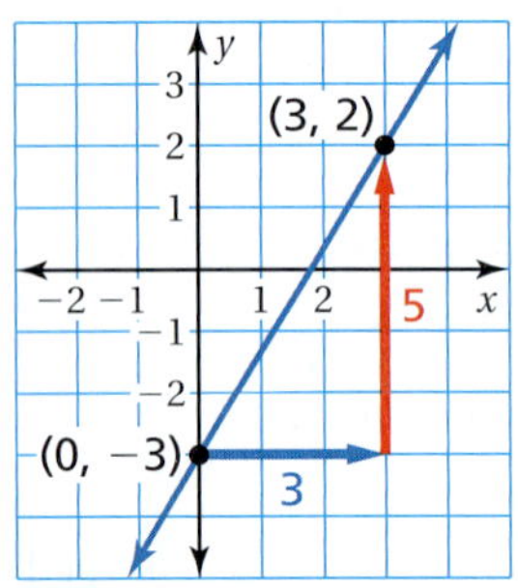

Find the slope and the *y*-intercept.

$$m = \frac{y_2 - y_1}{x_2 - x_1} = \frac{-3 - 2}{0 - 3} = \frac{-5}{-3}, \text{ or } \frac{5}{3}$$

Because the line crosses the *y*-axis at (0, −3), the *y*-intercept is −3.

slope y-intercept

So, the equation is $y = \frac{5}{3}x + (-3)$, or $y = \frac{5}{3}x - 3$.

On Your Own

Now You're Ready
Exercises 5–10

Write an equation of the line in slope-intercept form.

1.

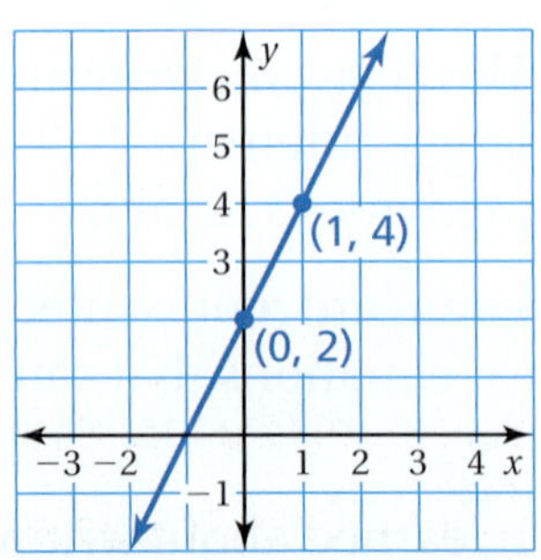

2.

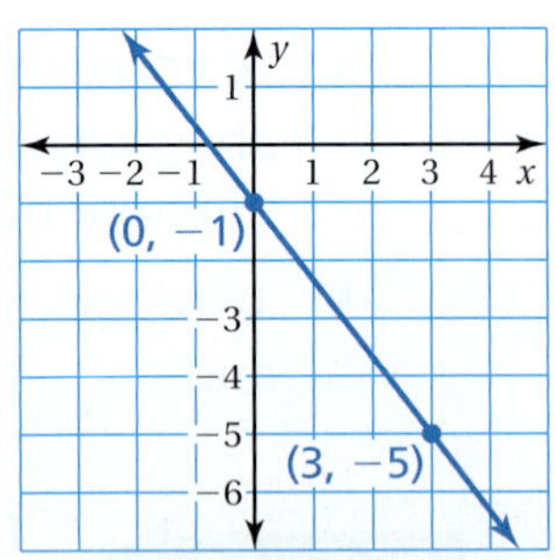

EXAMPLE 2 Writing an Equation

Which equation is shown in the graph?

(A) $y = -4$ (B) $y = -3$

(C) $y = 0$ (D) $y = -3x$

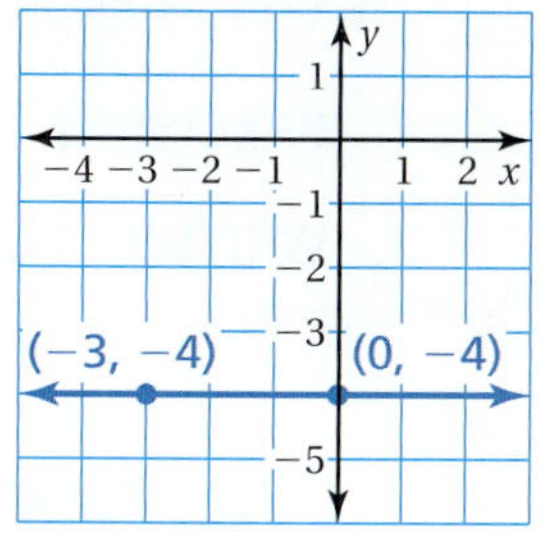

Remember

The graph of $y = a$ is a horizontal line that passes through $(0, a)$.

Find the slope and the y-intercept.

The line is horizontal, so the change in y is 0.

$$m = \frac{\text{change in } y}{\text{change in } x} = \frac{0}{3} = 0$$

Because the line crosses the y-axis at $(0, -4)$, the y-intercept is -4.

So, the equation is $y = 0x + (-4)$, or $y = -4$. The correct answer is (A).

EXAMPLE 3 Real-Life Application

The graph shows the distance remaining to complete a tunnel. (a) Write an equation that represents the distance y (in feet) remaining after x months. (b) How much time does it take to complete the tunnel?

Engineers used tunnel boring machines like the ones shown above to dig an extension of the Metro Gold Line in Los Angeles. The new tunnels are 1.7 miles long and 21 feet wide.

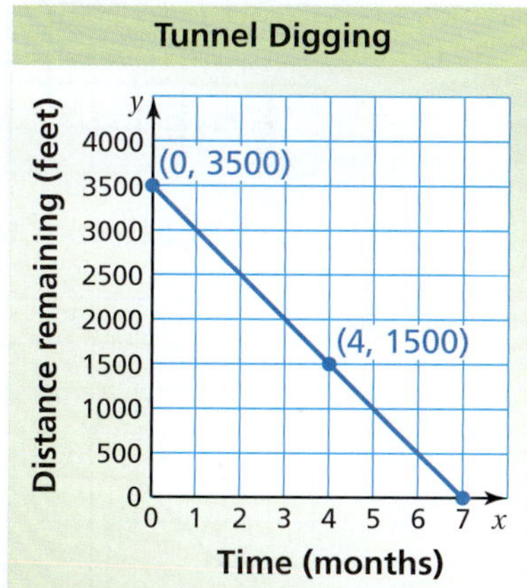

a. Find the slope and the y-intercept.

$$m = \frac{\text{change in } y}{\text{change in } x} = \frac{-2000}{4} = -500$$

Because the line crosses the y-axis at $(0, 3500)$, the y-intercept is 3500.

So, the equation is $y = -500x + 3500$.

b. The tunnel is complete when the distance remaining is 0 feet. So, find the value of x when $y = 0$.

$y = -500x + 3500$	Write the equation.
$0 = -500x + 3500$	Substitute 0 for y.
$-3500 = -500x$	Subtract 3500 from each side.
$7 = x$	Divide each side by -500.

It takes 7 months to complete the tunnel.

On Your Own

Now You're Ready Exercises 13–15

3. Write an equation of the line that passes through $(0, 5)$ and $(4, 5)$.

4. **WHAT IF?** In Example 3, the points are $(0, 3500)$ and $(5, 1500)$. How long does it take to complete the tunnel?

4.6 Exercises

Check It Out
Help with Homework
BigIdeasMath.com

Vocabulary and Concept Check

1. **PRECISION** Explain how to find the slope of a line given the intercepts of the line.
2. **WRITING** Explain how to write an equation of a line using its graph.

Practice and Problem Solving

Write an equation that represents each side of the figure.

3. 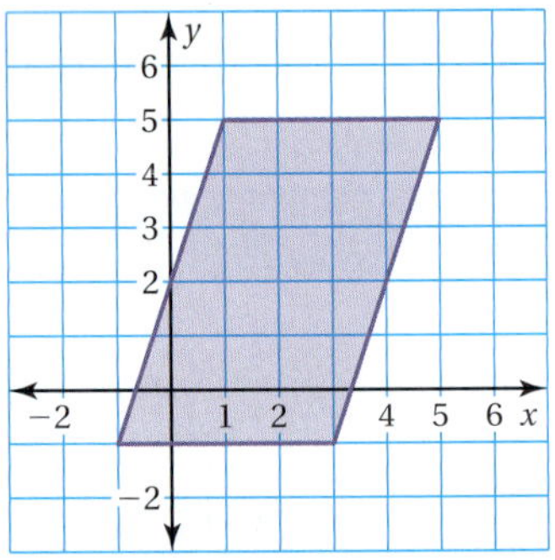

4. 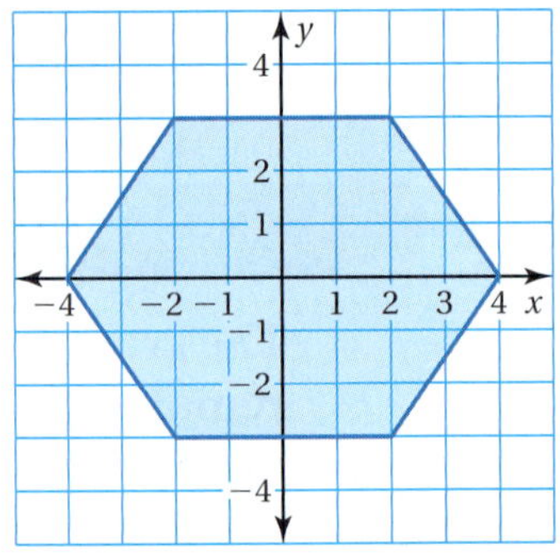

Write an equation of the line in slope-intercept form.

1 5. 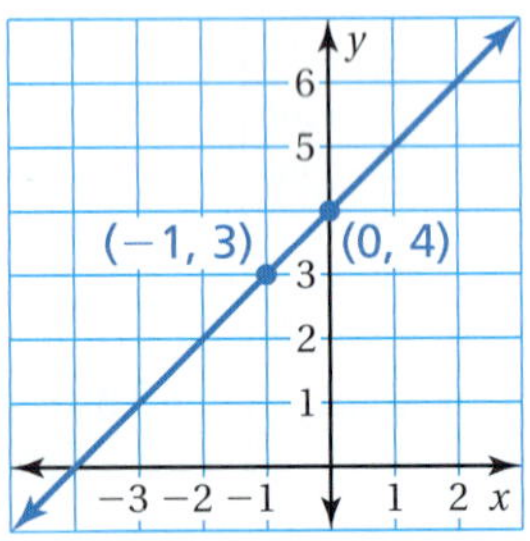

6. 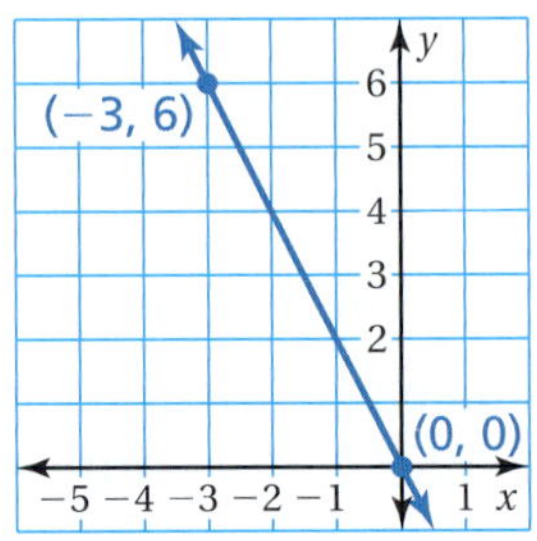

7. 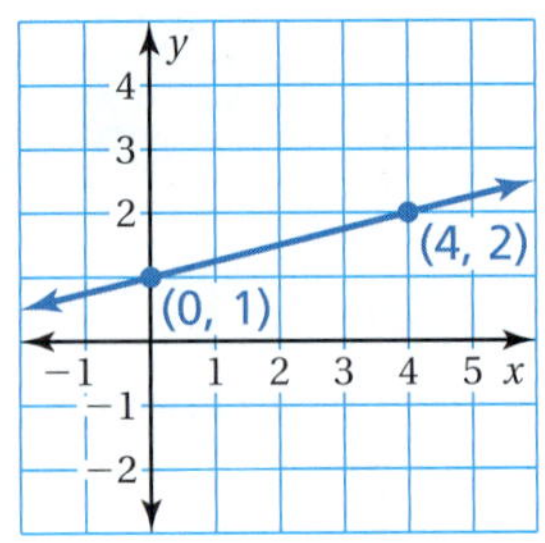

8. 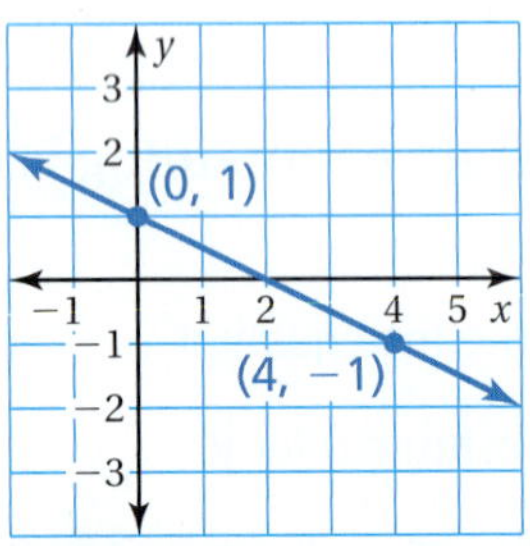

9. 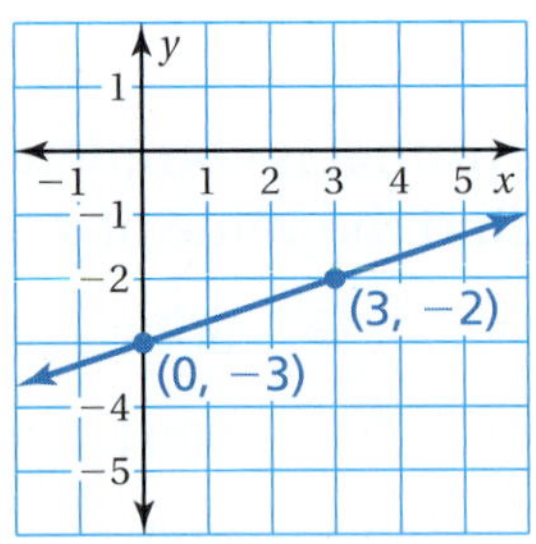

10. 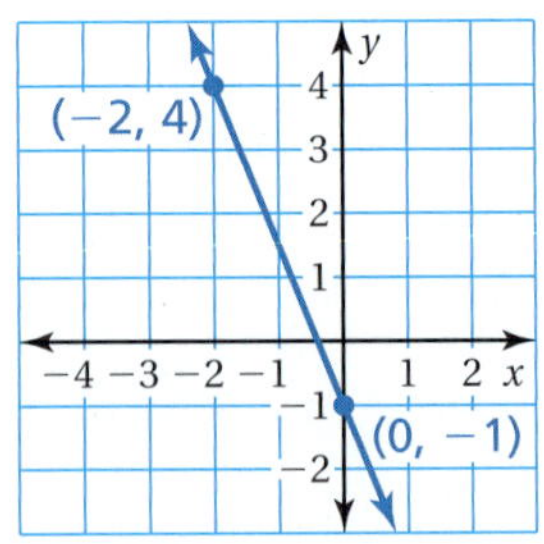

11. **ERROR ANALYSIS** Describe and correct the error in writing an equation of the line.

✗ $y = \frac{1}{2}x + 4$

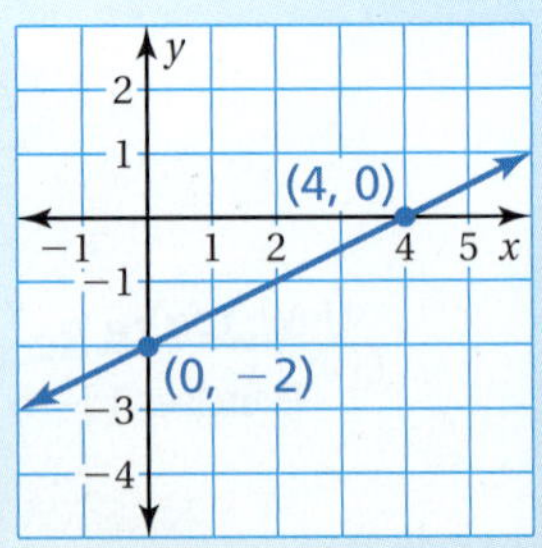

12. **BOA** A boa constrictor is 18 inches long at birth and grows 8 inches per year. Write an equation that represents the length y (in feet) of a boa constrictor that is x years old.

Write an equation of the line that passes through the points.

2 **13.** (2, 5), (0, 5) **14.** (−3, 0), (0, 0) **15.** (0, −2), (4, −2)

16. WALKATHON One of your friends gives you $10 for a charity walkathon. Another friend gives you an amount per mile. After 5 miles, you have raised $13.50 total. Write an equation that represents the amount y of money you have raised after x miles.

17. BRAKING TIME During each second of braking, an automobile slows by about 10 miles per hour.

a. Plot the points (0, 60) and (6, 0). What do the points represent?

b. Draw a line through the points. What does the line represent?

c. Write an equation of the line.

18. PAPER You have 500 sheets of notebook paper. After 1 week, you have 72% of the sheets left. You use the same number of sheets each week. Write an equation that represents the number y of pages remaining after x weeks.

19. Critical Thinking The palm tree on the left is 10 years old. The palm tree on the right is 8 years old. The trees grow at the same rate.

a. Estimate the height y (in feet) of each tree.

b. Plot the two points (x, y), where x is the age of each tree and y is the height of each tree.

c. What is the rate of growth of the trees?

d. Write an equation that represents the height of a palm tree in terms of its age.

Fair Game Review What you learned in previous grades & lessons

Plot the ordered pair in a coordinate plane. *(Skills Review Handbook)*

20. (1, 4) **21.** (−1, −2) **22.** (0, 1) **23.** (2, 7)

24. MULTIPLE CHOICE Which of the following statements is true? *(Section 4.4)*

Ⓐ The x-intercept is 5.

Ⓑ The x-intercept is −2.

Ⓒ The y-intercept is 5.

Ⓓ The y-intercept is −2.

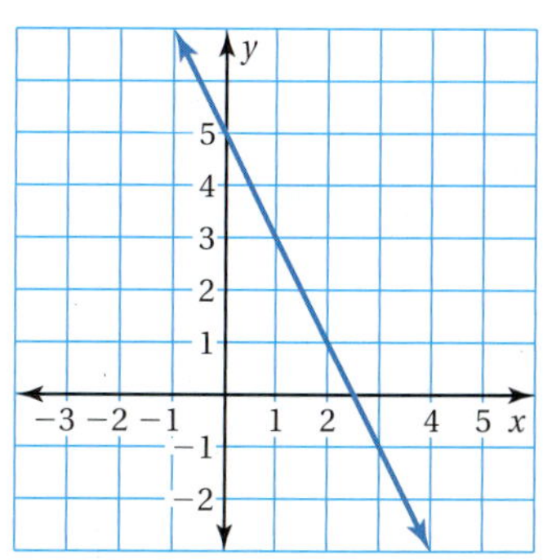

4.7 Writing Equations in Point-Slope Form

Essential Question How can you write an equation of a line when you are given the slope and a point on the line?

1 ACTIVITY: Writing Equations of Lines

Work with a partner.

- **Sketch the line that has the given slope and passes through the given point.**
- **Find the *y*-intercept of the line.**
- **Write an equation of the line.**

a. $m = -2$

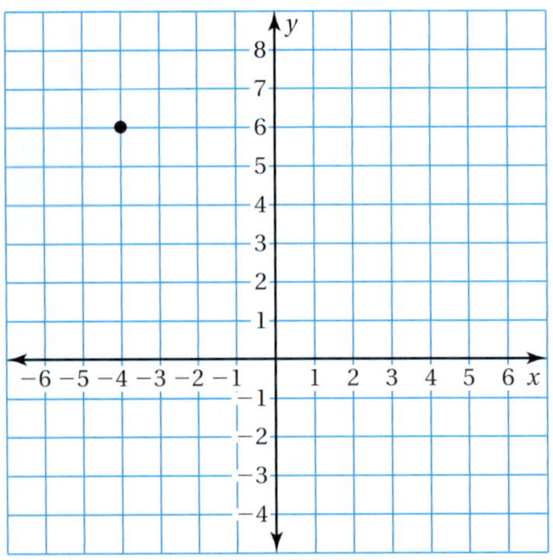

b. $m = \frac{1}{3}$

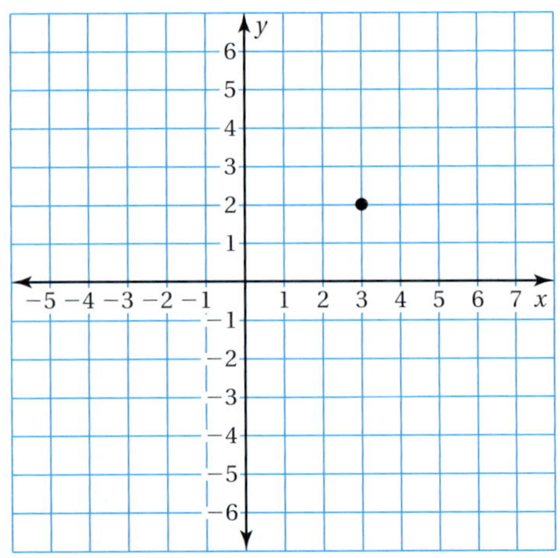

c. $m = -\frac{2}{3}$

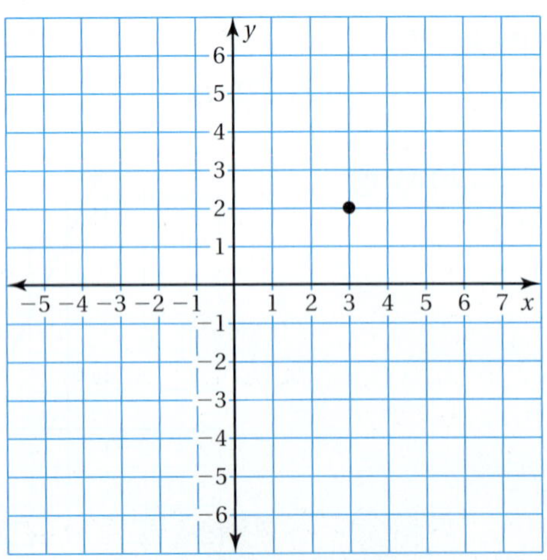

d. $m = \frac{5}{2}$

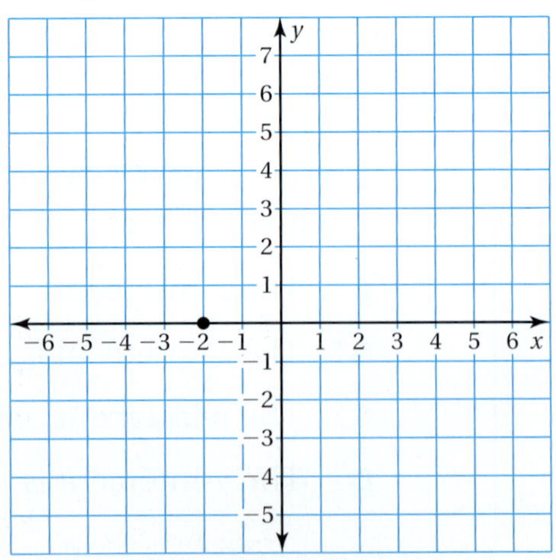

COMMON CORE

Writing Equations

In this lesson, you will

- write equations of lines using a slope and a point.
- write equations of lines using two points.

Preparing for Standard MACC.8.F.2.4

2 ACTIVITY: Deriving an Equation

Work with a partner.

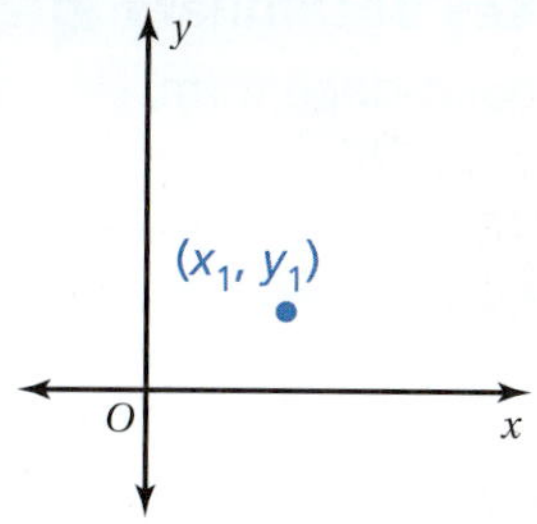

a. Draw a nonvertical line that passes through the point (x_1, y_1).

b. Plot another point on your line. Label this point as (x, y). This point represents any other point on the line.

c. Label the rise and the run of the line through the points (x_1, y_1) and (x, y).

d. The rise can be written as $y - y_1$. The run can be written as $x - x_1$. Explain why this is true.

e. Write an equation for the slope m of the line using the expressions from part (d).

f. Multiply each side of the equation by the expression in the denominator. Write your result. What does this result represent?

Math Practice 3

Construct Arguments

How does a graph help you derive an equation?

3 ACTIVITY: Writing an Equation

Work with a partner.

For 4 months, you saved \$25 a month. You now have \$175 in your savings account.

- Draw a graph that shows the balance in your account after t months.
- Use your result from Activity 2 to write an equation that represents the balance A after t months.

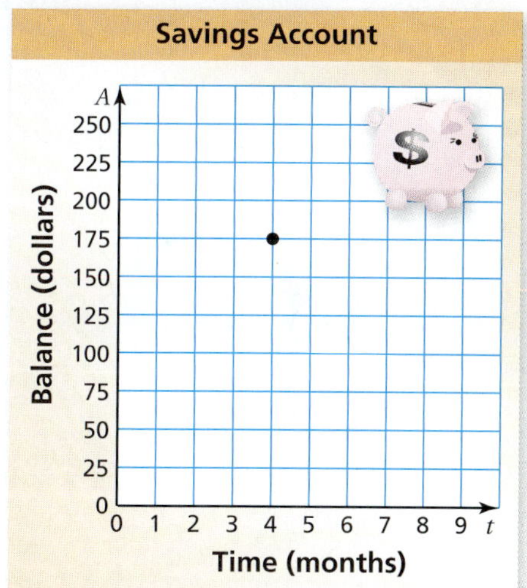

What Is Your Answer?

4. Redo Activity 1 using the equation you found in Activity 2. Compare the results. What do you notice?

5. Why do you think $y - y_1 = m(x - x_1)$ is called the *point-slope form* of the equation of a line? Why do you think it is important?

6. **IN YOUR OWN WORDS** How can you write an equation of a line when you are given the slope and a point on the line? Give an example that is different from those in Activity 1.

Practice Use what you learned about writing equations using a slope and a point to complete Exercises 3–5 on page 188.

4.7 Lesson

Check It Out
Lesson Tutorials
BigIdeasMath.com

Key Vocabulary
point-slope form, *p. 186*

Key Idea

Point-Slope Form

Words A linear equation written in the form $y - y_1 = m(x - x_1)$ is in **point-slope form**. The line passes through the point (x_1, y_1), and the slope of the line is m.

Algebra

$$y - y_1 = m(x - x_1)$$

slope: m

passes through (x_1, y_1)

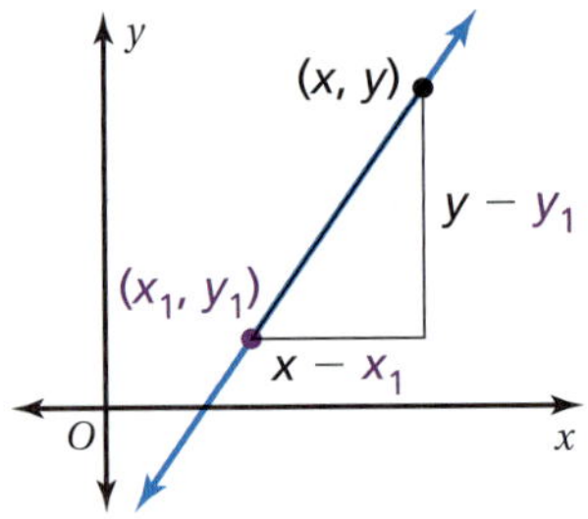

EXAMPLE 1 Writing an Equation Using a Slope and a Point

Write in point-slope form an equation of the line that passes through the point (−6, 1) with slope $\frac{2}{3}$.

$y - y_1 = m(x - x_1)$	Write the point-slope form.
$y - 1 = \frac{2}{3}[x - (-6)]$	Substitute $\frac{2}{3}$ for m, −6 for x_1, and 1 for y_1.
$y - 1 = \frac{2}{3}(x + 6)$	Simplify.

So, the equation is $y - 1 = \frac{2}{3}(x + 6)$.

Check Check that (−6, 1) is a solution of the equation.

$y - 1 = \frac{2}{3}(x + 6)$	Write the equation.
$1 - 1 \stackrel{?}{=} \frac{2}{3}(-6 + 6)$	Substitute.
$0 = 0$ ✓	Simplify.

On Your Own

Now You're Ready
Exercises 6–11

Write in point-slope form an equation of the line that passes through the given point and has the given slope.

1. $(1, 2);\ m = -4$ **2.** $(7, 0);\ m = 1$ **3.** $(-8, -5);\ m = -\frac{3}{4}$

Multi-Language Glossary at BigIdeasMath.com

EXAMPLE 2 Writing an Equation Using Two Points

Study Tip

You can use either of the given points to write the equation of the line.

Use $m = -2$ and $(5, -2)$.

$$y - (-2) = -2(x - 5)$$
$$y + 2 = -2x + 10$$
$$y = -2x + 8 \checkmark$$

Write in slope-intercept form an equation of the line that passes through the points (2, 4) and (5, −2).

Find the slope: $m = \dfrac{y_2 - y_1}{x_2 - x_1} = \dfrac{-2 - 4}{5 - 2} = \dfrac{-6}{3} = -2$

Then use the slope $m = -2$ and the point (2, 4) to write an equation of the line.

$y - y_1 = m(x - x_1)$	Write the point-slope form.
$y - 4 = -2(x - 2)$	Substitute −2 for m, 2 for x_1, and 4 for y_1.
$y - 4 = -2x + 4$	Distributive Property
$y = -2x + 8$	Write in slope-intercept form.

EXAMPLE 3 Real-Life Application

You finish parasailing and are being pulled back to the boat. After 2 seconds, you are 25 feet above the boat. (a) Write and graph an equation that represents your height *y* (in feet) above the boat after *x* seconds. (b) At what height were you parasailing?

a. You are being pulled down at the rate of 10 feet per second. So, the slope is −10. You are 25 feet above the boat after 2 seconds. So, the line passes through (2, 25). Use the point-slope form.

$y - 25 = -10(x - 2)$	Substitute for m, x_1, and y_1.
$y - 25 = -10x + 20$	Distributive Property
$y = -10x + 45$	Write in slope-intercept form.

So, the equation is $y = -10x + 45$.

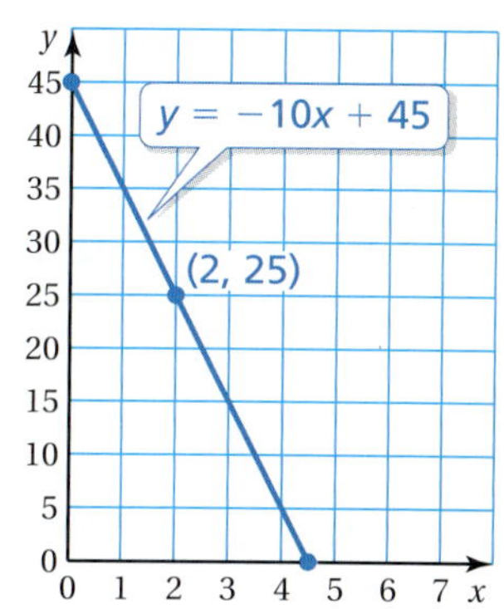

b. You start descending when $x = 0$. The y-intercept is 45. So, you were parasailing at a height of 45 feet.

On Your Own

Now You're Ready
Exercises 12–17

Write in slope-intercept form an equation of the line that passes through the given points.

4. (−2, 1), (3, −4) **5.** (−5, −5), (−3, 3) **6.** (−8, 6), (−2, 9)

7. WHAT IF? In Example 3, you are 35 feet above the boat after 2 seconds. Write and graph an equation that represents your height y (in feet) above the boat after x seconds.

4.7 Exercises

Check It Out
Help with Homework
BigIdeasMath.com

Vocabulary and Concept Check

1. **VOCABULARY** From the equation $y - 3 = -2(x + 1)$, identify the slope and a point on the line.

2. **WRITING** Describe how to write an equation of a line using (a) its slope and a point on the line and (b) two points on the line.

Practice and Problem Solving

Use the point-slope form to write an equation of the line with the given slope that passes through the given point.

3. $m = \frac{1}{2}$

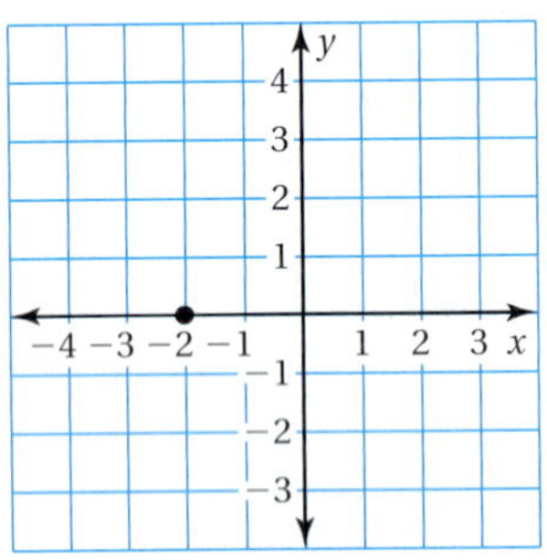

4. $m = -\frac{3}{4}$

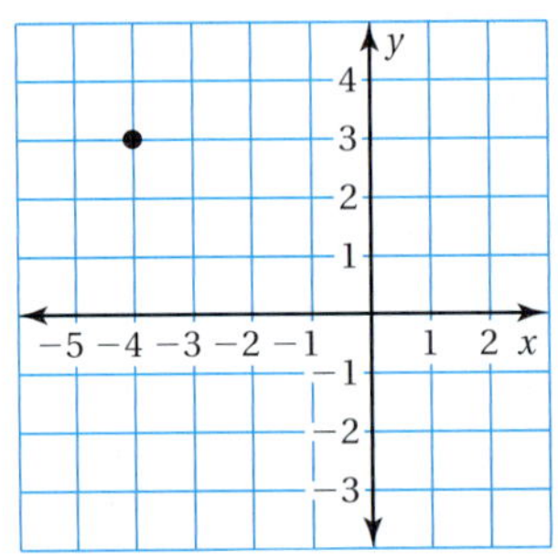

5. $m = -3$

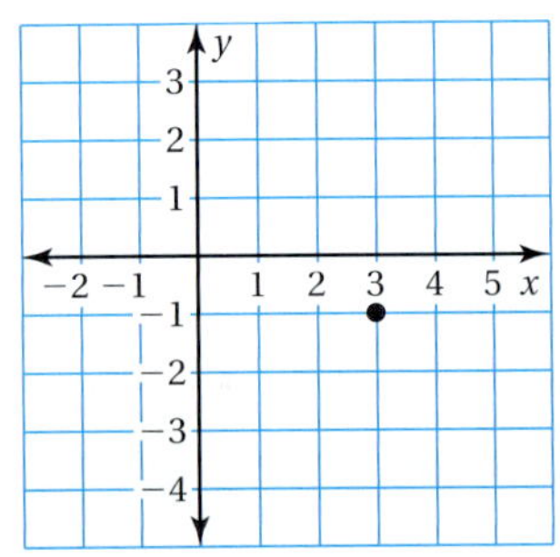

Write in point-slope form an equation of the line that passes through the given point and has the given slope.

① 6. $(3, 0);\ m = -\frac{2}{3}$

7. $(4, 8);\ m = \frac{3}{4}$

8. $(1, -3);\ m = 4$

9. $(7, -5);\ m = -\frac{1}{7}$

10. $(3, 3);\ m = \frac{5}{3}$

11. $(-1, -4);\ m = -2$

Write in slope-intercept form an equation of the line that passes through the given points.

② 12. $(-1, -1), (1, 5)$

13. $(2, 4), (3, 6)$

14. $(-2, 3), (2, 7)$

15. $(4, 1), (8, 2)$

16. $(-9, 5), (-3, 3)$

17. $(1, 2), (-2, -1)$

18. **CHEMISTRY** At 0 °C, the volume of a gas is 22 liters. For each degree the temperature T (in degrees Celsius) increases, the volume V (in liters) of the gas increases by $\frac{2}{25}$. Write an equation that represents the volume of the gas in terms of the temperature.

19. **CARS** After it is purchased, the value of a new car decreases \$4000 each year. After 3 years, the car is worth \$18,000.

a. Write an equation that represents the value V (in dollars) of the car x years after it is purchased.

b. What was the original value of the car?

20. **REASONING** Write an equation of a line that passes through the point (8, 2) that is (a) parallel and (b) perpendicular to the graph of the equation $y = 4x - 3$.

21. **CRICKETS** According to Dolbear's law, you can predict the temperature T (in degrees Fahrenheit) by counting the number x of chirps made by a snowy tree cricket in 1 minute. For each rise in temperature of 0.25°F, the cricket makes an additional chirp each minute.

a. A cricket chirps 40 times in 1 minute when the temperature is 50°F. Write an equation that represents the temperature in terms of the number of chirps in 1 minute.

b. You count 100 chirps in 1 minute. What is the temperature?

c. The temperature is 96°F. How many chirps would you expect the cricket to make?

Leaning Tower of Pisa

22. **WATERING CAN** You water the plants in your classroom at a constant rate. After 5 seconds, your watering can contains 58 ounces of water. Fifteen seconds later, the can contains 28 ounces of water.

a. Write an equation that represents the amount y (in ounces) of water in the can after x seconds.

b. How much water was in the can when you started watering the plants?

c. When is the watering can empty?

23. **Problem Solving** The Leaning Tower of Pisa in Italy was built between 1173 and 1350.

a. Write an equation for the yellow line.

b. The tower is 56 meters tall. How far off center is the top of the tower?

Fair Game Review What you learned in previous grades & lessons

Graph the linear equation. *(Section 4.4)*

24. $y = 4x$

25. $y = -2x + 1$

26. $y = 3x - 5$

27. **MULTIPLE CHOICE** What is the x-intercept of the equation $3x + 5y = 30$? *(Section 4.5)*

(A) -10 (B) -6 (C) 6 (D) 10

4.4–4.7 Quiz

Check It Out
Progress Check
BigIdeasMath.com

Find the slope and the y-intercept of the graph of the linear equation. *(Section 4.4)*

1. $y = \frac{1}{4}x - 8$

2. $y = -x + 3$

Find the x- and y-intercepts of the graph of the equation. *(Section 4.5)*

3. $3x - 2y = 12$

4. $x + 5y = 15$

Write an equation of the line in slope-intercept form. *(Section 4.6)*

5.

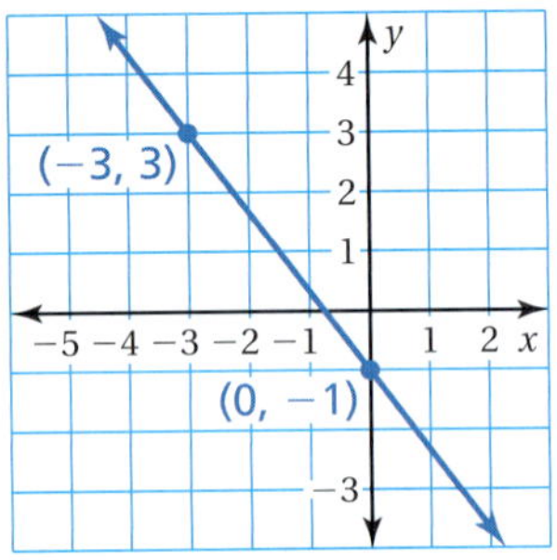

6.

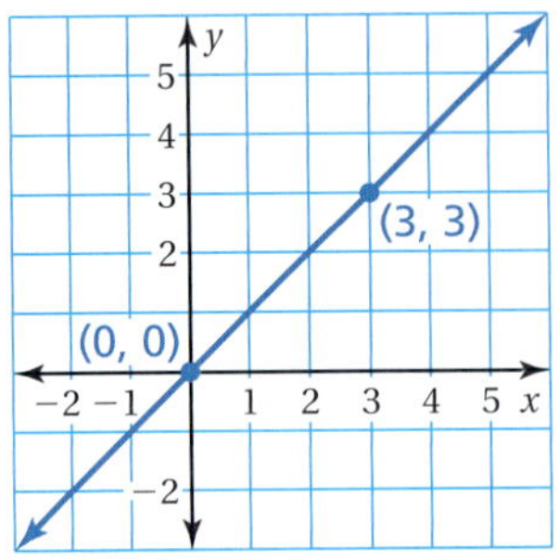

7.

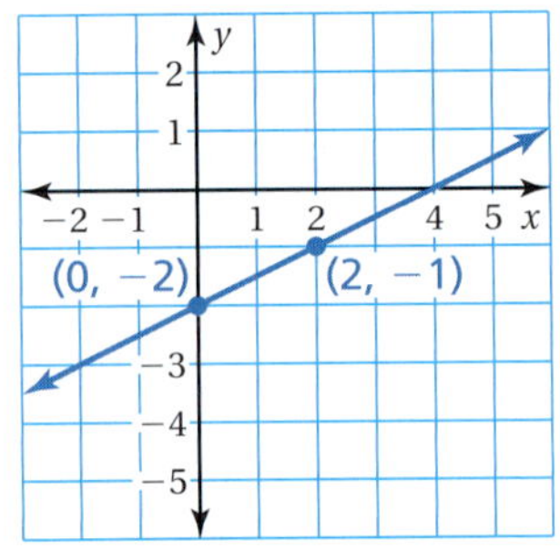

Write in point-slope form an equation of the line that passes through the given point and has the given slope. *(Section 4.7)*

8. $(1, 3);\ m = 2$

9. $(-3, -2);\ m = \frac{1}{3}$

10. $(-1, 4);\ m = -1$

11. $(8, -5);\ m = -\frac{1}{8}$

Write in slope-intercept form an equation of the line that passes through the given points. *(Section 4.7)*

12. $\left(0, -\frac{2}{3}\right) \left(-3, -\frac{2}{3}\right)$

13. $(4, 0), (0, 4)$

14. STATE FAIR The cost y (in dollars) of one person buying admission to a fair and going on x rides is $y = x + 12$. *(Section 4.4)*

a. Graph the equation.

b. Interpret the y-intercept and the slope.

15. PAINTING You used \$90 worth of paint for a school float. *(Section 4.5)*

a. Graph the equation $18x + 15y = 90$, where x is the number of gallons of blue paint and y is the number of gallons of white paint.

b. Interpret the intercepts.

16. CONSTRUCTION A construction crew is extending a highway sound barrier that is 13 miles long. The crew builds $\frac{1}{2}$ of a mile per week. Write an equation that represents the length y (in miles) of the barrier after x weeks. *(Section 4.6)*

4 Chapter Review

Check It Out
Vocabulary Help
BigIdeasMath.com

Review Key Vocabulary

linear equation *p. 144*
solution of a linear equation, *p. 144*
slope, *p. 150*
rise, *p. 150*
run, *p. 150*
x-intercept, *p. 168*
y-intercept, *p. 168*
slope-intercept form, *p. 168*
standard form, *p. 174*
point-slope form, *p. 186*

Review Examples and Exercises

4.1 Graphing Linear Equations *(pp. 142–147)*

Graph $y = 3x - 1$.

Step 1: Make a table of values.

x	$y = 3x - 1$	y	(x, y)
-2	$y = 3(-2) - 1$	-7	$(-2, -7)$
-1	$y = 3(-1) - 1$	-4	$(-1, -4)$
0	$y = 3(0) - 1$	-1	$(0, -1)$
1	$y = 3(1) - 1$	2	$(1, 2)$

Step 2: Plot the ordered pairs.

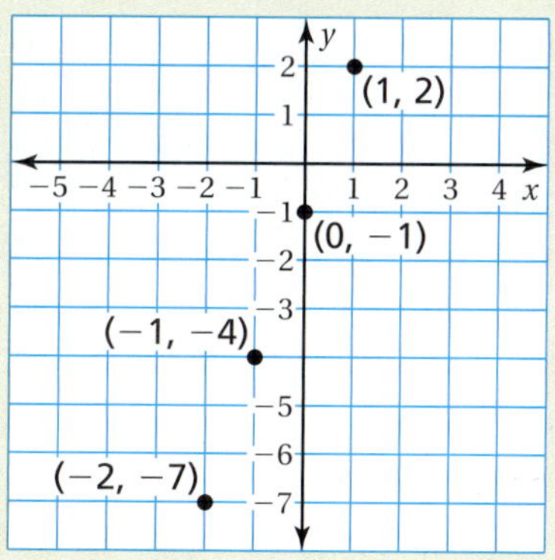

Step 3: Draw a line through the points.

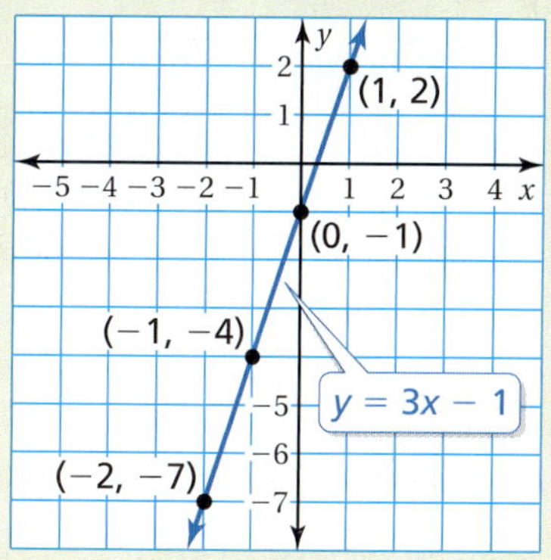

Exercises

Graph the linear equation.

1. $y = \frac{3}{5}x$

2. $y = -2$

3. $y = 9 - x$

4. $y = 1$

5. $y = \frac{2}{3}x + 2$

6. $x = -5$

4.2 Slope of a Line (pp. 148–157)

Find the slope of each line in the graph.

Red Line: $m = \dfrac{y_2 - y_1}{x_2 - x_1} = \dfrac{5 - (-3)}{2 - 2} = \dfrac{8}{0}$

∴ The slope of the red line is undefined.

Blue Line: $m = \dfrac{y_2 - y_1}{x_2 - x_1} = \dfrac{-1 - 2}{4 - (-3)} = \dfrac{-3}{7}$, or $-\dfrac{3}{7}$

Green Line: $m = \dfrac{y_2 - y_1}{x_2 - x_1} = \dfrac{4 - 4}{5 - 0} = \dfrac{0}{5}$, or 0

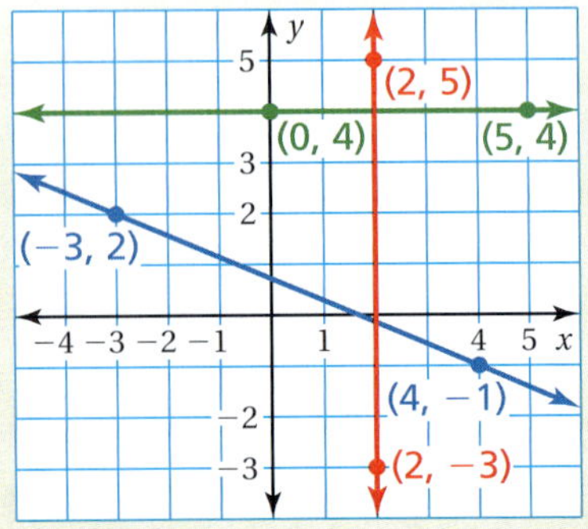

Exercises

The points in the table lie on a line. Find the slope of the line.

7.

x	0	1	2	3
y	−1	0	1	2

8.

x	−2	0	2	4
y	3	4	5	6

9. Are the lines $x = 2$ and $y = 4$ parallel? Are they perpendicular? Explain.

4.3 Graphing Proportional Relationships (pp. 158–163)

The cost y (in dollars) for x tickets to a movie is represented by the equation $y = 7x$. Graph the equation and interpret the slope.

The equation shows that the slope m is 7. So, the graph passes through (0, 0) and (1, 7).

Plot the points and draw a line through the points. Because negative values of x do not make sense in this context, graph in the first quadrant only.

∴ The slope indicates that the unit cost is $7 per ticket.

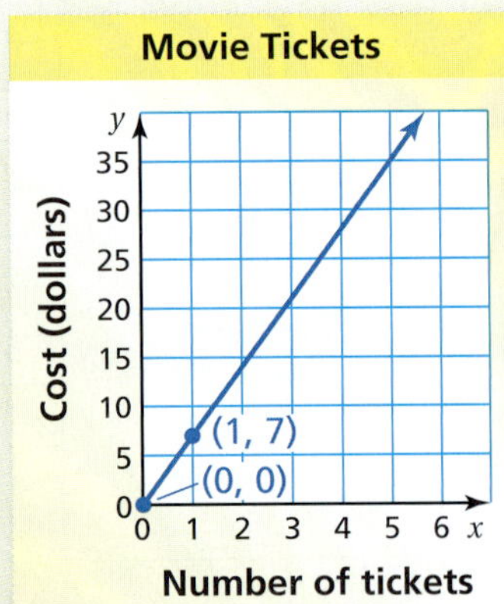

Exercises

10. **RUNNING** The number y of miles you run after x weeks is represented by the equation $y = 8x$. Graph the equation and interpret the slope.

11. **STUDYING** The number y of hours that you study after x days is represented by the equation $y = 1.5x$. Graph the equation and interpret the slope.

4.4 Graphing Linear Equations in Slope-Intercept Form *(pp. 166–171)*

Graph $y = 0.5x - 3$. Identify the x-intercept.

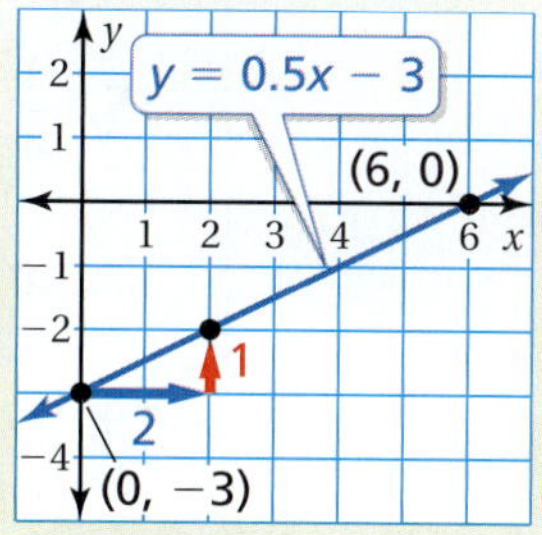

Step 1: Find the slope and the y-intercept.

$$y = 0.5x + (-3)$$

slope → 0.5; y-intercept → -3

Step 2: The y-intercept is -3. So, plot $(0, -3)$.

Step 3: Use the slope to find another point and draw the line.

$$m = \frac{\text{rise}}{\text{run}} = \frac{1}{2}$$

Plot the point that is 2 units right and 1 unit up from $(0, -3)$. Draw a line through the two points.

∴ The line crosses the x-axis at $(6, 0)$. So, the x-intercept is 6.

Exercises

Graph the linear equation. Identify the x-intercept. Use a graphing calculator to check your answer.

12. $y = 2x - 6$

13. $y = -4x + 8$

14. $y = -x - 8$

4.5 Graphing Linear Equations in Standard Form *(pp. 172–177)*

Graph $8x + 4y = 16$.

Step 1: Write the equation in slope-intercept form.

$8x + 4y = 16$	Write the equation.
$4y = -8x + 16$	Subtract $8x$ from each side.
$y = -2x + 4$	Divide each side by 4.

Step 2: Use the slope and the y-intercept to graph the equation.

$$y = -2x + 4$$

slope → -2; y-intercept → 4

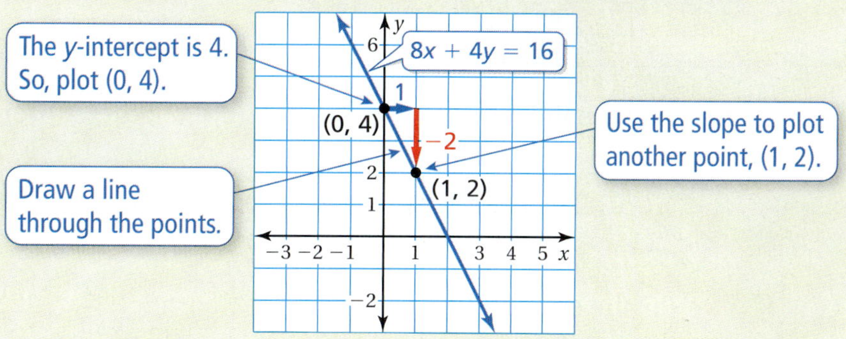

Exercises

Graph the linear equation.

15. $\frac{1}{4}x + y = 3$

16. $-4x + 2y = 8$

17. $x + 5y = 10$

18. $-\frac{1}{2}x + \frac{1}{8}y = \frac{3}{4}$

19. A dog kennel charges $30 per night to board your dog and $6 for each hour of playtime. The amount of money you spend is given by $30x + 6y = 180$, where x is the number of nights and y is the number of hours of playtime. Graph the equation and interpret the intercepts.

4.6 Writing Equations in Slope-Intercept Form *(pp. 178–183)*

Write an equation of the line in slope-intercept form.

a.

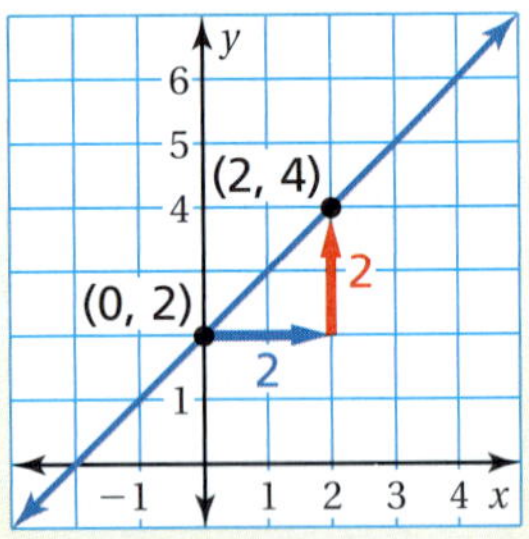

Find the slope and the y-intercept.

$$m = \frac{y_2 - y_1}{x_2 - x_1} = \frac{4 - 2}{2 - 0} = \frac{2}{2}, \text{ or } 1$$

Because the line crosses the y-axis at (0, 2), the y-intercept is 2.

slope | y-intercept

So, the equation is $y = 1x + 2$, or $y = x + 2$.

b.

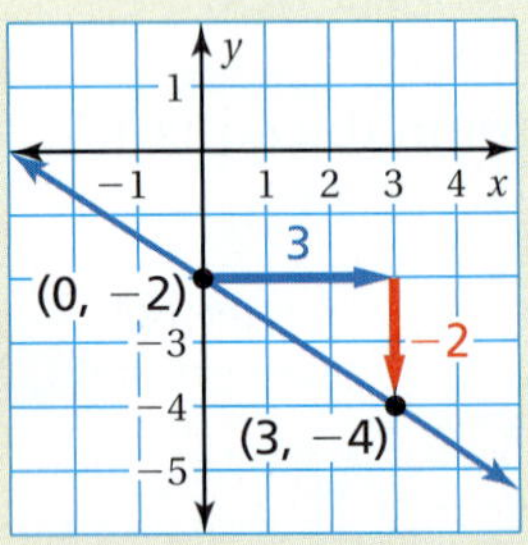

Find the slope and the y-intercept.

$$m = \frac{y_2 - y_1}{x_2 - x_1} = \frac{-4 - (-2)}{3 - 0} = \frac{-2}{3}, \text{ or } -\frac{2}{3}$$

Because the line crosses the y-axis at (0, −2), the y-intercept is −2.

slope | y-intercept

So, the equation is $y = -\frac{2}{3}x + (-2)$, or $y = -\frac{2}{3}x - 2$.

Exercises

Write an equation of the line in slope-intercept form.

20.

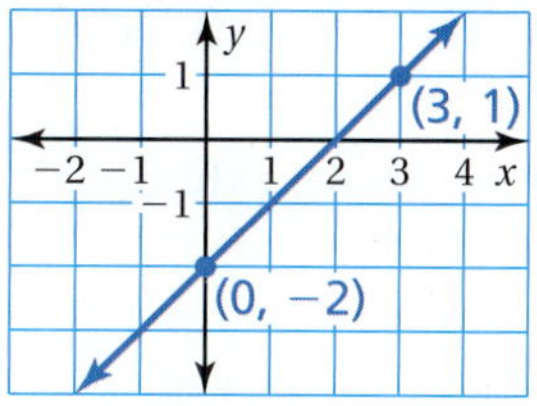

21.

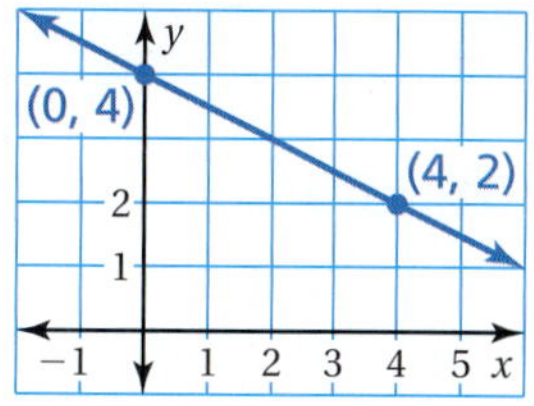

22.

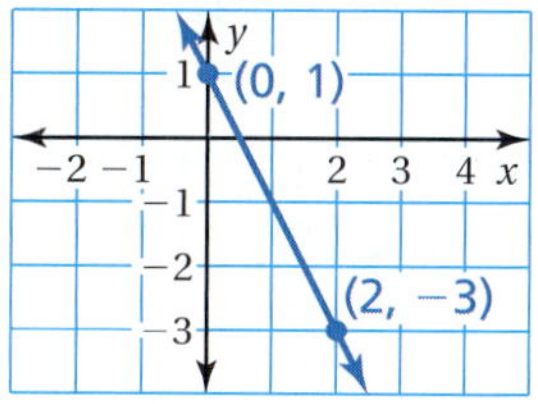

23.

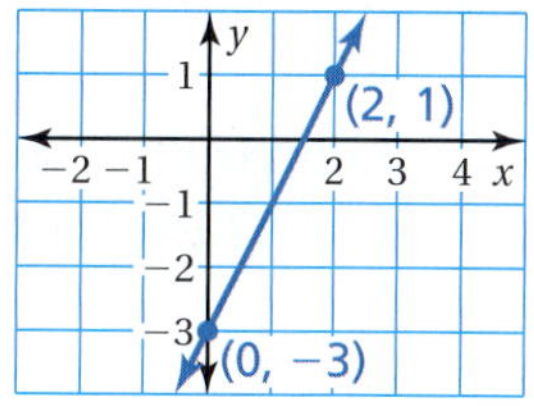

24. Write an equation of the line that passes through (0, 8) and (6, 8).

25. Write an equation of the line that passes through (0, −5) and (−5, −5).

4.7 Writing Equations in Point-Slope Form *(pp. 184–189)*

Write in slope-intercept form an equation of the line that passes through the points (2, 1) and (3, 5).

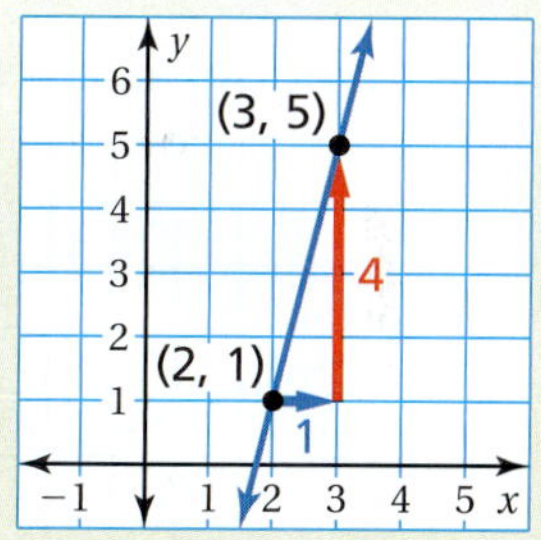

Find the slope.

$$m = \frac{y_2 - y_1}{x_2 - x_1} = \frac{5 - 1}{3 - 2} = \frac{4}{1}, \text{ or } 4$$

Then use the slope and one of the given points to write an equation of the line.

Use $m = 4$ and (2, 1).

$y - y_1 = m(x - x_1)$ — Write the point-slope form.

$y - 1 = 4(x - 2)$ — Substitute 4 for m, 2 for x_1, and 1 for y_1.

$y - 1 = 4x - 8$ — Distributive Property

$y = 4x - 7$ — Write in slope-intercept form.

So, the equation is $y = 4x - 7$.

Exercises

26. Write in point-slope form an equation of the line that passes through the point (4, 4) with slope 3.

27. Write in slope-intercept form an equation of the line that passes through the points (−4, 2) and (6, −3).

4 Chapter Test

Check It Out
Test Practice
BigIdeasMath.com

Find the slope and the y-intercept of the graph of the linear equation.

1. $y = 6x - 5$ **2.** $y = 20x + 15$ **3.** $y = -5x - 16$

4. $y - 1 = 3x + 8.4$ **5.** $y + 4.3 = 0.1x$ **6.** $-\frac{1}{2}x + 2y = 7$

Graph the linear equation.

7. $y = 2x + 4$ **8.** $y = -\frac{1}{2}x - 5$ **9.** $-3x + 6y = 12$

10. Which lines are parallel? Which lines are perpendicular? Explain.

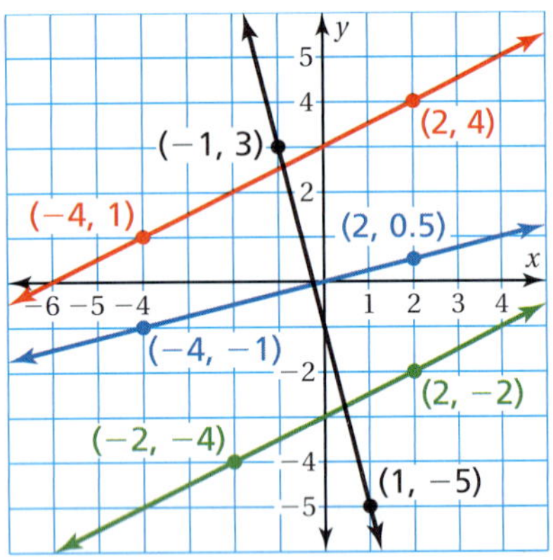

11. The points in the table lie on a line. Find the slope of the line.

x	y
−1	−4
0	−1
1	2
2	5

Write an equation of the line in slope-intercept form.

12.

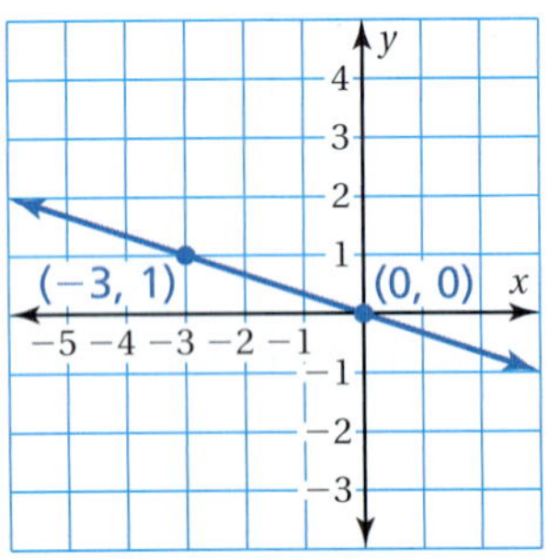

13.

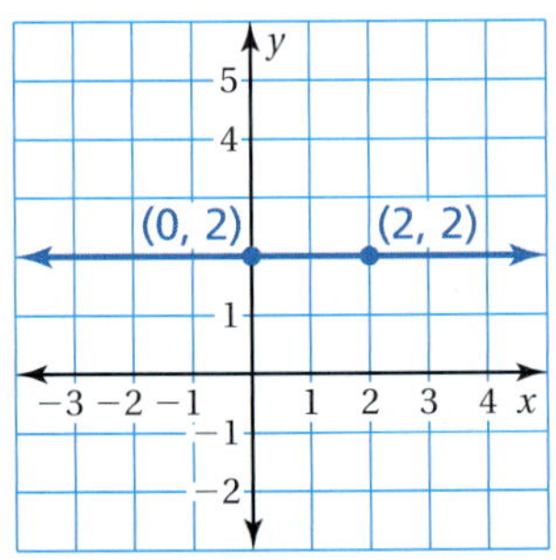

Write in slope-intercept form an equation of the line that passes through the given points.

14. $(-1, 5), (3, -3)$ **15.** $(-4, 1), (4, 3)$ **16.** $(-2, 5), (-1, 1)$

17. VOCABULARY The number y of new vocabulary words that you learn after x weeks is represented by the equation $y = 15x$.

a. Graph the equation and interpret the slope.

b. How many new vocabulary words do you learn after 5 weeks?

c. How many more vocabulary words do you learn after 6 weeks than after 4 weeks?

4 Standards Assessment

1. Which equation matches the line shown in the graph? *(MACC.8.EE.2.6)*

   A. $y = 2x - 2$

   B. $y = 2x + 1$

   C. $y = x - 2$

   D. $y = x + 1$

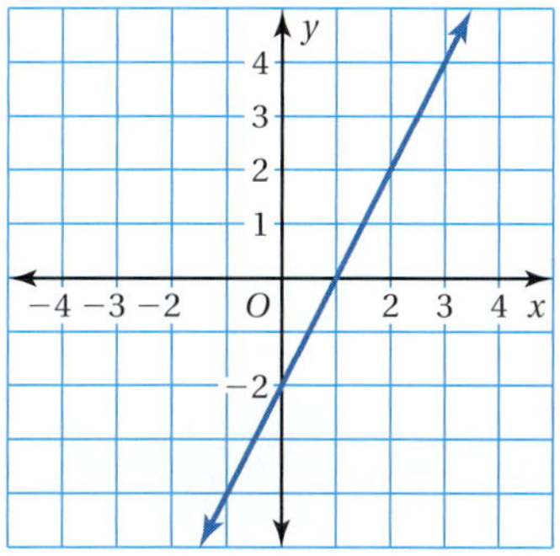

Test-Taking Strategy
Estimate the Answer

"Using estimation, you can see that there are about 300 hairs. So, it has to be C."

2. The equation $6x - 5y = 14$ is written in standard form. Which point lies on the graph of this equation? *(MACC.8.EE.2.6)*

   F. $(-4, -1)$

   G. $(-2, 4)$

   H. $(-1, -4)$

   I. $(4, -2)$

3. Which line has a slope of 0? *(MACC.8.EE.2.6)*

   A.

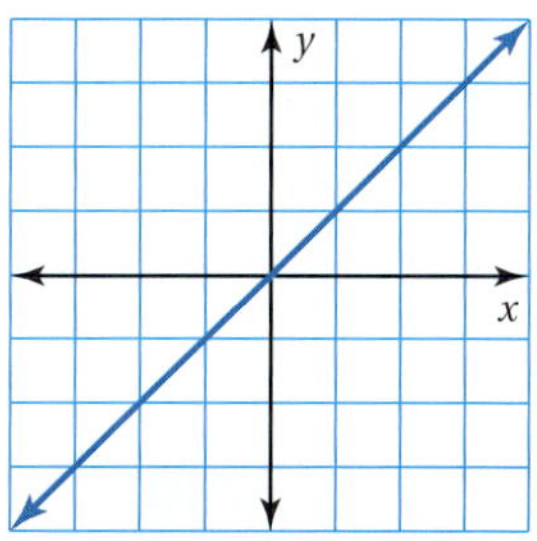

   B.

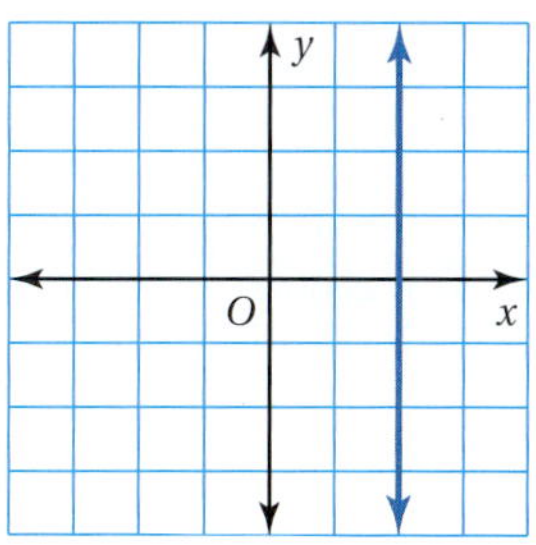

   C.

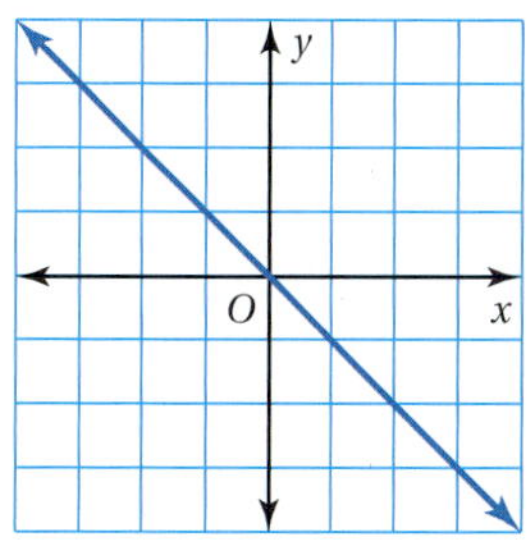

   D.

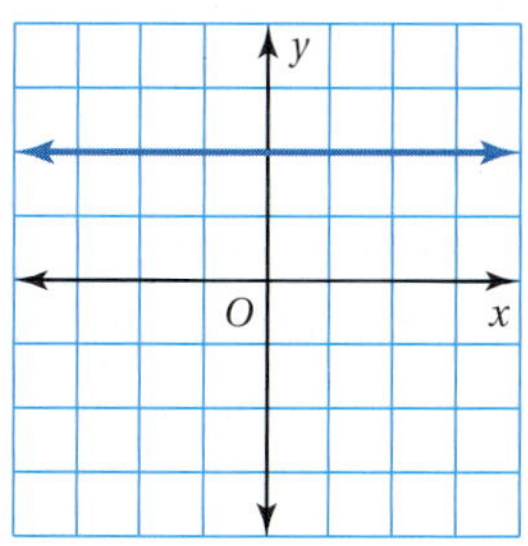

4. Which of the following is the equation of a line perpendicular to the line shown in the graph? *(MACC.8.EE.2.6)*

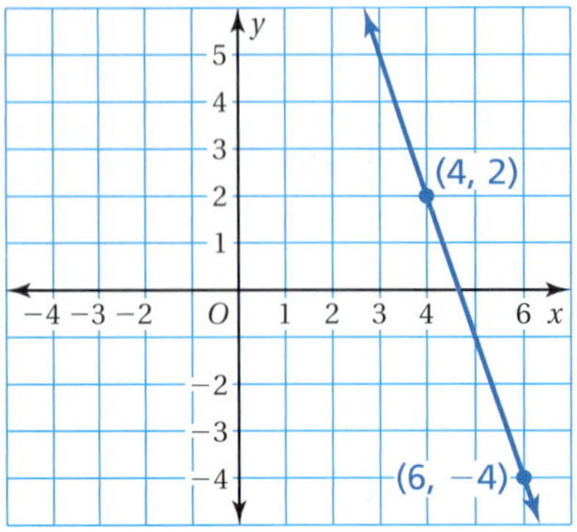

F. $y = 3x - 10$

G. $y = \frac{1}{3}x + 12$

H. $y = -3x + 5$

I. $y = -\frac{1}{3}x - 18$

5. What is the slope of the line that passes through the points $(2, -2)$ and $(8, 1)$? *(MACC.8.EE.2.6)*

6. A cell phone plan costs \$10 per month plus \$0.10 for each minute used. Last month, you spent \$18.50 using this plan. This can be modeled by the equation below, where m represents the number of minutes used.

$$0.1m + 10 = 18.5$$

How many minutes did you use last month? *(MACC.8.EE.3.7b)*

A. 8.4 min

B. 85 min

C. 185 min

D. 285 min

7. It costs \$40 to rent a car for one day. In addition, the rental agency charges you for each mile driven, as shown in the graph. *(MACC.8.EE.2.6)*

Think Solve Explain

Part A Determine the slope of the line joining the points on the graph.

Part B Explain what the slope represents.

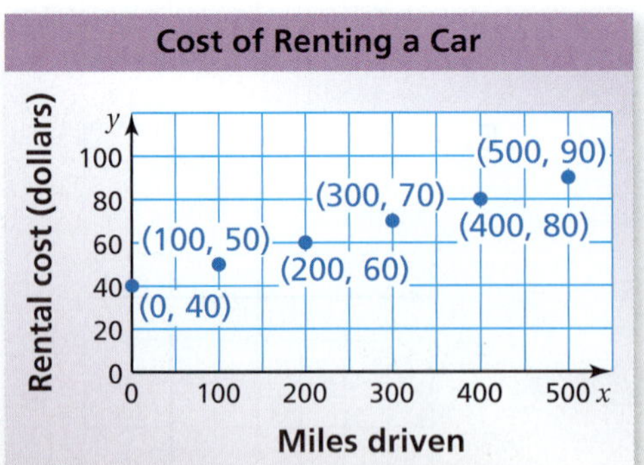

8. What value of x makes the equation below true? *(MACC.8.EE.3.7a)*

$$7 + 2x = 4x - 5$$

9. Trapezoid *KLMN* is graphed in the coordinate plane shown.

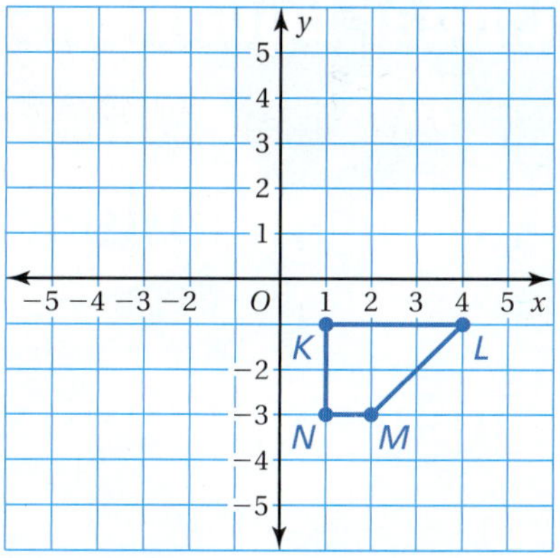

Rotate Trapezoid *KLMN* 90° clockwise about the origin. What are the coordinates of point M', the image of point M after the rotation? *(MACC.8.G.1.3)*

F. $(-3, -2)$

G. $(-2, -3)$

H. $(-2, 3)$

I. $(3, 2)$

10. Solve the formula $K = 3M - 7$ for M. *(MACC.8.EE.3.7b)*

A. $M = K + 7$

B. $M = \frac{K+7}{3}$

C. $M = \frac{K}{3} + 7$

D. $M = \frac{K-7}{3}$

11. What is the distance d across the canyon? *(MACC.8.G.1.5)*

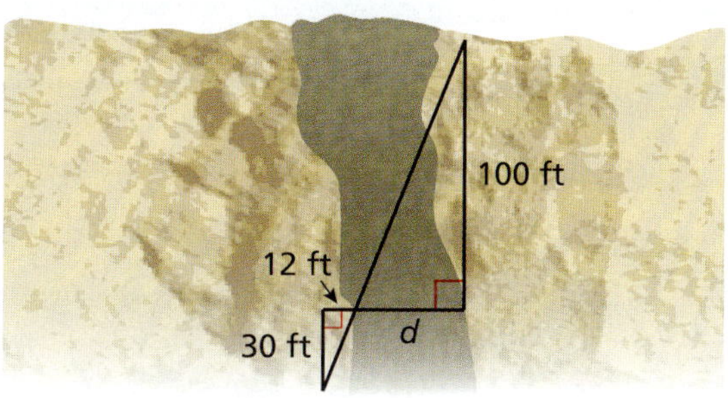

F. 3.6 ft

G. 12 ft

H. 40 ft

I. 250 ft

5 Systems of Linear Equations

"Can you graph a system of linear equations that shows the number of biscuits and treats that I am going to share with you?"

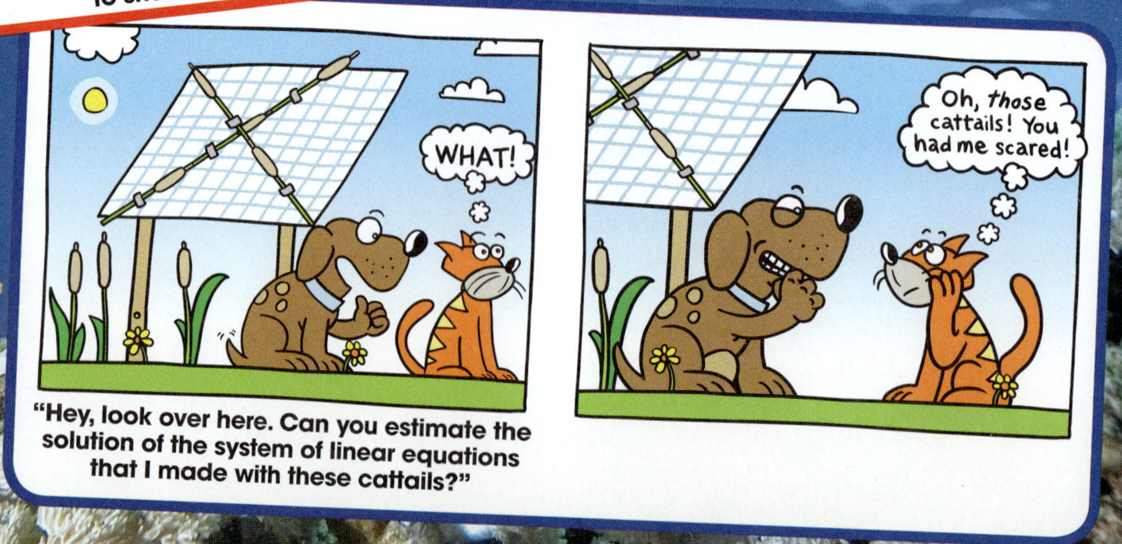

"Hey, look over here. Can you estimate the solution of the system of linear equations that I made with these cattails?"

What You Learned Before

"Hold your tail a bit lower."

Combining Like Terms (MACC.6.EE.1.3)

Example 1 Simplify each expression.

a. $4x + 7 + 5x - 2$

$4x + 7 + 5x - 2 = 4x + 5x + 7 - 2$ — Commutative Property of Addition

$= (4 + 5)x + 7 - 2$ — Distributive Property

$= 9x + 5$ — Simplify.

b. $z + z + z + z$

$z + z + z + z = 1z + 1z + 1z + 1z$ — Multiplication Property of One

$= (1 + 1 + 1 + 1)z$ — Distributive Property

$= 4z$ — Add coefficients.

Try It Yourself

Simplify the expression.

1. $5 + 4z - 2z$

2. $5(c + 8) + c + 3$

Solving Multi-Step Equations (MACC.8.EE.3.7b)

Example 2 Solve $4x - 2(3x + 1) = 16$.

$4x - 2(3x + 1) = 16$ — Write the equation.

$4x - 6x - 2 = 16$ — Distributive Property

$-2x - 2 = 16$ — Combine like terms.

$-2x = 18$ — Add 2 to each side.

$x = -9$ — Divide each side by -2.

The solution is $x = -9$.

Try It Yourself

Solve the equation. Check your solution.

3. $-5x + 8 = -7$

4. $7w + w - 15 = 17$

5. $-3(z - 8) + 10 = -5$

6. $2 = 10c - 4(2c - 9)$

5.1 Solving Systems of Linear Equations by Graphing

Essential Question How can you solve a system of linear equations?

1 ACTIVITY: Writing a System of Linear Equations

Work with a partner.

Your family starts a bed-and-breakfast. It spends \$500 fixing up a bedroom to rent. The cost for food and utilities is \$10 per night. Your family charges \$60 per night to rent the bedroom.

a. Write an equation that represents the costs.

Cost, C (in dollars) = \$10 per night • Number of nights, x + \$500

b. Write an equation that represents the revenue (income).

Revenue, R (in dollars) = \$60 per night • Number of nights, x

c. A set of two (or more) linear equations is called a **system of linear equations**. Write the system of linear equations for this problem.

2 ACTIVITY: Using a Table to Solve a System

Work with a partner. Use the cost and revenue equations from Activity 1 to find how many nights your family needs to rent the bedroom before recovering the cost of fixing up the bedroom. This is the *break-even point*.

a. Copy and complete the table.

x	0	1	2	3	4	5	6	7	8	9	10	11
C												
R												

b. How many nights does your family need to rent the bedroom before breaking even?

COMMON CORE

Systems of Equations

In this lesson, you will

- write and solve systems of linear equations by graphing.
- solve real-life problems.

Learning Standards
MACC.8.EE.3.8a
MACC.8.EE.3.8b
MACC.8.EE.3.8c

3 ACTIVITY: Using a Graph to Solve a System

Work with a partner.

a. Graph the cost equation from Activity 1.

b. In the same coordinate plane, graph the revenue equation from Activity 1.

c. Find the point of intersection of the two graphs. What does this point represent? How does this compare to the break-even point in Activity 2? Explain.

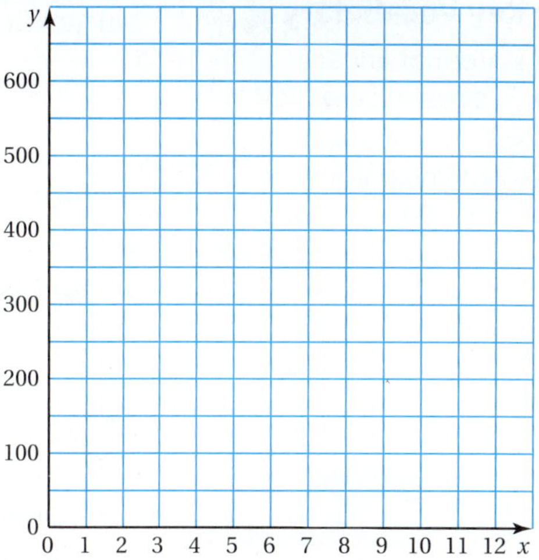

4 ACTIVITY: Using a Graphing Calculator

Work with a partner. Use a graphing calculator to solve the system.

$y = 10x + 500$ Equation 1

$y = 60x$ Equation 2

Math Practice 5

Use Technology to Explore

How do you decide the values for the viewing window of your calculator? What other viewing windows could you use?

a. Enter the equations into your calculator. Then graph the equations. What is an appropriate window?

b. On your graph, how can you determine which line is the graph of which equation? Label the equations on the graph shown.

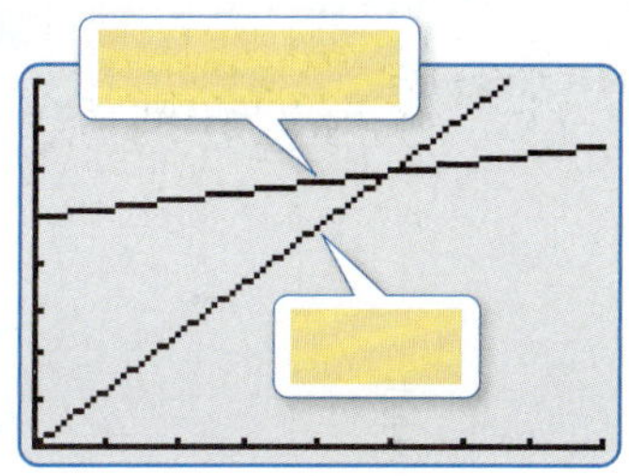

c. Visually estimate the point of intersection of the graphs.

d. To find the solution, use the *intersect* feature to find the point of intersection. The solution is (___ , ___).

What Is Your Answer?

5. **IN YOUR OWN WORDS** How can you solve a system of linear equations? How can you check your solution?

6. **CHOOSE TOOLS** Solve one of the systems by using a table, another system by sketching a graph, and the remaining system by using a graphing calculator. Explain why you chose each method.

a. $y = 4.3x + 1.2$
$y = -1.7x - 2.4$

b. $y = x$
$y = -2x + 9$

c. $y = -x - 5$
$y = 3x + 1$

Practice Use what you learned about systems of linear equations to complete Exercises 4–6 on page 206.

5.1 Lesson

Check It Out
Lesson Tutorials
BigIdeasMath.com

Key Vocabulary
system of linear equations, *p. 204*
solution of a system of linear equations, *p. 204*

A **system of linear equations** is a set of two or more linear equations in the same variables. An example is shown below.

$$y = x + 1 \qquad \text{Equation 1}$$
$$y = 2x - 7 \qquad \text{Equation 2}$$

A **solution of a system of linear equations** in two variables is an ordered pair that is a solution of each equation in the system. The solution of a system of linear equations is the point of intersection of the graphs of the equations.

Reading
A system of linear equations is also called a *linear system*.

Key Idea

Solving a System of Linear Equations by Graphing

Step 1: Graph each equation in the same coordinate plane.

Step 2: Estimate the point of intersection.

Step 3: Check the point from Step 2 by substituting for x and y in each equation of the original system.

EXAMPLE 1 Solving a System of Linear Equations by Graphing

Solve the system by graphing. $\quad$ $\mathbf{y = 2x + 5}$ Equation 1
$\qquad\qquad\qquad\qquad\qquad\quad$ $\mathbf{y = -4x - 1}$ Equation 2

Step 1: Graph each equation.

Step 2: Estimate the point of intersection. The graphs appear to intersect at $(-1, 3)$.

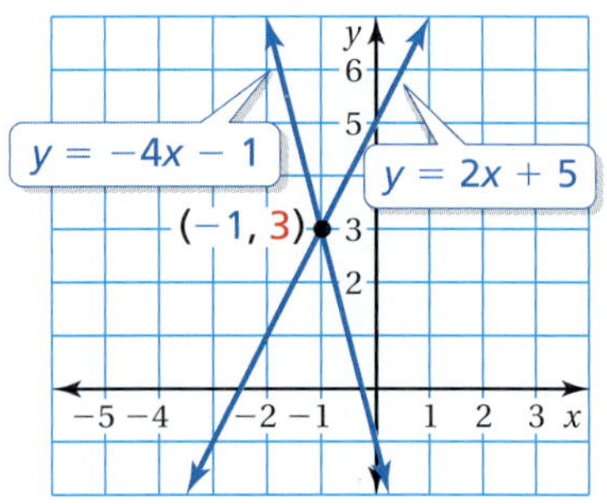

Step 3: Check the point from Step 2.

Equation 1	Equation 2
$y = 2x + 5$	$y = -4x - 1$
$3 \stackrel{?}{=} 2(-1) + 5$	$3 \stackrel{?}{=} -4(-1) - 1$
$3 = 3$ ✓	$3 = 3$ ✓

∴ The solution is $(-1, 3)$.

Check

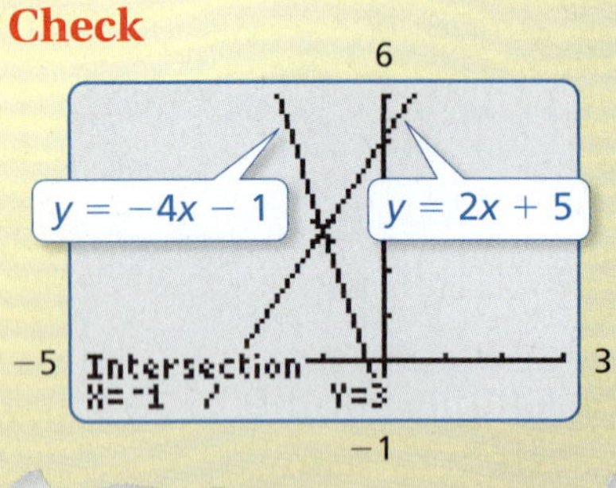

On Your Own

Now You're Ready
Exercises 10–12

Solve the system of linear equations by graphing.

1. $y = x - 1$
   $y = -x + 3$

2. $y = -5x + 14$
   $y = x - 10$

3. $y = x$
   $y = 2x + 1$

Multi-Language Glossary at BigIdeasMath.com

EXAMPLE 2 Real-Life Application

A kicker on a football team scores 1 point for making an extra point and 3 points for making a field goal. The kicker makes a total of 8 extra points and field goals in a game and scores 12 points. Write and solve a system of linear equations to find the number x of extra points and the number y of field goals.

Use a verbal model to write a system of linear equations.

Number of extra points, x	+	Number of field goals, y	=	Total number of kicks

Points per extra point	•	Number of extra points, x	+	Points per field goal	•	Number of field goals, y	=	Total number of points

The system is: $x + y = 8$ Equation 1

$x + 3y = 12$ Equation 2

Step 1: Graph each equation.

Step 2: Estimate the point of intersection. The graphs appear to intersect at (6, 2).

Step 3: Check your point from Step 2.

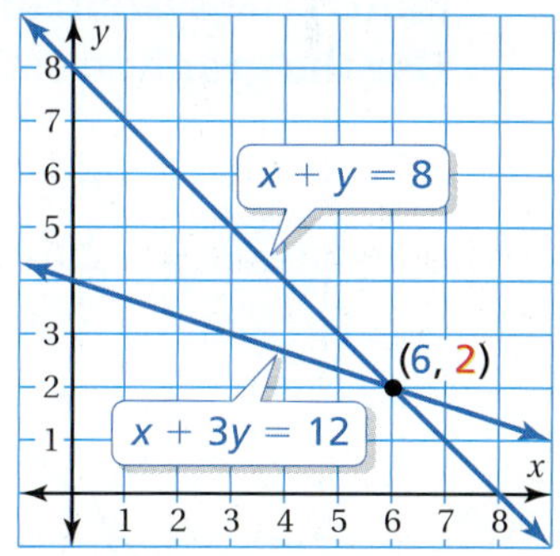

Equation 1	Equation 2
$x + y = 8$	$x + 3y = 12$
$6 + 2 \stackrel{?}{=} 8$	$6 + 3(2) \stackrel{?}{=} 12$
$8 = 8$ ✓	$12 = 12$ ✓

∴ The solution is (6, 2). So, the kicker made 6 extra points and 2 field goals.

Check

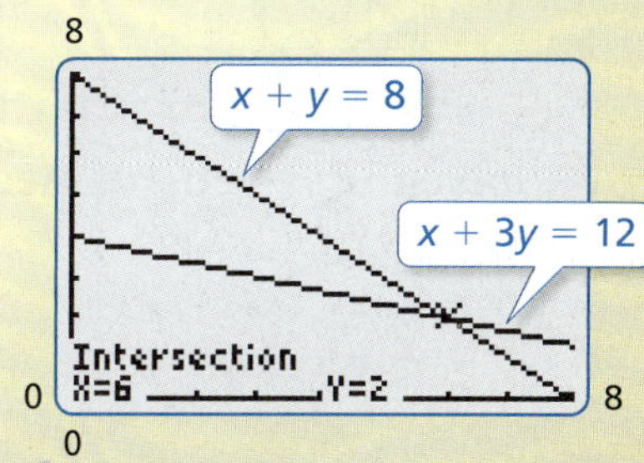

Study Tip

It may be easier to graph the equations in a system by rewriting the equations in slope-intercept form.

On Your Own

Now You're Ready
Exercises 13–15

Solve the system of linear equations by graphing.

4. $y = -4x - 7$
$x + y = 2$

5. $x - y = 5$
$-3x + y = -1$

6. $\frac{1}{2}x + y = -6$
$6x + 2y = 8$

7. **WHAT IF?** The kicker makes a total of 7 extra points and field goals and scores 17 points. Write and solve a system of linear equations to find the numbers of extra points and field goals.

5.1 Exercises

Check It Out
Help with Homework
BigIdeasMath.com

Vocabulary and Concept Check

1. **VOCABULARY** Do the equations $4x - 3y = 5$ and $7y + 2x = -8$ form a system of linear equations? Explain.
2. **WRITING** What does it mean to solve a system of equations?
3. **WRITING** You graph a system of linear equations, and the solution appears to be (3, 4). How can you verify that the solution is (3, 4)?

Practice and Problem Solving

Use a table to find the break-even point. Check your solution.

4. $C = 15x + 150$
$R = 45x$

5. $C = 24x + 80$
$R = 44x$

6. $C = 36x + 200$
$R = 76x$

Match the system of linear equations with the corresponding graph. Use the graph to estimate the solution. Check your solution.

7. $y = 1.5x - 2$
$y = -x + 13$

8. $y = x + 4$
$y = 3x - 1$

9. $y = \frac{2}{3}x - 3$
$y = -2x + 5$

A.

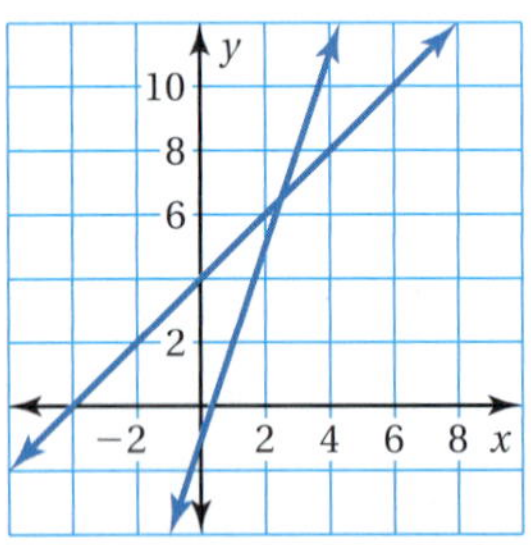

B.

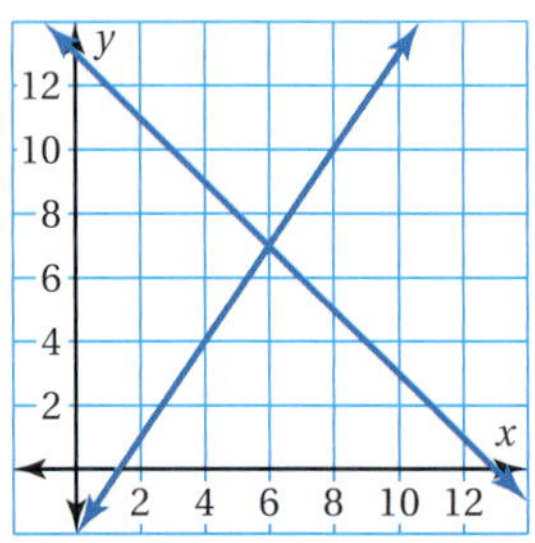

C.

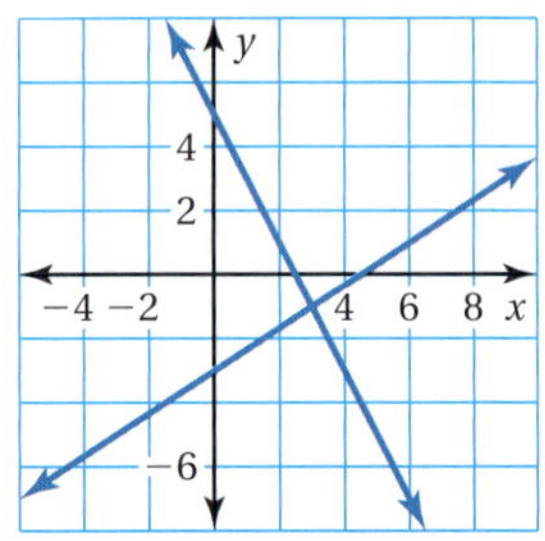

Solve the system of linear equations by graphing.

(1) **10.** $y = 2x + 9$
$y = 6 - x$

11. $y = -x - 4$
$y = \frac{3}{5}x + 4$

12. $y = 2x + 5$
$y = \frac{1}{2}x - 1$

(2) **13.** $x + y = 27$
$y = x + 3$

14. $y - x = 17$
$y = 4x + 2$

15. $x - y = 7$
$0.5x + y = 5$

16. **CARRIAGE RIDES** The cost C (in dollars) for the care and maintenance of a horse and carriage is $C = 15x + 2000$, where x is the number of rides.

a. Write an equation for the revenue R in terms of the number of rides.

b. How many rides are needed to break even?

Use a graphing calculator to solve the system of linear equations.

17. $2.2x + y = 12.5$
$1.4x - 4y = 1$

18. $2.1x + 4.2y = 14.7$
$-5.7x - 1.9y = -11.4$

19. $-1.1x - 5.5y = -4.4$
$0.8x - 3.2y = -11.2$

20. ERROR ANALYSIS Describe and correct the error in solving the system of linear equations.

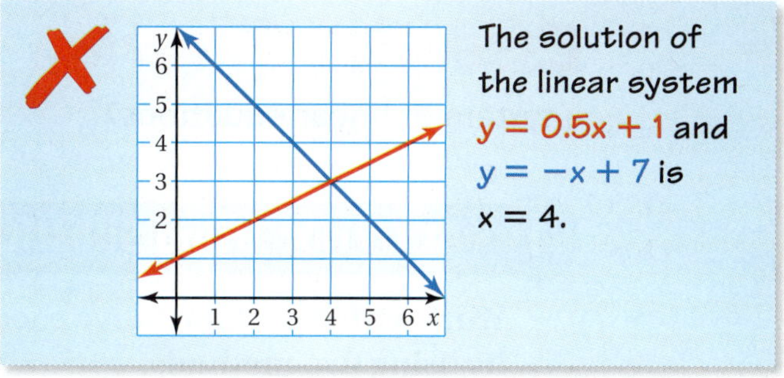

21. REASONING Is it possible for a system of two linear equations to have exactly two solutions? Explain your reasoning.

22. MODELING You have a total of 42 math and science problems for homework. You have 10 more math problems than science problems. How many problems do you have in each subject? Use a system of linear equations to justify your answer.

23. CANOE RACE You and your friend are in a canoe race. Your friend is a half mile in front of you and paddling 3 miles per hour. You are paddling 3.4 miles per hour.

a. You are 8.5 miles from the finish line. How long will it take you to catch up to your friend?

b. You both maintain your paddling rates for the remainder of the race. How far ahead of your friend will you be when you cross the finish line?

24. Critical Thinking Your friend is trying to grow her hair as long as her cousin's hair. The table shows their hair lengths (in inches) in different months.

Month	Friend's Hair (in.)	Cousin's Hair (in.)
3	4	7
8	6.5	9

a. Write a system of linear equations that represents this situation.

b. Will your friend's hair ever be as long as her cousin's hair? If so, in what month?

Fair Game Review What you learned in previous grades & lessons

Solve the equation. Check your solution. *(Section 1.2)*

25. $\frac{3}{4}c - \frac{1}{4}c + 3 = 7$

26. $5(2 - y) + y = -6$

27. $6x - 3(x + 8) = 9$

28. MULTIPLE CHOICE What is the slope of the line that passes through $(-2, -2)$ and $(3, -1)$? *(Section 4.2)*

Ⓐ -5 Ⓑ $-\frac{1}{5}$ Ⓒ $\frac{1}{5}$ Ⓓ 5

5.2 Solving Systems of Linear Equations by Substitution

Essential Question How can you use substitution to solve a system of linear equations?

1 ACTIVITY: Using Substitution to Solve a System

Work with a partner. Solve each system of linear equations by using two methods.

Method 1: Solve for x first.

Solve for x in one of the equations. Use the expression for x to find the solution of the system. Explain how you did it.

Method 2: Solve for y first.

Solve for y in one of the equations. Use the expression for y to find the solution of the system. Explain how you did it.

Is the solution the same using both methods?

a. $6x - y = 11$
$2x + 3y = 7$

b. $2x - 3y = -1$
$x - y = 1$

c. $3x + y = 5$
$5x - 4y = -3$

d. $5x - y = 2$
$3x - 6y = 12$

e. $x + y = -1$
$5x + y = -13$

f. $2x - 6y = -6$
$7x - 8y = 5$

2 ACTIVITY: Writing and Solving a System of Equations

Work with a partner.

a. Roll a pair of number cubes that have different colors. Then write the ordered pair shown by the number cubes. The ordered pair at the right is (3, 4).

b. Write a system of linear equations that has this ordered pair as its solution.

c. Exchange systems with your partner. Use one of the methods from Activity 1 to solve the system.

COMMON CORE

Systems of Equations

In this lesson, you will

- write and solve systems of linear equations by substitution.
- solve real-life problems.

Learning Standards
MACC.8.EE.3.8b
MACC.8.EE.3.8c

3 ACTIVITY: Solving a Secret Code

Math Practice 1

Check Progress

As you complete each system of equations, how do you know your answer is correct?

Work with a partner. Decode the quote by Archimedes.

__ __ __ __   __ __   __   __ __ __ __ __   __ __   __ __ __ __ __ ,

-8 -7 7 -5 -4 -5 -3 -2 -1 -3 0 -5 1 2 3 1 -3 4 5

__ __ __   __   __ __ __ __   __ __ __ __   __ __ __   __ __ __ __ __ .

-3 4 5 -7 6 -7 -1 -1 -4 2 7 -5 1 8 -5 -5 -3 9 1 8

(A, C) $x + y = -3$, $x - y = -3$

(D, E) $x + y = 0$, $x - y = 10$

(G, H) $x + y = 0$, $x - y = -16$

(I, L) $x + 2y = -9$, $2x - y = -13$

(M, N) $x + 2y = 4$, $2x - y = -12$

(O, P) $x + 2y = -2$, $2x - y = 6$

(R, S) $2x + y = 21$, $x - y = 6$

(T, U) $2x + y = -7$, $x - y = 10$

(V, W) $2x + y = 20$, $x - y = 1$

What Is Your Answer?

4. **IN YOUR OWN WORDS** How can you use substitution to solve a system of linear equations?

Practice Use what you learned about systems of linear equations to complete Exercises 4–6 on page 212.

5.2 Lesson

Check It Out
Lesson Tutorials
BigIdeasMath.com

Another way to solve systems of linear equations is to use substitution.

Key Idea

Solving a System of Linear Equations by Substitution

Step 1: Solve one of the equations for one of the variables.

Step 2: Substitute the expression from Step 1 into the other equation and solve for the other variable.

Step 3: Substitute the value from Step 2 into one of the original equations and solve.

EXAMPLE 1 Solving a System of Linear Equations by Substitution

Solve the system by substitution. $\mathbf{y = 2x - 4}$ Equation 1

$\mathbf{7x - 2y = 5}$ Equation 2

Step 1: Equation 1 is already solved for y.

Step 2: Substitute $2x - 4$ for y in Equation 2.

$7x - 2y = 5$	Equation 2
$7x - 2(2x - 4) = 5$	Substitute $2x - 4$ for y.
$7x - 4x + 8 = 5$	Distributive Property
$3x + 8 = 5$	Combine like terms.
$3x = -3$	Subtract 8 from each side.
$x = -1$	Divide each side by 3.

Step 3: Substitute -1 for x in Equation 1 and solve for y.

$y = 2x - 4$	Equation 1
$= 2(-1) - 4$	Substitute -1 for x.
$= -2 - 4$	Multiply.
$= -6$	Subtract.

The solution is $(-1, -6)$.

Check

Equation 1

$y = 2x - 4$

$-6 \stackrel{?}{=} 2(-1) - 4$

$-6 = -6$ ✓

Equation 2

$7x - 2y = 5$

$7(-1) - 2(-6) \stackrel{?}{=} 5$

$5 = 5$ ✓

On Your Own

Now You're Ready
Exercises 10–15

Solve the system of linear equations by substitution. Check your solution.

1. $y = 2x + 3$
   $y = 5x$
2. $4x + 2y = 0$
   $y = \frac{1}{2}x - 5$
3. $x = 5y + 3$
   $2x + 4y = -1$

EXAMPLE 2 Real-Life Application

You buy a total of 50 turkey burgers and veggie burgers for $90. You pay $2 per turkey burger and $1.50 per veggie burger. Write and solve a system of linear equations to find the number *x* of turkey burgers and the number *y* of veggie burgers you buy.

Use a verbal model to write a system of linear equations.

[Number of turkey burgers, *x*] + [Number of veggie burgers, *y*] = [Total number of burgers]

[Cost per turkey burger] • [Number of turkey burgers, *x*] + [Cost per veggie burger] • [Number of veggie burgers, *y*] = [Total cost]

The system is: $x + y = 50$ Equation 1

$2x + 1.5y = 90$ Equation 2

Step 1: Solve Equation 1 for x.

$x + y = 50$ Equation 1

$x = 50 - y$ Subtract y from each side.

Step 2: Substitute $50 - y$ for x in Equation 2.

$2x + 1.5y = 90$ Equation 2

$2(50 - y) + 1.5y = 90$ Substitute $50 - y$ for x.

$100 - 2y + 1.5y = 90$ Distributive Property

$-0.5y = -10$ Simplify.

$y = 20$ Divide each side by -0.5.

Step 3: Substitute 20 for y in Equation 1 and solve for x.

$x + y = 50$ Equation 1

$x + 20 = 50$ Substitute 20 for y.

$x = 30$ Subtract 20 from each side.

You buy 30 turkey burgers and 20 veggie burgers.

Study Tip

It is easiest to solve for a variable that has a coefficient of 1 or −1.

Check

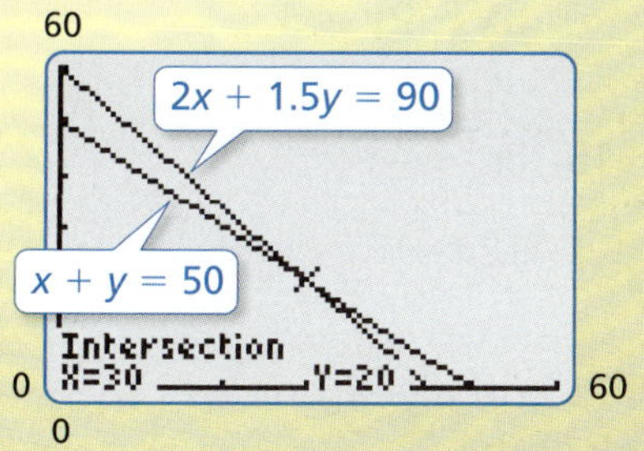

On Your Own

Now You're Ready
Exercises 18–20

4. You sell lemonade for $2 per cup and orange juice for $3 per cup. You sell a total of 100 cups for $240. Write and solve a system of linear equations to find the number of cups of lemonade and the number of cups of orange juice you sold.

5.2 Exercises

Check It Out
Help with Homework
BigIdeasMath.com

Vocabulary and Concept Check

1. **WRITING** Describe how to solve a system of linear equations by substitution.
2. **NUMBER SENSE** When solving a system of linear equations by substitution, how do you decide which variable to solve for in Step 1?
3. **REASONING** Does solving a system of linear equations by graphing give the same solution as solving by substitution? Explain your reasoning.

Practice and Problem Solving

Write a system of linear equations that has the ordered pair as its solution. Use a method from Activity 1 to solve the system.

4. 

5. 

6. 

Tell which equation you would choose to solve for one of the variables when solving the system by substitution. Explain your reasoning.

7. $2x + 3y = 5$
   $4x - y = 3$

8. $\frac{2}{3}x + 5y = -1$
   $x + 6y = 0$

9. $2x + 10y = 14$
   $5x - 9y = 1$

Solve the system of linear equations by substitution. Check your solution.

1 10. $y = x - 4$
   $y = 4x - 10$

11. $y = 2x + 5$
   $y = 3x - 1$

12. $x = 2y + 7$
   $3x - 2y = 3$

13. $4x - 2y = 14$
   $y = \frac{1}{2}x - 1$

14. $2x = y - 10$
   $x + 7 = y$

15. $8x - \frac{1}{3}y = 0$
   $12x + 3 = y$

16. **SCHOOL CLUBS** There are a total of 64 students in a drama club and a yearbook club. The drama club has 10 more students than the yearbook club.
    a. Write a system of linear equations that represents this situation.
    b. How many students are in the drama club? the yearbook club?

17. **THEATER** A drama club earns \$1040 from a production. It sells a total of 64 adult tickets and 132 student tickets. An adult ticket costs twice as much as a student ticket.
    a. Write a system of linear equations that represents this situation.
    b. What is the cost of each ticket?

Solve the system of linear equations by substitution. Check your solution.

2 **18.** $y - x = 0$
$2x - 5y = 9$

19. $x + 4y = 14$
$3x + 7y = 22$

20. $-2x - 5y = 3$
$3x + 8y = -6$

21. ERROR ANALYSIS Describe and correct the error in solving the system of linear equations.

✗

$2x + y = 5$	Equation 1	Step 1:	Step 2:
$3x - 2y = 4$	Equation 2	$2x + y = 5$	$2x + (-2x + 5) = 5$
		$y = -2x + 5$	$2x - 2x + 5 = 5$
			$5 = 5$

22. STRUCTURE The measure of the obtuse angle in the isosceles triangle is two and a half times the measure of one base angle. Write and solve a system of linear equations to find the measures of all the angles.

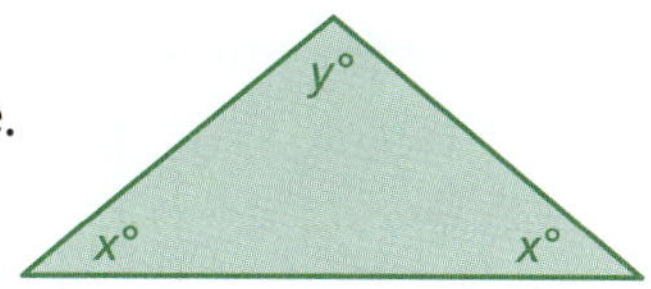

23. ANIMAL SHELTER An animal shelter has a total of 65 abandoned cats and dogs. The ratio of cats to dogs is 6 : 7. How many cats are in the shelter? How many dogs are in the shelter? Justify your answers.

24. NUMBER SENSE The sum of the digits of a two-digit number is 8. When the digits are reversed, the number increases by 36. Find the original number.

25. Repeated Reasoning A DJ has a total of 1075 dance, rock, and country songs on her system. The dance selection is three times the size of the rock selection. The country selection has 105 more songs than the rock selection. How many songs on the system are dance? rock? country?

Fair Game Review What you learned in previous grades & lessons

Write the equation in standard form. *(Section 4.5)*

26. $3x - 9 = 7y$

27. $8 - 5y = -2x$

28. $6x = y + 3$

29. MULTIPLE CHOICE Use the figure to find the measure of $\angle 2$. *(Section 3.1)*

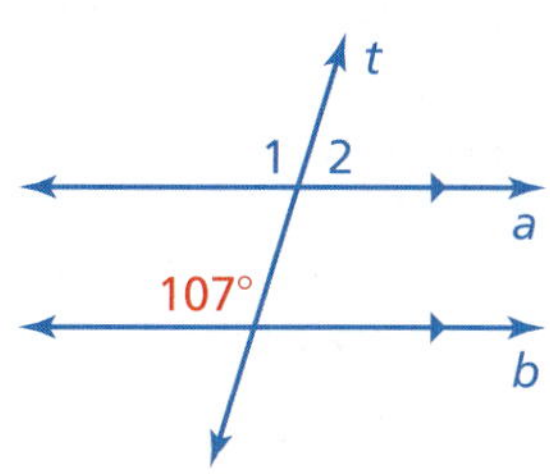

Ⓐ 17° Ⓑ 73°

Ⓒ 83° Ⓓ 107°

5 Study Help

Check It Out
Graphic Organizer
BigIdeasMath.com

You can use a **notetaking organizer** to write notes, vocabulary, and questions about a topic. Here is an example of a notetaking organizer for solving systems of linear equations by graphing.

Write important vocabulary or formulas in this space. →

system of linear equations solution of a system of linear equations	**Solving systems of linear equations by graphing** **Step 1:** Graph each equation. **Step 2:** Estimate the point of intersection. **Step 3:** Check the point from Step 2. Example: Solve the system. $y = x + 1$, $y = -2x - 2$

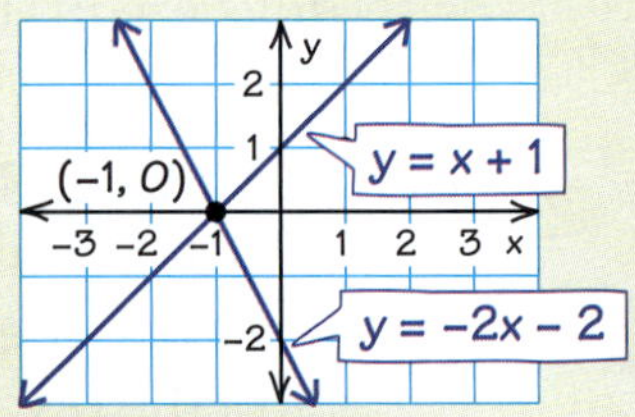

The solution is (–1, 0).

Will a system of linear equations always have a solution?

← Write your notes about the topic in this space.

← Write your questions about the topic in this space.

On Your Own

Make a notetaking organizer to help you study this topic.

1. solving systems of linear equations by substitution

After you complete this chapter, make notetaking organizers for the following topics.

2. solving systems of linear equations by elimination
3. solving systems of linear equations with no solution or infinitely many solutions
4. solving linear equations by graphing

"My notetaking organizer has me thinking about retirement when I won't have to fetch sticks anymore."

5.1–5.2 Quiz

Check It Out
Progress Check
BigIdeasMath.com

Match the system of linear equations with the corresponding graph. Use the graph to estimate the solution. Check your solution. *(Section 5.1)*

1. $y = x - 2$

$y = -2x + 1$

2. $y = x - 3$

$y = -\frac{1}{3}x + 1$

3. $y = \frac{1}{2}x - 2$

$y = 4x + 5$

A.

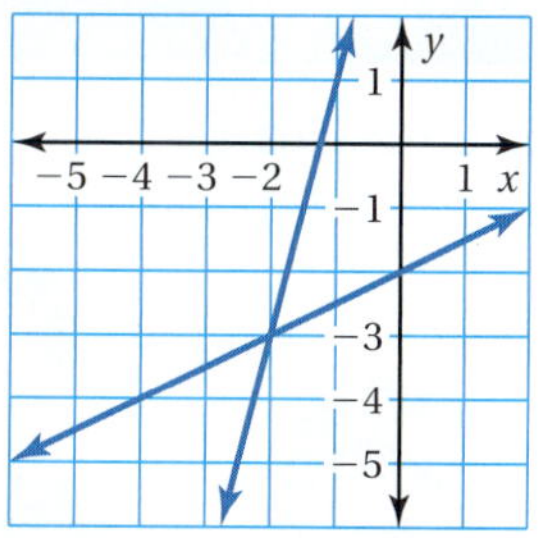

B.

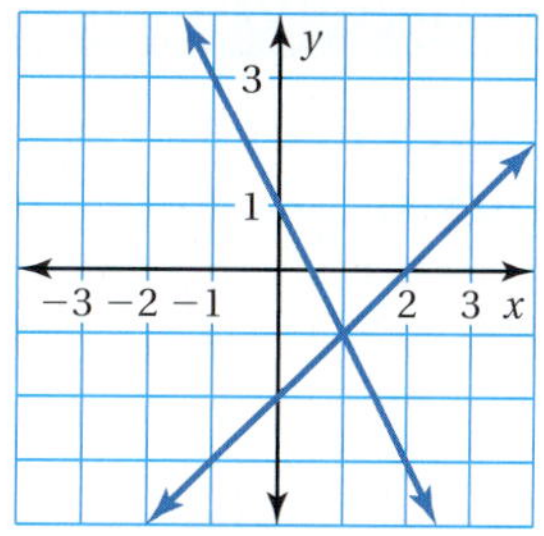

C.

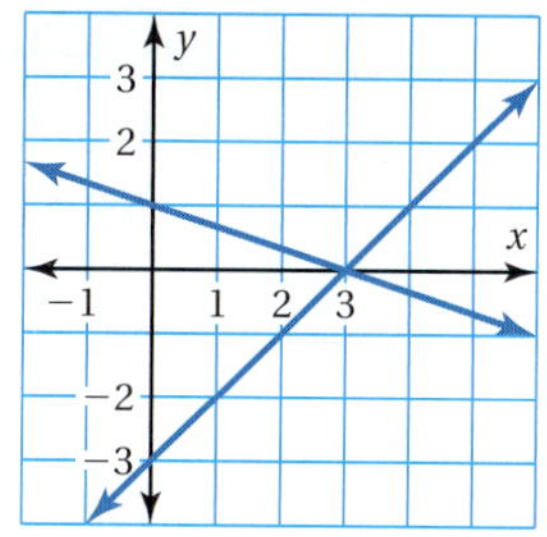

Solve the system of linear equations by graphing. *(Section 5.1)*

4. $y = 2x - 3$

$y = -x + 9$

5. $6x + y = -2$

$y = -3x + 1$

6. $4x + 2y = 2$

$3x = 4 - y$

Solve the system of linear equations by substitution. Check your solution. *(Section 5.2)*

7. $y = x - 8$

$y = 2x - 14$

8. $x = 2y + 2$

$2x - 5y = 1$

9. $x - 5y = 1$

$-2x + 9y = -1$

10. MOVIE CLUB Members of a movie rental club pay a \$15 annual membership fee and \$2 for new release movies. Nonmembers pay \$3 for new release movies. *(Section 5.1)*

a. Write a system of linear equations that represents this situation.

b. When is it beneficial to have a membership?

11. NUMBER SENSE The sum of two numbers is 38. The greater number is 8 more than the other number. Find each number. Use a system of linear equations to justify your answer. *(Section 5.1)*

12. VOLLEYBALL The length of a sand volleyball court is twice its width. The perimeter is 180 feet. Find the length and width of the sand volleyball court. *(Section 5.2)*

13. MEDICAL STAFF A hospital employs a total of 77 nurses and doctors. The ratio of nurses to doctors is 9 : 2. How many nurses are employed at the hospital? How many doctors are employed at the hospital? *(Section 5.2)*

5.3 Solving Systems of Linear Equations by Elimination

Essential Question How can you use elimination to solve a system of linear equations?

1 ACTIVITY: Using Elimination to Solve a System

Work with a partner. Solve each system of linear equations by using two methods.

Method 1: Subtract.

Subtract Equation 2 from Equation 1. What is the result? Explain how you can use the result to solve the system of equations.

Method 2: Add.

Add the two equations. What is the result? Explain how you can use the result to solve the system of equations.

Is the solution the same using both methods?

a. $2x + y = 4$
$2x - y = 0$

b. $3x - y = 4$
$3x + y = 2$

c. $x + 2y = 7$
$x - 2y = -5$

2 ACTIVITY: Using Elimination to Solve a System

Work with a partner.

$2x + y = 2$ Equation 1

$x + 5y = 1$ Equation 2

a. Can you add or subtract the equations to solve the system of linear equations? Explain.

b. Explain what property you can apply to Equation 1 in the system so that the y-coefficients are the same.

c. Explain what property you can apply to Equation 2 in the system so that the x-coefficients are the same.

d. You solve the system in part (b). Your partner solves the system in part (c). Compare your solutions.

e. Use a graphing calculator to check your solution.

COMMON CORE

Systems of Equations

In this lesson, you will

- write and solve systems of linear equations by elimination.
- solve real-life problems.

Learning Standards
MACC.8.EE.3.8b
MACC.8.EE.3.8c

3 ACTIVITY: Solving a Secret Code

Math Practice 1

Find Entry Points

What is the first thing you do to solve a system of linear equations? Why?

Work with a partner. Solve the puzzle to find the name of a famous mathematician who lived in Egypt around 350 A.D.

4	B	W	R	M	F	Y	K	N
3	O	J	A	S	I	D	X	Z
2	Q	P	C	E	G	B	T	J
1	M	R	C	Z	N	O	U	W
0	K	X	U	H	L	Y	S	Q
−1	F	E	A	S	W	K	R	M
−2	G	J	Z	N	H	V	D	G
−3	E	L	X	L	F	Q	O	B
	−3	−2	−1	0	1	2	3	4

□ □ □ □ □ □ □

1. $2x + y = 0$, $x - y = 3$
2. $x + y = 2$, $2x - 2y = 4$
3. $3x + 3y = 0$, $2x - 2y = -8$
4. $2x + y = -3$, $x - y = 0$
5. $x + y = 5$, $x - y = 1$
6. $x + y = 4$, $x - y = -2$
7. $x + 2y = 5$, $2x - y = -5$

What Is Your Answer?

4. **IN YOUR OWN WORDS** How can you use elimination to solve a system of linear equations?
5. **STRUCTURE** When can you add or subtract equations in a system to solve the system? When do you have to multiply first? Justify your answers with examples.
6. **LOGIC** In Activity 2, why can you multiply equations in the system by a constant and not change the solution of the system? Explain your reasoning.

Practice Use what you learned about systems of linear equations to complete Exercises 4–6 on page 221.

5.3 Lesson

Check It Out
Lesson Tutorials
BigIdeasMath.com

Key Idea

Solving a System of Linear Equations by Elimination

Step 1: Multiply, if necessary, one or both equations by a constant so at least 1 pair of like terms has the same or opposite coefficients.

Step 2: Add or subtract the equations to eliminate one of the variables.

Step 3: Solve the resulting equation for the remaining variable.

Step 4: Substitute the value from Step 3 into one of the original equations and solve.

EXAMPLE 1 Solving a System of Linear Equations by Elimination

Solve the system by elimination.

$x + 3y = -2$ Equation 1

$x - 3y = 16$ Equation 2

Step 1: The coefficients of the y-terms are already opposites.

Step 2: Add the equations.

$x + 3y = -2$ Equation 1

$x - 3y = 16$ Equation 2

$2x = 14$ Add the equations.

Step 3: Solve for x.

$2x = 14$ Equation from Step 2

$x = 7$ Divide each side by 2.

Step 4: Substitute 7 for x in one of the original equations and solve for y.

$x + 3y = -2$ Equation 1

$7 + 3y = -2$ Substitute 7 for x.

$3y = -9$ Subtract 7 from each side.

$y = -3$ Divide each side by 3.

∴ The solution is $(7, -3)$.

Study Tip

Because the coefficients of x are the same, you can also solve the system by subtracting in Step 2.

$x + 3y = -2$

$x - 3y = 16$

$6y = -18$

So, $y = -3$.

Check

Equation 1

$x + 3y = -2$

$7 + 3(-3) \stackrel{?}{=} -2$

$-2 = -2$ ✓

Equation 2

$x - 3y = 16$

$7 - 3(-3) \stackrel{?}{=} 16$

$16 = 16$ ✓

On Your Own

Now You're Ready
Exercises 7–12

Solve the system of linear equations by elimination. Check your solution.

1. $2x - y = 9$
   $4x + y = 21$

2. $-5x + 2y = 13$
   $5x + y = -1$

3. $3x + 4y = -6$
   $7x + 4y = -14$

EXAMPLE 2 Solving a System of Linear Equations by Elimination

Solve the system by elimination.

$$-6x + 5y = 25 \quad \text{Equation 1}$$
$$-2x - 4y = 14 \quad \text{Equation 2}$$

Step 1: Multiply Equation 2 by 3.

$-6x + 5y = 25$		$-6x + 5y = 25$	Equation 1
$-2x - 4y = 14$	**Multiply by 3.** →	$-6x - 12y = 42$	Revised Equation 2

Step 2: Subtract the equations.

$$-6x + 5y = 25 \quad \text{Equation 1}$$
$$-6x - 12y = 42 \quad \text{Revised Equation 2}$$
$$17y = -17 \quad \text{Subtract the equations.}$$

Step 3: Solve for y.

$$17y = -17 \quad \text{Equation from Step 2}$$
$$y = -1 \quad \text{Divide each side by 17.}$$

Step 4: Substitute -1 for y in one of the original equations and solve for x.

$$-2x - 4y = 14 \quad \text{Equation 2}$$
$$-2x - 4(-1) = 14 \quad \text{Substitute } -1 \text{ for } y.$$
$$-2x + 4 = 14 \quad \text{Multiply.}$$
$$-2x = 10 \quad \text{Subtract 4 from each side.}$$
$$x = -5 \quad \text{Divide each side by } -2.$$

The solution is $(-5, -1)$.

Check

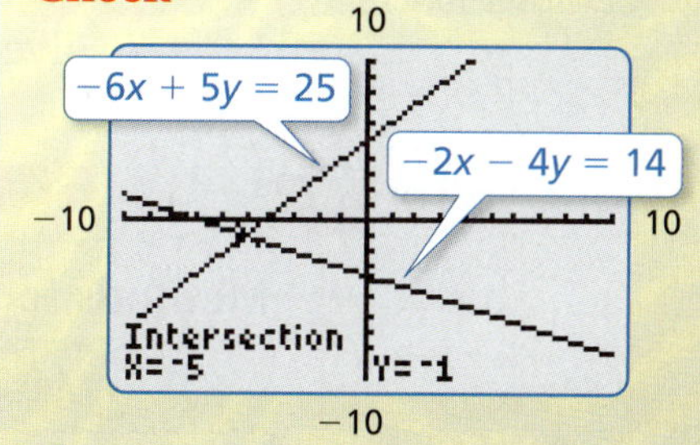

Study Tip

In Example 2, notice that you can also multiply Equation 2 by -3 and then add the equations.

On Your Own

Solve the system of linear equations by elimination. Check your solution.

4. $3x + y = 11$
   $6x + 3y = 24$

5. $4x - 5y = -19$
   $-x - 2y = 8$

6. $5y = 15 - 5x$
   $y = -2x + 3$

EXAMPLE 3 Real-Life Application

You buy 8 hostas and 15 daylilies for \$193. Your friend buys 3 hostas and 12 daylilies for \$117. Write and solve a system of linear equations to find the cost of each daylily.

Use a verbal model to write a system of linear equations.

Number of hostas	$\cdot$	Cost of each hosta, x	$+$	Number of daylilies	$\cdot$	Cost of each daylily, y	$=$	Total cost

The system is: $8x + 15y = 193$ — Equation 1 (You)

$3x + 12y = 117$ — Equation 2 (Your friend)

Step 1: To find the cost y of each daylily, eliminate the x-terms. Multiply Equation 1 by 3. Multiply Equation 2 by 8.

$8x + 15y = 193$ **Multiply by 3.** → $24x + 45y = 579$ — Revised Equation 1

$3x + 12y = 117$ **Multiply by 8.** → $24x + 96y = 936$ — Revised Equation 2

Step 2: Subtract the revised equations.

$$\begin{aligned} 24x + 45y &= 579 && \text{Revised Equation 1} \\ 24x + 96y &= 936 && \text{Revised Equation 2} \\ \hline -51y &= -357 && \text{Subtract the equations.} \end{aligned}$$

Step 3: Solving the equation $-51y = -357$ gives $y = 7$.

∴ So, each daylily costs \$7.

On Your Own

Now You're Ready
Exercises 16–21

7. A landscaper buys 4 peonies and 9 geraniums for \$190. Another landscaper buys 5 peonies and 6 geraniums for \$185. Write and solve a system of linear equations to find the cost of each peony.

Summary

Methods for Solving Systems of Linear Equations

Method	When to Use
Graphing *(Lesson 5.1)*	To estimate solutions
Substitution *(Lesson 5.2)*	When one of the variables in one of the equations has a coefficient of 1 or -1
Elimination *(Lesson 5.3)*	When at least 1 pair of like terms has the same or opposite coefficients
Elimination (Multiply First) *(Lesson 5.3)*	When one of the variables cannot be eliminated by adding or subtracting the equations

5.3 Exercises

Check It Out
Help with Homework
BigIdeasMath.com

Vocabulary and Concept Check

1. **WRITING** Describe how to solve a system of linear equations by elimination.
2. **NUMBER SENSE** When should you use multiplication to solve a system of linear equations by elimination?
3. **WHICH ONE DOESN'T BELONG?** Which system of equations does *not* belong with the other three? Explain your reasoning.

$3x + 3y = 3$ $2x - 3y = 7$	$-2x + y = 6$ $2x - 3y = -10$	$2x + 3y = 11$ $3x - 2y = 10$	$x + y = 5$ $3x - y = 3$

Practice and Problem Solving

Use a method from Activity 1 to solve the system.

4. $x + y = 3$
   $x - y = 1$
5. $-x + 3y = 0$
   $x + 3y = 12$
6. $3x + 2y = 3$
   $3x - 2y = -9$

Solve the system of linear equations by elimination. Check your solution.

(1) 7. $x + 3y = 5$
   $-x - y = -3$
8. $x - 2y = -7$
   $3x + 2y = 3$
9. $4x + 3y = -5$
   $-x + 3y = -10$
10. $2x + 7y = 1$
    $2x - 4y = 12$
11. $2x + 5y = 16$
    $3x - 5y = -1$
12. $3x - 2y = 4$
    $6x - 2y = -2$

13. **ERROR ANALYSIS** Describe and correct the error in solving the system of linear equations.

$5x + 2y = 9$ Equation 1
$3x - 2y = -1$ Equation 2
$2x \quad = 10$
$x = 5$
The solution is (5, −8).

14. **RAFFLE TICKETS** You and your friend are selling raffle tickets for a new laptop. You sell 14 more tickets than your friend sells. Together, you and your friend sell 58 tickets.
    a. Write a system of linear equations that represents this situation.
    b. How many tickets does each of you sell?

15. **JOGGING** You can jog around your block twice and the park once in 10 minutes. You can jog around your block twice and the park 3 times in 22 minutes.
    a. Write a system of linear equations that represents this situation.
    b. How long does it take you to jog around the park?

Solve the system of linear equations by elimination. Check your solution.

2 3 **16.** $2x - y = 0$
$3x - 2y = -3$

17. $x + 4y = 1$
$3x + 5y = 10$

18. $-2x + 3y = 7$
$5x + 8y = -2$

19. $3x + 3 = 3y$
$2x - 6y = 2$

20. $2x - 6 = 4y$
$7y = -3x + 9$

21. $5x = 4y + 8$
$3y = 3x - 3$

22. **ERROR ANALYSIS** Describe and correct the error in solving the system of linear equations.

> ✗
> $x + y = 1$ Equation 1 **Multiply by −5.** → $-5x + 5y = -5$
> $5x + 3y = -3$ Equation 2 → $5x + 3y = -3$
> $8y = -8$
> $y = -1$
>
> The solution is $(2, -1)$.

23. **REASONING** For what values of a and b should you solve the system by elimination?

a. $4x - y = 3$
$ax + 10y = 6$

b. $x - 7y = 6$
$-6x + by = 9$

Determine whether the line through the first pair of points intersects the line through the second pair of points. Explain.

24. Line 1: $(-2, 1), (2, 7)$
Line 2: $(-4, -1), (0, 5)$

25. Line 1: $(3, -2), (7, -1)$
Line 2: $(5, 2), (6, -2)$

26. **AIRPLANES** Two airplanes are flying to the same airport. Their positions are shown in the graph. Write a system of linear equations that represents this situation. Solve the system by elimination to justify your answer.

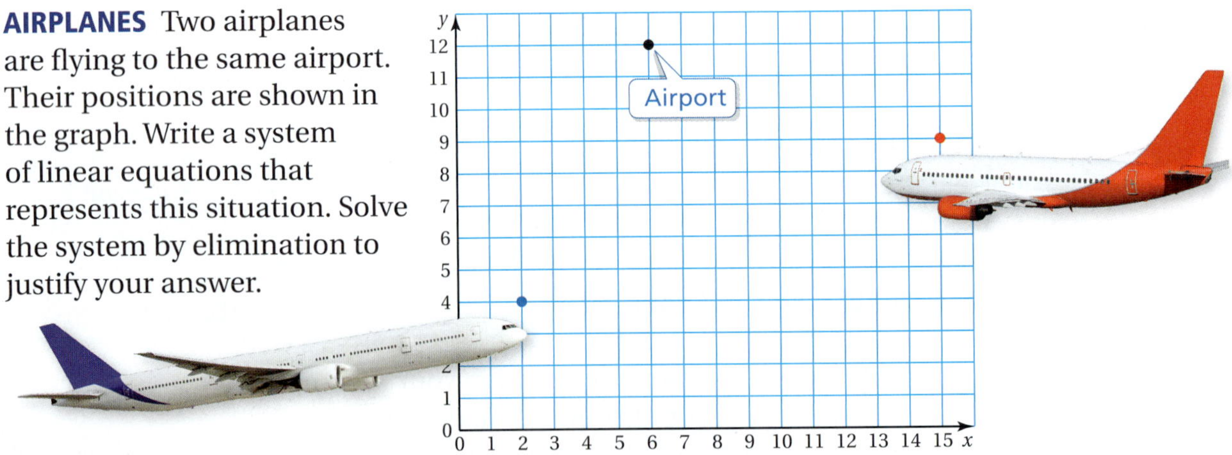

27. **TEST PRACTICE** The table shows the number of correct answers on a practice standardized test. You score 86 points on the test, and your friend scores 76 points.

	You	Your Friend
Multiple Choice	23	28
Short Response	10	5

a. Write a system of linear equations that represents this situation.

b. How many points is each type of question worth?

28. **LOGIC** You solve a system of equations in which x represents the number of adult tickets sold and y represents the number of student tickets sold. Can $(-6, 24)$ be the solution of the system? Explain your reasoning.

29. **VACATION** The table shows the activities of two tourists at a vacation resort. You want to go parasailing for 1 hour and horseback riding for 2 hours. How much do you expect to pay?

	Parasailing	Horseback Riding	Total Cost
Tourist 1	2 hours	5 hours	\$205
Tourist 2	3 hours	3 hours	\$240

30. **REASONING** The solution of a system of linear equations is $(2, -4)$. One equation in the system is $2x + y = 0$. Explain how you could find a second equation for the system. Then find a second equation. Solve the system by elimination to justify your answer.

31. **JEWELER** A metal alloy is a mixture of two or more metals. A jeweler wants to make 8 grams of 18-carat gold, which is 75% gold. The jeweler has an alloy that is 90% gold and an alloy that is 50% gold. How much of each alloy should the jeweler use?

32. **PROBLEM SOLVING** A powerboat takes 30 minutes to travel 10 miles downstream. The return trip takes 50 minutes. What is the speed of the current?

33. **Critical Thinking** Solve the system of equations by elimination.

$$2x - y + 3z = -1$$
$$x + 2y - 4z = -1$$
$$y - 2z = 0$$

Fair Game Review What you learned in previous grades & lessons

Decide whether the two equations are equivalent. *(Section 1.2 and Section 1.3)*

34. $4n + 1 = n - 8$

$3n = -9$

35. $2a + 6 = 12$

$a + 3 = 6$

36. $7v - \frac{3}{2} = 5$

$14v - 3 = 15$

37. **MULTIPLE CHOICE** Which line has the same slope as $y = \frac{1}{2}x - 3$? *(Section 4.4)*

(A) $y = -2x + 4$ (B) $y = 2x + 3$ (C) $y - 2x = 5$ (D) $2y - x = 7$

5.4 Solving Special Systems of Linear Equations

Essential Question Can a system of linear equations have no solution? Can a system of linear equations have many solutions?

1 ACTIVITY: Writing a System of Linear Equations

Work with a partner. Your cousin is 3 years older than you. You can represent your ages by two linear equations.

$y = t$ Your age

$y = t + 3$ Your cousin's age

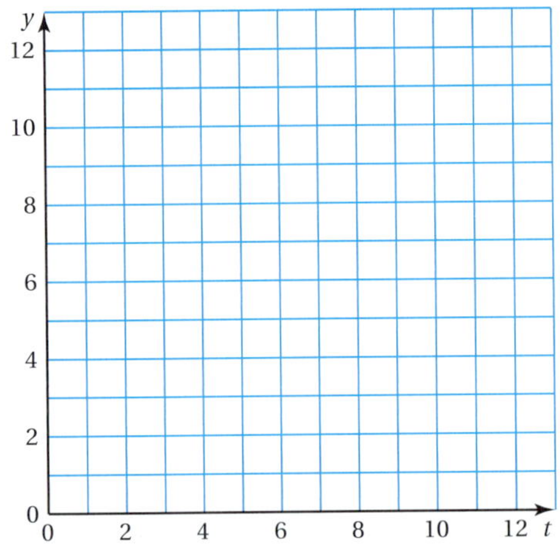

a. Graph both equations in the same coordinate plane.

b. What is the vertical distance between the two graphs? What does this distance represent?

c. Do the two graphs intersect? Explain what this means in terms of your age and your cousin's age.

2 ACTIVITY: Using a Table to Solve a System

Work with a partner. You invest \$500 for equipment to make dog backpacks. Each backpack costs you \$15 for materials. You sell each backpack for \$15.

a. Copy and complete the table for your cost C and your revenue R.

x	0	1	2	3	4	5	6	7	8	9	10
C											
R											

b. When will you break even? What is wrong?

COMMON CORE

Systems of Equations

In this lesson, you will

- solve systems of linear equations with no solution or infinitely many solutions.

Learning Standards
MACC.8.EE.3.8a
MACC.8.EE.3.8b
MACC.8.EE.3.8c

3 ACTIVITY: Using a Graph to Solve a Puzzle

Math Practice 4

Analyze Relationships

What do you know about the graphs of the two equations? How does this relate to the number of solutions?

Work with a partner. Let x and y be two numbers. Here are two clues about the values of x and y.

	Words	Equation
Clue 1:	y is 4 more than twice the value of x.	$y = 2x + 4$
Clue 2:	The difference of $3y$ and $6x$ is 12.	$3y - 6x = 12$

a. Graph both equations in the same coordinate plane.

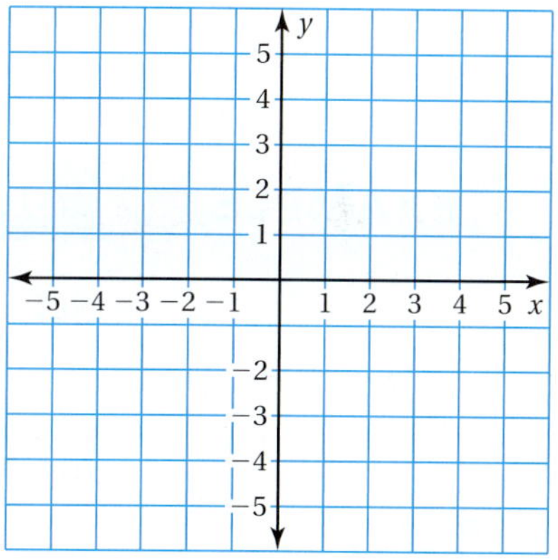

b. Do the two lines intersect? Explain.

c. What is the solution of the puzzle?

d. Use the equation $y = 2x + 4$ to complete the table.

x	0	1	2	3	4	5	6	7	8	9	10
y											

e. Does each solution in the table satisfy *both* clues?

f. What can you conclude? How many solutions does the puzzle have? How can you describe them?

What Is Your Answer?

4. IN YOUR OWN WORDS Can a system of linear equations have no solution? Can a system of linear equations have many solutions? Give examples to support your answers.

Practice Use what you learned about special systems of linear equations to complete Exercises 3 and 4 on page 228.

5.4 Lesson

Check It Out
Lesson Tutorials
BigIdeasMath.com

Key Idea

Solutions of Systems of Linear Equations

A system of linear equations can have *one solution*, *no solution*, or *infinitely many solutions*.

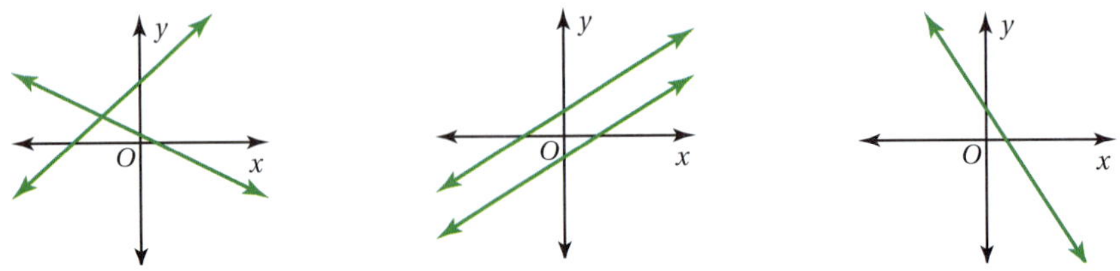

One solution	**No solution**	**Infinitely many solutions**
The lines intersect.	The lines are parallel.	The lines are the same.

EXAMPLE 1 Solving a System: No Solution

Solve the system. $y = 3x + 1$ Equation 1

$y = 3x - 3$ Equation 2

Method 1: Solve by graphing.

Graph each equation. The lines have the same slope and different y-intercepts. So, the lines are parallel.

Because parallel lines do not intersect, there is no point that is a solution of both equations.

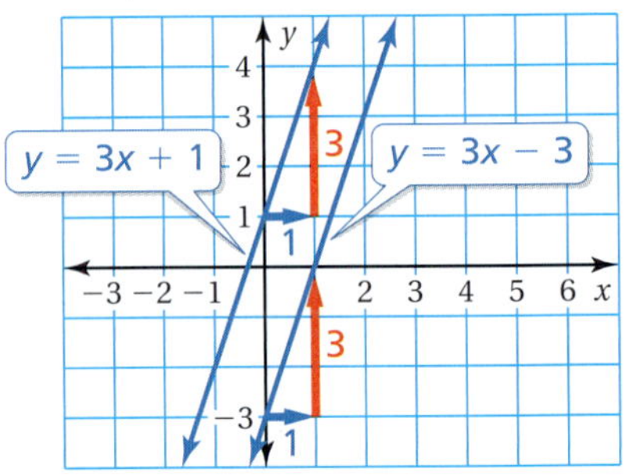

So, the system of linear equations has no solution.

Method 2: Solve by substitution.

Substitute $3x - 3$ for y in Equation 1.

$y = 3x + 1$ Equation 1

$3x - 3 = 3x + 1$ Substitute $3x - 3$ for y.

$-3 = 1$ ✗ Subtract $3x$ from each side.

The equation $-3 = 1$ is never true. So, the system of linear equations has no solution.

Study Tip

You can solve some linear systems by inspection. In Example 1, notice you can rewrite the system as

$-3x + y = 1$

$-3x + y = -3.$

This system has no solution because $-3x + y$ cannot be equal to 1 and -3 at the same time.

On Your Own

Now You're Ready
Exercises 8–10

Solve the system of linear equations. Check your solution.

1. $y = -x + 3$
$y = -x + 5$

2. $y = -5x - 2$
$5x + y = 0$

3. $x = 2y + 10$
$2x + 3y = -1$

EXAMPLE 2 Solving a System: Infinitely Many Solutions

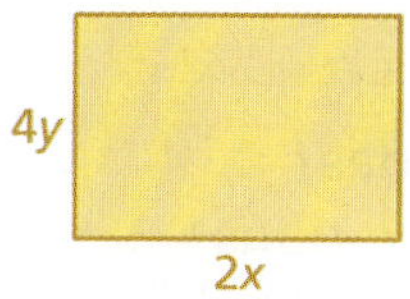

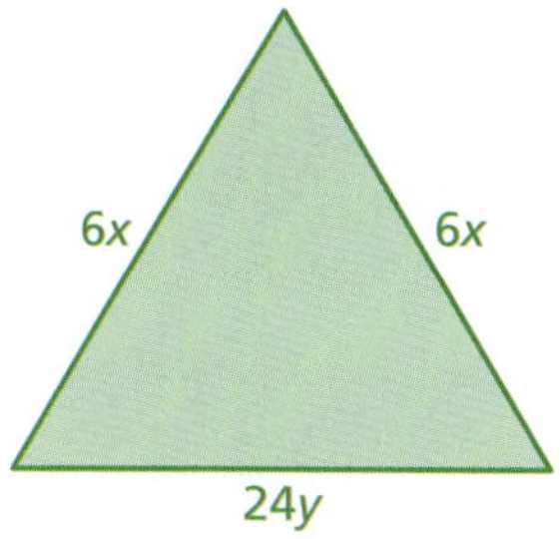

The perimeter of the rectangle is 36 units. The perimeter of the triangle is 108 units. Write and solve a system of linear equations to find the values of *x* and *y*.

Perimeter of Rectangle

$2(2x) + 2(4y) = 36$

$4x + 8y = 36$ Equation 1

Perimeter of Triangle

$6x + 6x + 24y = 108$

$12x + 24y = 108$ Equation 2

The system is: $4x + 8y = 36$ Equation 1

$12x + 24y = 108$ Equation 2

Method 1: Solve by graphing.

Graph each equation.

The lines have the same slope and the same *y*-intercept. So, the lines are the same.

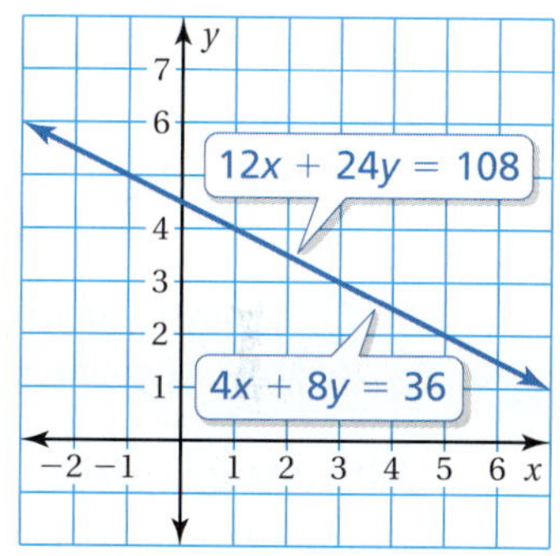

In this context, *x* and *y* must be positive. Because the lines are the same, all the points on the line in Quadrant I are solutions of both equations. So, the system of linear equations has infinitely many solutions.

Method 2: Solve by elimination.

Multiply Equation 1 by 3 and subtract the equations.

$4x + 8y = 36$	**Multiply by 3.**	$12x + 24y = 108$	Revised Equation 1
$12x + 24y = 108$		$12x + 24y = 108$	Equation 2
		$0 = 0$	Subtract.

The equation $0 = 0$ is always true. In this context, *x* and *y* must be positive. So, the solutions are all the points on the line $4x + 8y = 36$ in Quadrant I. The system of linear equations has infinitely many solutions.

On Your Own

Now You're Ready
Exercises 11–13

Solve the system of linear equations. Check your solution.

4. $x + y = 3$
$x - y = -3$

5. $2x + y = 5$
$4x + 2y = 0$

6. $2x - 4y = 10$
$-12x + 24y = -60$

7. **WHAT IF?** What happens to the solution in Example 2 if the perimeter of the rectangle is 54 units? Explain.

5.4 Exercises

Check It Out
Help with Homework
BigIdeasMath.com

Vocabulary and Concept Check

1. **WRITING** Describe the difference between the graph of a system of linear equations that has *no solution* and the graph of a system of linear equations that has *infinitely many solutions*.

2. **REASONING** When solving a system of linear equations algebraically, how do you know when the system has *no solution*? *infinitely many solutions*?

Practice and Problem Solving

Let x and y be two numbers. Find the solution of the puzzle.

3. y is $\frac{1}{3}$ more than 4 times the value of x.

   The difference of $3y$ and $12x$ is 1.

4. $\frac{1}{2}$ of x plus 3 is equal to y.

   x is 6 more than twice the value of y.

Without graphing, determine whether the system of linear equations has *one solution, infinitely many solutions,* or *no solution.* Explain your reasoning.

5. $y = 5x - 9$

   $y = 5x + 9$

6. $y = 6x + 2$

   $y = 3x + 1$

7. $y = 8x - 2$

   $y - 8x = -2$

Solve the system of linear equations. Check your solution.

① 8. $y = 2x - 2$

   $y = 2x + 9$

9. $y = 3x + 1$

   $-x + 2y = -3$

10. $y = \frac{\pi}{3}x + \pi$

    $-\pi x + 3y = -6\pi$

② 11. $y = -\frac{1}{6}x + 5$

    $x + 6y = 30$

12. $\frac{1}{3}x + y = 1$

    $2x + 6y = 6$

13. $-2x + y = 1.3$

    $2(0.5x - y) = 4.6$

14. **ERROR ANALYSIS** Describe and correct the error in solving the system of linear equations.

    ✗ $y = -2x + 4$

    $y = -2x + 6$

    The lines have the same slope, so, there are infinitely many solutions.

15. **PIG RACE** In a pig race, your pig gets a head start of 3 feet and is running at a rate of 2 feet per second. Your friend's pig is also running at a rate of 2 feet per second. A system of linear equations that represents this situation is $y = 2x + 3$ and $y = 2x$. Will your friend's pig catch up to your pig? Explain.

16. REASONING One equation in a system of linear equations has a slope of -3. The other equation has a slope of 4. How many solutions does the system have? Explain.

17. LOGIC How can you use the slopes and the y-intercepts of equations in a system of linear equations to determine whether the system has *one solution, infinitely many solutions,* or *no solution*? Explain your reasoning.

$$4x + 8y = 64$$
$$8x + 16y = 128$$

18. MONEY You and a friend both work two different jobs. The system of linear equations represents the total earnings for x hours worked at the first job and y hours worked at the second job. Your friend earns twice as much as you.

a. One week, both of you work 4 hours at the first job. How many hours do you and your friend work at the second job?

b. Both of you work the same number of hours at the second job. Compare the number of hours each of you works at the first job.

19. DOWNLOADS You download a digital album for \$10. Then you and your friend download the same number of individual songs for \$0.99 each. Write a system of linear equations that represents this situation. Will you and your friend spend the same amount of money? Explain.

20. REASONING Does the system shown *always, sometimes,* or *never* have no solution when $a = b$? $a \geq b$? $a < b$? Explain your reasoning.

$$y = ax + 1$$
$$y = bx + 4$$

21. SKIING The table shows the number of lift tickets and ski rentals sold to two different groups. Is it possible to determine how much each lift ticket costs? Justify your answer.

Group	1	2
Number of Lift Tickets	36	24
Number of Ski Rentals	18	12
Total Cost (dollars)	684	456

22. Precision Find the values of a and b so the system shown has the solution (2, 3). Does the system have any other solutions? Explain.

$$12x - 2by = 12$$
$$3ax - by = 6$$

Fair Game Review What you learned in previous grades & lessons

Write an equation of the line that passes through the given points. *(Section 4.7)*

23. (0, 0), (2, 6)　　**24.** (0, -3), (3, 3)　　**25.** (-6, 5), (0, 2)

26. MULTIPLE CHOICE What is the solution of $-2(y + 5) \leq 16$? *(Skills Review Handbook)*

Ⓐ $y \leq -13$　　Ⓑ $y \geq -13$　　Ⓒ $y \leq -3$　　Ⓓ $y \geq -3$

Extension 5.4 Solving Linear Equations by Graphing

Check It Out
Lesson Tutorials
BigIdeasMath.com

COMMON CORE

Systems of Equations

In this extension, you will

- solve linear equations by graphing a system of linear equations.
- solve real-life problems.

Applying Standards
MACC.8.EE.3.7
MACC.8.EE.3.8

Key Idea

Solving Equations Using Graphs

Step 1: To solve the equation $ax + b = cx + d$, write two linear equations.

$$ax + b = cx + d$$

$y = ax + b$ and $y = cx + d$

Step 2: Graph the system of linear equations. The x-value of the solution of the system of linear equations is the solution of the equation $ax + b = cx + d$.

EXAMPLE 1 Solving an Equation Using a Graph

Solve $x - 2 = -\frac{1}{2}x + 1$ using a graph. Check your solution.

Step 1: Write a system of linear equations using each side of the equation.

$$x - 2 = -\frac{1}{2}x + 1$$

$y = x - 2$ and $y = -\frac{1}{2}x + 1$

Step 2: Graph the system.

$$y = x - 2$$

$$y = -\frac{1}{2}x + 1$$

The graphs intersect at (2, 0).

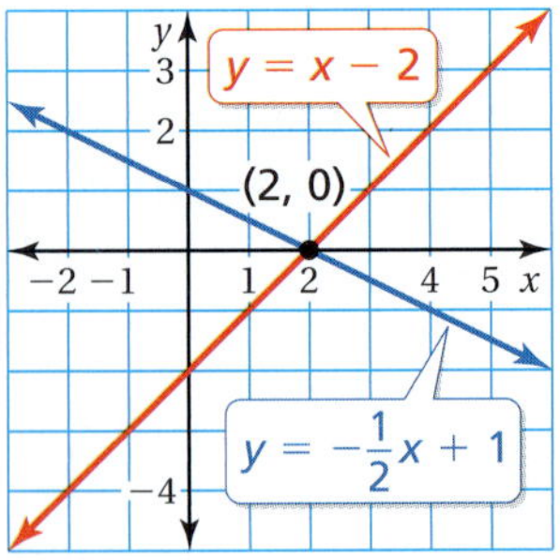

So, the solution of the equation is $x = 2$.

Check

$$x - 2 = -\frac{1}{2}x + 1$$

$$2 - 2 \stackrel{?}{=} -\frac{1}{2}(2) + 1$$

$$0 = 0 \checkmark$$

Practice

Use a graph to solve the equation. Check your solution.

1. $2x + 3 = 4$
2. $2x = x - 3$
3. $3x + 1 = 3x + 2$
4. $\frac{1}{3}x = x + 8$
5. $1.5x + 2 = 11 - 3x$
6. $3 - 2x = -2x + 3$
7. **STRUCTURE** Write an equation with variables on both sides that has no solution. How can you change the equation so that it has infinitely many solutions?

EXAMPLE 2 Real-Life Application

Plant A grows 0.6 inch per month. Plant B grows twice as fast.

a. **Use the model to write an equation.**

b. **After how many months x are the plants the same height?**

Growth rate	•	Months, x	+	Original height	=	Growth rate	•	Months, x	+	Original height

a. The equation is $0.6x + 12 = 1.2x + 9$.

b. Write a system of linear equations using each side of the equation. Then use a graphing calculator to graph the system.

$$0.6x + 12 = 1.2x + 9$$

$y = 0.6x + 12$ $\qquad$ $y = 1.2x + 9$

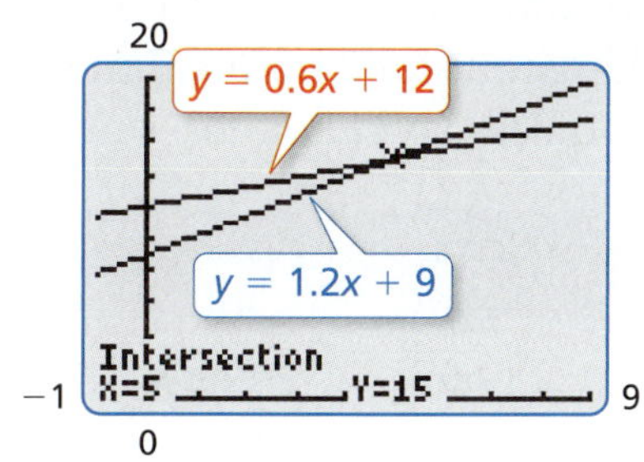

The solution of the system is (5, 15).

So, the plants are both 15 inches tall after 5 months.

Study Tip

One way to check your answer is to solve the equation algebraically as in Section 1.3.

$0.6x + 12 = 1.2x + 9$

$12 = 0.6x + 9$

$3 = 0.6x$

$5 = x$

Practice

Use a graph to solve the equation. Check your solution.

8. $6x - 2 = x + 11$ $\qquad$ **9.** $\frac{4}{3}x - 1 = \frac{2}{3}x + 6$ $\qquad$ **10.** $1.75x = 2.25x + 10.25$

11. **WHAT IF?** In Example 2, the growth rate of Plant A is 0.5 inch per month. After how many months x are the plants the same height?

5.3–5.4 Quiz

Check It Out
Progress Check
BigIdeasMath.com

Solve the system of linear equations by elimination. Check your solution. *(Section 5.3)*

1. $x + 2y = 4$
$-x - y = 2$

2. $2x - y = 1$
$x + 3y - 4 = 0$

3. $3x = -4y + 10$
$4x + 3y = 11$

4. $2x + 5y = 60$
$2x - 5y = -20$

Solve the system of linear equations. Check your solution. *(Section 5.4)*

5. $3x - 2y = 16$
$6x - 4y = 32$

6. $4y = x - 8$
$-\frac{1}{4}x + y = -1$

7. $-2x + y = -2$
$3x + y = 3$

8. $3x = \frac{1}{3}y + 2$
$9x - y = -6$

Use a graph to solve the equation. Check your solution. *(Section 5.4)*

9. $4x - 1 = 2x$

10. $-\frac{1}{2}x + 1 = -x + 1$

11. $1 - 3x = -3x + 2$

12. $1 - 5x = 3 - 7x$

13. TOURS The table shows the activities of two visitors at a park. You want to take the boat tour for 2 hours and the walking tour for 3 hours. Can you determine how much you will pay? Explain. *(Section 5.4)*

	Boat Tour	Walking Tour	Total Cost
Visitor 1	1 hour	2 hours	\$19
Visitor 2	1.5 hours	3 hours	\$28.50

14. RENTALS A business rents bicycles and in-line skates. Bicycle rentals cost \$25 per day, and in-line skate rentals cost \$20 per day. The business has 20 rentals today and makes \$455. *(Section 5.3)*

a. Write a system of linear equations that represents this situation.

b. How many bicycle rentals and in-line skate rentals did the business have today?

5 Chapter Review

Check It Out
Vocabulary Help
BigIdeasMath.com

Review Key Vocabulary

system of linear equations, *p. 204*

solution of a system of linear equations, *p. 204*

Review Examples and Exercises

5.1 Solving Systems of Linear Equations by Graphing *(pp. 202–207)*

Solve the system by graphing. $\mathbf{y = -2x}$ Equation 1
$\mathbf{y = 3x + 5}$ Equation 2

Step 1: Graph each equation.

Step 2: Estimate the point of intersection. The graphs appear to intersect at $(-1, 2)$.

Step 3: Check the point from Step 2.

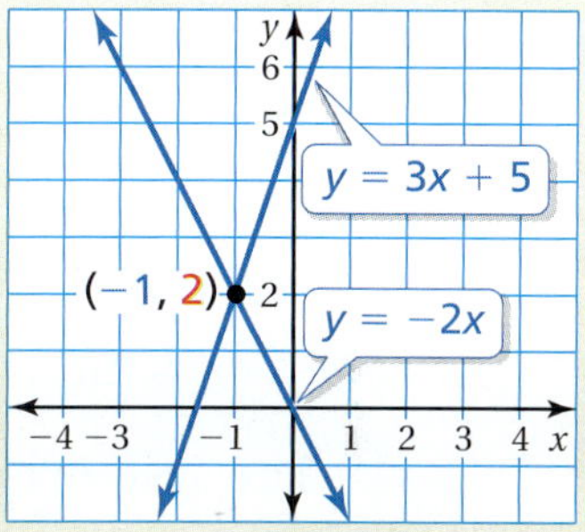

Equation 1	Equation 2
$y = -2x$	$y = 3x + 5$
$2 \stackrel{?}{=} -2(-1)$	$2 \stackrel{?}{=} 3(-1) + 5$
$2 = 2$ ✓	$2 = 2$ ✓

∴ The solution is $(-1, 2)$.

Exercises

Solve the system of linear equations by graphing.

1. $y = 2x - 3$
   $y = x + 2$
2. $y = -x + 4$
   $x + 3y = 0$
3. $x - y = -2$
   $2x - 3y = -2$

5.2 Solving Systems of Linear Equations by Substitution *(pp. 208–213)*

Solve the system by substitution. $\mathbf{x = 1 + y}$ Equation 1
$\mathbf{x + 3y = 13}$ Equation 2

Step 1: Equation 1 is already solved for x.

Step 2: Substitute $1 + y$ for x in Equation 2.

$1 + y + 3y = 13$ Substitute $1 + y$ for x.

$y = 3$ Solve for y.

Step 3: Substituting 3 for y in Equation 1 gives $x = 4$.

∴ The solution is $(4, 3)$.

Exercises

Solve the system of linear equations by substitution. Check your solution.

4. $y = -3x - 7$
   $y = x + 9$

5. $\frac{1}{2}x + y = -4$
   $y = 2x + 16$

6. $-x + 5y = 28$
   $x + 3y = 20$

5.3 Solving Systems of Linear Equations by Elimination (pp. 216–223)

You have a total of 5 quarters and dimes in your pocket. The value of the coins is \$0.80. Write and solve a system of linear equations to find the number *x* of dimes and the number *y* of quarters in your pocket.

Use a verbal model to write a system of linear equations.

Number of dimes, x + Number of quarters, y = Number of coins

Value of a dime • Number of dimes, x + Value of a quarter • Number of quarters, y = Total value

The system is $x + y = 5$ and $0.1x + 0.25y = 0.8$.

Step 1: Multiply Equation 2 by 10.

$x + y = 5$		$x + y = 5$	Equation 1
$0.1x + 0.25y = 0.8$	Multiply by 10.	$x + 2.5y = 8$	Revised Equation 2

Step 2: Subtract the equations.

$$x + y = 5 \quad \text{Equation 1}$$
$$x + 2.5y = 8 \quad \text{Revised Equation 2}$$
$$-1.5y = -3 \quad \text{Subtract the equations.}$$

Step 3: Solving the equation $-1.5y = -3$ gives $y = 2$.

Step 4: Substitute 2 for y in one of the original equations and solve for x.

$$x + y = 5 \quad \text{Equation 1}$$
$$x + 2 = 5 \quad \text{Substitute 2 for } y.$$
$$x = 3 \quad \text{Subtract 2 from each side.}$$

So, you have 3 dimes and 2 quarters in your pocket.

Exercises

7. **GIFT BASKET** A gift basket that contains jars of jam and packages of bread mix costs \$45. There are 8 items in the basket. Jars of jam cost \$6 each, and packages of bread mix cost \$5 each. Write and solve a system of linear equations to find the number of jars of jam and the number of packages of bread mix in the gift basket.

5.4 Solving Special Systems of Linear Equations (pp. 224–231)

a. Solve the system. $y = -5x - 8$ Equation 1

$y = -5x + 4$ Equation 2

Solve by substitution. Substitute $-5x + 4$ for y in Equation 1.

$y = -5x - 8$ Equation 1

$-5x + 4 = -5x - 8$ Substitute $-5x + 4$ for y.

$4 = -8$ ✗ Add $5x$ to each side.

The equation $4 = -8$ is never true. So, the system of linear equations has no solution.

b. Solve $x + 1 = \frac{1}{3}x + 3$ using a graph. Check your solution.

Step 1: Write a system of linear equations using each side of the equation.

$$x + 1 = \frac{1}{3}x + 3$$

$y = x + 1$ $y = \frac{1}{3}x + 3$

Step 2: Graph the system.

$y = x + 1$

$y = \frac{1}{3}x + 3$

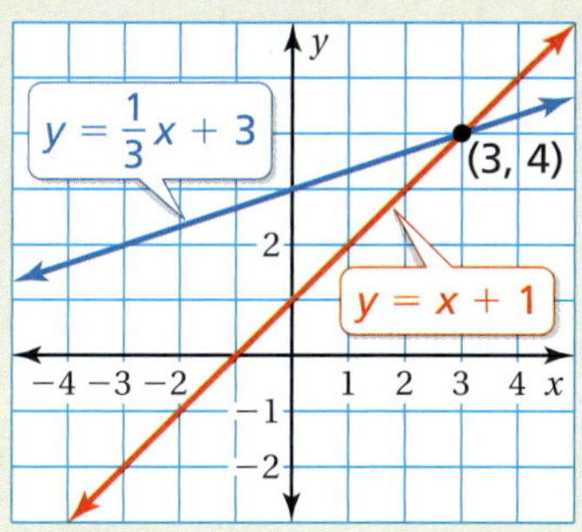

Check

$x + 1 = \frac{1}{3}x + 3$

$3 + 1 \stackrel{?}{=} \frac{1}{3}(3) + 3$

$4 = 4$ ✓

The graphs intersect at (3, 4).

So, the solution is $x = 3$.

Exercises

Solve the system of linear equations. Check your solution.

8. $x + 2y = -5$
$x - 2y = -5$

9. $3x - 2y = 1$
$9x - 6y = 3$

10. $8x - 2y = 16$
$-4x + y = 8$

11. Use a graph to solve $2x - 9 = 7x + 11$. Check your solution.

5 Chapter Test

Check It Out
Test Practice
BigIdeasMath.com

Solve the system of linear equations by graphing.

1. $y = 4 - x$
$y = x - 4$

2. $y = \frac{1}{2}x + 10$
$y = 4x - 4$

3. $y + x = 0$
$3y + 6x = -9$

Solve the system of linear equations by substitution. Check your solution.

4. $-3x + y = 2$
$-x + y - 4 = 0$

5. $x + y = 20$
$y = 2x - 1$

6. $x - y = 3$
$x + 2y = -6$

Solve the system of linear equations by elimination. Check your solution.

7. $2x + y = 3$
$x - y = 3$

8. $x + y = 12$
$3x = 2y + 6$

9. $-2x + y + 3 = 0$
$3x + 4y = -1$

Without graphing, determine whether the system of linear equations has *one solution*, *infinitely many solutions*, or *no solution*. Explain your reasoning.

10. $y = 4x + 8$
$y = 5x + 1$

11. $2y = 16x - 2$
$y = 8x - 1$

12. $y = -3x + 2$
$6x + 2y = 10$

Use a graph to solve the equation. Check your solution.

13. $\frac{1}{4}x - 4 = \frac{3}{4}x + 2$

14. $8x - 14 = -2x - 4$

15. **FRUIT** The price of 2 pears and 6 apples is \$14. The price of 3 pears and 9 apples is \$21. Can you determine the unit prices for pears and apples? Explain.

16. **BOUQUET** A bouquet of lilies and tulips has 12 flowers. Lilies cost \$3 each, and tulips cost \$2 each. The bouquet costs \$32. Write and solve a system of linear equations to find the number of lilies and tulips in the bouquet.

17. **DINNER** How much does it cost for 2 specials and 2 glasses of milk?

5 Standards Assessment

Test-Taking Strategy
Read Question Before Answering

"Take your time and read the question carefully before choosing your answer."

1. What is the solution of the system of equations shown below? *(MACC.8.EE.3.8b)*

$$y = -\frac{2}{3}x - 1$$
$$4x + 6y = -6$$

A. $\left(-\frac{3}{2}, 0\right)$ **C.** no solution

B. $(0, -1)$ **D.** infinitely many solutions

2. What is the slope of a line that is perpendicular to the line $y = -0.25x + 3$? *(MACC.8.EE.2.6)*

3. On the grid below, Rectangle *EFGH* is plotted and its vertices are labeled.

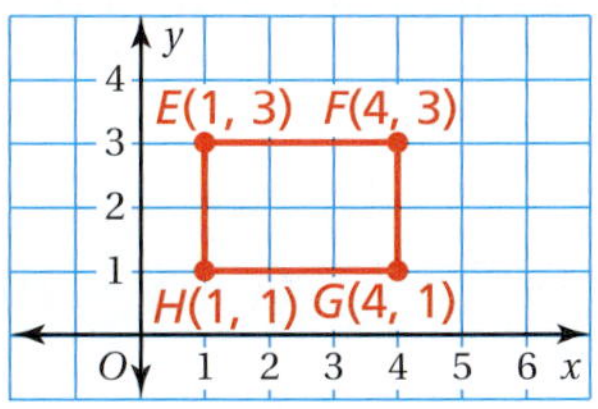

Which of the following shows Rectangle $E'F'G'H'$, the image of Rectangle *EFGH* after it is reflected in the x-axis? *(MACC.8.G.1.3)*

F.
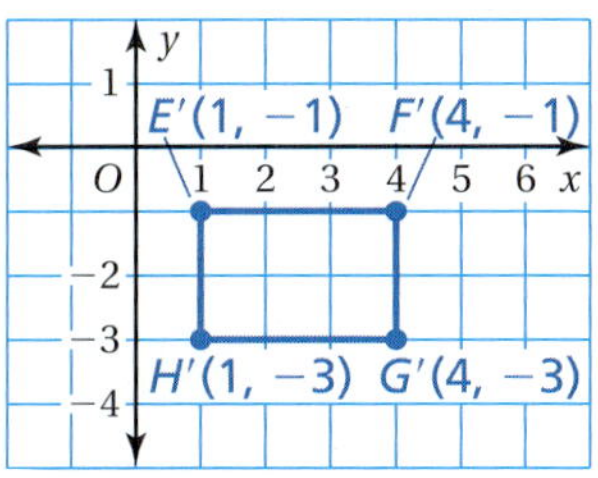

H.
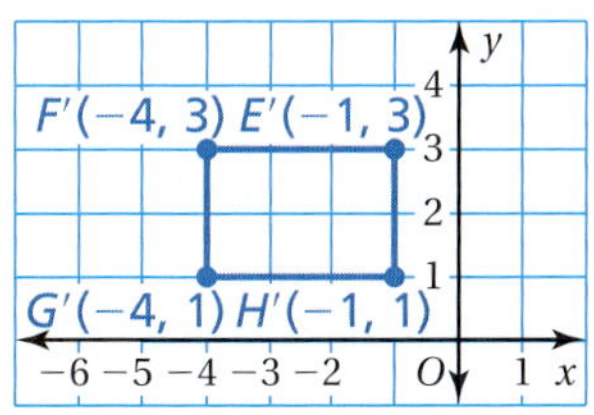

G.
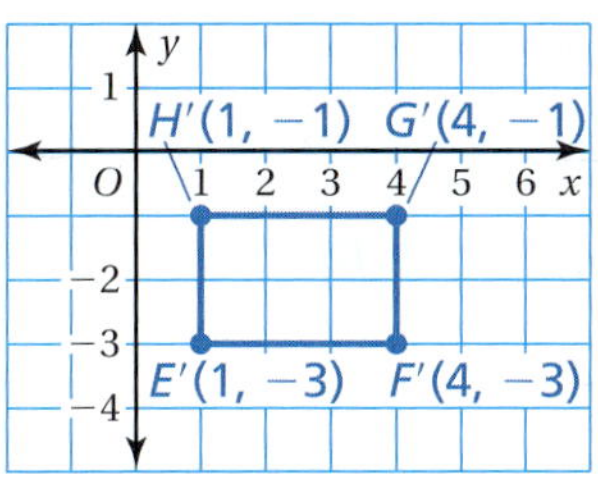

I.
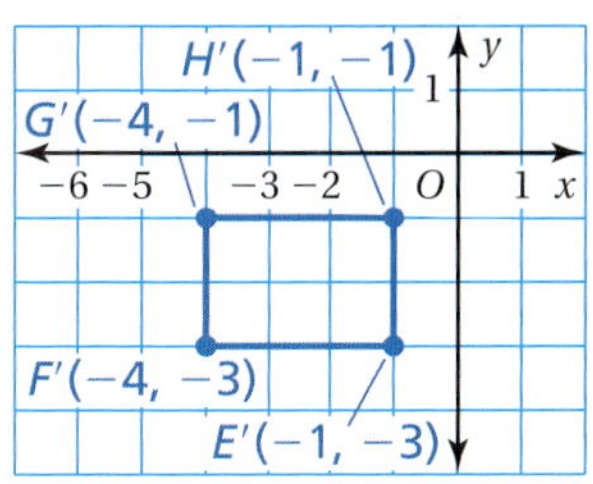

4. Which point is a solution of the system of equations shown below? *(MACC.8.EE.3.8b)*

$$x + 3y = 10$$
$$x = 2y - 5$$

A. (1, 3)

B. (3, 1)

C. (55, −15)

D. (−35, −15)

5. The graph of a system of two linear equations is shown. How many solutions does the system have? *(MACC.8.EE.3.8a)*

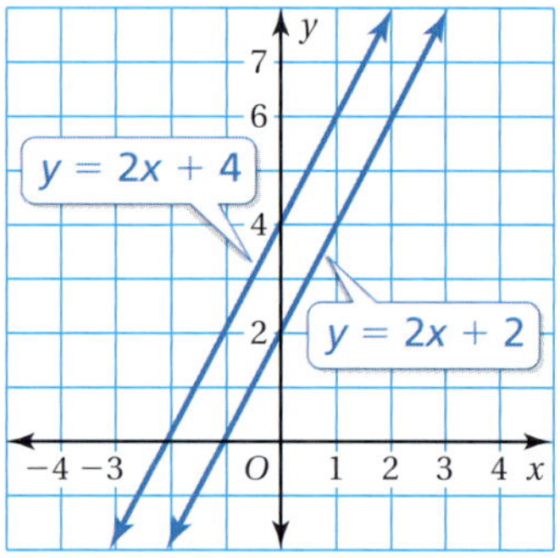

F. none

G. exactly one

H. exactly two

I. infinitely many

6. A scenic train ride has one price for adults and one price for children. One family of two adults and two children pays \$62 for the train ride. Another family of one adult and four children pays \$70. Which system of linear equations can you use to find the price x for an adult and the price y for a child? *(MACC.8.EE.3.8c)*

A. $2x + 2y = 70$
$x + 4y = 62$

B. $x + y = 62$
$x + y = 70$

C. $2x + 2y = 62$
$4x + y = 70$

D. $2x + 2y = 62$
$x + 4y = 70$

7. Which of the following is true about the graph of the linear equation $y = -7x + 5$? *(MACC.8.EE.2.6)*

F. The slope is 5, and the y-intercept is −7.

G. The slope is −5, and the y-intercept is −7.

H. The slope is −7, and the y-intercept is −5.

I. The slope is −7, and the y-intercept is 5.

8. What value of w makes the equation below true? *(MACC.8.EE.3.7b)*

$$7w - 3w = 2(3w + 11)$$

9. The graph of which equation is parallel to the line that passes through the points $(-1, 5)$ and $(4, 7)$? *(MACC.8.EE.2.6)*

A. $y = \frac{2}{3}x + 6$

B. $y = -\frac{5}{2}x + 4$

C. $y = \frac{2}{5}x + 1$

D. $y = \frac{5}{2}x - 1$

10. You buy 3 T-shirts and 2 pairs of shorts for \$42.50. Your friend buys 5 T-shirts and 3 pairs of shorts for \$67.50. Use a system of linear equations to find the cost of each T-shirt. Show your work and explain your reasoning. *(MACC.8.EE.3.8c)*

Think
Solve
Explain

11. The two figures have the same area. What is the value of y? *(MACC.8.EE.3.7b)*

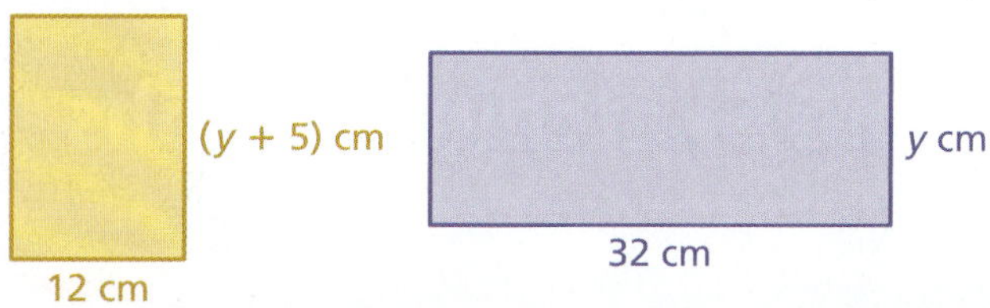

F. $\frac{1}{4}$

G. $\frac{15}{8}$

H. 3

I. 8

12. A system of two linear equations has infinitely many solutions. What can you conclude about the graphs of the two equations? *(MACC.8.EE.3.8a)*

A. The lines have the same slope and the same y-intercept.

B. The lines have the same slope and different y-intercepts.

C. The lines have different slopes and the same y-intercept.

D. The lines have different slopes and different y-intercepts.

13. The sum of one-third of a number and 10 is equal to 13. What is the number? *(MACC.8.EE.3.7b)*

F. $\frac{8}{3}$ **G.** 9 **H.** 29 **I.** 69

14. Solve the equation $4x + 7y = 16$ for x. *(MACC.8.EE.3.7b)*

A. $x = 4 + \frac{7}{4}y$

B. $x = 4 - \frac{7}{4}y$

C. $x = 4 + \frac{4}{7}y$

D. $x = 16 - 7y$

6 Functions

"Here's a math anagram."

"Here's another one."

"It is my treat-converter function machine. However many cat treats I input, the machine outputs TWICE that many dog biscuits. Isn't that cool?"

"Do you think the stripes in this shirt make me look too linear?"

What You Learned Before

Identifying Patterns (MACC.5.OA.2.3)

Example 1 Find the missing value in the table.

x	y
30	0
40	10
50	20
60	

Each y-value is 30 less than the x-value.

So, the missing value is $60 - 30 = 30$.

Try It Yourself

Find the missing value in the table.

1.

x	y
5	10
7	14
10	20
40	

2.

x	y
0.5	1
1.5	2
3	3.5
9.5	

3.

x	y
15	5
30	10
45	15
60	

Evaluating Algebraic Expressions (MACC.7.NS.1.3)

Example 2 Evaluate $2x - 12$ when $x = 5$.

$$2x - 12 = 2(5) - 12$$ Substitute 5 for x.

$$= 10 - 12$$ Using order of operations, multiply 2 and 5.

$$= 10 + (-12)$$ Add the opposite of 12.

$$= -2$$ Add.

Try It Yourself

Evaluate the expression when $y = 4$.

4. $-4y + 2$

5. $\frac{y}{2} - 8$

6. $-10 - 6y$

6.1 Relations and Functions

Essential Question How can you use a mapping diagram to show the relationship between two data sets?

1 ACTIVITY: Constructing Mapping Diagrams

Work with a partner. Copy and complete the mapping diagram.

a. Area A

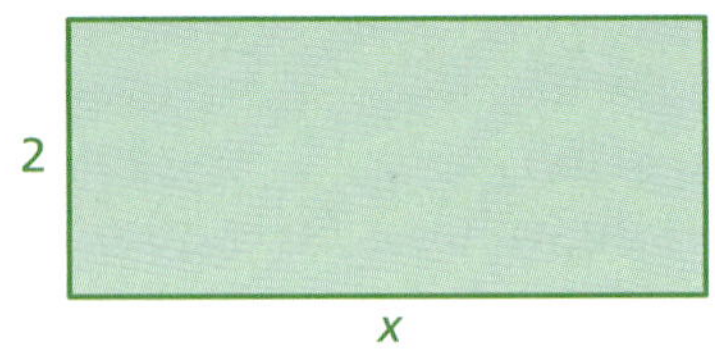

Input, x **Output, A**

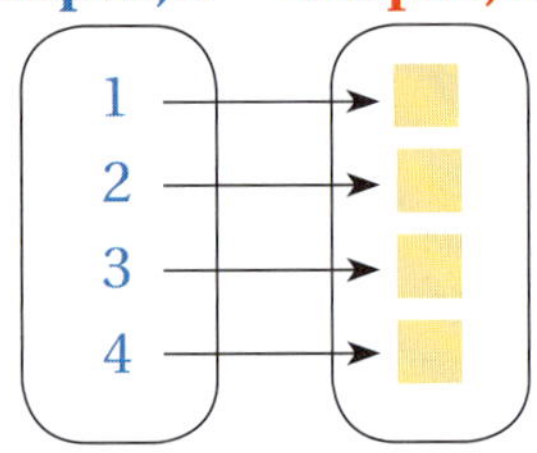

b. Perimeter P

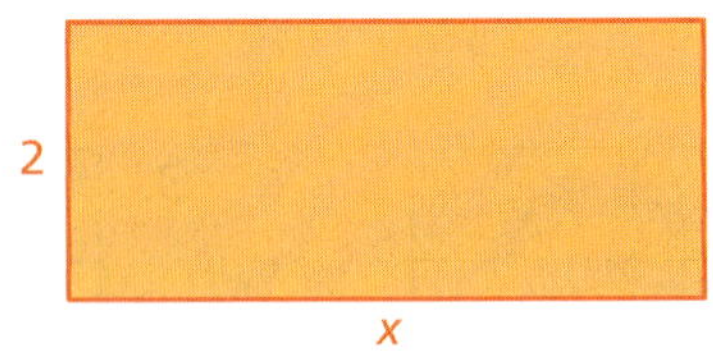

Input, x **Output, P**

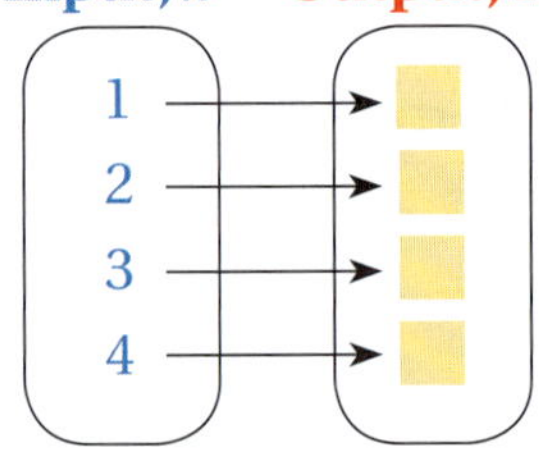

c. Circumference C

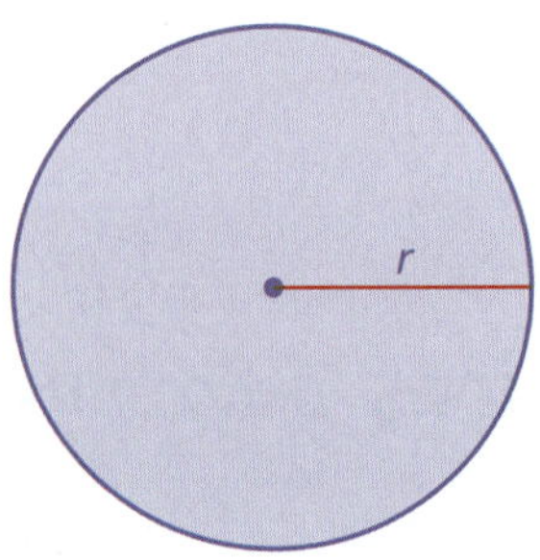

Input, r **Output, C**

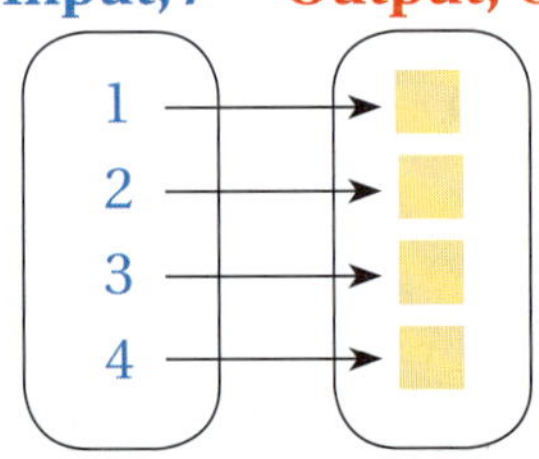

d. Volume V

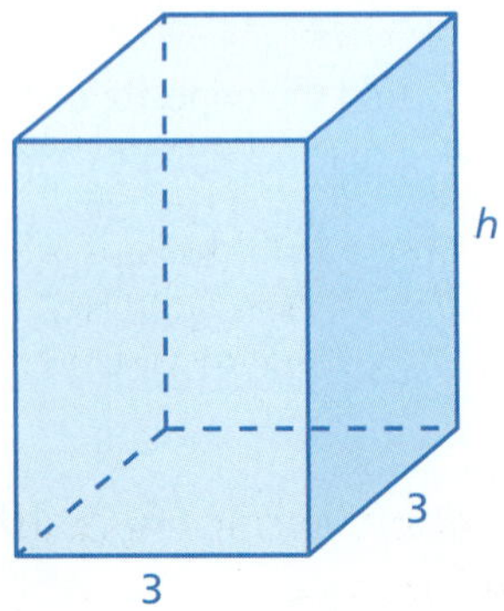

Input, h **Output, V**

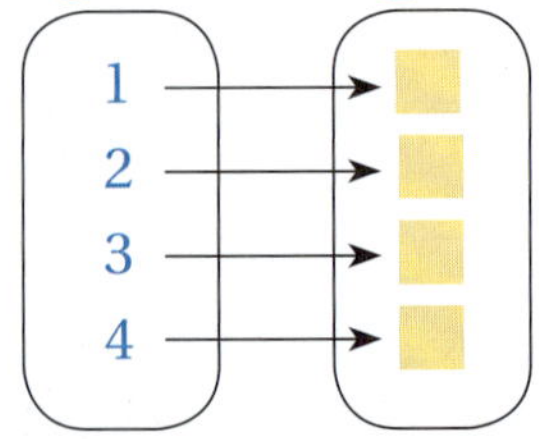

COMMON CORE

Functions

In this lesson, you will

- define relations and functions.
- determine whether relations are functions.
- describe patterns in mapping diagrams.

Learning Standard
MACC.8.F.1.1

2 ACTIVITY: Describing Situations

Math Practice 7

View as Components

What are the input values? Do any of the input values point to more than one output value? How does this help you describe a possible situation?

Work with a partner. How many outputs are assigned to each input? Describe a possible situation for each mapping diagram.

a.

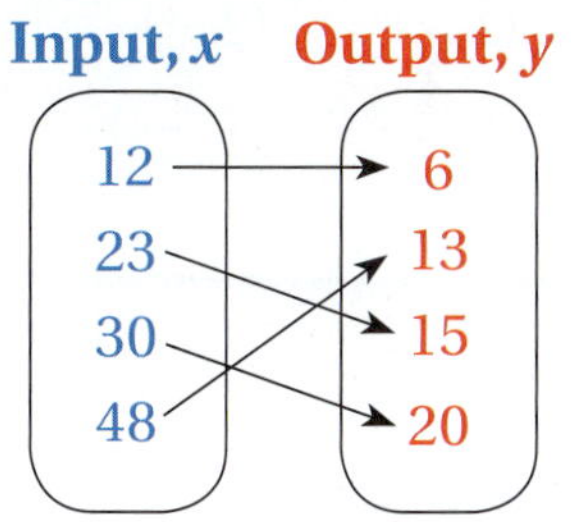

b.

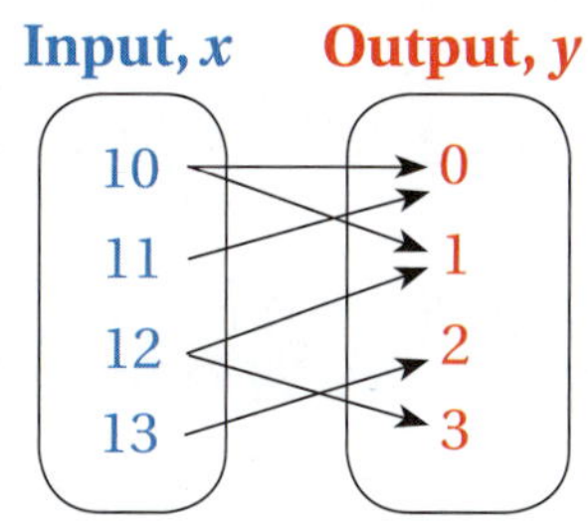

3 ACTIVITY: Interpreting Mapping Diagrams

Work with a partner. Describe the pattern in the mapping diagram. Copy and complete the diagram.

a.

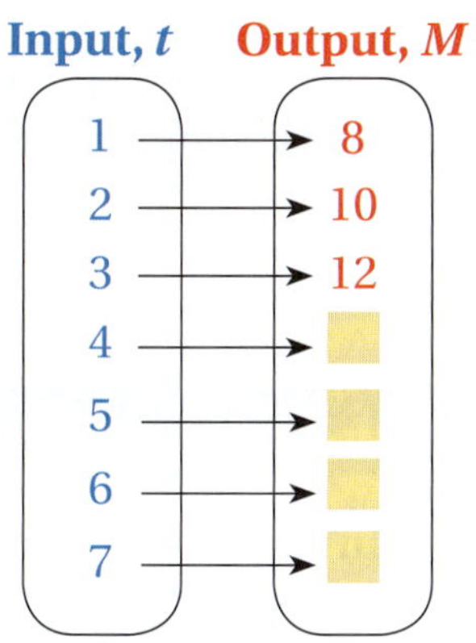

b.

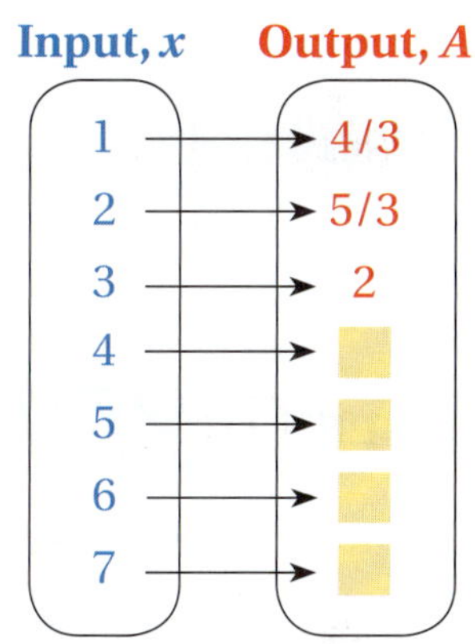

What Is Your Answer?

4. **IN YOUR OWN WORDS** How can you use a mapping diagram to show the relationship between two data sets?

"I made a mapping diagram." "It shows how I feel about my skateboard with each passing day."

Practice Use what you learned about mapping diagrams to complete Exercises 3–5 on page 246.

6.1 Lesson

Check It Out
Lesson Tutorials
BigIdeasMath.com

Key Vocabulary
input, *p. 244*
output, *p. 244*
relation, *p. 244*
mapping diagram, *p. 244*
function, *p. 245*

Ordered pairs can be used to show **inputs** and **outputs**.

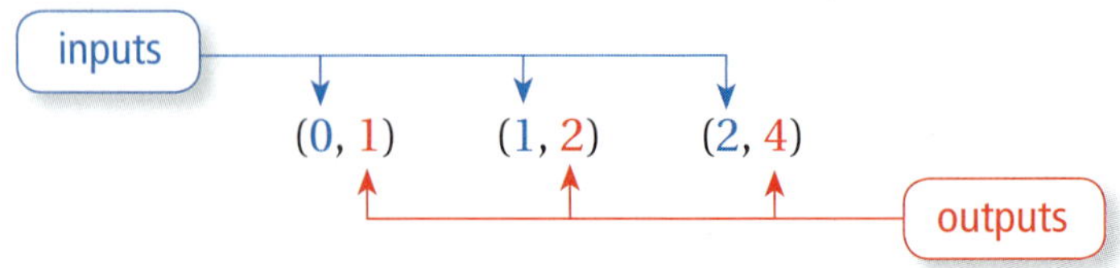

Key Idea

Relations and Mapping Diagrams

A **relation** pairs inputs with outputs. A relation can be represented by ordered pairs or a **mapping diagram**.

Ordered Pairs

(0, 1)

(1, 2)

(2, 4)

Mapping Diagram

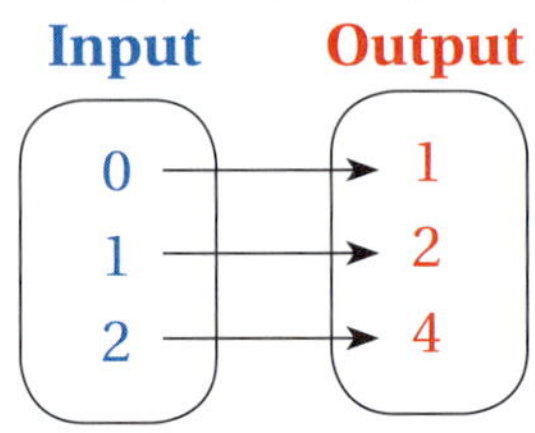

EXAMPLE 1 Listing Ordered Pairs of a Relation

List the ordered pairs shown in the mapping diagram.

a.

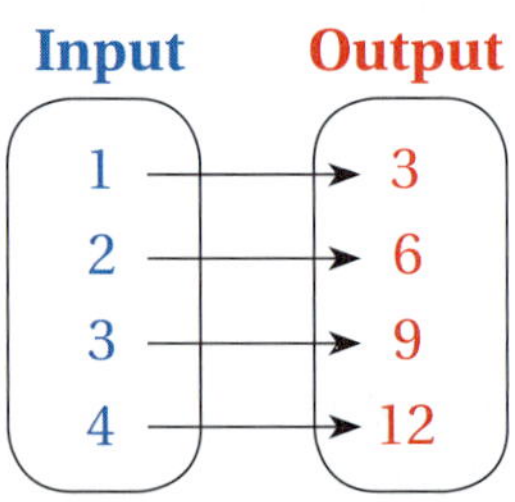

∴ The ordered pairs are (1, 3), (2, 6), (3, 9), and (4, 12).

b.

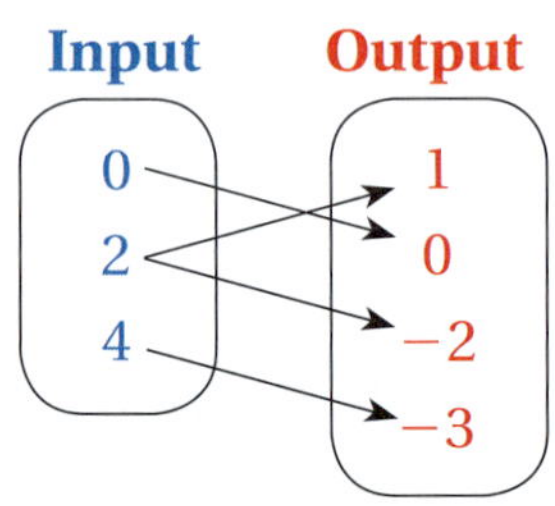

∴ The ordered pairs are (0, 0), (2, 1), (2, −2), and (4, −3).

On Your Own

Now You're Ready
Exercises 6–8

List the ordered pairs shown in the mapping diagram.

1.

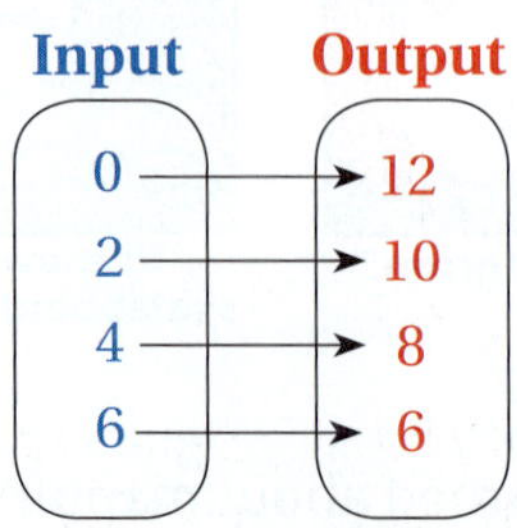

2.

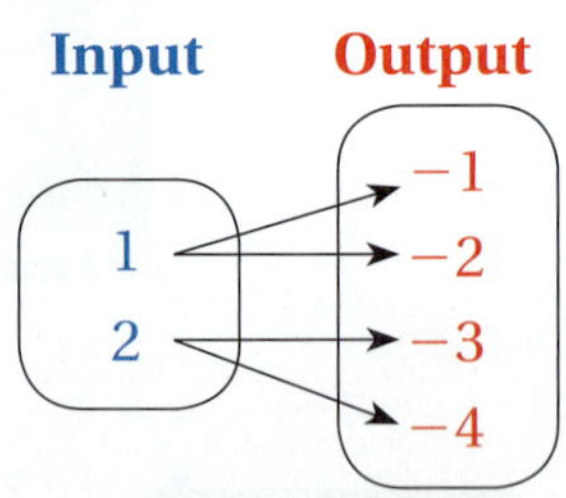

A relation that pairs each input with *exactly one* output is a **function**.

EXAMPLE 2 Determining Whether Relations Are Functions

Determine whether each relation is a function.

a.

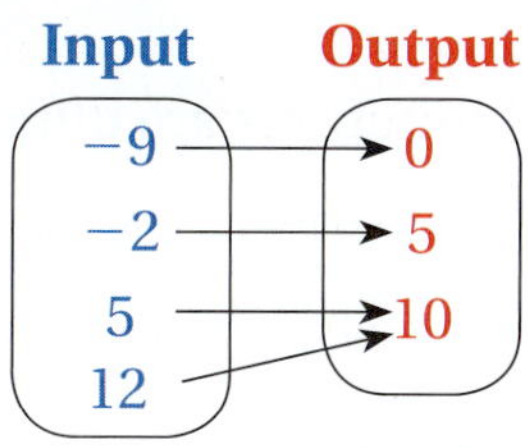

Each input has exactly one output. So, the relation is a function.

b.

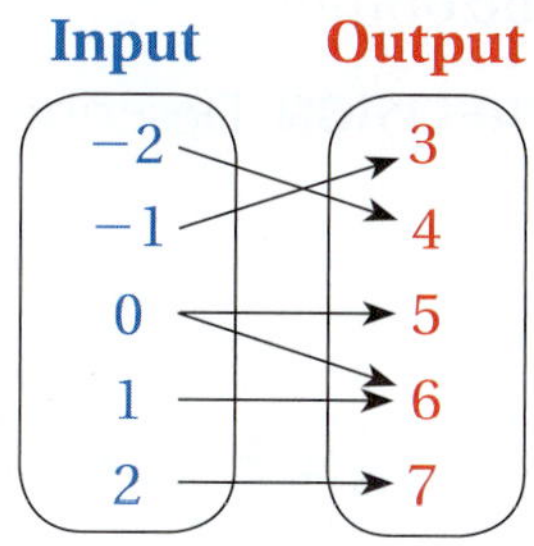

The input 0 has two outputs, 5 and 6. So, the relation is *not* a function.

EXAMPLE 3 Describing a Mapping Diagram

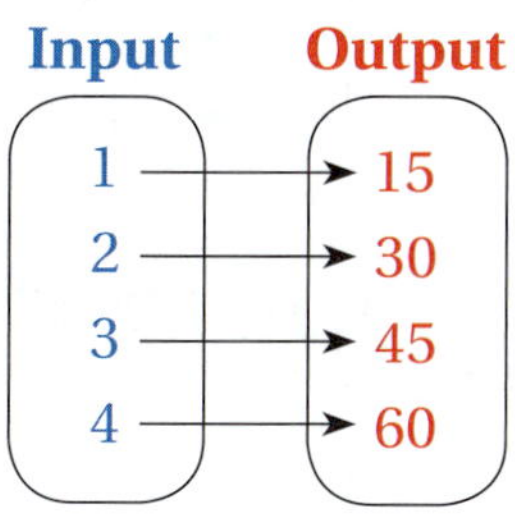

Consider the mapping diagram at the left.

a. Determine whether the relation is a function.

Each input has exactly one output.

So, the relation is a function.

b. Describe the pattern of inputs and outputs in the mapping diagram.

Look at the relationship between the inputs and the outputs.

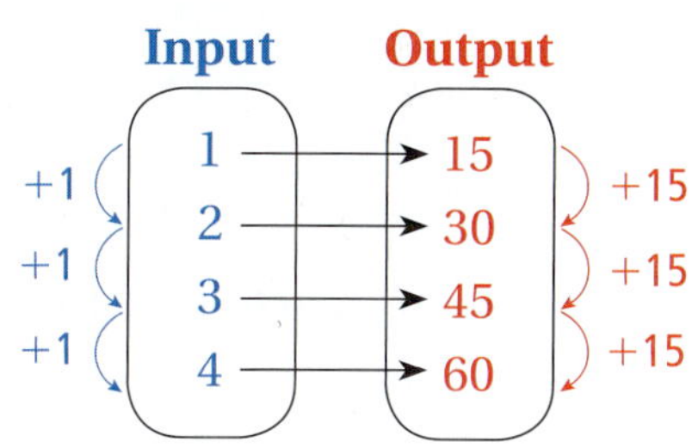

As each input increases by 1, the output increases by 15.

On Your Own

Now You're Ready
Exercises 9–11 and 13–15

Determine whether the relation is a function.

3.

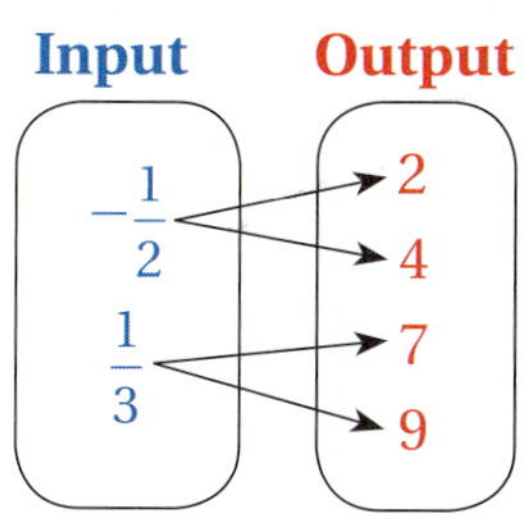

4.

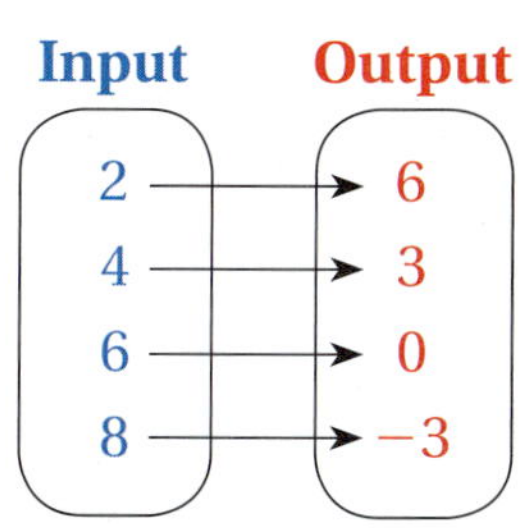

5. Describe the pattern of inputs and outputs in the mapping diagram in On Your Own 4.

6.1 Exercises

Check It Out
Help with Homework
BigIdeasMath.com

Vocabulary and Concept Check

1. **VOCABULARY** In an ordered pair, which number represents the input? the output?

2. **PRECISION** Describe how relations and functions are different.

Practice and Problem Solving

Describe the pattern in the mapping diagram. Copy and complete the diagram.

3. 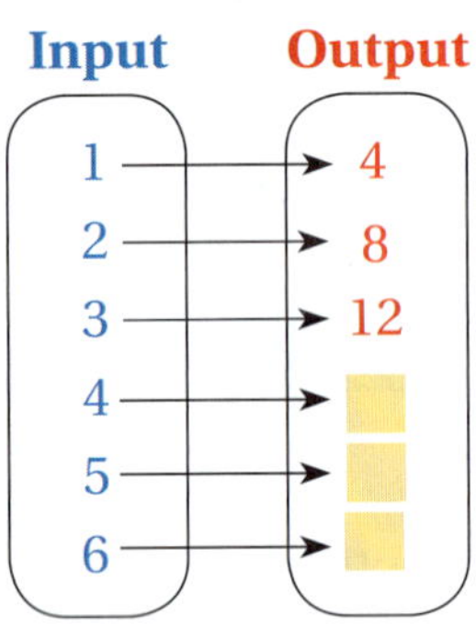

4. 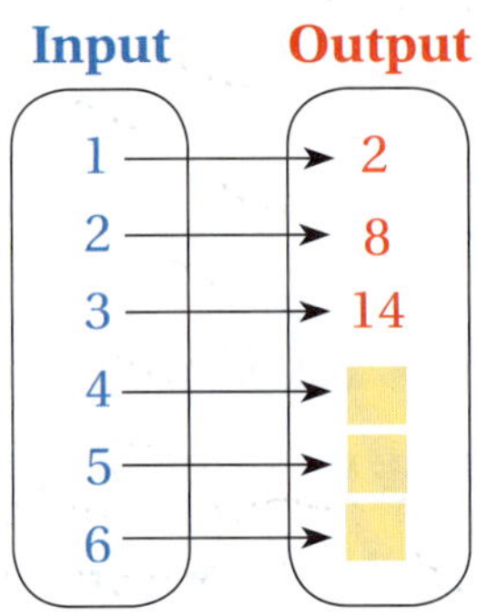

5. 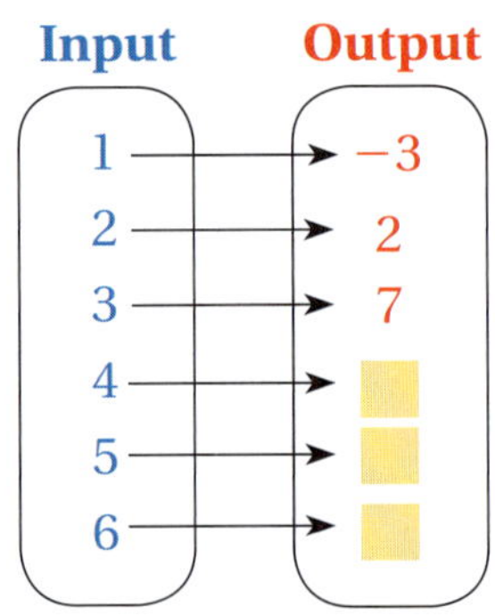

List the ordered pairs shown in the mapping diagram.

① 6. 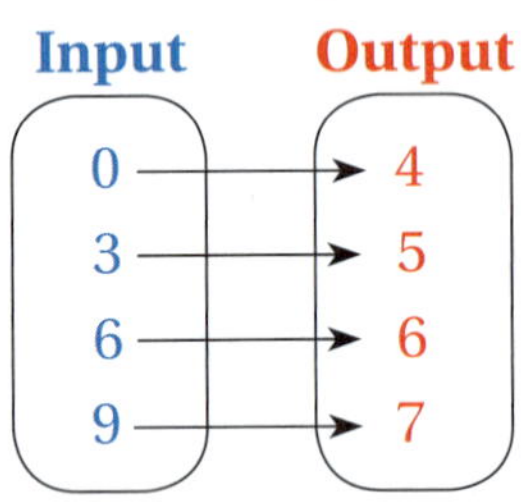

7. 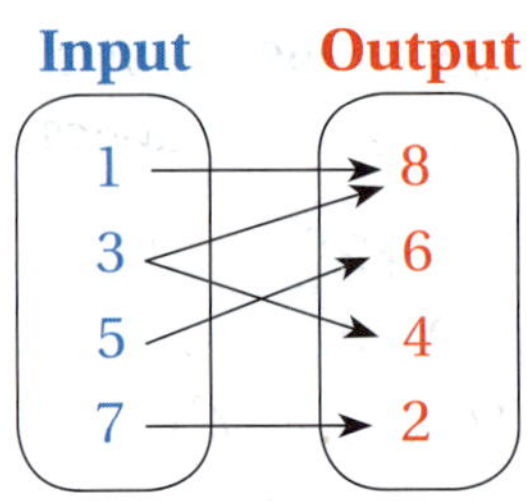

8. 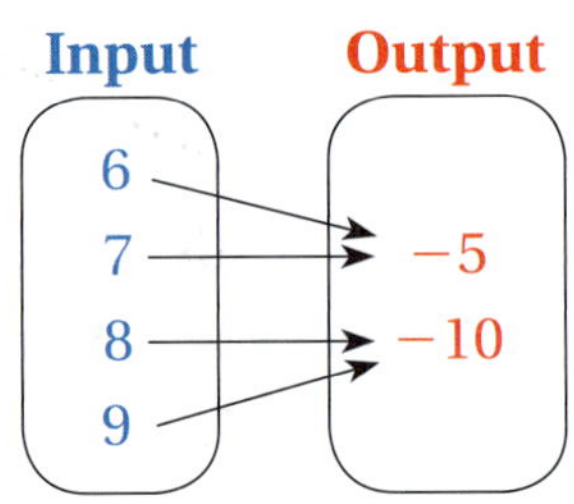

Determine whether the relation is a function.

② 9. 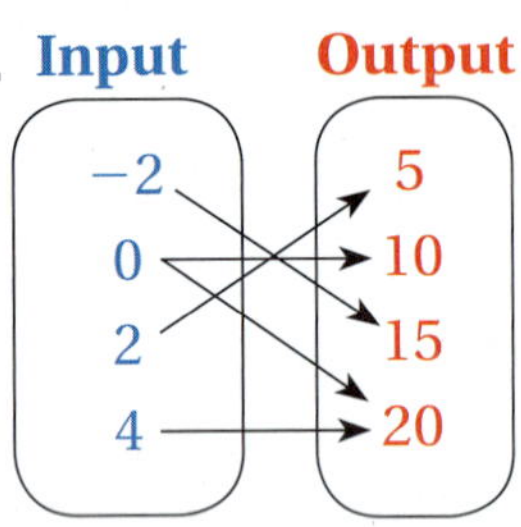

10. 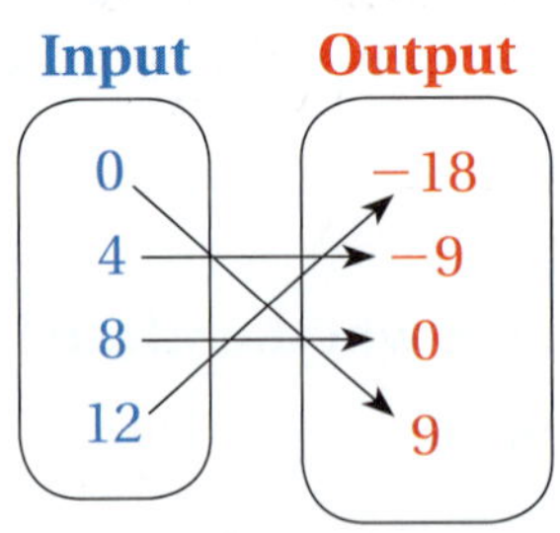

11. 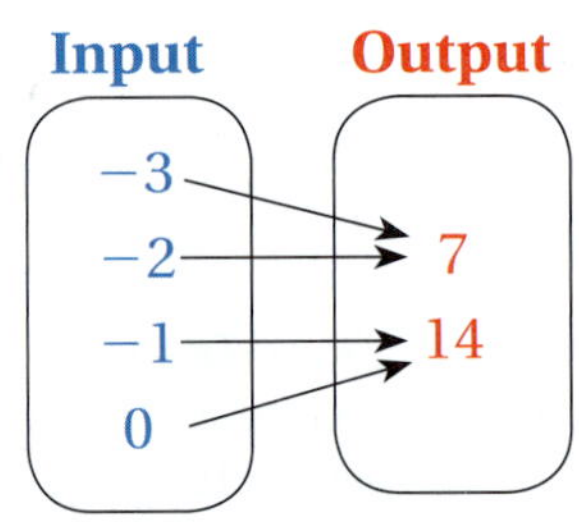

12. **ERROR ANALYSIS** Describe and correct the error in determining whether the relation is a function.

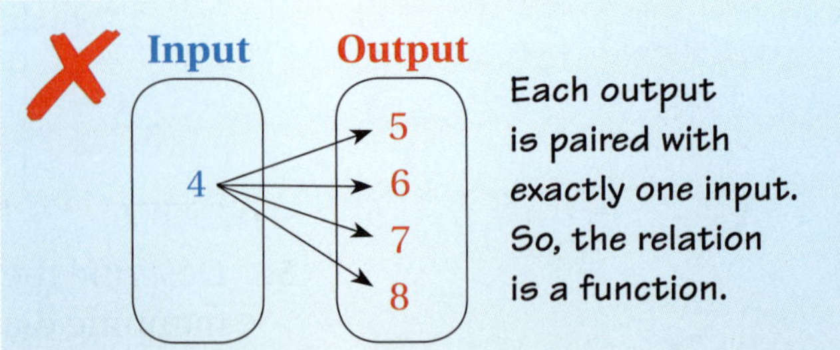

Draw a mapping diagram for the graph. Then describe the pattern of inputs and outputs.

3 **13.**

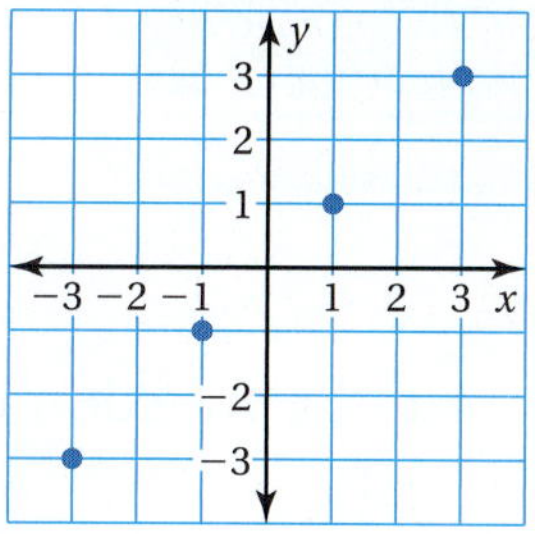

14.

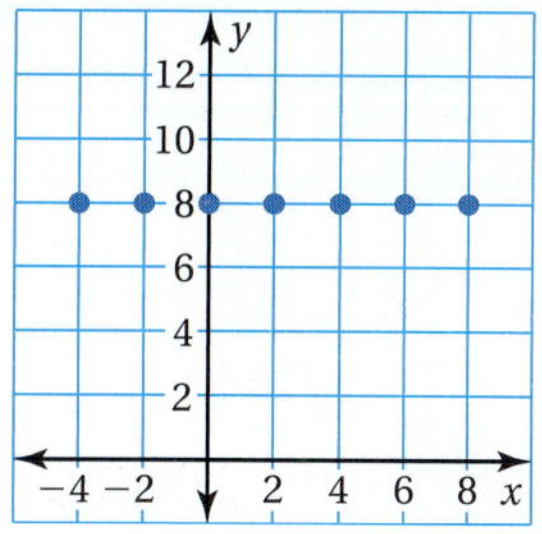

15.

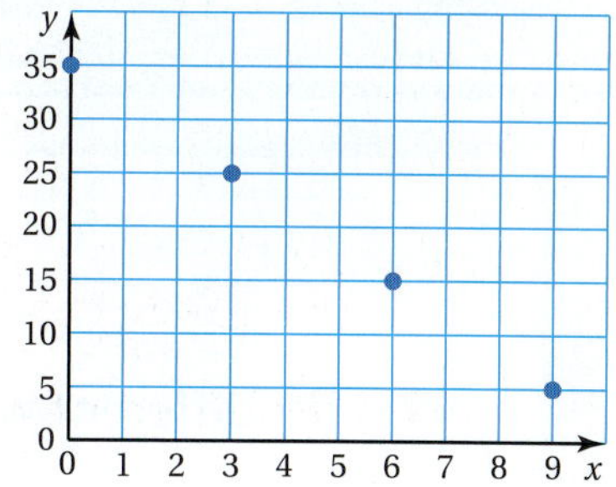

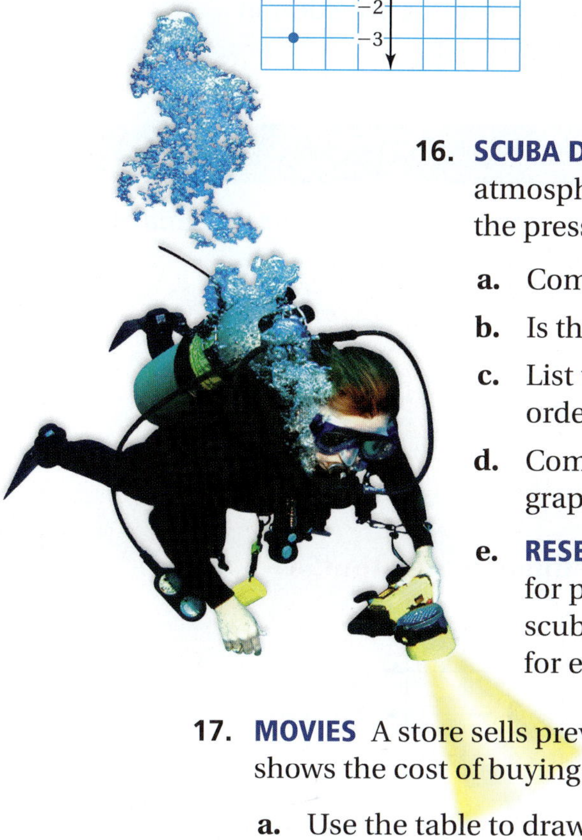

16. SCUBA DIVING The normal pressure at sea level is one atmosphere of pressure (1 ATM). As you dive below sea level, the pressure increases by 1 ATM for each 10 meters of depth.

a. Complete the mapping diagram.

b. Is the relation a function? Explain.

c. List the ordered pairs. Then plot the ordered pairs in a coordinate plane.

d. Compare the mapping diagram and graph. Which do you prefer? Why?

e. RESEARCH What are common depths for people who are just learning to scuba dive? What are common depths for experienced scuba divers?

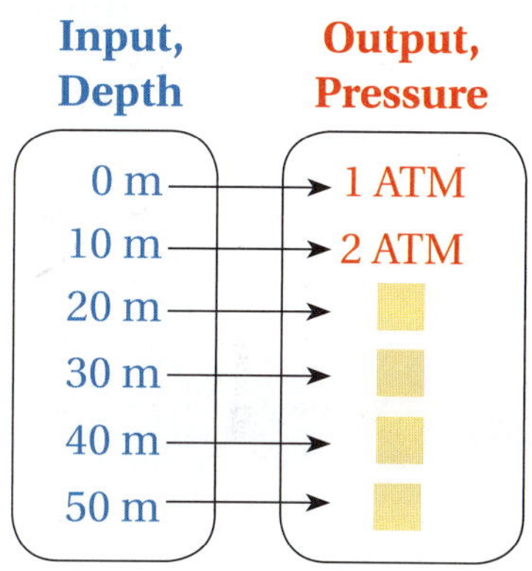

17. MOVIES A store sells previously viewed movies. The table shows the cost of buying 1, 2, 3, or 4 movies.

a. Use the table to draw a mapping diagram.

b. Is the relation a function? Explain.

c. Describe the pattern. How does the cost per movie change as you buy more movies?

Movies	Cost
1	\$10
2	\$18
3	\$24
4	\$28

18. Repeated Reasoning The table shows the outputs for several inputs. Use two methods to find the output for an input of 200.

Input, x	0	1	2	3	4
Output, y	25	30	35	40	45

Fair Game Review What you learned in previous grades & lessons

The coordinates of a point and its image are given. Is the reflection in the *x*-axis or *y*-axis? *(Section 2.3)*

19. $(3, -3) \rightarrow (-3, -3)$ **20.** $(-5, 1) \rightarrow (-5, -1)$ **21.** $(-2, -4) \rightarrow (-2, 4)$

22. MULTIPLE CHOICE Which word best describes two figures that have the same size and the same shape? *(Section 2.1)*

Ⓐ congruent Ⓑ dilation Ⓒ parallel Ⓓ similar

6.2 Representations of Functions

Essential Question How can you represent a function in different ways?

1 ACTIVITY: Describing a Function

Work with a partner. Copy and complete the mapping diagram for the area of the figure. Then write an equation that describes the function.

a.

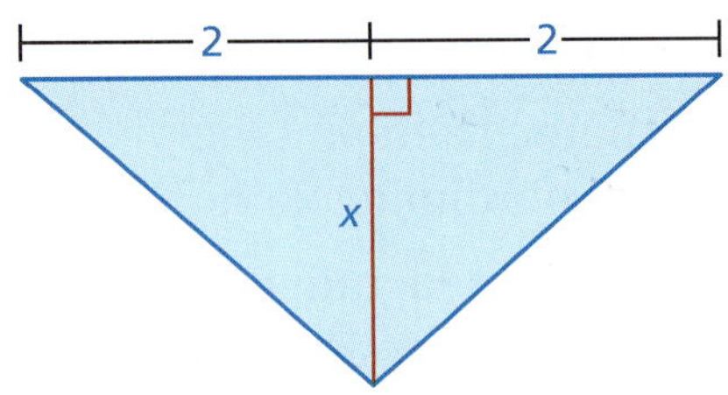

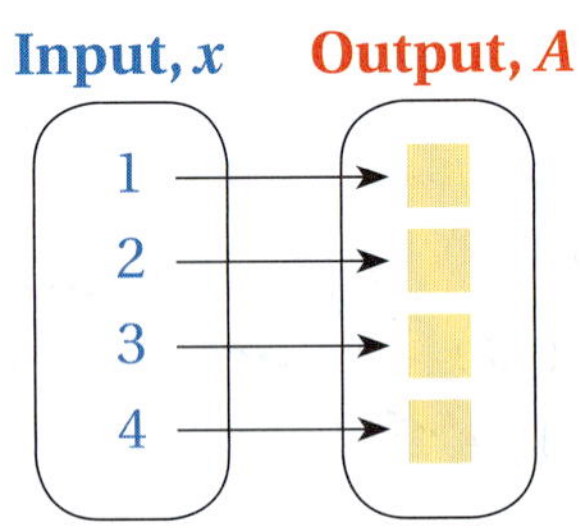

b.

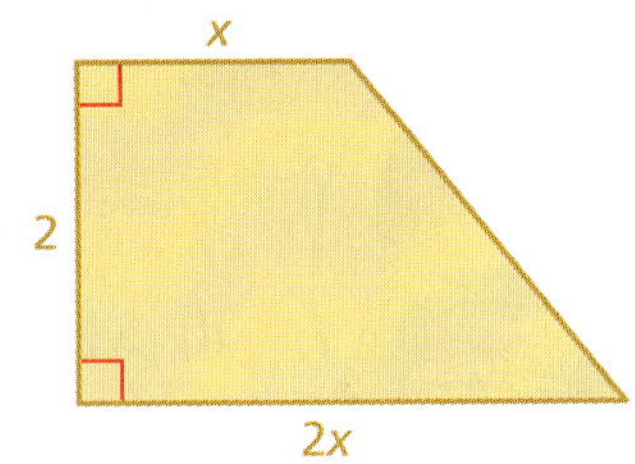

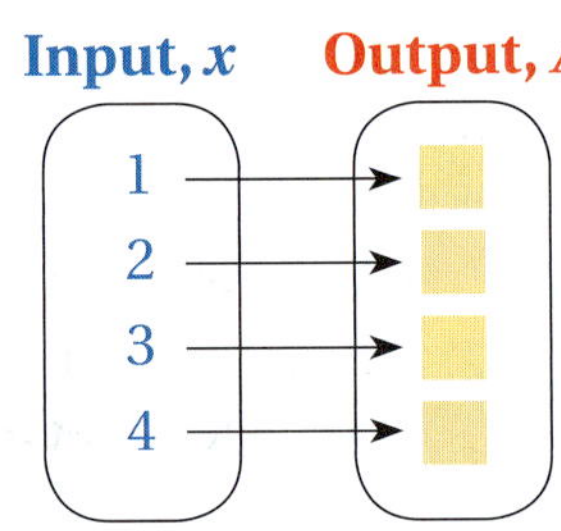

2 ACTIVITY: Using a Table

Work with a partner. Make a table that shows the pattern for the area, where the input is the figure number x and the output is the area A. Write an equation that describes the function. Then use your equation to find which figure has an area of 81 when the pattern continues.

a.

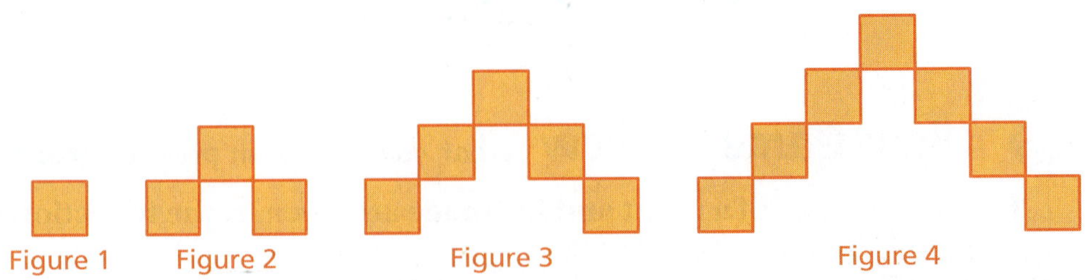

b.

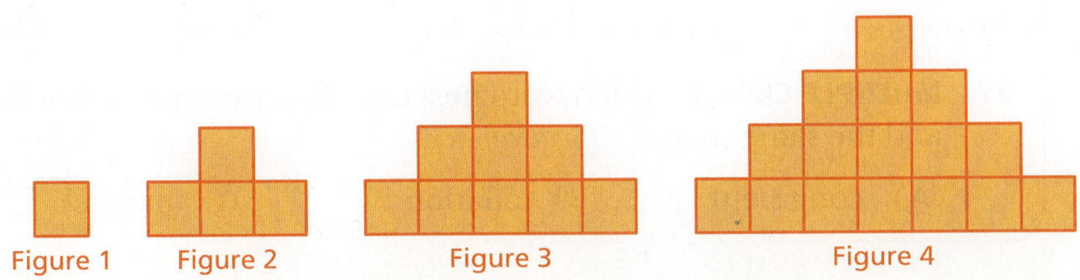

COMMON CORE

Functions

In this lesson, you will

- write function rules.
- use input-output tables to represent functions.
- use graphs to represent functions.

Learning Standard
MACC.8.F.1.1

3 ACTIVITY: Using a Graph

> **Math Practice 3**
> **Construct Arguments**
> How does the graph help you determine whether the statement is true?

Work with a partner. Graph the data. Use the graph to test the truth of each statement. If the statement is true, write an equation that shows how to obtain one measurement from the other measurement.

a. "You can find the horsepower of a race car engine if you know its volume in cubic inches."

Volume (cubic inches), x	200	350	350	500
Horsepower, y	375	650	250	600

b. "You can find the volume of a race car engine in cubic centimeters if you know its volume in cubic inches."

Volume (cubic inches), x	100	200	300
Volume (cubic centimeters), y	1640	3280	4920

4 ACTIVITY: Interpreting a Graph

Work with a partner. The table shows the average speeds of the winners of the Daytona 500. Graph the data. Can you use the graph to predict future winning speeds? Explain why or why not.

Year, x	2004	2005	2006	2007	2008	2009	2010	2011	2012
Speed (mi/h), y	156	135	143	149	153	133	137	130	140

What Is Your Answer?

5. **IN YOUR OWN WORDS** How can you represent a function in different ways?

Practice Use what you learned about representing functions to complete Exercises 4–6 on page 253.

6.2 Lesson

Check It Out
Lesson Tutorials
BigIdeasMath.com

Key Vocabulary
function rule, *p. 250*

Remember

An independent variable represents a quantity that can change freely. A dependent variable *depends* on the independent variable.

Key Idea

Functions as Equations

A **function rule** is an equation that describes the relationship between inputs (independent variable) and outputs (dependent variable).

EXAMPLE 1 Writing Function Rules

a. Write a function rule for "The output is five less than the input."

Words	The output	is	five less than	the input.
Equation	y	$=$	x	$-\ 5$

∴ A function rule is $y = x - 5$.

b. Write a function rule for "The output is the square of the input."

Words	The output	is	the square	of the input.
Equation	y	$=$	x^2	

∴ A function rule is $y = x^2$.

EXAMPLE 2 Evaluating a Function

What is the value of $y = 2x + 5$ when $x = 3$?

$y = 2x + 5$ — Write the equation.

$= 2(3) + 5$ — Substitute 3 for x.

$= 11$ — Simplify.

∴ When $x = 3$, $y = 11$.

On Your Own

Now You're Ready
Exercises 7–18

1. Write a function rule for "The output is one-fourth of the input."

Find the value of y when $x = 5$.

2. $y = 4x - 1$ 3. $y = 10x$ 4. $y = 7 - 3x$

Multi-Language Glossary at BigIdeasMath.com

Key Idea

Functions as Tables and Graphs

A function can be represented by an input-output table and by a graph. The table and graph below represent the function $y = x + 2$.

Input, x	Output, y	Ordered Pair, (x, y)
1	3	(1, 3)
2	4	(2, 4)
3	5	(3, 5)

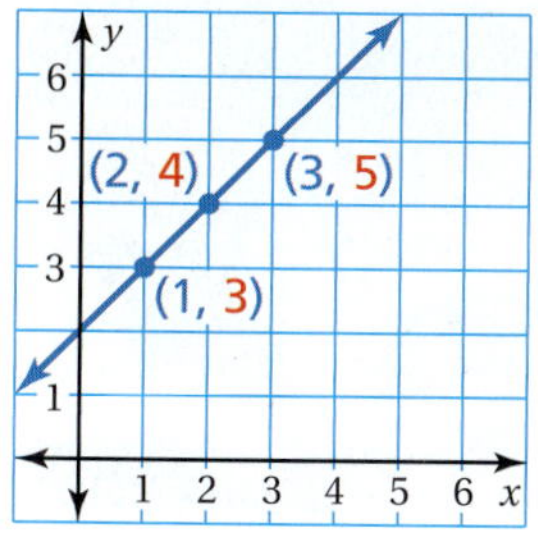

By drawing a line through the points, you graph *all* of the solutions of the function $y = x + 2$.

EXAMPLE 3 Graphing a Function

Graph the function $y = -2x + 1$ using inputs of $-1, 0, 1,$ and 2.

Make an input-output table.

Input, x	$-2x + 1$	Output, y	Ordered Pair, (x, y)
-1	$-2(-1) + 1$	3	$(-1, 3)$
0	$-2(0) + 1$	1	$(0, 1)$
1	$-2(1) + 1$	-1	$(1, -1)$
2	$-2(2) + 1$	-3	$(2, -3)$

Plot the ordered pairs and draw a line through the points.

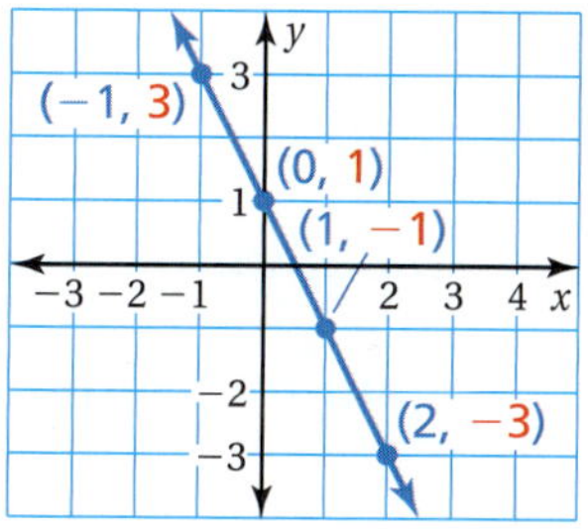

On Your Own

Now You're Ready
Exercises 19–24

Graph the function.

5. $y = x + 1$ **6.** $y = -3x$ **7.** $y = 3x + 2$

EXAMPLE 4 Real-Life Application

The number of pounds p of carbon dioxide produced by a car is 20 times the number of gallons g of gasoline used by the car. Write and graph a function that describes the relationship between g and p.

Write a function rule using the variables g and p.

Words The number of pounds of carbon dioxide is 20 times the number of gallons of gasoline used.

Equation $p = 20 \cdot g$

Make an input-output table that represents the function $p = 20g$.

Input, g	$20g$	Output, p	Ordered Pair, (g, p)
1	20(1)	20	(1, 20)
2	20(2)	40	(2, 40)
3	20(3)	60	(3, 60)

Plot the ordered pairs and draw a line through the points.

Because you cannot have a negative number of gallons, use only positive values of g.

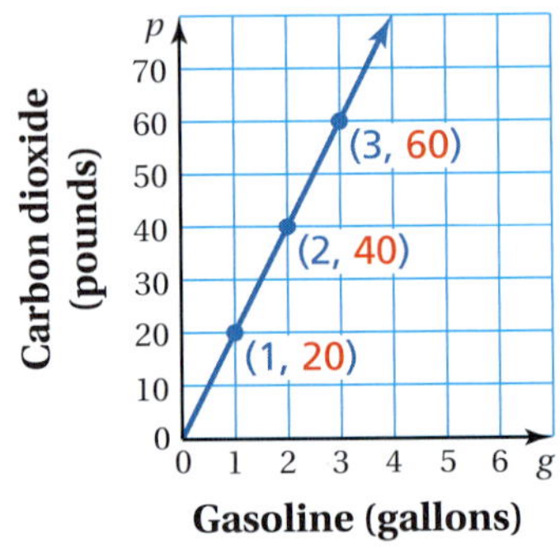

On Your Own

Now You're Ready
Exercise 26

8. **WHAT IF?** For a truck, p is 25 times g. Write and graph a function that describes the relationship between g and p.

Summary

Representations of Functions

Words An output is 2 more than the input.

Equation $y = x + 2$

Input-Output Table

Input, x	Output, y
−1	1
0	2
1	3
2	4

Mapping Diagram

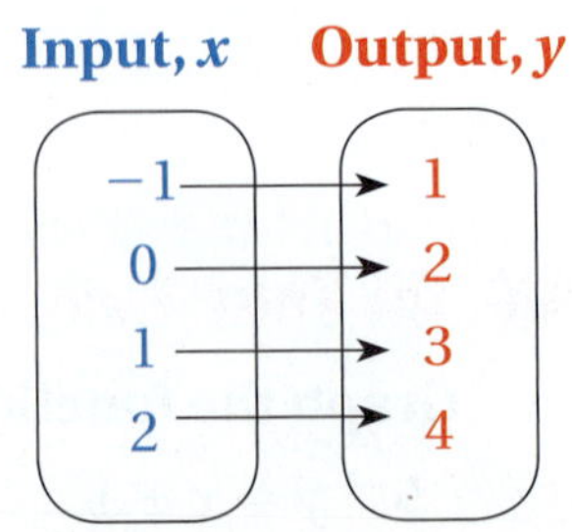

Graph

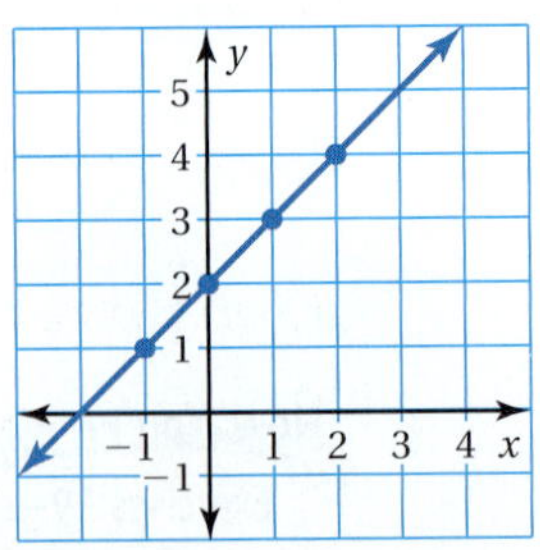

6.2 Exercises

Check It Out
Help with Homework
BigIdeasMath.com

Vocabulary and Concept Check

1. **VOCABULARY** Identify the input variable and the output variable for the function rule $y = 2x + 5$.
2. **WRITING** Describe five ways to represent a function.
3. **DIFFERENT WORDS, SAME QUESTION** Which is different? Find "both" answers.

What output is 4 more than twice the input 3?

What output is twice the sum of the input 3 and 4?

What output is the sum of 2 times the input 3 and 4?

What output is 4 increased by twice the input 3?

Practice and Problem Solving

Write an equation that describes the function.

4. Input, x Output, y

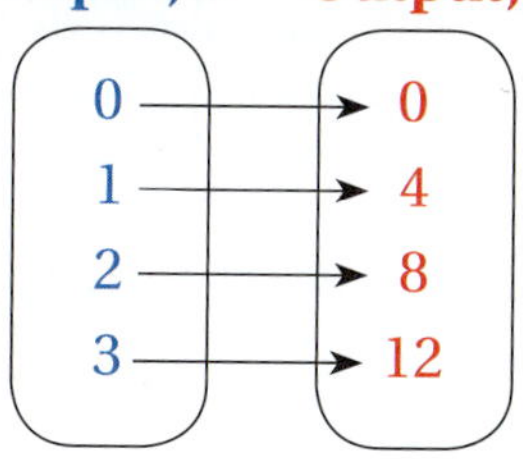

5.

Input, x	Output, y
1	8
2	9
3	10
4	11

6.

Input, x	Output, y
1	0
3	-2
5	-4
7	-6

Write a function rule for the statement.

(1) 7. The output is half of the input.
8. The output is eleven more than the input.
9. The output is three less than the input.
10. The output is the cube of the input.
11. The output is six times the input.
12. The output is one more than twice the input.

Find the value of y for the given value of x.

(2) 13. $y = x + 5;\ x = 3$
14. $y = 7x;\ x = -5$
15. $y = 1 - 2x;\ x = 9$
16. $y = 3x + 2;\ x = 0.5$
17. $y = 2x^3;\ x = 3$
18. $y = \frac{x}{2} + 9;\ x = -12$

Graph the function.

(3) 19. $y = x + 4$
20. $y = 2x$
21. $y = -5x + 3$
22. $y = \frac{x}{4}$
23. $y = \frac{3}{2}x + 1$
24. $y = 1 + 0.5x$

25. ERROR ANALYSIS Describe and correct the error in graphing the function represented by the input-output table.

Input, x	-4	-2	0	2
Output, y	-1	1	3	5

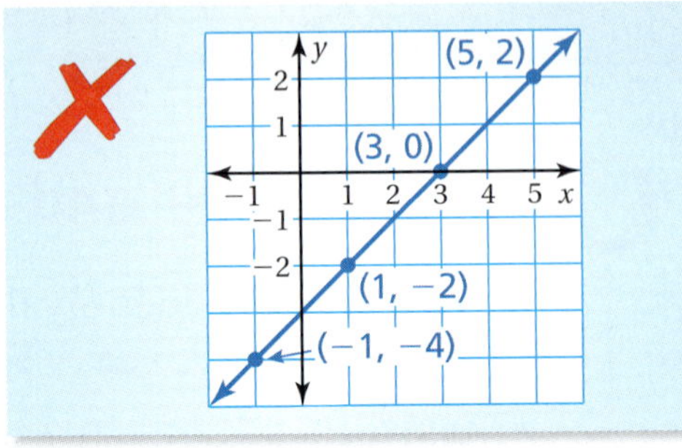

26. DOLPHIN A dolphin eats 30 pounds of fish per day.

a. Write and graph a function that relates the number of pounds p of fish that a dolphin eats in d days.

b. How many pounds of fish does a dolphin eat in 30 days?

Match the graph with the function it represents.

27.

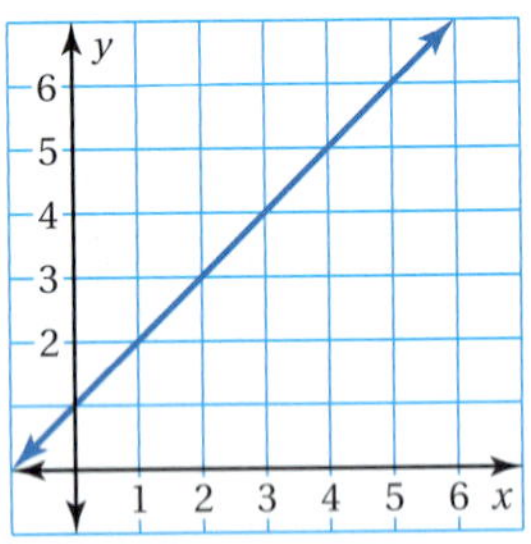

28.

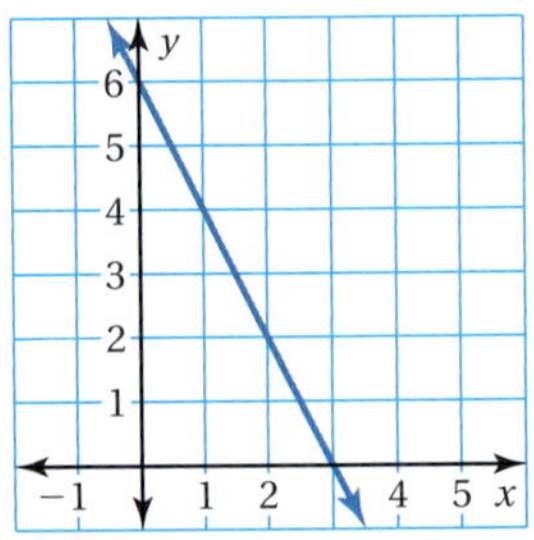

29.

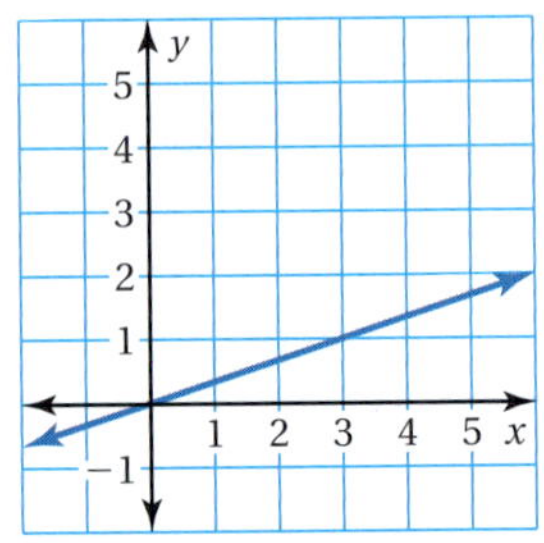

A. $y = \frac{x}{3}$ B. $y = x + 1$ C. $y = -2x + 6$

Find the value of *x* for the given value of *y*.

30. $y = 5x - 7;\ y = -22$ **31.** $y = 9 - 7x;\ y = 37$ **32.** $y = \frac{x}{4} - 7;\ y = 2$

33. BRACELETS You decide to make and sell bracelets. The cost of your materials is \$84. You charge \$3.50 for each bracelet.

a. Write a function that represents the profit P for selling b bracelets.

b. Which variable is independent? dependent? Explain.

c. You will *break even* when the cost of your materials equals your income. How many bracelets must you sell to break even?

34. SALE A furniture store is having a sale where everything is 40% off.

a. Write a function that represents the amount of discount d on an item with a regular price p.

b. Graph the function using the inputs 100, 200, 300, 400, and 500 for p.

c. You buy a bookshelf that has a regular price of \$85. What is the sale price of the bookshelf?

35. AIRBOAT TOURS You want to take a two-hour airboat tour.

a. Write a function that represents the cost G of a tour at Gator Tours.

b. Write a function that represents the cost S of a tour at Snake Tours.

c. Which is a better deal? Explain.

36. REASONING The graph of a function is a line that goes through the points (3, 2), (5, 8), and (8, y). What is the value of y?

37. CRITICAL THINKING Make a table where the independent variable is the side length of a square and the dependent variable is the *perimeter*. Make a second table where the independent variable is the side length of a square and the dependent variable is the *area*. Graph both functions in the same coordinate plane. Compare the functions and graphs.

38. Puzzle The blocks that form the diagonals of each square are shaded. Each block is one square unit. Find the "green area" of Square 20. Find the "green area" of Square 21. Explain your reasoning.

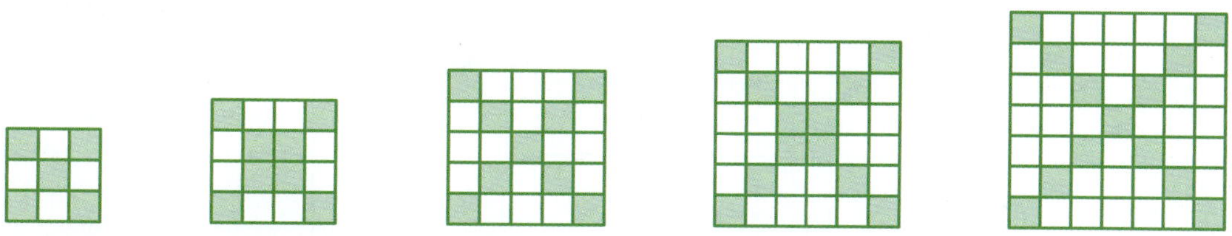

Fair Game Review What you learned in previous grades & lessons

Find the slope of the line. *(Section 4.2)*

39.

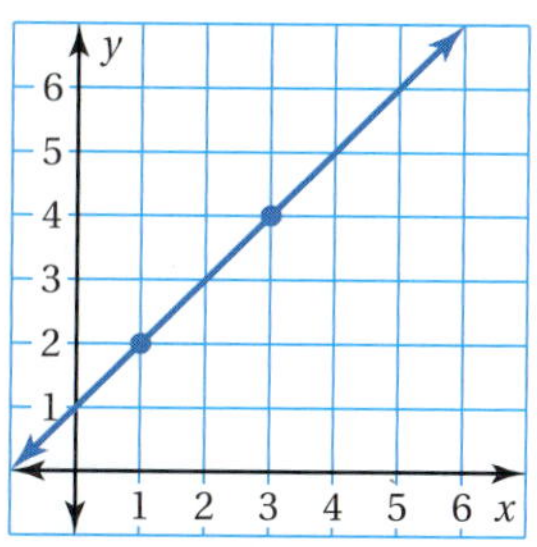

40.

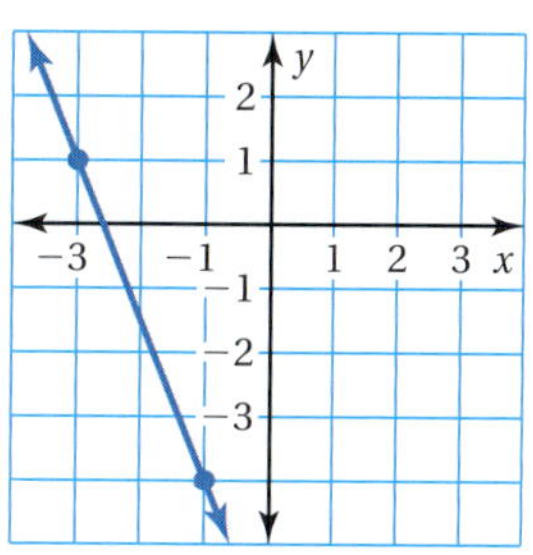

41.

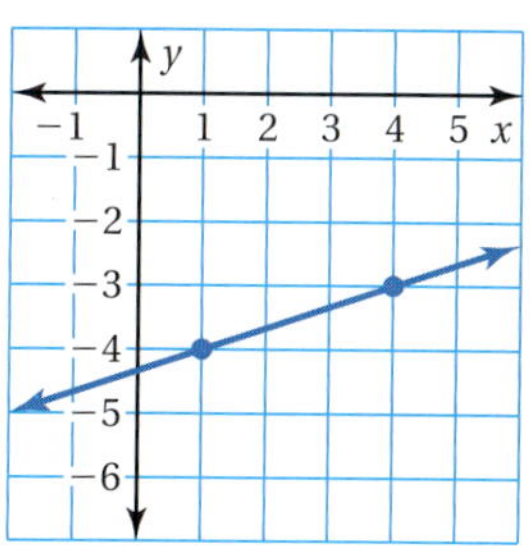

42. MULTIPLE CHOICE You want to volunteer for at most 20 hours each month. So far, you have volunteered for 7 hours this month. Which inequality represents the number of hours h you can volunteer for the rest of this month? *(Skills Review Handbook)*

Ⓐ $h \geq 13$ Ⓑ $h \geq 27$ Ⓒ $h \leq 13$ Ⓓ $h < 27$

6.3 Linear Functions

Essential Question How can you use a function to describe a linear pattern?

1 ACTIVITY: Finding Linear Patterns

Work with a partner.

- **Plot the points from the table in a coordinate plane.**
- **Write a linear equation for the function represented by the graph.**

a.

x	0	2	4	6	8
y	150	125	100	75	50

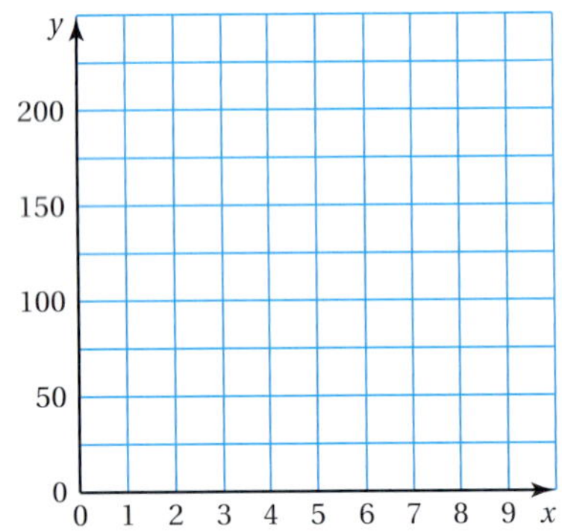

b.

x	4	6	8	10	12
y	15	20	25	30	35

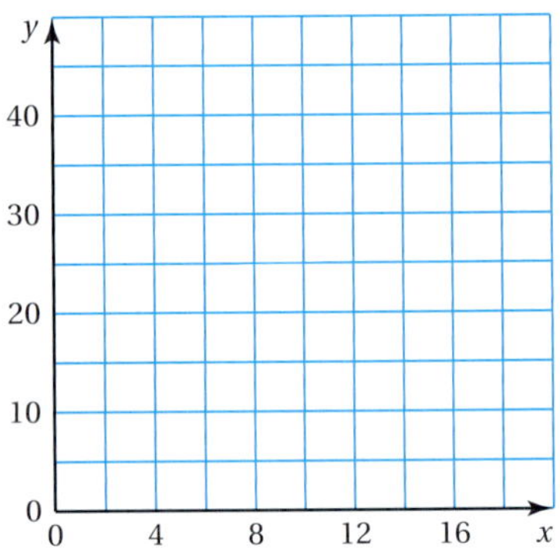

c.

x	−4	−2	0	2	4
y	4	6	8	10	12

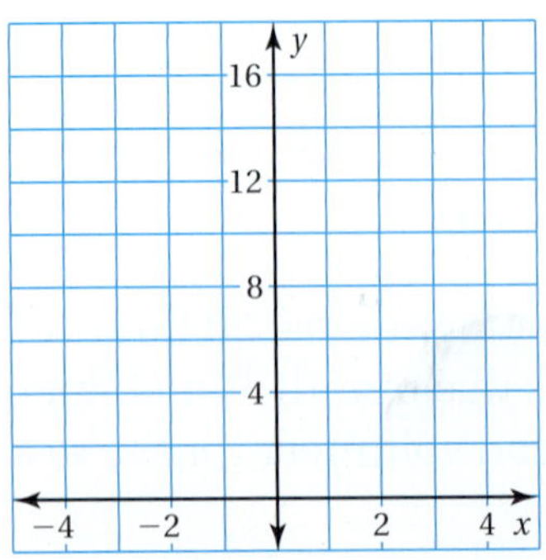

d.

x	−4	−2	0	2	4
y	1	0	−1	−2	−3

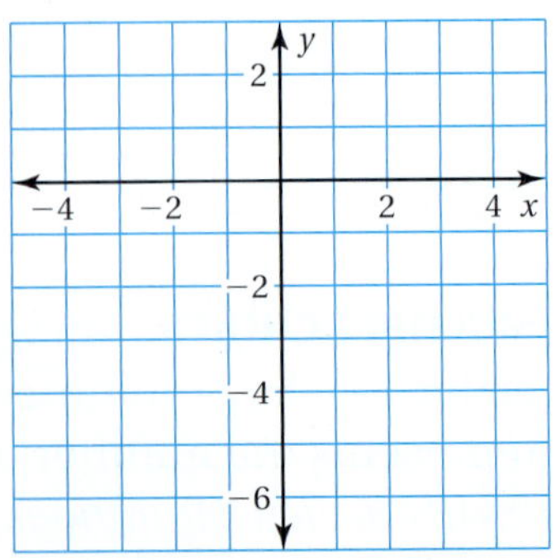

COMMON CORE

Functions

In this lesson, you will

- understand that the equation $y = mx + b$ defines a linear function.
- write linear functions using graphs or tables.
- compare linear functions.

Learning Standards
MACC.8.F.1.2
MACC.8.F.1.3
MACC.8.F.2.4

2 ACTIVITY: Finding Linear Patterns

Math Practice 6

Label Axes

How do you know what to label the axes? How does this help you accurately graph the data?

Work with a partner. The table shows a familiar linear pattern from geometry.

- **Write a function that relates y to x.**
- **What do the variables x and y represent?**
- **Graph the function.**

a.

x	1	2	3	4	5
y	2π	4π	6π	8π	10π

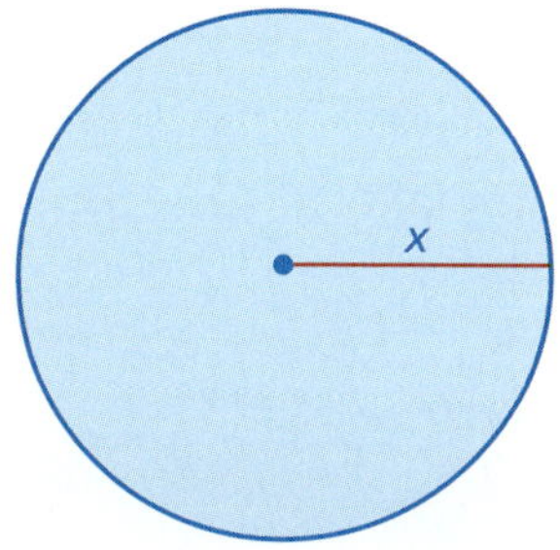

b.

x	1	2	3	4	5
y	10	12	14	16	18

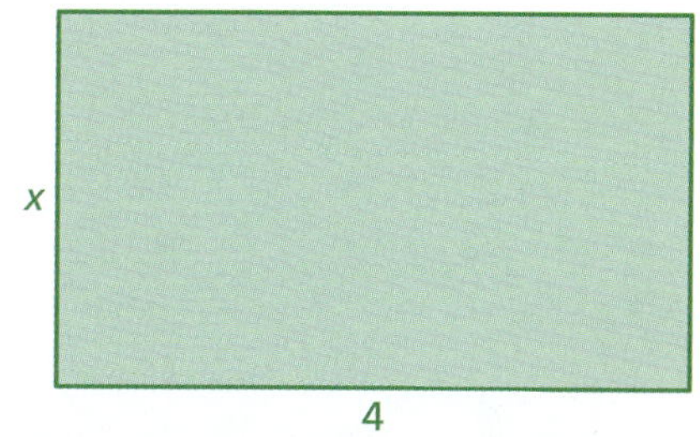

c.

x	1	2	3	4	5
y	5	6	7	8	9

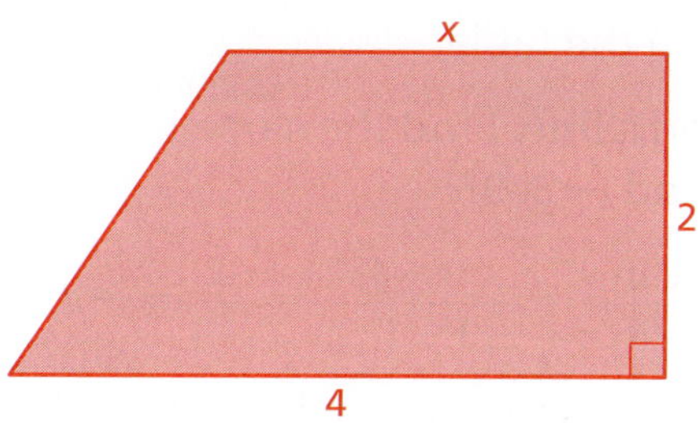

d.

x	1	2	3	4	5
y	28	40	52	64	76

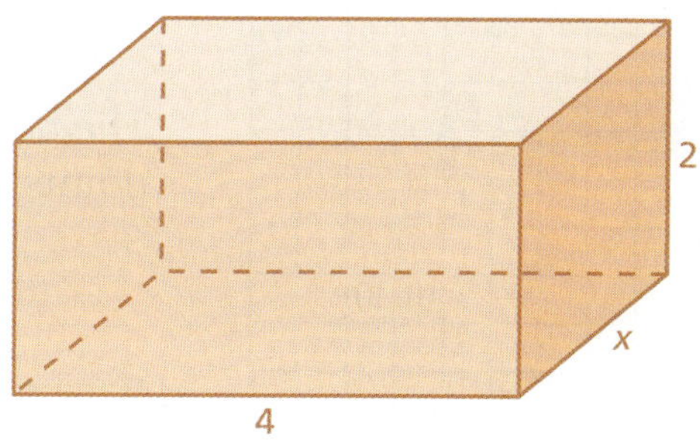

What Is Your Answer?

3. **IN YOUR OWN WORDS** How can you use a function to describe a linear pattern?

4. Describe the strategy you used to find the functions in Activities 1 and 2.

Practice Use what you learned about linear patterns to complete Exercises 3 and 4 on page 261.

6.3 Lesson

Check It Out
Lesson Tutorials
BigIdeasMath.com

Key Vocabulary
linear function, *p. 258*

A **linear function** is a function whose graph is a nonvertical line. A linear function can be written in the form $y = mx + b$, where m is the slope and b is the y-intercept.

EXAMPLE 1 Writing a Linear Function Using a Graph

Use the graph to write a linear function that relates y to x.

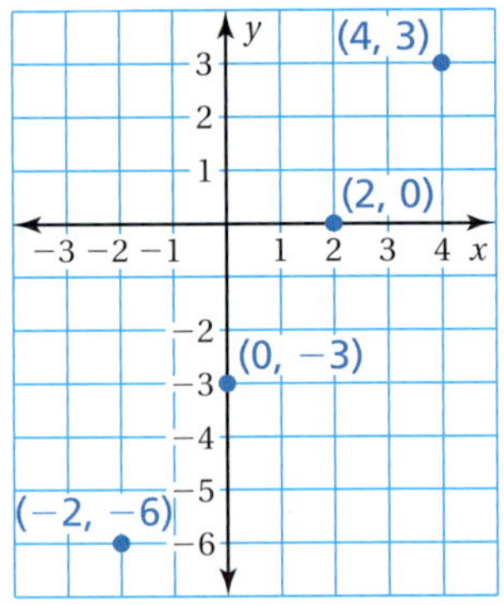

The points lie on a line. Find the slope by using the points (2, 0) and (4, 3).

$$m = \frac{\text{change in } y}{\text{change in } x} = \frac{3 - 0}{4 - 2} = \frac{3}{2}$$

Because the line crosses the y-axis at $(0, -3)$, the y-intercept is -3.

So, the linear function is $y = \frac{3}{2}x - 3$.

EXAMPLE 2 Writing a Linear Function Using a Table

Use the table to write a linear function that relates y to x.

x	−3	−2	−1	0
y	9	7	5	3

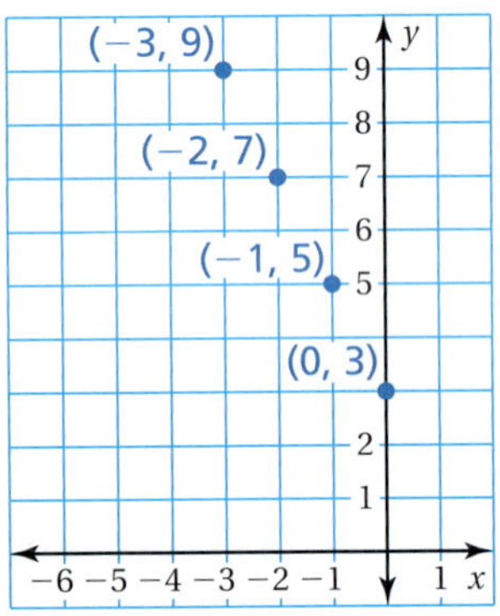

Plot the points in the table.

The points lie on a line. Find the slope by using the points $(-2, 7)$ and $(-3, 9)$.

$$m = \frac{\text{change in } y}{\text{change in } x} = \frac{9 - 7}{-3 - (-2)} = \frac{2}{-1} = -2$$

Because the line crosses the y-axis at $(0, 3)$, the y-intercept is 3.

So, the linear function is $y = -2x + 3$.

On Your Own

Now You're Ready
Exercises 5–10

Use the graph or table to write a linear function that relates y to x.

1. 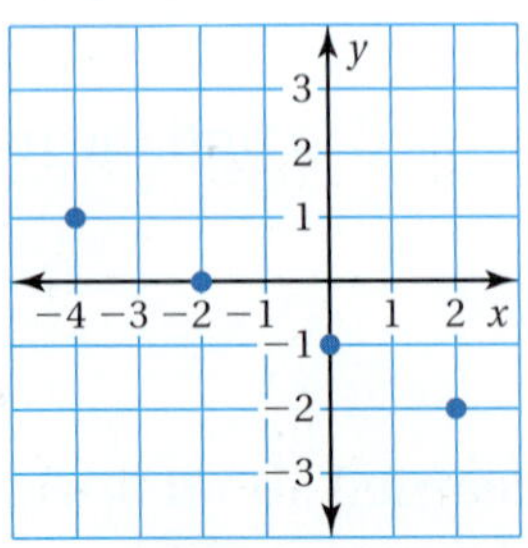

2.

x	−2	−1	0	1
y	2	2	2	2

Multi-Language Glossary at BigIdeasMath.com

EXAMPLE 3 Real-Life Application

Minutes, x	Height (thousands of feet), y
0	65
10	60
20	55
30	50

You are controlling an unmanned aerial vehicle (UAV) for surveillance. The table shows the height y (in thousands of feet) of the UAV x minutes after you start its descent from cruising altitude.

a. Write a linear function that relates y to x. Interpret the slope and the y-intercept.

You can write a linear function that relates the dependent variable y to the independent variable x because the table shows a constant rate of change. Find the slope by using the points (0, 65) and (10, 60).

$$m = \frac{\text{change in } y}{\text{change in } x} = \frac{60 - 65}{10 - 0} = \frac{-5}{10} = -0.5$$

Because the line crosses the y-axis at (0, 65), the y-intercept is 65.

So, the linear function is $y = -0.5x + 65$. The slope indicates that the height decreases 500 feet per minute. The y-intercept indicates that the descent begins at a cruising altitude of 65,000 feet.

Common Error

Make sure you consider the units when interpreting the slope and the y-intercept.

b. Graph the linear function.

Plot the points in the table and draw a line through the points.

Because time cannot be negative in this context, use only positive values of x.

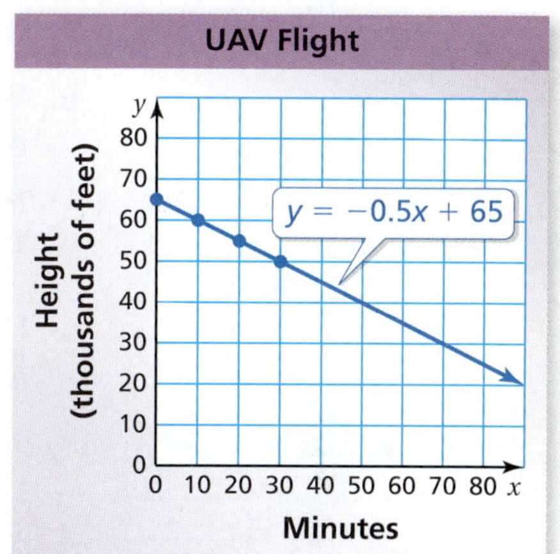

c. Find the height of the UAV when you stop the descent after 1 hour.

Because 1 hour = 60 minutes, find the value of y when $x = 60$.

$y = -0.5x + 65$ — Write the equation.

$= -0.5(60) + 65$ — Substitute 60 for x.

$= 35$ — Simplify.

So, the descent of the UAV stops at a height of 35,000 feet.

On Your Own

Now You're Ready
Exercises 11–13

3. **WHAT IF?** You double the rate of descent. Repeat parts (a)–(c).

EXAMPLE 4 Comparing Linear Functions

The earnings y (in dollars) of a nighttime employee working x hours are represented by the linear function $y = 7.5x + 30$. The table shows the earnings of a daytime employee.

Time (hours), x	1	2	3	4
Earnings (dollars), y	12.50	25.00	37.50	50.00

(+1, +1, +1 between times; +12.50, +12.50, +12.50 between earnings)

a. Which employee has a higher hourly wage?

Nighttime Employee

$y = 7.5x + 30$

The slope is 7.5.

The nighttime employee earns \$7.50 per hour.

Daytime Employee

$\dfrac{\text{change in earnings}}{\text{change in time}} = \dfrac{\$12.50}{1 \text{ hour}}$

The daytime employee earns \$12.50 per hour.

So, the daytime employee has a higher hourly wage.

b. Write a linear function that relates the daytime employee's earnings to the number of hours worked. In the same coordinate plane, graph the linear functions that represent the earnings of the two employees. Interpret the graphs.

Use a verbal model to write a linear function that represents the earnings of the daytime employee.

Earnings = Hourly wage • Hours worked

$y = 12.5x$

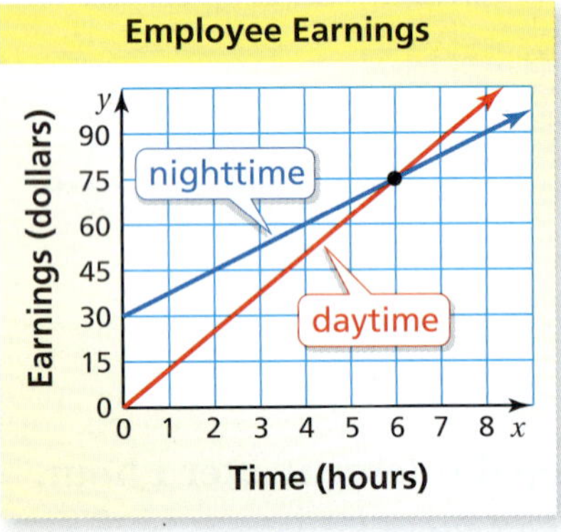

The graph shows that the daytime employee has a higher hourly wage but does not earn more money than the nighttime employee until each person has worked more than 6 hours.

On Your Own

Now You're Ready
Exercise 14

4. Manager A earns \$15 per hour and receives a \$50 bonus. The graph shows the earnings of Manager B.

   a. Which manager has a higher hourly wage?

   b. After how many hours does Manager B earn more money than Manager A?

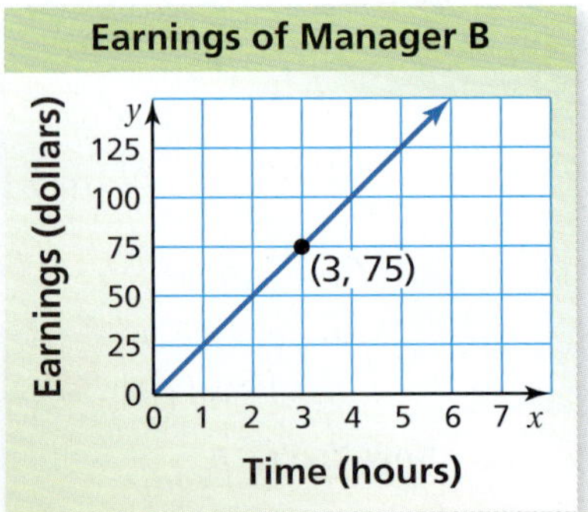

6.3 Exercises

Check It Out
Help with Homework
BigIdeasMath.com

Vocabulary and Concept Check

1. **STRUCTURE** Is $y = mx + b$ a linear function when $b = 0$? Explain.

2. **WRITING** Explain why the vertical line does not represent a linear function.

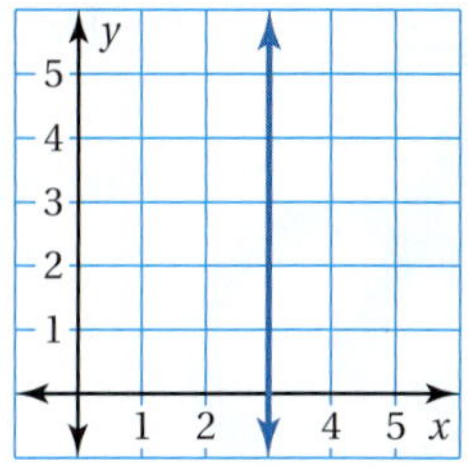

Practice and Problem Solving

The table shows a familiar linear pattern from geometry. Write a function that relates y to x. What do the variables x and y represent? Graph the function.

3.

x	1	2	3	4	5
y	π	2π	3π	4π	5π

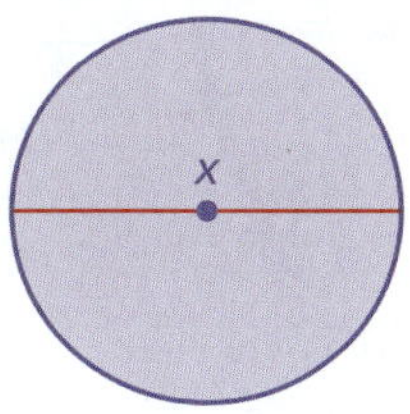

4.

x	1	2	3	4	5
y	2	4	6	8	10

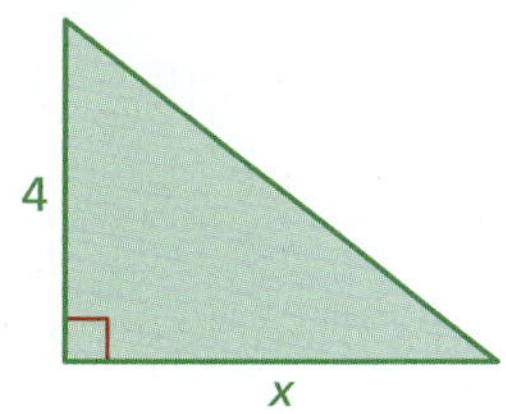

Use the graph or table to write a linear function that relates y to x.

1 2 5.

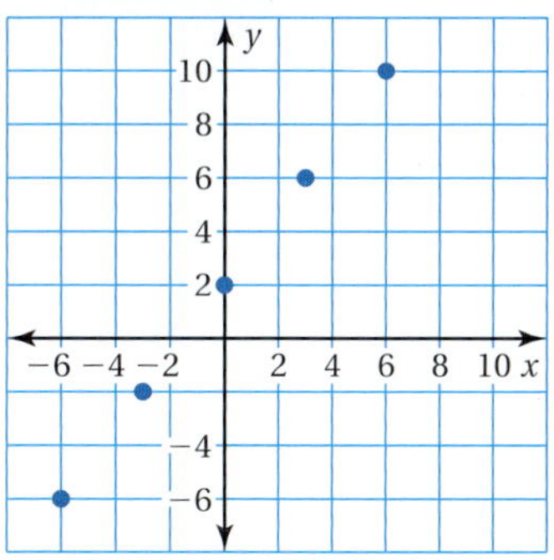

6.

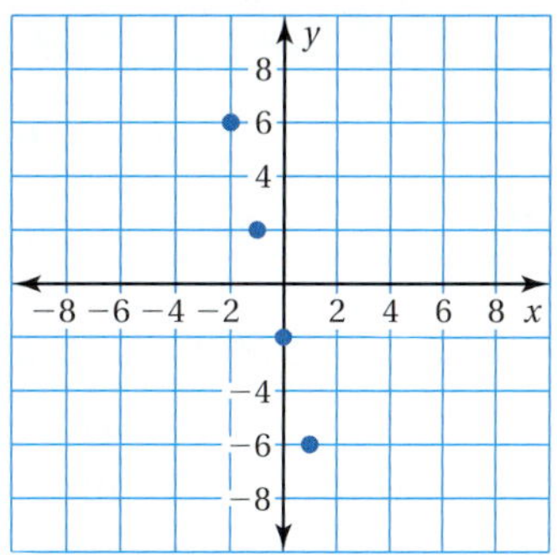

7.

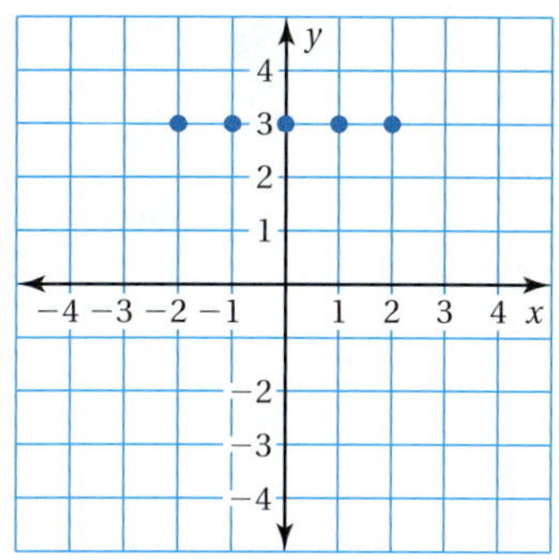

8.

x	−2	−1	0	1
y	−4	−2	0	2

9.

x	−8	−4	0	4
y	2	1	0	−1

10.

x	−3	0	3	6
y	3	5	7	9

3 11. **MOVIES** The table shows the cost y (in dollars) of renting x movies.

a. Which variable is independent? dependent?

b. Write a linear function that relates y to x. Interpret the slope.

c. Graph the linear function.

d. How much does it cost to rent three movies?

Number of Movies, x	0	1	2	4
Cost, y	0	3	6	12

12. **BIKE JUMPS** A *bunny hop* is a bike trick in which the rider brings both tires off the ground without using a ramp. The table shows the height y (in inches) of a bunny hop on a bike that weighs x pounds.

Weight (pounds), x	19	21	23
Height (inches), y	10.2	9.8	9.4

a. Write a linear function that relates the height of a bunny hop to the weight of the bike.

b. Graph the linear function.

c. What is the height of a bunny hop on a bike that weighs 21.5 pounds?

13. **BATTERY** The graph shows the percent y (in decimal form) of battery power remaining x hours after you turn on a laptop computer.

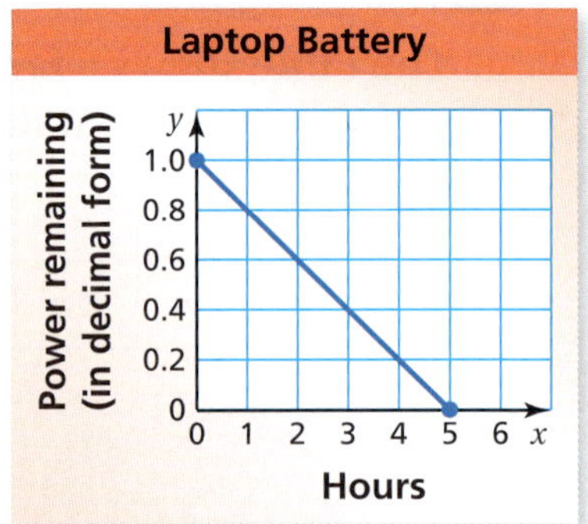

a. Write a linear function that relates y to x.

b. Interpret the slope, the x-intercept, and the y-intercept.

c. After how many hours is the battery power at 75%?

4 14. **RACE** You and a friend race each other. You give your friend a 50-foot head start. The distance y (in feet) your friend runs after x seconds is represented by the linear function $y = 14x + 50$. The table shows the distances you run.

Time (seconds), x	2	4	6	8
Distance (feet), y	38	76	114	152

a. Who runs at a faster rate? What is that rate?

b. Write a linear function that relates your distance to the number of seconds. In the same coordinate plane, graph the linear functions that represent the distances of you and your friend.

c. For what distances will you win the race? Explain.

15. **CALORIES** The number of calories burned y after x minutes of kayaking is represented by the linear function $y = 4.5x$. The graph shows the calories burned by hiking.

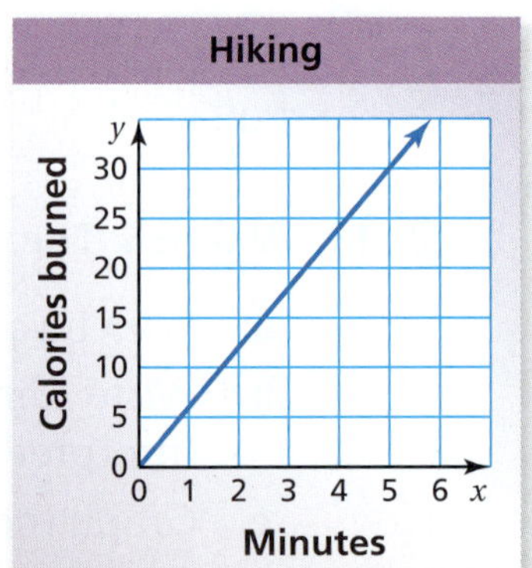

a. Which activity burns more calories per minute?

b. How many more calories are burned by doing the activity in part (a) than the other activity for 45 minutes?

16. SAVINGS You and your friend are saving money to buy bicycles that cost \$175 each. The amount y (in dollars) you save after x weeks is represented by the equation $y = 5x + 45$. The graph shows your friend's savings.

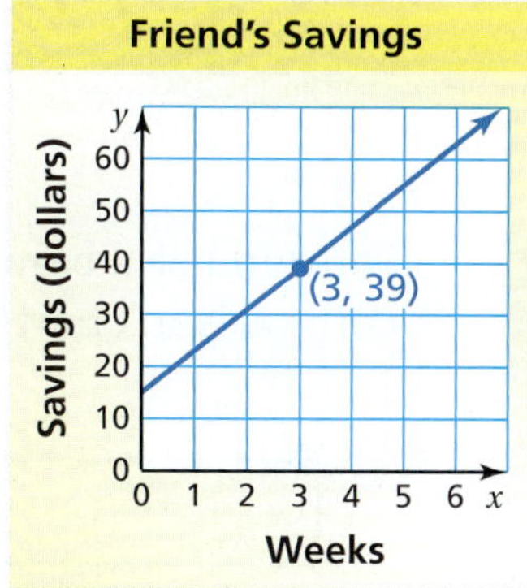

a. Who has more money to start? Who saves more per week?

b. Who can buy a bicycle first? Explain.

17. REASONING Can the graph of a linear function be a horizontal line? Explain your reasoning.

Years of Education, x	Annual Salary, y
0	28
2	40
4	52
6	64
10	88

18. SALARY The table shows a person's annual salary y (in thousands of dollars) after x years of education beyond high school.

a. Graph the data. Then describe the pattern.

b. What is the annual salary of the person after 8 years of education beyond high school?

c. Find the annual salary of a person with 30 years of education. Do you think this situation makes sense? Explain.

19. Problem Solving The Heat Index is calculated using the relative humidity and the temperature. For every 1 degree increase in the temperature from 94°F to 98°F at 75% relative humidity, the Heat Index rises 4°F.

a. On a summer day, the relative humidity is 75%, the temperature is 94°F, and the Heat Index is 122°F. Construct a table that relates the temperature t to the Heat Index H. Start the table at 94°F and end it at 98°F.

b. Identify the independent and dependent variables.

c. Write a linear function that represents this situation.

d. Estimate the Heat Index when the temperature is 100°F.

Fair Game Review What you learned in previous grades & lessons

Solve the equation. *(Section 1.1)*

20. $b - 1.6 \div 4 = -3$

21. $w + |-2.8| = 4.3$

22. $\frac{3}{4} = y - \frac{1}{5}(8)$

23. MULTIPLE CHOICE Which of the following describes the translation from the red figure to the blue figure? *(Section 2.2)*

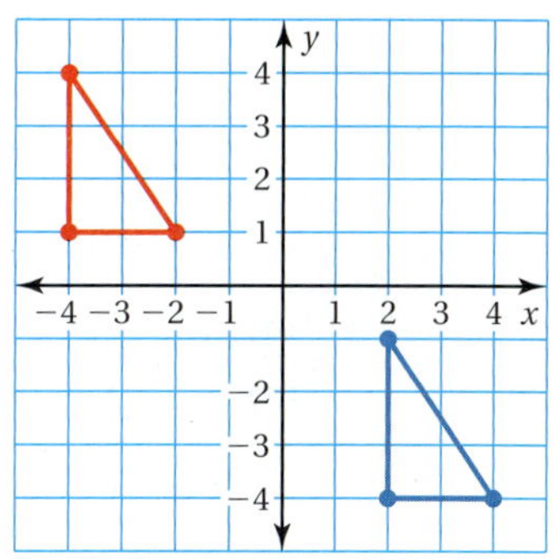

Ⓐ $(x - 6, y + 5)$ Ⓑ $(x - 5, y + 6)$

Ⓒ $(x + 6, y - 5)$ Ⓓ $(x + 5, y - 6)$

6 Study Help

Check It Out
Graphic Organizer
BigIdeasMath.com

You can use a **comparison chart** to compare two topics. Here is an example of a comparison chart for relations and functions.

	Relations	Functions
Definition	A relation pairs inputs with outputs.	A relation that pairs each input with *exactly one* output is a function.
Ordered pairs	(1, 0) (3, −1) (3, 6) (7, 14)	(1, 0) (2, −1) (5, 7) (6, 20)
Mapping diagram	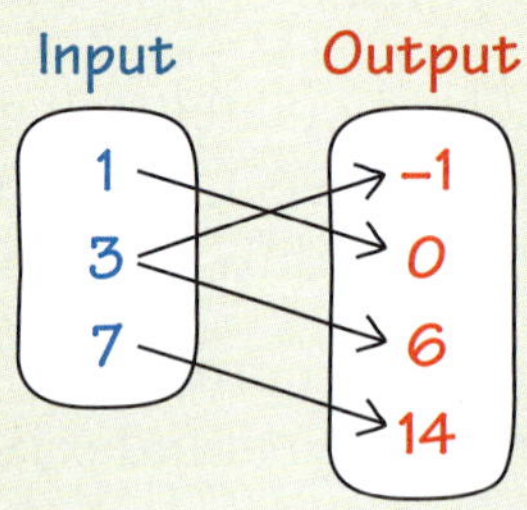	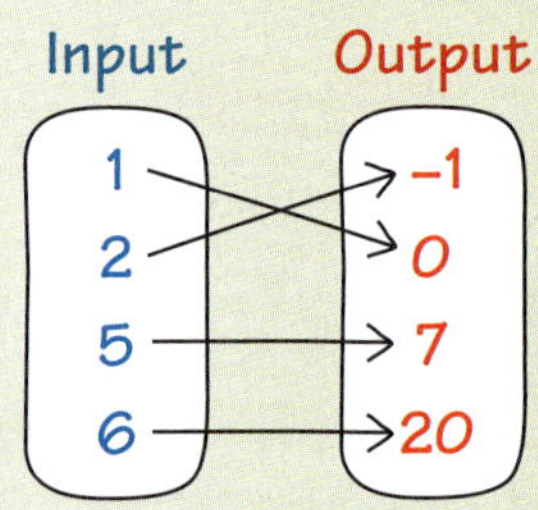

On Your Own

Make comparison charts to help you study and compare these topics.

1. functions as tables and functions as graphs
2. linear functions with positive slopes and linear functions with negative slopes

After you complete this chapter, make comparison charts for the following topics.

3. linear functions and nonlinear functions
4. graphs with numerical values on the axes and graphs without numerical values on the axes

"Creating a comparison chart causes canines to crystalize concepts."

6.1–6.3 Quiz

Check It Out
Progress Check
BigIdeasMath.com

List the ordered pairs shown in the mapping diagram. Then determine whether the relation is a function. *(Section 6.1)*

1.

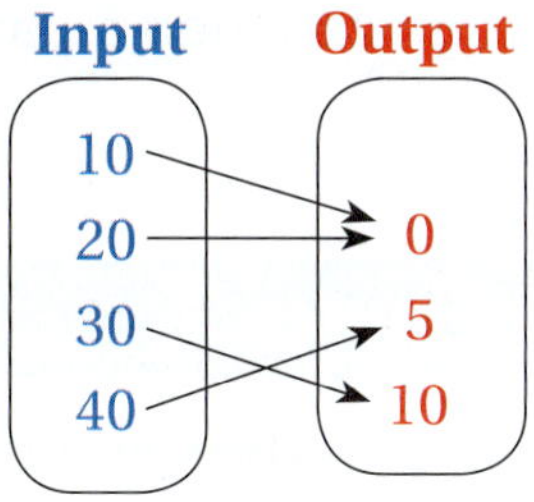

2.

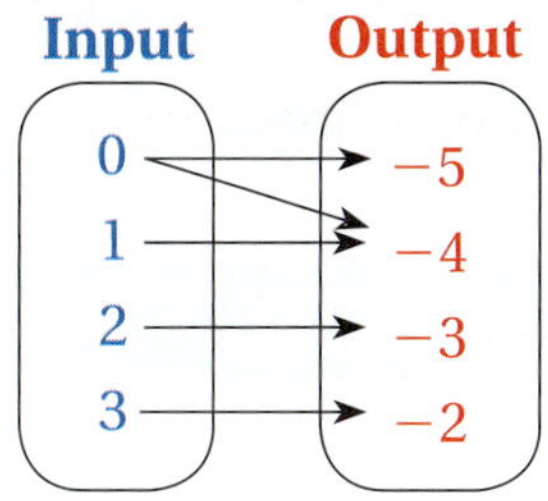

Find the value of *y* for the given value of *x*. *(Section 6.2)*

3. $y = 10x;\ x = -3$

4. $y = 6 - 2x;\ x = 11$

5. $y = 4x + 5;\ x = \frac{1}{2}$

Graph the function. *(Section 6.2)*

6. $y = x - 10$

7. $y = 2x + 3$

8. $y = \frac{x}{2}$

Use the graph or table to write a linear function that relates *y* to *x*. *(Section 6.3)*

9.

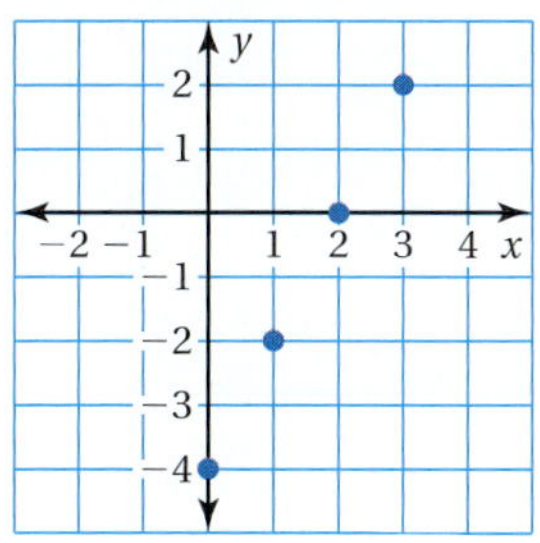

10.

x	*y*
−3	−3
0	−1
3	1
6	3

11. PUPPIES The table shows the ages of four puppies and their weights. Use the table to draw a mapping diagram. *(Section 6.1)*

Age (weeks)	Weight (oz)
3	11
4	85
6	85
10	480

12. MUSIC An online music store sells songs for $0.90 each. *(Section 6.2)*

a. Write a function that you can use to find the cost C of buying s songs.

b. What is the cost of buying 5 songs?

13. ADVERTISING The table shows the revenue R (in millions of dollars) of a company when it spends A (in millions of dollars) on advertising. *(Section 6.3)*

Advertising, *A*	Revenue, *R*
0	2
2	6
4	10
6	14
8	18

a. Write and graph a linear function that relates the revenue to the advertising cost.

b. What is the revenue of the company when it spends $15 million on advertising?

6.4 Comparing Linear and Nonlinear Functions

Essential Question How can you recognize when a pattern in real life is linear or nonlinear?

1 ACTIVITY: Finding Patterns for Similar Figures

Work with a partner. Copy and complete each table for the sequence of similar rectangles. Graph the data in each table. Decide whether each pattern is linear or nonlinear.

a. Perimeters of similar rectangles

x	1	2	3	4	5
P					

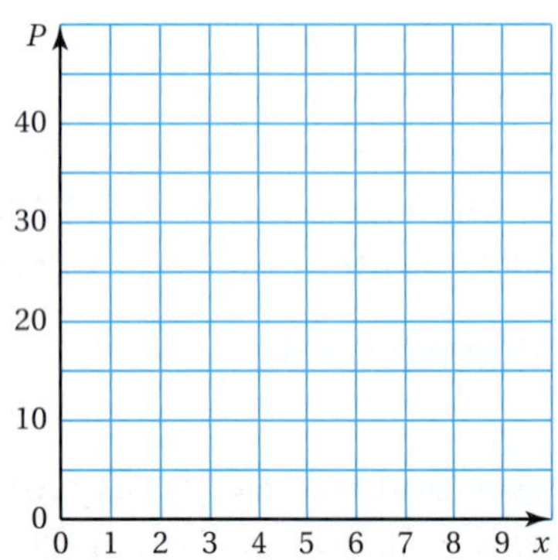

b. Areas of similar rectangles

x	1	2	3	4	5
A					

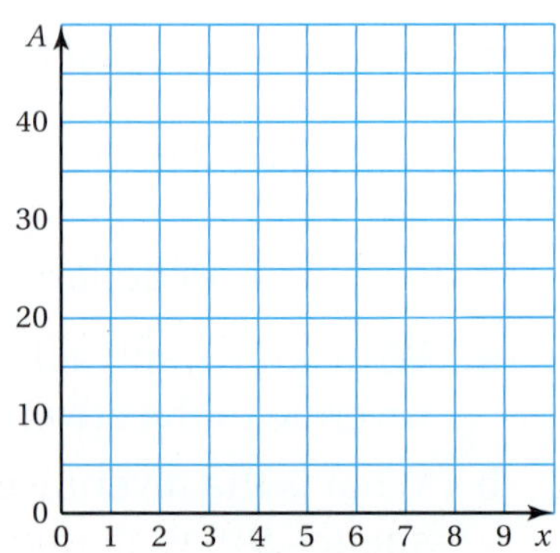

COMMON CORE

Functions

In this lesson, you will

- identify linear and nonlinear functions from tables or graphs.
- compare linear and nonlinear functions.

Learning Standard
MACC.8.F.1.3

2 ACTIVITY: Comparing Linear and Nonlinear Functions

Work with a partner. Each table shows the height *h* (in feet) of a falling object at *t* seconds.

- **Graph the data in each table.**
- **Decide whether each graph is linear or nonlinear.**
- **Compare the two falling objects. Which one has an increasing speed?**

a. Falling parachute jumper

t	0	1	2	3	4
h	300	285	270	255	240

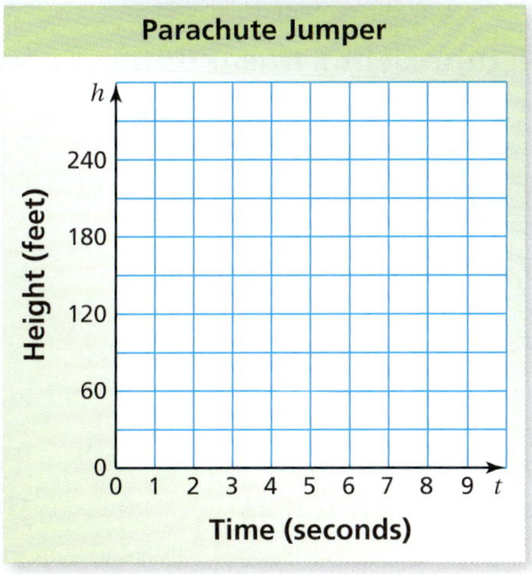

b. Falling bowling ball

t	0	1	2	3	4
h	300	284	236	156	44

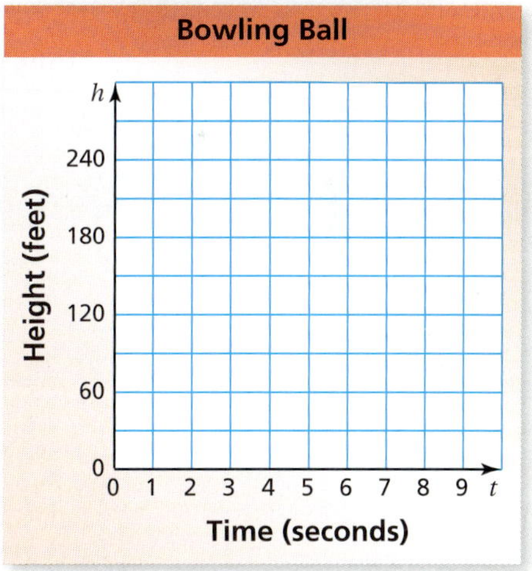

Math Practice 4

Apply Mathematics

What will the graph look like for an object that has a constant speed? an increasing speed? Explain.

What Is Your Answer?

3. **IN YOUR OWN WORDS** How can you recognize when a pattern in real life is linear or nonlinear? Describe two real-life patterns: one that is linear and one that is nonlinear. Use patterns that are different from those described in Activities 1 and 2.

Practice Use what you learned about comparing linear and nonlinear functions to complete Exercises 3–6 on page 270.

6.4 Lesson

Key Vocabulary
nonlinear function, *p. 268*

The graph of a linear function shows a constant rate of change. A **nonlinear function** does not have a constant rate of change. So, its graph is *not* a line.

EXAMPLE 1 Identifying Functions from Tables

Does the table represent a *linear* or *nonlinear* function? Explain.

a.

+3 +3 +3

x	3	6	9	12
y	40	32	24	16

−8 −8 −8

As x increases by 3, y decreases by 8. The rate of change is constant. So, the function is linear.

b.

+2 +2 +2

x	1	3	5	7
y	2	11	33	88

+9 +22 +55

As x increases by 2, y increases by different amounts. The rate of change is *not* constant. So, the function is nonlinear.

Study Tip
A constant rate of change describes a quantity that changes by equal amounts over equal intervals.

EXAMPLE 2 Identifying Functions from Graphs

Does the graph represent a *linear* or *nonlinear* function? Explain.

a.

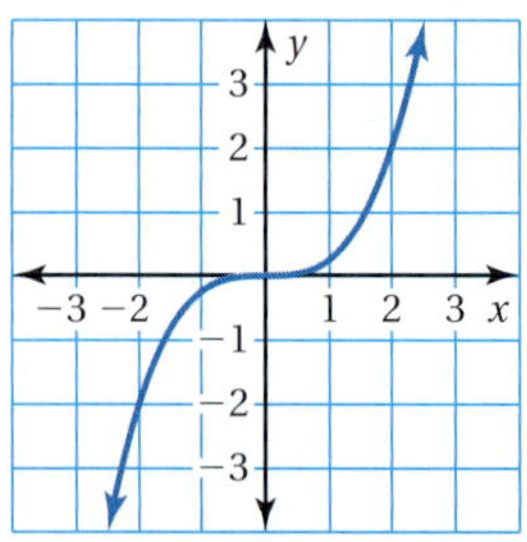

The graph is *not* a line. So, the function is nonlinear.

b.

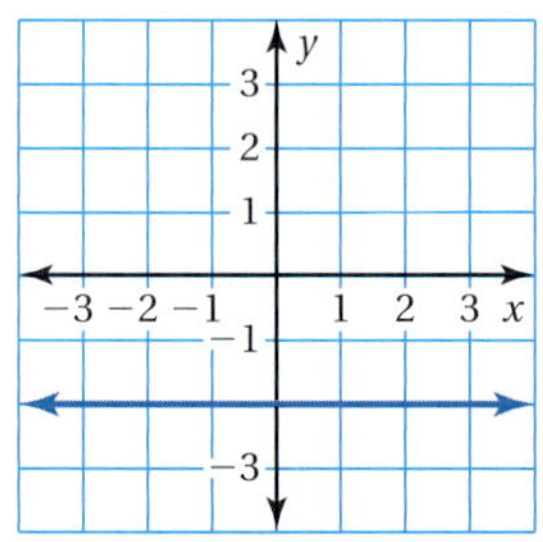

The graph is a line. So, the function is linear.

On Your Own

Now You're Ready
Exercises 7–10

Does the table or graph represent a *linear* or *nonlinear* function? Explain.

1.

x	y
0	25
7	20
14	15
21	10

2.

x	y
2	8
4	4
6	0
8	−4

3.

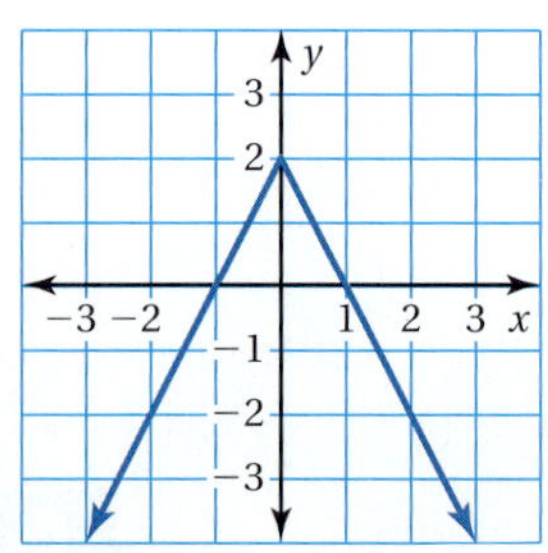

EXAMPLE 3 Identifying a Nonlinear Function

Which equation represents a *nonlinear* function?

Ⓐ $y = 4.7$ Ⓑ $y = \pi x$

Ⓒ $y = \frac{4}{x}$ Ⓓ $y = 4(x - 1)$

You can rewrite the equations $y = 4.7$, $y = \pi x$, and $y = 4(x - 1)$ in slope-intercept form. So, they are linear functions.

You cannot rewrite the equation $y = \frac{4}{x}$ in slope-intercept form. So, it is a nonlinear function.

∴ The correct answer is Ⓒ.

EXAMPLE 4 Real-Life Application

Account A earns simple interest. Account B earns compound interest. The table shows the balances for 5 years. Graph the data and compare the graphs.

Year, t	Account A Balance	Account B Balance
0	\$100	\$100
1	\$110	\$110
2	\$120	\$121
3	\$130	\$133.10
4	\$140	\$146.41
5	\$150	\$161.05

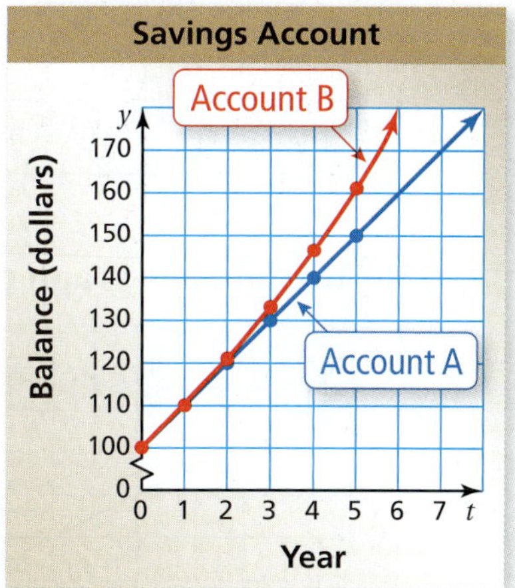

Both graphs show that the balances are positive and increasing.

The balance of Account A has a constant rate of change of \$10. So, the function representing the balance of Account A is linear.

The balance of Account B increases by different amounts each year. Because the rate of change is not constant, the function representing the balance of Account B is nonlinear.

On Your Own

Now You're Ready
Exercises 12–14

Does the equation represent a *linear* or *nonlinear* function? Explain.

4. $y = x + 5$ **5.** $y = \frac{4x}{3}$ **6.** $y = 1 - x^2$

6.4 Exercises

Vocabulary and Concept Check

1. **VOCABULARY** Describe how linear functions and nonlinear functions are different.

2. **WHICH ONE DOESN'T BELONG?** Which equation does *not* belong with the other three? Explain your reasoning.

$5y = 2x$ $\qquad y = \frac{2}{5}x$ $\qquad 10y = 4x$ $\qquad 5xy = 2$

Practice and Problem Solving

Graph the data in the table. Decide whether the graph is *linear* or *nonlinear*.

3.

x	0	1	2	3
y	4	8	12	16

4.

x	1	2	3	4
y	1	2	6	24

5.

x	6	5	4	3
y	21	15	10	6

6.

x	−1	0	1	2
y	−7	−3	1	5

Does the table or graph represent a *linear* or *nonlinear* function? Explain.

1 2 7.

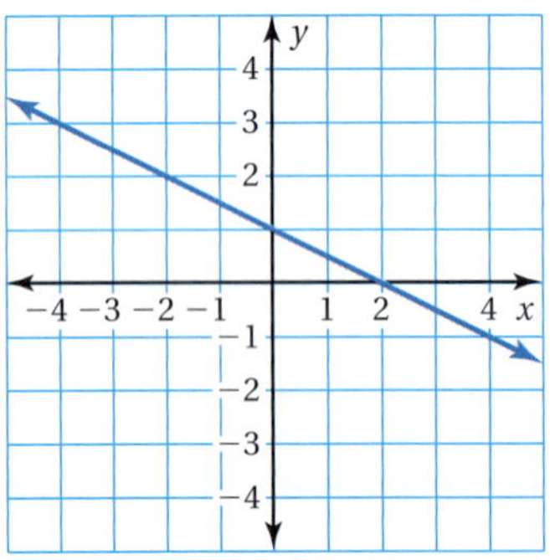

8.

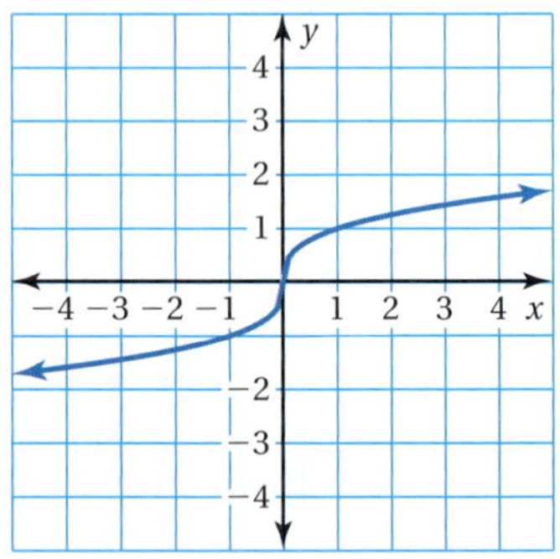

9.

x	5	11	17	23
y	7	11	15	19

10.

x	−3	−1	1	3
y	9	1	1	9

11. **VOLUME** The table shows the volume V (in cubic feet) of a cube with an edge length of x feet. Does the table represent a linear or nonlinear function? Explain.

Edge Length, *x*	1	2	3	4	5	6	7	8
Volume, *V*	1	8	27	64	125	216	343	512

Does the equation represent a *linear* or *nonlinear* function? Explain.

3 **12.** $2x + 3y = 7$ **13.** $y + x = 4x + 5$ **14.** $y = \frac{8}{x^2}$

15. **LIGHT** The frequency y (in terahertz) of a light wave is a function of its wavelength x (in nanometers). Does the table represent a linear or nonlinear function? Explain.

Color	Red	Yellow	Green	Blue	Violet
Wavelength, x	660	595	530	465	400
Frequency, y	454	504	566	645	749

16. **MODELING** The table shows the cost y (in dollars) of x pounds of sunflower seeds.

Pounds, x	Cost, y
2	2.80
3	?
4	5.60

a. What is the missing y-value that makes the table represent a linear function?

b. Write a linear function that represents the cost y of x pounds of seeds. Interpret the slope.

c. Does the function have a maximum value? Explain your reasoning.

17. **TREES** Tree A is 5 feet tall and grows at a rate of 1.5 feet per year. The table shows the height h (in feet) of Tree B after x years.

Years, x	Height, h
0	5
1	11
4	17
9	23

a. Does the table represent a linear or nonlinear function? Explain.

b. Which tree is taller after 10 years? Explain.

18. Number Sense The ordered pairs represent a function.

$(0, -1), (1, 0), (2, 3), (3, 8)$, and $(4, 15)$

a. Graph the ordered pairs and describe the pattern. Is the function linear or nonlinear?

b. Write an equation that represents the function.

Fair Game Review What you learned in previous grades & lessons

The vertices of a figure are given. Draw the figure and its image after a dilation with the given scale factor *k*. Identify the type of dilation. *(Section 2.7)*

19. $A(-3, 1), B(-1, 3), C(-1, 1); k = 3$

20. $J(-8, -4), K(2, -4), L(6, -10), M(-8, -10); k = \frac{1}{4}$

21. **MULTIPLE CHOICE** What is the value of x? *(Section 3.3)*

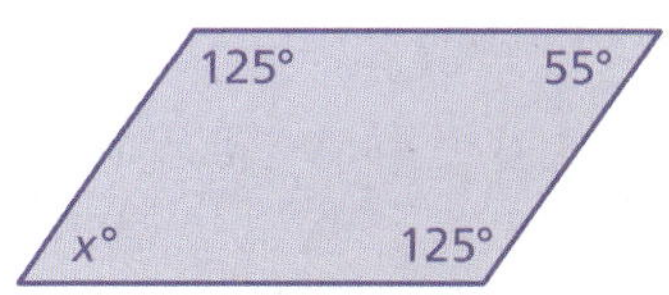

Ⓐ 25 Ⓑ 35

Ⓒ 55 Ⓓ 125

6.5 Analyzing and Sketching Graphs

Essential Question How can you use a graph to represent relationships between quantities without using numbers?

1 ACTIVITY: Interpreting a Graph

Work with a partner. Use the graph shown.

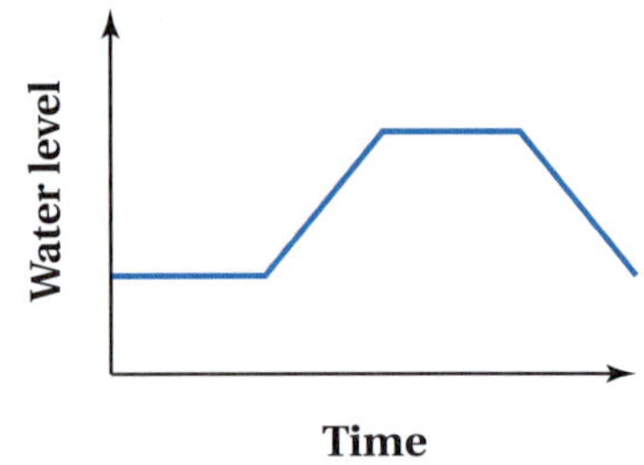

a. How is this graph different from the other graphs you have studied?

b. Write a short paragraph that describes how the water level changes over time.

c. What situation can this graph represent?

2 ACTIVITY: Matching Situations to Graphs

Work with a partner. You are riding your bike. Match each situation with the appropriate graph. Explain your reasoning.

A.

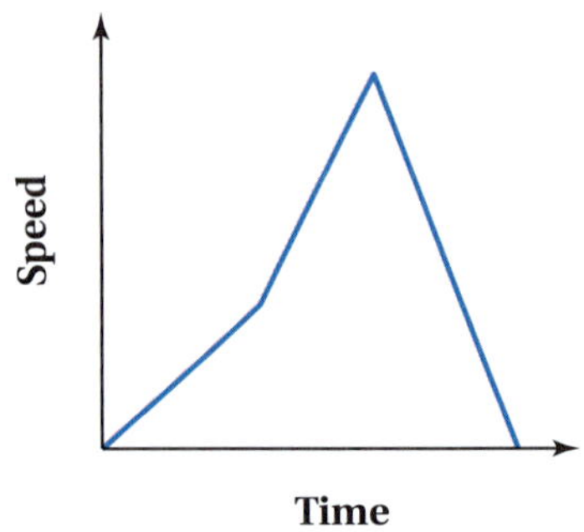

B.

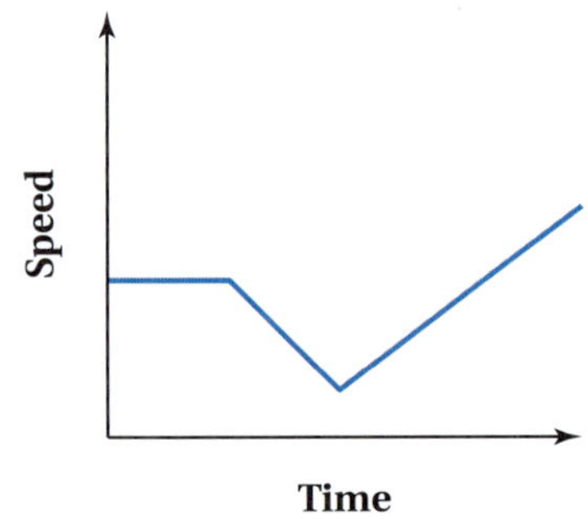

C.

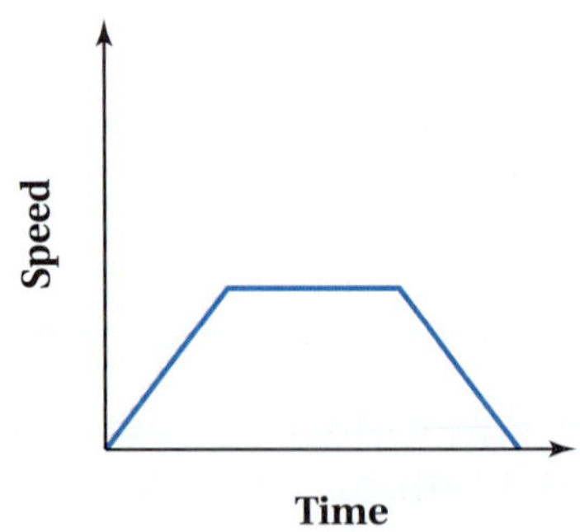

D.

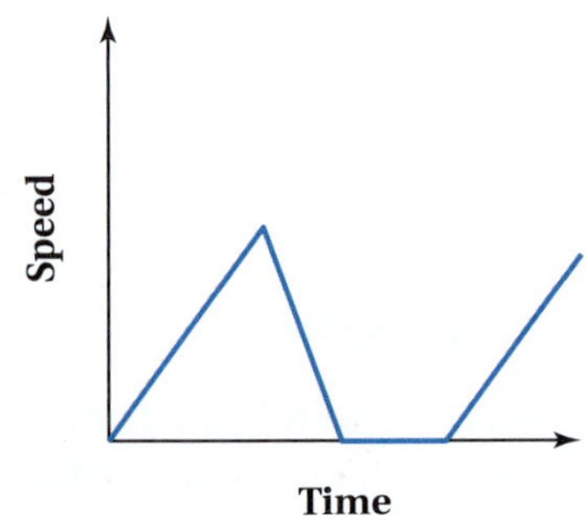

a. You gradually increase your speed, then ride at a constant speed along a bike path. You then slow down until you reach your friend's house.

b. You gradually increase your speed, then go down a hill. You then quickly come to a stop at an intersection.

c. You gradually increase your speed, then stop at a store for a couple of minutes. You then continue to ride, gradually increasing your speed.

d. You ride at a constant speed, then go up a hill. Once on top of the hill, you gradually increase your speed.

COMMON CORE

Functions

In this lesson, you will

- analyze the relationship between two quantities using graphs.
- sketch graphs to represent the relationship between two quantities.

Learning Standard
MACC.8.F.2.5

③ ACTIVITY: Comparing Graphs

Work with a partner. The graphs represent the heights of a rocket and a weather balloon after they are launched.

a. How are the graphs similar? How are they different? Explain.

b. Compare the steepness of each graph.

c. Which graph do you think represents the height of the rocket? Explain.

Graph A

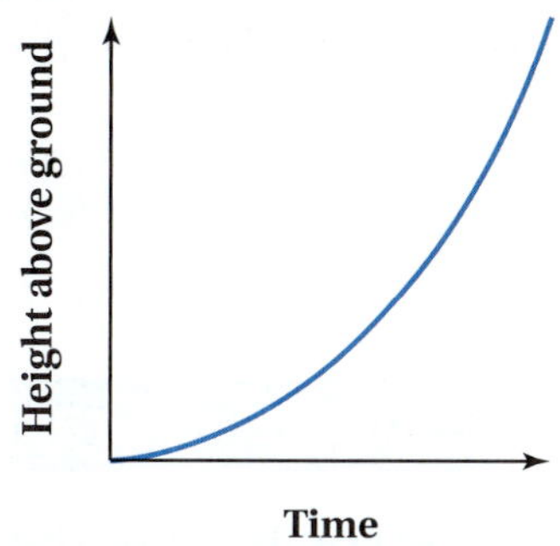

Graph B

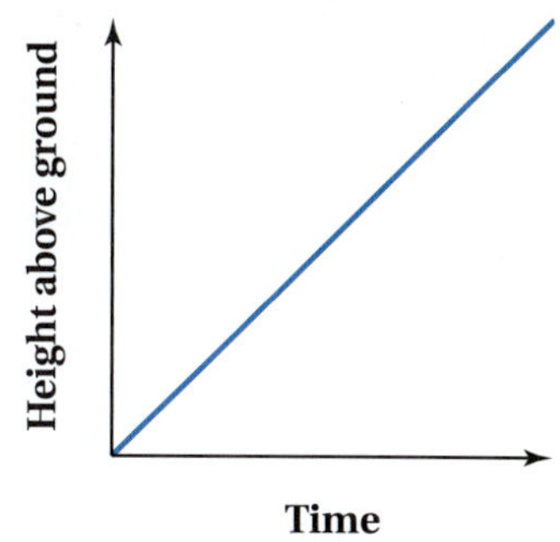

④ ACTIVITY: Comparing Graphs

Math Practice 1

Consider Similar Problems

How is this activity similar to the previous activity?

Work with a partner. The graphs represent the speeds of two cars. One car is approaching a stop sign. The other car is approaching a yield sign.

a. How are the graphs similar? How are they different? Explain.

b. Compare the steepness of each graph.

c. Which graph do you think represents the car approaching a stop sign? Explain.

Graph A

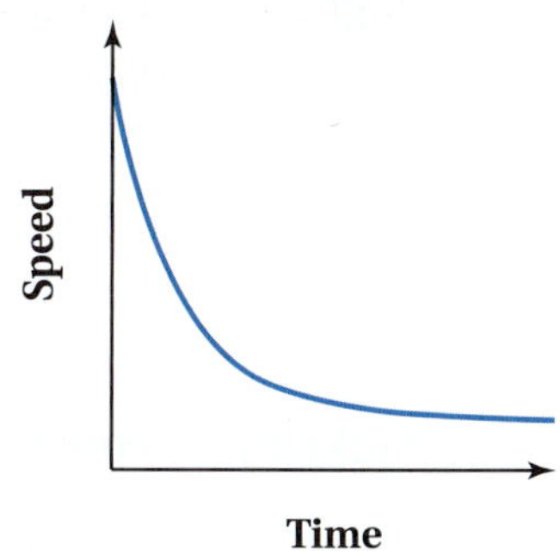

Graph B

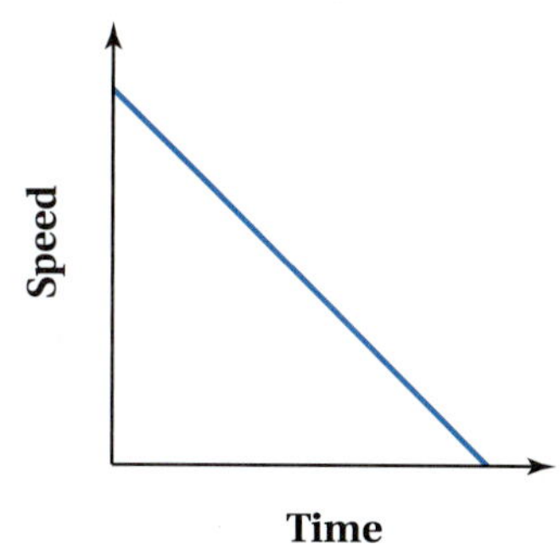

What Is Your Answer?

5. **IN YOUR OWN WORDS** How can you use a graph to represent relationships between quantities without using numbers?

6. Describe a possible situation represented by the graph shown.

7. Sketch a graph similar to the graphs in Activities 1 and 2. Exchange graphs with a classmate and describe a possible situation represented by the graph. Discuss the results.

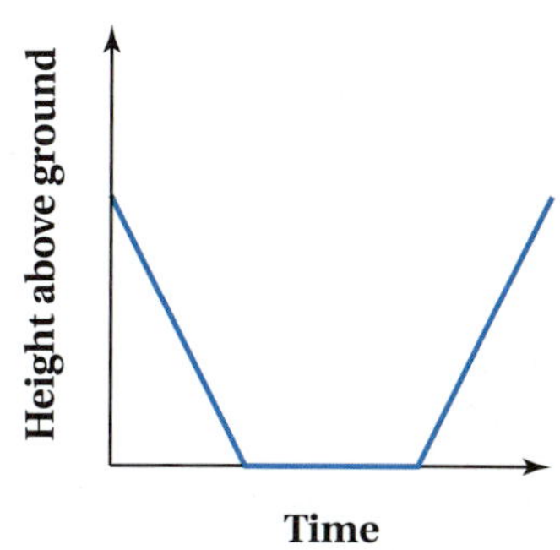

Practice Use what you learned about analyzing and sketching graphs to complete Exercises 7–9 on page 276.

6.5 Lesson

Check It Out
Lesson Tutorials
BigIdeasMath.com

Graphs can show the relationship between quantities without using specific numbers on the axes.

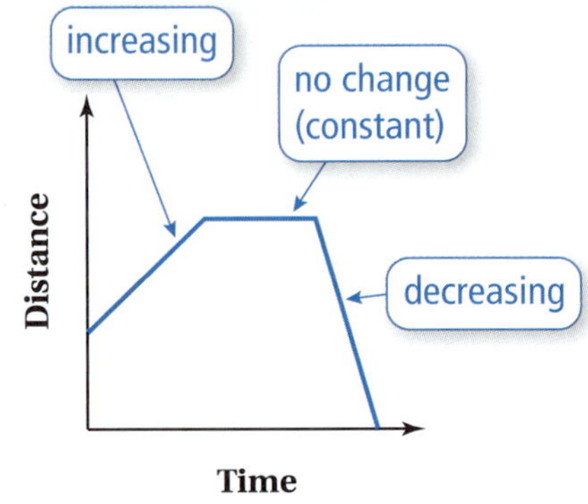

EXAMPLE 1 Analyzing Graphs

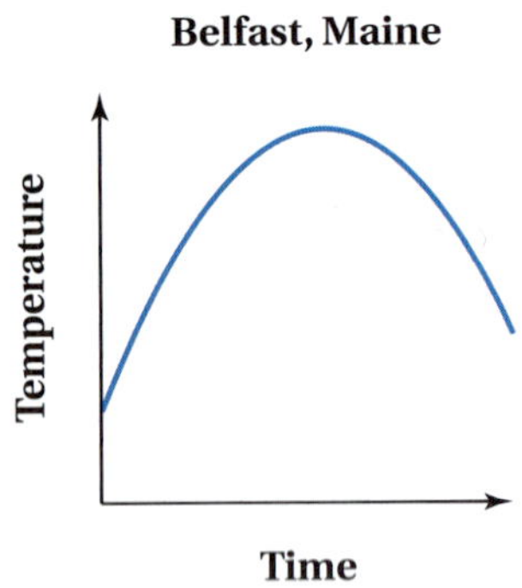

The graphs show the temperatures throughout the day in two cities.

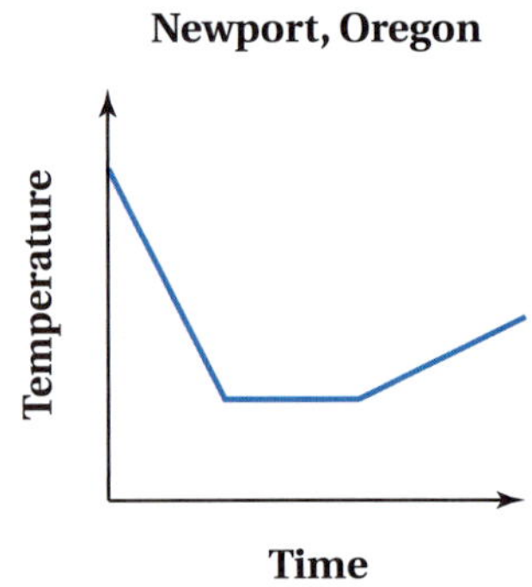

a. Describe the change in temperature in each city.

Belfast: The temperature increases at the beginning of the day. Then the temperature begins to decrease at a faster and faster rate for the rest of the day.

Newport: The temperature decreases at a constant rate at the beginning of the day. Then the temperature stays the same for a while before increasing at a constant rate for the rest of the day.

b. Make three comparisons from the graphs.

Three possible comparisons follow:

- Both graphs show increasing and decreasing temperatures.
- Both graphs are nonlinear, but the graph of the temperatures in Newport consists of three linear sections.
- In Belfast, it was warmer at the end of the day than at the beginning. In Newport, it was colder at the end of the day than at the beginning.

Study Tip

The comparisons given in Example 1(b) are sample answers. You can make many other correct comparisons.

On Your Own

Now You're Ready
Exercises 7–12

1. The graphs show the paths of two birds diving to catch fish.
   a. Describe the path of each bird.
   b. Make three comparisons from the graphs.

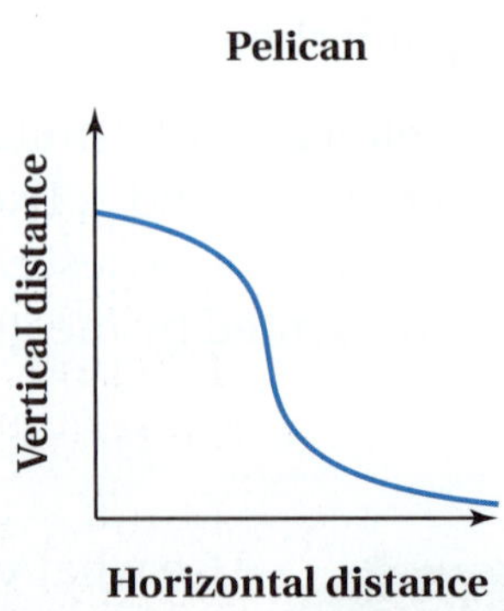

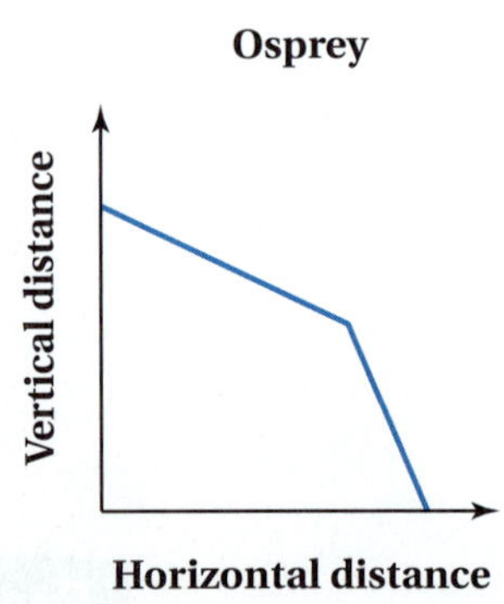

You can sketch graphs showing relationships between quantities that are described verbally.

EXAMPLE 2 Sketching Graphs

Sketch a graph that represents each situation.

a. A stopped subway train gains speed at a constant rate until it reaches its maximum speed. It travels at this speed for a while, and then slows down at a constant rate until coming to a stop at the next station.

Step 1: Draw the axes. Label the vertical axis "Speed" and the horizontal axis "Time."

Step 2: Sketch the graph.

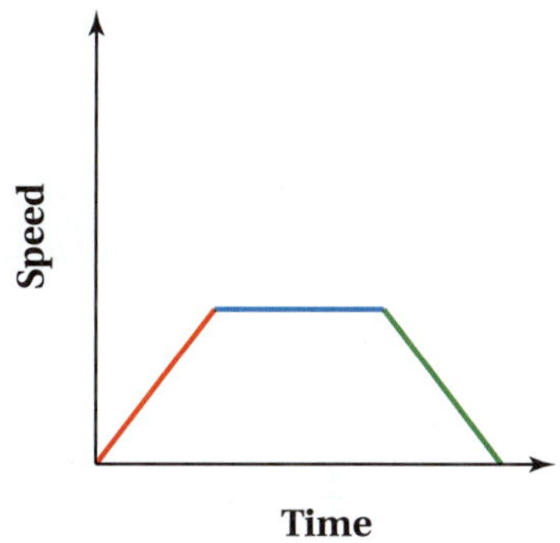

Words	Graph
A stopped subway train gains speed at a constant rate . . .	increasing line segment starting at the origin
until it reaches its maximum speed. It travels at this speed for a while, . . .	horizontal line segment
and then slows down at a constant rate until coming to a stop at the next station.	decreasing line segment ending at the horizontal axis

b. As television size increases, the price increases at an increasing rate.

Step 1: Draw the axes. Label the vertical axis "Price" and the horizontal axis "TV size."

Step 2: Sketch the graph.

The price *increases at an increasing rate*. So, the graph is nonlinear and becomes steeper and steeper as the TV size increases.

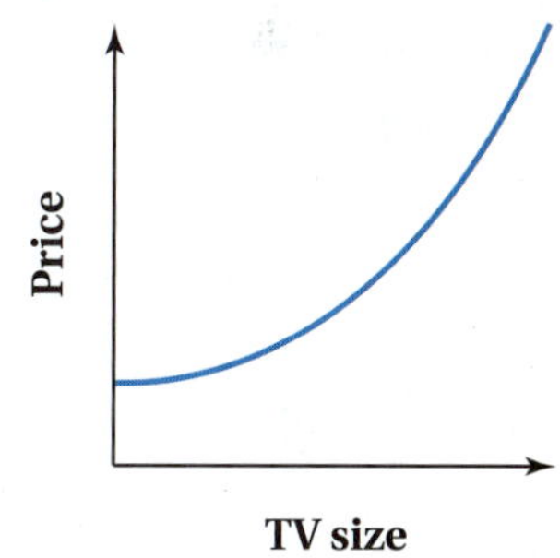

On Your Own

Now You're Ready
Exercises 15–18

Sketch a graph that represents the situation.

2. A fully charged battery loses its charge at a constant rate until it has no charge left. You plug it in and recharge it fully. Then it loses its charge at a constant rate until it has no charge left.

3. As the available quantity of a product increases, the price decreases at a decreasing rate.

6.5 Exercises

Vocabulary and Concept Check

MATCHING **Match the verbal description with the part of the graph it describes.**

1. stays the same
2. slowly decreases at a constant rate
3. slowly increases at a constant rate
4. increases at an increasing rate
5. quickly decreases at a constant rate
6. quickly increases at a constant rate

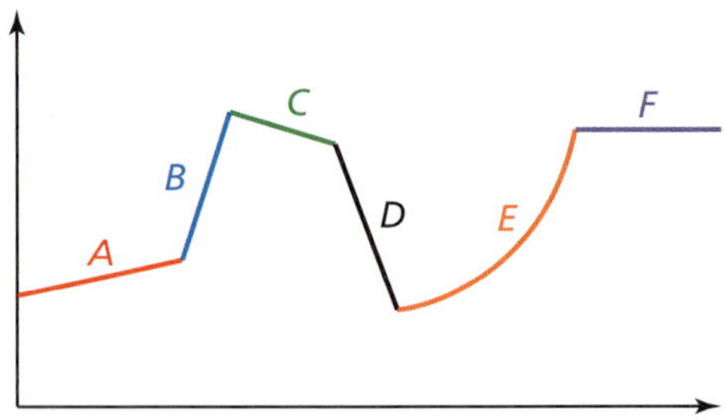

Practice and Problem Solving

Describe the relationship between the two quantities.

1 7.

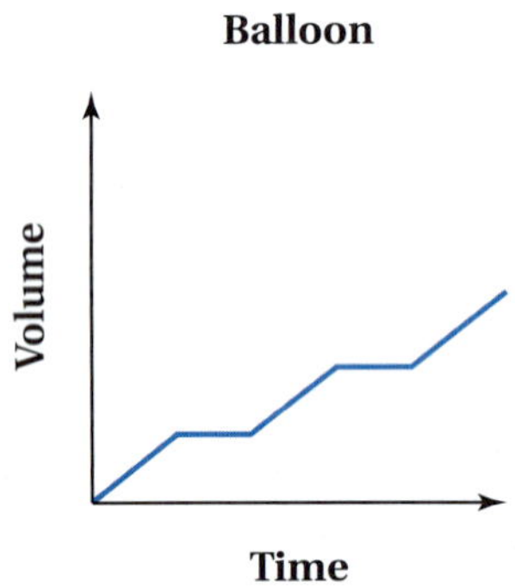

8.

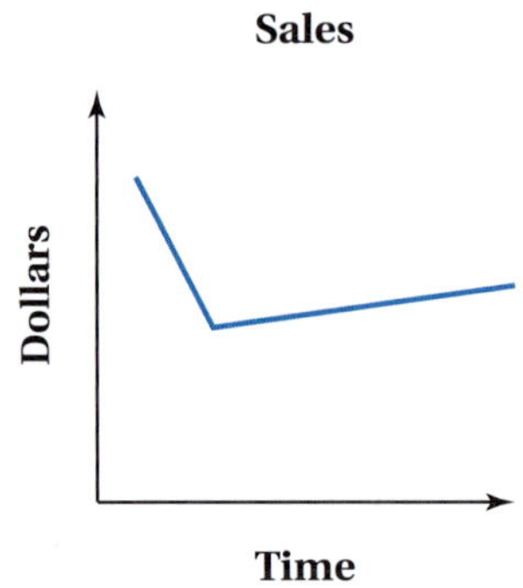

9.

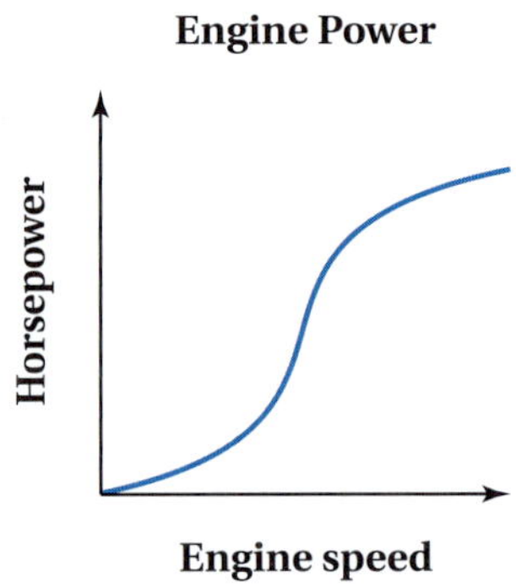

10.

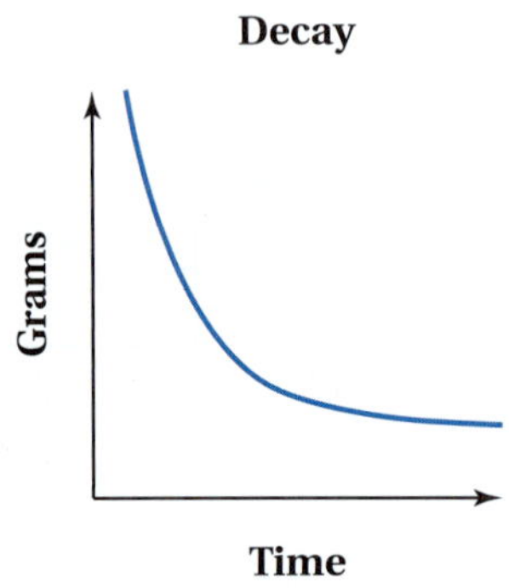

11.

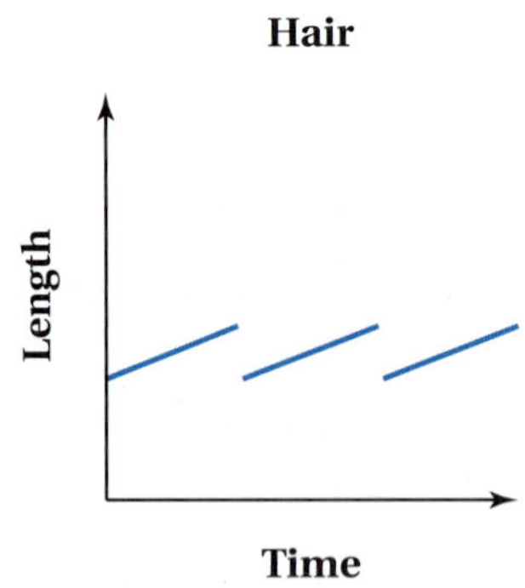

12.

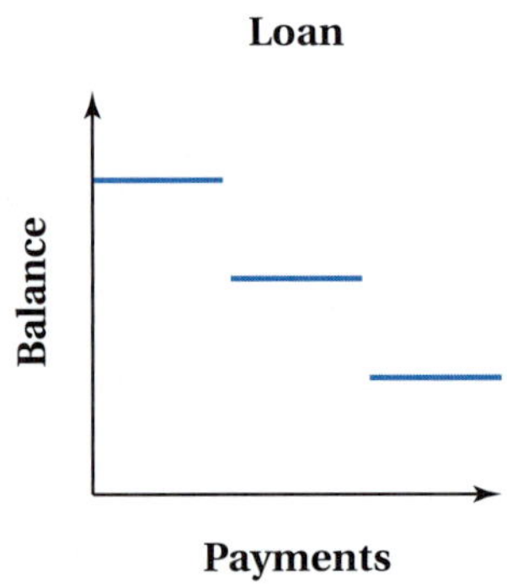

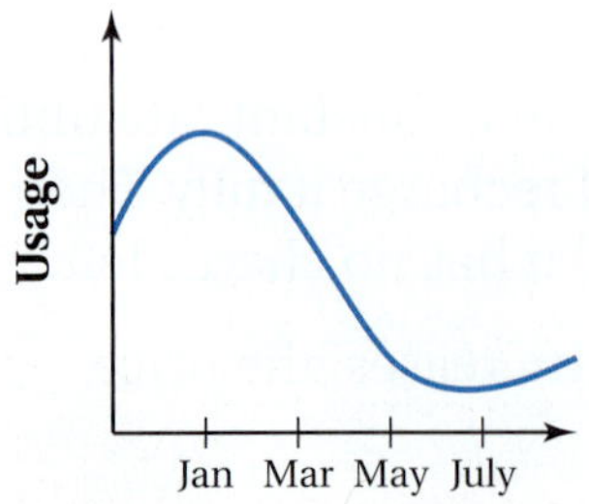

13. **NATURAL GAS** The graph shows the natural gas usage for a house.

    a. Describe the change in usage from January to March.

    b. Describe the change in usage from March to May.

14. **REASONING** The graph shows two bowlers' averages during a bowling season.

a. Describe each bowler's performance.

b. Who had a greater average most of the season? Who had a greater average at the end of the season?

Sketch a graph that represents the situation.

2 15. The value of a car depreciates. The value decreases quickly at first and then more slowly.

16. The distance from the ground changes as your friend swings on a swing.

17. The value of a rare coin increases at an increasing rate.

18. You are typing at a constant rate. You pause to think about your next paragraph, and then you resume typing at the same constant rate.

19. Economics You can use a *supply and demand model* to understand how the price of a product changes in a market. The *supply curve* of a particular product represents the quantity suppliers will produce at various prices. The *demand curve* for the product represents the quantity consumers are willing to buy at various prices.

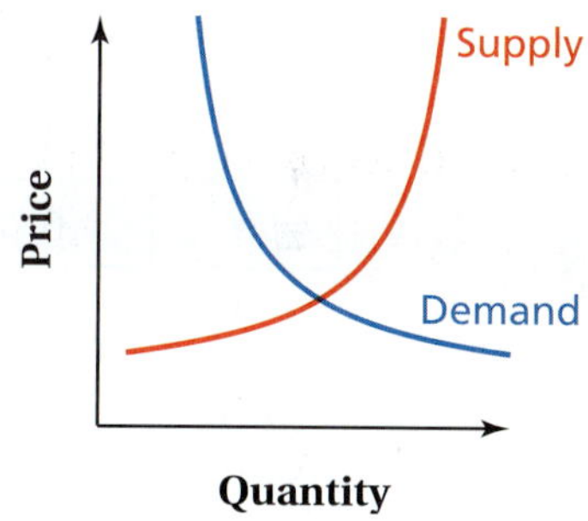

a. Describe and interpret each curve.

b. Which part of the graph represents a surplus? a shortage? Explain your reasoning.

c. The curves intersect at the *equilibrium point*, which is where the quantity produced equals the quantity demanded. Suppose that demand for a product suddenly increases, causing the entire demand curve to shift to the right. What happens to the equilibrium point?

Fair Game Review What you learned in previous grades & lessons

Solve the system of linear equations by graphing. *(Section 5.1)*

20. $y = x + 2$
$y = -x - 4$

21. $x - y = 3$
$-2x + y = -5$

22. $3x + 2y = 2$
$5x - 3y = -22$

23. **MULTIPLE CHOICE** Which triangle is a rotation of Triangle D? *(Section 2.4)*

Ⓐ Triangle A

Ⓑ Triangle B

Ⓒ Triangle C

Ⓓ none

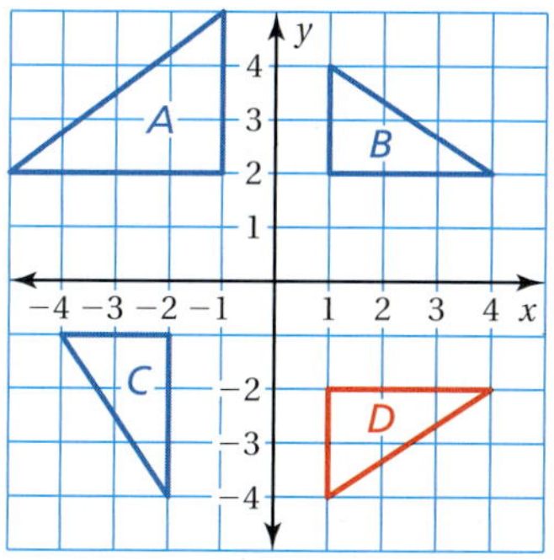

6.4–6.5 Quiz

Check It Out
Progress Check
BigIdeasMath.com

Does the table or graph represent a *linear* or *nonlinear* function? Explain. *(Section 6.4)*

1.

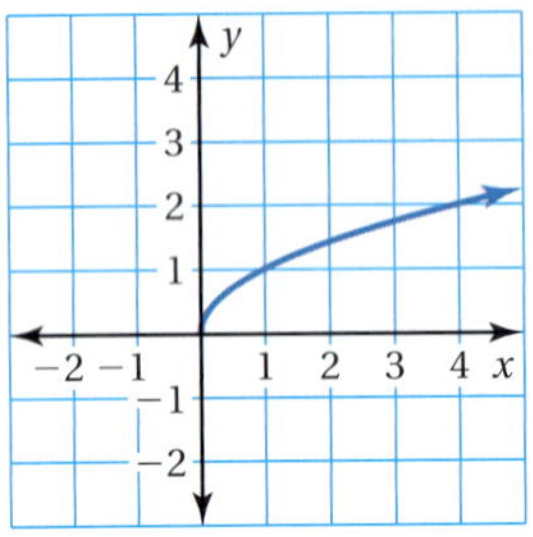

2.

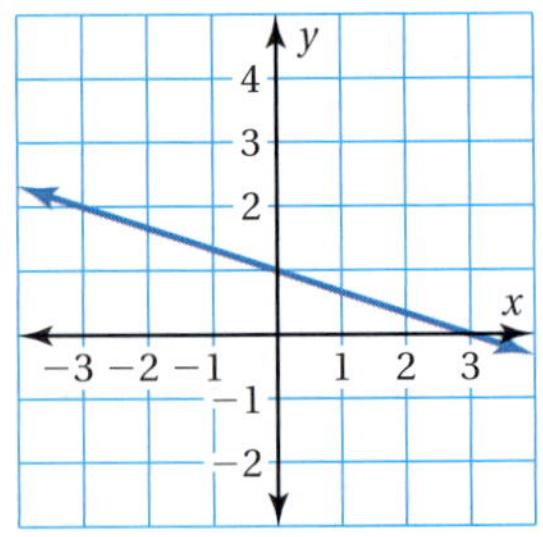

3.

x	*y*
0	3
3	0
6	3
9	6

4.

x	*y*
−1	3
1	7
3	11
5	15

5. CHICKEN SALAD The equation $y = 7.9x$ represents the cost y (in dollars) of buying x pounds of chicken salad. Does this equation represent a linear or nonlinear function? Explain. *(Section 6.4)*

6. HEIGHTS The graphs show the heights of two people over time. *(Section 6.5)*

a. Describe the change in height of each person.

b. Make three comparisons from the graphs.

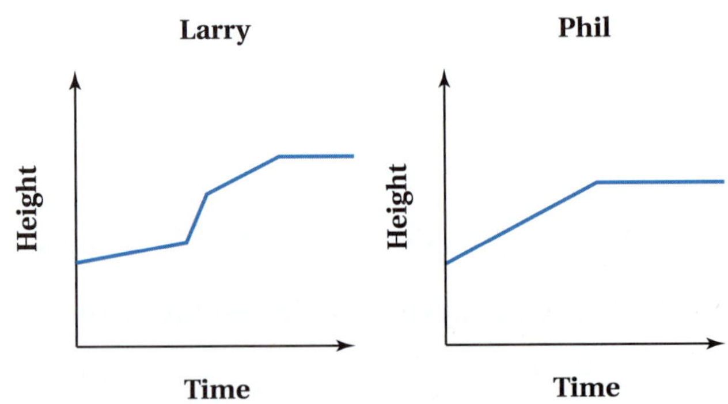

You are snowboarding down a hill. Sketch a graph that represents the situation. *(Section 6.5)*

7. You gradually increase your speed at a constant rate over time but fall about halfway down the hill. You take a short break, then get up, and gradually increase your speed again.

8. You gradually increase your speed at a constant rate over time. You come to a steep section of the hill and rapidly increase your speed at a constant rate. You then decrease your speed at a constant rate until you come to a stop.

Check It Out
Vocabulary Help
BigIdeasMath.com

Review Key Vocabulary

input, *p. 244*
output, *p. 244*
relation, *p. 244*
mapping diagram, *p. 244*
function, *p. 245*
function rule, *p. 250*
linear function, *p. 258*
nonlinear function, *p. 268*

Review Examples and Exercises

6.1 Relations and Functions *(pp. 242–247)*

Determine whether the relation is a function.

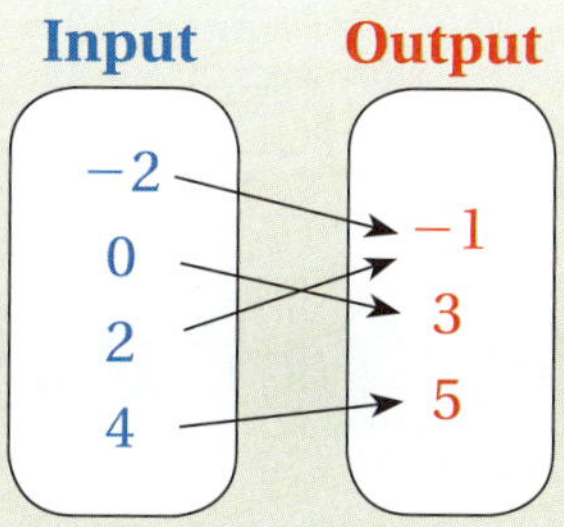

Each input has exactly one output.

So, the relation is a function.

Exercises

Determine whether the relation is a function.

1.

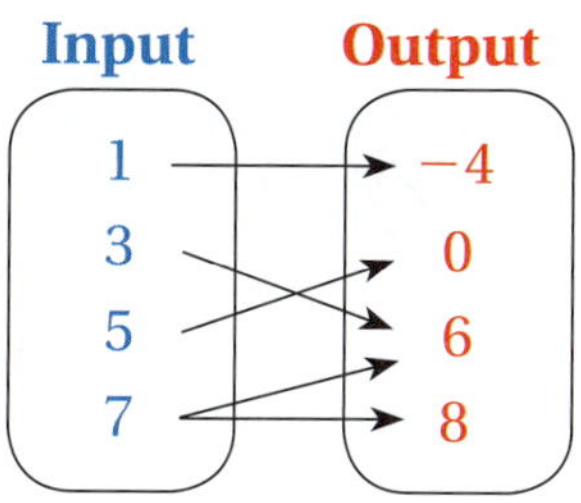

2.

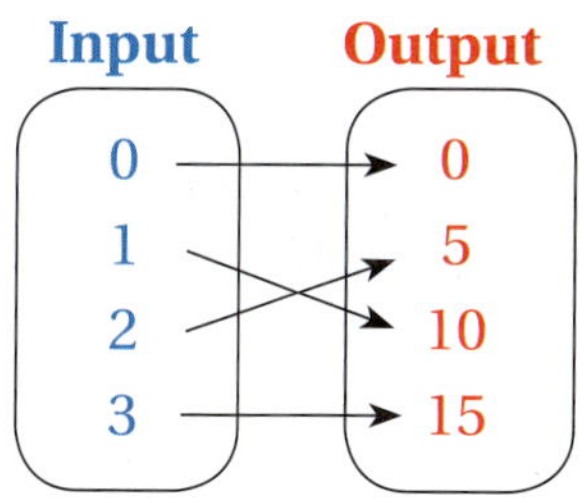

6.2 Representations of Functions *(pp. 248–255)*

Graph the function $y = x - 1$ using inputs of −1, 0, 1, and 2.

Make an input-output table.

Input, x	$x - 1$	Output, y	Ordered Pair, (x, y)
−1	−1 − 1	−2	(−1, −2)
0	0 − 1	−1	(0, −1)
1	1 − 1	0	(1, 0)
2	2 − 1	1	(2, 1)

Plot the ordered pairs and draw a line through the points.

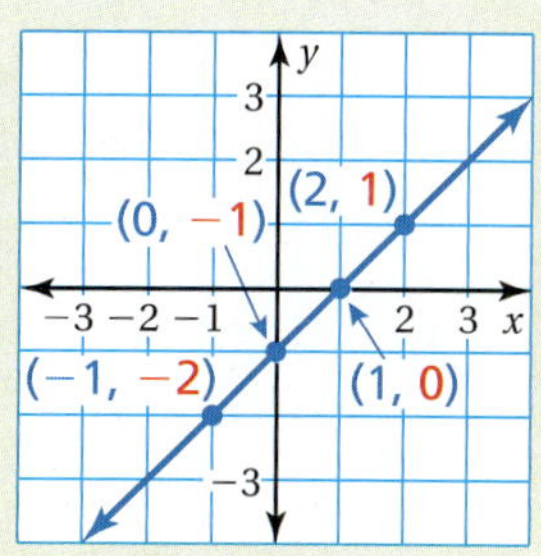

Exercises

Find the value of y for the given value of x.

3. $y = 2x - 3;\ x = -4$

4. $y = 2 - 9x;\ x = \frac{2}{3}$

5. $y = \frac{x}{3} + 5;\ x = 6$

Graph the function.

6. $y = x + 3$

7. $y = -5x$

8. $y = 3 - 3x$

6.3 Linear Functions (pp. 256–263)

Use the graph to write a linear function that relates y to x.

The points lie on a line. Find the slope by using the points (1, 1) and (2, 3).

$$m = \frac{\text{change in } y}{\text{change in } x} = \frac{3-1}{2-1} = \frac{2}{1} = 2$$

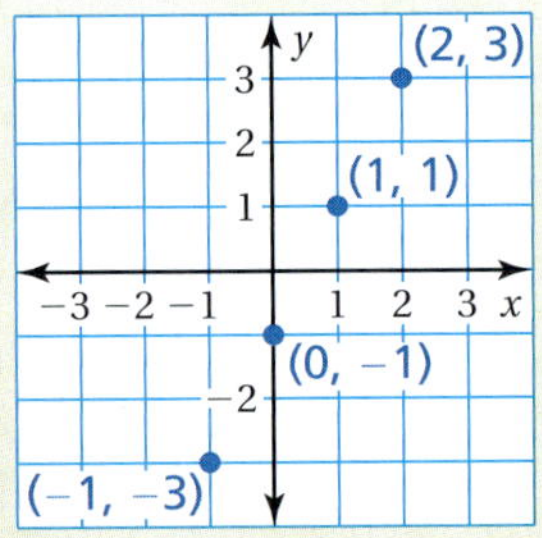

Because the line crosses the y-axis at $(0, -1)$, the y-intercept is -1.

So, the linear function is $y = 2x - 1$.

Exercises

Use the graph or table to write a linear function that relates y to x.

9.

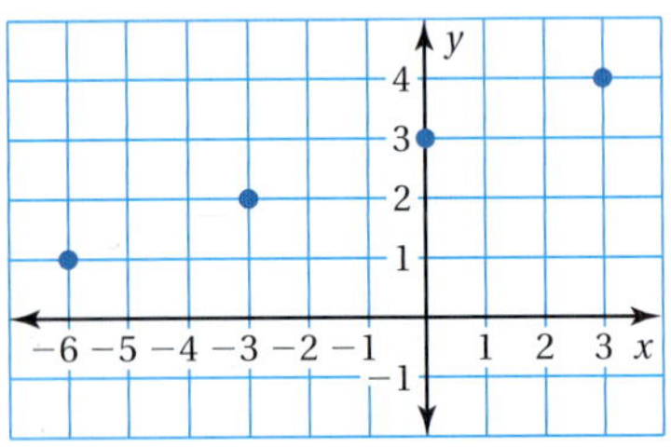

10.

x	−2	0	2	4
y	−7	−7	−7	−7

6.4 Comparing Linear and Nonlinear Functions (pp. 266–271)

Does the table represent a *linear* or *nonlinear* function? Explain.

a.

+2 +2 +2

x	0	2	4	6
y	0	1	4	9

+1 +3 +5

As x increases by 2, y increases by different amounts. The rate of change is *not* constant. So, the function is nonlinear.

b.

+5 +5 +5

x	0	5	10	15
y	50	40	30	20

−10 −10 −10

As x increases by 5, y decreases by 10. The rate of change is constant. So, the function is linear.

Exercises

Does the table represent a *linear* or *nonlinear* function? Explain.

11.

x	3	6	9	12
y	1	10	19	28

12.

x	1	3	5	7
y	3	1	1	3

6.5 Analyzing and Sketching Graphs *(pp. 272–277)*

The graphs show the populations of two cities over several years.

a. Describe the change in population in each city.

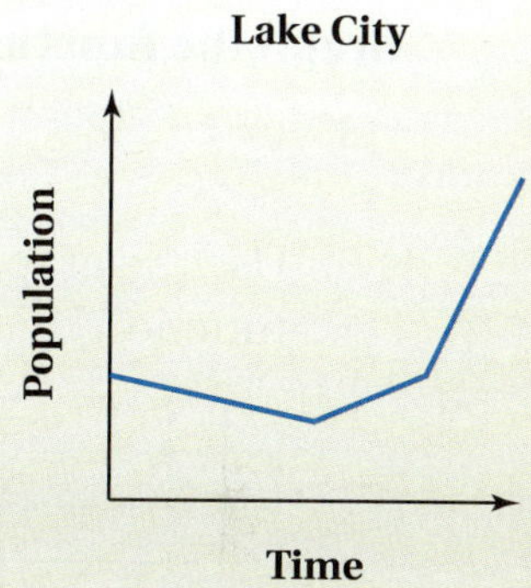

Lake City: The population gradually decreases at a constant rate, then gradually increases at a constant rate. Then the population rapidly increases at a constant rate.

Gold Point: The population rapidly increases at a constant rate. Then the population stays the same for a short period of time before gradually decreasing at a constant rate.

b. Make three comparisons from the graphs.

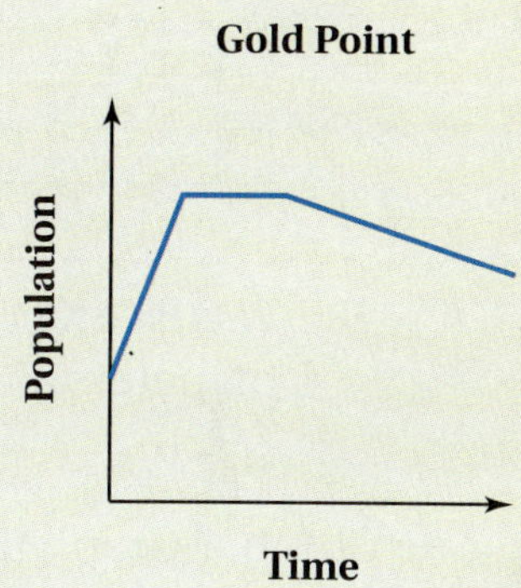

- Both graphs show increasing and decreasing populations.
- Both graphs are nonlinear, but both graphs consist of three linear sections.
- Both populations at the end of the time period are greater than the populations at the beginning of the time period.

Exercises

13. SALES The graphs show the sales of two companies.

a. Describe the sales of each company.

b. Make three comparisons from the graphs.

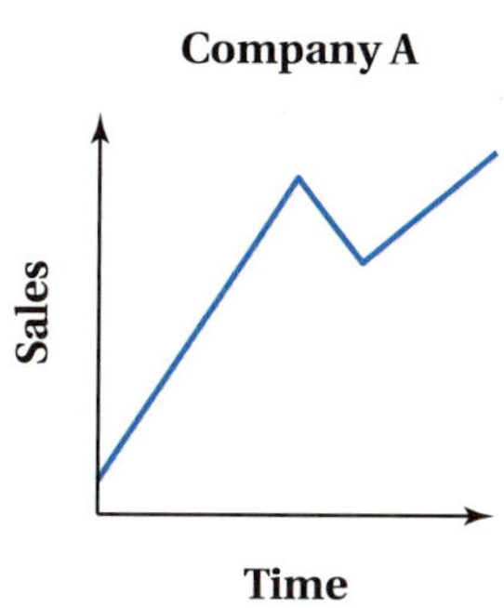

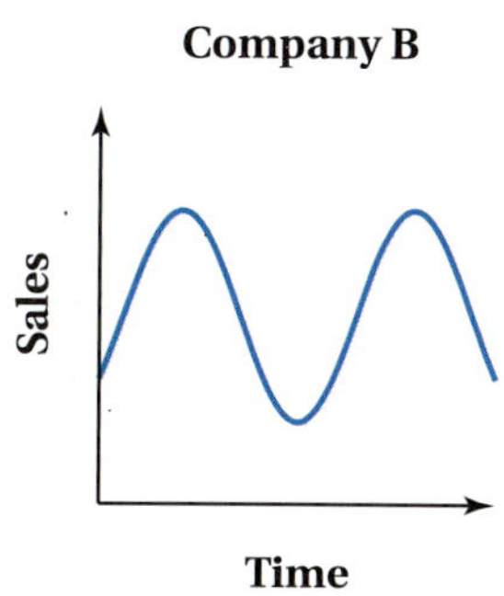

Sketch a graph that represents the situation.

14. You climb up a climbing wall. You gradually climb halfway up the wall at a constant rate, then stop and take a break. You then climb to the top of the wall at a constant rate.

15. The price of a stock steadily increases at a constant rate for several months before the stock market crashes. The price then quickly decreases at a constant rate.

6 Chapter Test

Check It Out
Test Practice
BigIdeasMath.com

Determine whether the relation is a function.

1.

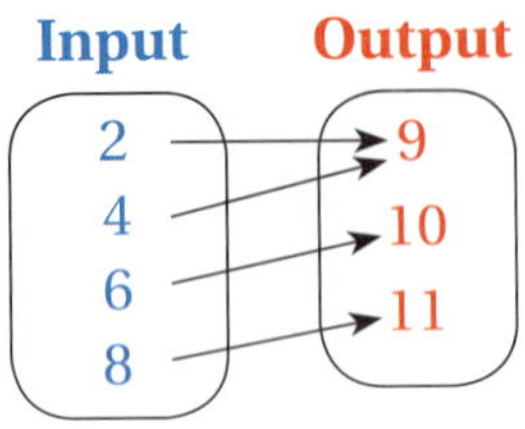

2.

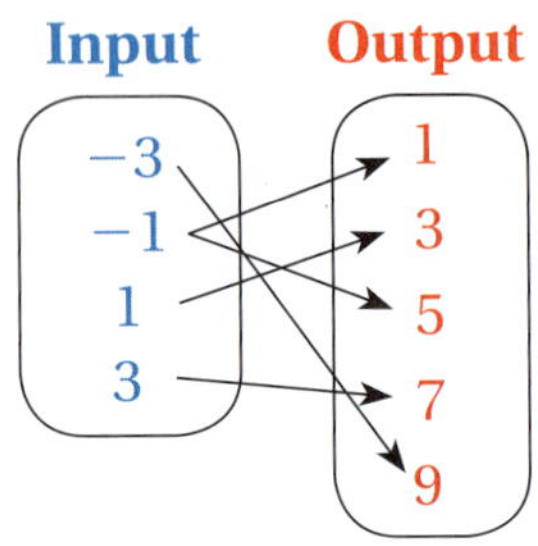

Graph the function.

3. $y = x + 8$

4. $y = 1 - 3x$

5. $y = x - 4$

6. Use the graph to write a linear function that relates y to x.

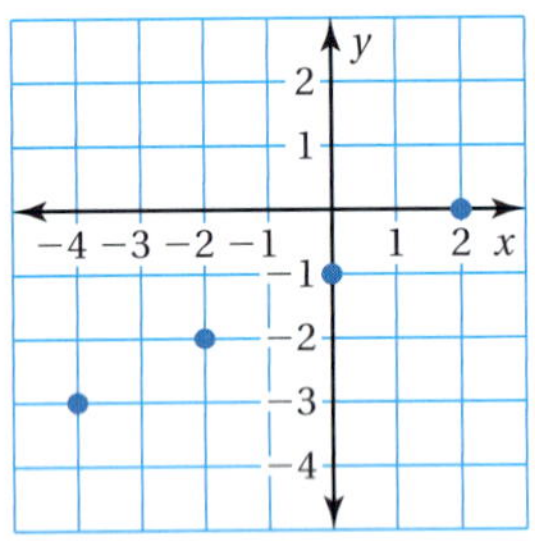

7. Does the table represent a *linear* or *nonlinear* function? Explain.

x	0	2	4	6
y	8	0	−8	−16

8. WATER SKI The table shows the number of meters a water skier travels in x minutes.

Minutes, x	1	2	3	4	5
Meters, y	600	1200	1800	2400	3000

a. Write a function that relates x to y.

b. Graph the linear function.

c. At this rate, how many *kilometers* would the water skier travel in 12 minutes?

9. STOCKS The graphs show the prices of two stocks during one day.

a. Describe the prices of each stock.

b. Make three comparisons from the graphs.

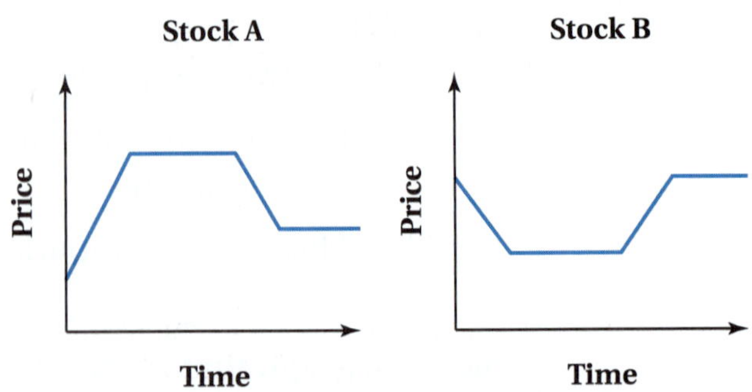

10. RACE You are competing in a race. You begin the race by increasing your speed at a constant rate. You then run at a constant speed until you get a cramp and have to stop. You wait until your cramp goes away before you start gradually increasing your speed again at a constant rate. Sketch a graph that represents the situation.

6 Standards Assessment

1. What is the slope of the line shown in the graph below? *(MACC.8.EE.2.6)*

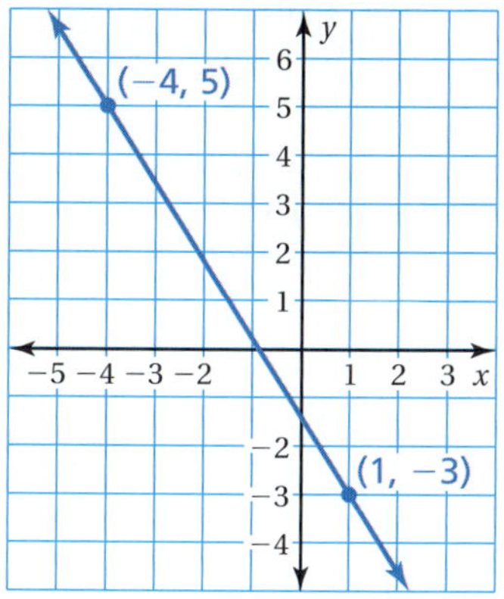

A. $-\frac{8}{3}$

B. $-\frac{8}{5}$

C. $-\frac{2}{3}$

D. $-\frac{2}{5}$

Test-Taking Strategy
Work Backwards

"**Work backwards** by trying 1, 2, 3, and 4. You will see that 3(4) = 12. So, D is correct."

2. Which value of a makes the equation below true? *(MACC.8.EE.3.7b)*

$$24 = \frac{a}{3} - 9$$

F. 5

G. 11

H. 45

I. 99

3. A mapping diagram is shown.

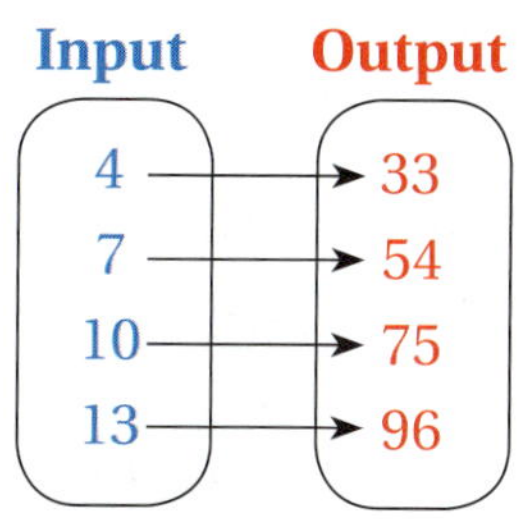

What number belongs in the box below so that the equation will correctly describe the function represented by the mapping diagram? *(MACC.8.F.1.1)*

$$y = \square x + 5$$

4. What is the solution of the system of linear equations shown below? *(MACC.8.EE.3.8b)*

$$y = 2x - 1$$

$$y = 3x + 5$$

A. (−13, −6)

B. (−6, −13)

C. (−13, 6)

D. (−6, 13)

5. A system of two linear equations has no solution. What can you conclude about the graphs of the two equations? *(MACC.8.EE.3.8a)*

F. The lines have the same slope and the same y-intercept.

G. The lines have the same slope and different y-intercepts.

H. The lines have different slopes and the same y-intercept.

I. The lines have different slopes and different y-intercepts.

6. Which graph shows a nonlinear function? *(MACC.8.F.1.3)*

A.

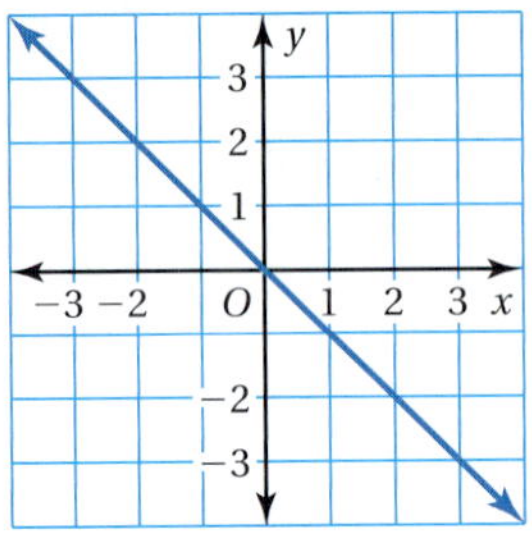

C.

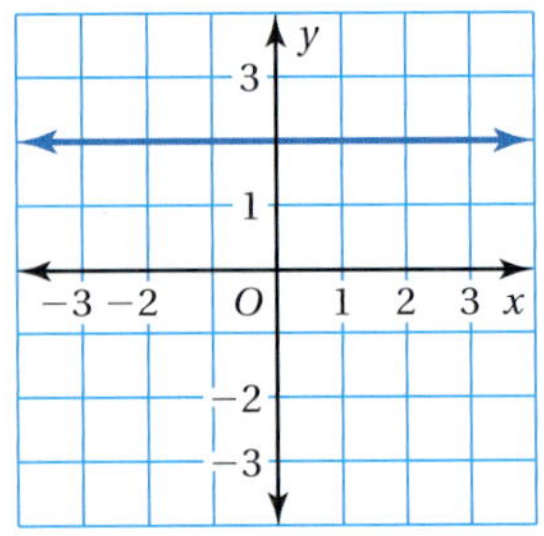

B.

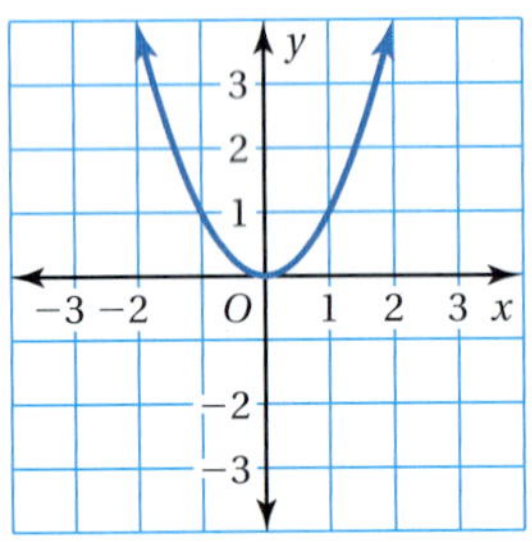

D.

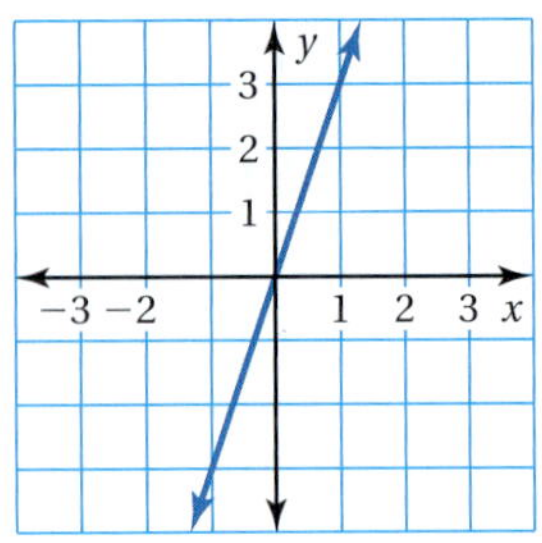

7. What is the value of x? *(MACC.8.G.1.5)*

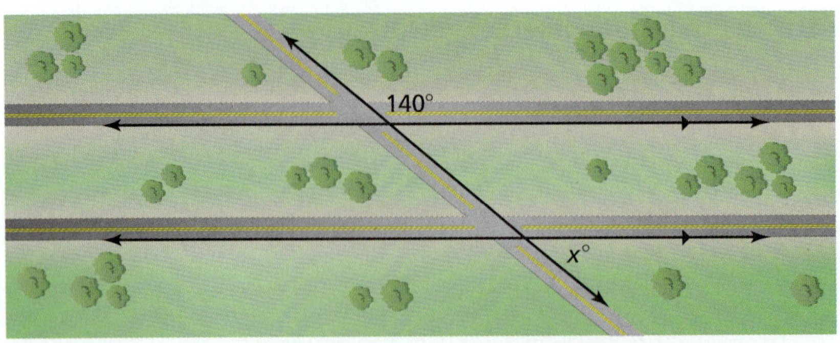

F. 40

G. 50

H. 140

I. 220

8. The tables show the sales (in millions of dollars) for two companies over a 5-year period. Examine the data in the tables. *(MACC.8.F.1.3)*

Think
Solve
Explain

Year	1	2	3	4	5
Sales	2	4	6	8	10

Year	1	2	3	4	5
Sales	1	1	2	3	5

Part A Does the first table show a linear function? Explain your reasoning.

Part B Does the second table show a linear function? Explain your reasoning.

9. The equations $y = -x + 4$ and $y = \frac{1}{2}x - 8$ form a system of linear equations. The table below shows the y-value for each equation at six different values of x. *(MACC.8.EE.3.8a)*

x	0	2	4	6	8	10
$y = -x + 4$	4	2	0	−2	−4	−6
$y = \frac{1}{2}x - 8$	−8	−7	−6	−5	−4	−3

What can you conclude from the table?

A. The system has one solution, when $x = 0$.

B. The system has one solution, when $x = 4$.

C. The system has one solution, when $x = 8$.

D. The system has no solution.

10. In the diagram below, Triangle ABC is a dilation of Triangle DEF. What is the value of x? *(MACC.8.G.1.4)*

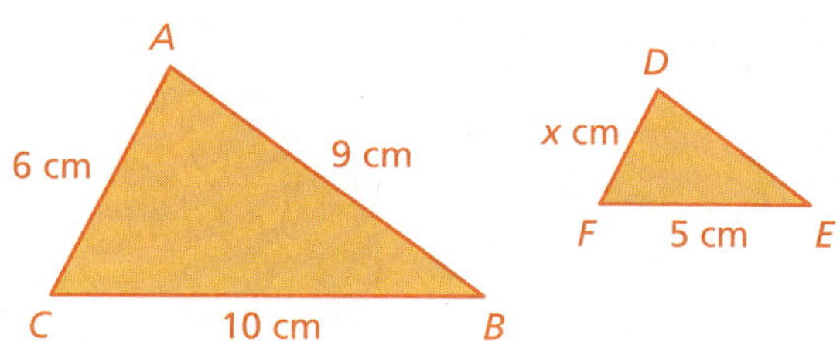

7 Real Numbers and the Pythagorean Theorem

What You Learned Before

"I just remember that the sum of the squares of the shorter sides is equal to the square of the longest side."

Comparing Decimals (MACC.5.NBT.1.3b)

Complete the number sentence with <, >, or =.

Example 1 **1.1 ▢ 1.01**

Because $\frac{110}{100}$ is greater than $\frac{101}{100}$, 1.1 is greater than 1.01.

So, $1.1 > 1.01$.

Example 2 **−0.3 ▢ −0.003**

Because $-\frac{300}{1000}$ is less than $-\frac{3}{1000}$, -0.3 is less than -0.003.

So, $-0.3 < -0.003$.

Example 3 **Find three decimals that make the number sentence $-5.12 >$ ▢ true.**

Any decimal less than -5.12 will make the sentence true.

Sample answer: -10.1, -9.05, -8.25

Try It Yourself

Complete the number sentence with <, >, or =.

1. 2.10 ▢ 2.1 **2.** -4.5 ▢ -4.25 **3.** π ▢ 3.2

Find three decimals that make the number sentence true.

4. $-0.01 \leq$ ▢ **5.** $1.75 >$ ▢ **6.** $0.75 \geq$ ▢

Using Order of Operations (MACC.7.NS.1.1, MACC.7.NS.1.2)

Example 4 **Evaluate $8^2 \div (32 \div 2) - 2(3 - 5)$.**

First:	Parentheses	$8^2 \div (32 \div 2) - 2(3 - 5) = 8^2 \div 16 - 2(-2)$
Second:	Exponents	$= 64 \div 16 - 2(-2)$
Third:	Multiplication and Division (from left to right)	$= 4 + 4$
Fourth:	Addition and Subtraction (from left to right)	$= 8$

Try It Yourself

Evaluate the expression.

7. $15\left(\frac{12}{3}\right) - 7^2 - 2 \cdot 7$ **8.** $3^2 \cdot 4 \div 18 + 30 \cdot 6 - 1$ **9.** $-1 + \left(\frac{4}{2}(6 - 1)\right)^2$

7.1 Finding Square Roots

Essential Question How can you find the dimensions of a square or a circle when you are given its area?

When you multiply a number by itself, you square the number.

Symbol for squaring is the exponent 2. → $4^2 = 4 \cdot 4$

$= 16$ 4 squared is 16.

To "undo" this, take the *square root* of the number.

Symbol for square root is a *radical sign*, $\sqrt{\ }$. → $\sqrt{16} = \sqrt{4^2} = 4$ The square root of 16 is 4.

1 ACTIVITY: Finding Square Roots

Work with a partner. Use a square root symbol to write the side length of the square. Then find the square root. Check your answer by multiplying.

a. Sample: $s = \sqrt{121} = 11$ ft

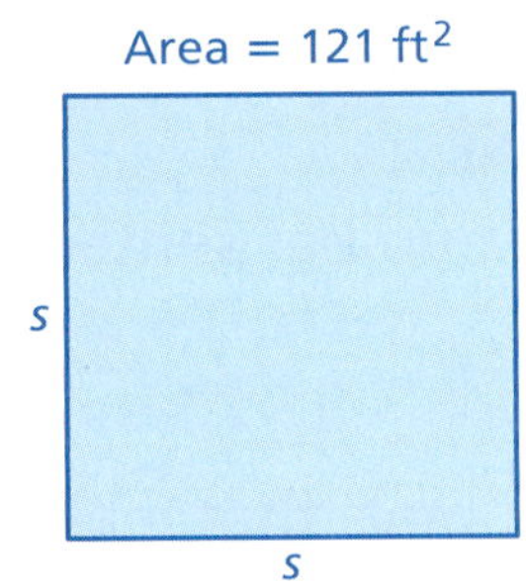

Check

$$\begin{array}{r} 11 \\ \times\ 11 \\ \hline 11 \\ 110 \\ \hline 121 \end{array}$$ ✓

∴ The side length of the square is 11 feet.

b.

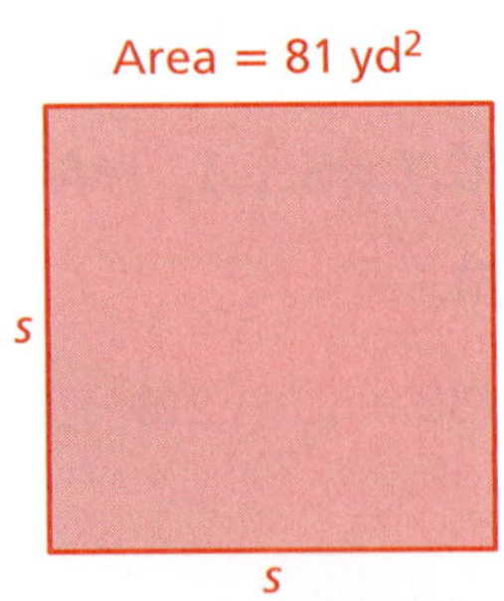

c.

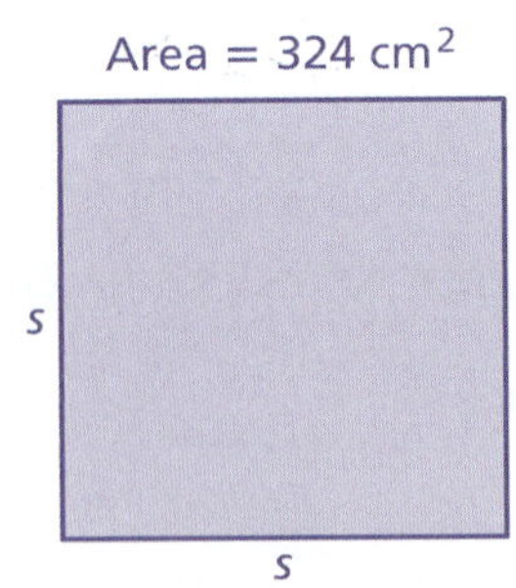

d.

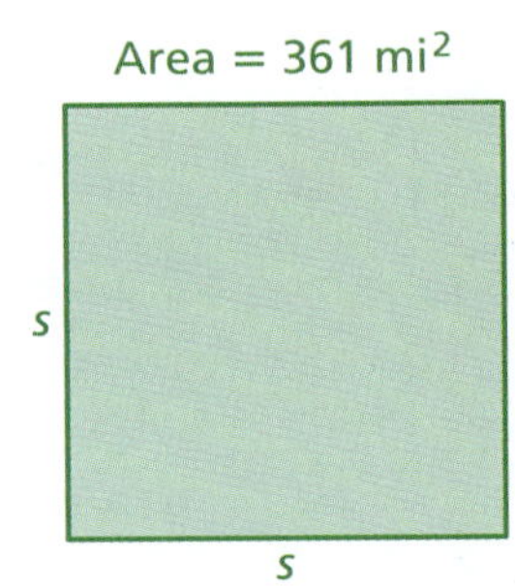

e.

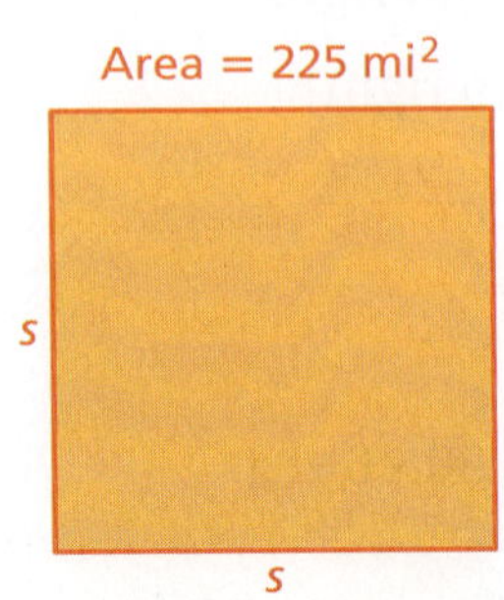

f.

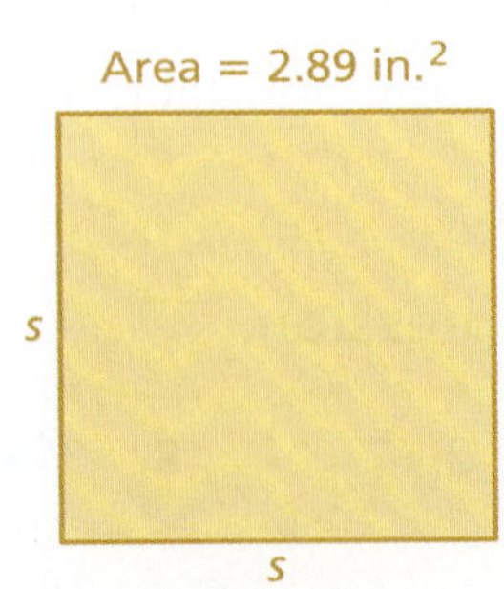

g.

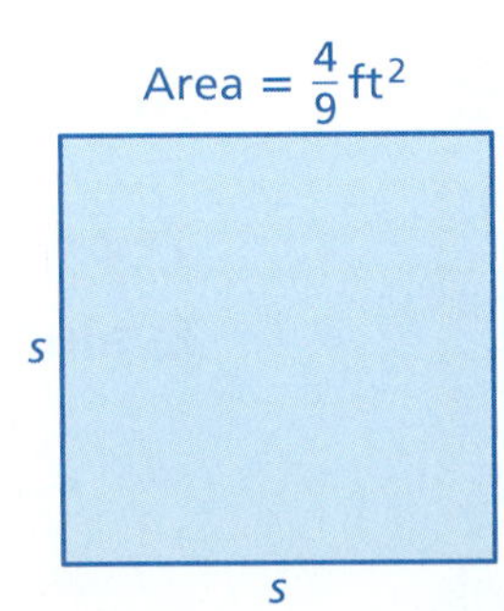

COMMON CORE

Square Roots

In this lesson, you will

- find square roots of perfect squares.
- evaluate expressions involving square roots.
- use square roots to solve equations.

Learning Standard
MACC.8.EE.1.2

2 ACTIVITY: Using Square Roots

Work with a partner. Find the radius of each circle.

a.

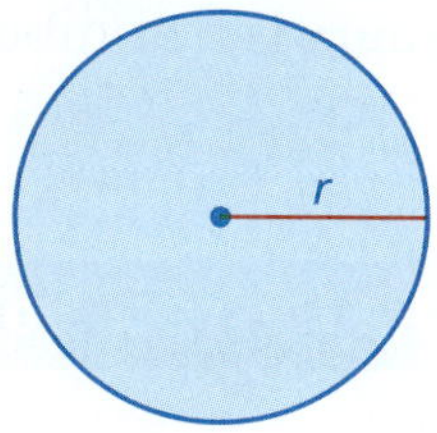

Area = 36π in.^{2}

b.

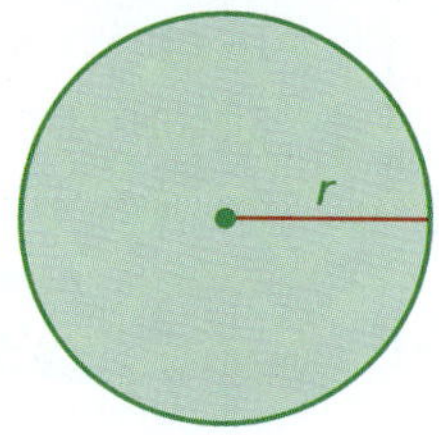

Area = π yd^2

c.

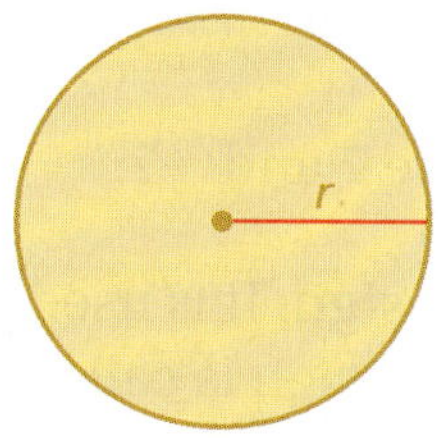

Area = 0.25π ft^2

d.

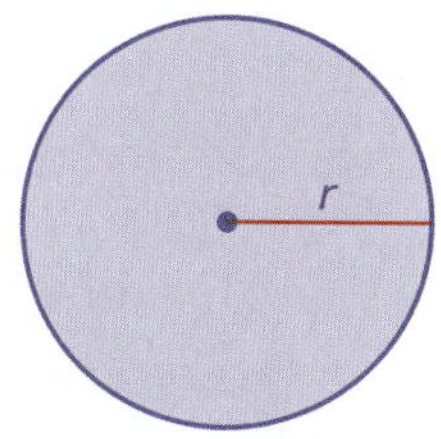

Area = $\frac{9}{16}\pi$ m^2

3 ACTIVITY: The Period of a Pendulum

Math Practice 6

Calculate Accurately

How can you use the graph to help you determine whether you calculated the values of *T* correctly?

Work with a partner.

The period of a pendulum is the time (in seconds) it takes the pendulum to swing back *and* forth.

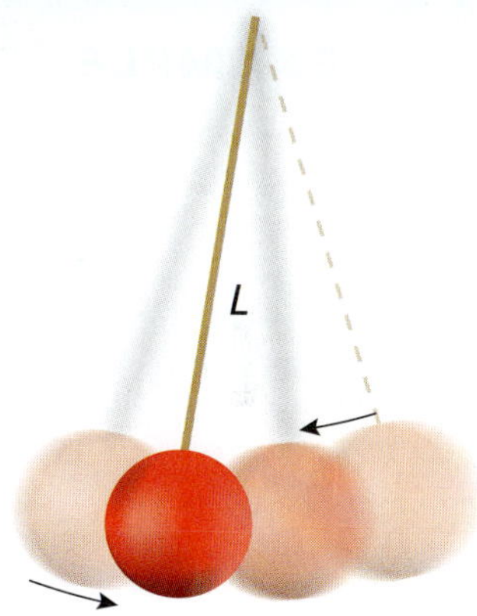

The period T is represented by $T = 1.1\sqrt{L}$, where L is the length of the pendulum (in feet).

Copy and complete the table. Then graph the function. Is the function linear?

L	1.00	1.96	3.24	4.00	4.84	6.25	7.29	7.84	9.00
T									

What Is Your Answer?

4. **IN YOUR OWN WORDS** How can you find the dimensions of a square or a circle when you are given its area? Give an example of each. How can you check your answers?

Practice Use what you learned about finding square roots to complete Exercises 4–6 on page 292.

7.1 Lesson

Check It Out
Lesson Tutorials
BigIdeasMath.com

Key Vocabulary
square root, *p. 290*
perfect square, *p. 290*
radical sign, *p. 290*
radicand, *p. 290*

A **square root** of a number is a number that, when multiplied by itself, equals the given number. Every positive number has a positive *and* a negative square root. A **perfect square** is a number with integers as its square roots.

EXAMPLE 1 Finding Square Roots of a Perfect Square

Find the two square roots of 49.

$7 \cdot 7 = 49$ and $(-7) \cdot (-7) = 49$

So, the square roots of 49 are 7 and -7.

Study Tip

Zero has one square root, which is 0.

The symbol $\sqrt{}$ is called a **radical sign**. It is used to represent a square root. The number under the radical sign is called the **radicand**.

Positive Square Root, $\sqrt{}$	Negative Square Root, $-\sqrt{}$	Both Square Roots, $\pm\sqrt{}$
$\sqrt{16} = 4$	$-\sqrt{16} = -4$	$\pm\sqrt{16} = \pm 4$

EXAMPLE 2 Finding Square Roots

Find the square root(s).

a. $\sqrt{25}$

Because $5^2 = 25$, $\sqrt{25} = \sqrt{5^2} = 5$.

$\sqrt{25}$ represents the *positive* square root.

b. $-\sqrt{\frac{9}{16}}$

Because $\left(\frac{3}{4}\right)^2 = \frac{9}{16}$, $-\sqrt{\frac{9}{16}} = -\sqrt{\left(\frac{3}{4}\right)^2} = -\frac{3}{4}$.

$-\sqrt{\frac{9}{16}}$ represents the *negative* square root.

c. $\pm\sqrt{2.25}$

$\pm\sqrt{2.25}$ represents both the *positive* and the *negative* square roots.

Because $1.5^2 = 2.25$, $\pm\sqrt{2.25} = \pm\sqrt{1.5^2} = 1.5$ and -1.5.

On Your Own

Now You're Ready
Exercises 7–18

Find the two square roots of the number.

1. 36
2. 100
3. 121

Find the square root(s).

4. $-\sqrt{1}$
5. $\pm\sqrt{\frac{4}{25}}$
6. $\sqrt{12.25}$

Squaring a positive number and finding a square root are inverse operations. You can use this relationship to evaluate expressions and solve equations involving squares.

EXAMPLE 3 Evaluating Expressions Involving Square Roots

Evaluate each expression.

a. $5\sqrt{36} + 7 = 5(6) + 7$ — Evaluate the square root.

$= 30 + 7$ — Multiply.

$= 37$ — Add.

b. $\frac{1}{4} + \sqrt{\frac{18}{2}} = \frac{1}{4} + \sqrt{9}$ — Simplify.

$= \frac{1}{4} + 3$ — Evaluate the square root.

$= 3\frac{1}{4}$ — Add.

c. $(\sqrt{81})^2 - 5 = 81 - 5$ — Evaluate the power using inverse operations.

$= 76$ — Subtract.

EXAMPLE 4 Real-Life Application

The area of a crop circle is 45,216 square feet. What is the radius of the crop circle? Use 3.14 for π.

$A = \pi r^2$ — Write the formula for the area of a circle.

$45{,}216 \approx 3.14r^2$ — Substitute 45,216 for A and 3.14 for π.

$14{,}400 = r^2$ — Divide each side by 3.14.

$\sqrt{14{,}400} = \sqrt{r^2}$ — Take positive square root of each side.

$120 = r$ — Simplify.

∴ The radius of the crop circle is about 120 feet.

On Your Own

Now You're Ready
Exercises 20–27

Evaluate the expression.

7. $12 - 3\sqrt{25}$ **8.** $\sqrt{\frac{28}{7}} + 2.4$ **9.** $15 - (\sqrt{4})^2$

10. The area of a circle is 2826 square feet. Write and solve an equation to find the radius of the circle. Use 3.14 for π.

7.1 Exercises

Check It Out
Help with Homework
BigIdeasMath.com

Vocabulary and Concept Check

1. **VOCABULARY** Is 26 a perfect square? Explain.
2. **REASONING** Can the square of an integer be a negative number? Explain.
3. **NUMBER SENSE** Does $\sqrt{256}$ represent the positive square root of 256, the negative square root of 256, or both? Explain.

Practice and Problem Solving

Find the dimensions of the square or circle. Check your answer.

4. Area = 441 cm^2

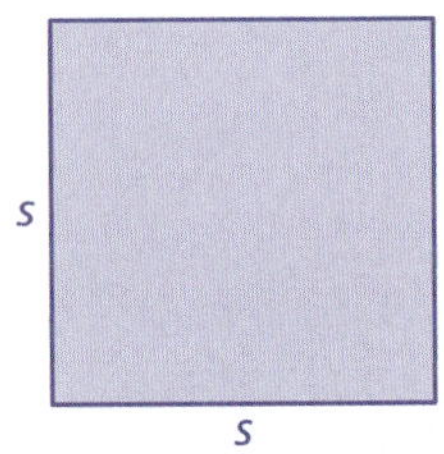

5. Area = 1.69 km^2

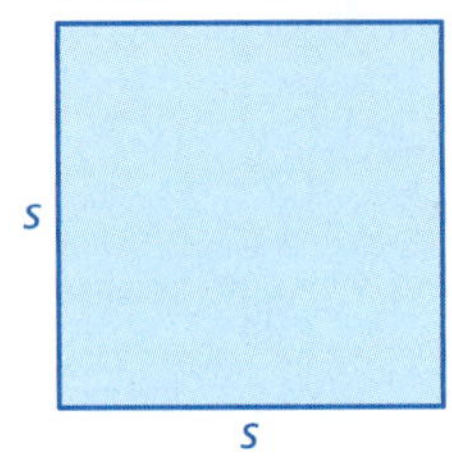

6. Area = 64π in.^{2}

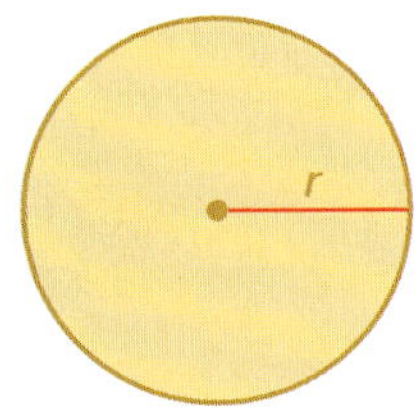

Find the two square roots of the number.

(1) 7. 9   8. 64   9. 4   10. 144

Find the square root(s).

(2) 11. $\sqrt{625}$ 12. $\pm\sqrt{196}$ 13. $\pm\sqrt{\frac{1}{961}}$ 14. $-\sqrt{\frac{9}{100}}$

15. $\pm\sqrt{4.84}$ 16. $\sqrt{7.29}$ 17. $-\sqrt{361}$ 18. $-\sqrt{2.25}$

19. **ERROR ANALYSIS** Describe and correct the error in finding the square roots.

$$\pm\sqrt{\frac{1}{4}} = \frac{1}{2}$$ ✗

Evaluate the expression.

(3) 20. $(\sqrt{9})^2 + 5$ 21. $28 - (\sqrt{144})^2$ 22. $3\sqrt{16} - 5$ 23. $10 - 4\sqrt{\frac{1}{16}}$

24. $\sqrt{6.76} + 5.4$ 25. $8\sqrt{8.41} + 1.8$ 26. $2\left(\sqrt{\frac{80}{5}} - 5\right)$ 27. $4\left(\sqrt{\frac{147}{3}} + 3\right)$

28. **NOTEPAD** The area of the base of a square notepad is 2.25 square inches. What is the length of one side of the base of the notepad?

$A = 25\pi$ mm^2

29. **CRITICAL THINKING** There are two square roots of 25. Why is there only one answer for the radius of the button?

Copy and complete the statement with <, >, or =.

30. $\sqrt{81} \;\square\; 8$ **31.** $0.5 \;\square\; \sqrt{0.25}$ **32.** $\frac{3}{2} \;\square\; \sqrt{\frac{25}{4}}$

33. SAILBOAT The area of a sail is $40\frac{1}{2}$ square feet. The base and the height of the sail are equal. What is the height of the sail (in feet)?

34. REASONING Is the product of two perfect squares always a perfect square? Explain your reasoning.

35. ENERGY The kinetic energy K (in joules) of a falling apple is represented by $K = \frac{v^2}{2}$, where v is the speed of the apple (in meters per second). How fast is the apple traveling when the kinetic energy is 32 joules?

36. PRECISION The areas of the two watch faces have a ratio of 16 : 25.

a. What is the ratio of the radius of the smaller watch face to the radius of the larger watch face?

b. What is the radius of the larger watch face?

37. WINDOW The cost C (in dollars) of making a square window with a side length of n inches is represented by $C = \frac{n^2}{5} + 175$. A window costs \$355. What is the length (in feet) of the window?

38. Geometry The area of the triangle is represented by the formula $A = \sqrt{s(s-21)(s-17)(s-10)}$, where s is equal to half the perimeter. What is the height of the triangle?

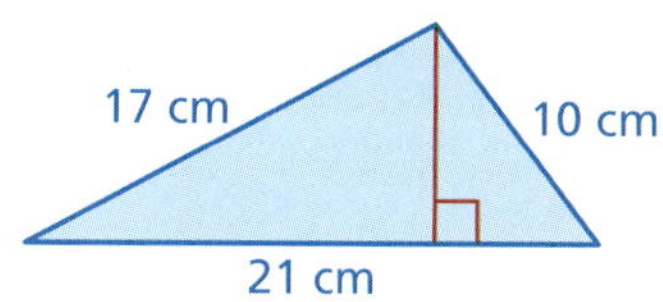

Fair Game Review What you learned in previous grades & lessons

Write in slope-intercept form an equation of the line that passes through the given points. *(Section 4.7)*

39. (2, 4), (5, 13) **40.** (−1, 7), (3, −1) **41.** (−5, −2), (5, 4)

42. MULTIPLE CHOICE What is the value of x? *(Section 3.2)*

Ⓐ 41 Ⓑ 44

Ⓒ 88 Ⓓ 134

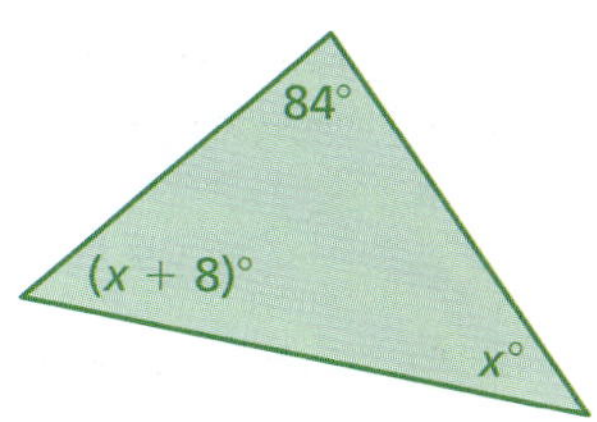

7.2 Finding Cube Roots

Essential Question How is the cube root of a number different from the square root of a number?

When you multiply a number by itself twice, you cube the number.

Symbol for cubing is the exponent 3. → $4^3 = 4 \cdot 4 \cdot 4$

$= 64$ 4 cubed is 64.

To "undo" this, take the *cube root* of the number.

Symbol for cube root is $\sqrt[3]{\ }$. → $\sqrt[3]{64} = \sqrt[3]{4^3} = 4$ The cube root of 64 is 4.

1 ACTIVITY: Finding Cube Roots

Work with a partner. Use a cube root symbol to write the edge length of the cube. Then find the cube root. Check your answer by multiplying.

a. Sample:

$s = \sqrt[3]{343} = \sqrt[3]{7^3} = 7$ inches

Volume = 343 in.^{3}

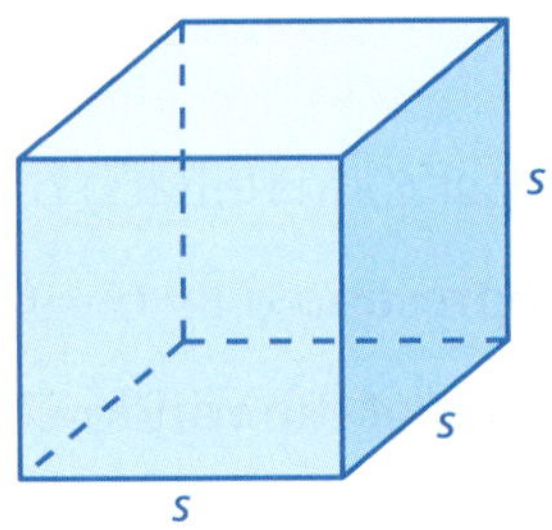

Check

$7 \cdot 7 \cdot 7 = 49 \cdot 7$

$= 343$ ✓

The edge length of the cube is 7 inches.

b. Volume = 27 ft^3

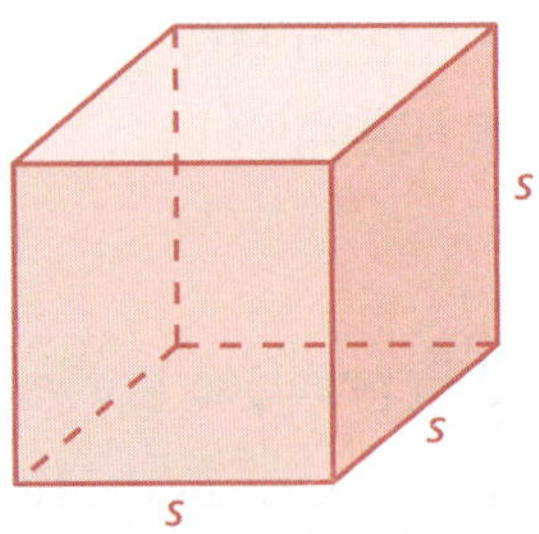

c. Volume = 125 m^3

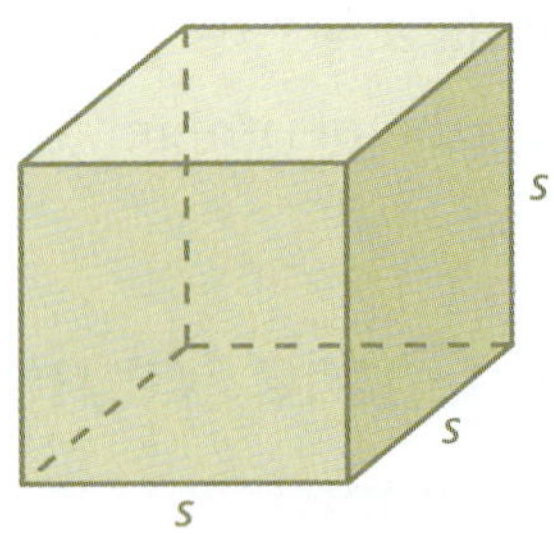

d. Volume = 0.001 cm^3

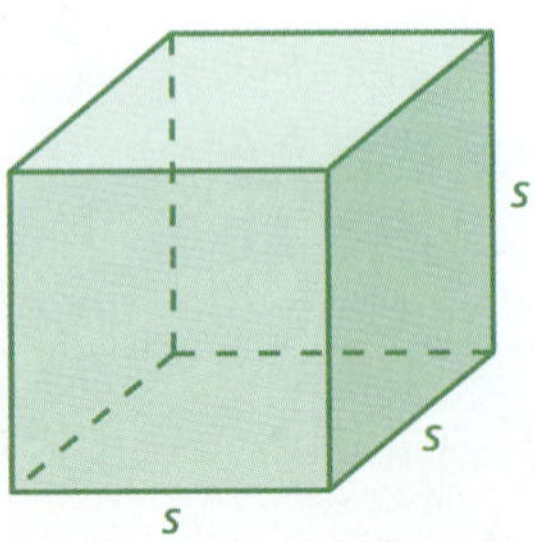

e. Volume = $\frac{1}{8}$ yd^3

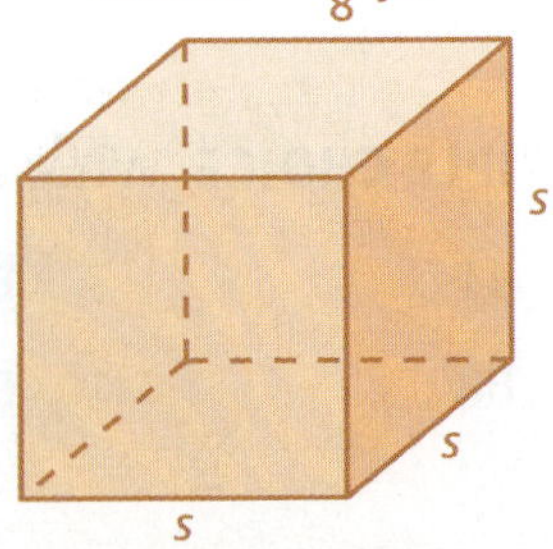

COMMON CORE

Cube Roots

In this lesson, you will

- find cube roots of perfect cubes.
- evaluate expressions involving cube roots.
- use cube roots to solve equations.

Learning Standard MACC.8.EE.1.2

2 ACTIVITY: Using Prime Factorizations to Find Cube Roots

Math Practice 7

View as Components

When writing the prime factorizations in Activity 2, how many times do you expect to see each factor? Why?

Work with a partner. Write the prime factorization of each number. Then use the prime factorization to find the cube root of the number.

a. 216

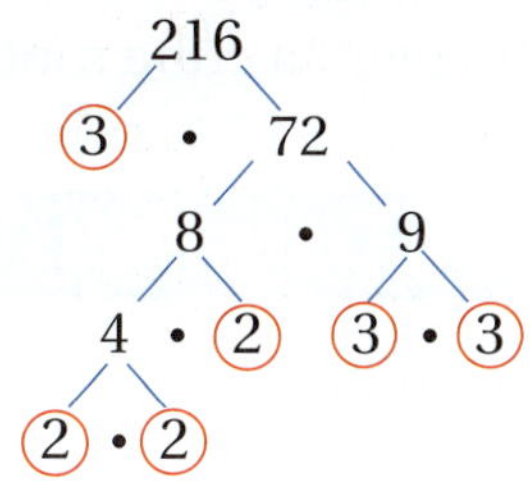

$216 = 3 \cdot 2 \cdot 3 \cdot 3 \cdot 2 \cdot 2$ Prime factorization

$= (3 \cdot \square) \cdot (3 \cdot \square) \cdot (3 \cdot \square)$ Commutative Property of Multiplication

$= \square \cdot \square \cdot \square$ Simplify.

The cube root of 216 is $\square$.

b. 1000

c. 3375

d. **STRUCTURE** Does this procedure work for every number? Explain why or why not.

What Is Your Answer?

3. Complete each statement using *positive* or *negative*.

a. A positive number times a positive number is a ________ number.

b. A negative number times a negative number is a ________ number.

c. A positive number multiplied by itself twice is a ________ number.

d. A negative number multiplied by itself twice is a ________ number.

4. **REASONING** Can a negative number have a cube root? Give an example to support your explanation.

5. **IN YOUR OWN WORDS** How is the cube root of a number different from the square root of a number?

6. Give an example of a number whose square root and cube root are equal.

7. A cube has a volume of 13,824 cubic meters. Use a calculator to find the edge length.

Practice Use what you learned about cube roots to complete Exercises 3–5 on page 298.

7.2 Lesson

Check It Out
Lesson Tutorials
BigIdeasMath.com

Key Vocabulary
cube root, *p. 296*
perfect cube, *p. 296*

A **cube root** of a number is a number that, when multiplied by itself, and then multiplied by itself again, equals the given number. A **perfect cube** is a number that can be written as the cube of an integer. The symbol $\sqrt[3]{}$ is used to represent a cube root.

EXAMPLE 1 Finding Cube Roots

Find each cube root.

a. $\sqrt[3]{8}$

Because $2^3 = 8$, $\sqrt[3]{8} = \sqrt[3]{2^3} = 2$.

b. $\sqrt[3]{-27}$

Because $(-3)^3 = -27$, $\sqrt[3]{-27} = \sqrt[3]{(-3)^3} = -3$.

c. $\sqrt[3]{\frac{1}{64}}$

Because $\left(\frac{1}{4}\right)^3 = \frac{1}{64}$, $\sqrt[3]{\frac{1}{64}} = \sqrt[3]{\left(\frac{1}{4}\right)^3} = \frac{1}{4}$.

Cubing a number and finding a cube root are inverse operations. You can use this relationship to evaluate expressions and solve equations involving cubes.

EXAMPLE 2 Evaluating Expressions Involving Cube Roots

Evaluate each expression.

a. $2\sqrt[3]{-216} - 3 = 2(-6) - 3$ Evaluate the cube root.

$= -12 - 3$ Multiply.

$= -15$ Subtract.

b. $\left(\sqrt[3]{125}\right)^3 + 21 = 125 + 21$ Evaluate the power using inverse operations.

$= 146$ Add.

On Your Own

Now You're Ready
Exercises 6–17

Find the cube root.

1. $\sqrt[3]{1}$
2. $\sqrt[3]{-343}$
3. $\sqrt[3]{-\frac{27}{1000}}$

Evaluate the expression.

4. $18 - 4\sqrt[3]{8}$
5. $\left(\sqrt[3]{-64}\right)^3 + 43$
6. $5\sqrt[3]{512} - 19$

Multi-Language Glossary at BigIdeasMath.com

EXAMPLE 3 Evaluating an Algebraic Expression

Evaluate $\frac{x}{4} + \sqrt[3]{\frac{x}{3}}$ when $x = 192$.

$$\frac{x}{4} + \sqrt[3]{\frac{x}{3}} = \frac{192}{4} + \sqrt[3]{\frac{192}{3}} \qquad \text{Substitute 192 for } x.$$

$$= 48 + \sqrt[3]{64} \qquad \text{Simplify.}$$

$$= 48 + 4 \qquad \text{Evaluate the cube root.}$$

$$= 52 \qquad \text{Add.}$$

On Your Own

Now You're Ready
Exercises 18–20

Evaluate the expression for the given value of the variable.

7. $\sqrt[3]{8y} + y, y = 64$

8. $2b - \sqrt[3]{9b}, b = -3$

EXAMPLE 4 Real-Life Application

Find the surface area of the baseball display case.

The baseball display case is in the shape of a cube. Use the formula for the volume of a cube to find the edge length s.

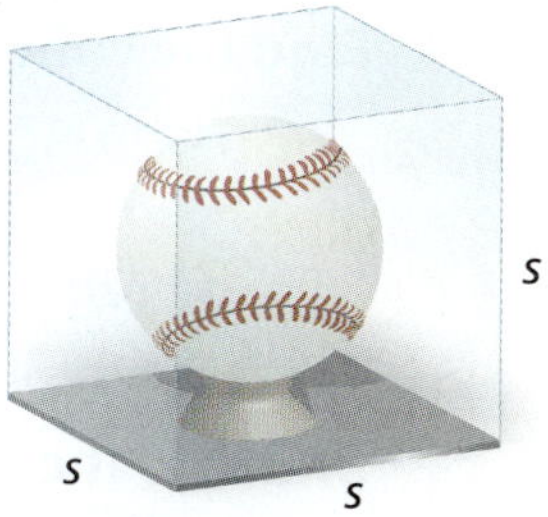

Volume = 125 in.^{3}

Remember

The volume V of a cube with edge length s is given by $V = s^3$. The surface area S is given by $S = 6s^2$.

$$V = s^3 \qquad \text{Write formula for volume.}$$

$$125 = s^3 \qquad \text{Substitute 125 for } V.$$

$$\sqrt[3]{125} = \sqrt[3]{s^3} \qquad \text{Take the cube root of each side.}$$

$$5 = s \qquad \text{Simplify.}$$

The edge length is 5 inches. Use a formula to find the surface area of the cube.

$$S = 6s^2 \qquad \text{Write formula for surface area.}$$

$$= 6(5)^2 \qquad \text{Substitute 5 for } s.$$

$$= 150 \qquad \text{Simplify.}$$

So, the surface area of the baseball display case is 150 square inches.

On Your Own

9. The volume of a music box that is shaped like a cube is 512 cubic centimeters. Find the surface area of the music box.

7.2 Exercises

Check It Out
Help with Homework
BigIdeasMath.com

Vocabulary and Concept Check

1. **VOCABULARY** Is 25 a perfect cube? Explain.
2. **REASONING** Can the cube of an integer be a negative number? Explain.

Practice and Problem Solving

Find the edge length of the cube.

3. Volume = 125,000 in.^{3}

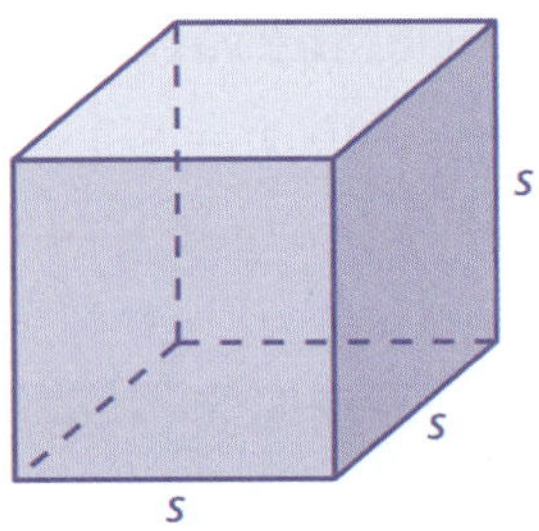

4. Volume = $\frac{1}{27}$ ft^3

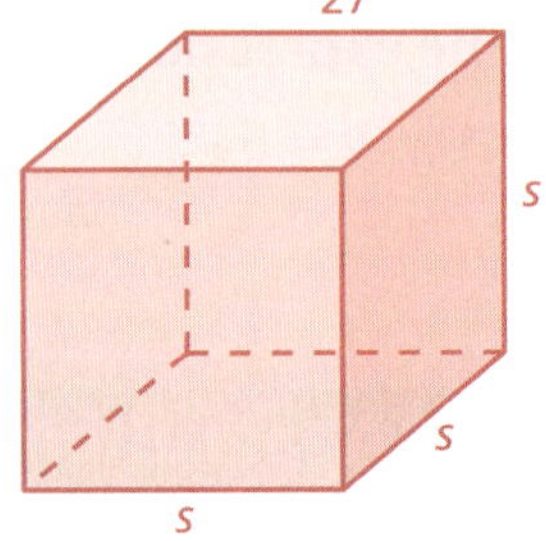

5. Volume = 0.064 m^3

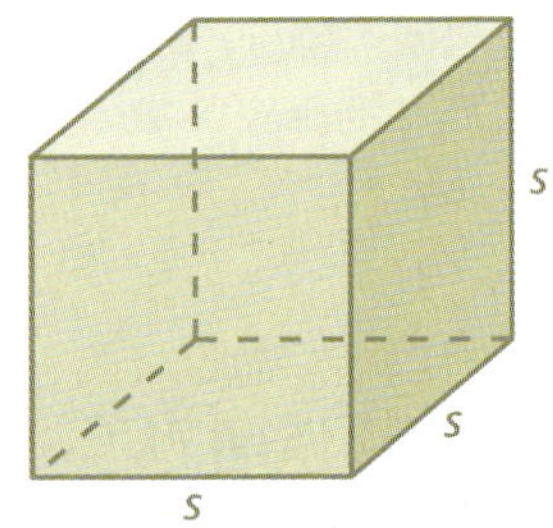

Find the cube root.

1 6. $\sqrt[3]{729}$

7. $\sqrt[3]{-125}$

8. $\sqrt[3]{-1000}$

9. $\sqrt[3]{1728}$

10. $\sqrt[3]{-\frac{1}{512}}$

11. $\sqrt[3]{\frac{343}{64}}$

Evaluate the expression.

2 12. $18 - \left(\sqrt[3]{27}\right)^3$

13. $\left(\sqrt[3]{-\frac{1}{8}}\right)^3 + 3\frac{3}{4}$

14. $5\sqrt[3]{729} - 24$

15. $\frac{1}{4} - 2\sqrt[3]{-\frac{1}{216}}$

16. $54 + \sqrt[3]{-4096}$

17. $4\sqrt[3]{8000} - 6$

Evaluate the expression for the given value of the variable.

3 18. $\sqrt[3]{\frac{n}{4}} + \frac{n}{10}$, $n = 500$

19. $\sqrt[3]{6w} - w$, $w = 288$

20. $2d + \sqrt[3]{-45d}$, $d = 75$

21. **STORAGE CUBE** The volume of a plastic storage cube is 27,000 cubic centimeters. What is the edge length of the storage cube?

22. **ICE SCULPTURE** The volume of a cube of ice for an ice sculpture is 64,000 cubic inches.
    a. What is the edge length of the cube of ice?
    b. What is the surface area of the cube of ice?

Copy and complete the statement with <, >, or =.

23. $-\frac{1}{4}$ ▭ $\sqrt[3]{-\frac{8}{125}}$ **24.** $\sqrt[3]{0.001}$ ▭ 0.01 **25.** $\sqrt[3]{64}$ ▭ $\sqrt{64}$

26. DRAG RACE The estimated velocity v (in miles per hour) of a car at the end of a drag race is $v = 234\sqrt[3]{\frac{p}{w}}$, where p is the horsepower of the car and w is the weight (in pounds) of the car. A car has a horsepower of 1311 and weighs 2744 pounds. Find the velocity of the car at the end of a drag race. Round your answer to the nearest whole number.

27. NUMBER SENSE There are three numbers that are their own cube roots. What are the numbers?

28. LOGIC Each statement below is true for square roots. Determine whether the statement is also true for cube roots. Explain your reasoning and give an example to support your explanation.

a. You cannot find the square root of a negative number.

b. Every positive number has a positive square root and a negative square root.

29. GEOMETRY The pyramid has a volume of 972 cubic inches. What are the dimensions of the pyramid?

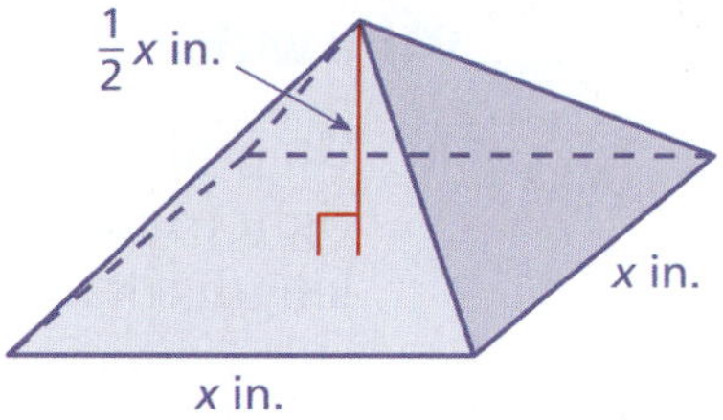

30. RATIOS The ratio $125 : x$ is equivalent to the ratio $x^2 : 125$. What is the value of x?

Critical Thinking **Solve the equation.**

31. $(3x + 4)^3 = 2197$ **32.** $(8x^3 - 9)^3 = 5832$ **33.** $((5x - 16)^3 - 4)^3 = 216{,}000$

Fair Game Review What you learned in previous grades & lessons

Evaluate the expression. *(Skills Review Handbook)*

34. $3^2 + 4^2$ **35.** $8^2 + 15^2$ **36.** $13^2 - 5^2$ **37.** $25^2 - 24^2$

38. MULTIPLE CHOICE Which linear function is shown by the table? *(Section 6.3)*

x	1	2	3	4
y	4	7	10	13

Ⓐ $y = \frac{1}{3}x + 1$ Ⓑ $y = 4x$ Ⓒ $y = 3x + 1$ Ⓓ $y = \frac{1}{4}x$

7.3 The Pythagorean Theorem

Essential Question How are the lengths of the sides of a right triangle related?

Pythagoras was a Greek mathematician and philosopher who discovered one of the most famous rules in mathematics. In mathematics, a rule is called a **theorem**. So, the rule that Pythagoras discovered is called the Pythagorean Theorem.

Pythagoras
(c. 570–c. 490 B.C.)

1 ACTIVITY: Discovering the Pythagorean Theorem

Work with a partner.

a. On grid paper, draw any right triangle. Label the lengths of the two shorter sides a and b.

b. Label the length of the longest side c.

c. Draw squares along each of the three sides. Label the areas of the three squares a^2, b^2, and c^2.

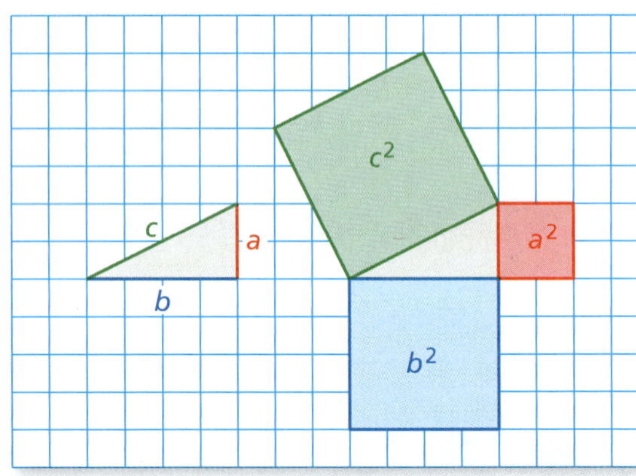

d. Cut out the three squares. Make eight copies of the right triangle and cut them out. Arrange the figures to form two identical larger squares.

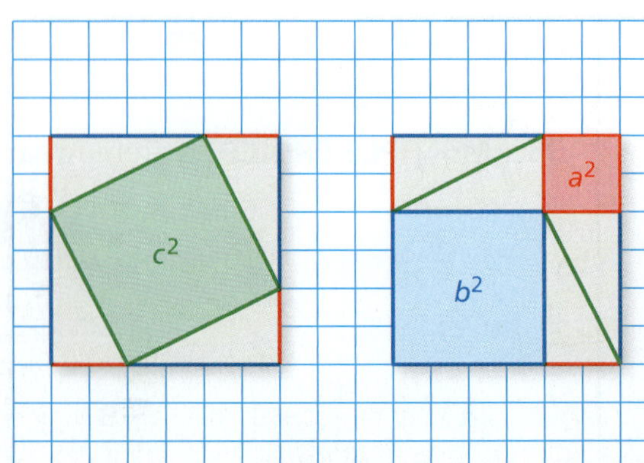

e. **MODELING** The Pythagorean Theorem describes the relationship among a^2, b^2, and c^2. Use your result from part (d) to write an equation that describes this relationship.

COMMON CORE

Pythagorean Theorem

In this lesson, you will

- provide geometric proof of the Pythagorean Theorem.
- use the Pythagorean Theorem to find missing side lengths of right triangles.
- solve real-life problems.

Learning Standards
MACC.8.EE.1.2
MACC.8.G.2.6
MACC.8.G.2.7
MACC.8.G.2.8

2 ACTIVITY: Using the Pythagorean Theorem in Two Dimensions

Work with a partner. Use a ruler to measure the longest side of each right triangle. Verify the result of Activity 1 for each right triangle.

a.

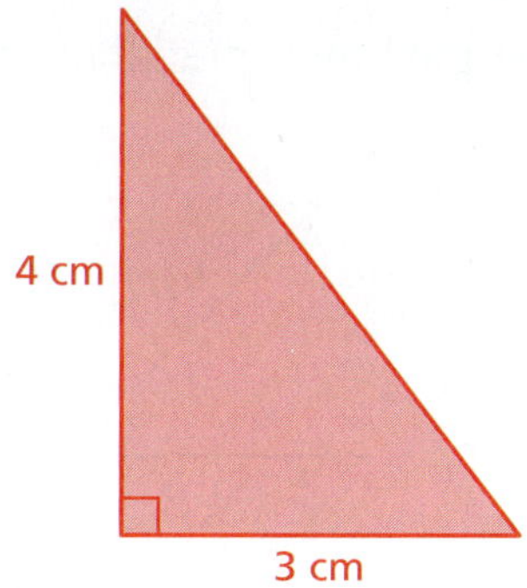

b.

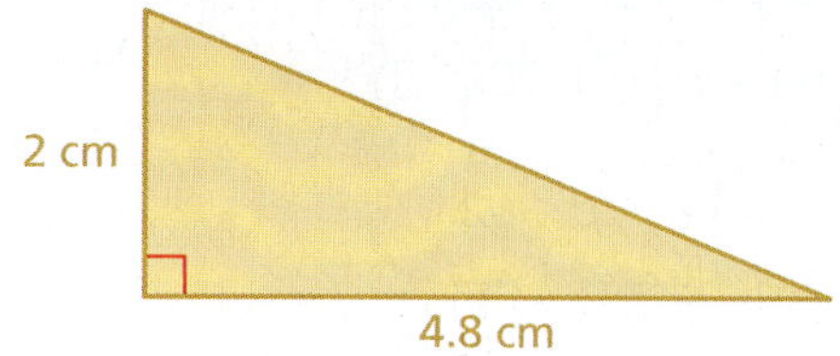

c.

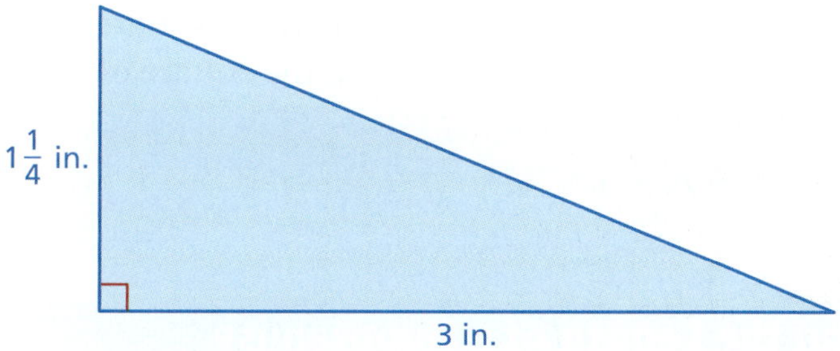

d.

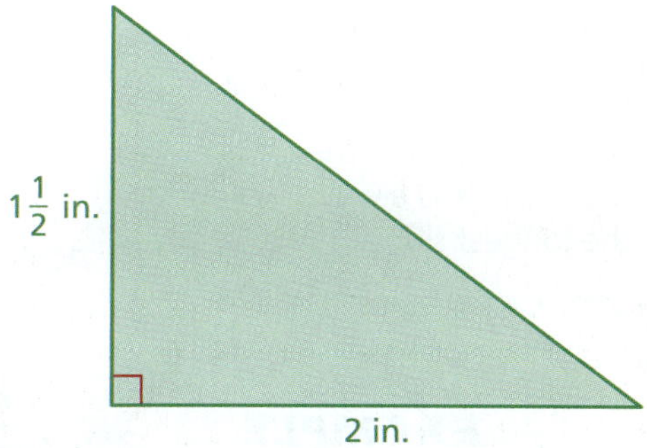

3 ACTIVITY: Using the Pythagorean Theorem in Three Dimensions

Math Practice 3

Use Definitions

How can you use what you know about the Pythagorean Theorem to describe the procedure for finding the length of the guy wire?

Work with a partner. A guy wire attached 24 feet above ground level on a telephone pole provides support for the pole.

a. PROBLEM SOLVING Describe a procedure that you could use to find the length of the guy wire without directly measuring the wire.

b. Find the length of the wire when it meets the ground 10 feet from the base of the pole.

What Is Your Answer?

4. **IN YOUR OWN WORDS** How are the lengths of the sides of a right triangle related? Give an example using whole numbers.

Practice Use what you learned about the Pythagorean Theorem to complete Exercises 3 and 4 on page 304.

Check It Out
Lesson Tutorials
BigIdeasMath.com

Key Vocabulary
theorem, *p. 300*
legs, *p. 302*
hypotenuse, *p. 302*
Pythagorean Theorem, *p. 302*

Key Ideas

Sides of a Right Triangle

The sides of a right triangle have special names.

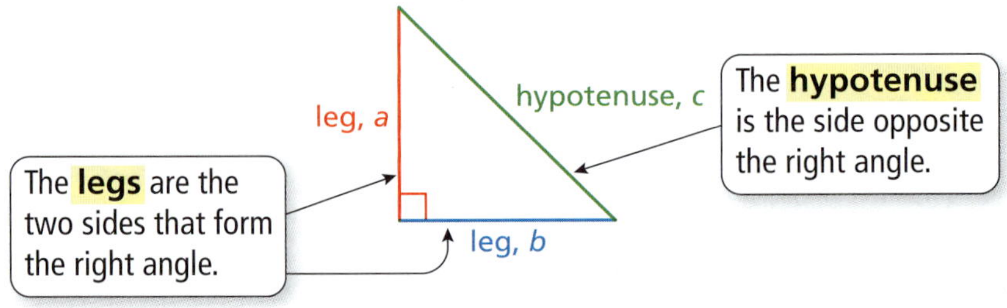

The Pythagorean Theorem

Words In any right triangle, the sum of the squares of the lengths of the legs is equal to the square of the length of the hypotenuse.

Algebra $a^2 + b^2 = c^2$

Study Tip
In a right triangle, the legs are the shorter sides and the hypotenuse is always the longest side.

EXAMPLE 1 Finding the Length of a Hypotenuse

Find the length of the hypotenuse of the triangle.

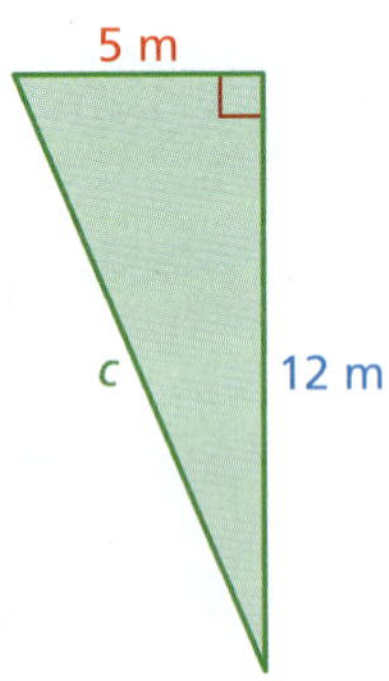

$a^2 + b^2 = c^2$	Write the Pythagorean Theorem.
$5^2 + 12^2 = c^2$	Substitute 5 for *a* and 12 for *b*.
$25 + 144 = c^2$	Evaluate powers.
$169 = c^2$	Add.
$\sqrt{169} = \sqrt{c^2}$	Take positive square root of each side.
$13 = c$	Simplify.

∴ The length of the hypotenuse is 13 meters.

On Your Own

Now You're Ready
Exercises 3 and 4

Find the length of the hypotenuse of the triangle.

1.

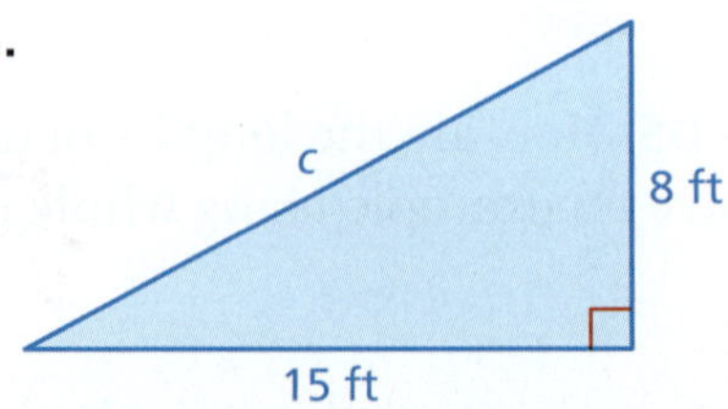

2.

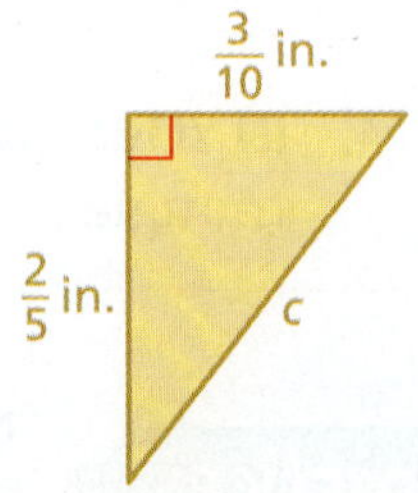

EXAMPLE 2 Finding the Length of a Leg

Find the missing length of the triangle.

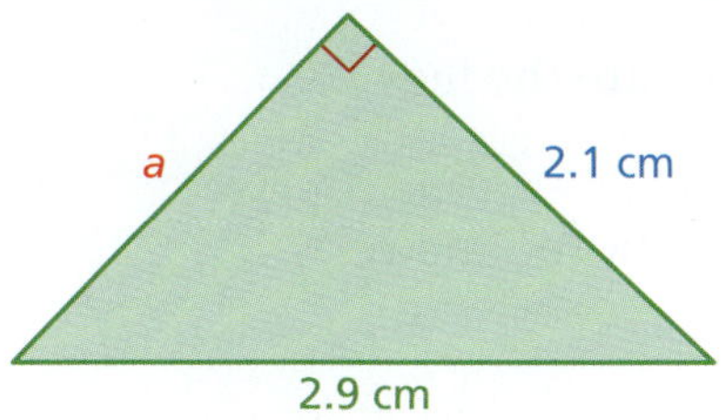

$a^2 + b^2 = c^2$	Write the Pythagorean Theorem.
$a^2 + 2.1^2 = 2.9^2$	Substitute 2.1 for *b* and 2.9 for *c*.
$a^2 + 4.41 = 8.41$	Evaluate powers.
$a^2 = 4$	Subtract 4.41 from each side.
$a = 2$	Take positive square root of each side.

∴ The missing length is 2 centimeters.

EXAMPLE 3 Real-Life Application

You are playing capture the flag. You are 50 yards north and 20 yards east of your team's base. The other team's base is 80 yards north and 60 yards east of your base. How far are you from the other team's base?

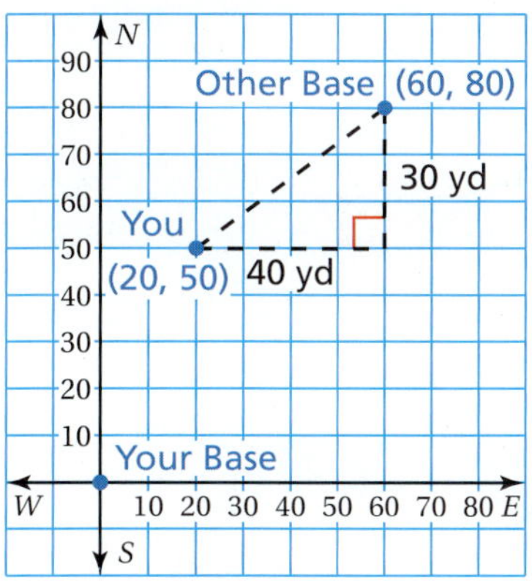

Step 1: Draw the situation in a coordinate plane. Let the origin represent your team's base. From the descriptions, you are at (20, 50) and the other team's base is at (60, 80).

Step 2: Draw a right triangle with a hypotenuse that represents the distance between you and the other team's base. The lengths of the legs are 30 yards and 40 yards.

Step 3: Use the Pythagorean Theorem to find the length of the hypotenuse.

$a^2 + b^2 = c^2$	Write the Pythagorean Theorem.
$30^2 + 40^2 = c^2$	Substitute 30 for a and 40 for b.
$900 + 1600 = c^2$	Evaluate powers.
$2500 = c^2$	Add.
$50 = c$	Take positive square root of each side.

∴ So, you are 50 yards from the other team's base.

On Your Own

Now You're Ready
Exercises 5–8

Find the missing length of the triangle.

3.

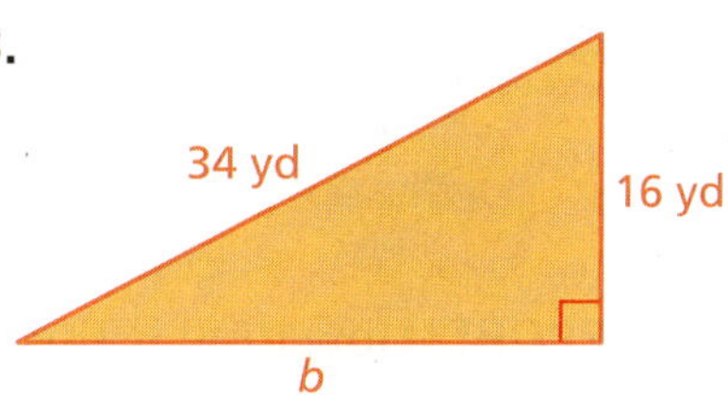

4.

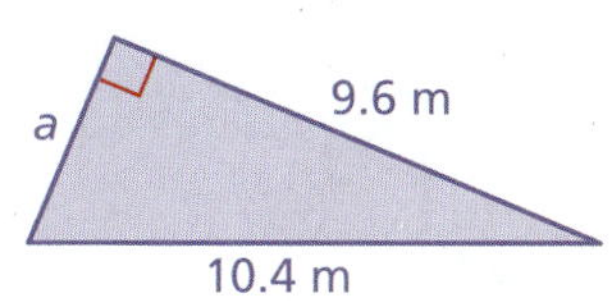

5. In Example 3, what is the distance between the bases?

7.3 Exercises

Check It Out
Help with Homework
BigIdeasMath.com

Vocabulary and Concept Check

1. **VOCABULARY** In a right triangle, how can you tell which sides are the legs and which side is the hypotenuse?

2. **DIFFERENT WORDS, SAME QUESTION** Which is different? Find "both" answers.

Which side is the hypotenuse?

Which side is the longest?

Which side is a leg?

Which side is opposite the right angle?

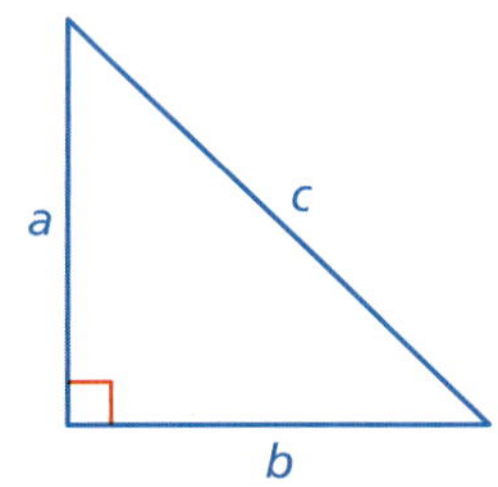

Practice and Problem Solving

Find the missing length of the triangle.

3.
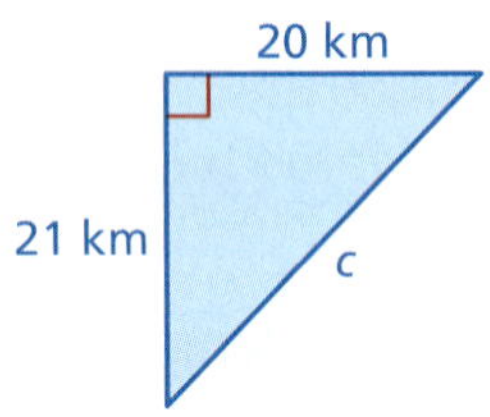

4.
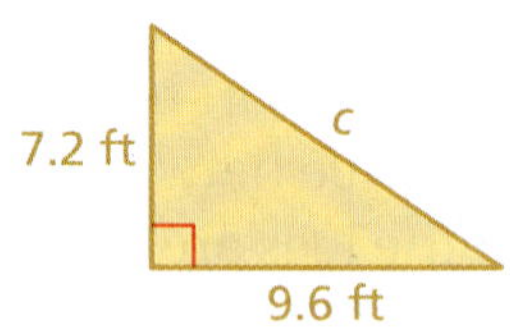

5.
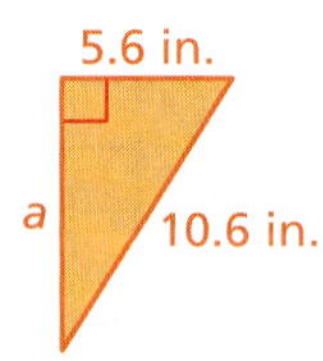

6.
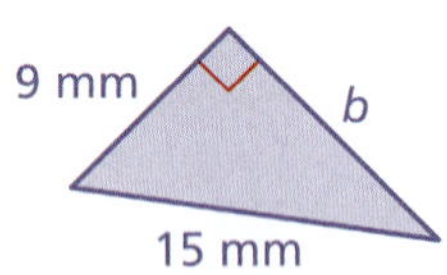

7.
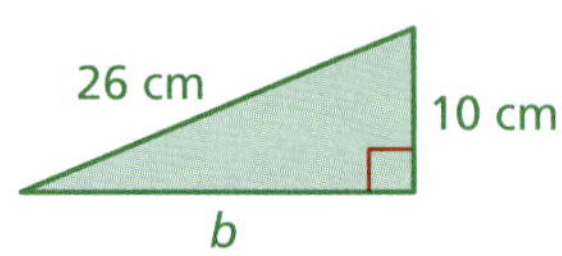

8.
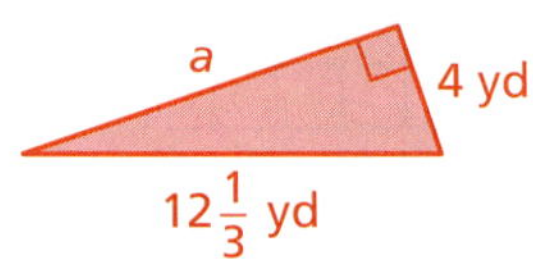

9. **ERROR ANALYSIS** Describe and correct the error in finding the missing length of the triangle.

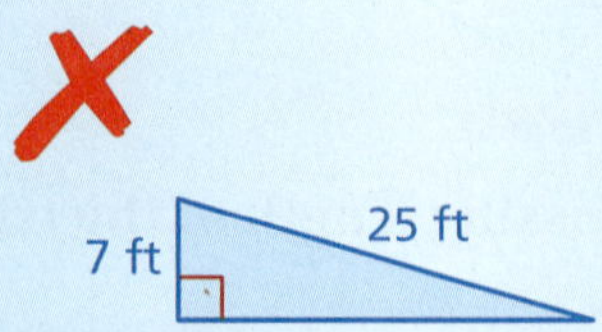

$$a^2 + b^2 = c^2$$
$$7^2 + 25^2 = c^2$$
$$674 = c^2$$
$$\sqrt{674} = c$$

10. **TREE SUPPORT** How long is the wire that supports the tree?

Find the missing length of the figure.

11.

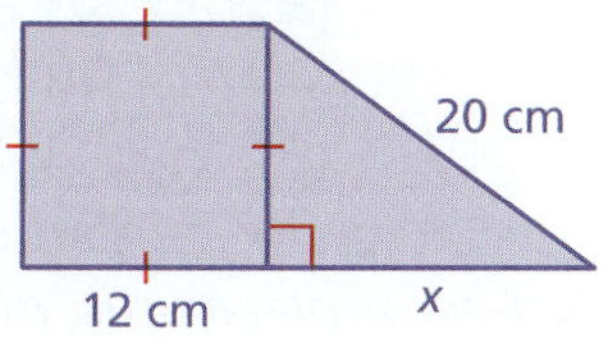

12.

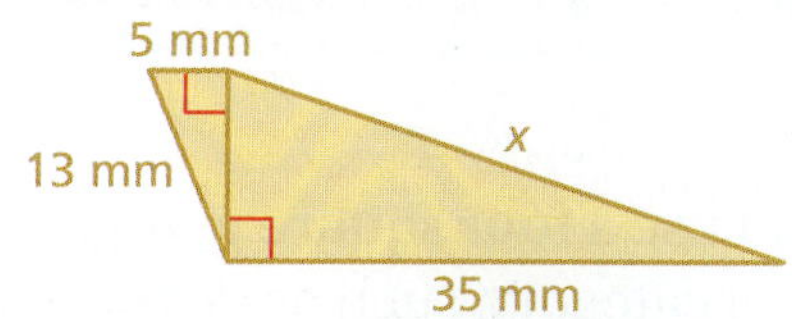

13. GOLF The figure shows the location of a golf ball after a tee shot. How many feet from the hole is the ball?

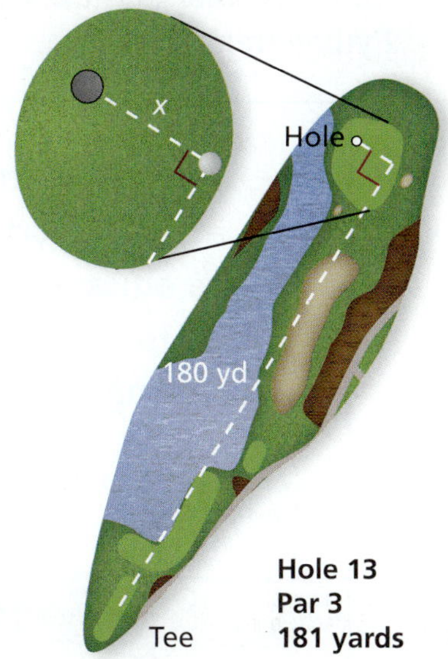

14. TENNIS A tennis player asks the referee a question. The sound of the player's voice travels only 30 feet. Can the referee hear the question? Explain.

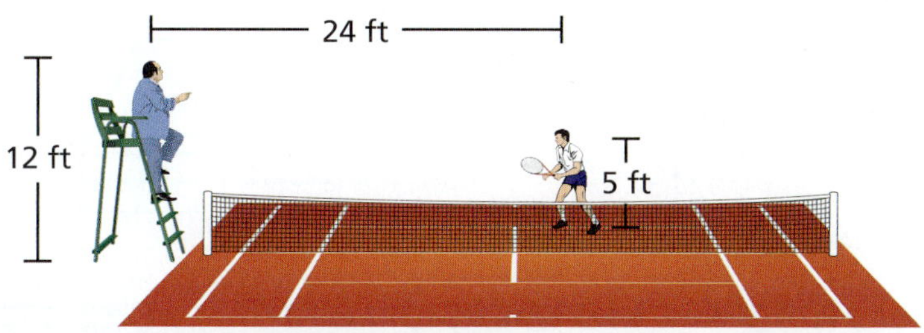

15. PROJECT Measure the length, width, and height of a rectangular room. Use the Pythagorean Theorem to find length *BC* and length *AB*.

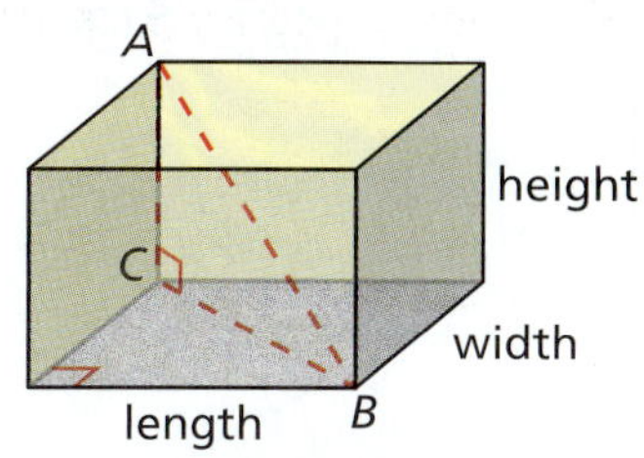

16. ALGEBRA The legs of a right triangle have lengths of 28 meters and 21 meters. The hypotenuse has a length of $5x$ meters. What is the value of x?

17. SNOWBALLS You and a friend stand back-to-back. You run 20 feet forward, then 15 feet to your right. At the same time, your friend runs 16 feet forward, then 12 feet to her right. She stops and hits you with a snowball.

a. Draw the situation in a coordinate plane.

b. How far does your friend throw the snowball?

18. Precision A box has a length of 6 inches, a width of 8 inches, and a height of 24 inches. Can a cylindrical rod with a length of 63.5 centimeters fit in the box? Explain your reasoning.

Fair Game Review What you learned in previous grades & lessons

Find the square root(s). *(Section 7.1)*

19. $\pm\sqrt{36}$ **20.** $-\sqrt{121}$ **21.** $\sqrt{169}$ **22.** $-\sqrt{225}$

23. MULTIPLE CHOICE What is the solution of the system of linear equations $y = 4x + 1$ and $2x + y = 13$? *(Section 5.2)*

Ⓐ $x = 1, y = 5$ Ⓑ $x = 5, y = 3$ Ⓒ $x = 2, y = 9$ Ⓓ $x = 9, y = 2$

7 Study Help

Check It Out
Graphic Organizer
BigIdeasMath.com

You can use a **four square** to organize information about a topic. Each of the four squares can be a category, such as *definition, vocabulary, example, non-example, words, algebra, table, numbers, visual, graph,* or *equation*. Here is an example of a four square for the Pythagorean Theorem.

Pythagorean Theorem

Definition

In any right triangle, the sum of the squares of the lengths of the legs is equal to the square of the length of the hypotenuse.

Visual

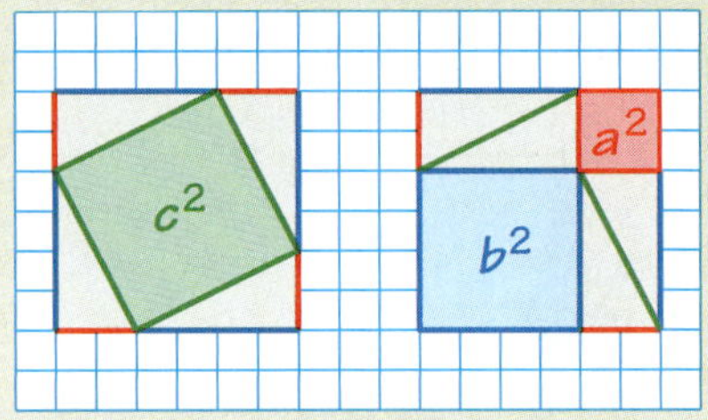

Algebra

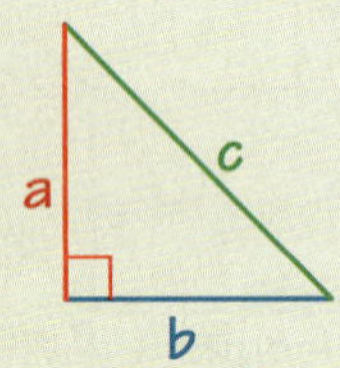

$a^2 + b^2 = c^2$

Example

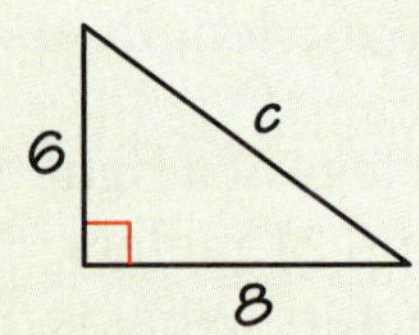

$$6^2 + 8^2 = c^2$$
$$36 + 64 = c^2$$
$$100 = c^2$$
$$10 = c$$

On Your Own

Make four squares to help you study these topics.

1. square roots
2. cube roots

After you complete this chapter, make four squares for the following topics.

3. irrational numbers
4. real numbers
5. converse of the Pythagorean Theorem
6. distance formula

"I'm taking a survey for my **four square**. How many fleas do you have?"

7.1–7.3 Quiz

Check It Out
Progress Check
BigIdeasMath.com

Find the square root(s). *(Section 7.1)*

1. $-\sqrt{4}$

2. $\sqrt{\frac{16}{25}}$

3. $\pm\sqrt{6.25}$

Find the cube root. *(Section 7.2)*

4. $\sqrt[3]{64}$

5. $\sqrt[3]{-216}$

6. $\sqrt[3]{-\frac{343}{1000}}$

Evaluate the expression. *(Section 7.1 and Section 7.2)*

7. $3\sqrt{49} + 5$

8. $10 - 4\sqrt{16}$

9. $\frac{1}{4} + \sqrt{\frac{100}{4}}$

10. $\left(\sqrt[3]{-27}\right)^3 + 61$

11. $15 + 3\sqrt[3]{125}$

12. $2\sqrt[3]{-729} - 5$

Find the missing length of the triangle. *(Section 7.3)*

13.

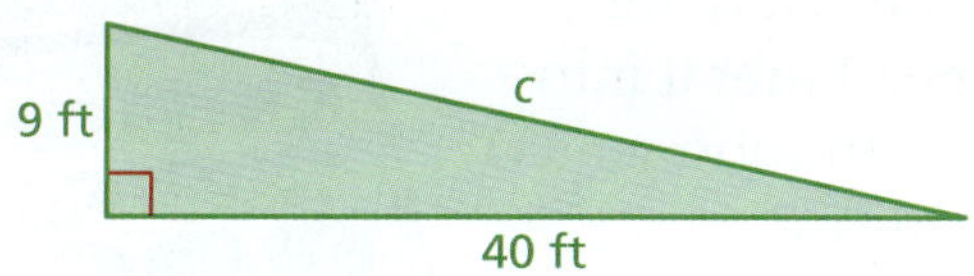

14.

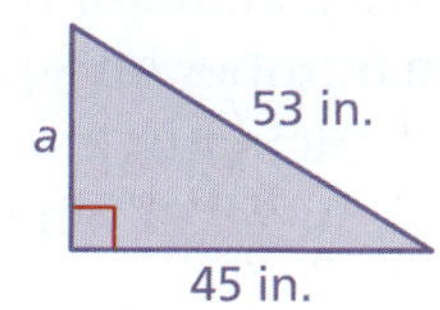

15.

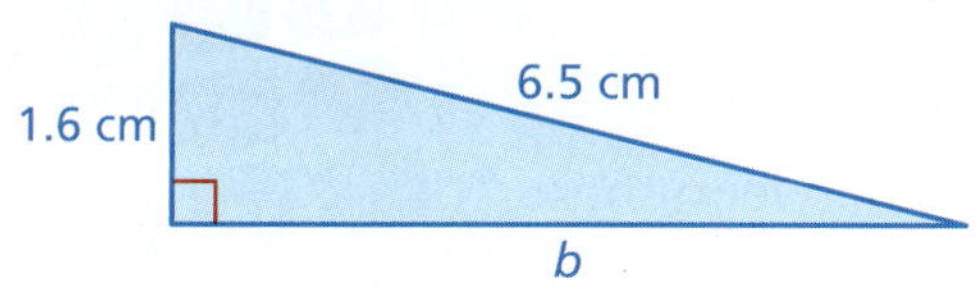

16.

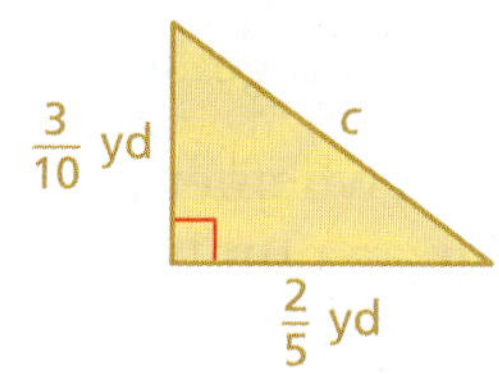

17. POOL The area of a circular pool cover is 314 square feet. Write and solve an equation to find the diameter of the pool cover. Use 3.14 for π. *(Section 7.1)*

18. PACKAGE A cube-shaped package has a volume of 5832 cubic inches. What is the edge length of the package? *(Section 7.2)*

19. FABRIC You are cutting a rectangular piece of fabric in half along the diagonal. The fabric measures 28 inches wide and $1\frac{1}{4}$ yards long. What is the length (in inches) of the diagonal? *(Section 7.3)*

7.4 Approximating Square Roots

Essential Question How can you find decimal approximations of square roots that are not rational?

1 ACTIVITY: Approximating Square Roots

Work with a partner. Archimedes was a Greek mathematician, physicist, engineer, inventor, and astronomer. He tried to find a rational number whose square is 3. Two that he tried were $\frac{265}{153}$ and $\frac{1351}{780}$.

a. Are either of these numbers equal to $\sqrt{3}$? Explain.

b. Use a calculator to approximate $\sqrt{3}$. Write the number on a piece of paper. Enter it into the calculator and square it. Then subtract 3. Do you get 0? What does this mean?

c. The value of $\sqrt{3}$ is between which two integers?

d. Tell whether the value of $\sqrt{3}$ is between the given numbers. Explain your reasoning.

1.7 and 1.8 | 1.72 and 1.73 | 1.731 and 1.732

2 ACTIVITY: Approximating Square Roots Geometrically

Work with a partner. Refer to the square on the number line below.

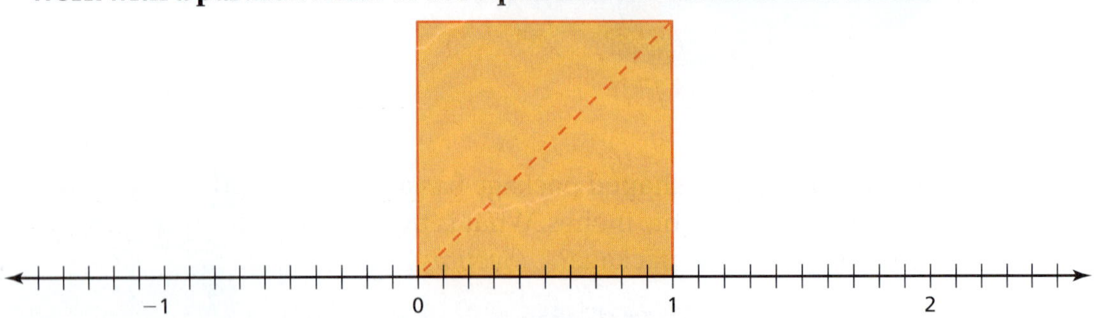

a. What is the length of the diagonal of the square?

b. Copy the square and its diagonal onto a piece of transparent paper. Rotate it about zero on the number line so that the diagonal aligns with the number line. Use the number line to estimate the length of the diagonal.

c. **STRUCTURE** How do you think your answers in parts (a) and (b) are related?

COMMON CORE

Square Roots

In this lesson, you will

- define irrational numbers.
- approximate square roots.
- approximate values of expressions involving irrational numbers.

Learning Standards
MACC.8.NS.1.1
MACC.8.NS.1.2
MACC.8.EE.1.2

3 ACTIVITY: Approximating Square Roots Geometrically

Math Practice 5

Recognize Usefulness of Tools

Why is the Pythagorean Theorem a useful tool when approximating a square root?

Work with a partner.

a. Use grid paper and the given scale to draw a horizontal line segment 1 unit in length. Label this segment *AC*.

b. Draw a vertical line segment 2 units in length. Label this segment *DC*.

c. Set the point of a compass on *A*. Set the compass to 2 units. Swing the compass to intersect segment *DC*. Label this intersection as *B*.

d. Use the Pythagorean Theorem to find the length of segment *BC*.

e. Use the grid paper to approximate $\sqrt{3}$ to the nearest tenth.

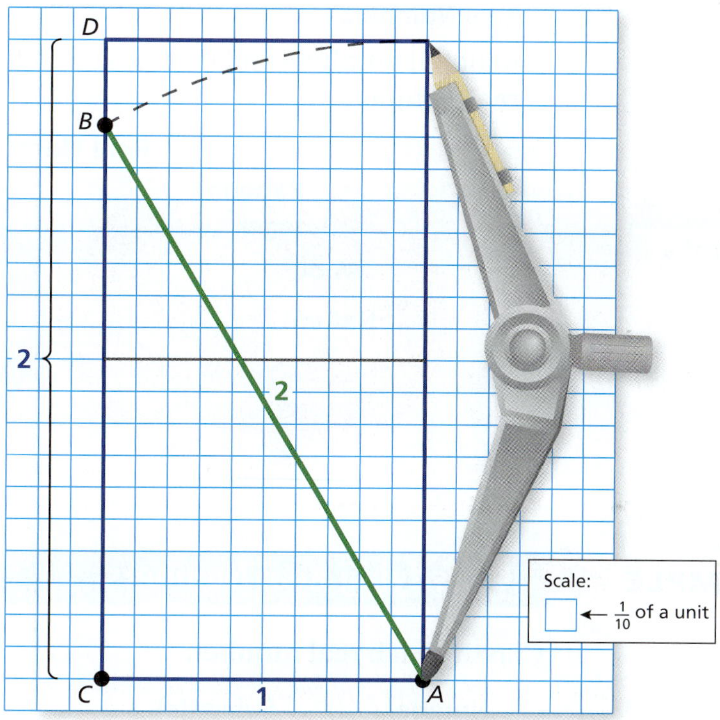

What Is Your Answer?

4. Compare your approximation in Activity 3 with your results from Activity 1.

5. Repeat Activity 3 for a triangle in which segment *AC* is 2 units and segment *BA* is 3 units. Use the Pythagorean Theorem to find the length of segment *BC*. Use the grid paper to approximate $\sqrt{5}$ to the nearest tenth.

6. **IN YOUR OWN WORDS** How can you find decimal approximations of square roots that are not rational?

Practice Use what you learned about approximating square roots to complete Exercises 5–8 on page 313.

7.4 Lesson

Check It Out
Lesson Tutorials
BigIdeasMath.com

Key Vocabulary
irrational number, *p. 310*
real numbers, *p. 310*

A rational number is a number that can be written as the ratio of two integers. An **irrational number** cannot be written as the ratio of two integers.

- The square root of any whole number that is not a perfect square is irrational. The cube root of any integer that is not a perfect cube is irrational.
- The decimal form of an irrational number neither terminates nor repeats.

Key Idea

Real Numbers

Rational numbers and irrational numbers together form the set of **real numbers**.

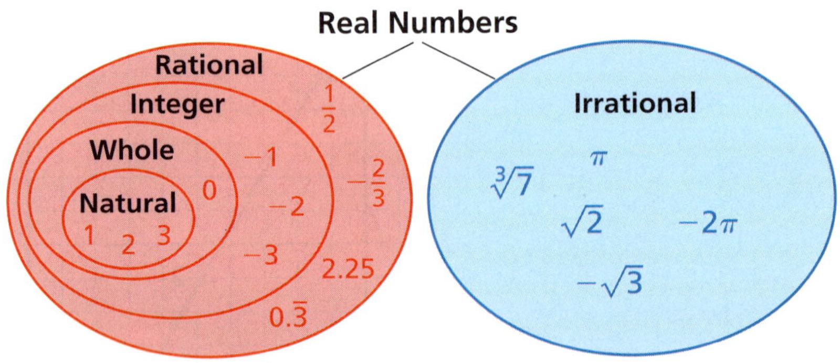

Remember
The decimal form of a rational number either terminates or repeats.

EXAMPLE 1 Classifying Real Numbers

Classify each real number.

	Number	Subset(s)	Reasoning
a.	$\sqrt{12}$	Irrational	12 is not a perfect square.
b.	$-0.\overline{25}$	Rational	$-0.\overline{25}$ is a repeating decimal.
c.	$-\sqrt{9}$	Integer, Rational	$-\sqrt{9}$ is equal to -3.
d.	$\frac{72}{4}$	Natural, Whole, Integer, Rational	$\frac{72}{4}$ is equal to 18.
e.	π	Irrational	The decimal form of π neither terminates nor repeats.

Study Tip
When classifying a real number, list all the subsets in which the number belongs.

On Your Own

Classify the real number.

1. 0.121221222 . . . **2.** $-\sqrt{196}$ **3.** $\sqrt[3]{2}$

Now You're Ready
Exercises 9–16

EXAMPLE 2 Approximating a Square Root

Estimate $\sqrt{71}$ to the nearest (a) integer and (b) tenth.

a. Make a table of numbers whose squares are close to 71.

Number	7	8	9	10
Square of Number	49	64	81	100

The table shows that 71 is between the perfect squares 64 and 81. Because 71 is closer to 64 than to 81, $\sqrt{71}$ is closer to 8 than to 9.

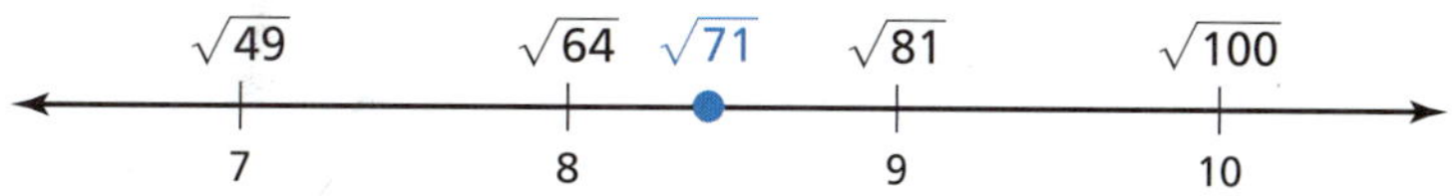

So, $\sqrt{71} \approx 8$.

b. Make a table of numbers between 8 and 9 whose squares are close to 71.

Number	8.3	8.4	8.5	8.6
Square of Number	68.89	70.56	72.25	73.96

Because 71 is closer to 70.56 than to 72.25, $\sqrt{71}$ is closer to 8.4 than to 8.5.

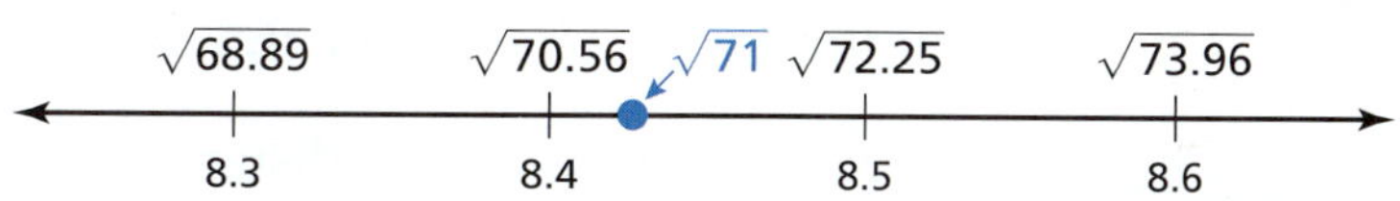

So, $\sqrt{71} \approx 8.4$.

Study Tip

You can continue the process shown in Example 2 to approximate square roots using more decimal places.

On Your Own

Now You're Ready
Exercises 20–25

Estimate the square root to the nearest (a) integer and (b) tenth.

4. $\sqrt{8}$ **5.** $-\sqrt{13}$ **6.** $-\sqrt{24}$ **7.** $\sqrt{110}$

EXAMPLE 3 Comparing Real Numbers

Which is greater, $\sqrt{5}$ or $2\frac{2}{3}$?

Estimate $\sqrt{5}$ to the nearest integer. Then graph the numbers on a number line.

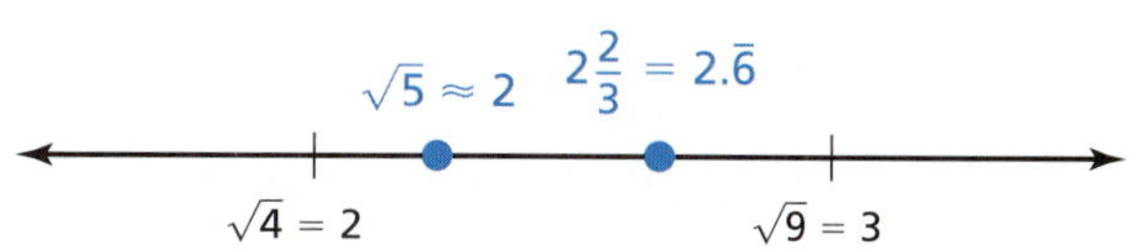

$2\frac{2}{3}$ is to the right of $\sqrt{5}$. So, $2\frac{2}{3}$ is greater.

EXAMPLE 4 Approximating the Value of an Expression

The radius of a circle with area A is approximately $\sqrt{\frac{A}{3}}$. The area of a circular mouse pad is 51 square inches. Estimate its radius to the nearest integer.

$$\sqrt{\frac{A}{3}} = \sqrt{\frac{51}{3}}$$ Substitute 51 for A.

$$= \sqrt{17}$$ Divide.

The nearest perfect square less than 17 is 16. The nearest perfect square greater than 17 is 25.

Because 17 is closer to 16 than to 25, $\sqrt{17}$ is closer to 4 than to 5.

So, the radius is about 4 inches.

EXAMPLE 5 Real-Life Application

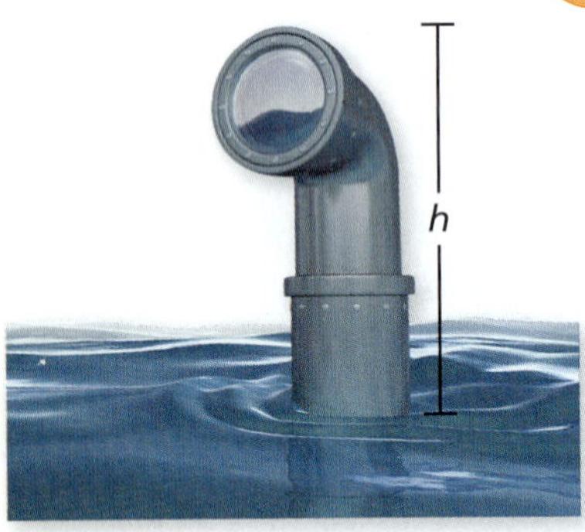

The distance (in nautical miles) you can see with a periscope is $1.17\sqrt{h}$, where h is the height of the periscope above the water. Can you see twice as far with a periscope that is 6 feet above the water than with a periscope that is 3 feet above the water? Explain.

Use a calculator to find the distances.

3 Feet Above Water		*6 Feet Above Water*
$1.17\sqrt{h} = 1.17\sqrt{3}$	Substitute for h.	$1.17\sqrt{h} = 1.17\sqrt{6}$
≈ 2.03	Use a calculator.	≈ 2.87

You can see $\frac{2.87}{2.03} \approx 1.41$ times farther with the periscope that is 6 feet above the water than with the periscope that is 3 feet above the water.

No, you cannot see twice as far with the periscope that is 6 feet above the water.

On Your Own

Now You're Ready
Exercises 26–31

Which number is greater? Explain.

8. $4\frac{1}{5}, \sqrt{23}$ **9.** $\sqrt{10}, -\sqrt{5}$ **10.** $-\sqrt{2}, -2$

11. The area of a circular mouse pad is 64 square inches. Estimate its radius to the nearest integer.

12. In Example 5, you use a periscope that is 10 feet above the water. Can you see farther than 4 nautical miles? Explain.

7.4 Exercises

Check It Out
Help with Homework
BigIdeasMath.com

Vocabulary and Concept Check

1. **VOCABULARY** How are rational numbers and irrational numbers different?
2. **WRITING** Describe a method of approximating $\sqrt{32}$.
3. **VOCABULARY** What are real numbers? Give three examples.
4. **WHICH ONE DOESN'T BELONG?** Which number does *not* belong with the other three? Explain your reasoning.

$-\frac{11}{12}$	25.075	$\sqrt{8}$	$-3.\overline{3}$

Practice and Problem Solving

Tell whether the rational number is a reasonable approximation of the square root.

5. $\frac{559}{250}, \sqrt{5}$
6. $\frac{3021}{250}, \sqrt{11}$
7. $\frac{678}{250}, \sqrt{28}$
8. $\frac{1677}{250}, \sqrt{45}$

Classify the real number.

1 9. 0
10. $\sqrt[3]{343}$
11. $\frac{\pi}{6}$
12. $-\sqrt{81}$
13. -1.125
14. $\frac{52}{13}$
15. $\sqrt[3]{-49}$
16. $\sqrt{15}$

17. **ERROR ANALYSIS** Describe and correct the error in classifying the number.

✗ $\sqrt{144}$ is irrational.

18. **SCRAPBOOKING** You cut a picture into a right triangle for your scrapbook. The lengths of the legs of the triangle are 4 inches and 6 inches. Is the length of the hypotenuse a rational number? Explain.

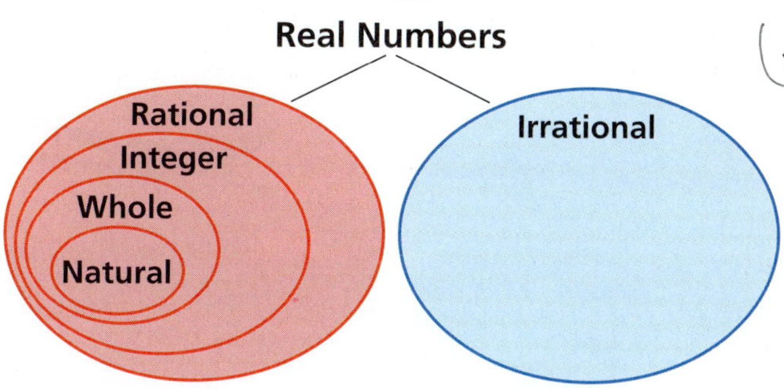

19. **VENN DIAGRAM** Place each number in the correct area of the Venn Diagram.
    a. the last digit of your phone number
    b. the square root of any prime number
    c. the ratio of the circumference of a circle to its diameter

Estimate the square root to the nearest (a) integer and (b) tenth.

2 20. $\sqrt{46}$
21. $\sqrt{685}$
22. $-\sqrt{61}$
23. $-\sqrt{105}$
24. $\sqrt{\frac{27}{4}}$
25. $-\sqrt{\frac{335}{2}}$

Which number is greater? Explain.

3 **26.** $\sqrt{20}$, 10 **27.** $\sqrt{15}$, -3.5 **28.** $\sqrt{133}$, $10\frac{3}{4}$

29. $\frac{2}{3}$, $\sqrt{\frac{16}{81}}$ **30.** $-\sqrt{0.25}$, -0.25 **31.** $-\sqrt{182}$, $-\sqrt{192}$

Use the graphing calculator screen to determine whether the statement is *true* or *false*.

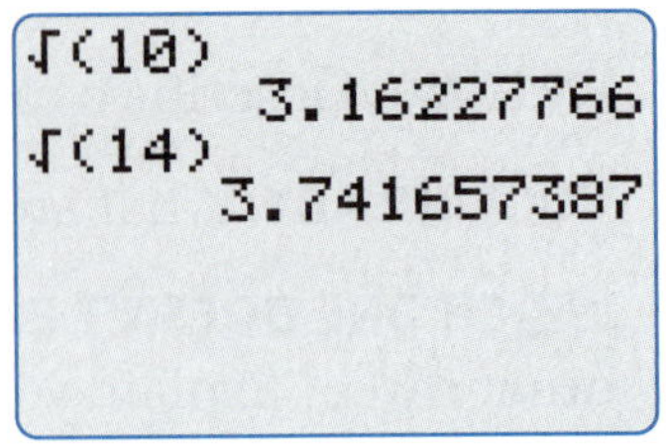

32. To the nearest tenth, $\sqrt{10} = 3.1$.

33. The value of $\sqrt{14}$ is between 3.74 and 3.75.

34. $\sqrt{10}$ lies between 3.1 and 3.16 on a number line.

35. FOUR SQUARE The area of a four square court is 66 square feet. Estimate the side length s to the nearest tenth of a foot.

36. CHECKERS A checkers board is 8 squares long and 8 squares wide. The area of each square is 14 square centimeters. Estimate the perimeter of the checkers board to the nearest tenth of a centimeter.

Approximate the length of the diagonal of the square or rectangle to the nearest tenth.

37.

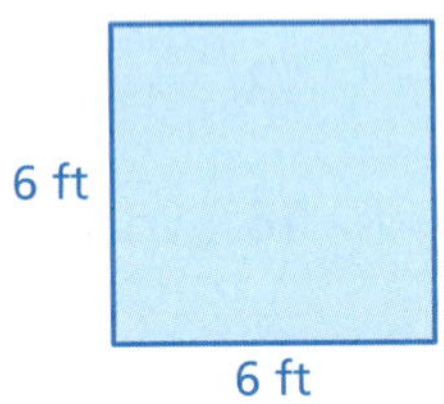

38.

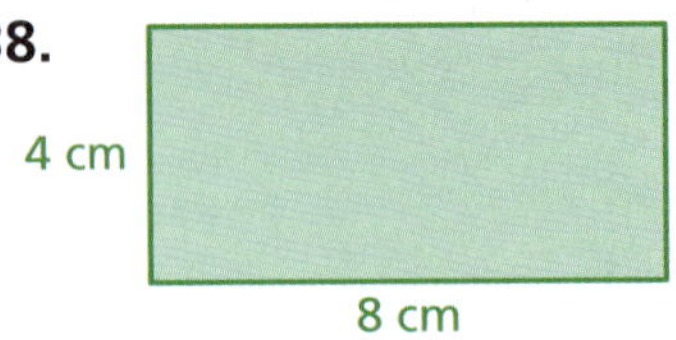

39.

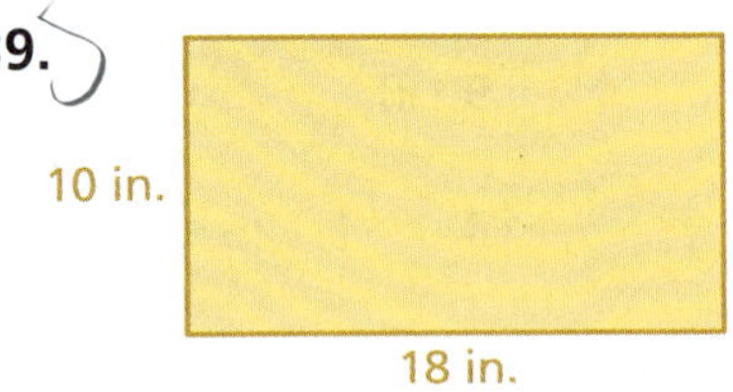

40. WRITING Explain how to continue the method in Example 2 to estimate $\sqrt{71}$ to the nearest hundredth.

41. REPEATED REASONING Describe a method that you can use to estimate a cube root to the nearest tenth. Use your method to estimate $\sqrt[3]{14}$ to the nearest tenth.

42. RADIO SIGNAL The maximum distance (in nautical miles) that a radio transmitter signal can be sent is represented by the expression $1.23\sqrt{h}$, where h is the height (in feet) above the transmitter.

Estimate the maximum distance x (in nautical miles) between the plane that is receiving the signal and the transmitter. Round your answer to the nearest tenth.

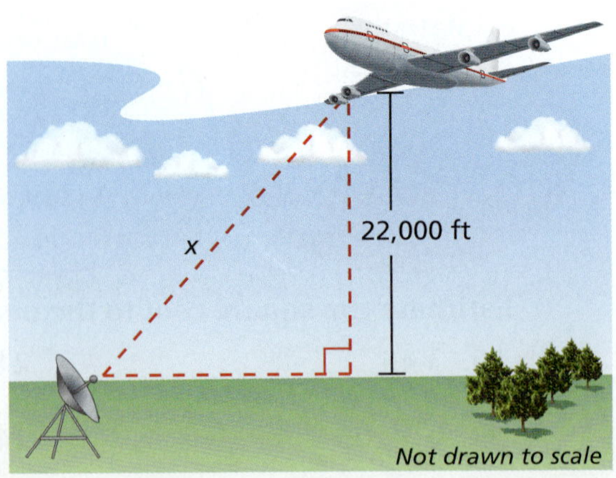

43. **OPEN-ENDED** Find two numbers a and b that satisfy the diagram.

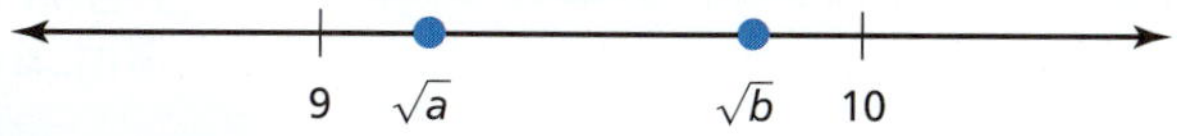

Estimate the square root to the nearest tenth.

44. $\sqrt{0.39}$ 45. $\sqrt{1.19}$ 46. $\sqrt{1.52}$

47. **ROLLER COASTER** The speed s (in meters per second) of a roller-coaster car is approximated by the equation $s = 3\sqrt{6r}$, where r is the radius of the loop. Estimate the speed of a car going around the loop. Round your answer to the nearest tenth.

48. **STRUCTURE** Is $\sqrt{\frac{1}{4}}$ a rational number? Is $\sqrt{\frac{3}{16}}$ a rational number? Explain.

49. **WATER BALLOON** The time t (in seconds) it takes a water balloon to fall d meters is represented by the equation $t = \sqrt{\frac{d}{4.9}}$. Estimate the time it takes the balloon to fall to the ground from a window that is 14 meters above the ground. Round your answer to the nearest tenth.

50. Number Sense Determine if the statement is *sometimes*, *always*, or *never* true. Explain your reasoning and give an example of each.

a. A rational number multiplied by a rational number is rational.

b. A rational number multiplied by an irrational number is rational.

c. An irrational number multiplied by an irrational number is rational.

Fair Game Review What you learned in previous grades & lessons

Find the missing length of the triangle. *(Section 7.3)*

51.

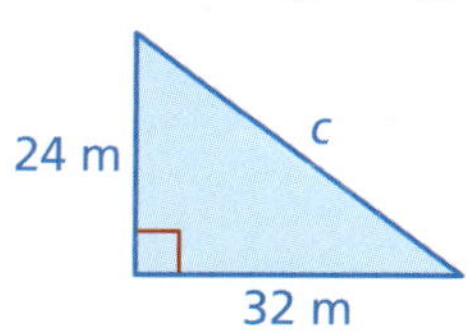

52.

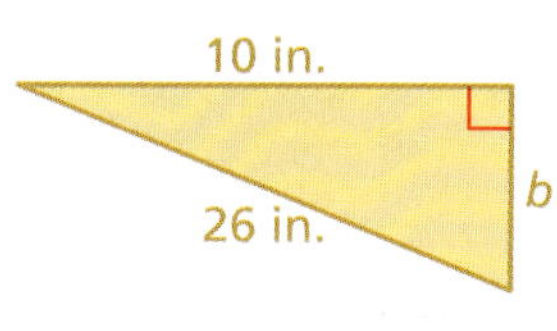

53.

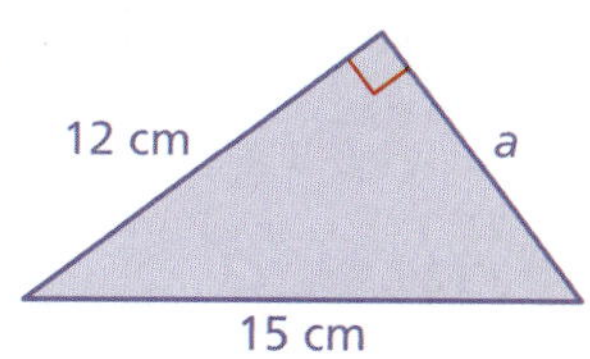

54. **MULTIPLE CHOICE** What is the ratio (red to blue) of the corresponding side lengths of the similar triangles? *(Section 2.5)*

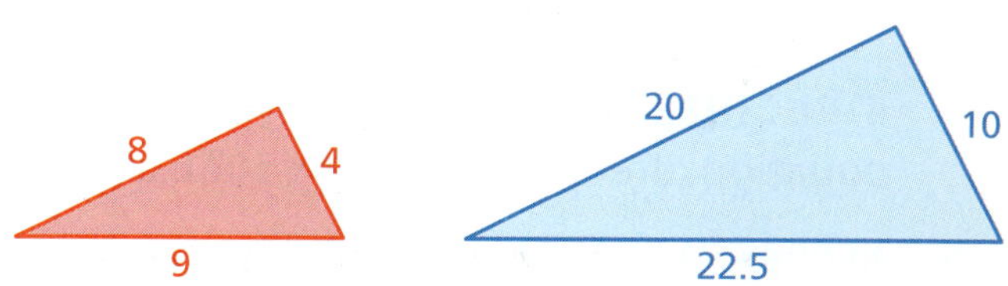

Ⓐ 1 : 3 Ⓑ 5 : 2

Ⓒ 3 : 4 Ⓓ 2 : 5

Extension 7.4 Repeating Decimals

Check It Out
Lesson Tutorials
BigIdeasMath.com

You have written terminating decimals as fractions. Because repeating decimals are rational numbers, you can also write repeating decimals as fractions.

Key Idea

Writing a Repeating Decimal as a Fraction

Let a variable x equal the repeating decimal d.

Step 1: Write the equation $x = d$.

Step 2: Multiply each side of the equation by 10^n to form a new equation, where n is the number of repeating digits.

Step 3: Subtract the original equation from the new equation.

Step 4: Solve for x.

COMMON CORE

Rational Numbers

In this extension, you will

- write a repeating decimal as a fraction.

Learning Standard
MACC.8.NS.1.1

EXAMPLE 1 Writing a Repeating Decimal as a Fraction (1 Digit Repeats)

Write $0.\overline{4}$ as a fraction in simplest form.

Let $x = 0.\overline{4}$.

$x = 0.\overline{4}$	Step 1: Write the equation.
$10 \cdot x = 10 \cdot 0.\overline{4}$	Step 2: There is 1 repeating digit, so multiply each side by $10^1 = 10$.
$10x = 4.\overline{4}$	Simplify.
$-(x = 0.\overline{4})$	Step 3: Subtract the original equation.
$9x = 4$	Simplify.
$x = \frac{4}{9}$	Step 4: Solve for x.

So, $0.\overline{4} = \frac{4}{9}$.

Check

$$\begin{array}{r} 0.44\ldots \\ 9\,\overline{)\,4.00} \\ \underline{3\,6} \\ 40 \\ \underline{36} \\ 40 \end{array}$$ ✓

Practice

Write the decimal as a fraction or a mixed number.

1. $0.\overline{1}$
2. $-0.\overline{5}$
3. $-1.\overline{2}$
4. $5.\overline{8}$
5. **STRUCTURE** In Example 1, why can you subtract the original equation from the new equation after multiplying by 10? Explain why these two steps are performed.
6. **REPEATED REASONING** Compare the repeating decimals and their equivalent fractions in Exercises 1–4. Describe the pattern. Use the pattern to explain how to write a repeating decimal as a fraction when only the tenths digit repeats.

EXAMPLE 2 Writing a Repeating Decimal as a Fraction (1 Digit Repeats)

Write $-0.2\overline{3}$ as a fraction in simplest form.

Let $x = -0.2\overline{3}$.

$x = -0.2\overline{3}$ — Step 1: Write the equation.

$10 \cdot x = 10 \cdot (-0.2\overline{3})$ — Step 2: There is 1 repeating digit, so multiply each side by $10^1 = 10$.

$10x = -2.\overline{3}$ — Simplify.

$-(x = -0.2\overline{3})$ — Step 3: Subtract the original equation.

$9x = -2.1$ — Simplify.

$x = \frac{-2.1}{9}$ — Step 4: Solve for x.

So, $-0.2\overline{3} = \frac{-2.1}{9} = -\frac{21}{90} = -\frac{7}{30}$.

Check

```
-7/30
     -.2333333333
```

EXAMPLE 3 Writing a Repeating Decimal as a Fraction (2 Digits Repeat)

Write $1.\overline{25}$ as a mixed number.

Let $x = 1.\overline{25}$.

$x = 1.\overline{25}$ — Step 1: Write the equation.

$100 \cdot x = 100 \cdot 1.\overline{25}$ — Step 2: There are 2 repeating digits, so multiply each side by $10^2 = 100$.

$100x = 125.\overline{25}$ — Simplify.

$-(x = 1.\overline{25})$ — Step 3: Subtract the original equation.

$99x = 124$ — Simplify.

$x = \frac{124}{99}$ — Step 4: Solve for x.

So, $1.\overline{25} = \frac{124}{99} = 1\frac{25}{99}$.

Check

```
124/99
      1.252525253
```

Practice

Write the decimal as a fraction or a mixed number.

7. $-0.4\overline{3}$ **8.** $2.0\overline{6}$ **9.** $0.\overline{27}$ **10.** $-4.\overline{50}$

11. **REPEATED REASONING** Find a pattern in the fractional representations of repeating decimals in which only the tenths and hundredths digits repeat. Use the pattern to explain how to write $9.\overline{04}$ as a mixed number.

7.5 Using the Pythagorean Theorem

Essential Question In what other ways can you use the Pythagorean Theorem?

The *converse* of a statement switches the hypothesis and the conclusion.

Statement:	Converse of the statement:
If p, then q.	If q, then p.

1 ACTIVITY: Analyzing Converses of Statements

Work with a partner. Write the converse of the true statement. Determine whether the converse is *true* or *false*. If it is true, justify your reasoning. If it is false, give a counterexample.

a. If $a = b$, then $a^2 = b^2$.

b. If $a = b$, then $a^3 = b^3$.

c. If one figure is a translation of another figure, then the figures are congruent.

d. If two triangles are similar, then the triangles have the same angle measures.

Is the converse of a true statement always true? always false? Explain.

2 ACTIVITY: The Converse of the Pythagorean Theorem

Work with a partner. The converse of the Pythagorean Theorem states: "If the equation $a^2 + b^2 = c^2$ is true for the side lengths of a triangle, then the triangle is a right triangle."

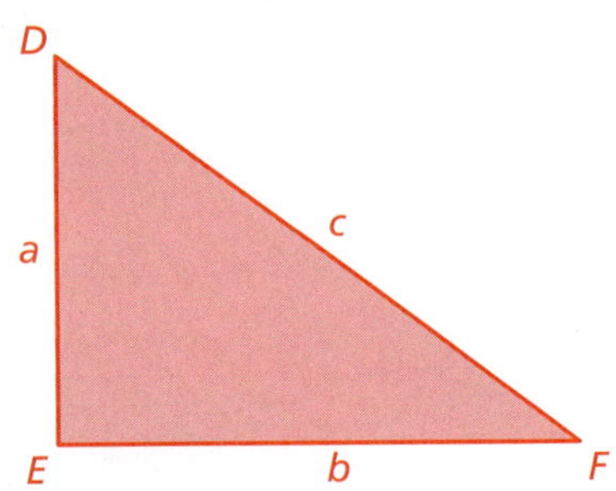

a. Do you think the converse of the Pythagorean Theorem is *true* or *false*? How could you use deductive reasoning to support your answer?

b. Consider $\triangle DEF$ with side lengths a, b, and c, such that $a^2 + b^2 = c^2$. Also consider $\triangle JKL$ with leg lengths a and b, where $\angle K = 90°$.

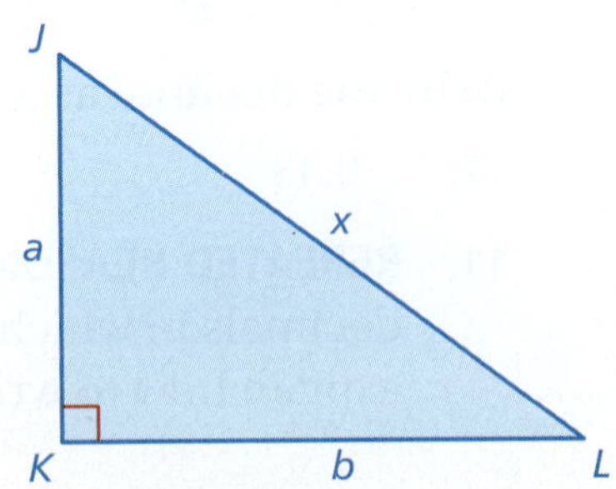

- What does the Pythagorean Theorem tell you about $\triangle JKL$?
- What does this tell you about c and x?
- What does this tell you about $\triangle DEF$ and $\triangle JKL$?
- What does this tell you about $\angle E$?
- What can you conclude?

COMMON CORE

Pythagorean Theorem

In this lesson, you will

- use the converse of the Pythagorean Theorem to identify right triangles.
- use the Pythagorean Theorem to find distances in a coordinate plane.
- solve real-life problems.

Learning Standards
MACC.8.EE.1.2
MACC.8.G.2.6
MACC.8.G.2.7
MACC.8.G.2.8

3 ACTIVITY: Developing the Distance Formula

Work with a partner. Follow the steps below to write a formula that you can use to find the distance between any two points in a coordinate plane.

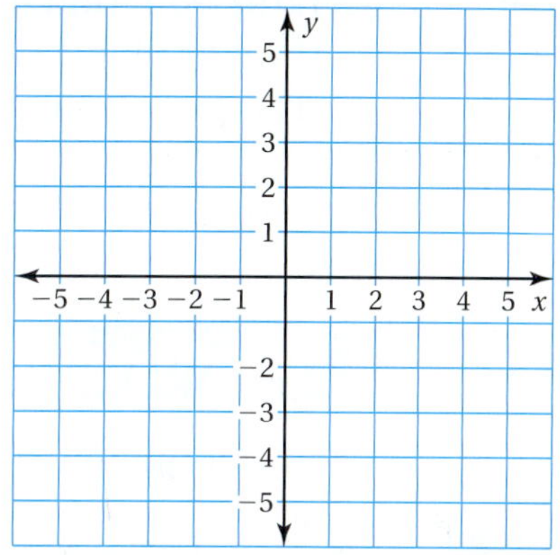

> **Math Practice 6**
>
> **Communicate Precisely**
>
> What steps can you take to make sure that you have written the distance formula accurately?

Step 1: Choose two points in the coordinate plane that do not lie on the same horizontal or vertical line. Label the points (x_1, y_1) and (x_2, y_2).

Step 2: Draw a line segment connecting the points. This will be the hypotenuse of a right triangle.

Step 3: Draw horizontal and vertical line segments from the points to form the legs of the right triangle.

Step 4: Use the x-coordinates to write an expression for the length of the horizontal leg.

Step 5: Use the y-coordinates to write an expression for the length of the vertical leg.

Step 6: Substitute the expressions for the lengths of the legs into the Pythagorean Theorem.

Step 7: Solve the equation in Step 6 for the hypotenuse c.

What does the length of the hypotenuse tell you about the two points?

What Is Your Answer?

4. **IN YOUR OWN WORDS** In what other ways can you use the Pythagorean Theorem?
5. What kind of real-life problems do you think the converse of the Pythagorean Theorem can help you solve?

Practice Use what you learned about the converse of a true statement to complete Exercises 3 and 4 on page 322.

7.5 Lesson

Check It Out
Lesson Tutorials
BigIdeasMath.com

Key Vocabulary
distance formula, *p. 320*

Key Ideas

Converse of the Pythagorean Theorem

If the equation $a^2 + b^2 = c^2$ is true for the side lengths of a triangle, then the triangle is a right triangle.

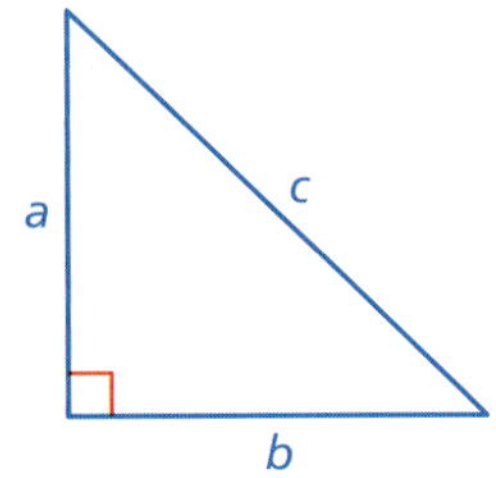

EXAMPLE 1 Identifying a Right Triangle

Tell whether each triangle is a right triangle.

a.

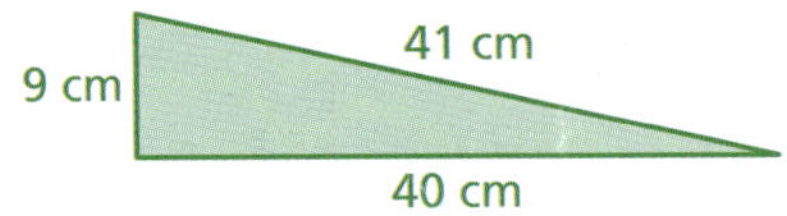

$$a^2 + b^2 = c^2$$
$$9^2 + 40^2 \stackrel{?}{=} 41^2$$
$$81 + 1600 \stackrel{?}{=} 1681$$
$$1681 = 1681 \checkmark$$

It *is* a right triangle.

b.

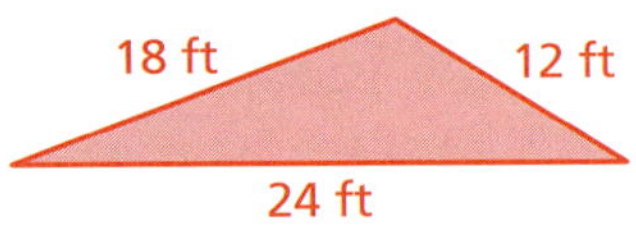

$$a^2 + b^2 = c^2$$
$$12^2 + 18^2 \stackrel{?}{=} 24^2$$
$$144 + 324 \stackrel{?}{=} 576$$
$$468 \neq 576 \text{ ✗}$$

It is *not* a right triangle.

Study Tip

A *Pythagorean triple* is a set of three positive integers *a*, *b*, and *c*, where $a^2 + b^2 = c^2$.

Common Error

When using the converse of the Pythagorean Theorem, always substitute the length of the longest side for *c*.

On Your Own

Tell whether the triangle with the given side lengths is a right triangle.

1. 28 in., 21 in., 20 in.
2. 1.25 mm, 1 mm, 0.75 mm

Now You're Ready
Exercises 5–10

On page 319, you used the Pythagorean Theorem to develop the *distance formula*. You can use the **distance formula** to find the distance between any two points in a coordinate plane.

Key Idea

Distance Formula

The distance d between any two points (x_1, y_1) and (x_2, y_2) is given by the formula

$$d = \sqrt{(x_2 - x_1)^2 + (y_2 - y_1)^2}.$$

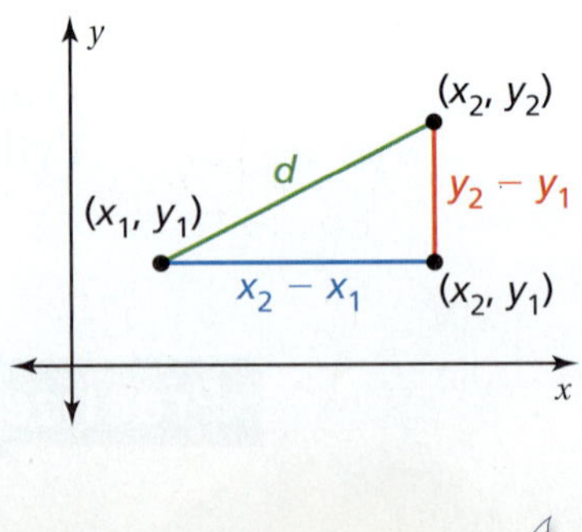

EXAMPLE 2 Finding the Distance Between Two Points

Find the distance between (1, 5) and (−4, −2).

Let $(x_1, y_1) = (1, 5)$ and $(x_2, y_2) = (-4, -2)$.

$$d = \sqrt{(x_2 - x_1)^2 + (y_2 - y_1)^2}$$ Write the distance formula.

$$= \sqrt{(-4 - 1)^2 + (-2 - 5)^2}$$ Substitute.

$$= \sqrt{(-5)^2 + (-7)^2}$$ Simplify.

$$= \sqrt{25 + 49}$$ Evaluate powers.

$$= \sqrt{74}$$ Add.

EXAMPLE 3 Real-Life Application

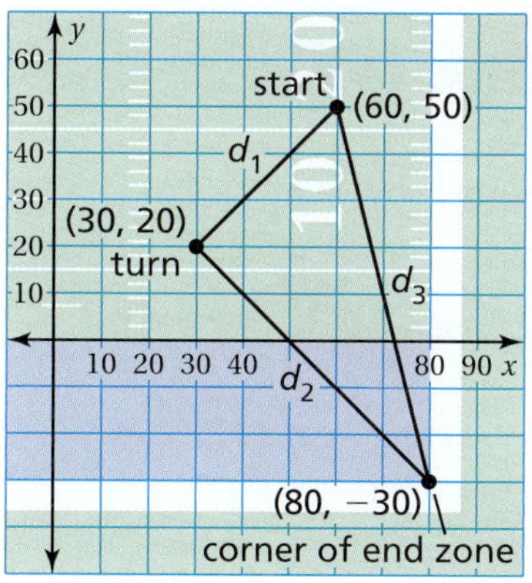

You design a football play in which a player runs down the field, makes a 90° turn, and runs to the corner of the end zone. Your friend runs the play as shown. Did your friend make a 90° turn? Each unit of the grid represents 10 feet.

Use the distance formula to find the lengths of the three sides.

$$d_1 = \sqrt{(60 - 30)^2 + (50 - 20)^2} = \sqrt{30^2 + 30^2} = \sqrt{1800} \text{ feet}$$

$$d_2 = \sqrt{(80 - 30)^2 + (-30 - 20)^2} = \sqrt{50^2 + (-50)^2} = \sqrt{5000} \text{ feet}$$

$$d_3 = \sqrt{(80 - 60)^2 + (-30 - 50)^2} = \sqrt{20^2 + (-80)^2} = \sqrt{6800} \text{ feet}$$

Use the converse of the Pythagorean Theorem to determine if the side lengths form a right triangle.

$$(\sqrt{1800})^2 + (\sqrt{5000})^2 \stackrel{?}{=} (\sqrt{6800})^2$$

$$1800 + 5000 \stackrel{?}{=} 6800$$

$$6800 = 6800 \checkmark$$

The sides form a right triangle.

So, your friend made a 90° turn.

On Your Own

Now You're Ready
Exercises 11–16

Find the distance between the two points.

3. (0, 0), (4, 5)
4. (7, −3), (9, 6)
5. (−2, −3), (−5, 1)
6. **WHAT IF?** In Example 3, your friend made the turn at (20, 10). Did your friend make a 90° turn?

7.5 Exercises

Check It Out
Help with Homework
BigIdeasMath.com

Vocabulary and Concept Check

1. **WRITING** Describe two ways to find the distance between two points in a coordinate plane.

2. **WHICH ONE DOESN'T BELONG?** Which set of numbers does *not* belong with the other three? Explain your reasoning.

3, 6, 8	6, 8, 10	5, 12, 13	7, 24, 25

Practice and Problem Solving

Write the converse of the true statement. Determine whether the converse is *true* or *false*. If it is true, justify your reasoning. If it is false, give a counterexample.

3. If a is an odd number, then a^2 is odd.

4. If $ABCD$ is a square, then $ABCD$ is a parallelogram.

Tell whether the triangle with the given side lengths is a right triangle.

1 5. 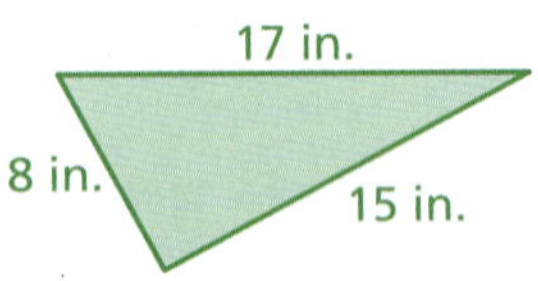

6. 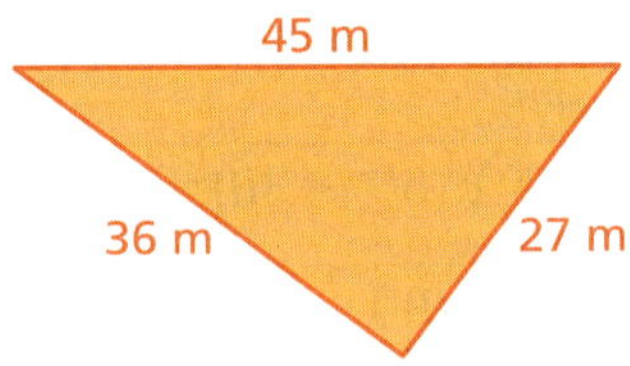

7. 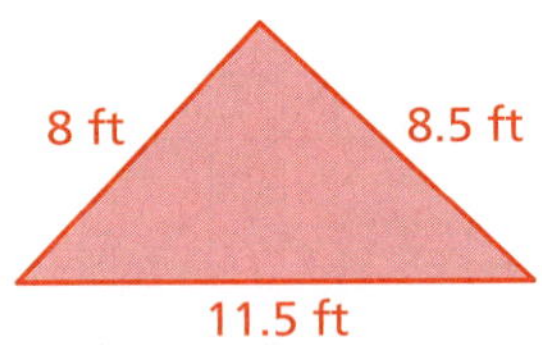

8. 14 mm, 19 mm, 23 mm

9. $\frac{9}{10}$ mi, $1\frac{1}{5}$ mi, $1\frac{1}{2}$ mi

10. 1.4 m, 4.8 m, 5 m

Find the distance between the two points.

2 11. (1, 2), (7, 6)

12. (4, −5), (−1, 7)

13. (2, 4), (7, 2)

14. (−1, −3), (1, 3)

15. (−6, −7), (0, 0)

16. (12, 5), (−12, −2)

17. **ERROR ANALYSIS** Describe and correct the error in finding the distance between the points (−3, −2) and (7, 4).

✗
$$d = \sqrt{[7-(-3)]^2 - [4-(-2)]^2}$$
$$= \sqrt{100 - 36}$$
$$= \sqrt{64} = 8$$

18. **CONSTRUCTION** A post and beam frame for a shed is shown in the diagram. Does the brace form a right triangle with the post and beam? Explain.

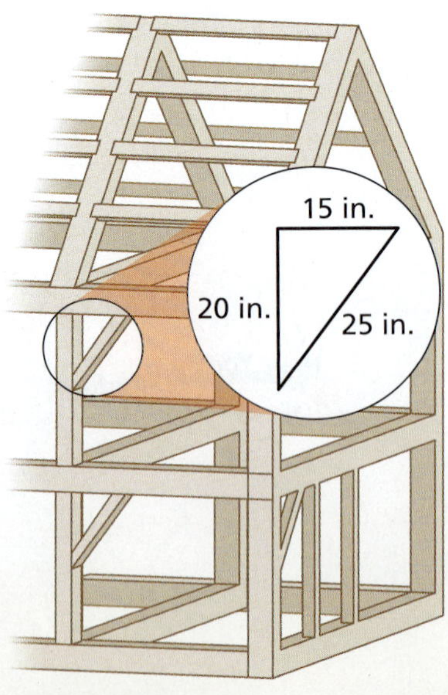

Tell whether a triangle with the given side lengths is a right triangle.

19. $\sqrt{63}, 9, 12$

20. $4, \sqrt{15}, 6$

21. $\sqrt{18}, \sqrt{24}, \sqrt{42}$

22. REASONING Plot the points $(-1, 3)$, $(4, -2)$, and $(1, -5)$ in a coordinate plane. Are the points the vertices of a right triangle? Explain.

23. GEOCACHING You spend the day looking for hidden containers in a wooded area using a Global Positioning System (GPS). You park your car on the side of the road, and then locate Container 1 and Container 2 before going back to the car. Does your path form a right triangle? Explain. Each unit of the grid represents 10 yards.

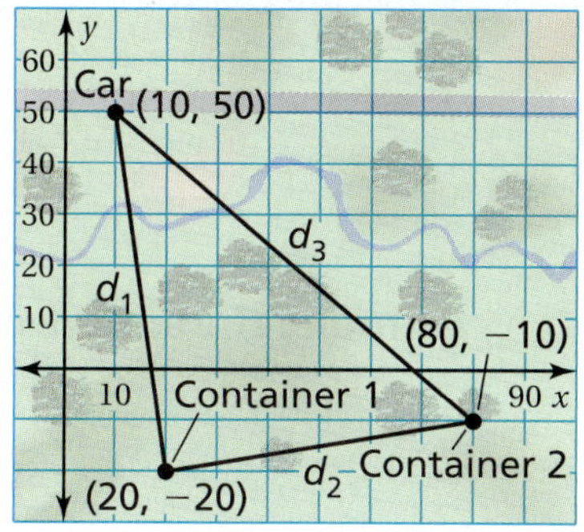

24. REASONING Your teacher wants the class to find the distance between the two points (2, 4) and (9, 7). You use (2, 4) for (x_1, y_1), and your friend uses (9, 7) for (x_1, y_1). Do you and your friend obtain the same result? Justify your answer.

25. AIRPORT Which plane is closer to the base of the airport tower? Explain.

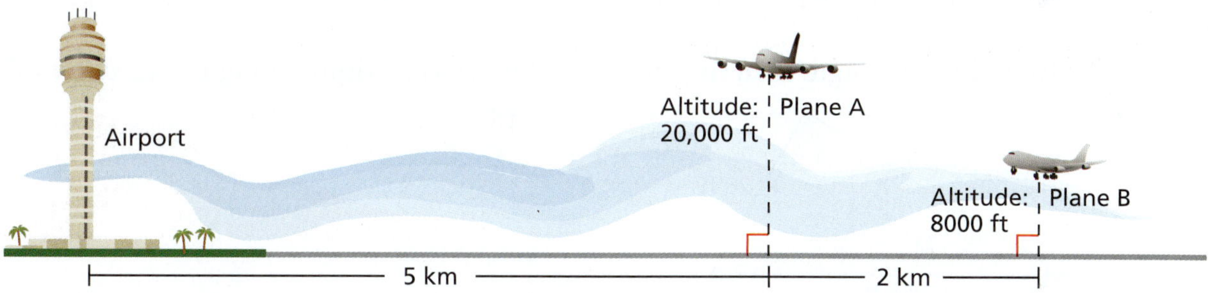

Not drawn to scale

26. Structure Consider the two points (x_1, y_1) and (x_2, y_2) in the coordinate plane. How can you find the point (x_m, y_m) located in the middle of the two given points? Justify your answer using the distance formula.

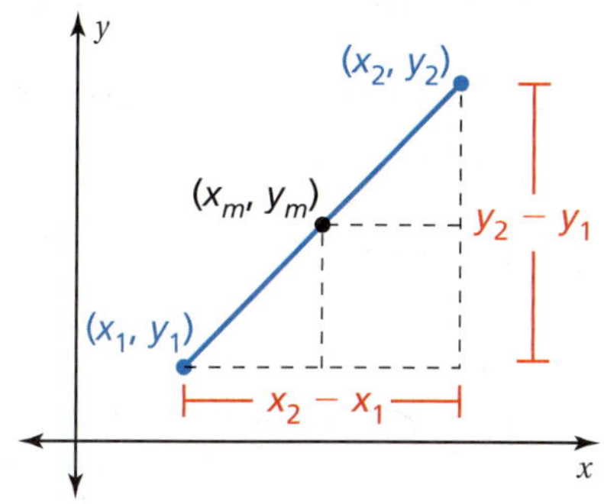

Fair Game Review What you learned in previous grades & lessons

Find the mean, median, and mode of the data. *(Skills Review Handbook)*

27. 12, 9, 17, 15, 12, 13

28. 21, 32, 16, 27, 22, 19, 10

29. 67, 59, 34, 71, 59

30. MULTIPLE CHOICE What is the sum of the interior angle measures of an octagon? *(Section 3.3)*

Ⓐ 720° Ⓑ 1080° Ⓒ 1440° Ⓓ 1800°

7.4–7.5 Quiz

Check It Out
Progress Check
BigIdeasMath.com

Classify the real number. *(Section 7.4)*

1. $-\sqrt{225}$

2. $-1\frac{1}{9}$

3. $\sqrt{41}$

4. $\sqrt{17}$

Estimate the square root to the nearest (a) integer and (b) tenth. *(Section 7.4)*

5. $\sqrt{38}$

6. $-\sqrt{99}$

7. $\sqrt{172}$

8. $\sqrt{115}$

Which number is greater? Explain. *(Section 7.4)*

9. $\sqrt{11}, 3\frac{3}{5}$

10. $\sqrt{1.44}, 1.1\overline{8}$

Write the decimal as a fraction or a mixed number. *(Section 7.4)*

11. $0.\overline{7}$

12. $-1.\overline{63}$

Tell whether the triangle with the given side lengths is a right triangle. *(Section 7.5)*

13.

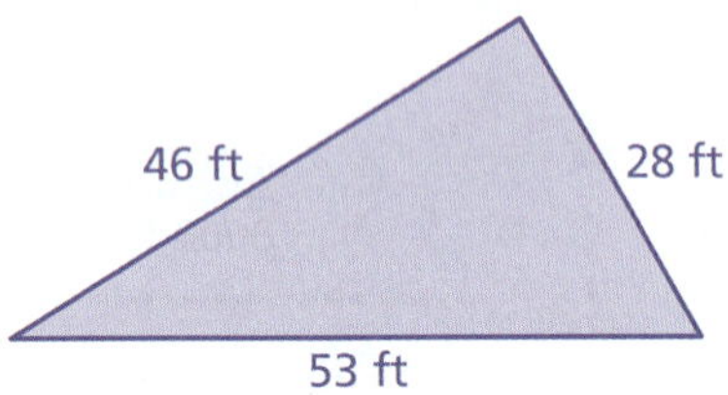

14.

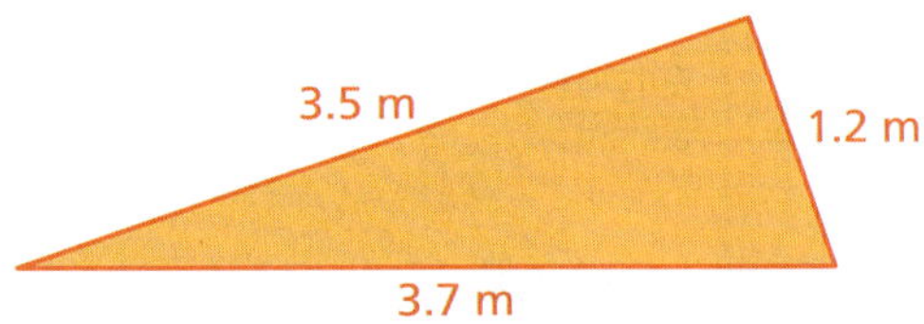

Find the distance between the two points. *(Section 7.5)*

15. $(-3, -1), (-1, -5)$

16. $(-4, 2), (5, 1)$

17. $(1, -2), (4, -5)$

18. $(-1, 1), (7, 4)$

19. $(-6, 5), (-4, -6)$

20. $(-1, 4), (1, 3)$

Use the figure to answer Exercises 21–24. Round your answer to the nearest tenth. *(Section 7.5)*

21. How far is the cabin from the peak?

22. How far is the fire tower from the lake?

23. How far is the lake from the peak?

24. You are standing at $(-5, -6)$. How far are you from the lake?

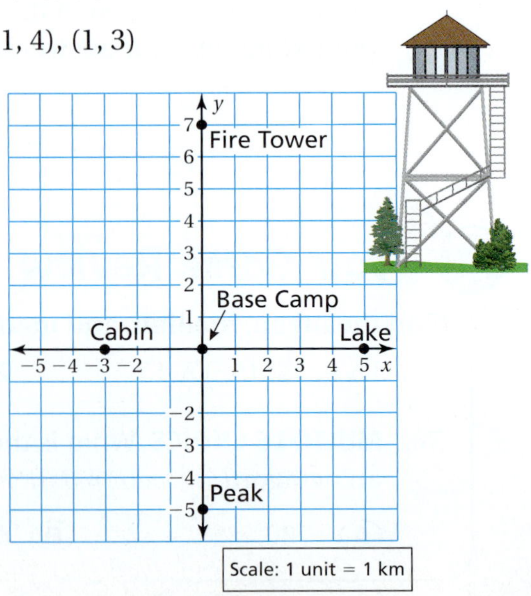

7 Chapter Review

Check It Out
Vocabulary Help
BigIdeasMath.com

Review Key Vocabulary

square root, *p. 290*
perfect square, *p. 290*
radical sign, *p. 290*
radicand, *p. 290*
cube root, *p. 296*
perfect cube, *p. 296*
theorem, *p. 300*
legs, *p. 302*
hypotenuse, *p. 302*
Pythagorean Theorem, *p. 302*
irrational number, *p. 310*
real numbers, *p. 310*
distance formula, *p. 320*

Review Examples and Exercises

7.1 Finding Square Roots *(pp. 288–293)*

Find $-\sqrt{36}$.

$-\sqrt{36}$ represents the *negative* square root.

Because $6^2 = 36$, $-\sqrt{36} = -\sqrt{6^2} = -6$.

Exercises

Find the square root(s).

1. $\sqrt{1}$ **2.** $-\sqrt{\dfrac{9}{25}}$ **3.** $\pm\sqrt{1.69}$

Evaluate the expression.

4. $15 - 4\sqrt{36}$ **5.** $\sqrt{\dfrac{54}{6}} + \dfrac{2}{3}$ **6.** $10\left(\sqrt{81} - 12\right)$

7.2 Finding Cube Roots *(pp. 294–299)*

Find $\sqrt[3]{\dfrac{125}{216}}$.

Because $\left(\dfrac{5}{6}\right)^3 = \dfrac{125}{216}$, $\sqrt[3]{\dfrac{125}{216}} = \sqrt[3]{\left(\dfrac{5}{6}\right)^3} = \dfrac{5}{6}$.

Exercises

Find the cube root.

7. $\sqrt[3]{729}$ **8.** $\sqrt[3]{\dfrac{64}{343}}$ **9.** $\sqrt[3]{-\dfrac{8}{27}}$

Evaluate the expression.

10. $\sqrt[3]{27} - 16$ **11.** $25 + 2\sqrt[3]{-64}$ **12.** $3\sqrt[3]{-125} - 27$

7.3 The Pythagorean Theorem (pp. 300–305)

Find the length of the hypotenuse of the triangle.

$a^2 + b^2 = c^2$	Write the Pythagorean Theorem.
$7^2 + 24^2 = c^2$	Substitute.
$49 + 576 = c^2$	Evaluate powers.
$625 = c^2$	Add.
$\sqrt{625} = \sqrt{c^2}$	Take positive square root of each side.
$25 = c$	Simplify.

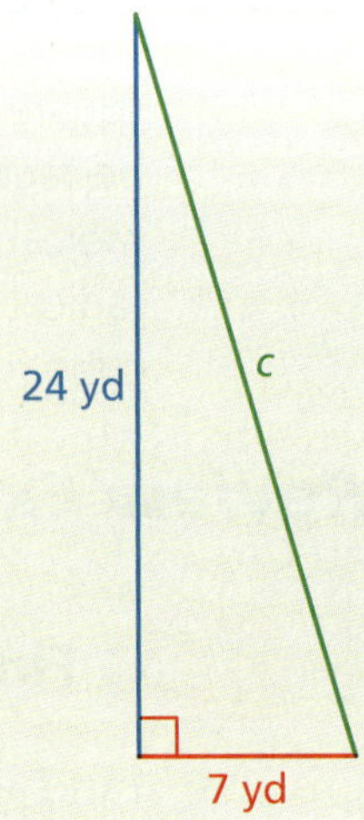

∴ The length of the hypotenuse is 25 yards.

Exercises

Find the missing length of the triangle.

13.

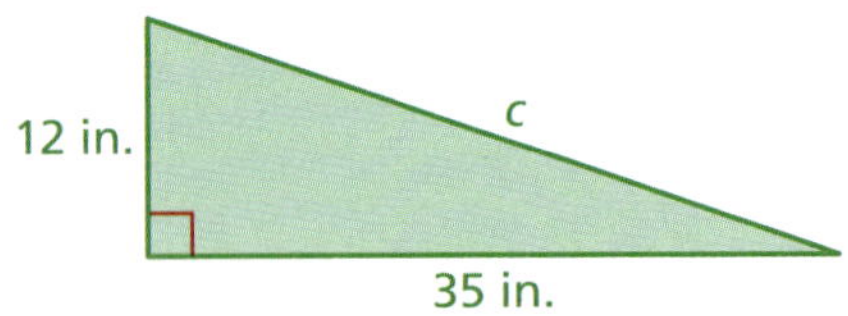

14.

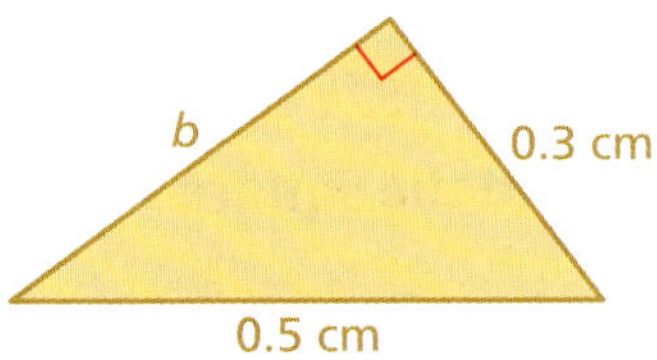

7.4 Approximating Square Roots (pp. 308–317)

a. Classify $\sqrt{19}$.

∴ The number $\sqrt{19}$ is irrational because 19 is not a perfect square.

b. Estimate $\sqrt{34}$ to the nearest integer.

Make a table of numbers whose squares are close to the radicand, 34.

Number	4	5	6	7
Square of Number	16	25	36	49

The table shows that 34 is between the perfect squares 25 and 36. Because 34 is closer to 36 than to 25, $\sqrt{34}$ is closer to 6 than to 5.

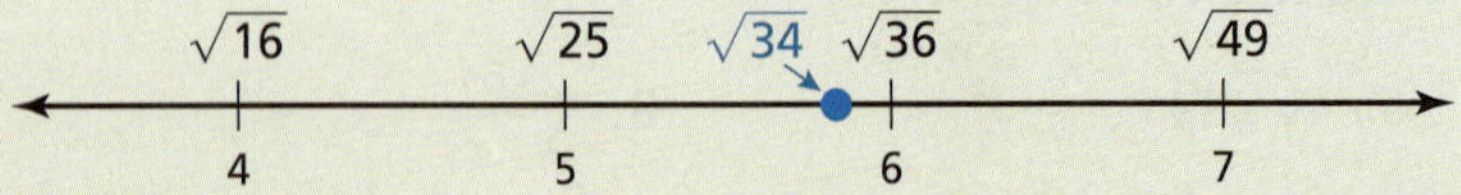

∴ So, $\sqrt{34} \approx 6$.

Exercises

Classify the real number.

15. $0.81\overline{5}$ **16.** $\sqrt{101}$ **17.** $\sqrt{4}$

Estimate the square root to the nearest (a) integer and (b) tenth.

18. $\sqrt{14}$ **19.** $\sqrt{90}$ **20.** $\sqrt{175}$

Write the decimal as a fraction.

21. $0.\overline{8}$ **22.** $0.\overline{36}$ **23.** $-1.\overline{6}$

7.5 Using the Pythagorean Theorem *(pp. 318–323)*

a. Is the triangle formed by the rope and the tent a right triangle?

$$a^2 + b^2 = c^2$$
$$64^2 + 48^2 \stackrel{?}{=} 80^2$$
$$4096 + 2304 \stackrel{?}{=} 6400$$
$$6400 = 6400 \checkmark$$

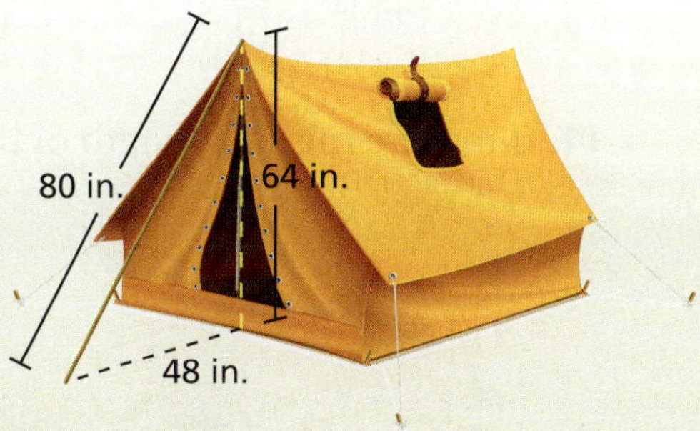

∴ It *is* a right triangle.

b. Find the distance between (−3, 1) and (4, 7).

Let $(x_1, y_1) = (-3, 1)$ and $(x_2, y_2) = (4, 7)$.

$d = \sqrt{(x_2 - x_1)^2 + (y_2 - y_1)^2}$ Write the distance formula.

$= \sqrt{[4 - (-3)]^2 + (7 - 1)^2}$ Substitute.

$= \sqrt{7^2 + 6^2}$ Simplify.

$= \sqrt{49 + 36}$ Evaluate powers.

$= \sqrt{85}$ Add.

Exercises

Tell whether the triangle is a right triangle.

24.

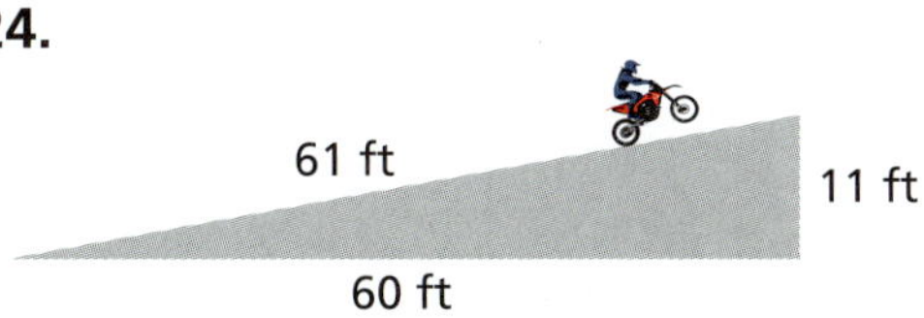

25.

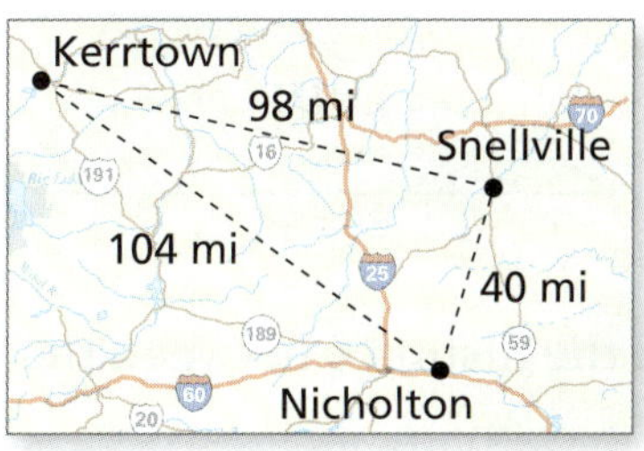

Find the distance between the two points.

26. $(-2, -5), (3, 5)$ **27.** $(-4, 7), (4, 0)$

7 Chapter Test

Check It Out
Test Practice
BigIdeasMath.com

Find the square root(s).

1. $-\sqrt{1600}$
2. $\sqrt{\frac{25}{49}}$
3. $\pm\sqrt{\frac{100}{9}}$

Find the cube root.

4. $\sqrt[3]{-27}$
5. $\sqrt[3]{\frac{8}{125}}$
6. $\sqrt[3]{-\frac{729}{64}}$

Evaluate the expression.

7. $12 + 8\sqrt{16}$
8. $\frac{1}{2} + \sqrt{\frac{72}{2}}$
9. $\left(\sqrt[3]{-125}\right)^3 + 75$
10. $50\sqrt[3]{\frac{512}{1000}} + 14$

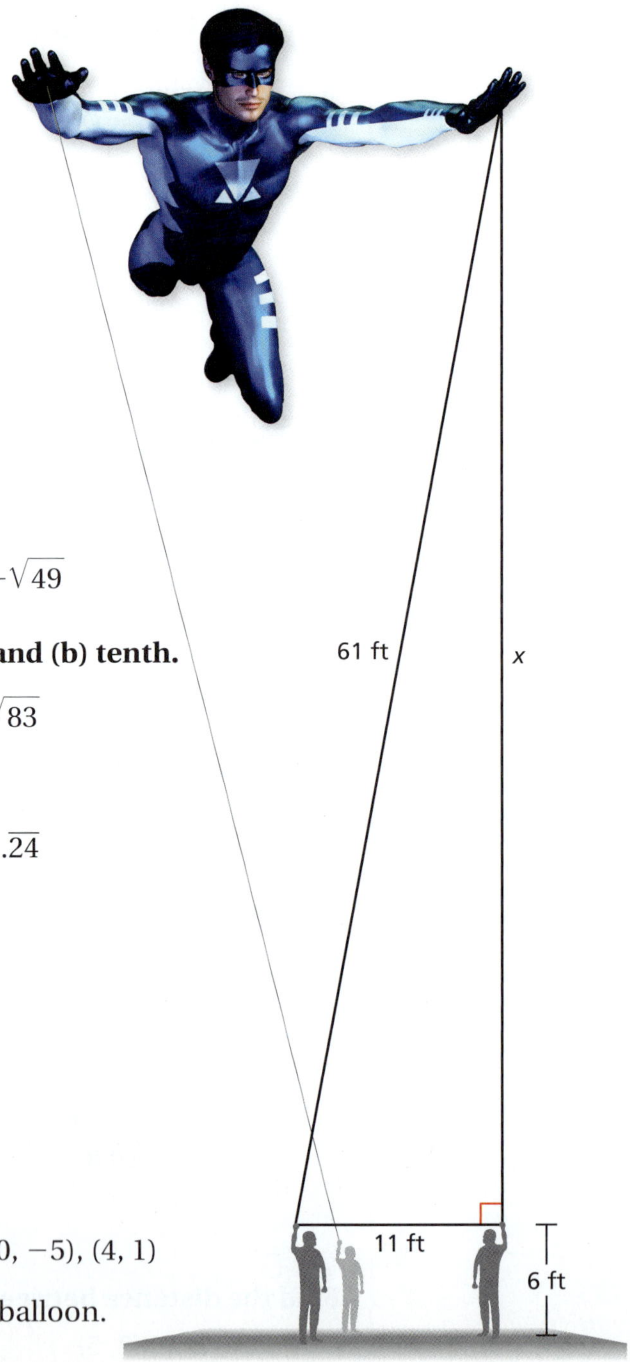

11. Find the missing length of the triangle.

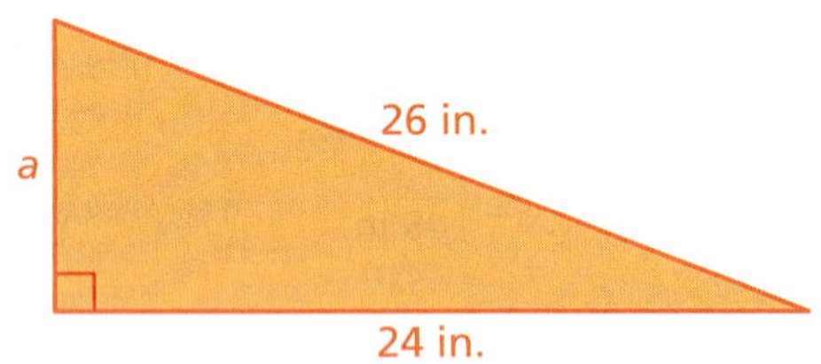

Classify the real number.

12. 16π
13. $-\sqrt{49}$

Estimate the square root to the nearest (a) integer and (b) tenth.

14. $\sqrt{58}$
15. $\sqrt{83}$

Write the decimal as a fraction or a mixed number.

16. $-0.\overline{3}$
17. $1.\overline{24}$

18. Tell whether the triangle is a right triangle.

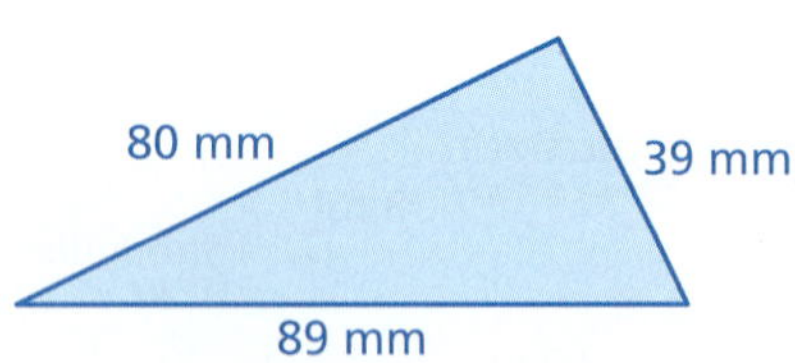

Find the distance between the two points.

19. $(-2, 3)$, $(6, 9)$
20. $(0, -5)$, $(4, 1)$
21. **SUPERHERO** Find the altitude of the superhero balloon.

7 Standards Assessment

Test-Taking Strategy
Answer Easy Questions First

"Scan the test and **answer the easy questions first**. You know the square root of 4 is 2."

1. The period T of a pendulum is the time, in seconds, it takes the pendulum to swing back and forth. The period can be found using the formula $T = 1.1\sqrt{L}$, where L is the length, in feet, of the pendulum. A pendulum has a length of 4 feet. Find its period. *(MACC.8.EE.1.2)*

A. 5.1 sec **C.** 3.1 sec

B. 4.4 sec **D.** 2.2 sec

2. Which parallelogram is a dilation of parallelogram *JKLM*? (Figures not drawn to scale.) *(MACC.8.G.1.4)*

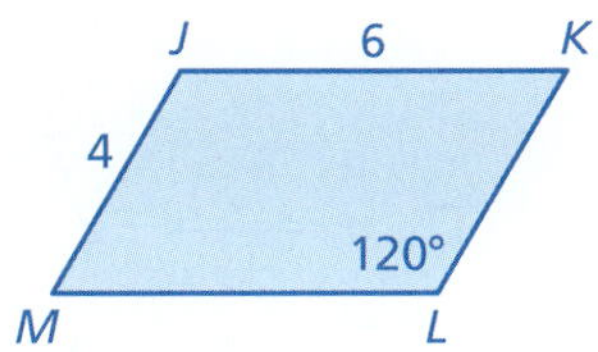

F.

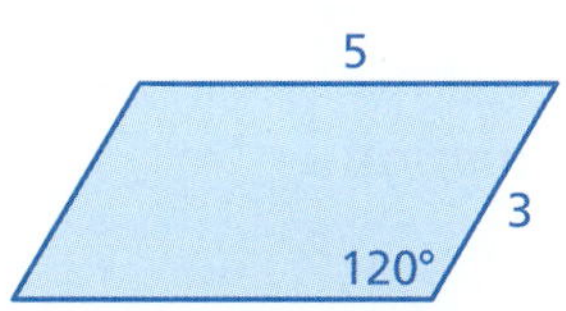

H.

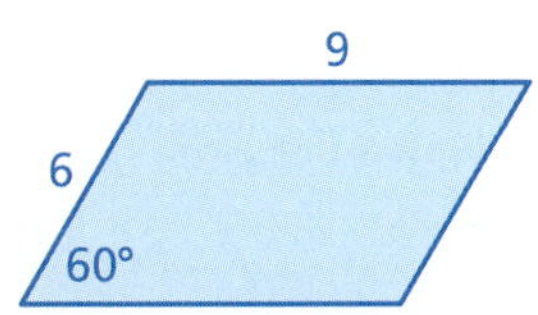

G.

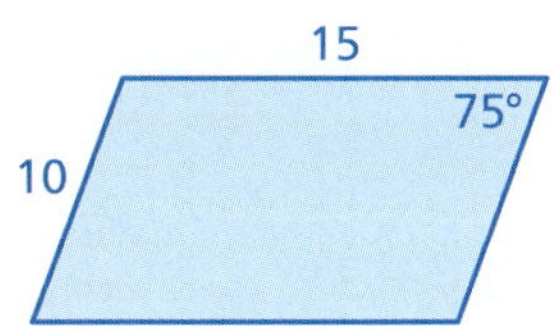

I.

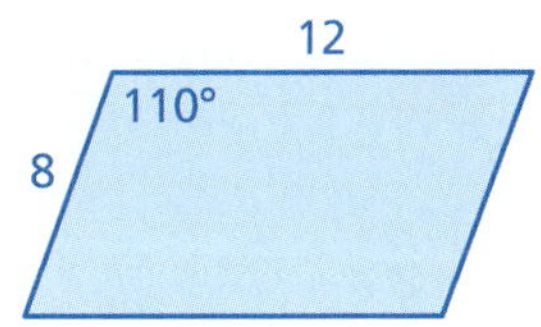

3. Which equation represents a linear function? *(MACC.8.F.1.3)*

A. $y = x^2$ **C.** $xy = 1$

B. $y = \frac{2}{x}$ **D.** $x + y = 1$

4. Which linear function matches the line shown in the graph? *(MACC.8.F.2.4)*

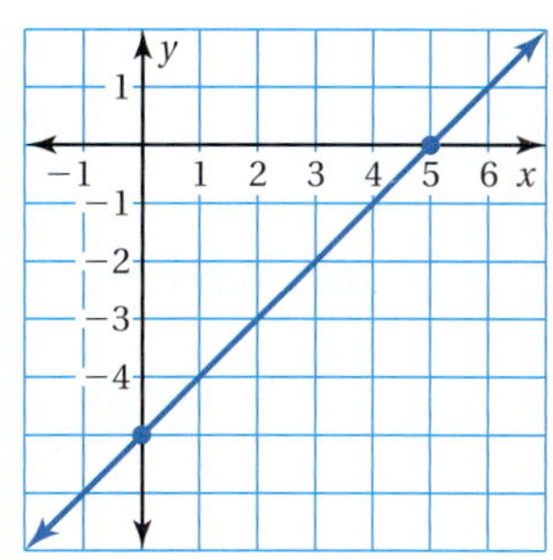

F. $y = x - 5$ **H.** $y = -x - 5$

G. $y = x + 5$ **I.** $y = -x + 5$

5. A football field is 40 yards wide and 120 yards long. Find the distance between opposite corners of the football field. Show your work and explain your reasoning. *(MACC.8.G.2.7)*

Think Solve Explain

6. A computer consultant charges \$50 plus \$40 for each hour she works. The consultant charged \$650 for one job. This can be represented by the equation below, where h represents the number of hours worked.

$$40h + 50 = 650$$

How many hours did the consultant work? *(MACC.8.EE.3.7b)*

7. You can use the formula below to find the sum S of the interior angle measures of a polygon with n sides. Solve the formula for n. *(MACC.8.EE.3.7b)*

$$S = 180(n - 2)$$

A. $n = 180(S - 2)$

B. $n = \frac{S}{180} + 2$

C. $n = \frac{S}{180} - 2$

D. $n = \frac{S}{180} + \frac{1}{90}$

8. The table below shows a linear pattern. Which linear function relates y to x? *(MACC.8.F.1.1)*

x	1	2	3	4	5
y	4	2	0	−2	−4

F. $y = 2x + 2$

G. $y = 4x$

H. $y = -2x + 2$

I. $y = -2x + 6$

9. An airplane flies from City 1 at (0, 0) to City 2 at (33, 56) and then to City 3 at (23, 32). What is the total number of miles it flies? Each unit of the coordinate grid represents 1 mile. *(MACC.8.G.2.8)*

10. What is the missing length of the right triangle shown? *(MACC.8.G.2.7)*

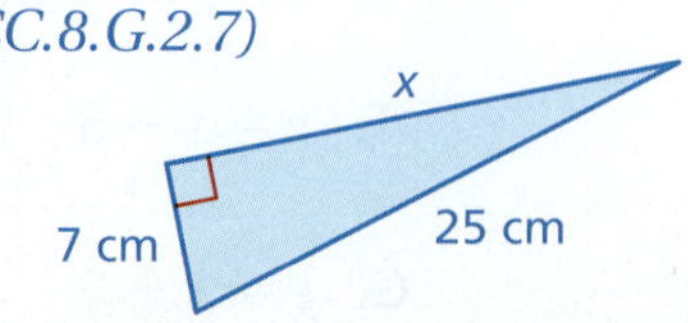

A. 16 cm

B. 18 cm

C. 24 cm

D. $\sqrt{674}$ cm

11. A system of linear equations is shown in the coordinate plane below. What is the solution for this system? *(MACC.8.EE.3.8a)*

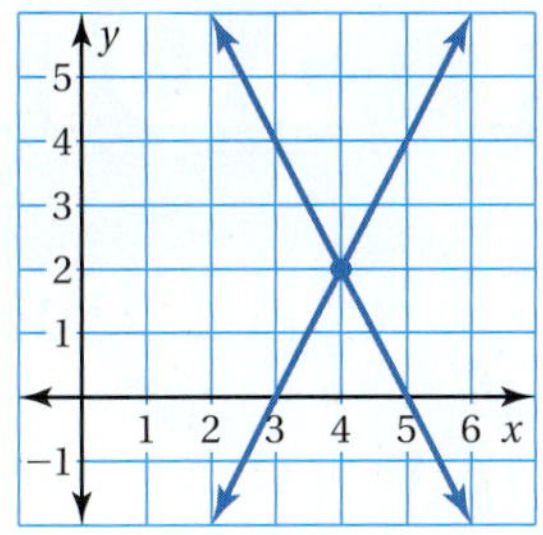

F. (0, 10)

G. (3, 0)

H. (4, 2)

I. (5, 0)

12. In the diagram, lines ℓ and m are parallel. Which angle has the same measure as $\angle 1$? *(MACC.8.G.1.5)*

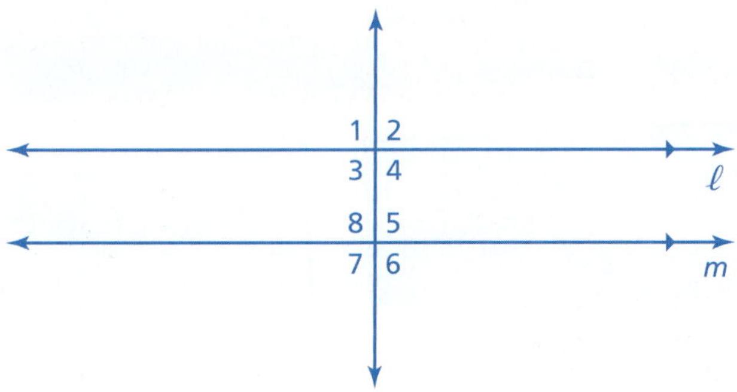

A. $\angle 2$

B. $\angle 5$

C. $\angle 7$

D. $\angle 8$

13. Which graph represents the linear equation $y = -2x - 2$? *(MACC.8.EE.2.6)*

F.

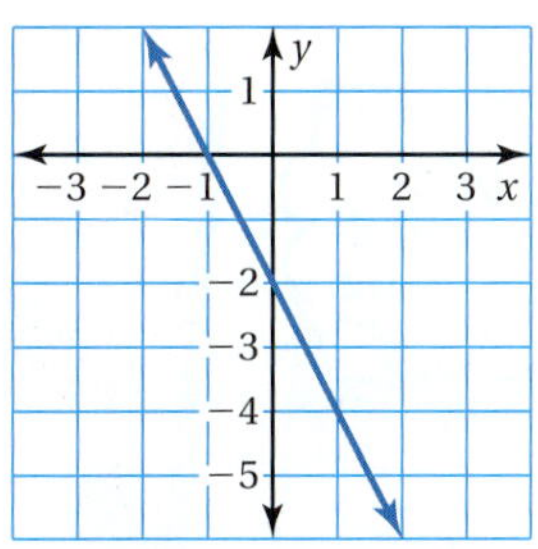

H.

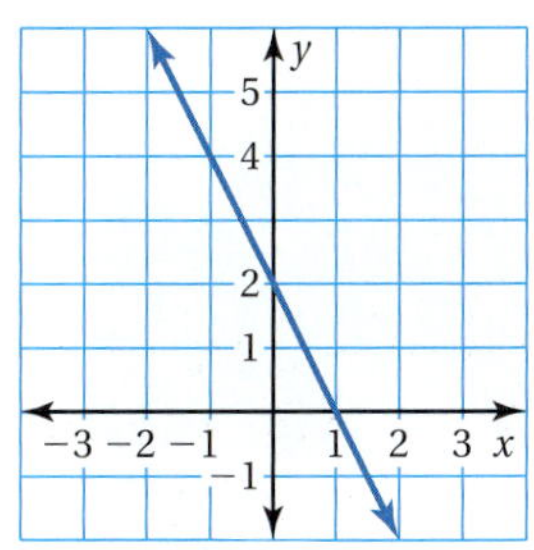

G.

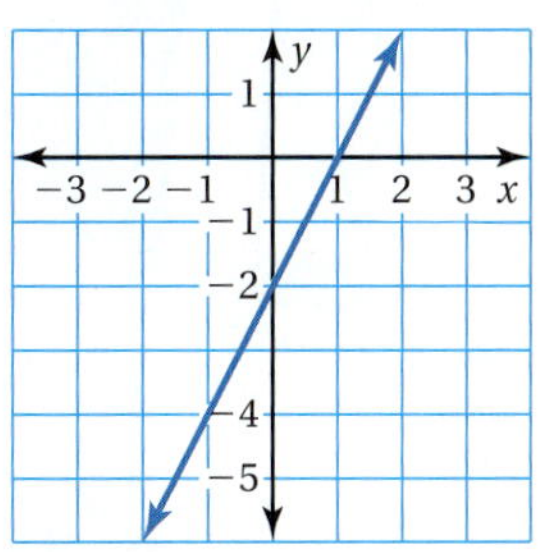

I.

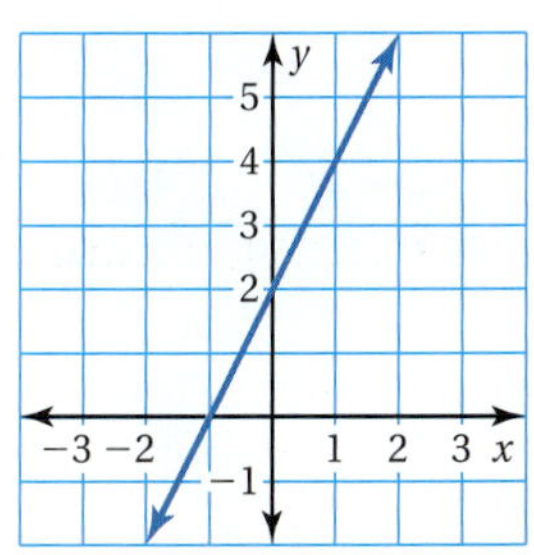

8 Volume and Similar Solids

"Dear Sir: Why do you sell dog food in tall cans and sell cat food in short cans?"

"Neither of these shapes is the optimal use of surface area when compared to volume."

"Do you know why the volume of a cone is one-third the volume of a cylinder with the same height and base?"

What You Learned Before

Finding the Area of a Composite Figure (MACC.7.G.2.6)

Example 1 Find the area of the figure.

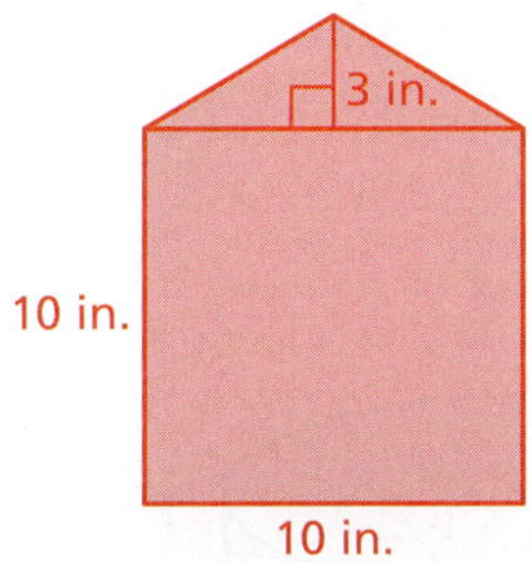

Area = Area of square + Area of triangle

$$A = s^2 + \frac{1}{2}bh$$

$$= 10^2 + \left(\frac{1}{2} \cdot 10 \cdot 3\right)$$

$$= 100 + 15$$

$$= 115 \text{ in.}^2$$

Try It Yourself

Find the area of the figure.

1.

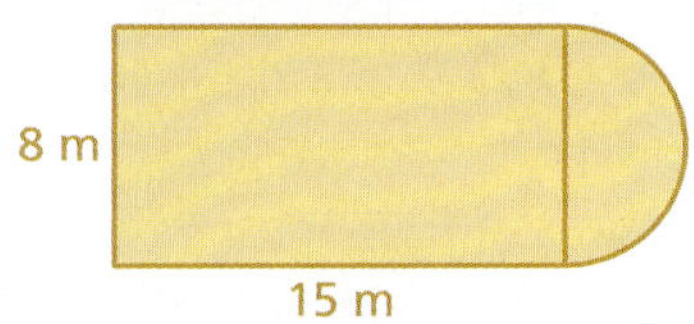

2.

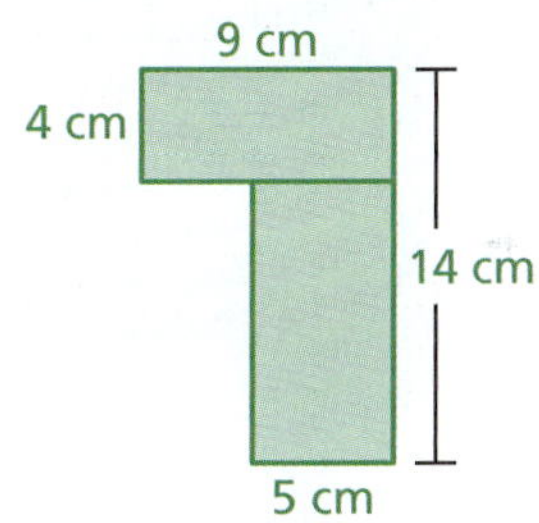

Finding the Areas of Circles (MACC.7.G.2.4)

Example 2 Find the area of the circle.

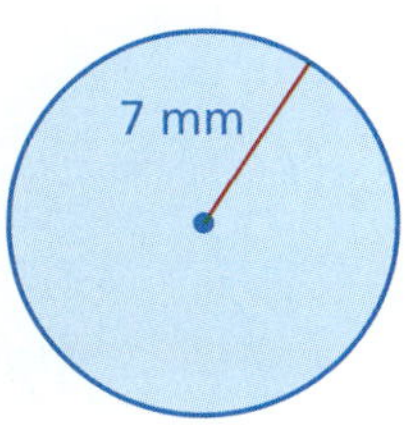

$$A = \pi r^2$$

$$\approx \frac{22}{7} \cdot 7^2$$

$$= \frac{22}{7} \cdot 49$$

$$= 154 \text{ mm}^2$$

Example 3 Find the area of the circle.

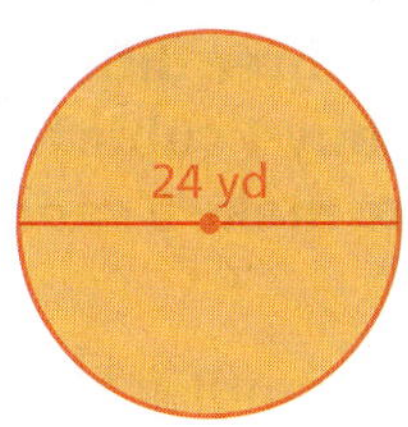

$$A = \pi r^2$$

$$\approx 3.14 \cdot 12^2$$

$$= 3.14 \cdot 144$$

$$= 452.16 \text{ yd}^2$$

Try It Yourself

Find the area of the circle.

3.

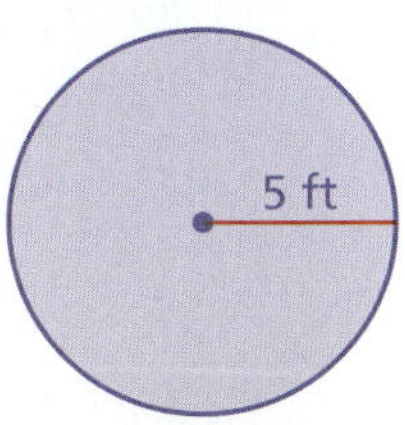

4.

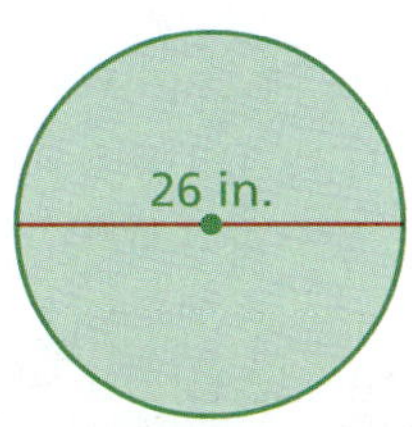

5.

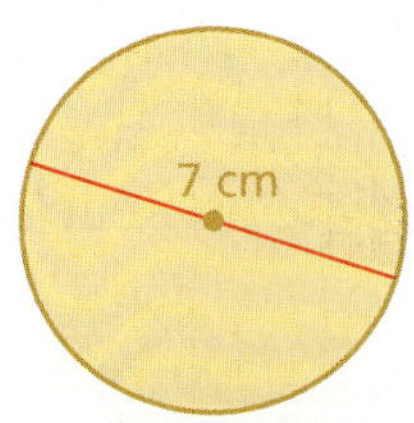

8.1 Volumes of Cylinders

Essential Question How can you find the volume of a cylinder?

1 ACTIVITY: Finding a Formula Experimentally

Work with a partner.

a. Find the area of the face of a coin.

b. Find the volume of a stack of a dozen coins.

c. Write a formula for the volume of a cylinder.

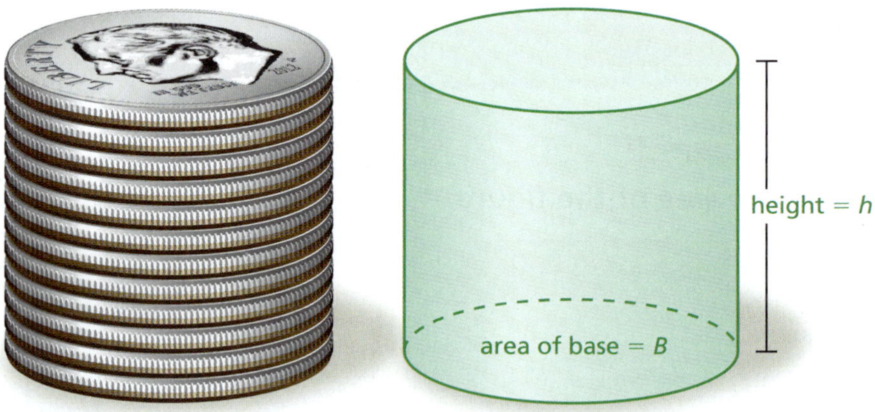

2 ACTIVITY: Making a Business Plan

Work with a partner. You are planning to make and sell three different sizes of cylindrical candles. You buy 1 cubic foot of candle wax for $20 to make 8 candles of each size.

a. Design the candles. What are the dimensions of each size of candle?

b. You want to make a profit of $100. Decide on a price for each size of candle.

c. Did you set the prices so that they are proportional to the volume of each size of candle? Why or why not?

COMMON CORE

Geometry

In this lesson, you will

- find the volumes of cylinders.
- find the heights of cylinders given the volumes.
- solve real-life problems.

Learning Standard
MACC.8.G.3.9

3 ACTIVITY: Science Experiment

Work with a partner. Use the diagram to describe how you can find the volume of a small object.

4 ACTIVITY: Comparing Cylinders

Math Practice 1

Consider Similar Problems

How can you use the results of Activity 1 to find the volumes of the cylinders?

Work with a partner.

a. Just by looking at the two cylinders, which one do you think has the greater volume? Explain your reasoning.

b. Find the volume of each cylinder. Was your prediction in part (a) correct? Explain your reasoning.

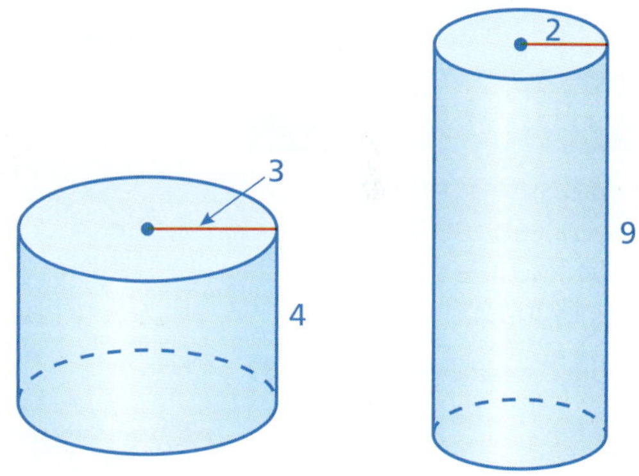

What Is Your Answer?

5. **IN YOUR OWN WORDS** How can you find the volume of a cylinder?

6. Compare your formula for the volume of a cylinder with the formula for the volume of a prism. How are they the same?

"Here's how I remember how to find the volume of <u>any</u> prism or cylinder." "Base times tall, will fill 'em all."

Practice Use what you learned about the volumes of cylinders to complete Exercises 3–5 on page 338.

Check It Out
Lesson Tutorials
BigIdeasMath.com

Key Idea

Volume of a Cylinder

Words The volume V of a cylinder is the product of the area of the base and the height of the cylinder.

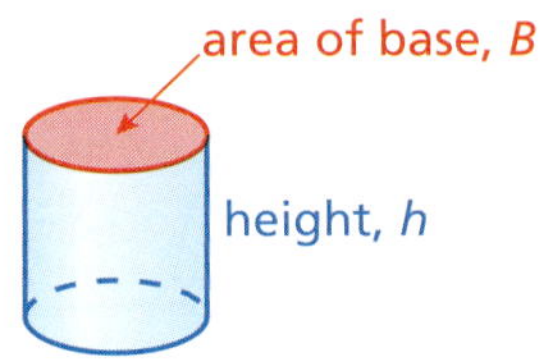

Algebra $V = Bh$

Area of base → B; Height of cylinder → h

EXAMPLE 1 Finding the Volume of a Cylinder

Find the volume of the cylinder. Round your answer to the nearest tenth.

$V = Bh$ Write formula for volume.

$= \pi(3)^2(6)$ Substitute.

$= 54\pi \approx 169.6$ Use a calculator.

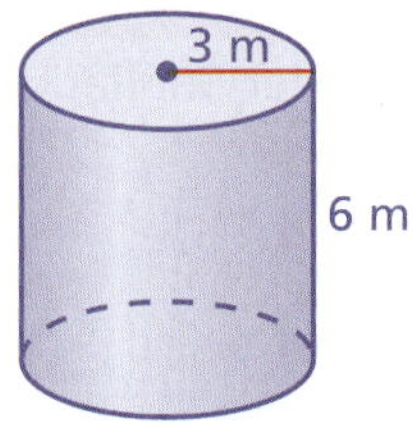

The volume is about 169.6 cubic meters.

Study Tip

Because $B = \pi r^2$, you can use $V = \pi r^2 h$ to find the volume of a cylinder.

EXAMPLE 2 Finding the Height of a Cylinder

Find the height of the cylinder. Round your answer to the nearest whole number.

The diameter is 10 inches. So, the radius is 5 inches.

$V = Bh$ Write formula for volume.

$314 = \pi(5)^2(h)$ Substitute.

$314 = 25\pi h$ Simplify.

$4 \approx h$ Divide each side by 25π.

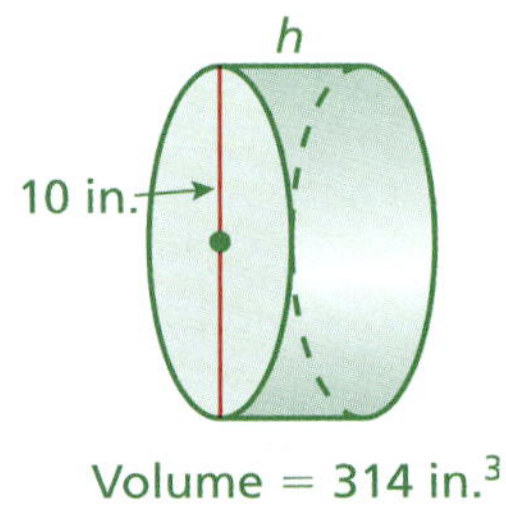

The height is about 4 inches.

On Your Own

Now You're Ready
Exercises 3–11 and 13–15

Find the volume V or height h of the cylinder. Round your answer to the nearest tenth.

1.

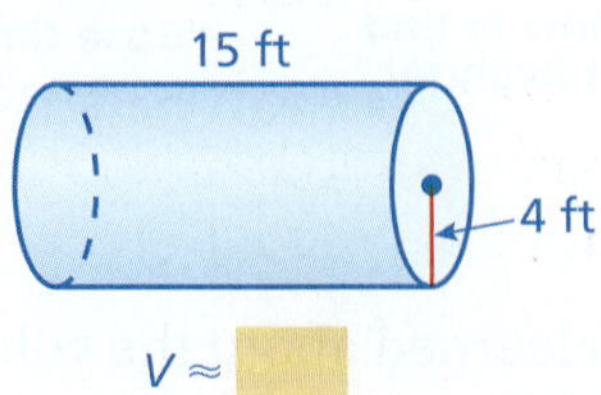

2.

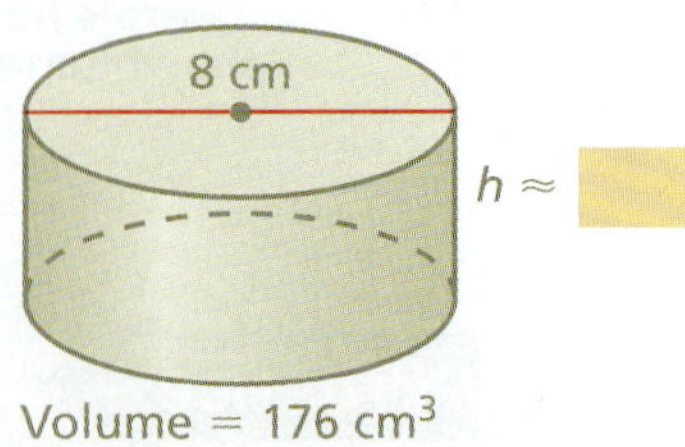

EXAMPLE 3 Real-Life Application

How much salsa is missing from the jar?

The empty space in the jar is a cylinder with a height of $10 - 4 = 6$ centimeters and a radius of 5 centimeters.

$V = Bh$	Write formula for volume.
$= \pi(5)^2(6)$	Substitute.
$= 150\pi \approx 471$	Use a calculator.

So, about 471 cubic centimeters of salsa are missing from the jar.

EXAMPLE 4 Real-Life Application

About how many gallons of water does the watercooler bottle contain? ($1 \text{ ft}^3 \approx 7.5$ gal)

Ⓐ 5.3 gallons Ⓑ 10 gallons Ⓒ 17 gallons Ⓓ 40 gallons

Find the volume of the cylinder. The diameter is 1 foot. So, the radius is 0.5 foot.

$V = Bh$	Write formula for volume.
$= \pi(0.5)^2(1.7)$	Substitute.
$= 0.425\pi \approx 1.3352$	Use a calculator.

So, the bottle contains about 1.3352 cubic feet of water. To find the number of gallons it contains, multiply by the conversion factor $\frac{7.5 \text{ gal}}{1 \text{ ft}^3}$.

$$1.3352 \cancel{\text{ft}^3} \times \frac{7.5 \text{ gal}}{1 \cancel{\text{ft}^3}} \approx 10 \text{ gal}$$

The watercooler bottle contains about 10 gallons of water. So, the correct answer is Ⓑ.

On Your Own

Now You're Ready
Exercise 12

3. **WHAT IF?** In Example 3, the height of the salsa in the jar is 5 centimeters. How much salsa is missing from the jar?

4. A cylindrical water tower has a diameter of 15 meters and a height of 5 meters. About how many gallons of water can the tower contain? ($1 \text{ m}^3 \approx 264$ gal)

8.1 Exercises

Check It Out
Help with Homework
BigIdeasMath.com

Vocabulary and Concept Check

1. **DIFFERENT WORDS, SAME QUESTION** Which is different? Find "both" answers.

How much does it take to fill the cylinder?

What is the capacity of the cylinder?

How much does it take to cover the cylinder?

How much does the cylinder contain?

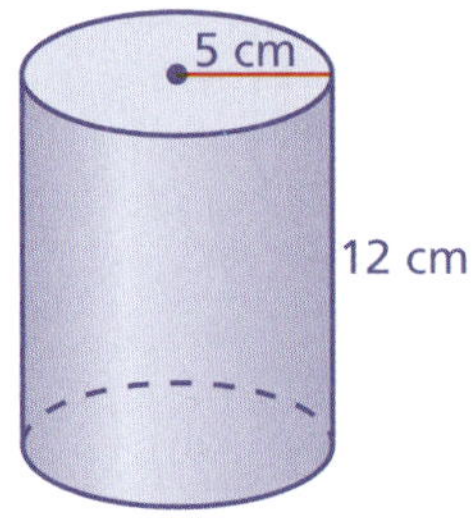

2. **REASONING** Without calculating, which of the solids has the greater volume? Explain.

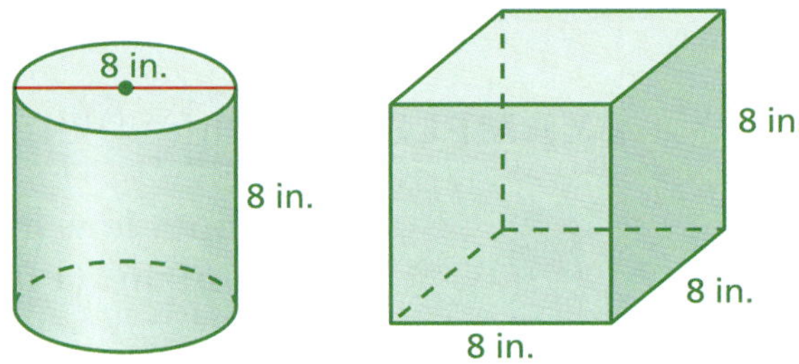

Practice and Problem Solving

Find the volume of the cylinder. Round your answer to the nearest tenth.

(1) 3.

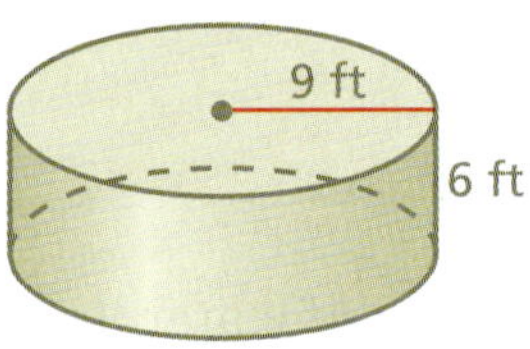

4.

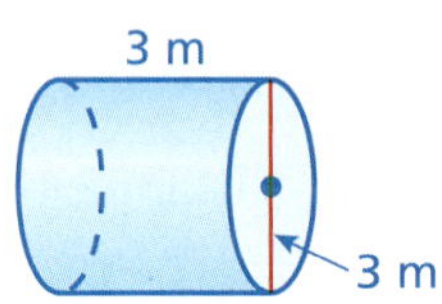

5.

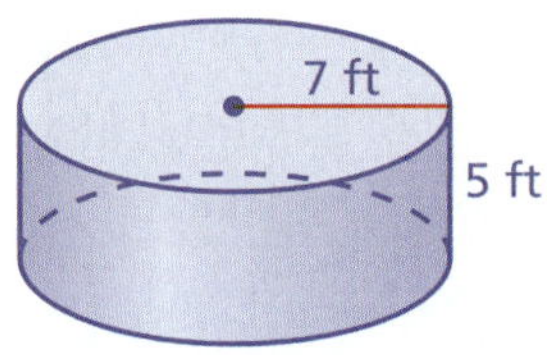

6.

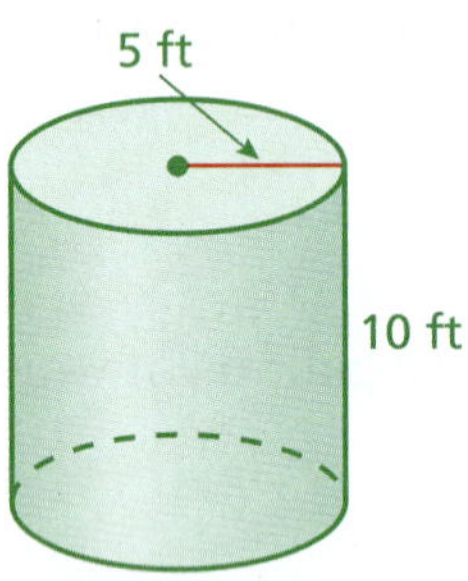

7.

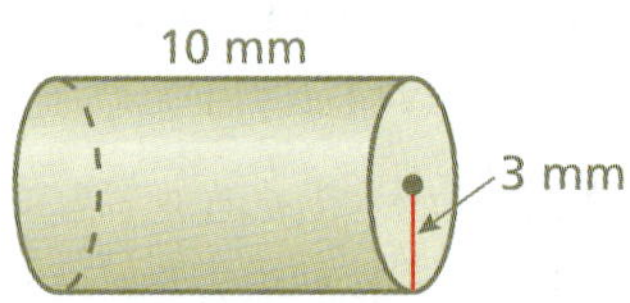

8.

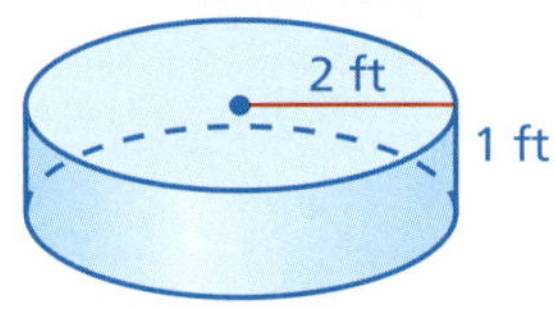

9.

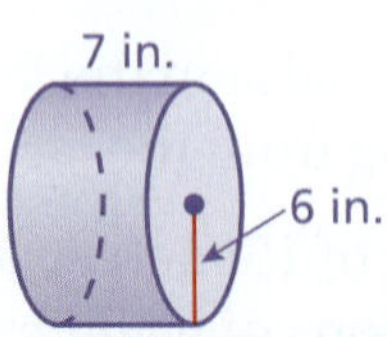

10.

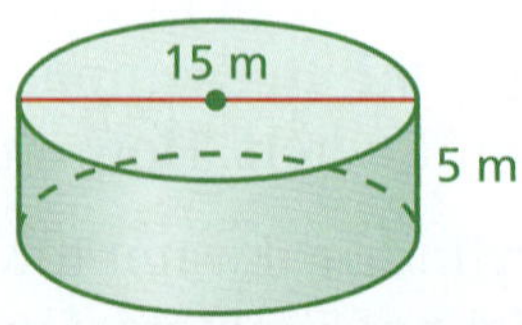

11.

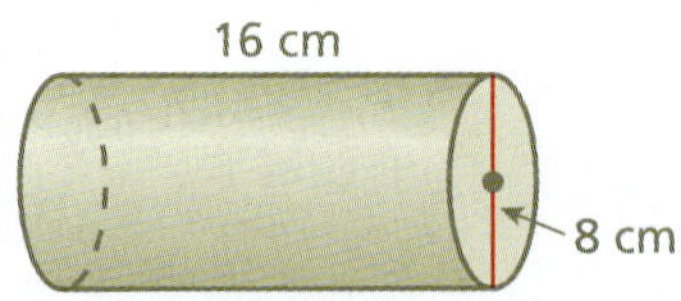

(4) 12. **SWIMMING POOL** A cylindrical swimming pool has a diameter of 16 feet and a height of 4 feet. About how many gallons of water can the pool contain? Round your answer to the nearest whole number. ($1 \text{ ft}^3 \approx 7.5$ gal)

Find the missing dimension of the cylinder. Round your answer to the nearest whole number.

2 **13.** Volume = 250 ft^3

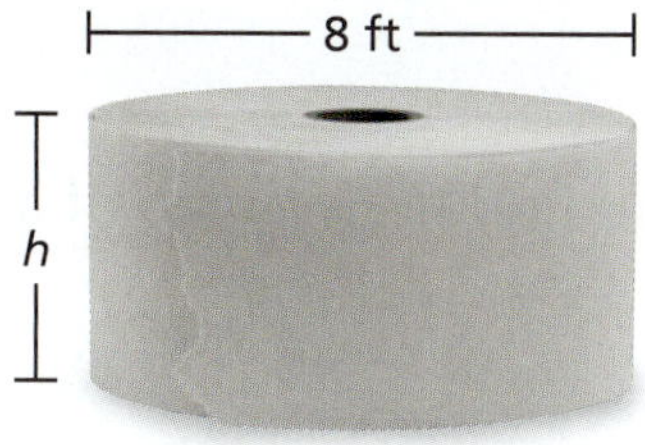

14. Volume = $10{,}000\pi \text{ in.}^3$

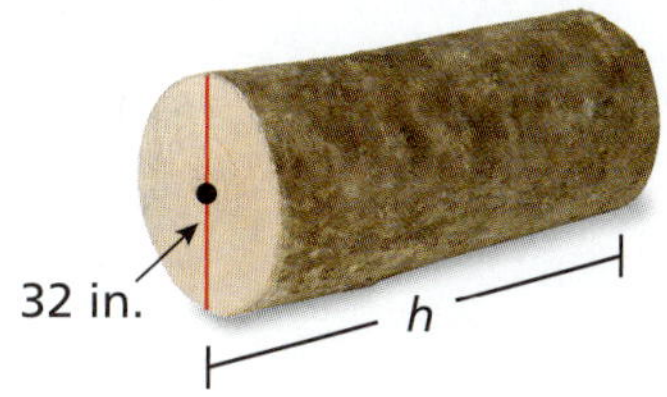

15. Volume = $600{,}000 \text{ cm}^3$

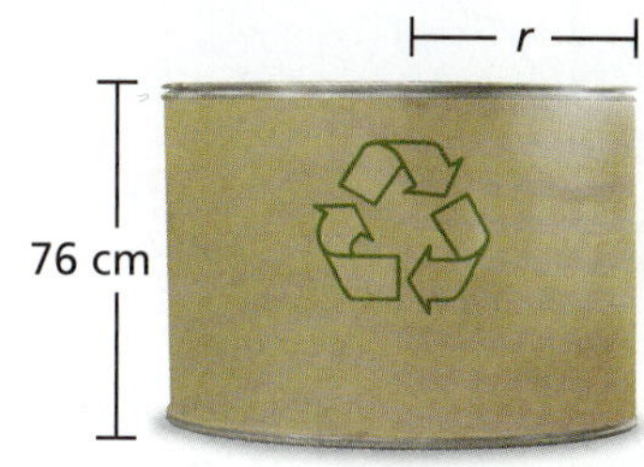

16. CRITICAL THINKING How does the volume of a cylinder change when its diameter is halved? Explain.

Round hay bale

17. MODELING A traditional "square" bale of hay is actually in the shape of a rectangular prism. Its dimensions are 2 feet by 2 feet by 4 feet. How many square bales contain the same amount of hay as one large "round" bale?

18. ROAD ROLLER A tank on a road roller is filled with water to make the roller heavy. The tank is a cylinder that has a height of 6 feet and a radius of 2 feet. One cubic foot of water weighs 62.5 pounds. Find the weight of the water in the tank.

19. VOLUME A cylinder has a surface area of 1850 square meters and a radius of 9 meters. Estimate the volume of the cylinder to the nearest whole number.

20. Problem Solving Water flows at 2 feet per second through a pipe with a diameter of 8 inches. A cylindrical tank with a diameter of 15 feet and a height of 6 feet collects the water.

a. What is the volume, in cubic inches, of water flowing out of the pipe every second?

b. What is the height, in inches, of the water in the tank after 5 minutes?

c. How many minutes will it take to fill 75% of the tank?

Fair Game Review What you learned in previous grades & lessons

Tell whether the triangle with the given side lengths is a right triangle. *(Section 7.5)*

21. 20 m, 21 m, 29 m

22. 1 in., 2.4 in., 2.6 in.

23. 5.6 ft, 8 ft, 10.6 ft

24. MULTIPLE CHOICE Which ordered pair is the solution of the linear system $3x + 4y = -10$ and $2x - 4y = 0$? *(Section 5.3)*

Ⓐ $(-6, 2)$ Ⓑ $(2, -6)$ Ⓒ $(-2, -1)$ Ⓓ $(-1, -2)$

8.2 Volumes of Cones

Essential Question How can you find the volume of a cone?

You already know how the volume of a pyramid relates to the volume of a prism. In this activity, you will discover how the volume of a cone relates to the volume of a cylinder.

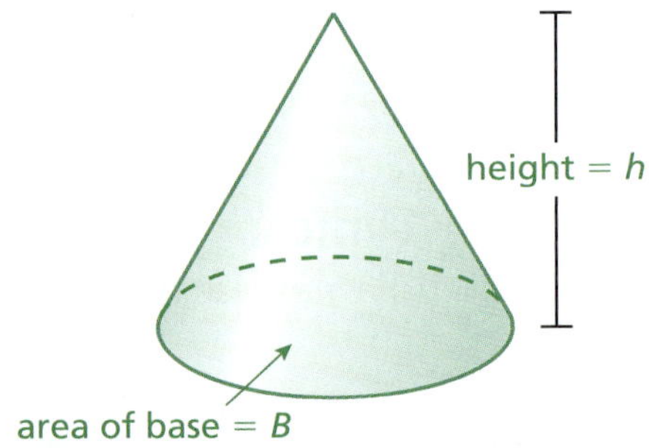

1 ACTIVITY: Finding a Formula Experimentally

Work with a partner. Use a paper cup that is shaped like a cone.

- Estimate the height of the cup.
- Trace the top of the cup on a piece of paper. Find the diameter of the circle.
- Use these measurements to draw a net for a cylinder with the same base and height as the paper cup.
- Cut out the net. Then fold and tape it to form an open cylinder.
- Fill the paper cup with rice. Then pour the rice into the cylinder. Repeat this until the cylinder is full. How many cones does it take to fill the cylinder?
- Use your result to write a formula for the volume of a cone.

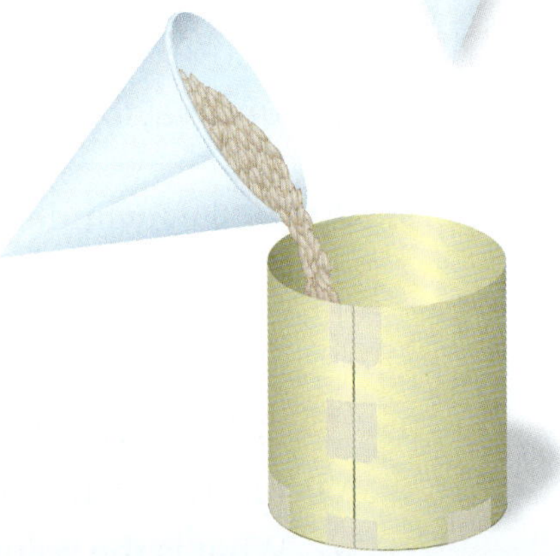

2 ACTIVITY: Summarizing Volume Formulas

Work with a partner. You can remember the volume formulas for prisms, cylinders, pyramids, and cones with just two concepts.

Volumes of Prisms and Cylinders

Volume = Area of base × ______

Volumes of Pyramids and Cones

Volume = ______ Volume of prism or cylinder with same base and height

Make a list of all the formulas you need to remember to find the area of a base. Talk about strategies for remembering these formulas.

COMMON CORE

Geometry

In this lesson, you will

- find the volumes of cones.
- find the heights of cones given the volumes.
- solve real-life problems.

Learning Standard
MACC.8.G.3.9

3 ACTIVITY: Volumes of Oblique Solids

Work with a partner. Think of a stack of paper. When you adjust the stack so that the sides are oblique (slanted), do you change the volume of the stack? If the volume of the stack does not change, then the formulas for volumes of right solids also apply to oblique solids.

Math Practice 2

Use Equations

What equation would you use to find the volume of the oblique solid? Explain.

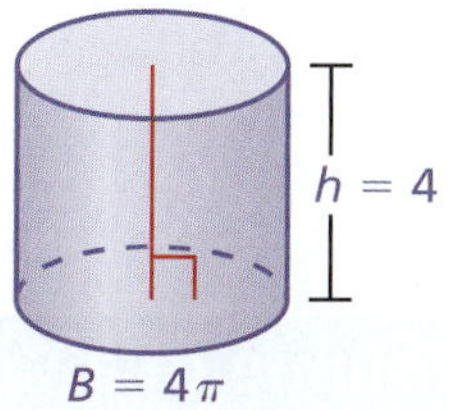

Right cylinder

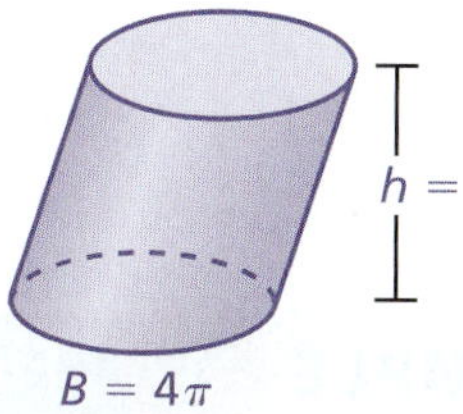

Oblique cylinder

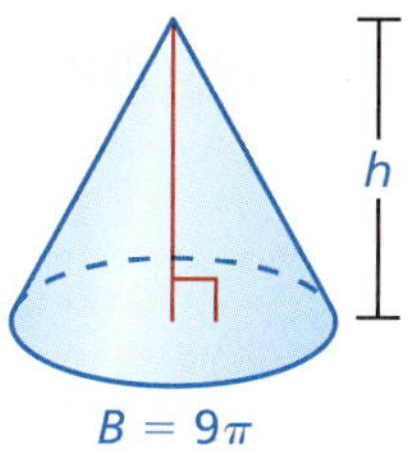

Right cone

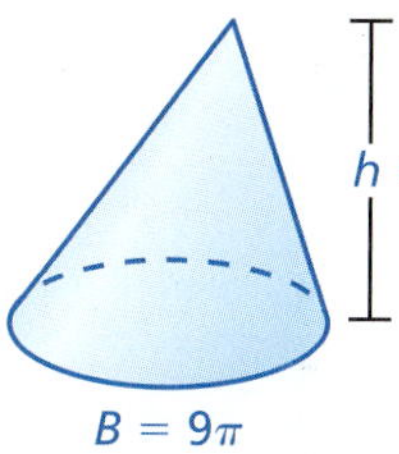

Oblique cone

What Is Your Answer?

4. **IN YOUR OWN WORDS** How can you find the volume of a cone?
5. Describe the intersection of the plane and the cone. Then explain how to find the volume of each section of the solid.

a.

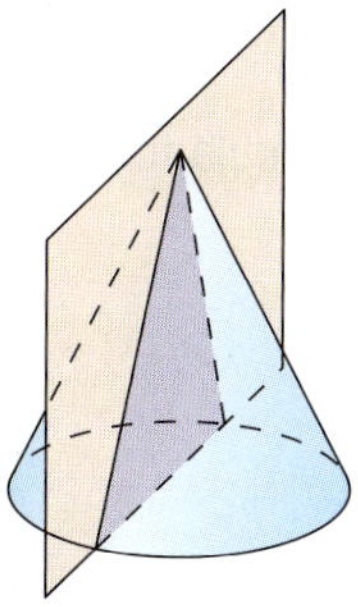

b.

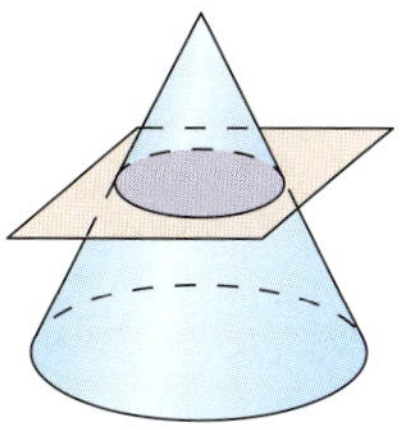

Practice Use what you learned about the volumes of cones to complete Exercises 4–6 on page 344.

8.2 Lesson

Check It Out
Lesson Tutorials
BigIdeasMath.com

Key Idea

Study Tip

The *height* of a cone is the perpendicular distance from the base to the vertex.

Volume of a Cone

Words The volume V of a cone is one-third the product of the area of the base and the height of the cone.

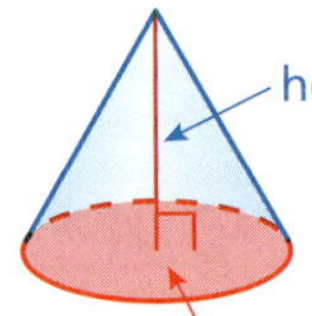

Algebra $V = \frac{1}{3}Bh$

Area of base (B)

Height of cone (h)

EXAMPLE 1 Finding the Volume of a Cone

Study Tip

Because $B = \pi r^2$, you can use $V = \frac{1}{3}\pi r^2 h$ to find the volume of a cone.

Find the volume of the cone. Round your answer to the nearest tenth.

The diameter is 4 meters. So, the radius is 2 meters.

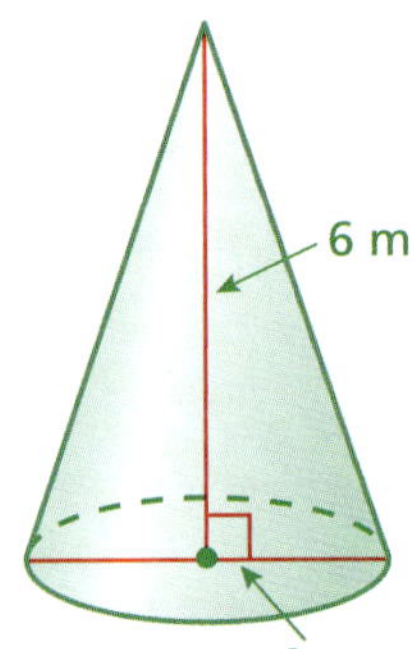

$V = \frac{1}{3}Bh$ — Write formula for volume.

$= \frac{1}{3}\pi(2)^2(6)$ — Substitute.

$= 8\pi \approx 25.1$ — Use a calculator.

The volume is about 25.1 cubic meters.

EXAMPLE 2 Finding the Height of a Cone

Find the height of the cone. Round your answer to the nearest tenth.

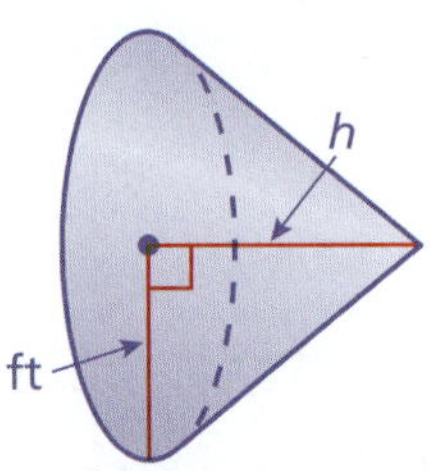

$V = \frac{1}{3}Bh$ — Write formula for volume.

$956 = \frac{1}{3}\pi(9)^2(h)$ — Substitute.

$956 = 27\pi h$ — Simplify.

$11.3 \approx h$ — Divide each side by 27π.

The height is about 11.3 feet.

On Your Own

Now You're Ready
Exercises 4–12 and 15–17

Find the volume V or height h of the cone. Round your answer to the nearest tenth.

1.

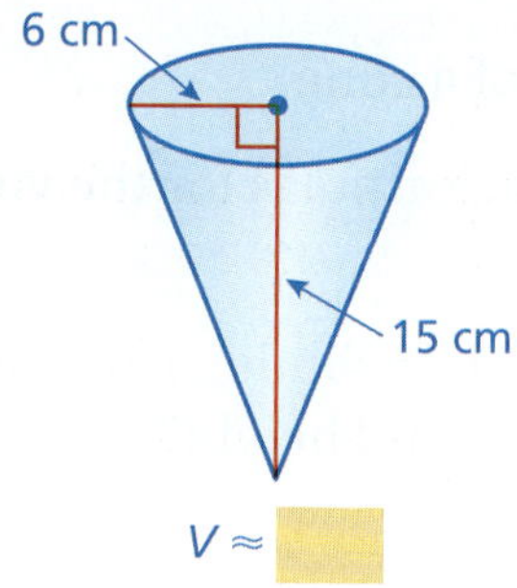

2.

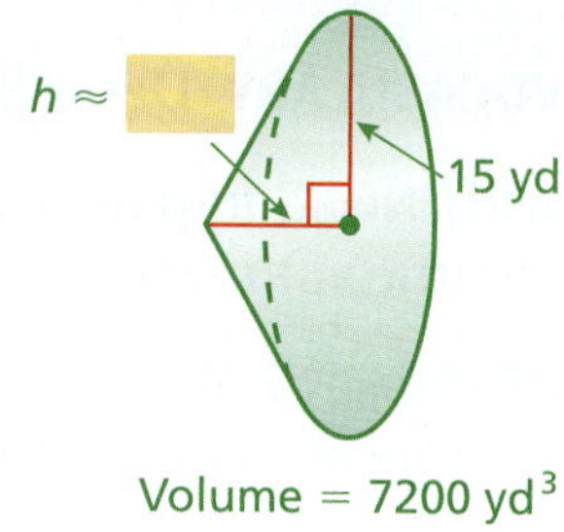

EXAMPLE 3 Real-Life Application

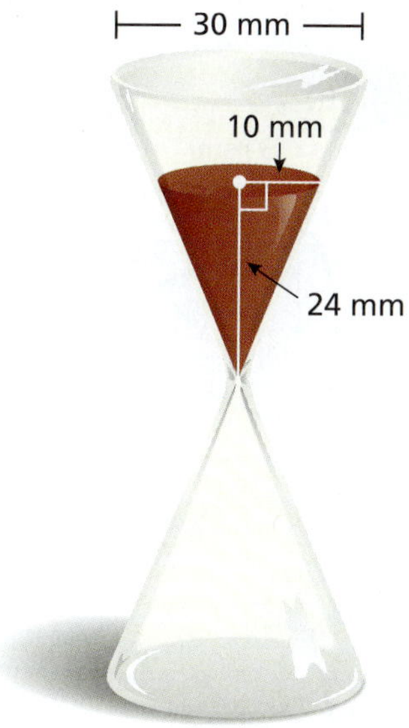

You must answer a trivia question before the sand in the timer falls to the bottom. The sand falls at a rate of 50 cubic millimeters per second. How much time do you have to answer the question?

Use the formula for the volume of a cone to find the volume of the sand in the timer.

$$V = \frac{1}{3}Bh \qquad \text{Write formula for volume.}$$

$$= \frac{1}{3}\pi(10)^2(24) \qquad \text{Substitute.}$$

$$= 800\pi \approx 2513 \qquad \text{Use a calculator.}$$

The volume of the sand is about 2513 cubic millimeters. To find the amount of time you have to answer the question, multiply the volume by the rate at which the sand falls.

$$2513 \text{ mm}^3 \times \frac{1 \text{ sec}}{50 \text{ mm}^3} = 50.26 \text{ sec}$$

So, you have about 50 seconds to answer the question.

On Your Own

3. **WHAT IF?** The sand falls at a rate of 60 cubic millimeters per second. How much time do you have to answer the question?

4. **WHAT IF?** The height of the sand in the timer is 12 millimeters, and the radius is 5 millimeters. How much time do you have to answer the question?

8.2 Exercises

Check It Out
Help with Homework
BigIdeasMath.com

Vocabulary and Concept Check

1. **VOCABULARY** Describe the height of a cone.
2. **WRITING** Compare and contrast the formulas for the volume of a pyramid and the volume of a cone.
3. **REASONING** You know the volume of a cylinder. How can you find the volume of a cone with the same base and height?

Practice and Problem Solving

Find the volume of the cone. Round your answer to the nearest tenth.

1 **4.**

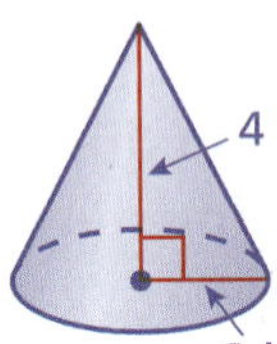

5.

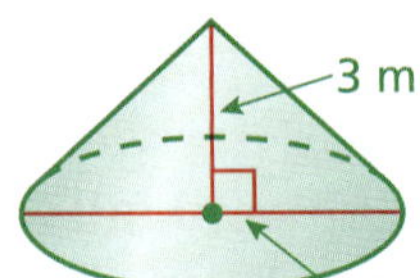

6.

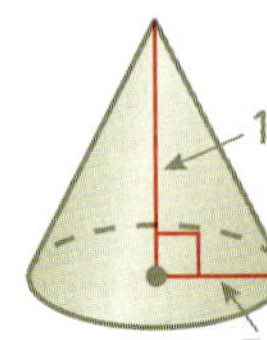

7.

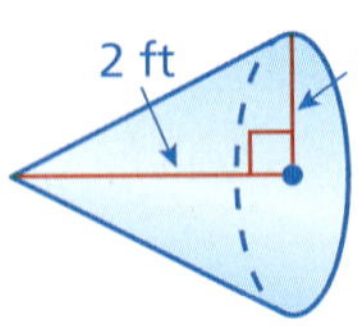

8.

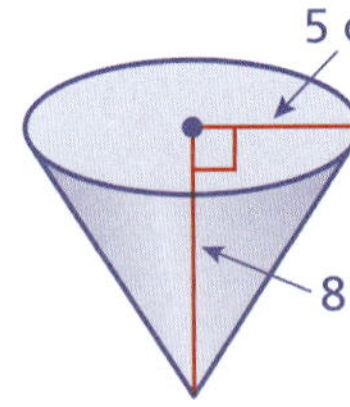

9.

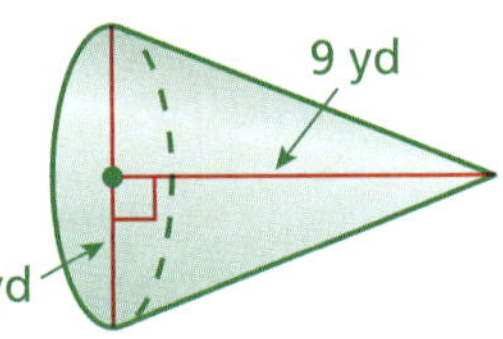

10.

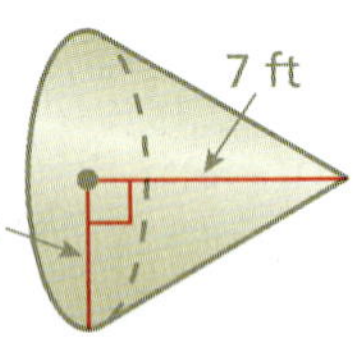

11.

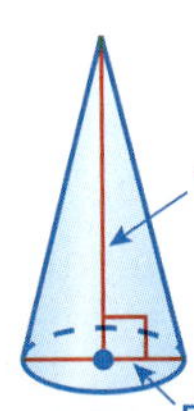

12.

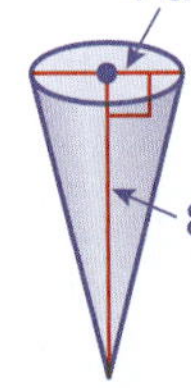

13. **ERROR ANALYSIS** Describe and correct the error in finding the volume of the cone.

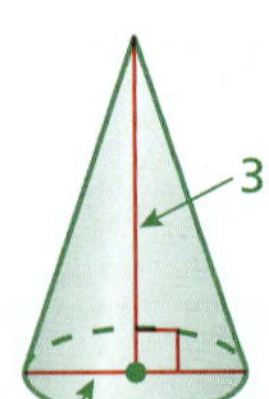

✗
$$V = \frac{1}{3}Bh$$
$$= \frac{1}{3}(\pi)(2)^2(3)$$
$$= 4\pi \text{ m}^3$$

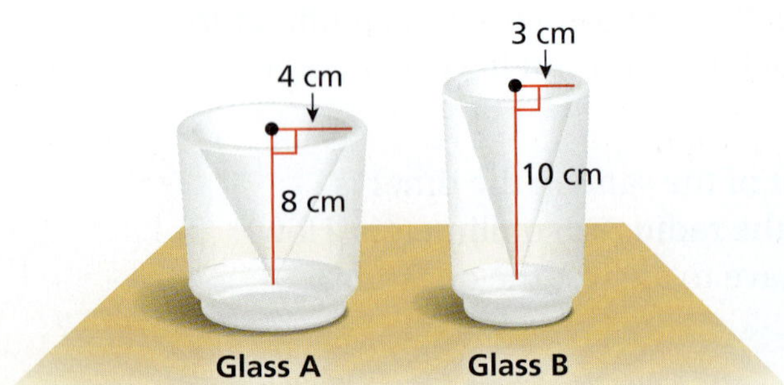

14. **GLASS** The inside of each glass is shaped like a cone. Which glass can hold more liquid? How much more?

Find the missing dimension of the cone. Round your answer to the nearest tenth.

2 **15.** Volume = $\frac{1}{18}\pi$ ft^3

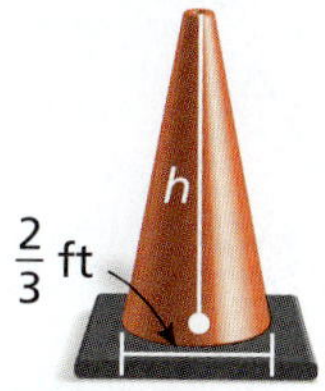

16. Volume = 225 cm^3

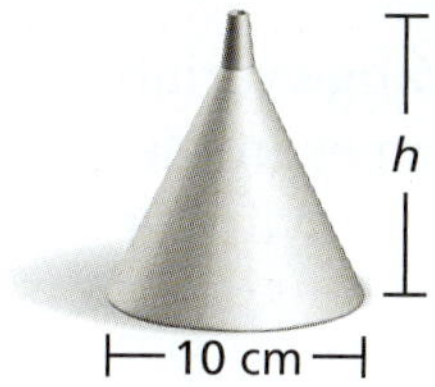

17. Volume = 3.6 in.^{3}

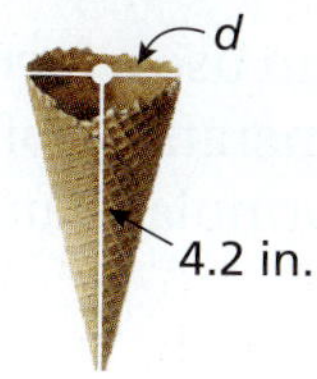

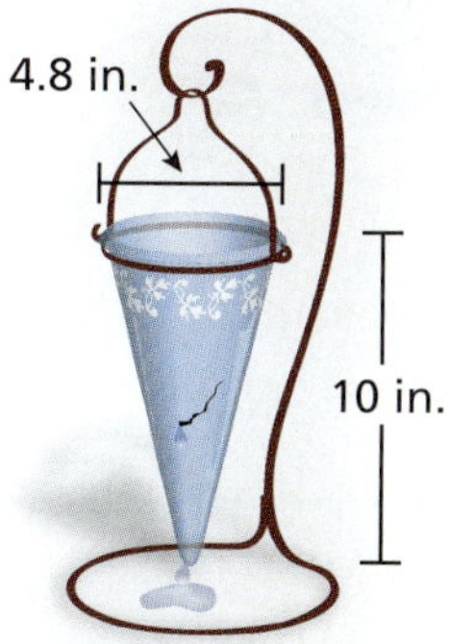

18. **REASONING** The volume of a cone is 20π cubic meters. What is the volume of a cylinder with the same base and height?

19. **VASE** Water leaks from a crack in a vase at a rate of 0.5 cubic inch per minute. How long does it take for 20% of the water to leak from a full vase?

20. **LEMONADE STAND** You have 10 gallons of lemonade to sell. (1 gal ≈ 3785 cm^3)

a. Each customer uses one paper cup. How many paper cups will you need?

b. The cups are sold in packages of 50. How many packages should you buy?

c. How many cups will be left over if you sell 80% of the lemonade?

21. **STRUCTURE** The cylinder and the cone have the same volume. What is the height of the cone?

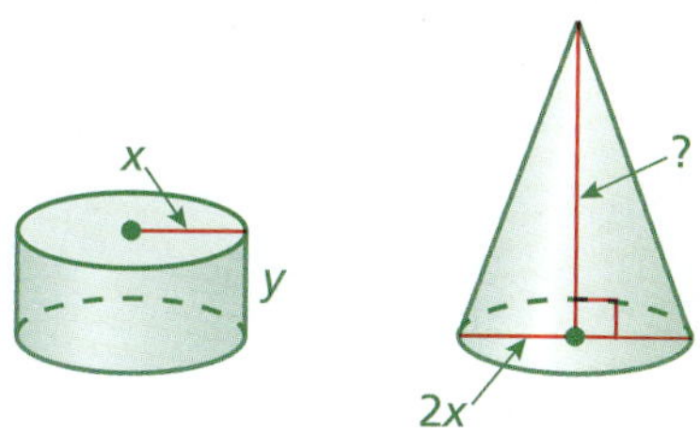

22. Critical Thinking In Example 3, you use a different timer with the same dimensions. The sand in this timer has a height of 30 millimeters. How much time do you have to answer the question?

Fair Game Review What you learned in previous grades & lessons

The vertices of a figure are given. Rotate the figure as described. Find the coordinates of the image. *(Section 2.4)*

23. $A(-1, 1)$, $B(2, 3)$, $C(2, 1)$
90° counterclockwise about vertex A

24. $E(-4, 1)$, $F(-3, 3)$, $G(-2, 3)$, $H(-1, 1)$
180° about the origin

25. **MULTIPLE CHOICE** $\triangle ABC \sim \triangle XYZ$ by a scale factor of 3. How many times greater is the area of $\triangle XYZ$ than the area of $\triangle ABC$? *(Section 2.6)*

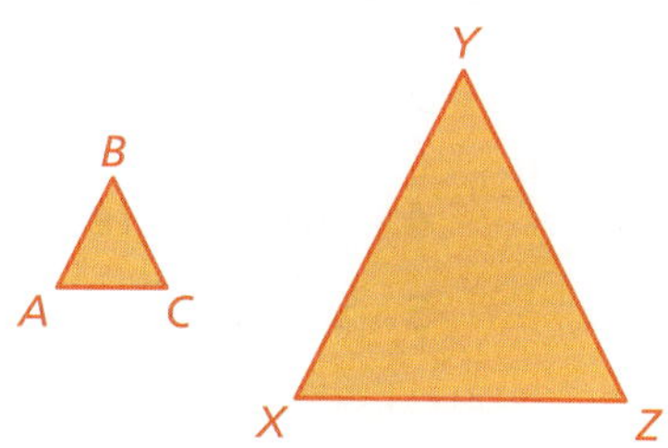

Ⓐ $\frac{1}{9}$ Ⓑ $\frac{1}{3}$

Ⓒ 3 Ⓓ 9

8 Study Help

Check It Out
Graphic Organizer
BigIdeasMath.com

You can use a **formula triangle** to arrange variables and operations of a formula. Here is an example of a formula triangle for the volume of a cylinder.

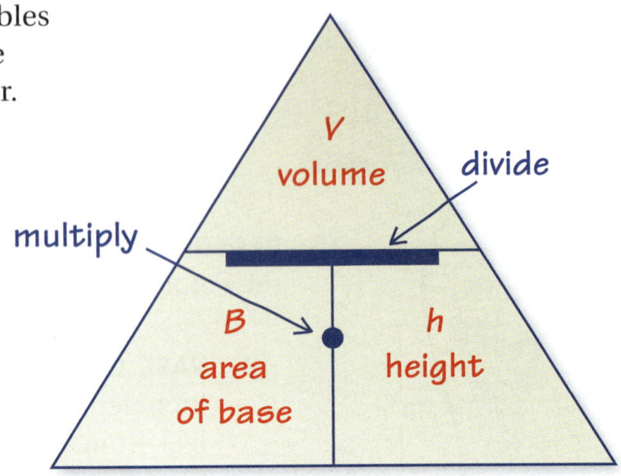

To find an unknown variable, use the other variables and the operation between them. For example, to find the area B of the base, cover up the B. Then you can see that you divide the volume V by the height h.

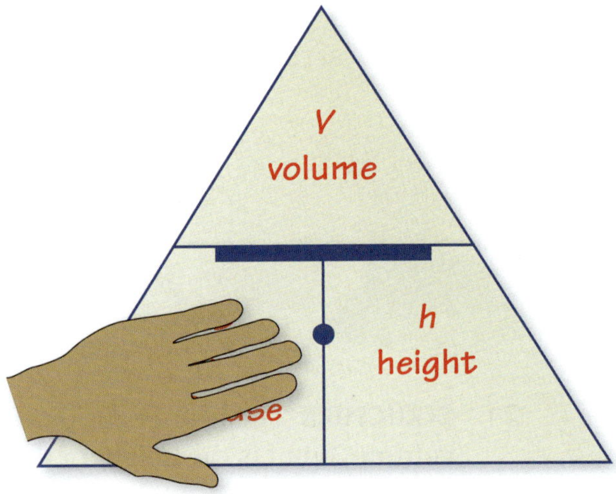

On Your Own

Make a formula triangle to help you study this topic. (*Hint:* Your formula triangle may have a different form than what is shown in the example.)

1. volume of a cone

After you complete this chapter, make formula triangles for the following topics.

2. volume of a sphere
3. volume of a composite solid
4. surface areas of similar solids
5. volumes of similar solids

"See how a formula triangle works? Cover any variable and you get its formula."

8.1–8.2 Quiz

Check It Out
Progress Check
BigIdeasMath.com

Find the volume of the solid. Round your answer to the nearest tenth. *(Section 8.1 and Section 8.2)*

1.

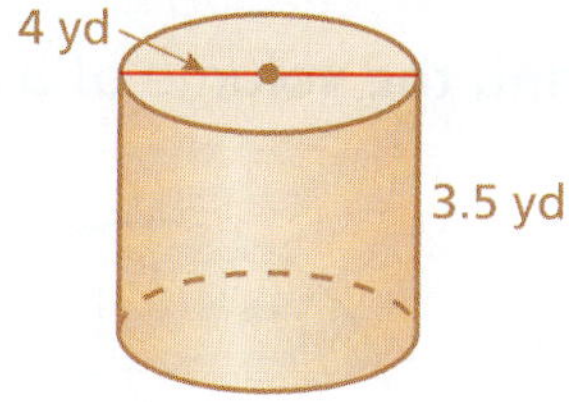

2.

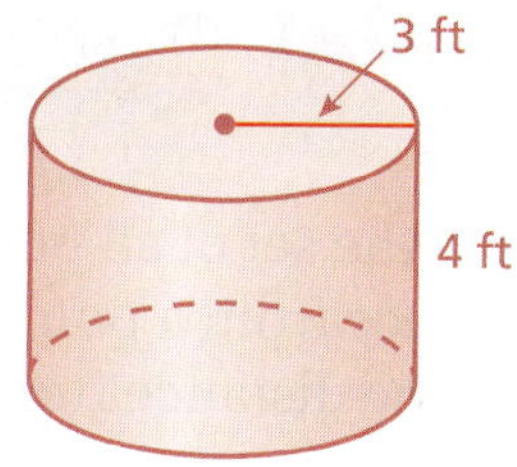

3.

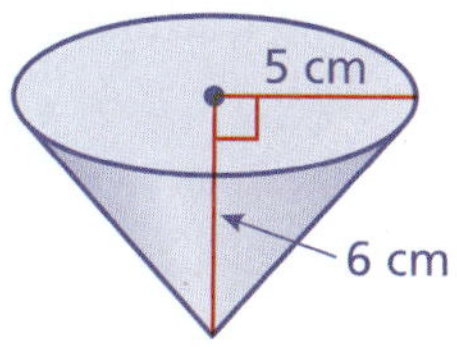

4.

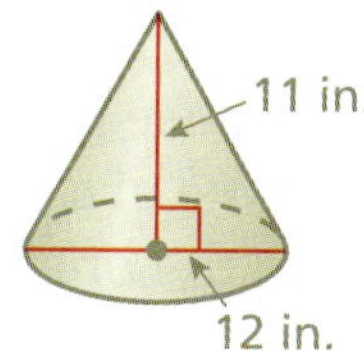

Find the missing dimension of the solid. Round your answer to the nearest tenth. *(Section 8.1 and Section 8.2)*

5.

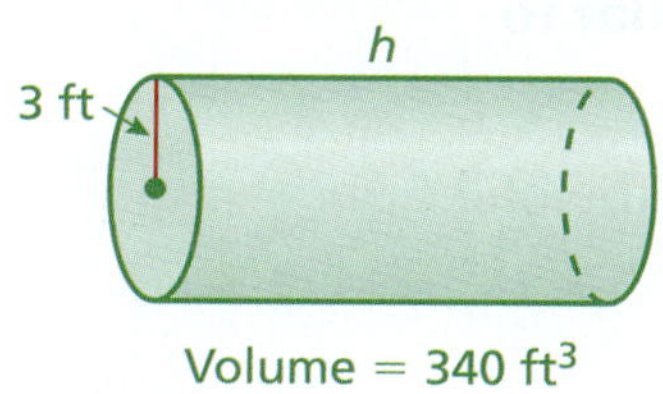

6.

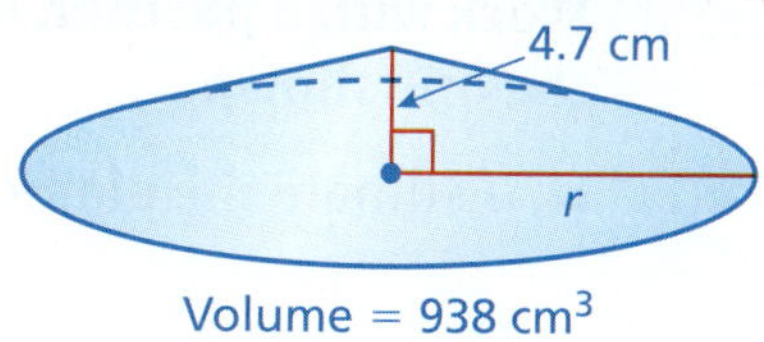

7. **PAPER CONE** The paper cone can hold 84.78 cubic centimeters of water. What is the height of the cone? *(Section 8.2)*

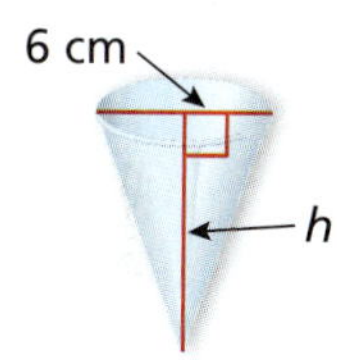

8. **GEOMETRY** Triple both dimensions of the cylinder. How many times greater is the volume of the new cylinder than the volume of the original cylinder? *(Section 8.1)*

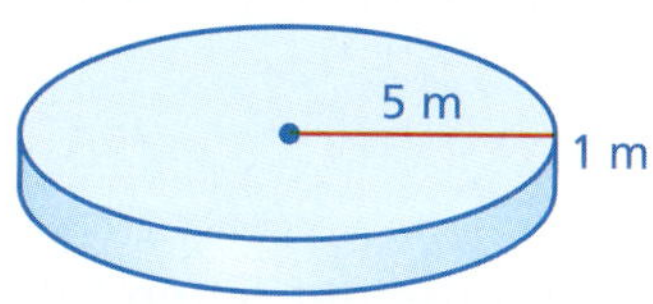

9. **SAND ART** There are 42.39 cubic inches of blue sand and 28.26 cubic inches of red sand in the cylindrical container. How many cubic inches of white sand are in the container? *(Section 8.1)*

10. **JUICE CAN** You are buying two cylindrical cans of juice. Each can holds the same amount of juice. What is the height of Can B? *(Section 8.1)*

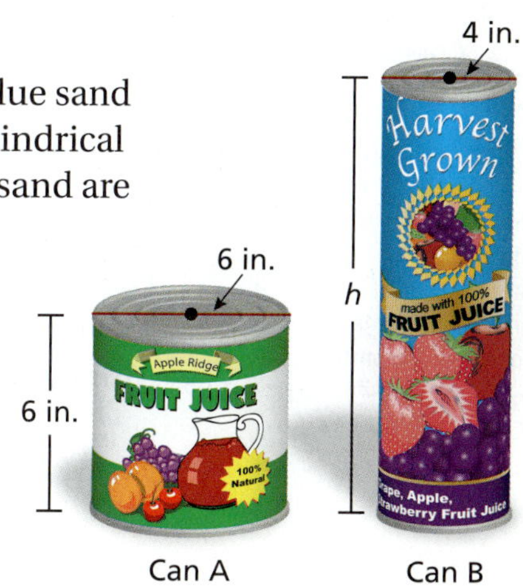

8.3 Volumes of Spheres

Essential Question How can you find the volume of a sphere?

A **sphere** is the set of all points in space that are the same distance from a point called the *center*. The *radius r* is the distance from the center to any point on the sphere.

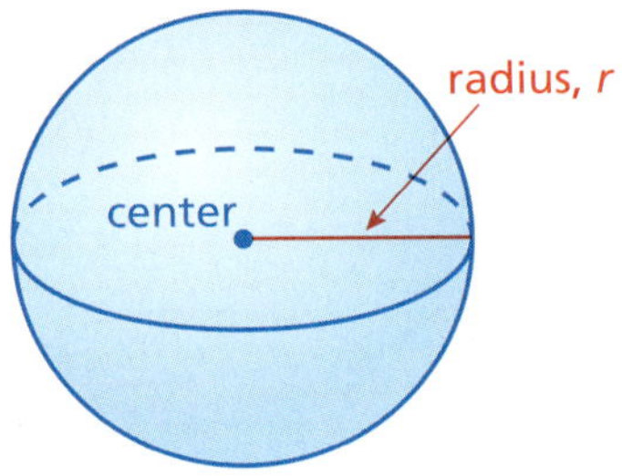

A sphere is different from the other solids you have studied so far because it does not have a base. To discover the volume of a sphere, you can use an activity similar to the one in the previous section.

1 ACTIVITY: Exploring the Volume of a Sphere

Work with a partner. Use a plastic ball similar to the one shown.

- Estimate the diameter and the radius of the ball.

- Use these measurements to draw a net for a cylinder with a diameter and a height equal to the diameter of the ball. How is the height *h* of the cylinder related to the radius *r* of the ball? Explain.

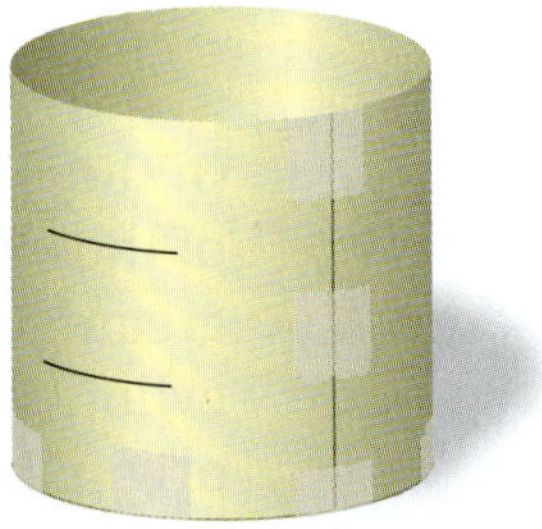

- Cut out the net. Then fold and tape it to form an open cylinder. Make two marks on the cylinder that divide it into thirds, as shown.

- Cover the ball with aluminum foil or tape. Leave one hole open. Fill the ball with rice. Then pour the rice into the cylinder. What fraction of the cylinder is filled with rice?

COMMON CORE

Geometry

In this lesson, you will
- find the volumes of spheres.
- find the radii of spheres given the volumes.
- solve real-life problems.

Learning Standard
MACC.8.G.3.9

2 ACTIVITY: Deriving the Formula for the Volume of a Sphere

Work with a partner. Use the results from Activity 1 and the formula for the volume of a cylinder to complete the steps.

> **Math Practice 4**
>
> **Analyze Relationships**
>
> What is the relationship between the volume of a sphere and the volume of a cylinder? How does this help you derive a formula for the volume of a sphere?

$V = \pi r^2 h$	Write formula for volume of a cylinder.
$= \dfrac{\square}{\square} \pi r^2 h$	Multiply by $\dfrac{\square}{\square}$ because the volume of a sphere is $\dfrac{\square}{\square}$ of the volume of the cylinder.
$= \dfrac{\square}{\square} \pi r^2 \square$	Substitute $\square$ for h.
$= \dfrac{\square}{\square} \pi \square$	Simplify.

3 ACTIVITY: Deriving the Formula for the Volume of a Sphere

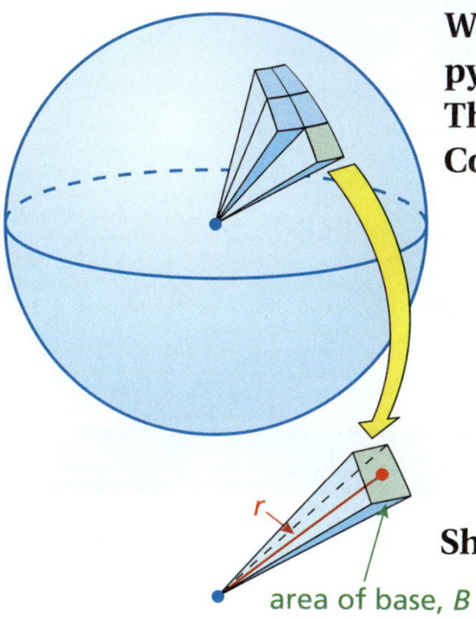

Work with a partner. Imagine filling the inside of a sphere with n small pyramids. The vertex of each pyramid is at the center of the sphere. The height of each pyramid is approximately equal to r, as shown. Complete the steps. (The surface area of a sphere is equal to $4\pi r^2$.)

$V = \dfrac{1}{3}Bh$	Write formula for volume of a pyramid.
$= n\dfrac{1}{3}B \square$	Multiply by the number of small pyramids n and substitute $\square$ for h.
$= \dfrac{1}{3}(4\pi r^2) \square$	$4\pi r^2 \approx n \cdot \square$

Show how this result is equal to the result in Activity 2.

What Is Your Answer?

4. **IN YOUR OWN WORDS** How can you find the volume of a sphere?

5. Describe the intersection of the plane and the sphere. Then explain how to find the volume of each section of the solid.

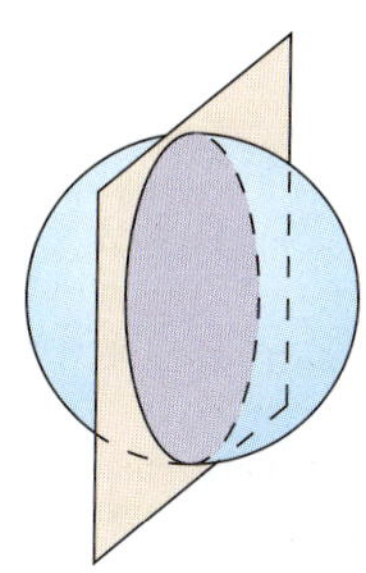

Practice Use what you learned about the volumes of spheres to complete Exercises 3–5 on page 352.

Check It Out
Lesson Tutorials
BigIdeasMath.com

Key Vocabulary
sphere, *p. 348*
hemisphere, *p. 351*

Key Idea

Volume of a Sphere

Words The volume V of a sphere is the product of $\frac{4}{3}\pi$ and the cube of the radius of the sphere.

Algebra $V = \frac{4}{3}\pi r^3$ ← Cube of radius of sphere

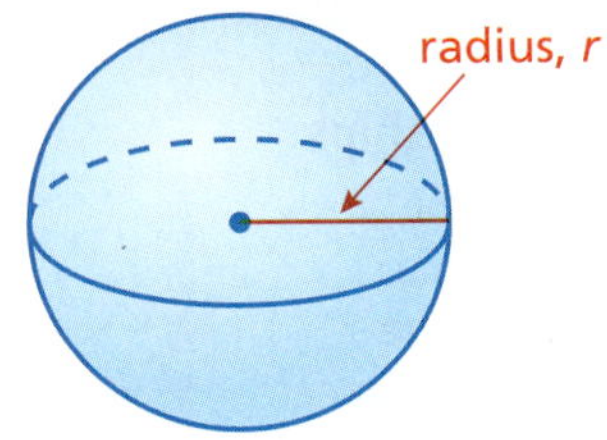

EXAMPLE 1 Finding the Volume of a Sphere

Find the volume of the sphere. Round your answer to the nearest tenth.

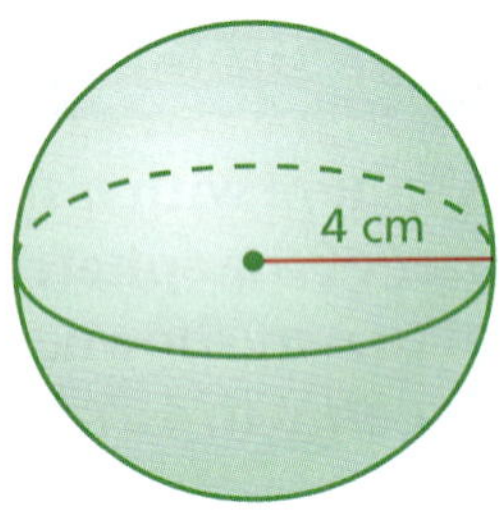

$V = \frac{4}{3}\pi r^3$ — Write formula for volume.

$= \frac{4}{3}\pi (4)^3$ — Substitute 4 for r.

$= \frac{256}{3}\pi$ — Simplify.

≈ 268.1 — Use a calculator.

The volume is about 268.1 cubic centimeters.

EXAMPLE 2 Finding the Radius of a Sphere

Find the radius of the sphere.

Volume = 288π in.^{3}

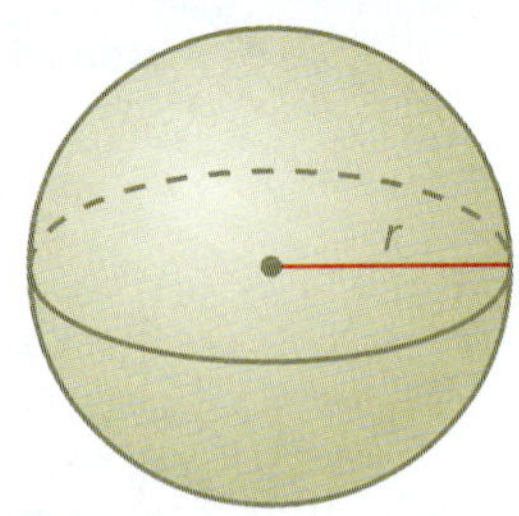

$V = \frac{4}{3}\pi r^3$ — Write formula.

$288\pi = \frac{4}{3}\pi r^3$ — Substitute.

$288\pi = \frac{4\pi}{3} r^3$ — Multiply.

$\frac{3}{4\pi} \cdot 288\pi = \frac{3}{4\pi} \cdot \frac{4\pi}{3} r^3$ — Multiplication Property of Equality

$216 = r^3$ — Simplify.

$6 = r$ — Take the cube root of each side.

The radius is 6 inches.

Multi-Language Glossary at BigIdeasMath.com

On Your Own

Now You're Ready
Exercises 3–11

Find the volume V or radius r of the sphere. Round your answer to the nearest tenth, if necessary.

1.

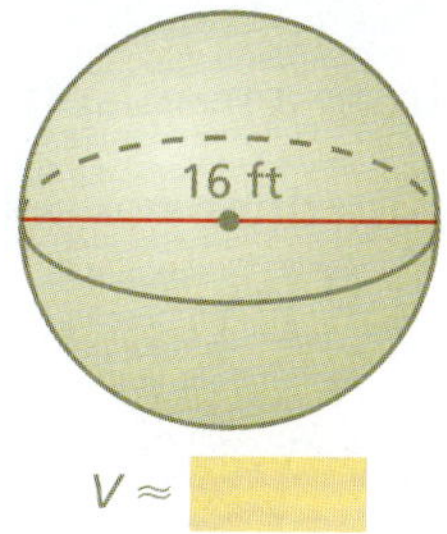

2.

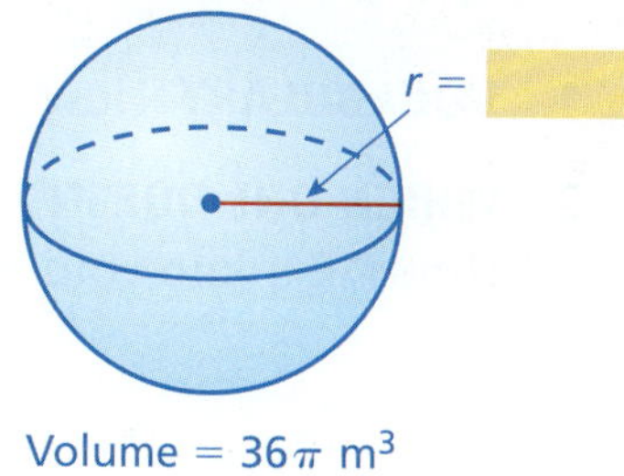

EXAMPLE 3 Finding the Volume of a Composite Solid

A hemisphere is one-half of a sphere. The top of the silo is a hemisphere with a radius of 12 feet. What is the volume of the silo? Round your answer to the nearest thousand.

The silo is made up of a cylinder and a hemisphere. Find the volume of each solid.

Cylinder

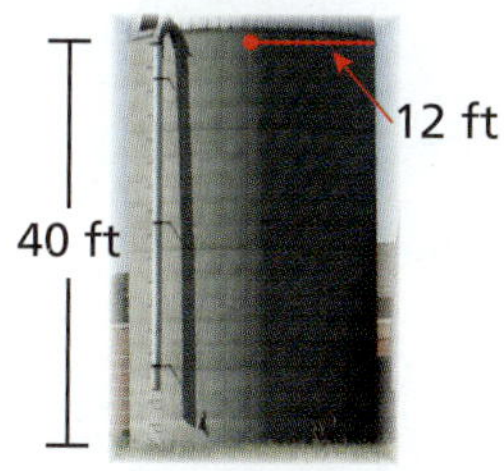

$V = Bh$

$= \pi(12)^2(40)$

$= 5760\pi$

Hemisphere

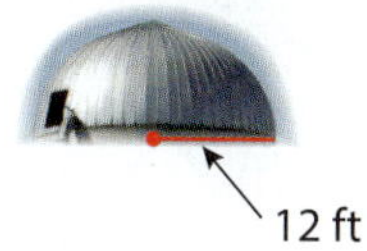

$V = \frac{1}{2} \cdot \frac{4}{3}\pi r^3$

$= \frac{1}{2} \cdot \frac{4}{3}\pi(12)^3$

$= 1152\pi$

So, the volume is $5760\pi + 1152\pi = 6912\pi \approx 22{,}000$ cubic feet.

Study Tip

In Example 3, the height of the cylindrical part of the silo is the difference of the silo height and the radius of the hemisphere.

52 − 12 = 40 ft

On Your Own

Now You're Ready
Exercises 14–16

Find the volume of the composite solid. Round your answer to the nearest tenth.

3.

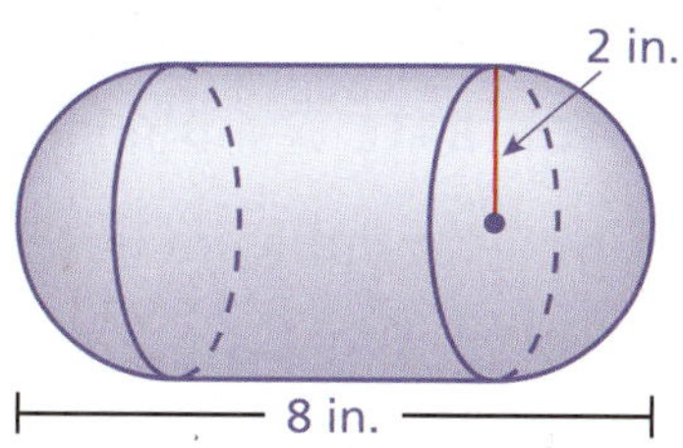

4.

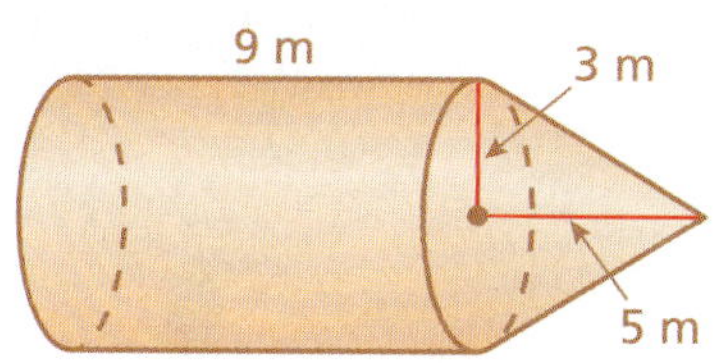

8.3 Exercises

Check It Out
Help with Homework
BigIdeasMath.com

Vocabulary and Concept Check

1. **VOCABULARY** How is a sphere different from a hemisphere?

2. **WHICH ONE DOESN'T BELONG?** Which figure does *not* belong with the other three? Explain your reasoning.

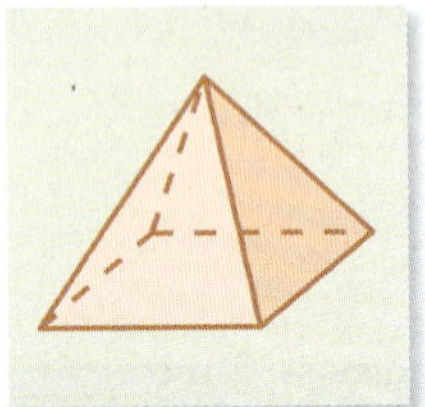
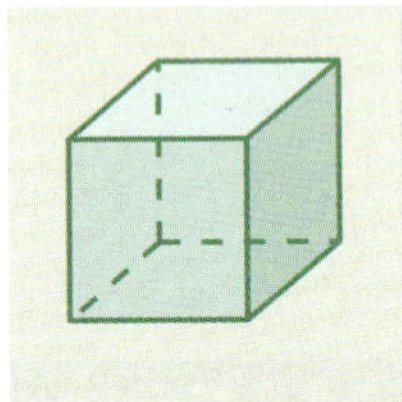
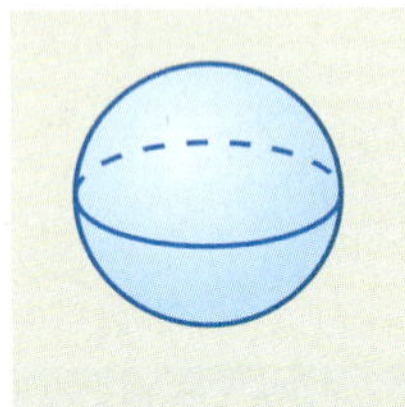
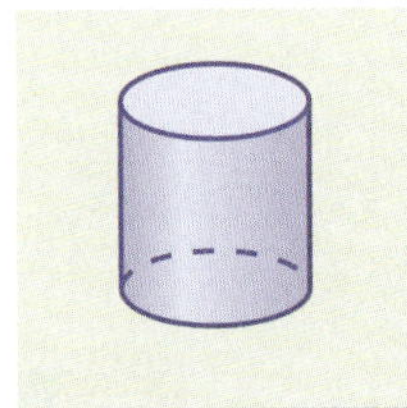

Practice and Problem Solving

Find the volume of the sphere. Round your answer to the nearest tenth.

1 3.
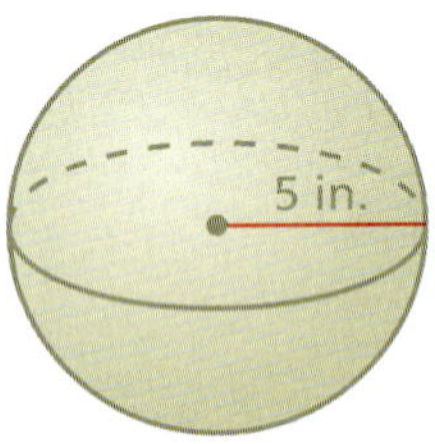

4.
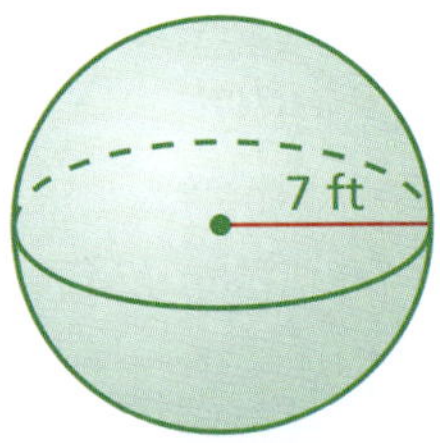

5.
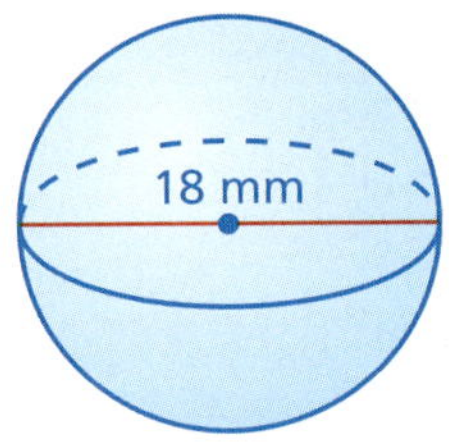

6.
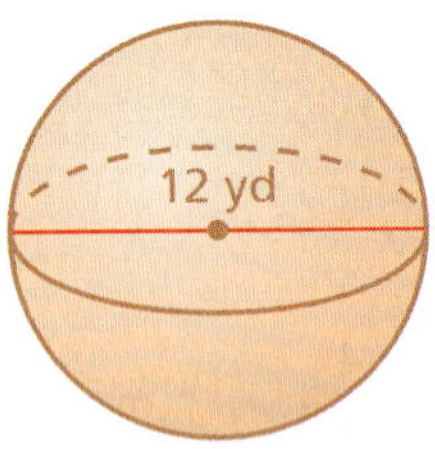

7.
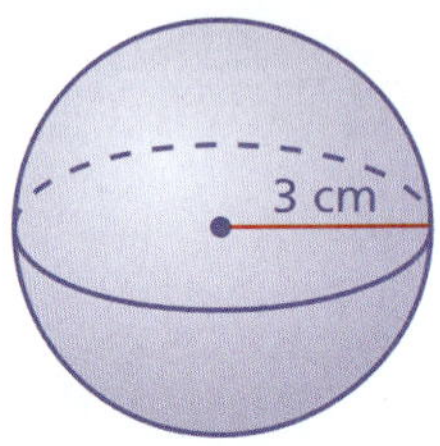

8.
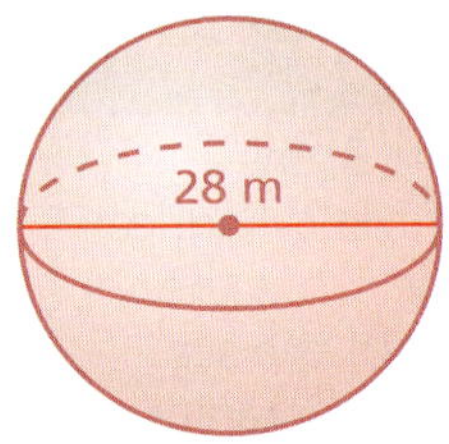

Find the radius of the sphere with the given volume.

2 9. Volume $= 972\pi\ \text{mm}^3$

10. Volume $= 4.5\pi\ \text{cm}^3$

11. Volume $= 121.5\pi\ \text{ft}^3$

12. **GLOBE** The globe of the Moon has a radius of 10 inches. Find the volume of the globe. Round your answer to the nearest whole number.

13. **SOFTBALL** A softball has a volume of $\frac{125}{6}\pi$ cubic inches. Find the radius of the softball.

Find the volume of the composite solid. Round your answer to the nearest tenth.

③ **14.**

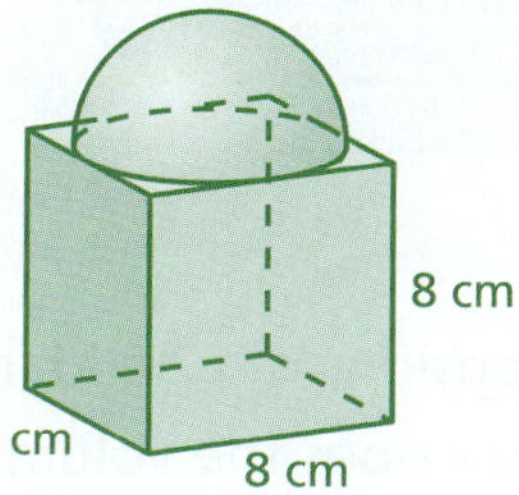

15.

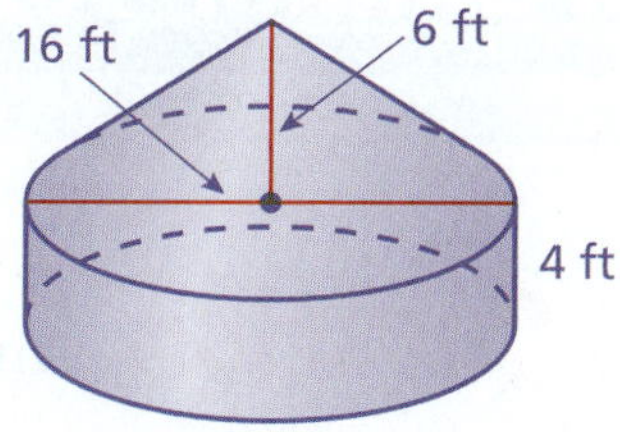

16.

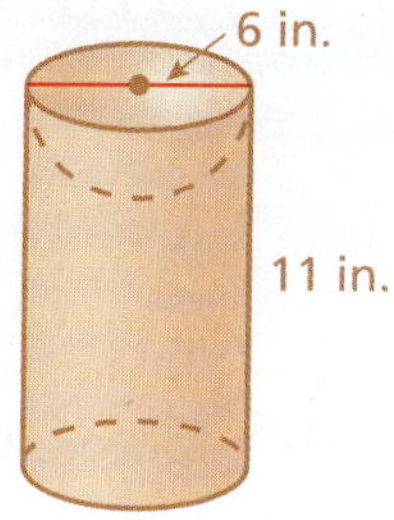

17. **REASONING** A sphere and a right cylinder have the same radius and volume. Find the radius r in terms of the height h of the cylinder.

18. **PACKAGING** A cylindrical container of three rubber balls has a height of 18 centimeters and a diameter of 6 centimeters. Each ball in the container has a radius of 3 centimeters. Find the amount of space in the container that is not occupied by rubber balls. Round your answer to the nearest whole number.

Volume = 4500π in.^{3}

19. **BASKETBALL** The basketball shown is packaged in a box that is in the shape of a cube. The edge length of the box is equal to the diameter of the basketball. What is the surface area and the volume of the box?

20. **Logic** Your friend says that the volume of a sphere with radius r is four times the volume of a cone with radius r. When is this true? Justify your answer.

Fair Game Review What you learned in previous grades & lessons

The blue figure is a dilation of the red figure. Identify the type of dilation and find the scale factor. *(Section 2.7)*

21.

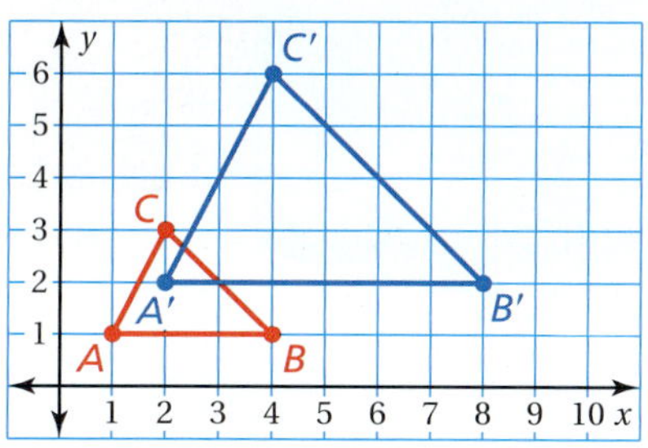

22.

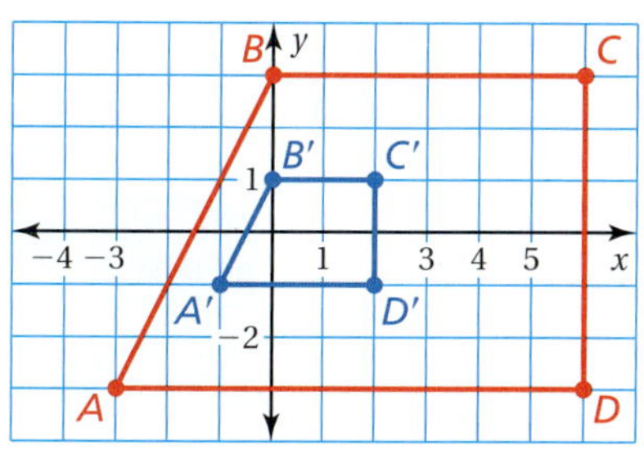

23. **MULTIPLE CHOICE** A person who is 5 feet tall casts a 6-foot-long shadow. A nearby flagpole casts a 30-foot-long shadow. What is the height of the flagpole? *(Section 3.4)*

Ⓐ 25 ft Ⓑ 29 ft Ⓒ 36 ft Ⓓ 40 ft

8.4 Surface Areas and Volumes of Similar Solids

Essential Question When the dimensions of a solid increase by a factor of k, how does the surface area change? How does the volume change?

1 ACTIVITY: Comparing Surface Areas and Volumes

Work with a partner. Copy and complete the table. Describe the pattern. Are the dimensions proportional? Explain your reasoning.

a.

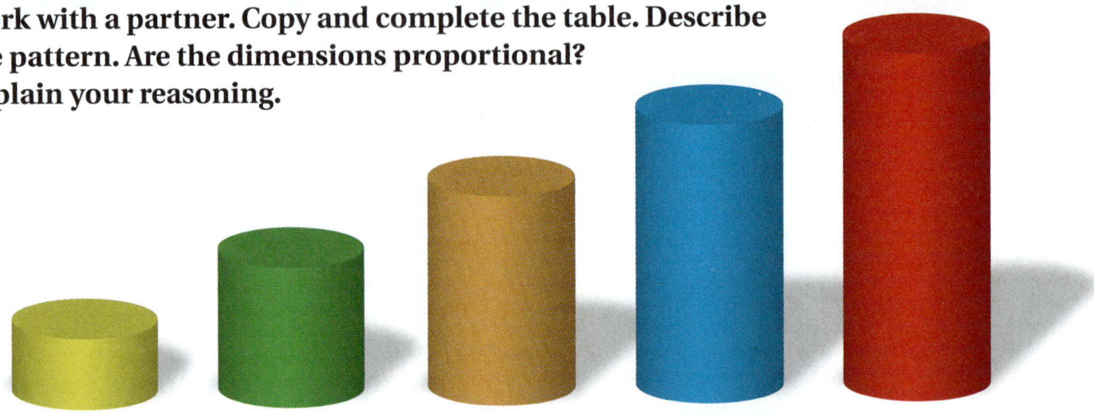

Radius	1	1	1	1	1
Height	1	2	3	4	5
Surface Area					
Volume					

b.

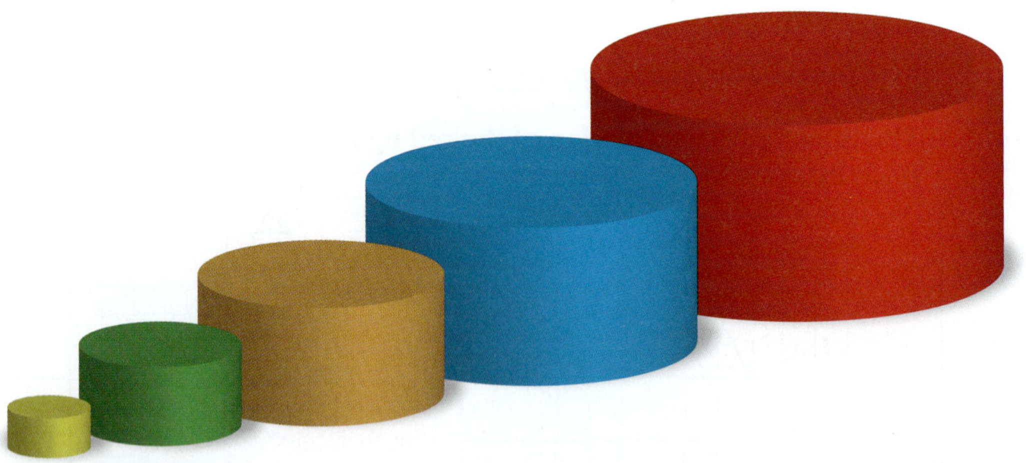

Radius	1	2	3	4	5
Height	1	2	3	4	5
Surface Area					
Volume					

COMMON CORE

Geometry

In this lesson, you will
- identify similar solids.
- use properties of similar solids to find missing measures.
- understand the relationship between surface areas of similar solids.
- understand the relationship between volumes of similar solids.
- solve real-life problems.

Applying Standard
MACC.8.G.3.9

2 ACTIVITY: Comparing Surface Areas and Volumes

Work with a partner. Copy and complete the table. Describe the pattern. Are the dimensions proportional? Explain.

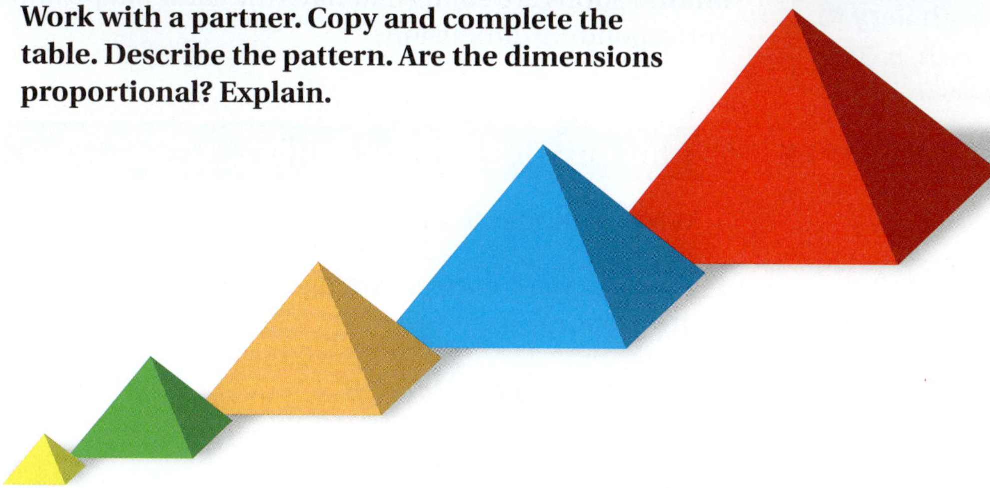

> **Math Practice 8**
> **Repeat Calculations**
> Which calculations are repeated? How does this help you describe the pattern?

Base Side	6	12	18	24	30
Height	4	8	12	16	20
Slant Height	5	10	15	20	25
Surface Area					
Volume					

What Is Your Answer?

3. **IN YOUR OWN WORDS** When the dimensions of a solid increase by a factor of k, how does the surface area change?

4. **IN YOUR OWN WORDS** When the dimensions of a solid increase by a factor of k, how does the volume change?

5. **REPEATED REASONING** All the dimensions of a prism increase by a factor of 5.

   a. How many times greater is the surface area? Explain.

   5   10   25   125

   b. How many times greater is the volume? Explain.

   5   10   25   125

Practice Use what you learned about surface areas and volumes of similar solids to complete Exercise 3 on page 359.

8.4 Lesson

Check It Out
Lesson Tutorials
BigIdeasMath.com

Key Vocabulary
similar solids, *p. 356*

Similar solids are solids that have the same shape and proportional corresponding dimensions.

EXAMPLE 1 Identifying Similar Solids

Which cylinder is similar to Cylinder A?

Cylinder B

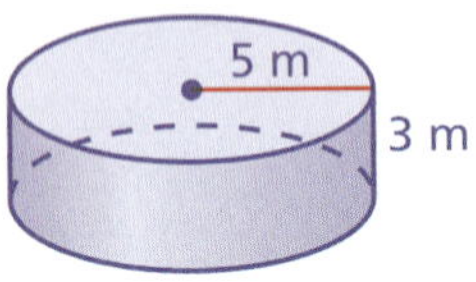

Cylinder C

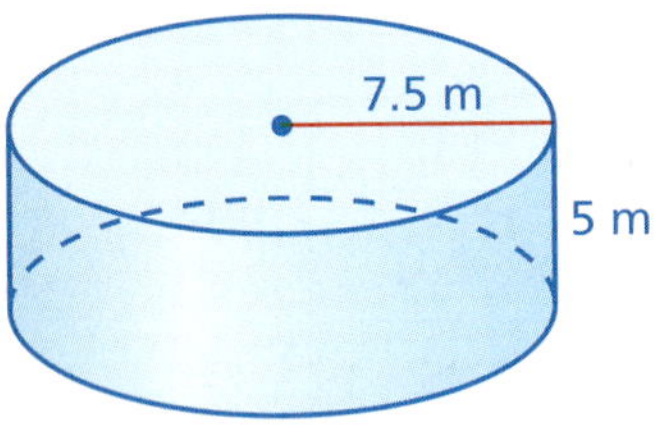

Cylinder A

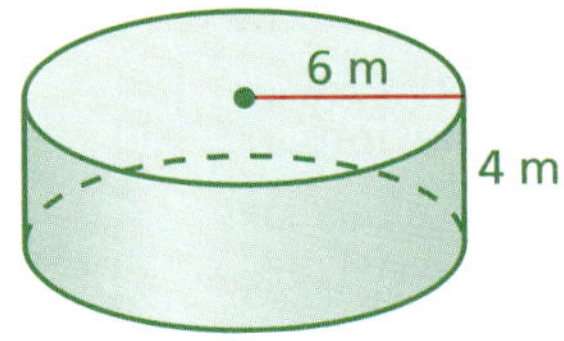

Check to see if corresponding dimensions are proportional.

Cylinder A and Cylinder B

$$\frac{\text{Height of A}}{\text{Height of B}} = \frac{4}{3} \qquad \frac{\text{Radius of A}}{\text{Radius of B}} = \frac{6}{5}$$

Not proportional

Cylinder A and Cylinder C

$$\frac{\text{Height of A}}{\text{Height of C}} = \frac{4}{5} \qquad \frac{\text{Radius of A}}{\text{Radius of C}} = \frac{6}{7.5} = \frac{4}{5}$$

Proportional

So, Cylinder C is similar to Cylinder A.

EXAMPLE 2 Finding Missing Measures in Similar Solids

The cones are similar. Find the missing slant height ℓ.

Cone X

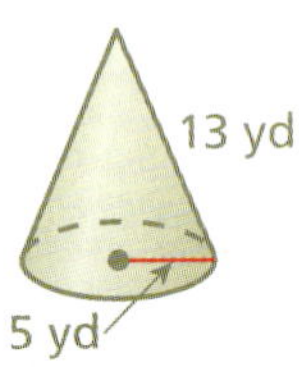

Cone Y

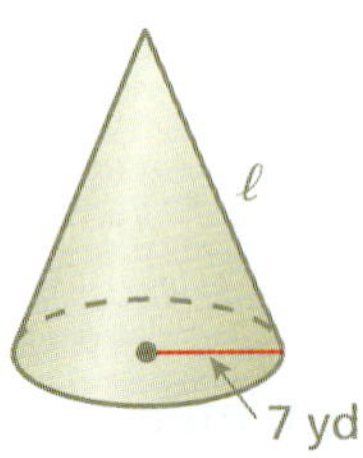

$$\frac{\text{Radius of X}}{\text{Radius of Y}} = \frac{\text{Slant height of X}}{\text{Slant height of Y}}$$

$\frac{5}{7} = \frac{13}{\ell}$ — Substitute.

$5\ell = 91$ — Cross Products Property

$\ell = 18.2$ — Divide each side by 5.

The slant height is 18.2 yards.

On Your Own

Now You're Ready
Exercises 4–9

1. Cylinder D has a radius of 7.5 meters and a height of 4.5 meters. Which cylinder in Example 1 is similar to Cylinder D?
2. The prisms at the right are similar. Find the missing width and length.

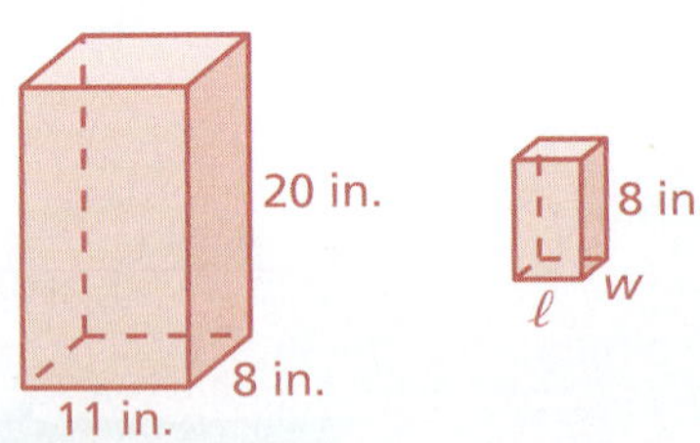

Key Ideas

Linear Measures

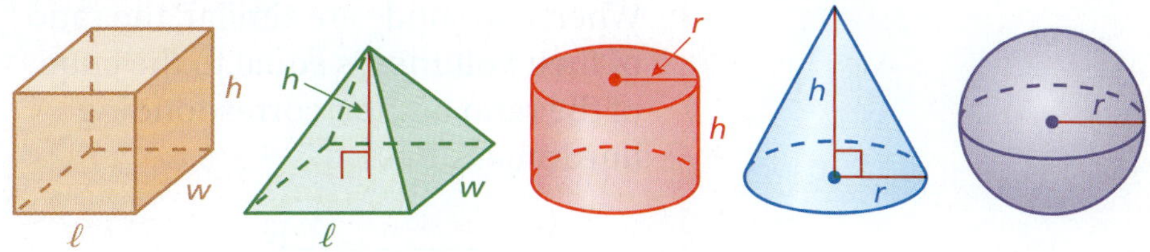

Surface Areas of Similar Solids

When two solids are similar, the ratio of their surface areas is equal to the square of the ratio of their corresponding linear measures.

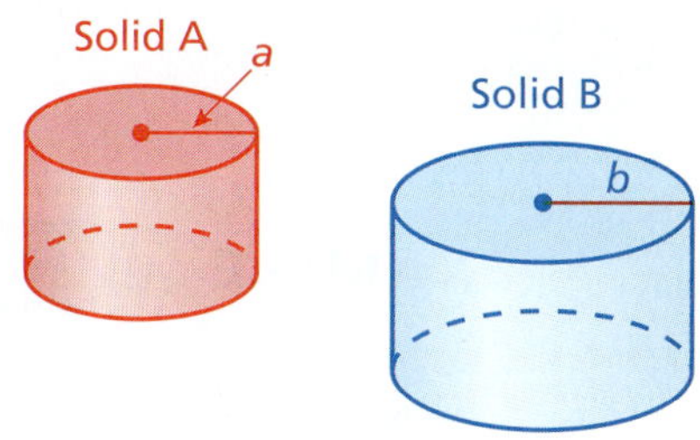

$$\frac{\text{Surface Area of A}}{\text{Surface Area of B}} = \left(\frac{a}{b}\right)^2$$

EXAMPLE 3 Finding Surface Area

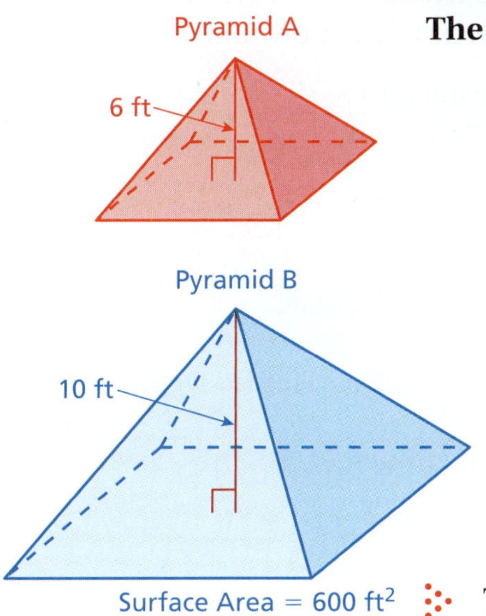

The pyramids are similar. What is the surface area of Pyramid A?

$$\frac{\text{Surface Area of A}}{\text{Surface Area of B}} = \left(\frac{\text{Height of A}}{\text{Height of B}}\right)^2$$

$\frac{S}{600} = \left(\frac{6}{10}\right)^2$ Substitute.

$\frac{S}{600} = \frac{36}{100}$ Evaluate.

$\frac{S}{600} \cdot 600 = \frac{36}{100} \cdot 600$ Multiplication Property of Equality

$S = 216$ Simplify.

∴ The surface area of Pyramid A is 216 square feet.

On Your Own

The solids are similar. Find the surface area of the red solid. Round your answer to the nearest tenth.

3.

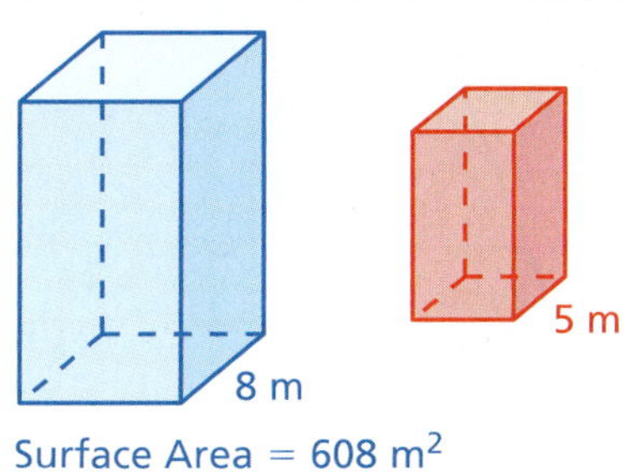

4.

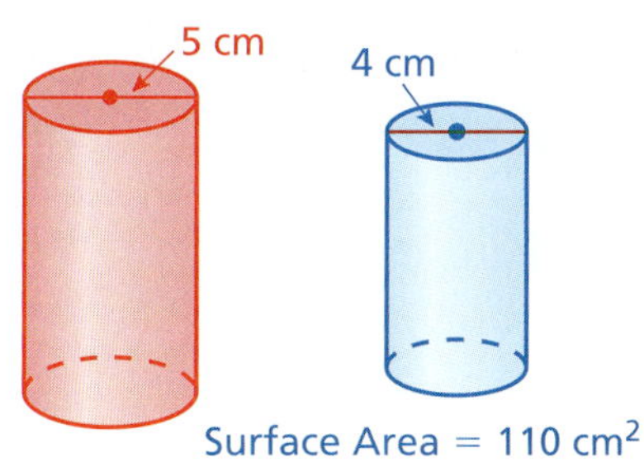

Key Idea

Volumes of Similar Solids

When two solids are similar, the ratio of their volumes is equal to the cube of the ratio of their corresponding linear measures.

$$\frac{\text{Volume of A}}{\text{Volume of B}} = \left(\frac{a}{b}\right)^3$$

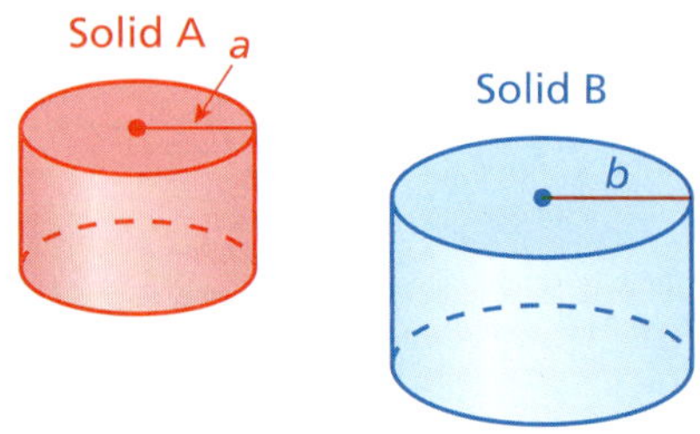

EXAMPLE 4 Finding Volume

Original Tank

Volume = 2000 ft^3

The dimensions of the touch tank at an aquarium are doubled. What is the volume of the new touch tank?

Ⓐ 150 ft^3 Ⓑ 4000 ft^3
Ⓒ 8000 ft^3 Ⓓ 16,000 ft^3

The dimensions are doubled, so the ratio of the dimensions of the original tank to the dimensions of the new tank is 1 : 2.

$$\frac{\text{Original volume}}{\text{New volume}} = \left(\frac{\text{Original dimension}}{\text{New dimension}}\right)^3$$

$$\frac{2000}{V} = \left(\frac{1}{2}\right)^3 \quad \text{Substitute.}$$

$$\frac{2000}{V} = \frac{1}{8} \quad \text{Evaluate.}$$

$$16{,}000 = V \quad \text{Cross Products Property}$$

∴ The volume of the new tank is 16,000 cubic feet. So, the correct answer is Ⓓ.

Study Tip

When the dimensions of a solid are multiplied by k, the surface area is multiplied by k^2 and the volume is multiplied by k^3.

On Your Own

Now You're Ready Exercises 10–13

The solids are similar. Find the volume of the red solid. Round your answer to the nearest tenth.

5.

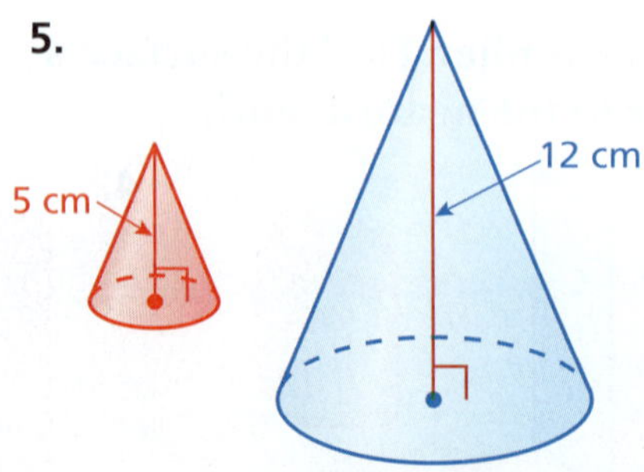

6.

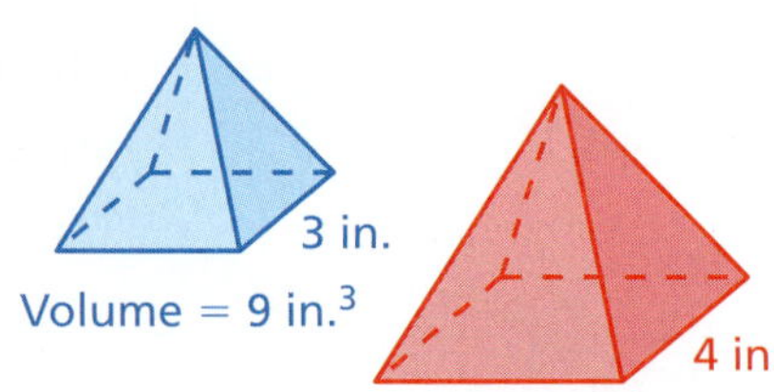

8.4 Exercises

Check It Out
Help with Homework
BigIdeasMath.com

Vocabulary and Concept Check

1. **VOCABULARY** What are similar solids?

2. **OPEN-ENDED** Draw two similar solids and label their corresponding linear measures.

Practice and Problem Solving

3. **NUMBER SENSE** All the dimensions of a cube increase by a factor of $\frac{3}{2}$.

   a. How many times greater is the surface area? Explain.

   b. How many times greater is the volume? Explain.

Determine whether the solids are similar.

1 4.

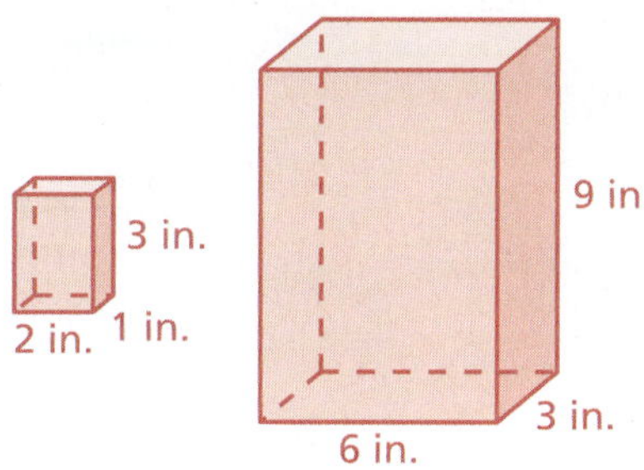

5.

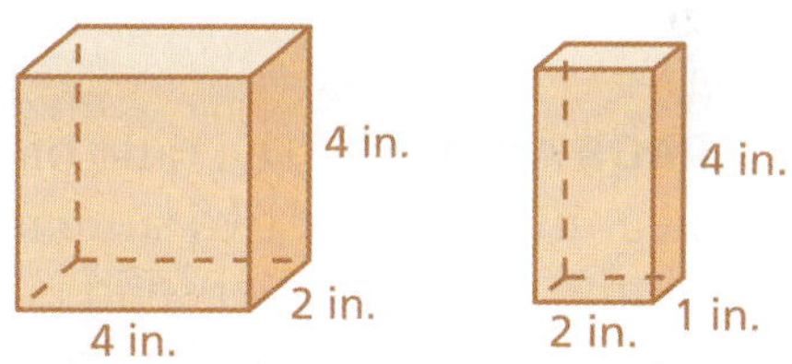

6.

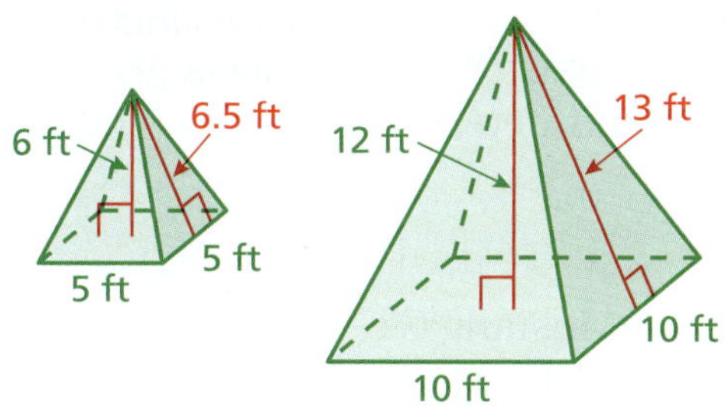

7.

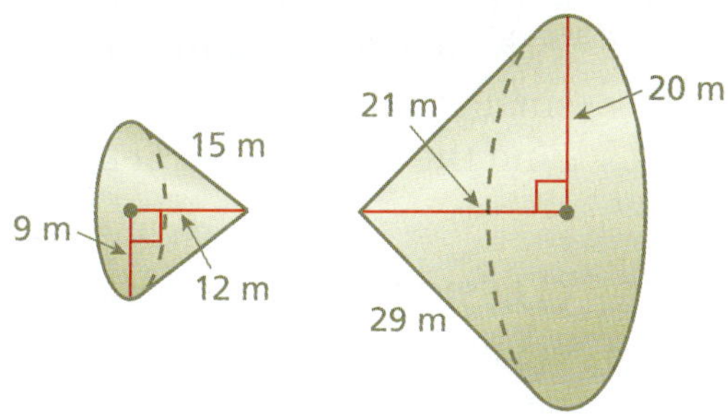

The solids are similar. Find the missing dimension(s).

2 8.

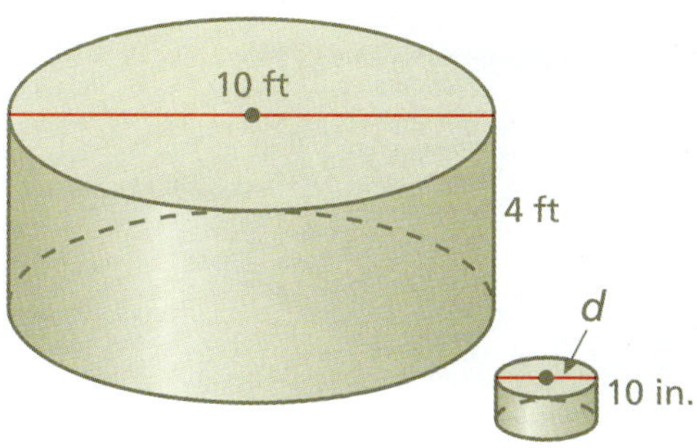

9.

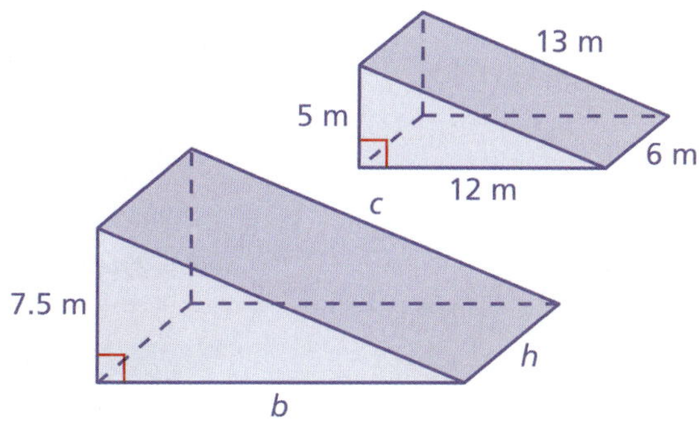

The solids are similar. Find the surface area S or volume V of the red solid. Round your answer to the nearest tenth.

③ ④ **10.**

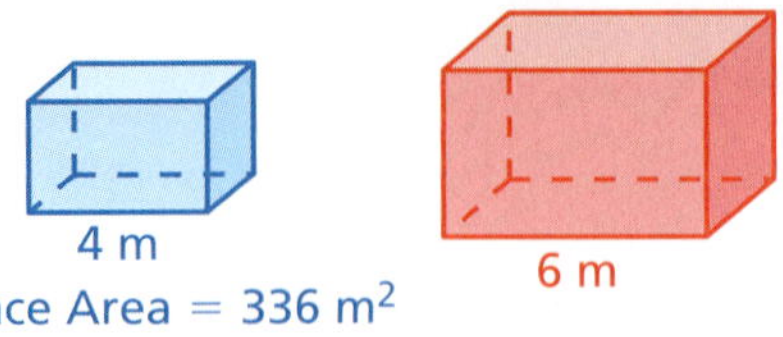

11.

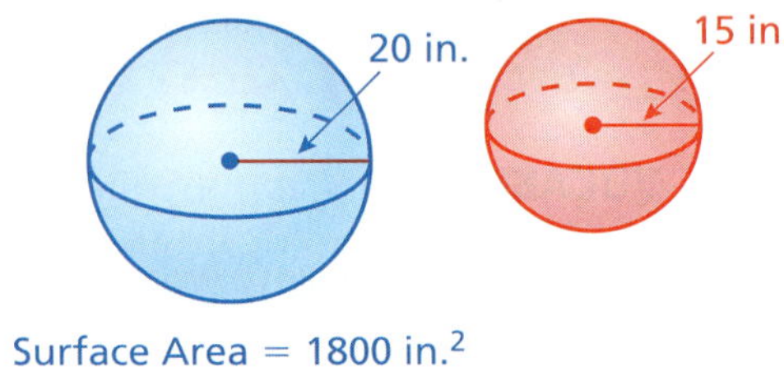

12.

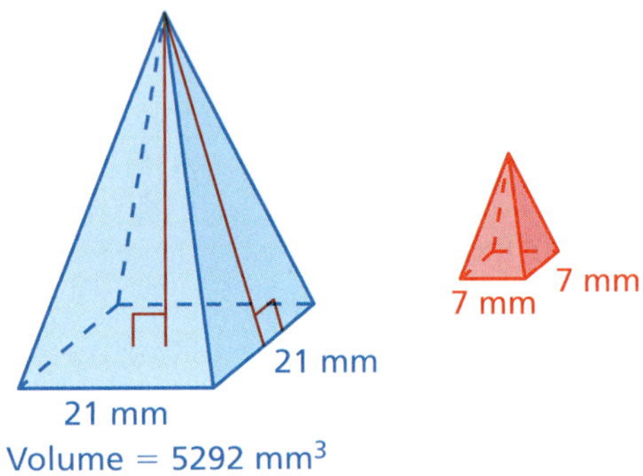

13.

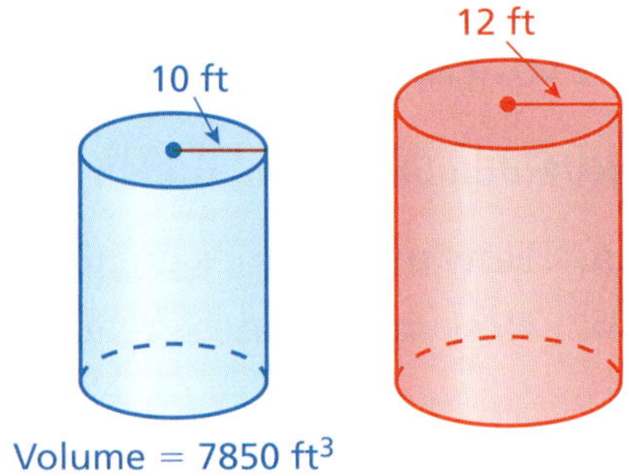

14. ERROR ANALYSIS The ratio of the corresponding linear measures of two similar solids is 3 : 5. The volume of the smaller solid is 108 cubic inches. Describe and correct the error in finding the volume of the larger solid.

✗

$$\frac{108}{V} = \left(\frac{3}{5}\right)^2$$

$$\frac{108}{V} = \frac{9}{25}$$

$$300 = V$$

The volume of the larger solid is 300 cubic inches.

15. MIXED FRUIT The ratio of the corresponding linear measures of two similar cans of fruit is 4 to 7. The smaller can has a surface area of 220 square centimeters. Find the surface area of the larger can.

16. CLASSIC MUSTANG The volume of a 1968 Ford Mustang GT engine is 390 cubic inches. Which scale model of the Mustang has the greater engine volume, a 1 : 18 scale model or a 1 : 24 scale model? How much greater is it?

17. **MARBLE STATUE** You have a small marble statue of Wolfgang Mozart. It is 10 inches tall and weighs 16 pounds. The original statue is 7 feet tall.

   a. Estimate the weight of the original statue. Explain your reasoning.

   b. If the original statue were 20 feet tall, how much would it weigh?

Wolfgang Mozart

18. **REPEATED REASONING** The largest doll is 7 inches tall. Each of the other dolls is 1 inch shorter than the next larger doll. Make a table that compares the surface areas and the volumes of the seven dolls.

19. **Precision** You and a friend make paper cones to collect beach glass. You cut out the largest possible three-fourths circle from each piece of paper.

   a. Are the cones similar? Explain your reasoning.

   b. Your friend says that because your sheet of paper is twice as large, your cone will hold exactly twice the volume of beach glass. Is this true? Explain your reasoning.

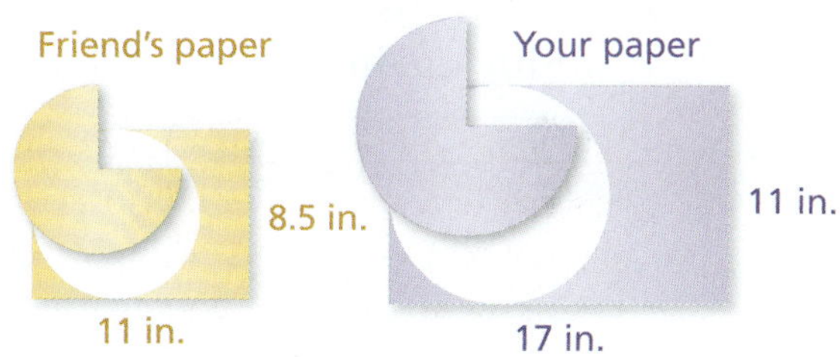

Fair Game Review What you learned in previous grades & lessons

Draw the figure and its reflection in the x-axis. Identify the coordinates of the image. *(Section 2.3)*

20. $A(1, 1)$, $B(3, 4)$, $C(4, 2)$

21. $J(-3, 0)$, $K(-4, 3)$, $L(-1, 4)$

22. **MULTIPLE CHOICE** Which system of linear equations has no solution? *(Section 5.4)*

   Ⓐ $y = 4x + 1$, $y = -4x + 1$

   Ⓑ $y = 2x - 7$, $y = 2x + 7$

   Ⓒ $3x + y = 1$, $6x + 2y = 2$

   Ⓓ $5x + y = 3$, $x + 5y = 15$

8.3–8.4 Quiz

Find the volume of the sphere. Round your answer to the nearest tenth. *(Section 8.3)*

1.

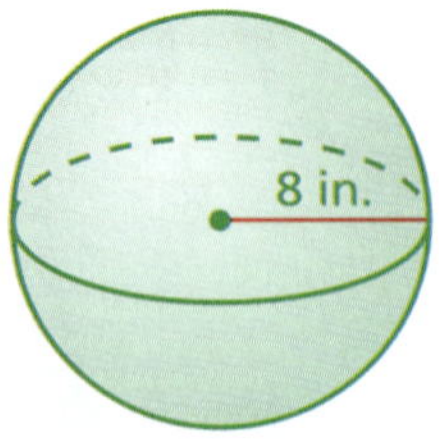

2.

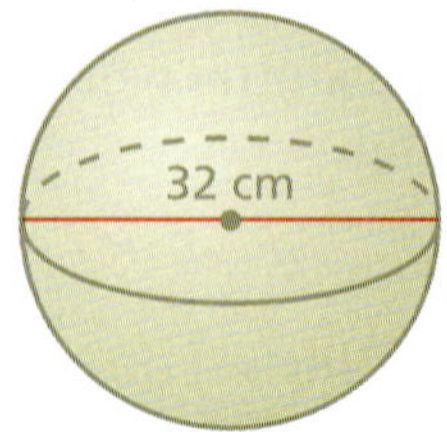

Find the radius of the sphere with the given volume. *(Section 8.3)*

3. Volume $= 4500\pi$ yd^3

4. Volume $= \frac{32}{3}\pi$ ft^3

5. Find the volume of the composite solid. Round your answer to the nearest tenth. *(Section 8.3)*

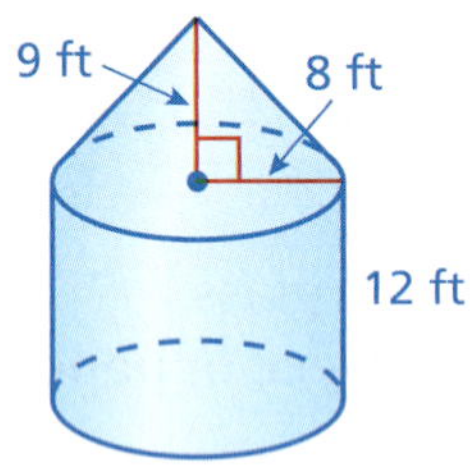

6. Determine whether the solids are similar. *(Section 8.4)*

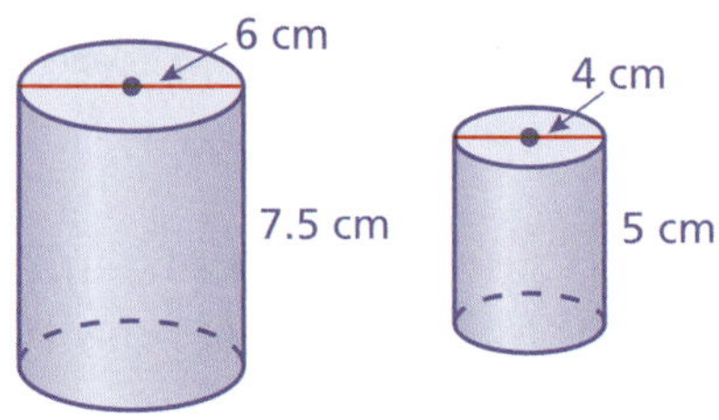

7. The prisms are similar. Find the missing width and height. *(Section 8.4)*

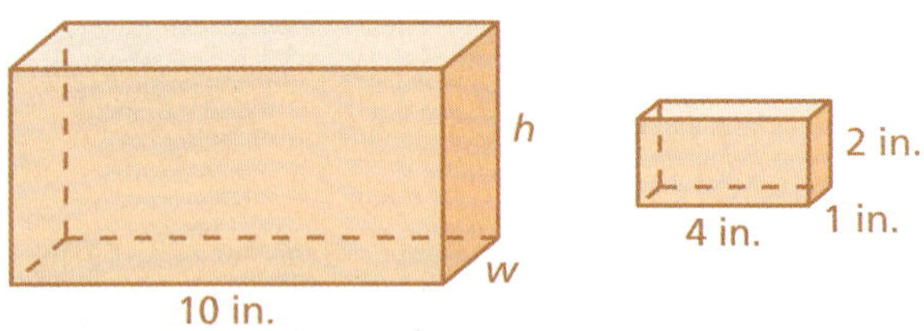

8. The solids are similar. Find the surface area of the red solid. *(Section 8.4)*

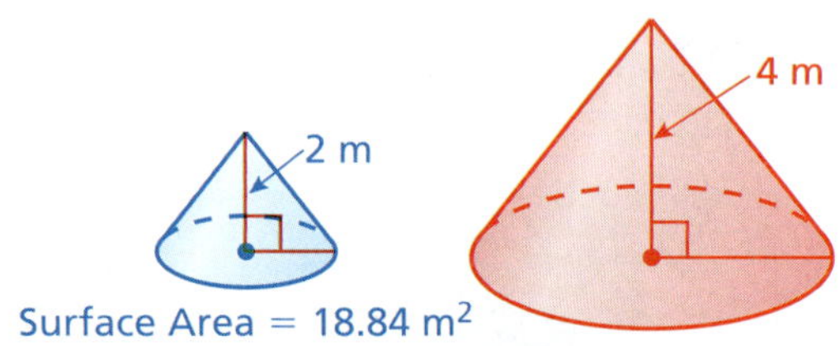

9. HAMSTER A hamster toy is in the shape of a sphere. What is the volume of the toy? Round your answer to the nearest whole number. *(Section 8.3)*

10. JEWELRY BOXES The ratio of the corresponding linear measures of two similar jewelry boxes is 2 to 3. The larger box has a volume of 162 cubic inches. Find the volume of the smaller jewelry box. *(Section 8.4)*

11. ARCADE You win a token after playing an arcade game. What is the volume of the gold ring? Round your answer to the nearest tenth. *(Section 8.3)*

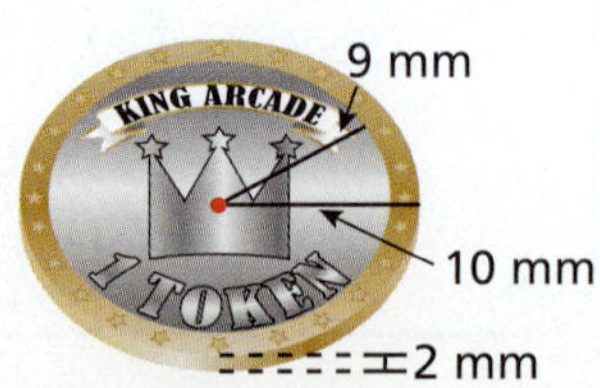

8 Chapter Review

Check It Out
Vocabulary Help
BigIdeasMath.com

Review Key Vocabulary

sphere, *p. 348* | hemisphere, *p. 351* | similar solids, *p. 356*

Review Examples and Exercises

8.1 Volumes of Cylinders *(pp. 334–339)*

Find the volume of the cylinder. Round your answer to the nearest tenth.

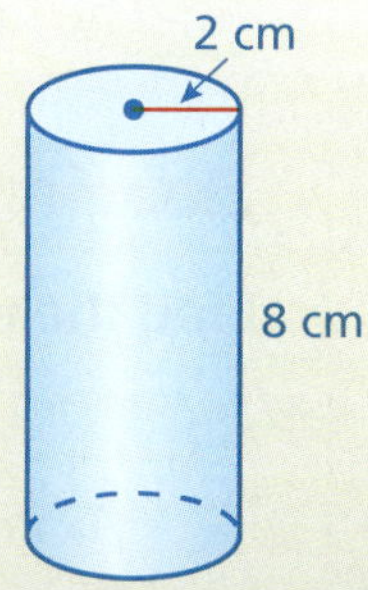

$V = Bh$ Write formula for volume.

$= \pi(2)^2(8)$ Substitute.

$= 32\pi \approx 100.5$ Use a calculator.

∴ The volume is about 100.5 cubic centimeters.

Exercises

Find the volume of the cylinder. Round your answer to the nearest tenth.

1.

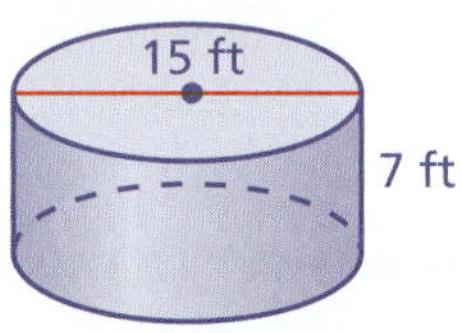

2.

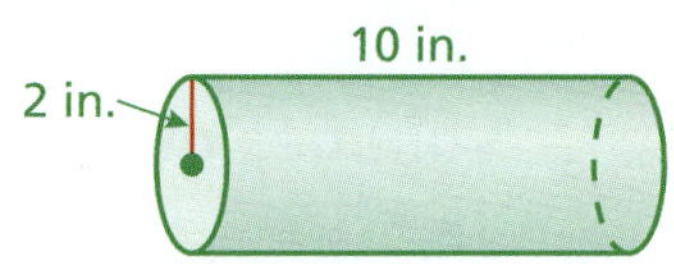

3.

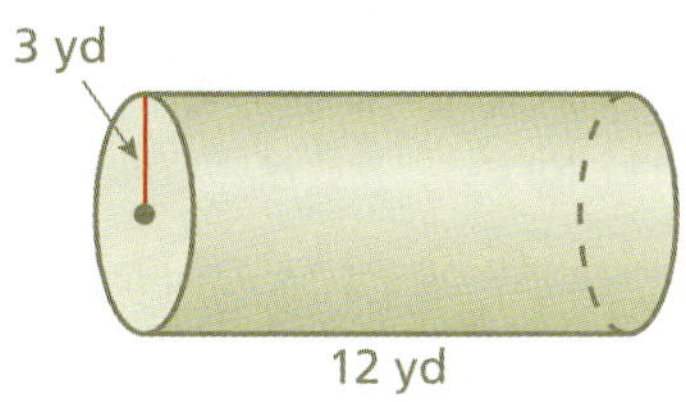

4.

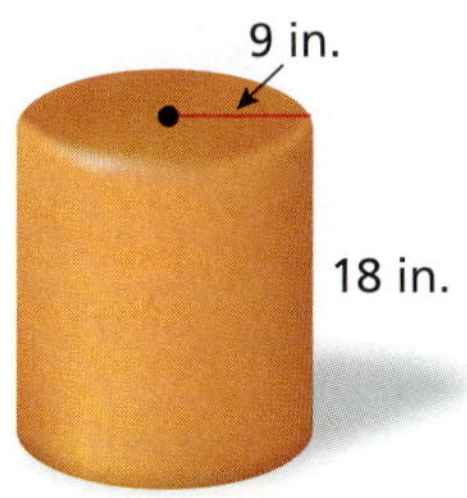

Find the missing dimension of the cylinder. Round your answer to the nearest whole number.

5. Volume $= 25 \text{ in.}^3$

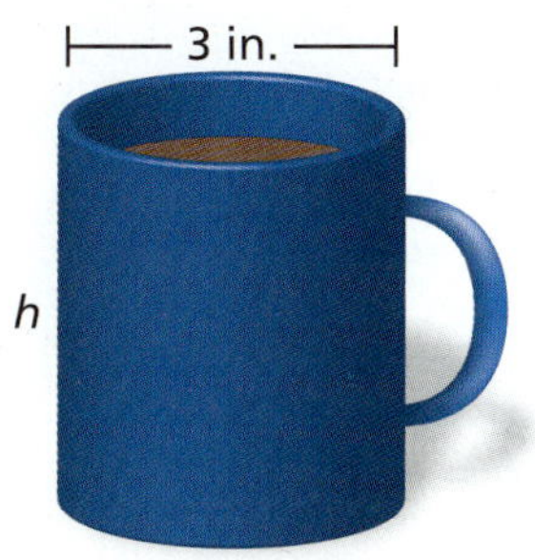

6. Volume $= 7599 \text{ m}^3$

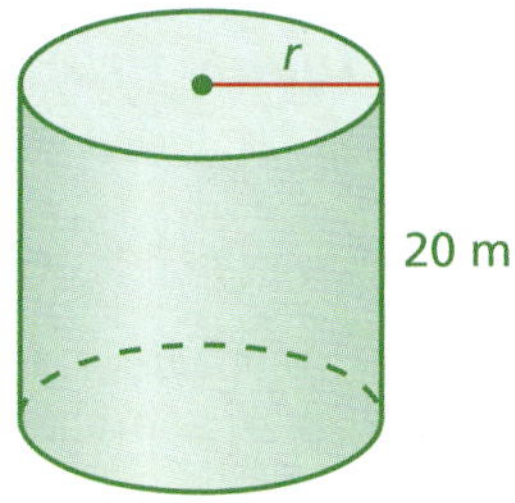

8.2 Volumes of Cones (pp. 340–345)

Find the height of the cone. Round your answer to the nearest tenth.

$V = \frac{1}{3}Bh$	Write formula for volume.
$900 = \frac{1}{3}\pi(6)^2(h)$	Substitute.
$900 = 12\pi h$	Simplify.
$23.9 \approx h$	Divide each side by 12π.

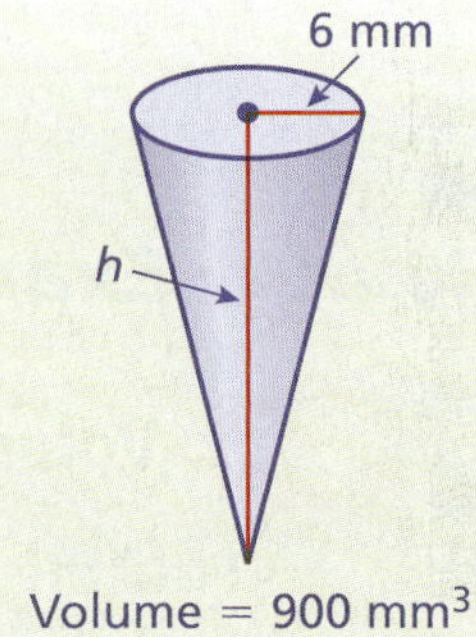

The height is about 23.9 millimeters.

Exercises

Find the volume V or height h of the cone. Round your answer to the nearest tenth.

7.

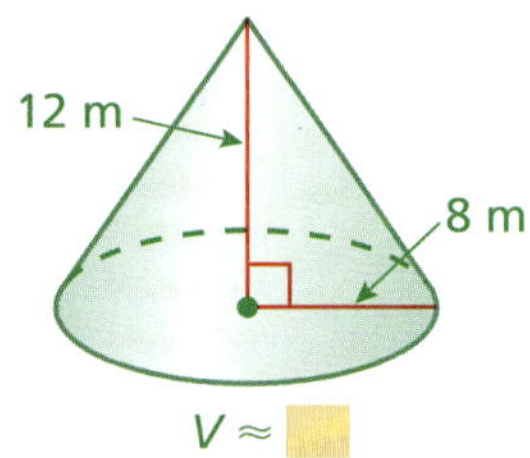

8.

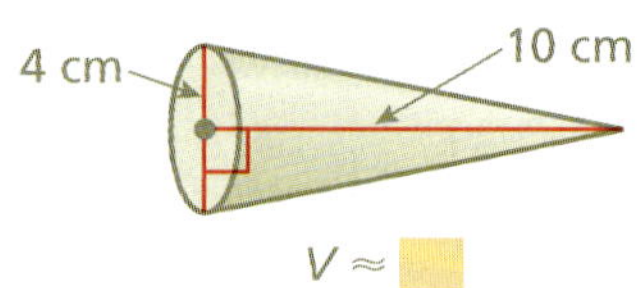

9.

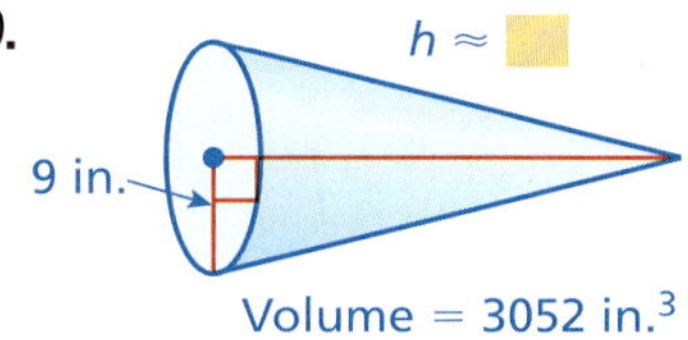

8.3 Volumes of Spheres (pp. 348–353)

a. Find the volume of the sphere. Round your answer to the nearest tenth.

$V = \frac{4}{3}\pi r^3$	Write formula for volume.
$= \frac{4}{3}\pi(11)^3$	Substitute 11 for r.
$= \frac{5324}{3}\pi$	Simplify.
≈ 5575.3	Use a calculator.

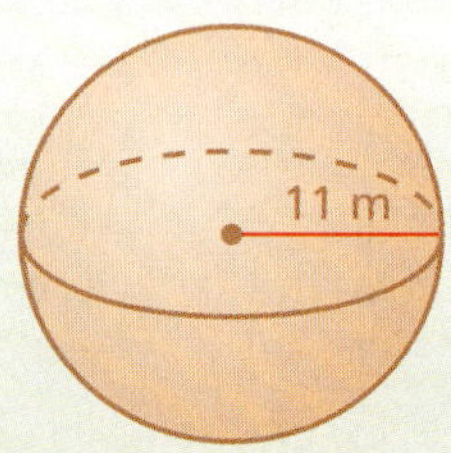

The volume is about 5575.3 cubic meters.

b. Find the volume of the composite solid. Round your answer to the nearest tenth.

Square Prism	*Cylinder*
$V = Bh$	$V = Bh$
$= (12)(12)(9)$	$= \pi(5)^2(9)$
$= 1296$	$= 225\pi \approx 706.9$

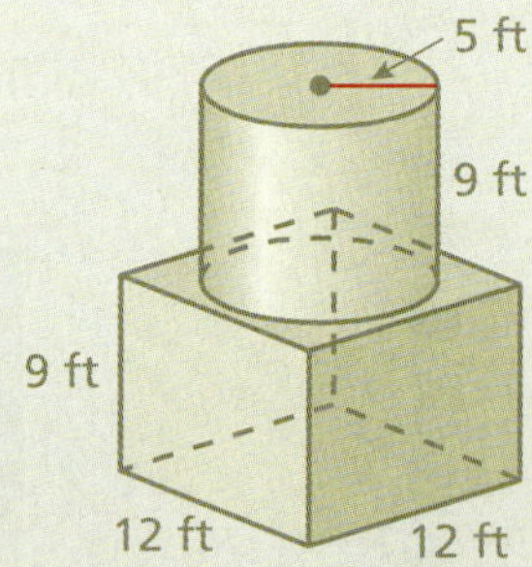

So, the volume is about $1296 + 706.9 = 2002.9$ cubic feet.

Exercises

Find the volume V or radius r of the sphere. Round your answer to the nearest tenth, if necessary.

10.

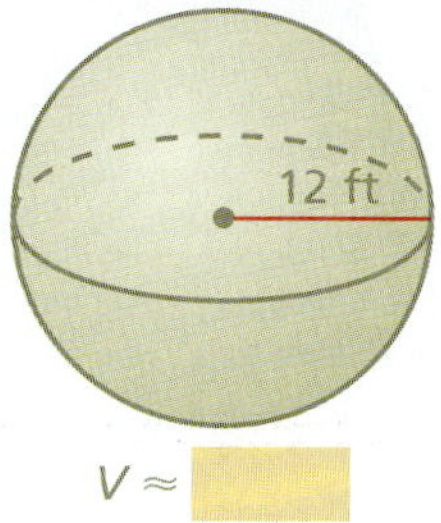

11.

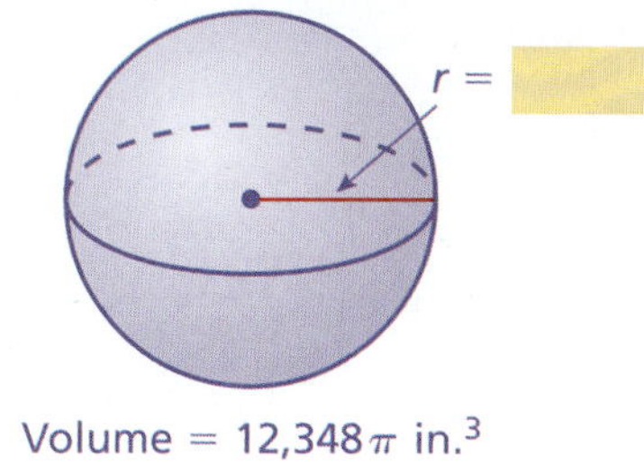

Find the volume of the composite solid. Round your answer to the nearest tenth.

12.

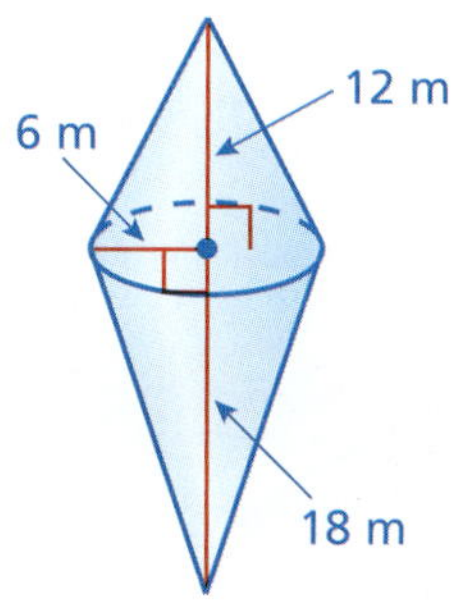

13.

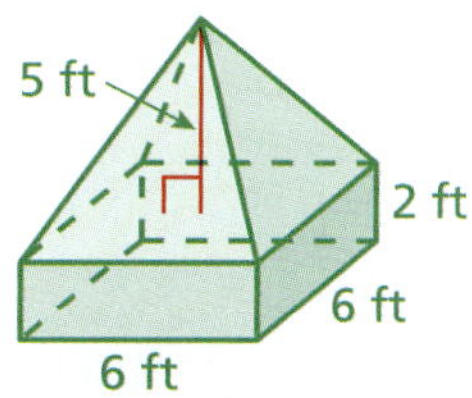

14.

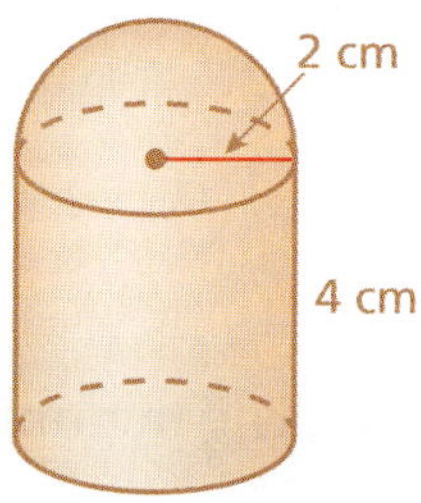

8.4 Surface Areas and Volumes of Similar Solids *(pp. 354–361)*

The cones are similar. What is the volume of the red cone? Round your answer to the nearest tenth.

$$\frac{\text{Volume of A}}{\text{Volume of B}} = \left(\frac{\text{Height of A}}{\text{Height of B}}\right)^3$$

$$\frac{V}{157} = \left(\frac{4}{6}\right)^3 \quad \text{Substitute.}$$

$$\frac{V}{157} = \frac{64}{216} \quad \text{Evaluate.}$$

$$V \approx 46.5 \quad \text{Solve for } V.$$

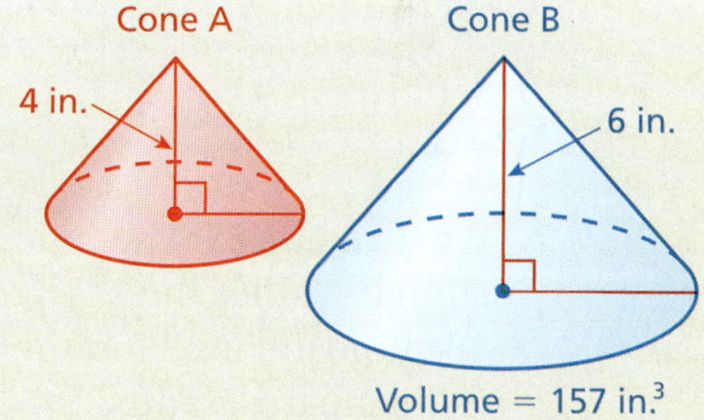

∴ The volume is about 46.5 cubic inches.

Exercises

The solids are similar. Find the surface area S or volume V of the red solid. Round your answer to the nearest tenth.

15.

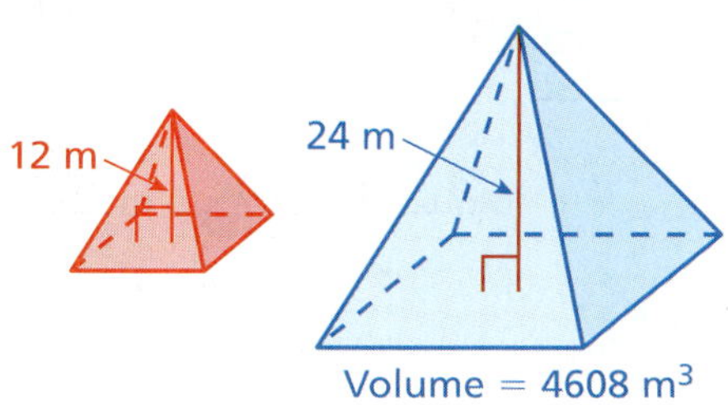

16.

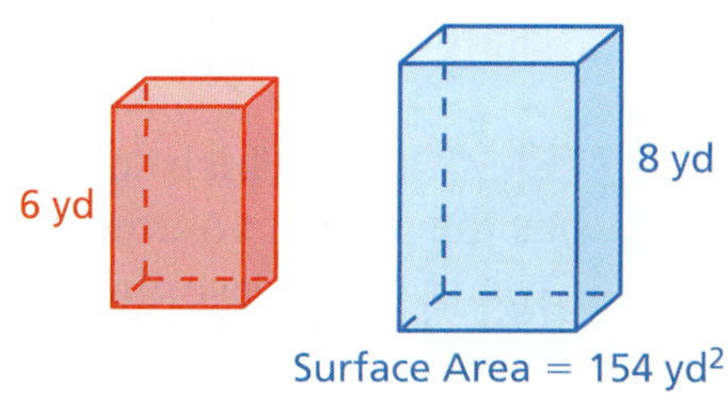

8 Chapter Test

Check It Out
Test Practice
BigIdeasMath.com

Find the volume of the solid. Round your answer to the nearest tenth.

1.

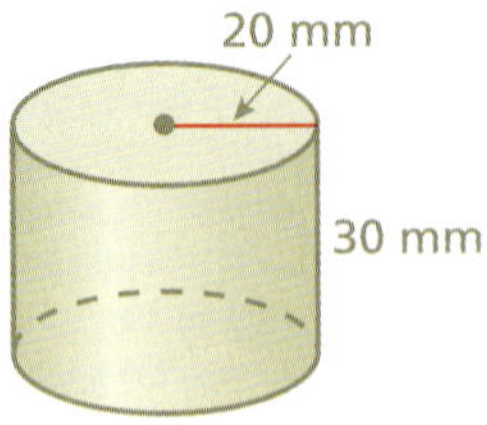

2.

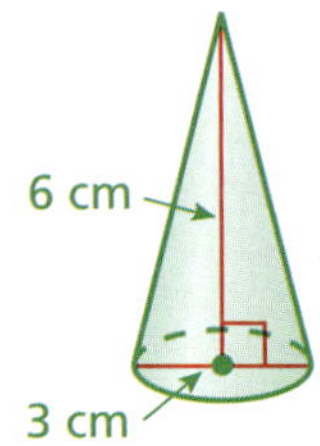

3.

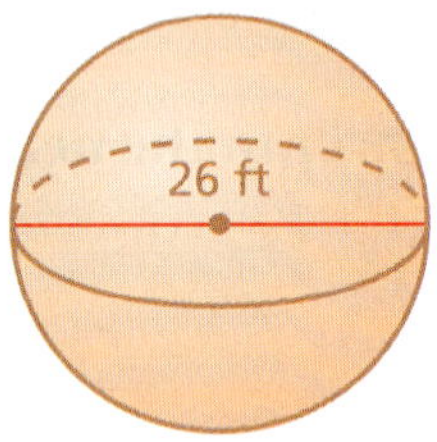

4.

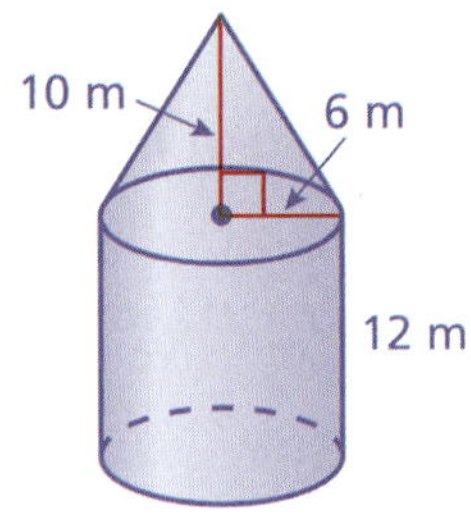

5. The pyramids are similar.

a. Find the missing dimension.

b. Find the surface area of the red pyramid.

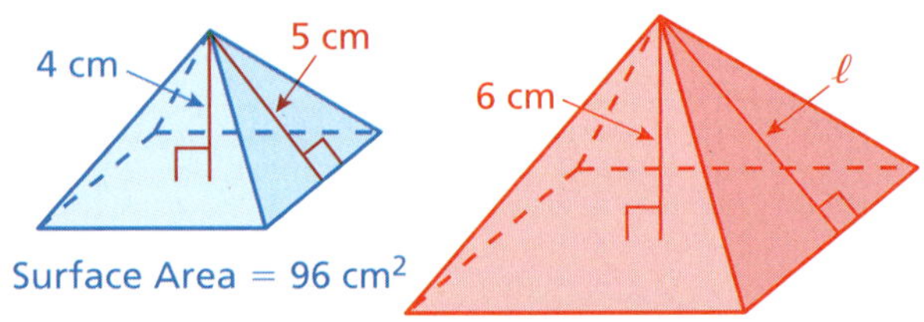

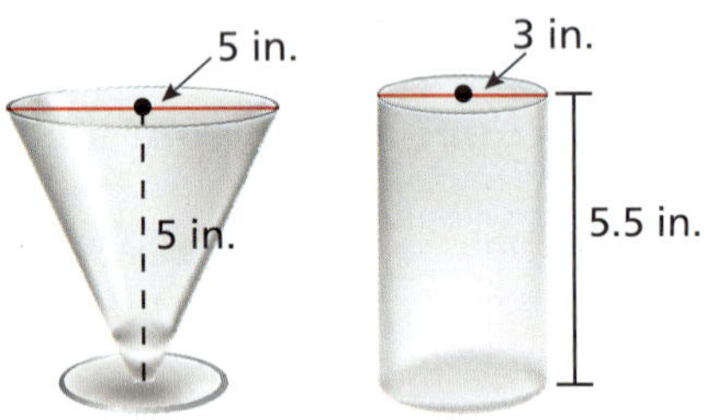

6. SMOOTHIES You are making smoothies. You will use either the cone-shaped glass or the cylindrical glass. Which glass holds more? About how much more?

7. WAFFLE CONES The ratio of the corresponding linear measures of two similar waffle cones is 3 to 4. The smaller cone has a volume of about 18 cubic inches. Find the volume of the larger cone. Round your answer to the nearest tenth.

8. OPEN-ENDED Draw two different composite solids that have the same volume but different surface areas. Explain your reasoning.

9. MILK Glass A has a diameter of 3.5 inches and a height of 4 inches. Glass B has a radius of 1.5 inches and a height of 5 inches. Which glass can hold more milk?

10. REASONING Which solid has the greater volume? Explain your reasoning.

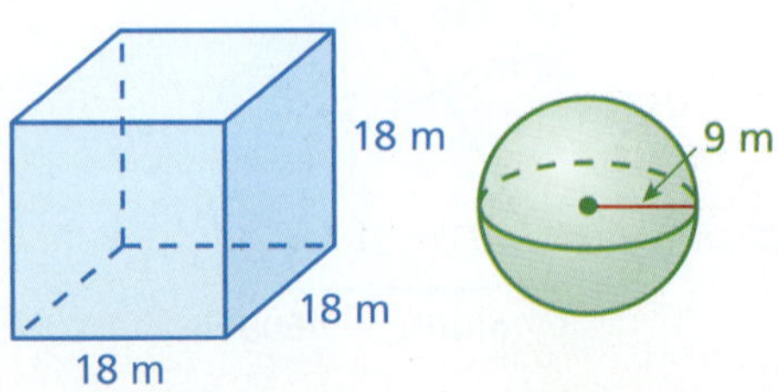

8 Standards Assessment

Test-Taking Strategy
After Answering Easy Questions, Relax

"After answering the easy questions, relax and try the harder ones. For this, $\pi r^2h = 2\pi$. So, it's A."

1. What value of w makes the equation below true? *(MACC.8.EE.3.7b)*

$$\frac{w}{3} = 3(w - 1) - 1$$

A. $\frac{1}{2}$ **C.** $\frac{5}{4}$

B. $\frac{3}{4}$ **D.** $\frac{3}{2}$

2. A right circular cone and its dimensions are shown below.

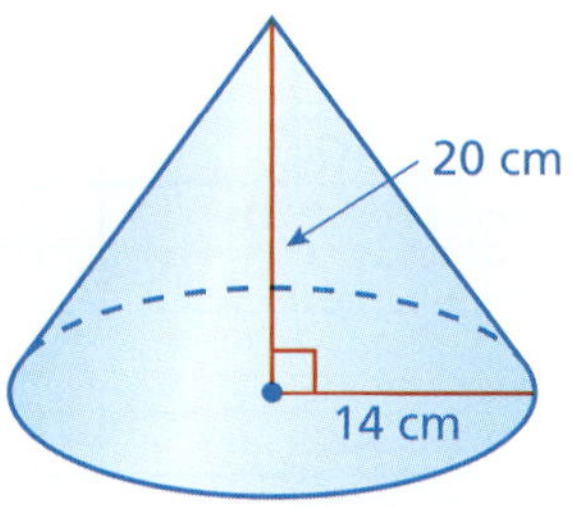

What is the volume of the right circular cone? $\left(\text{Use } \frac{22}{7} \text{ for } \pi.\right)$ *(MACC.8.G.3.9)*

F. $1{,}026\frac{2}{3}$ cm^3 **H.** $4{,}106\frac{2}{3}$ cm^3

G. $3{,}080$ cm^3 **I.** $12{,}320$ cm^3

3. Patricia solved the equation in the box shown.

What should Patricia do to correct the error that she made? *(MACC.8.EE.3.7b)*

$$-\frac{3}{2}(8x - 10) = -20$$
$$8x - 10 = -20\left(-\frac{3}{2}\right)$$
$$8x - 10 = 30$$
$$8x - 10 + 10 = 30 + 10$$
$$8x = 40$$
$$\frac{8x}{8} = \frac{40}{8}$$
$$x = 5$$

A. Add 10 to -20.

B. Distribute $-\frac{3}{2}$ to get $-12x - 15$.

C. Multiply both sides by $-\frac{2}{3}$ instead of $-\frac{3}{2}$.

D. Multiply both sides by $\frac{3}{2}$ instead of $-\frac{3}{2}$.

4. On the grid below, Rectangle *EFGH* is plotted and its vertices are labeled.

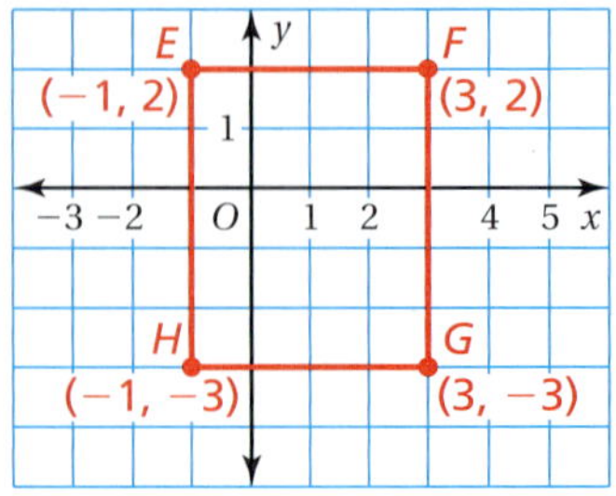

Which of the following shows Rectangle *E′F′G′H′*, the image of Rectangle *EFGH* after it is reflected in the *x*-axis? *(MACC.8.G.1.3)*

F.

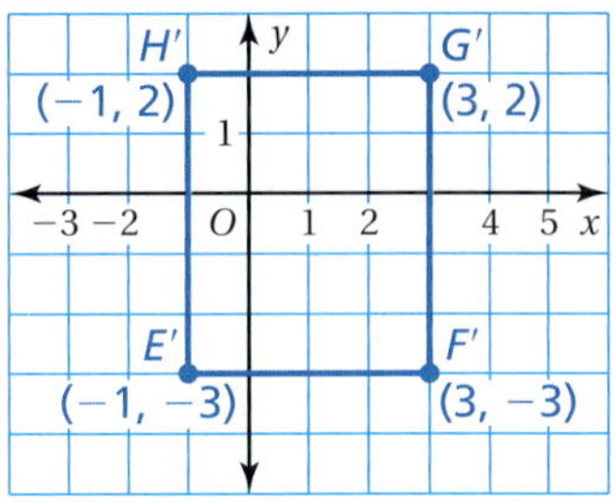

H.

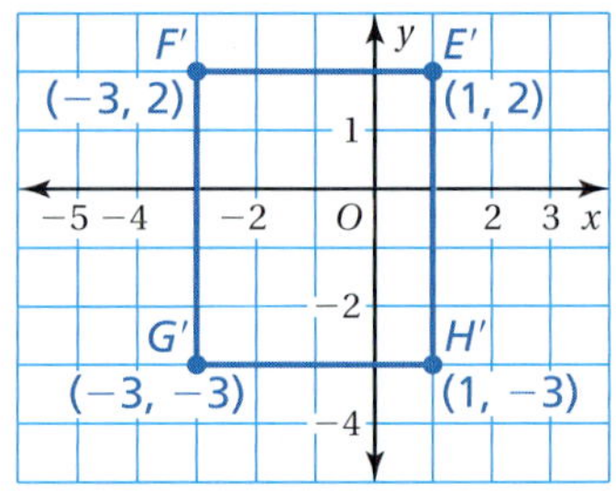

G.

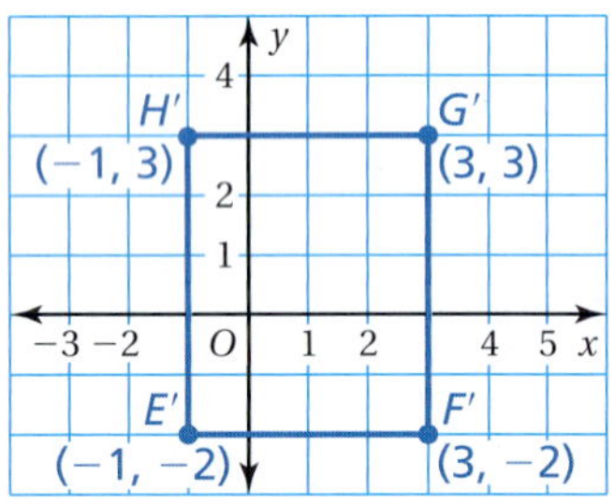

I.

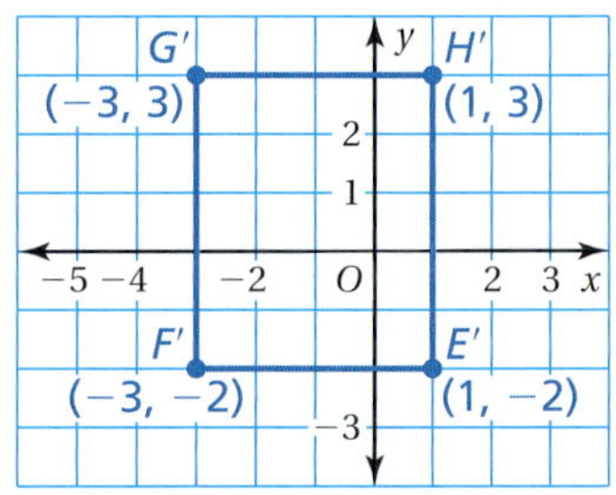

5. List the ordered pairs shown in the mapping diagram below. *(MACC.8.F.1.1)*

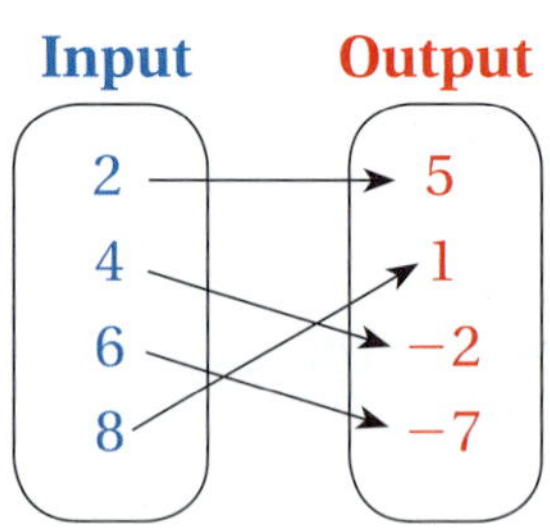

A. (2, 5), (4, −2), (6, −7), (8, 1)

B. (2, −7), (4, −2), (6, 1), (8, 5)

C. (2, 5), (4, 1), (6, −2), (8, −7)

D. (5, 2), (−2, 4), (−7, 6), (1, 8)

6. The temperature fell from 54 degrees Fahrenheit to 36 degrees Fahrenheit over a 6-hour period. The temperature fell by the same number of degrees each hour. How many degrees Fahrenheit did the temperature fall each hour? *(MACC.8.EE.3.7b)*

7. Solve the formula below for I. *(MACC.8.EE.3.7b)*

$$A = P + PI$$

F. $I = A - 2P$

G. $I = \frac{A}{P} - P$

H. $I = A - \frac{P}{P}$

I. $I = \frac{A - P}{P}$

8. A right circular cylinder has a volume of 1296 cubic inches. If you divide the radius of the cylinder by 12, what would be the volume, in cubic inches, of the smaller cylinder? *(MACC.8.G.3.9)*

9. Which graph represents a linear function? *(MACC.8.F.1.3)*

A.

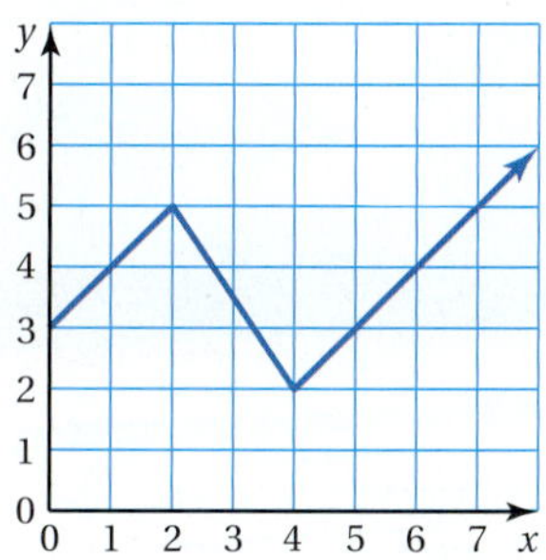

C.

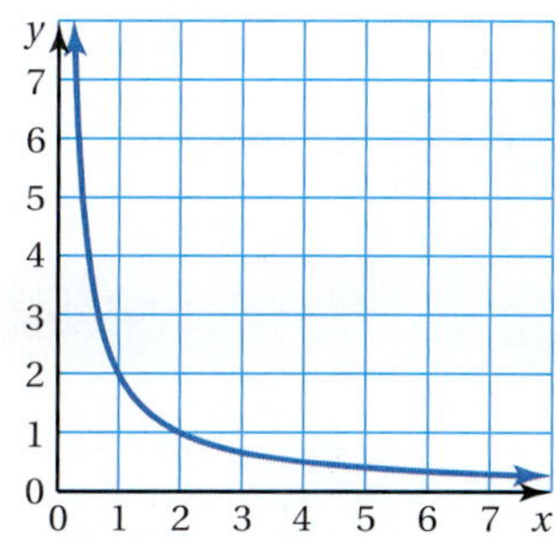

B.

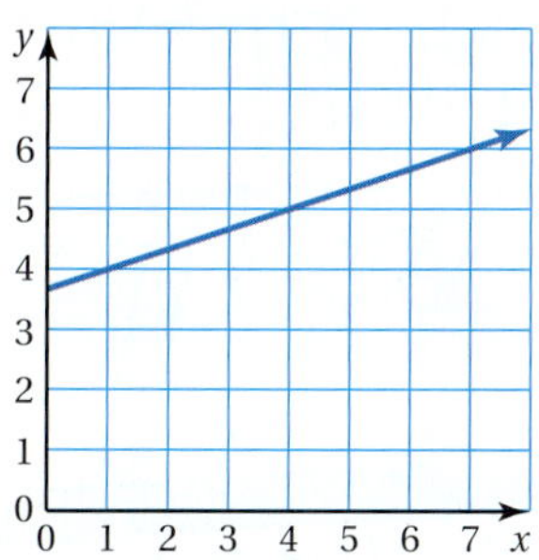

D.

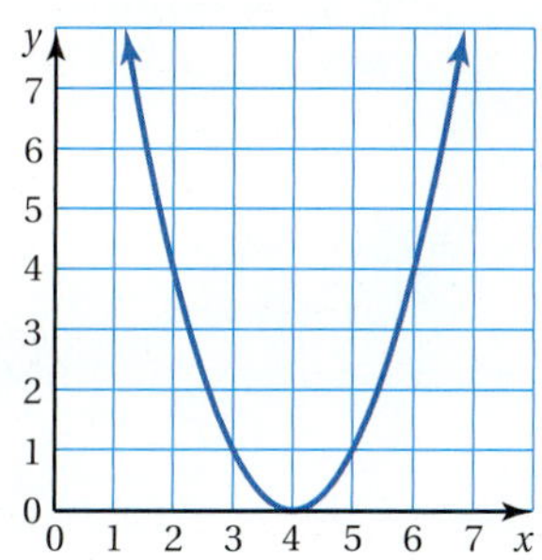

10. The figure below is a diagram for making a tin lantern.

Think
Solve
Explain

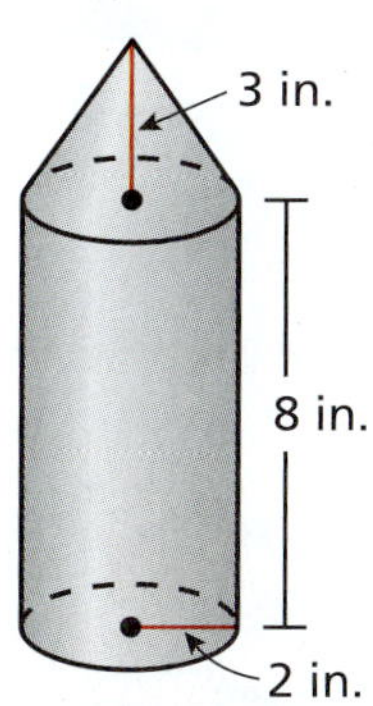

The figure consists of a right circular cylinder without its top base and a right circular cone without its base. What is the volume, in cubic inches, of the entire lantern? Show your work and explain your reasoning. (Use 3.14 for π.) *(MACC.8.G.3.9)*

9 Data Analysis and Displays

What You Learned Before

"Here's an interesting survey about favorite dog toys."

Plotting Points (MACC.6.NS.3.6c)

Example 1 **Plot (a) (−3, 2) and (b) (4, −2.5) in a coordinate plane. Describe the location of each point.**

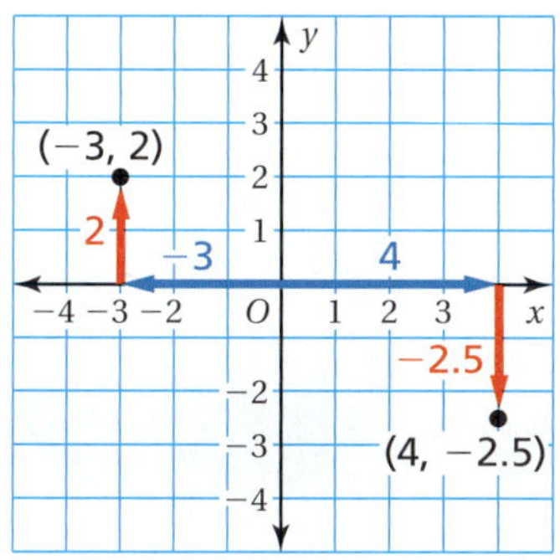

a. Start at the origin. Move 3 units left and 2 units up. Then plot the point.

The point is in Quadrant II.

b. Start at the origin. Move 4 units right and −2.5 units down. Then plot the point.

The point is in Quadrant IV.

Try It Yourself

Plot the ordered pair in a coordinate plane. Describe the location of the point.

1. $(1, 3)$ **2.** $(-2, 4)$ **3.** $(1, -3.5)$ **4.** $\left(-1\frac{3}{4}, -2\frac{1}{4}\right)$

Writing an Equation Using Two Points (MACC.8.F.2.4)

Example 2 **Write in slope-intercept form an equation of the line that passes through the points (4, 2) and (−1, −8).**

Find the slope:

$$m = \frac{y_2 - y_1}{x_2 - x_1} = \frac{-8 - 2}{-1 - 4} = \frac{-10}{-5} = 2$$

Then use the slope $m = 2$ and the point $(4, 2)$ to write an equation of the line.

$y - y_1 = m(x - x_1)$ Write the point-slope form.

$y - 2 = 2(x - 4)$ Substitute 2 for m, 4 for x_1, and 2 for y_1.

$y - 2 = 2x - 8$ Distributive Property

$y = 2x - 6$ Write in slope-intercept form.

Try It Yourself

Write in slope-intercept form an equation of the line that passes through the given points.

5. $(-1, 4), (3, 8)$ **6.** $(0, -1), (-8, -2)$ **7.** $(6, 8), (3, -9)$

9.1 Scatter Plots

Essential Question How can you construct and interpret a scatter plot?

1 ACTIVITY: Constructing a Scatter Plot

Work with a partner. The weights x (in ounces) and circumferences C (in inches) of several sports balls are shown.

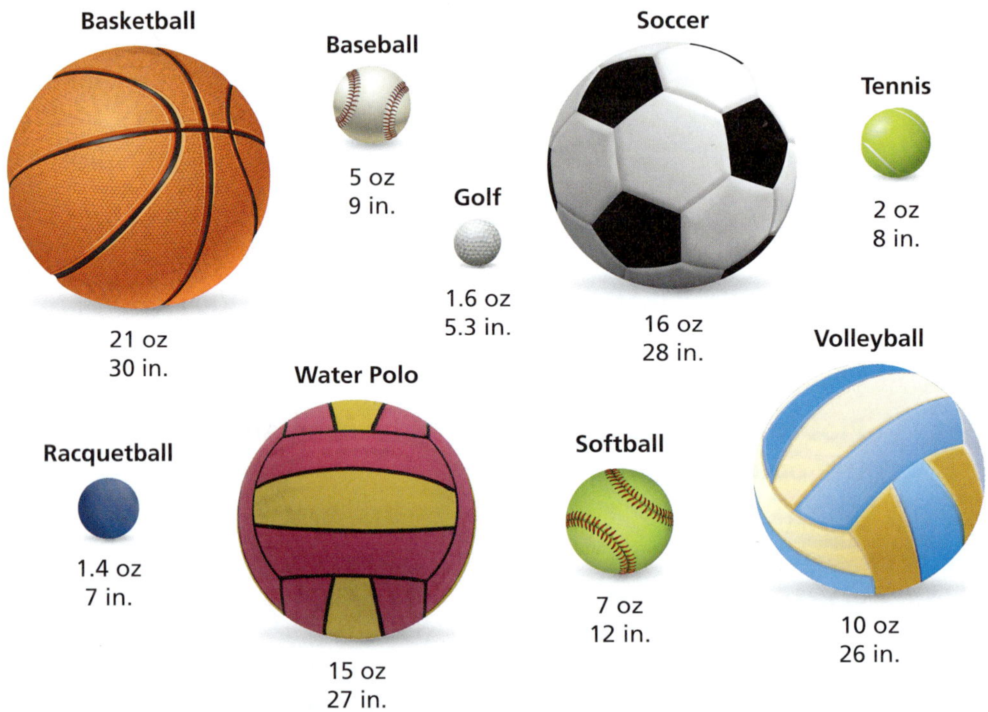

a. Choose a scale for the horizontal axis and the vertical axis of the coordinate plane shown.

b. Write the weight x and circumference C of each ball as an ordered pair. Then plot the ordered pairs in the coordinate plane.

c. Describe the relationship between weight and circumference. Are any of the points close together?

d. In general, do you think you can describe this relationship as *positive* or *negative*? *linear* or *nonlinear*? Explain.

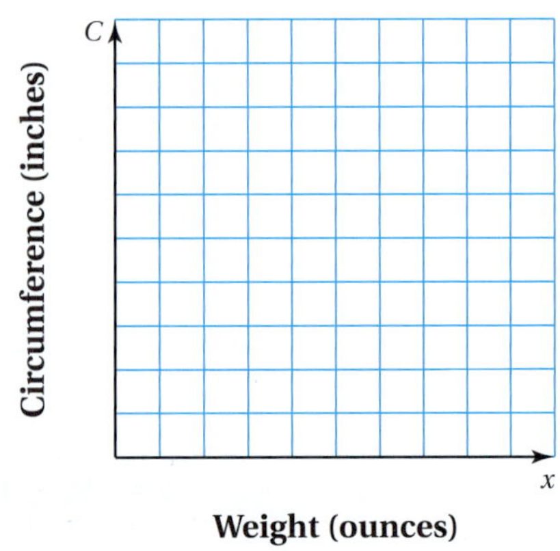

e. A bowling ball has a weight of 225 ounces and a circumference of 27 inches. Describe the location of the ordered pair that represents this data point in the coordinate plane. How does this point compare to the others? Explain your reasoning.

COMMON CORE

Data Analysis

In this lesson, you will

- construct and interpret scatter plots.
- describe patterns in scatter plots.

Learning Standard
MACC.8.SP.1.1

2 ACTIVITY: Constructing a Scatter Plot

Math Practice 5

Recognize Usefulness of Tools

How do you know when a scatter plot is a useful tool for making a prediction?

Work with a partner. The table shows the number of absences and the final grade for each student in a sample.

Absences	Final Grade
0	95
3	88
2	90
5	83
7	79
9	70
4	85
1	94
10	65
8	75

a. Write the ordered pairs from the table. Then plot them in a coordinate plane.

b. Describe the relationship between absences and final grade. How is this relationship similar to the relationship between weight and circumference in Activity 1? How is it different?

c. **MODELING** A student has been absent 6 days. Use the data to predict the student's final grade. Explain how you found your answer.

3 ACTIVITY: Identifying Scatter Plots

Work with a partner. Match the data sets with the most appropriate scatter plot. Explain your reasoning.

a. month of birth and birth weight for infants at a day care

b. quiz score and test score of each student in a class

c. age and value of laptop computers

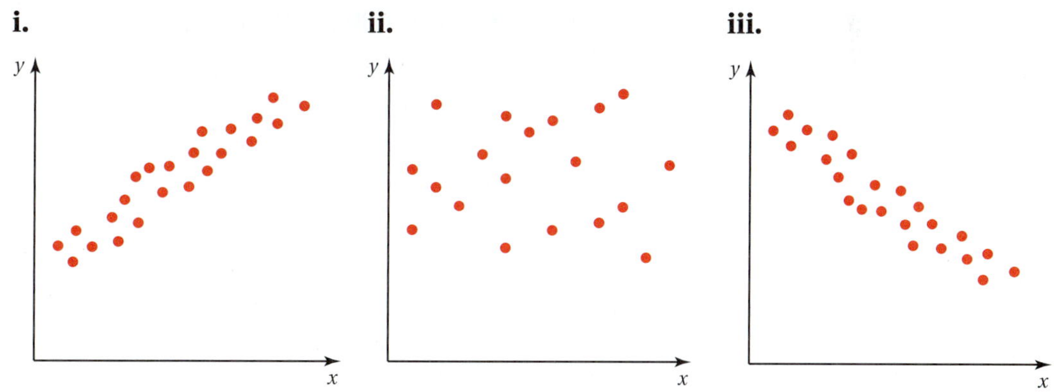

What Is Your Answer?

4. How would you define the term *scatter plot*?

5. **IN YOUR OWN WORDS** How can you construct and interpret a scatter plot?

Practice Use what you learned about scatter plots to complete Exercise 7 on page 376.

9.1 Lesson

Check It Out
Lesson Tutorials
BigIdeasMath.com

Key Vocabulary
scatter plot, *p. 374*

Key Idea

Scatter Plot

A **scatter plot** is a graph that shows the relationship between two data sets. The two sets of data are graphed as ordered pairs in a coordinate plane.

EXAMPLE 1 Interpreting a Scatter Plot

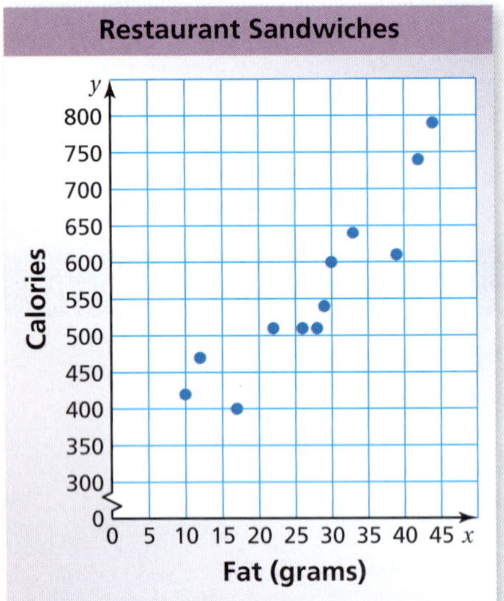

The scatter plot at the left shows the amounts of fat (in grams) and the numbers of calories in 12 restaurant sandwiches.

a. How many calories are in the sandwich that contains 17 grams of fat?

Draw a horizontal line from the point that has an x-value of 17. It crosses the y-axis at 400.

So, the sandwich has 400 calories.

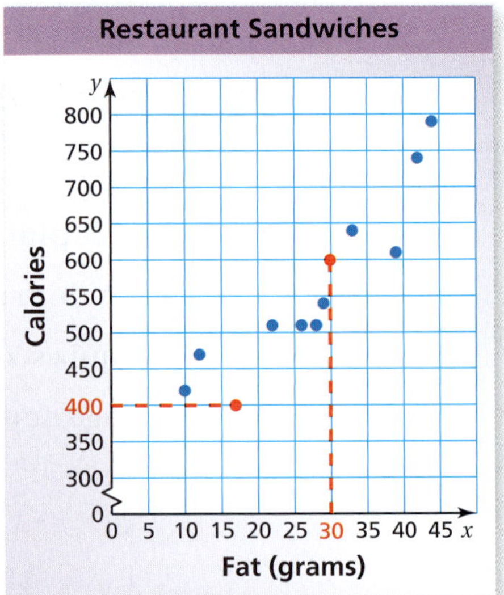

b. How many grams of fat are in the sandwich that contains 600 calories?

Draw a vertical line from the point that has a y-value of 600. It crosses the x-axis at 30.

So, the sandwich has 30 grams of fat.

c. What tends to happen to the number of calories as the number of grams of fat increases?

Looking at the graph, the plotted points go up from left to right.

So, as the number of grams of fat increases, the number of calories increases.

On Your Own

Now You're Ready
Exercises 8 and 9

1. **WHAT IF?** A sandwich has 650 calories. Based on the scatter plot in Example 1, how many grams of fat would you expect the sandwich to have? Explain your reasoning.

Multi-Language Glossary at BigIdeasMath.com

A scatter plot can show that a relationship exists between two data sets.

Positive Linear Relationship

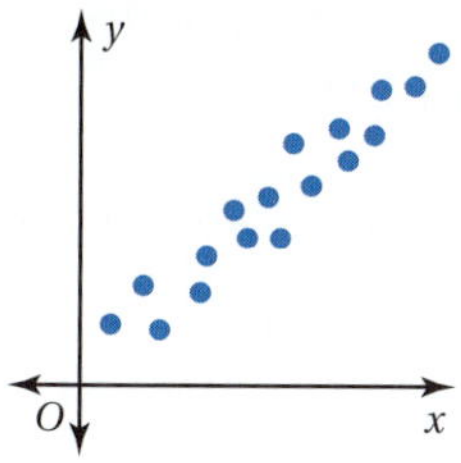

The points lie close to a line. As x increases, y increases.

Negative Linear Relationship

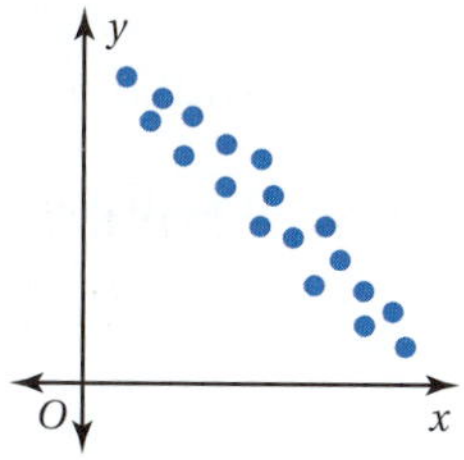

The points lie close to a line. As x increases, y decreases.

Nonlinear Relationship

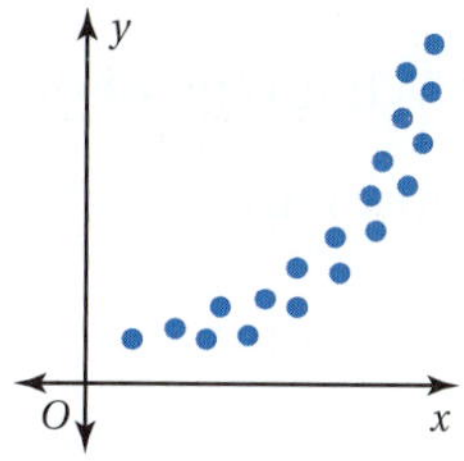

The points lie in the shape of a curve.

No Relationship

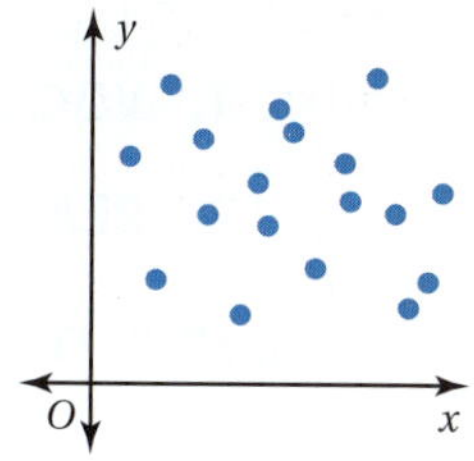

The points show no pattern.

EXAMPLE 2 Identifying Relationships

Describe the relationship between the data. Identify any outliers, gaps, or clusters.

a. television size and price

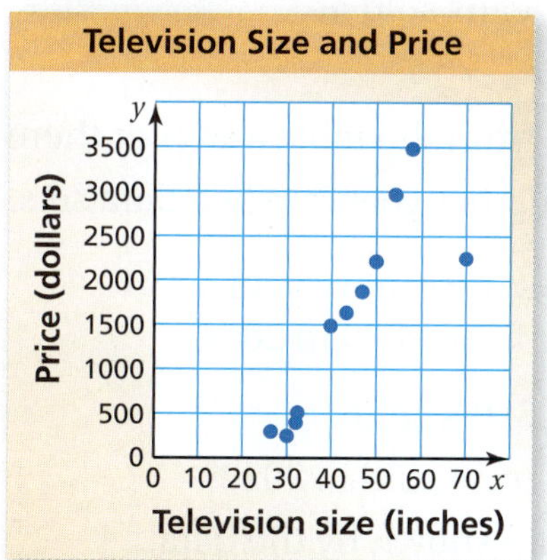

The points appear to lie close to a line. As x increases, y increases.

So, the scatter plot shows a positive linear relationship. There is an outlier at (70, 2250), a cluster of data under \$500, and a gap in the data from \$500 to \$1500.

b. age and number of pets owned

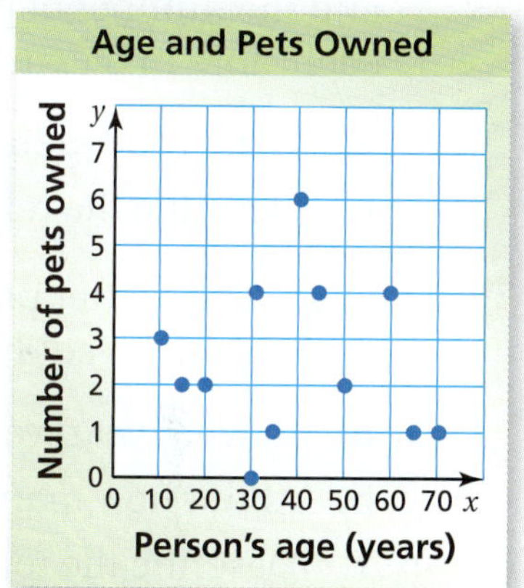

The points show no pattern.

So, the scatter plot shows no relationship. There are no obvious outliers, gaps, or clusters in the data.

On Your Own

Now You're Ready
Exercises 10–12

2. Make a scatter plot of the data and describe the relationship between the data. Identify any outliers, gaps, or clusters.

Study Time (min), x	30	20	60	90	45	10	30	75	120	80
Test Score, y	80	74	92	97	85	62	83	90	70	91

9.1 Exercises

Check It Out
Help with Homework
BigIdeasMath.com

Vocabulary and Concept Check

1. **VOCABULARY** What type of data do you need to make a scatter plot? Explain.
2. **REASONING** How can you identify an outlier in a scatter plot?

LOGIC Describe the relationship you would expect between the data. Explain.

3. shoe size of a student and the student's IQ
4. time since a train's departure and the distance to its destination
5. height of a bouncing ball and the time since it was dropped
6. number of toppings on a pizza and the price of the pizza

Practice and Problem Solving

7. **JEANS** The table shows the average price (in dollars) of jeans sold at different stores and the number of pairs of jeans sold at each store in one month.

Average Price	22	40	28	35	46
Number Sold	152	94	134	110	81

   a. Write the ordered pairs from the table and plot them in a coordinate plane.
   b. Describe the relationship between the two data sets.

1 8. **SUVS** The scatter plot shows the numbers of sport utility vehicles sold in a city from 2009 to 2014.

   a. In what year were 1000 SUVs sold?
   b. About how many SUVs were sold in 2013?
   c. Describe the relationship shown by the data.

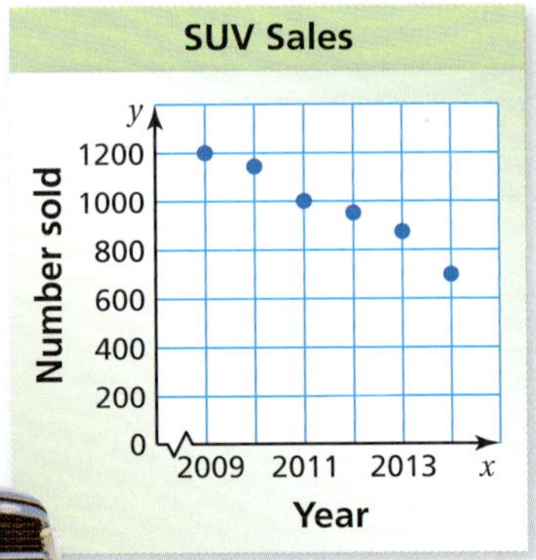

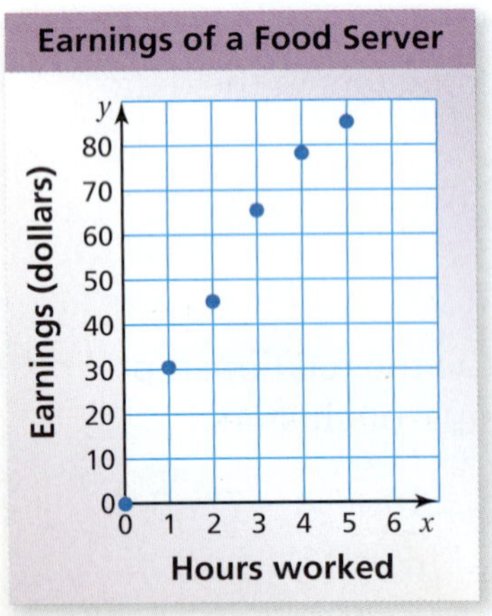

9. **EARNINGS** The scatter plot shows the total earnings (wages and tips) of a food server during one day.

   a. About how many hours must the server work to earn $70?
   b. About how much did the server earn for 5 hours of work?
   c. Describe the relationship shown by the data.

Describe the relationship between the data. Identify any outliers, gaps, or clusters.

2 **10.**

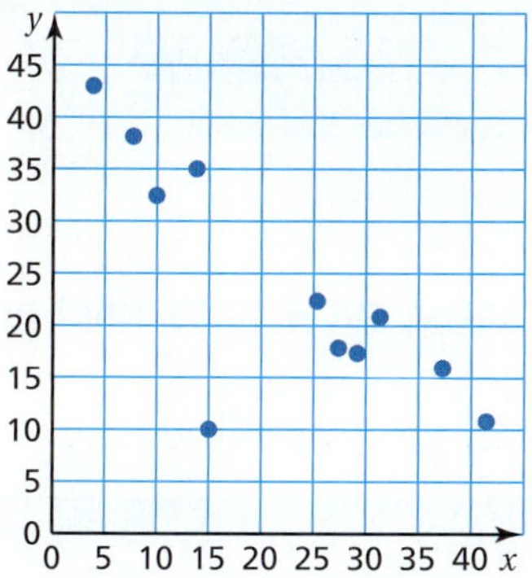

11.

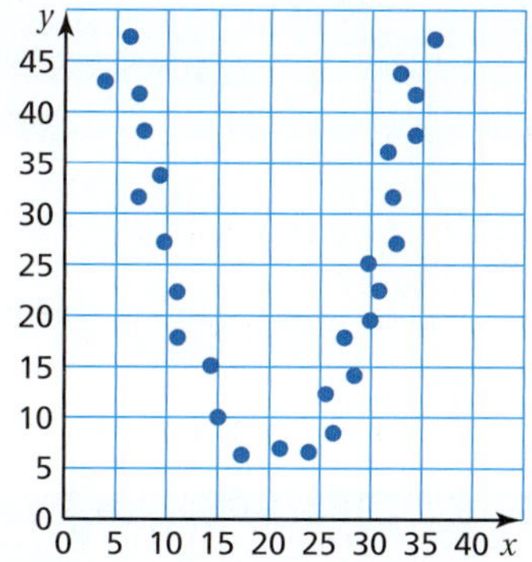

12.

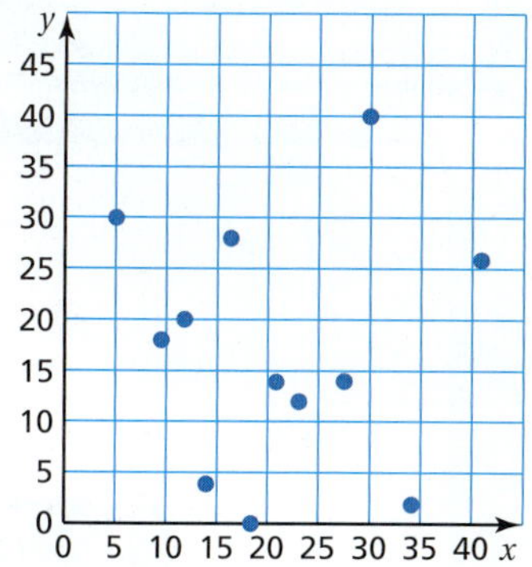

13. **HONEY** The table shows the average price per pound for honey in the United States from 2009 to 2012. What type of relationship do the data show?

Year, x	2009	2010	2011	2012
Average Price per Pound, y	\$4.65	\$4.85	\$5.15	\$5.53

14. **TEST SCORES** The scatter plot shows the numbers of minutes spent studying and the test scores for a science class. (a) What type of relationship do the data show? (b) Interpret the relationship.

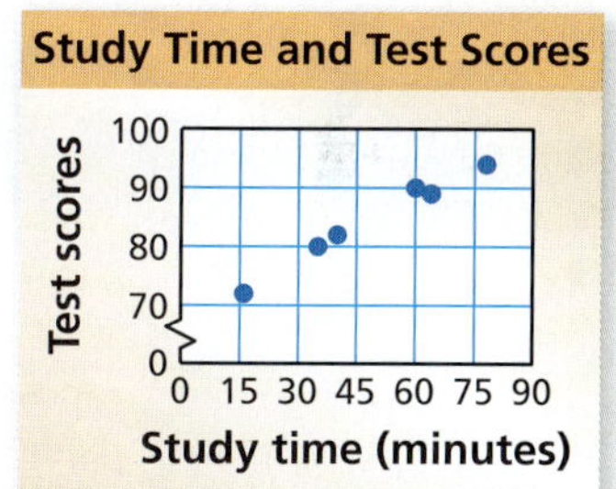

15. **OPEN-ENDED** Describe a set of real-life data that has a negative linear relationship.

16. **PROBLEM SOLVING** The table shows the memory capacities (in gigabytes) and prices (in dollars) of 7-inch tablet computers at a store. (a) Make a scatter plot of the data. Then describe the relationship between the data. (b) Identify any outliers, gaps, or clusters. Explain why you think they exist.

Memory (GB), x	8	16	4	32	4	16	4	8	16	8	16	8
Price (dollars), y	200	230	120	250	100	200	90	160	150	180	220	150

17. **Reasoning** Sales of sunglasses and beach towels at a store show a positive linear relationship in the summer. Does this mean that the sales of one item *cause* the sales of the other item to increase? Explain.

Fair Game Review What you learned in previous grades & lessons

Use a graph to solve the equation. Check your solution. *(Section 5.4)*

18. $5x = 2x + 6$

19. $7x + 3 = 9x - 13$

20. $\frac{2}{3}x = -\frac{1}{3}x - 4$

21. **MULTIPLE CHOICE** When graphing a proportional relationship represented by $y = mx$, which point is not on the graph? *(Section 4.3)*

Ⓐ (0, 0) Ⓑ $(0, m)$ Ⓒ $(1, m)$ Ⓓ $(2, 2m)$

9.2 Lines of Fit

Essential Question How can you use data to predict an event?

1 ACTIVITY: Representing Data by a Linear Equation

Work with a partner. You have been working on a science project for 8 months. Each month, you measured the length of a baby alligator.

The table shows your measurements.

September (month 0) | April (month 7)

Month, x	0	1	2	3	4	5	6	7
Length (in.), y	22.0	22.5	23.5	25.0	26.0	27.5	28.5	29.5

Use the following steps to predict the baby alligator's length next September.

a. Graph the data in the table.

b. Draw a line that you think best approximates the points.

c. Write an equation for your line.

d. **MODELING** Use the equation to predict the baby alligator's length next September.

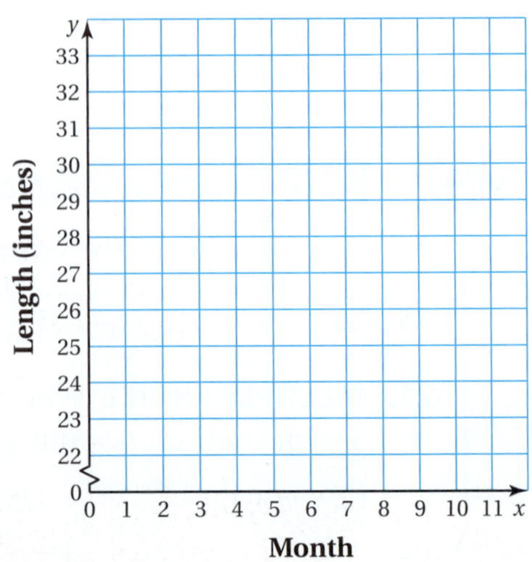

COMMON CORE

Data Analysis

In this lesson, you will

- find lines of fit.
- use lines of fit to solve problems.

Learning Standards
MACC.8.SP.1.1
MACC.8.SP.1.2
MACC.8.SP.1.3

2 ACTIVITY: Representing Data by a Linear Equation

Work with a partner. You are a biologist and study bat populations.

You are asked to predict the number of bats that will be living in an abandoned mine after 3 years.

To start, you find the number of bats that have been living in the mine during the past 8 years.

The table shows the results of your research.

Math Practice 4

Use a Graph

How can you draw a line that "fits" the collection of points? How should the points be positioned around the line?

7 years ago → 0; this year → 7

Year, x	0	1	2	3	4	5	6	7
Bats (thousands), y	327	306	299	270	254	232	215	197

Use the following steps to predict the number of bats that will be living in the mine after 3 years.

a. Graph the data in the table.

b. Draw a line that you think best approximates the points.

c. Write an equation for your line.

d. **MODELING** Use the equation to predict the number of bats in 3 years.

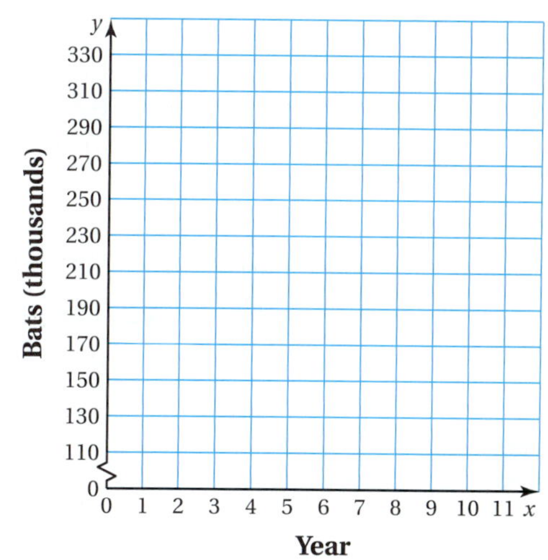

What Is Your Answer?

3. **IN YOUR OWN WORDS** How can you use data to predict an event?

4. **MODELING** Use the Internet or some other reference to find data that appear to have a linear pattern. List the data in a table, and then graph the data. Use an equation that is based on the data to predict a future event.

Practice Use what you learned about lines of fit to complete Exercise 4 on page 382.

9.2 Lesson

Check It Out
Lesson Tutorials
BigIdeasMath.com

Key Vocabulary
line of fit, *p. 380*
line of best fit, *p. 381*

A **line of fit** is a line drawn on a scatter plot close to most of the data points. It can be used to estimate data on a graph.

EXAMPLE 1 Finding a Line of Fit

Month, x	Depth (feet), y
0	20
1	19
2	15
3	13
4	11
5	10
6	8
7	7
8	5

The table shows the depth of a river x months after a monsoon season ends. (a) Make a scatter plot of the data and draw a line of fit. (b) Write an equation of the line of fit. (c) Interpret the slope and the y-intercept of the line of fit. (d) Predict the depth in month 9.

a. Plot the points in a coordinate plane. The scatter plot shows a negative linear relationship. Draw a line that is close to the data points. Try to have as many points above the line as below it.

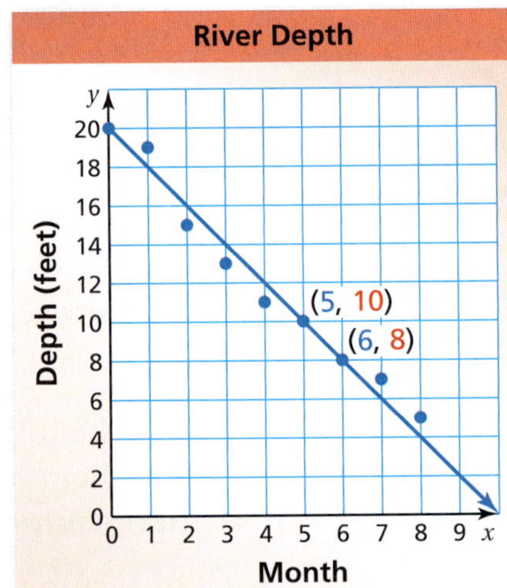

b. The line passes through (5, 10) and (6, 8).

$$\text{slope} = \frac{\text{rise}}{\text{run}} = \frac{-2}{1} = -2$$

Because the line crosses the y-axis at (0, 20), the y-intercept is 20.

So, an equation of the line of fit is $y = -2x + 20$.

c. The slope is -2, and the y-intercept is 20. So, the depth of the river is 20 feet at the end of the monsoon season and decreases by about 2 feet per month.

d. To predict the depth in month 9, substitute 9 for x in the equation of the line of fit.

$$y = -2x + 20 = -2(9) + 20 = 2$$

The depth in month 9 should be about 2 feet.

Study Tip
A line of fit does not need to pass through any of the data points.

On Your Own

Now You're Ready
Exercises 5 and 6

1. The table shows the numbers of people who have attended a festival over an 8-year period. (a) Make a scatter plot of the data and draw a line of fit. (b) Write an equation of the line of fit. (c) Interpret the slope and the y-intercept of the line of fit. (d) Predict the number of people who will attend the festival in year 10.

Year, x	1	2	3	4	5	6	7	8
Attendance, y	420	500	650	900	1100	1500	1750	2400

Multi-Language Glossary at BigIdeasMath.com

Study Tip

You know how to use two points to find an equation of a line of fit. When finding an equation of the line of best fit, every point in the data set is used.

Graphing calculators use a method called *linear regression* to find a precise line of fit called a **line of best fit**. This line best models a set of data. A calculator often gives a value r called the *correlation coefficient*. This value tells whether the correlation is positive or negative, and how closely the equation models the data. Values of r range from -1 to 1. When r is close to 1 or -1, there is a strong correlation between the variables. As r gets closer to 0, the correlation becomes weaker.

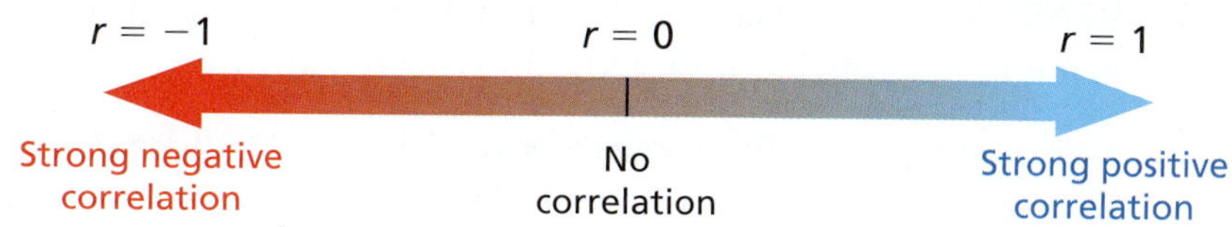

EXAMPLE 2 Finding a Line of Best Fit Using Technology

The table shows the worldwide movie ticket sales y (in billions of dollars) from 2000 to 2011, where $x = 0$ represents the year 2000. Use a graphing calculator to find an equation of the line of best fit. Identify and interpret the correlation coefficient.

Year, x	0	1	2	3	4	5	6	7	8	9	10	11
Ticket Sales, y	16	17	20	20	25	23	26	26	28	29	32	33

Step 1: Enter the data from the table into your calculator.

Step 2: Use the *linear regression* feature.

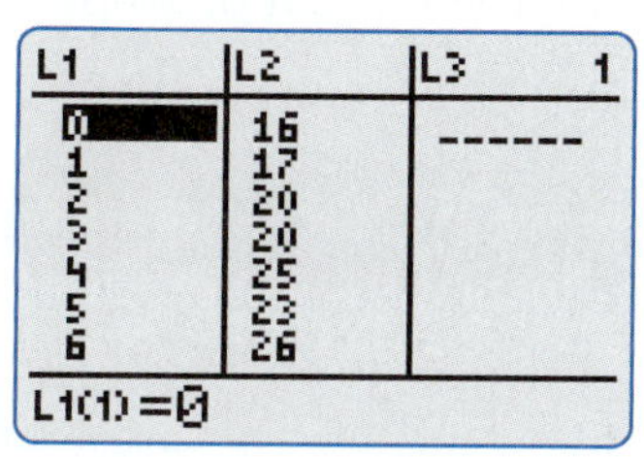

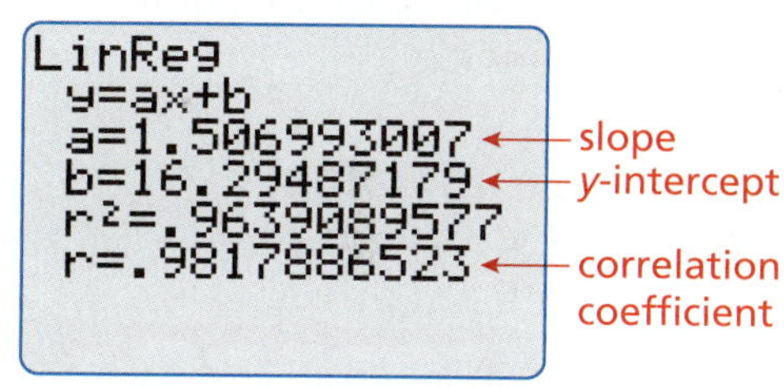

Study Tip

The slope of 1.5 indicates that sales are increasing by about \$1.5 billion each year. The *y*-intercept of 16 represents the ticket sales of \$16 billion for 2000.

An equation of the line of best fit is $y = 1.5x + 16$. The correlation coefficient is about 0.982. This means that the relationship between years and ticket sales is a strong positive correlation and that the equation closely models the data.

Check Use a graphing calculator to make a scatter plot and graph the line of best fit.

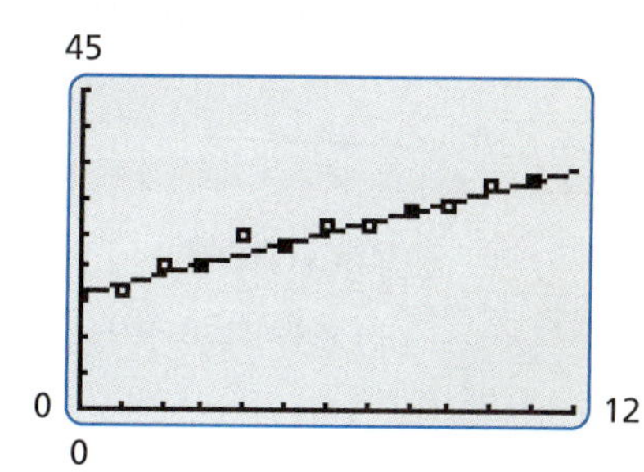

On Your Own

Now You're Ready Exercises 8–10

2. Use a graphing calculator to find an equation of the line of best fit for the data in Example 1. Identify and interpret the correlation coefficient.

9.2 Exercises

Check It Out
Help with Homework
BigIdeasMath.com

Vocabulary and Concept Check

1. **WRITING** Explain why a line of fit is helpful when analyzing data.
2. **REASONING** Tell whether the line drawn on the graph is a good fit for the data. Explain your reasoning.
3. **NUMBER SENSE** Which correlation coefficient indicates a stronger relationship: -0.98 or 0.91? Explain.

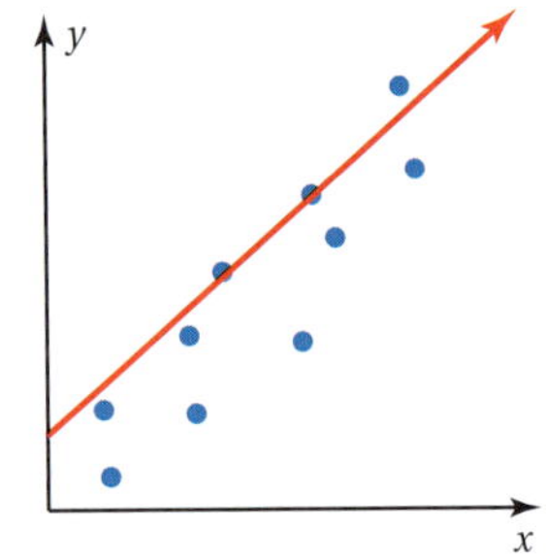

Practice and Problem Solving

4. **BLUEBERRIES** The table shows the weights y of x pints of blueberries.

Number of Pints, x	0	1	2	3	4	5
Weight (pounds), y	0	0.8	1.50	2.20	3.0	3.75

   a. Graph the data in the table.
   b. Draw a line that you think best approximates the points.
   c. Write an equation for your line.
   d. Use the equation to predict the weight of 10 pints of blueberries.
   e. Blueberries cost $2.25 per pound. How much do 10 pints of blueberries cost?

1 5. **HOT CHOCOLATE** The table shows the daily high temperature (°F) and the number of hot chocolates sold at a coffee shop for eight randomly selected days.

Temperature (°F), x	30	36	44	51	60	68	75	82
Hot Chocolates, y	45	43	36	35	30	27	23	17

   a. Make a scatter plot of the data and draw a line of fit.
   b. Write an equation of the line of fit.
   c. Interpret the slope and the y-intercept of the line of fit.
   d. Predict the number of hot chocolates sold when the high temperature is 20°F.

6. **VACATION** The table shows the distance you are away from home over a 6-hour period of your vacation.

   a. Make a scatter plot of the data and draw a line of fit.
   b. Write an equation of the line of fit.
   c. About how many miles per hour do you travel?
   d. About how far were you from home when you started?
   e. Predict the distance from home in 7 hours.

Hours, x	Distance (miles), y
1	62
2	123
3	188
4	228
5	280
6	344

7. **REASONING** A data set has no relationship. Is it possible to find a line of fit for the data? Explain.

(2) 8. **AMUSEMENT PARK** The table shows the attendance y (in thousands) at an amusement park from 2004 to 2013, where $x = 4$ represents the year 2004. Use a graphing calculator to find an equation of the line of best fit. Identify and interpret the correlation coefficient.

Year, x	4	5	6	7	8	9	10	11	12	13
Attendance (thousands), y	850	845	828	798	800	792	785	781	775	760

9. **SNOWSTORM** The table shows the total snow depth y (in inches) on the ground during a snowstorm x hours after it began. Use a graphing calculator to find an equation of the line of best fit. Identify and interpret the correlation coefficient. Use your equation to estimate how much snow was on the ground before the snowstorm began.

Hours, x	1	2	3	4	5	6	7	8
Snow Depth (inches), y	5	6	6.75	7.75	8.5	9.5	10.5	11.5

10. **TEXTING** The table shows the numbers y (in billions) of text messages sent from 2006 to 2011, where $x = 6$ represents the year 2006.

Year, x	Text Messages (billions), y
6	113
7	241
8	601
9	1360
10	1806
11	2206

a. Use a graphing calculator to find an equation of the line of best fit. Identify and interpret the correlation coefficient.

b. Interpret the slope of the line of best fit. Does the y-intercept make sense for this problem? Explain.

c. Predict the number of text messages sent in 2015.

11. **Modeling** The table shows the height y (in feet) of a baseball x seconds after it was hit.

Seconds, x	Height (feet), y
0	3
0.5	39
1	67
1.5	87
2	99

a. Use a graphing calculator to find an equation of the line of best fit. Identify and interpret the correlation coefficient.

b. Predict the height after 5 seconds.

c. The actual height after 5 seconds is about 3 feet. Why do you think this is different from your prediction?

Fair Game Review What you learned in previous grades & lessons

Write the decimal as a fraction or a mixed number. *(Section 7.4)*

12. $0.\overline{2}$ 13. $-2.\overline{7}$ 14. $-1.4\overline{6}$ 15. $0.\overline{81}$

16. **MULTIPLE CHOICE** Which expression represents the volume of a sphere with radius r? *(Section 8.3)*

(A) $\frac{1}{3}\pi r^2 h$ (B) $\pi r^2 h$ (C) $4\pi r^2$ (D) $\frac{4}{3}\pi r^3$

Check It Out
Graphic Organizer
BigIdeasMath.com

You can use an **information frame** to help you organize and remember concepts. Here is an example of an information frame for scatter plots.

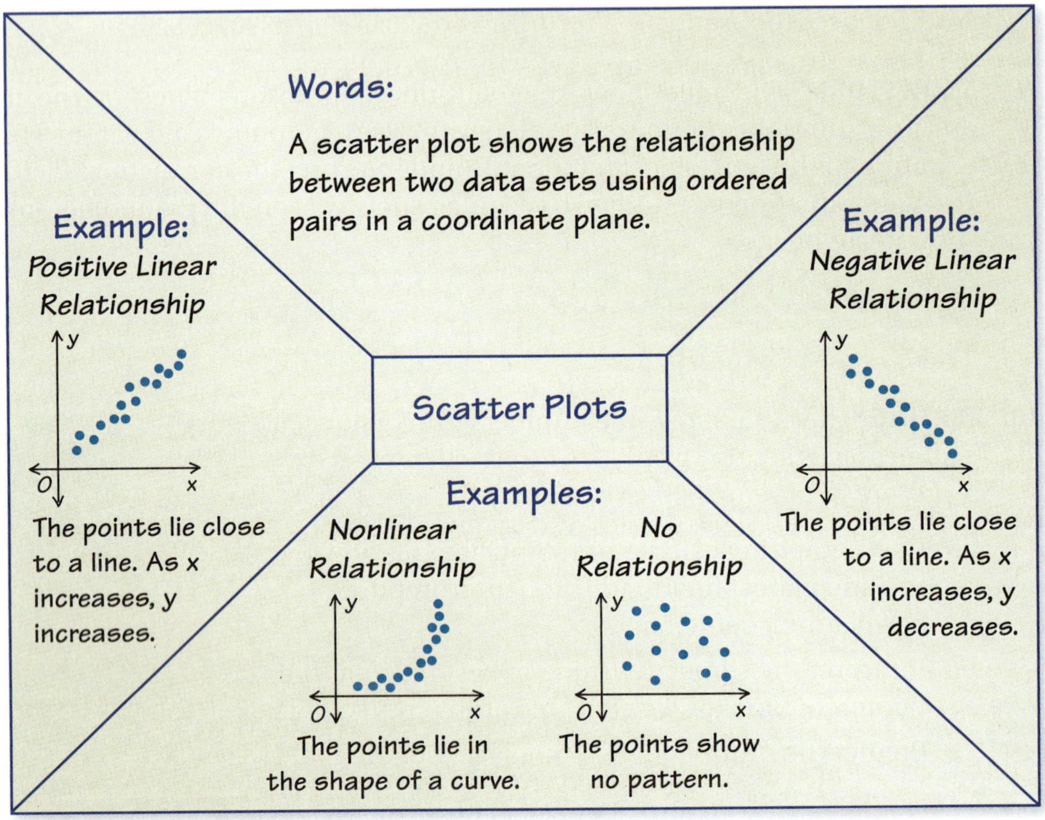

On Your Own

Make an information frame to help you study this topic.

1. lines of fit

After you complete this chapter, make information frames for the following topics.

2. two-way tables
3. data displays

"Dear Teacher, I am emailing my information frame showing the characteristics of circles."

9.1–9.2 Quiz

Check It Out
Progress Check
BigIdeasMath.com

1. **CHARITY** The scatter plot shows the amount of money donated to a charity from 2007 to 2012. *(Section 9.1)*

   a. In what year did the charity receive $150,000?

   b. How much did the charity receive in 2010?

   c. Describe the relationship shown by the data.

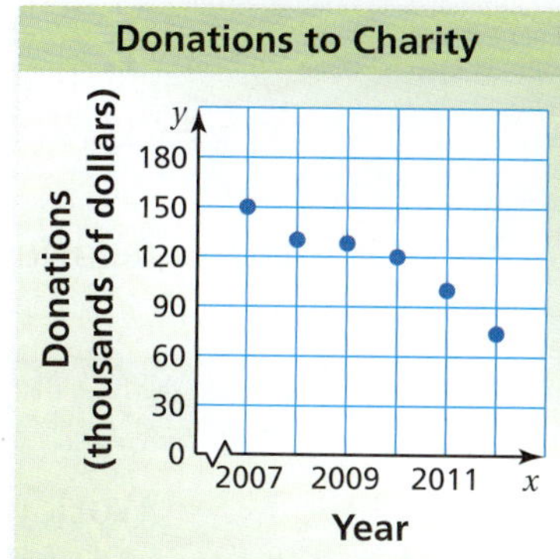

Describe the relationship between the data. Identify any outliers, gaps, or clusters. *(Section 9.1)*

2.

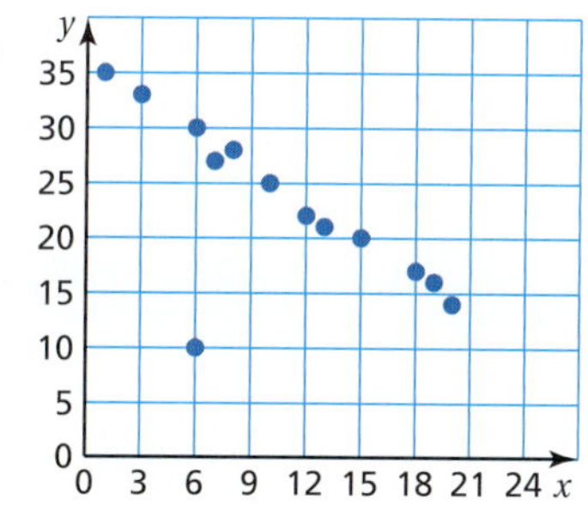

3.

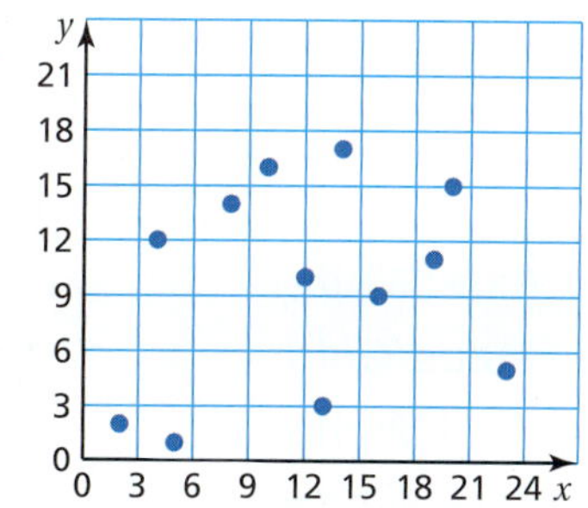

4.

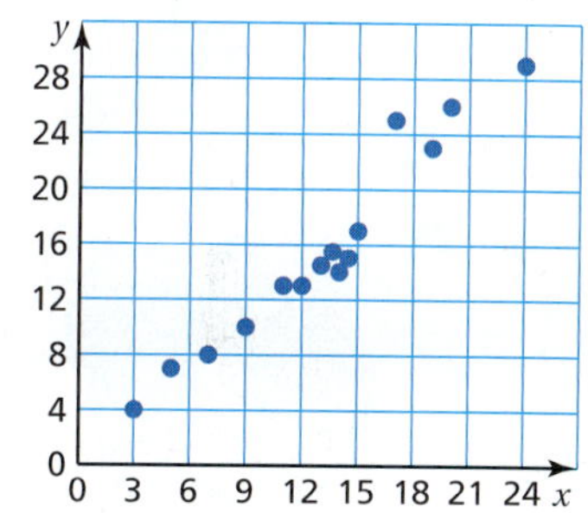

Year, x	Millions of Customers, y
0	12
1	18
2	21
3	28
4	33
5	38
6	42
7	48

5. **CUSTOMERS SERVED** The table shows the numbers of customers (in millions) served by a restaurant chain over an 8-year period. *(Section 9.1 and Section 9.2)*

   a. Make a scatter plot of the data. Then describe the relationship between the data. Identify any outliers, gaps, or clusters.

   b. Use a graphing calculator to find the equation of the line of best fit for the data. Identify and interpret the correlation coefficient.

6. **CATS** An animal shelter opens in December. The table shows the number of cats adopted from the shelter each month from January to September. *(Section 9.2)*

Month	1	2	3	4	5	6	7	8	9
Cats	3	6	7	11	13	14	15	18	19

   a. Make a scatter plot of the data and draw a line of fit.

   b. Write an equation of the line of fit.

   c. Interpret the slope and the y-intercept of the line of fit.

   d. Predict how many cats will be adopted in October.

9.3 Two-Way Tables

Essential Question How can you read and make a two-way table?

Two categories of data can be displayed in a *two-way table*.

1 ACTIVITY: Reading a Two-Way Table

Work with a partner. You are the manager of a sports shop. The two-way table shows the numbers of soccer T-shirts that your shop has left in stock at the end of the season.

		T-Shirt Size					
		S	M	L	XL	XXL	Total
Color	Blue/White	5	4	1	0	2	
	Blue/Gold	3	6	5	2	0	
	Red/White	4	2	4	1	3	
	Black/White	3	4	1	2	1	
	Black/Gold	5	2	3	0	2	
	Total						65

a. Complete the totals for the rows and columns.

b. Are there any black-and-gold XL T-shirts in stock? Justify your answer.

c. The numbers of T-shirts you ordered at the beginning of the season are shown below. Complete the two-way table.

		T-Shirt Size					
		S	M	L	XL	XXL	Total
Color	Blue/White	5	6	7	6	5	
	Blue/Gold	5	6	7	6	5	
	Red/White	5	6	7	6	5	
	Black/White	5	6	7	6	5	
	Black/Gold	5	6	7	6	5	
	Total						

d. **REASONING** How would you alter the numbers of T-shirts you order for next season? Explain your reasoning.

COMMON CORE

Data Analysis

In this lesson, you will

- read two-way tables.
- make and interpret two-way tables.

Learning Standard
MACC.8.SP.1.4

2 ACTIVITY: Analyzing Data

Math Practice 3

Construct Arguments

What are the advantages of using a table instead of a graph to analyze data?

Work with a partner. The three-dimensional two-way table shows information about the numbers of hours students at a high school work at part-time jobs during the school year.

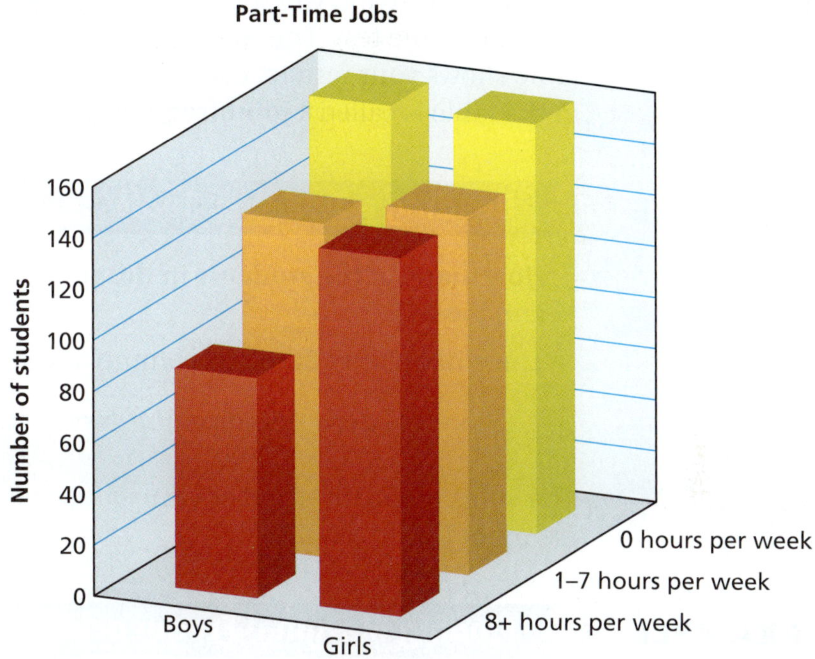

a. Make a two-way table showing the data. Use estimation to find the entries in your table.

b. Write two observations you can make that summarize the data in your table.

c. **REASONING** A newspaper article claims that more boys than girls drop out of high school to work full-time. Do the data support this claim? Explain your reasoning.

What Is Your Answer?

3. **IN YOUR OWN WORDS** How can you read and make a two-way table?

4. Find a real-life data set that you can represent by a two-way table. Then make a two-way table for the data set.

Practice Use what you learned about two-way tables to complete Exercises 3–6 on page 390.

9.3 Lesson

Key Vocabulary
two-way table, *p. 388*
joint frequency, *p. 388*
marginal frequency, *p. 388*

A **two-way table** displays two categories of data collected from the same source.

You randomly survey students in your school about their grades on the last test and whether they studied for the test. The two-way table shows your results. Each entry in the table is called a **joint frequency**.

		Student	
		Studied	Did Not Study
Grade	Passed	21	2
	Failed	1	6

joint frequency

EXAMPLE 1 Reading a Two-Way Table

How many of the students in the survey above studied for the test and passed?

The entry in the "Studied" column and "Passed" row is 21.

So, 21 of the students in the survey studied for the test and passed.

The sums of the rows and columns in a two-way table are called **marginal frequencies**.

EXAMPLE 2 Finding Marginal Frequencies

Find and interpret the marginal frequencies for the survey above.

Create a new column and a new row for the sums. Then add the entries.

		Student		
		Studied	Did Not Study	Total
Grade	Passed	21	2	23
	Failed	1	6	7
	Total	22	8	30

23 students passed.
7 students failed.
30 students were surveyed.
22 students studied.
8 students did not study.

On Your Own

Now You're Ready
Exercises 5–8

1. You randomly survey students in a cafeteria about their plans for a football game and a school dance. The two-way table shows your results.
   a. How many students will attend the dance but not the football game?
   b. Find and interpret the marginal frequencies for the survey.

		Football Game	
		Attend	Not Attend
Dance	Attend	35	5
	Not Attend	16	20

EXAMPLE 3 Making a Two-Way Table

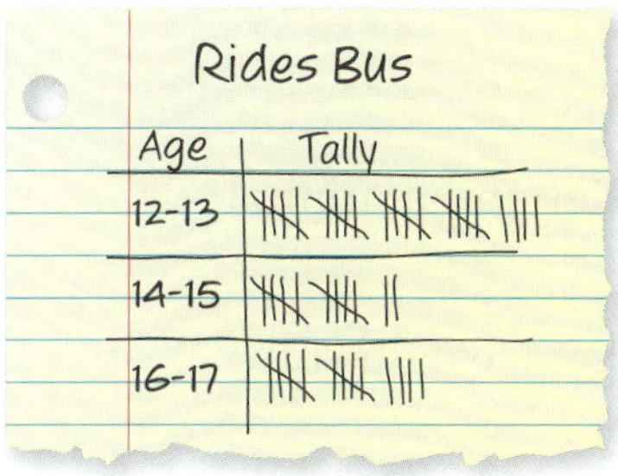

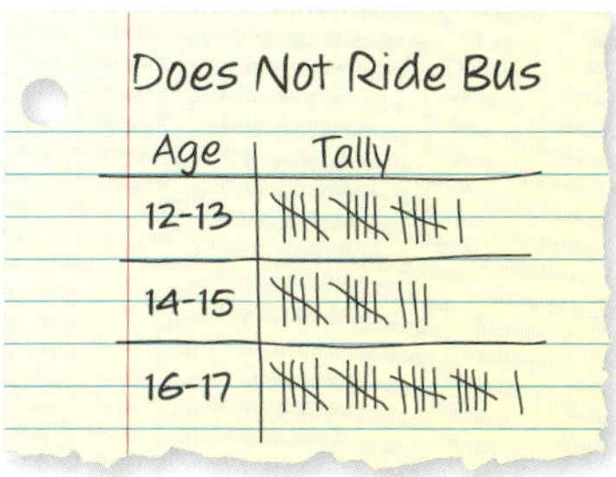

You randomly survey students between the ages of 12 and 17 about whether they ride the bus to school. The results are shown in the tally sheets. Make a two-way table that includes the marginal frequencies.

The two categories for the table are the ages and whether or not they ride the bus. Use the tally sheets to calculate each joint frequency. Then add to find each marginal frequency.

		Age			
		12–13	14–15	16–17	Total
Student	Rides Bus	24	12	14	50
Student	Does Not Ride Bus	16	13	21	50
	Total	40	25	35	100

EXAMPLE 4 Finding a Relationship in a Two-Way Table

Use the two-way table in Example 3.

a. For each age group, what percent of the students in the survey ride the bus to school? do not ride the bus to school? Organize the results in a two-way table. Explain what one of the entries represents.

		Age		
		12–13	14–15	16–17
Student	Rides Bus	60%	48%	40%
Student	Does Not Ride Bus	40%	52%	60%

$\frac{14}{35} = 0.4$

So, 40% of the 16- and 17-year-old students in the survey ride the bus to school.

b. Does the table in part (a) show a relationship between age and whether students ride the bus to school? Explain.

∴ Yes, the table shows that as age increases, students are less likely to ride the bus to school.

On Your Own

Now You're Ready
Exercise 10

2. You randomly survey students in a school about whether they buy a school lunch or pack a lunch. Your results are shown.

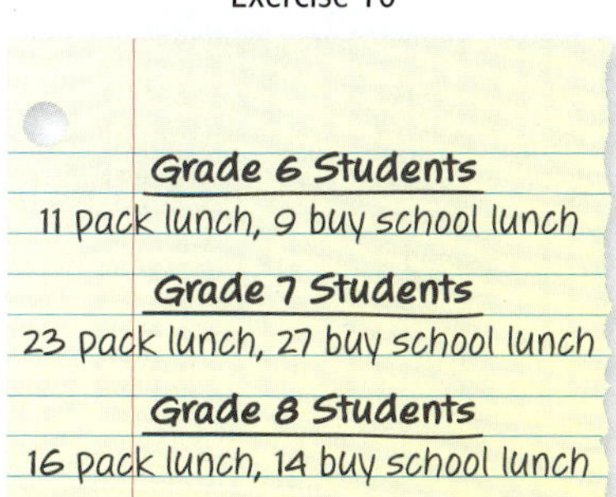

   a. Make a two-way table that includes the marginal frequencies.

   b. For each grade level, what percent of the students in the survey pack a lunch? buy a school lunch? Organize the results in a two-way table. Explain what one of the entries represents.

   c. Does the table in part (b) show a relationship between grade level and lunch choice? Explain.

9.3 Exercises

Check It Out
Help with Homework
BigIdeasMath.com

Vocabulary and Concept Check

1. **VOCABULARY** Explain the relationship between joint frequencies and marginal frequencies.

2. **OPEN-ENDED** Describe how you can use a two-way table to organize data you collect from a survey.

Practice and Problem Solving

You randomly survey students about participating in their class's yearly fundraiser. You display the two categories of data in the two-way table.

3. Find the total of each row.

4. Find the total of each column.

1 5. How many female students will be participating in the fundraiser?

6. How many male students will *not* be participating in the fundraiser?

		Fundraiser	
		No	Yes
Gender	Female	22	51
	Male	30	29

Find and interpret the marginal frequencies.

2 7.

		School Play	
		Attend	Not Attend
Class	Junior	41	30
	Senior	52	23

8.

		Cell Phone Minutes	
		Limited	Unlimited
Text Plan	Limited	78	0
	Unlimited	175	15

9. **GOALS** You randomly survey students in your school. You ask what is most important to them: grades, popularity, or sports. You display your results in the two-way table.

   a. How many 7th graders chose sports? How many 8th graders chose grades?

   b. Find and interpret the marginal frequencies for the survey.

   c. What percent of students in the survey are 6th graders who chose popularity?

		Goal		
		Grades	Popularity	Sports
Grade	6th	31	18	23
	7th	39	16	19
	8th	42	6	17

③ ④ 10. SAVINGS You randomly survey people in your neighborhood about whether they have at least \$1000 in savings. The results are shown in the tally sheets.

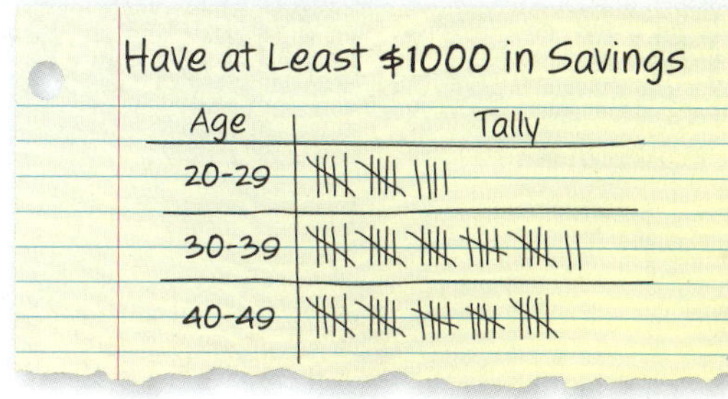

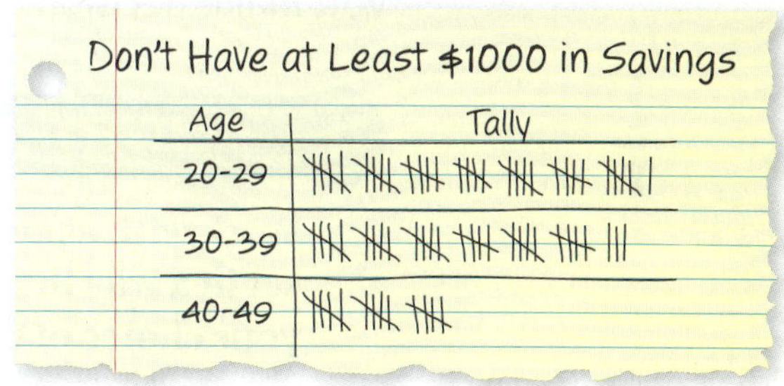

a. Make a two-way table that includes the marginal frequencies.

b. For each age group, what percent of the people have at least \$1000 in savings? do not have at least \$1000 in savings? Organize the results in a two-way table.

c. Does the table in part (b) show a relationship between age and whether people have at least \$1000 in savings? Explain.

11. EYE COLOR You randomly survey students in your school about the color of their eyes. The results are shown in the tables.

Eye Color of Males Surveyed		
Green	Blue	Brown
5	16	27

Eye Color of Females Surveyed		
Green	Blue	Brown
3	19	18

a. Make a two-way table.

b. Find and interpret the marginal frequencies for the survey.

c. For each eye color, what percent of the students in the survey are male? female? Organize the results in a two-way table. Explain what two of the entries represent.

12. REASONING Use the information from Exercise 11. For each gender, what percent of the students in the survey have green eyes? blue eyes? brown eyes? Organize the results in a two-way table. Explain what two of the entries represent.

13. Precision What percent of students in the survey in Exercise 11 are either female or have green eyes? What percent of students in the survey are males who do not have green eyes? Find and explain the sum of these two percents.

Fair Game Review What you learned in previous grades & lessons

Write an equation of the line that passes through the points. *(Section 4.6)*

14. (0, 1), (−2, −5) **15.** (0, −2), (3, 13) **16.** (−4, 1), (0, 3)

17. MULTIPLE CHOICE Which equation does not represent a linear function? *(Section 6.4)*

Ⓐ $y = 4x$ Ⓑ $xy = 8$ Ⓒ $y = -3$ Ⓓ $6x + 5y = -2$

9.4 Choosing a Data Display

Essential Question How can you display data in a way that helps you make decisions?

1 ACTIVITY: Displaying Data

Work with a partner. Analyze and display each data set in a way that best describes the data. Explain your choice of display.

a. **ROADKILL** A group of schools in New England participated in a 2-month study. They reported 3962 dead animals.

Birds: 307   Mammals: 2746
Amphibians: 145   Reptiles: 75
Unknown: 689

b. **BLACK BEAR ROADKILL** The data below show the numbers of black bears killed on a state's roads from 1993 to 2012.

1993:	30	2000:	47	2007:	99
1994:	37	2001:	49	2008:	129
1995:	46	2002:	61	2009:	111
1996:	33	2003:	74	2010:	127
1997:	43	2004:	88	2011:	141
1998:	35	2005:	82	2012:	135
1999:	43	2006:	109		

c. **RACCOON ROADKILL** A 1-week study along a 4-mile section of road found the following weights (in pounds) of raccoons that had been killed by vehicles.

13.4	14.8	17.0	12.9
21.3	21.5	16.8	14.8
15.2	18.7	18.6	17.2
18.5	9.4	19.4	15.7
14.5	9.5	25.4	21.5
17.3	19.1	11.0	12.4
20.4	13.6	17.5	18.5
21.5	14.0	13.9	19.0

d. What do you think can be done to minimize the number of animals killed by vehicles?

COMMON CORE
Data Analysis
In this lesson, you will
- choose appropriate data displays.
- identify and analyze misleading data displays.

Applying Standard
MACC.8.SP.1.1

2 ACTIVITY: Statistics Project

Math Practice 4

Use a Graph

How can you use a graph to represent the data you have gathered for your report? What does the graph tell you about the data?

ENDANGERED SPECIES PROJECT **Use the Internet or some other reference to write a report about an animal species that is (or has been) endangered. Include graphical displays of the data you have gathered.**

Sample: Florida Key Deer

In 1939, Florida banned the hunting of Key deer. The numbers of Key deer fell to about 100 in the 1940s.

In 1947, public sentiment was stirred by 11-year-old Glenn Allen from Miami. Allen organized Boy Scouts and others in a letter-writing campaign that led to the establishment of the National Key Deer Refuge in 1957. The approximately 8600-acre refuge includes 2280 acres of designated wilderness.

The Key Deer Refuge has increased the population of Key deer. A recent study estimated the total Key deer population to be approximately 800.

About half of Key deer deaths are due to vehicles.

One of two Key deer wildlife underpasses on Big Pine Key

What Is Your Answer?

3. **IN YOUR OWN WORDS** How can you display data in a way that helps you make decisions? Use the Internet or some other reference to find examples of the following types of data displays.

- Bar graph
- Circle graph
- Scatter plot
- Stem-and-leaf plot
- Box-and-whisker plot

Practice Use what you learned about choosing data displays to complete Exercise 3 on page 397.

9.4 Lesson

Check It Out
Lesson Tutorials
BigIdeasMath.com

Key Idea

Data Display	What does it do?	
Pictograph	shows data using pictures	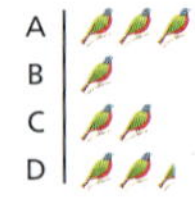
Bar Graph	shows data in specific categories	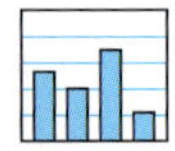
Circle Graph	shows data as parts of a whole	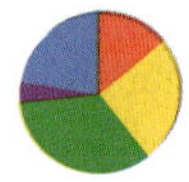
Line Graph	shows how data change over time	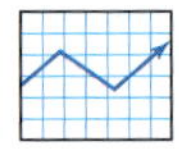
Histogram	shows frequencies of data values in intervals of the same size	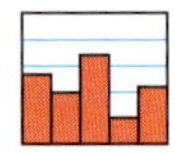
Stem-and-Leaf Plot	orders numerical data and shows how they are distributed	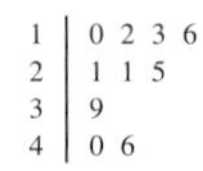
Box-and-Whisker Plot	shows the variability of a data set by using quartiles	
Dot Plot	shows the number of times each value occurs in a data set	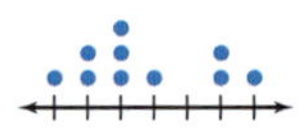
Scatter Plot	shows the relationship between two data sets by using ordered pairs in a coordinate plane	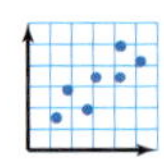

EXAMPLE 1 Choosing an Appropriate Data Display

Choose an appropriate data display for the situation. Explain your reasoning.

a. the number of students in a marching band each year

A line graph shows change over time. So, a line graph is an appropriate data display.

b. a comparison of people's shoe sizes and their heights

You want to compare two different data sets. So, a scatter plot is an appropriate data display.

On Your Own

Now You're Ready
Exercises 4–7

Choose an appropriate data display for the situation. Explain your reasoning.

1. the population of the United States divided into age groups

2. the percents of students in your school who play basketball, football, soccer, or lacrosse

EXAMPLE 2 Identifying an Appropriate Data Display

You record the number of hits for your school's new website for 5 months. Tell whether the data display is appropriate for representing how the number of hits changed during the 5 months. Explain your reasoning.

Month	Hits
August	250
September	320
October	485
November	650
December	925

a.

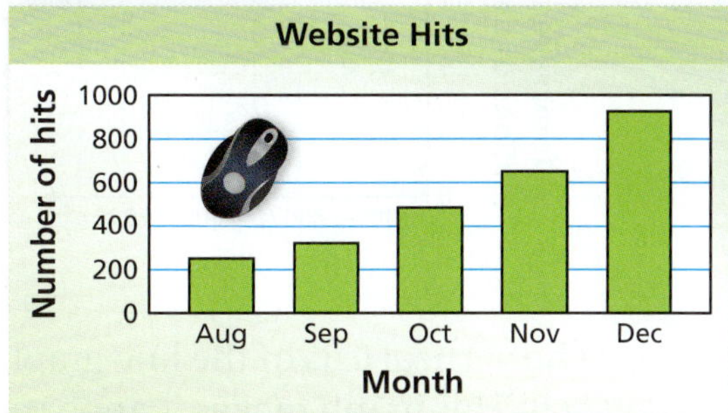

The bar graph shows the number of hits for each month. So, it is an appropriate data display.

b.

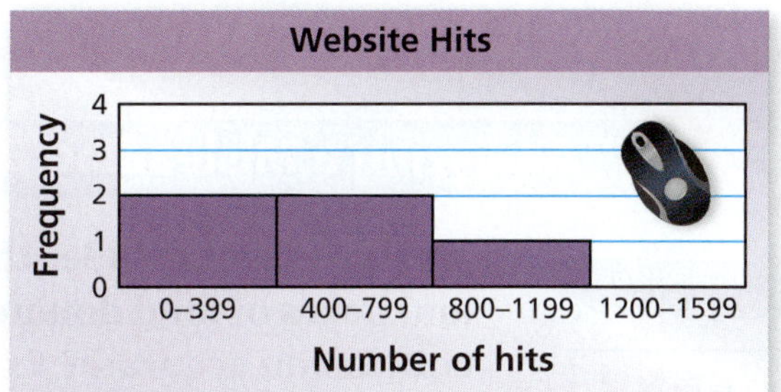

The histogram does not show the number of hits for each month or how the number of hits changes over time. So, it is *not* an appropriate data display.

c.

The line graph shows how the number of hits changes over time. So, it is an appropriate data display.

On Your Own

Now You're Ready
Exercises 8 and 9

Tell whether the data display is appropriate for representing the data in Example 2. Explain your reasoning.

3. dot plot **4.** circle graph **5.** stem-and-leaf plot

EXAMPLE 3 Identifying a Misleading Data Display

Which line graph is misleading? Explain.

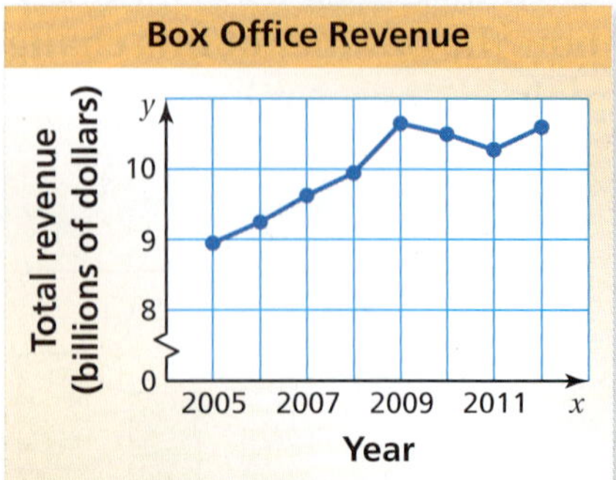

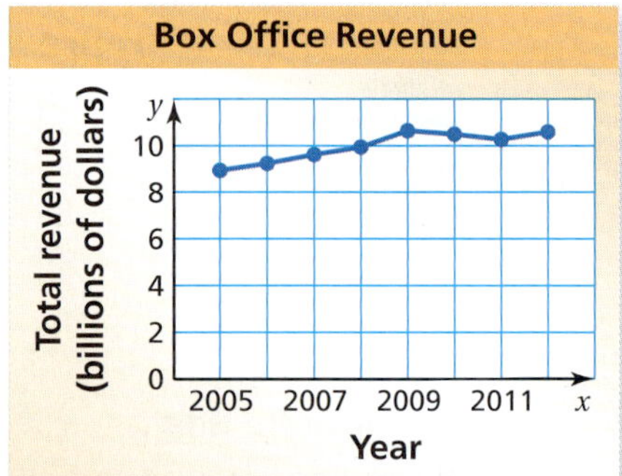

The vertical axis of the line graph on the left has a break (⩘) and begins at 8. This graph makes it appear that the total revenue increased rapidly from 2005 to 2009. The graph on the right has an unbroken axis. It is more honest and shows that the total revenue increased slowly.

∴ So, the graph on the left is misleading.

EXAMPLE 4 Analyzing a Misleading Data Display

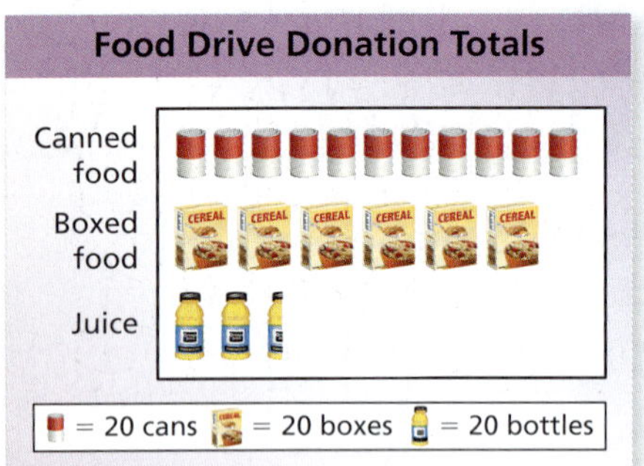

A volunteer concludes that the numbers of cans of food and boxes of food donated were about the same. Is this conclusion accurate? Explain.

Each icon represents the same number of items. Because the box icon is larger than the can icon, it looks like the number of boxes is about the same as the number of cans. But the number of boxes is actually about half of the number of cans.

∴ So, the conclusion is not accurate.

On Your Own

Now You're Ready
Exercises 11–14

Explain why the data display is misleading.

6.

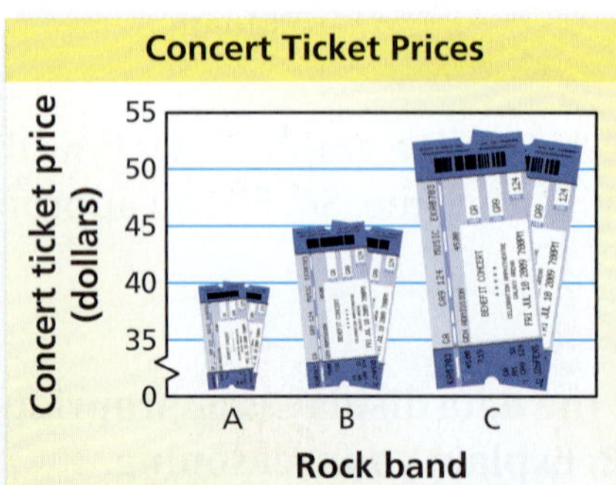

7.

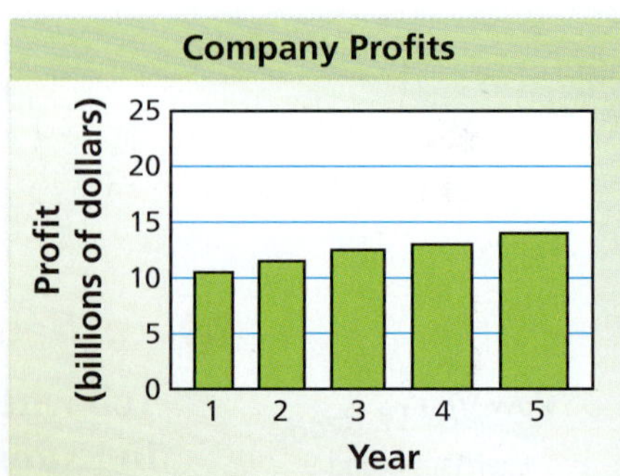

9.4 Exercises

Check It Out
Help with Homework
BigIdeasMath.com

Vocabulary and Concept Check

1. **REASONING** Can more than one display be appropriate for a data set? Explain.
2. **OPEN-ENDED** Describe how a histogram can be misleading.

Practice and Problem Solving

3. Analyze and display the data in a way that best describes the data. Explain your choice of display.

Notebooks Sold in One Week				
192 red	170 green	203 black	183 pink	230 blue
165 yellow	210 purple	250 orange	179 white	218 other

Choose an appropriate data display for the situation. Explain your reasoning.

1 4. a student's test scores and how the scores are spread out
5. the distance a person drives each month
6. the outcome of rolling a number cube
7. homework problems assigned each day

2 8. **LIFEGUARD** The table shows how many hours you worked as a lifeguard from May to August. Tell whether the data display is appropriate for representing how the number of hours worked changed during the 4 months. Explain your reasoning.

Lifeguard Schedule	
Month	**Hours Worked**
May	40
June	80
July	160
August	120

a.
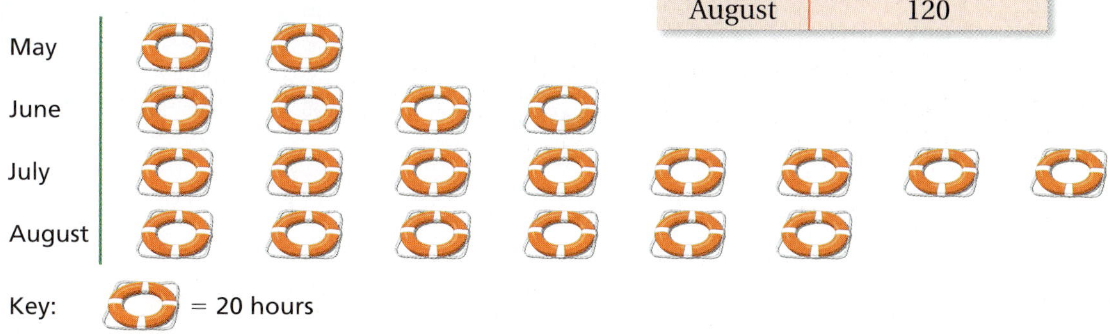

b.
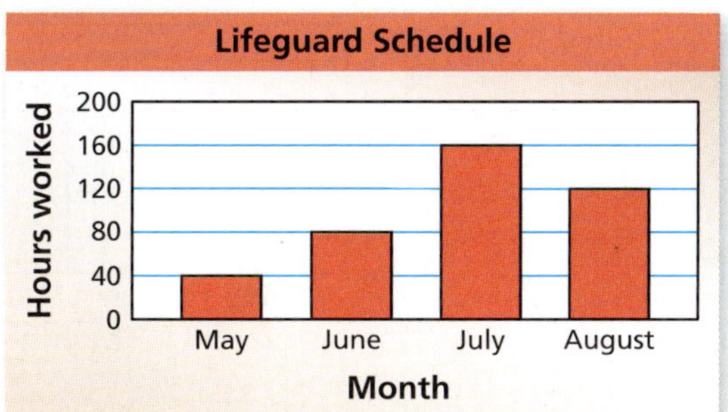

9. **FAVORITE SUBJECT** A survey asked 800 students to choose their favorite subject. The results are shown in the table. Tell whether the data display is appropriate for representing the portion of students who prefer math. Explain your reasoning.

Favorite School Subject	
Subject	**Number of Students**
Science	200
Math	160
Literature	240
Social Studies	120
Other	80

a.

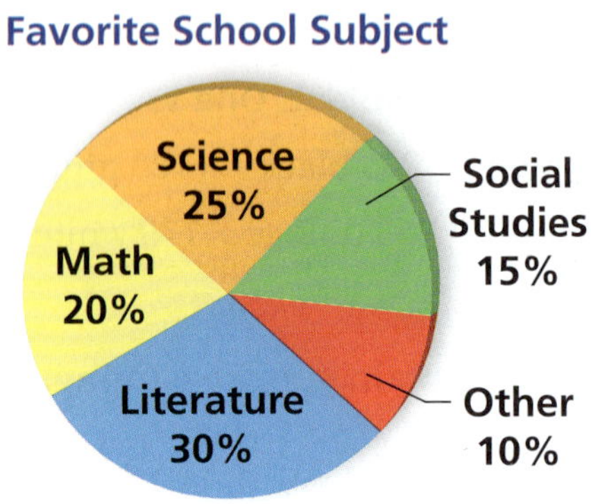

b.

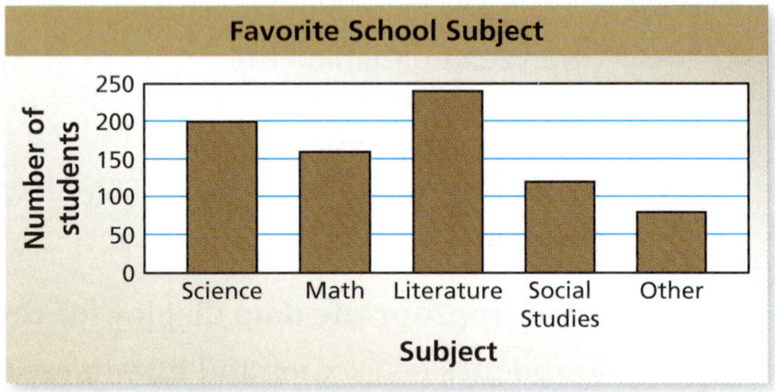

10. **WRITING** When should you use a histogram instead of a bar graph to display data? Use an example to support your answer.

Explain why the data display is misleading.

3 4 11.

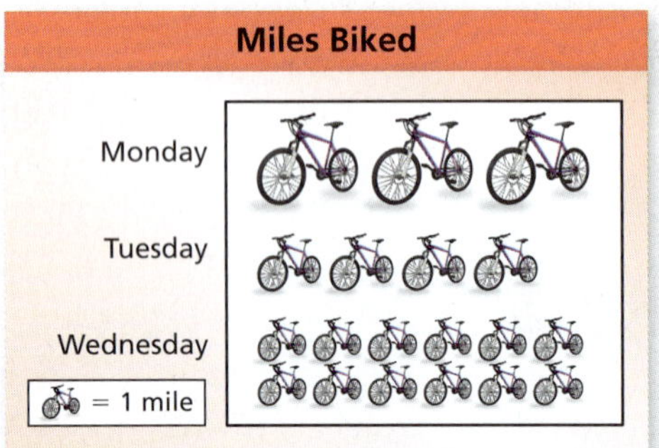

12.

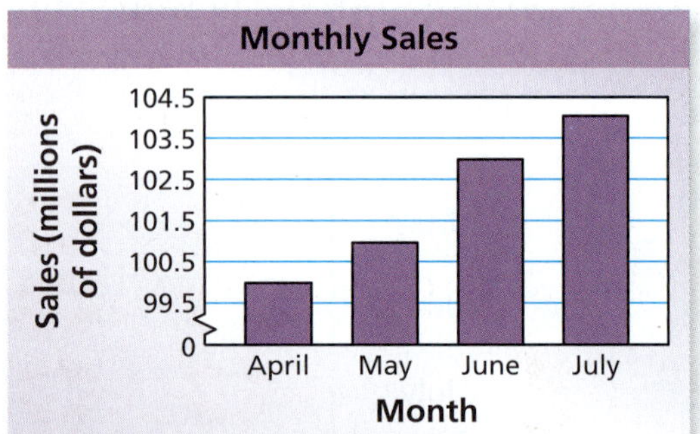

13.

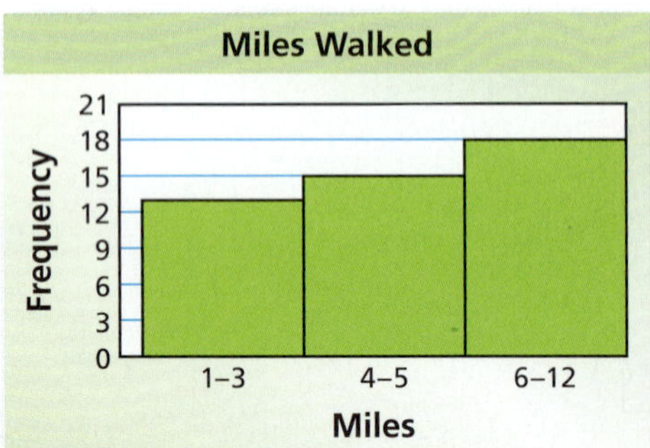

14.

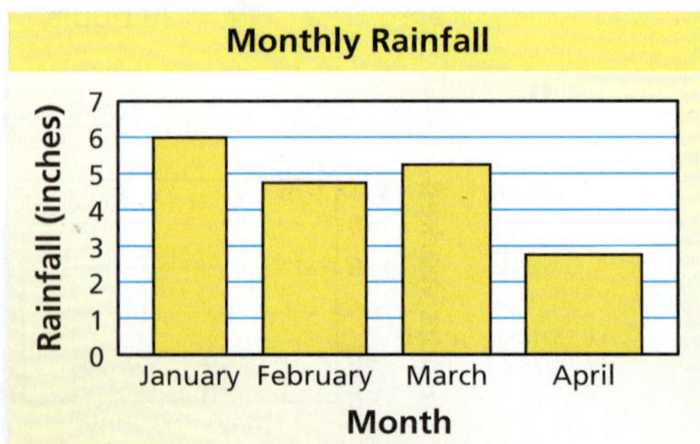

15. **VEGETABLES** A nutritionist wants to use a data display to show the favorite vegetables of the students at a school. Choose an appropriate data display for the situation. Explain your reasoning.

16. **CHEMICALS** A scientist gathers data about a decaying chemical compound. The results are shown in the scatter plot. Is the data display misleading? Explain.

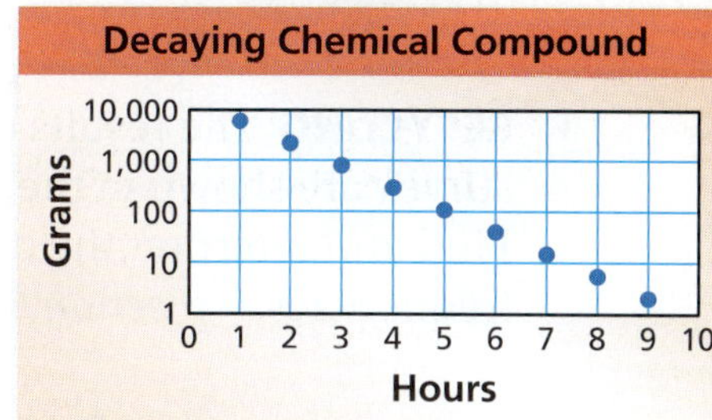

17. **REASONING** What type of data display is appropriate for showing the mode of a data set?

18. **SPORTS** A survey asked 100 students to choose their favorite sports. The results are shown in the circle graph.

a. Explain why the graph is misleading.

b. What type of data display would be more appropriate for the data? Explain.

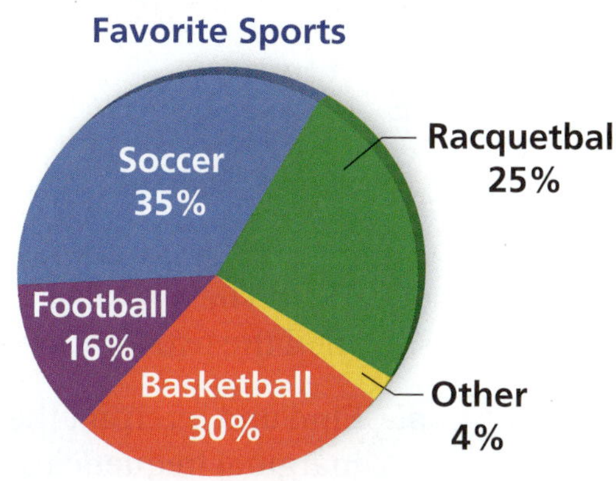

19. **Structure** With the help of computers, mathematicians have computed and analyzed billions of digits of the irrational number π. One of the things they analyze is the frequency of each of the numbers 0 through 9. The table shows the frequency of each number in the first 100,000 digits of π.

a. Display the data in a bar graph.

b. Display the data in a circle graph.

c. Which data display is more appropriate? Explain.

d. Describe the distribution.

Number	0	1	2	3	4	5	6	7	8	9
Frequency	9999	10,137	9908	10,025	9971	10,026	10,029	10,025	9978	9902

Fair Game Review What you learned in previous grades & lessons

Estimate the square root to the nearest (a) integer and (b) tenth. *(Section 7.4)*

20. $\sqrt{20}$ 21. $-\sqrt{74}$ 22. $\sqrt{140}$

23. **MULTIPLE CHOICE** What is 20% of 25% of 400? *(Skills Review Handbook)*

Ⓐ 20 Ⓑ 200 Ⓒ 240 Ⓓ 380

9.3–9.4 Quiz

Check It Out
Progress Check
BigIdeasMath.com

1. RECYCLING The results of a recycling survey are shown in the two-way table. Find and interpret the marginal frequencies. *(Section 9.3)*

		Recycle	
		Yes	**No**
Gender	**Female**	28	9
	Male	24	14

2. MUSIC The results of a music survey are shown in the two-way table. Find and interpret the marginal frequencies. *(Section 9.3)*

		Jazz	
		Likes	**Dislikes**
Country	**Likes**	26	14
	Dislikes	17	8

3. ELECTION The results of a voting survey are shown in the two-way table. *(Section 9.3)*

		Voter's Age		
		18–34	**35–64**	**65+**
Candidate	**Smith**	36	25	6
	Jackson	12	32	24

a. Find and interpret the marginal frequencies.

b. For each age group, what percent of voters prefer Smith? prefer Jackson? Organize your results in a two-way table.

c. Does your table in part (b) show a relationship between age and candidate preference? Explain.

Choose an appropriate data display for the situation. Explain your reasoning. *(Section 9.4)*

4. the percent of band students in each section of instruments

5. a company's profit for each week

6. TURTLES The tables show the weights (in pounds) of turtles caught in two ponds. Which type of data display would you use for this information? Explain. *(Section 9.4)*

Pond A			
12	13	15	6
7	8	12	7

Pond B			
9	12	5	8
12	15	16	19

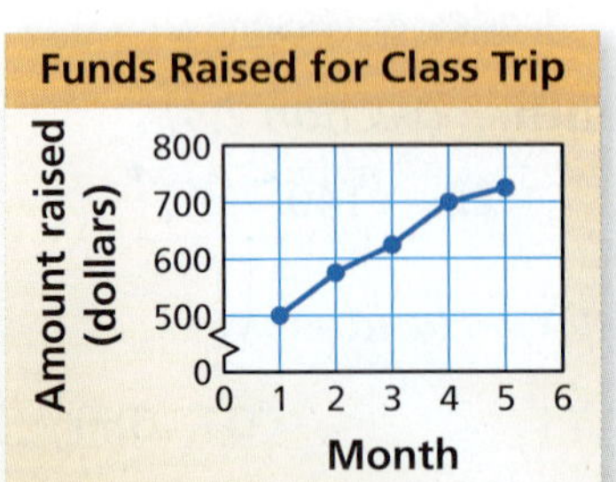

7. FUNDRAISER The line graph shows the amount of money that the eighth-grade students at a school raised each month to pay for a class trip. Is the graph misleading? Explain. *(Section 9.4)*

9 Chapter Review

Check It Out
Vocabulary Help
BigIdeasMath.com

Review Key Vocabulary

scatter plot, *p. 374*
line of fit, *p. 380*
line of best fit, *p. 381*
two-way table, *p. 388*
joint frequency, *p. 388*
marginal frequency, *p. 388*

Review Examples and Exercises

9.1 Scatter Plots *(pp. 372–377)*

Your school is ordering custom T-shirts. The scatter plot shows the number of T-shirts ordered and the cost per shirt. What tends to happen to the cost per shirt as the number of T-shirts ordered increases?

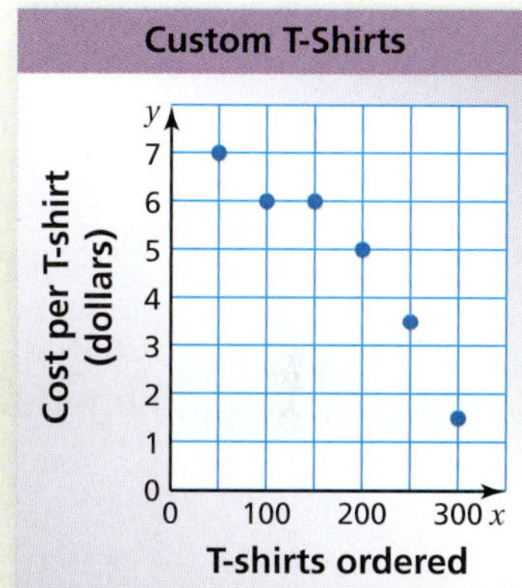

Looking at the graph, the plotted points go down from left to right.

So, as the number of T-shirts ordered increases, the cost per shirt decreases.

Exercises

1. **MIGRATION** The scatter plot shows the number of geese that migrated to a park each season.

   a. In what year did 270 geese migrate?

   b. How many geese migrated in 2010?

   c. Describe the relationship shown by the data.

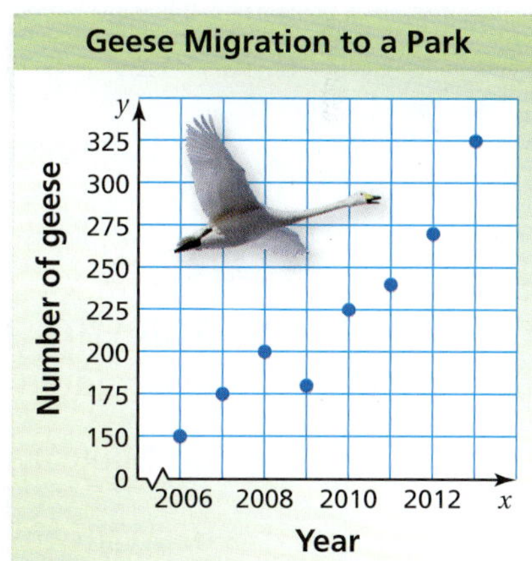

Describe the relationship between the data. Identify any outliers, gaps, or clusters.

2.

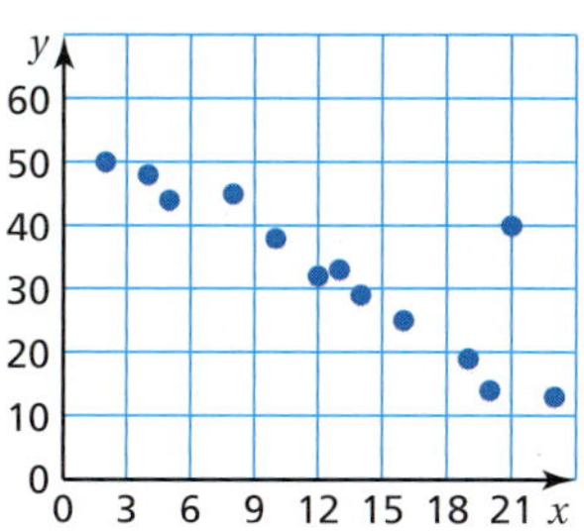

3.

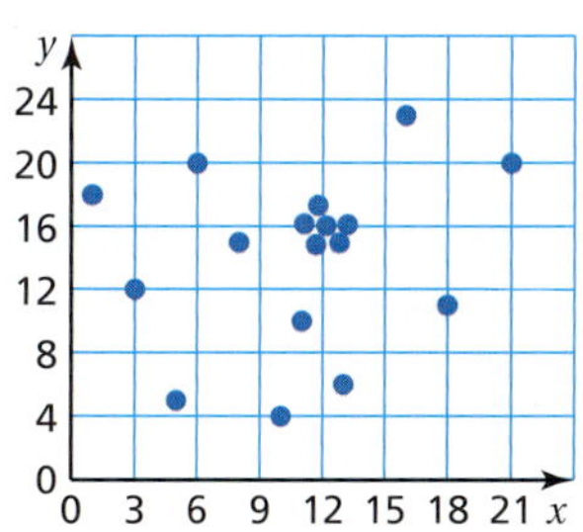

4.

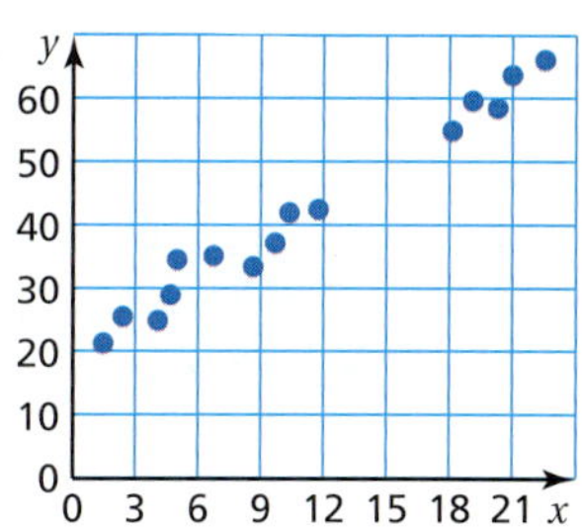

9.2 Lines of Fit (pp. 378–383)

The table shows the revenue (in millions of dollars) for a company over an 8-year period. (a) Make a scatter plot of the data and draw a line of fit. (b) Write an equation of the line of fit. (c) Interpret the slope and the *y*-intercept of the line of fit. (d) Predict what the revenue will be in year 9.

Year, *x*	1	2	3	4	5	6	7	8
Revenue (millions of dollars), *y*	20	35	46	56	68	82	92	108

a. Plot the points in a coordinate plane. The scatter plot shows a positive linear relationship. Draw a line that is close to the data points.

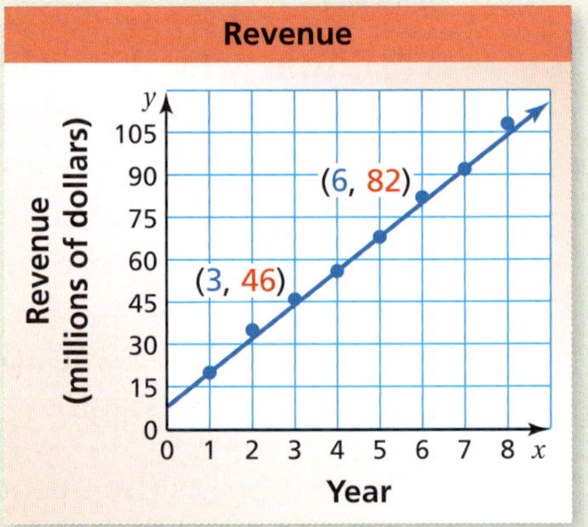

b. slope $= \frac{\text{rise}}{\text{run}} = \frac{36}{3} = 12$

Because the line crosses the *y*-axis at (0, 8), the *y*-intercept is 8.

So, an equation of the line of fit is $y = 12x + 8$.

c. The slope is 12. So, the revenue increased by about \$12 million each year. The *y*-intercept is 8. So, you can estimate that the revenue was \$8 million in the year before this 8-year period.

d. $y = 12x + 8 = 12(9) + 8 = 116$

The revenue in year 9 will be about \$116 million.

Exercises

5. **STUDENTS** The table shows the number of students at a middle school over a 10-year period.

   a. Make a scatter plot of the data and draw a line of fit.

   b. Write an equation of the line of fit.

   c. Interpret the slope and the *y*-intercept of the line of fit.

   d. Predict the number of students in year 11.

Year, *x*	Number of Students, *y*
1	492
2	507
3	520
4	535
5	550
6	562
7	577
8	591
9	604
10	618

6. **LINE OF BEST FIT** Use a graphing calculator to find an equation of the line of best fit for the data in Exercise 5. Identify and interpret the correlation coefficient.

9.3 Two-Way Tables (pp. 386–391)

You randomly survey students in your school about whether they liked a recent school play. The results are shown. Make a two-way table that includes the marginal frequencies. What percent of the students surveyed liked the play?

Male Students
48 likes, 12 dislikes

Female Students
56 likes, 14 dislikes

Of the 130 students surveyed, 104 students liked the play.

Because $\frac{104}{130} = 0.8$, 80% of the students in the survey liked the play.

		Student		
		Liked	**Did Not Like**	**Total**
Gender	**Male**	48	12	60
	Female	56	14	70
	Total	104	26	130

Exercises

You randomly survey people at a mall about whether they like the new food court. The results are shown.

Teenagers
96 likes, 4 dislikes

Adults
21 likes, 79 dislikes

Senior Citizens
18 likes, 82 dislikes

7. Make a two-way table that includes the marginal frequencies.
8. For each group, what percent of the people surveyed like the food court? dislike the food court? Organize your results in a two-way table.
9. Does your table in Exercise 8 show a relationship between age and whether people like the food court?

9.4 Choosing a Data Display (pp. 392–399)

Choose an appropriate data display for the situation. Explain your reasoning.

a. the percent of votes that each candidate received in an election

A circle graph shows data as parts of a whole. So, a circle graph is an appropriate data display.

b. the distribution of the ages of U.S. presidents

A stem-and-leaf plot orders numerical data and shows how they are distributed. So, a stem-and-leaf plot is an appropriate data display.

Exercises

Choose an appropriate data display for the situation. Explain your reasoning.

10. the number of pairs of shoes sold by a store each week
11. a comparison of the heights of brothers and sisters

9 Chapter Test

Check It Out
Test Practice
BigIdeasMath.com

1. **POPULATION** The graph shows the population (in millions) of the United States from 1960 to 2010.

   a. In what year was the population of the United States about 180 million?

   b. What was the approximate population of the United States in 1990?

   c. Describe the trend shown by the data.

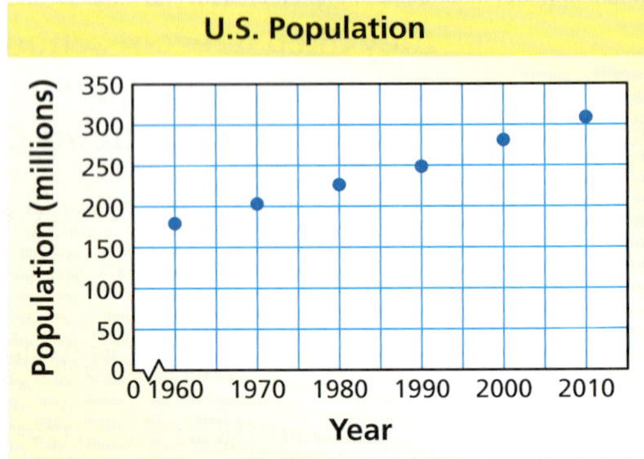

2. **WEIGHT** The table shows the weight of a baby over several months.

   a. Make a scatter plot of the data and draw a line of fit.

   b. Write an equation of the line of fit.

   c. Interpret the slope and the y-intercept of the line of fit.

   d. Predict how much the baby will weigh at 7 months.

Age (months)	Weight (pounds)
1	8
2	9.25
3	11.75
4	13
5	14.5
6	16

		Nonfiction	
		Likes	Dislikes
Fiction	Likes	26	20
	Dislikes	22	2

3. **READING** You randomly survey students at your school about what type of books they like to read. The two-way table shows your results. Find and interpret the marginal frequencies.

Choose an appropriate data display for the situation. Explain your reasoning.

4. magazine sales grouped by price

5. the distance a person hikes each week

6. **SAT** The table shows the numbers y of students (in thousands) who took the SAT from 2006 to 2010, where $x = 6$ represents the year 2006. Use a graphing calculator to find an equation of the line of best fit. Identify and interpret the correlation coefficient.

Year, x	6	7	8	9	10
Number of Students, y	1466	1495	1519	1530	1548

7. **RECYCLING** You randomly survey shoppers at a supermarket about whether they use reusable bags. Of 60 male shoppers, 15 use reusable bags. Of 110 female shoppers, 60 use reusable bags. Organize your results in a two-way table. Include the marginal frequencies.

9 Standards Assessment

1. What is the volume of the trash bin? *(MACC.8.G.3.9)*

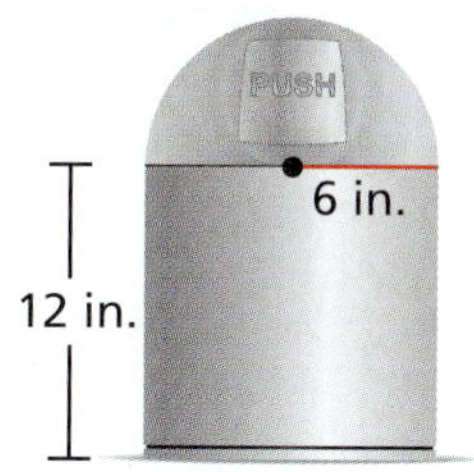

A. 288π in.^{3}

B. 576π in.^{3}

C. 648π in.^{3}

D. 720π in.^{3}

Test-Taking Strategy

Read All Choices Before Answering

"Reading all choices before answering can sometimes point out the obvious answer!"

2. The diagram below shows parallel lines cut by a transversal. Which angle is the corresponding angle for $\angle 6$? *(MACC.8.G.1.5)*

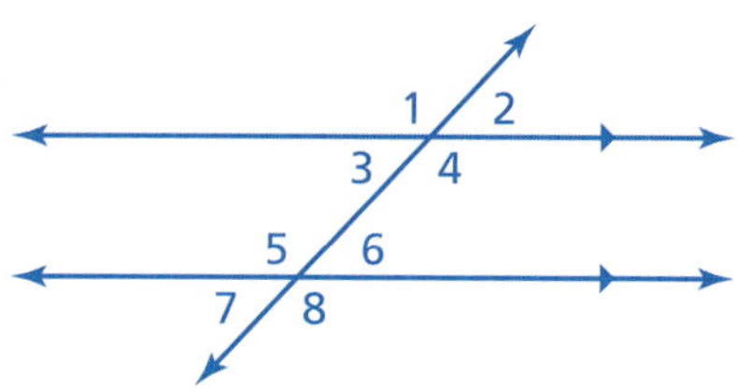

F. $\angle 2$

G. $\angle 3$

H. $\angle 4$

I. $\angle 8$

3. You randomly survey students in your school. You ask whether they have jobs. You display your results in the two-way table. How many male students do *not* have a job? *(MACC.8.SP.1.4)*

		Job	
		Yes	No
Gender	Male	27	12
	Female	31	17

4. Which scatter plot shows a negative relationship between x and y? *(MACC.8.SP.1.1)*

A.

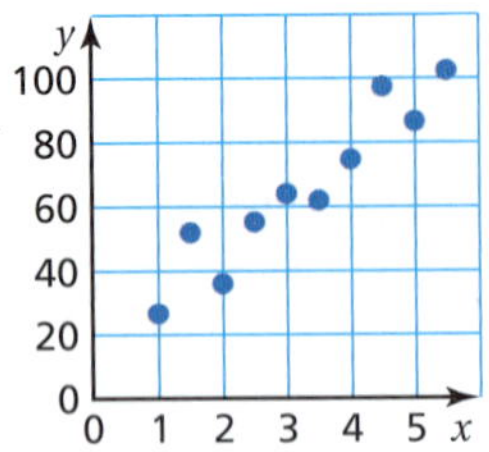

B.

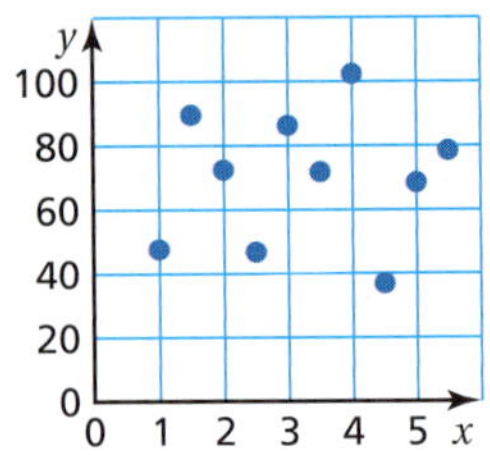

C.

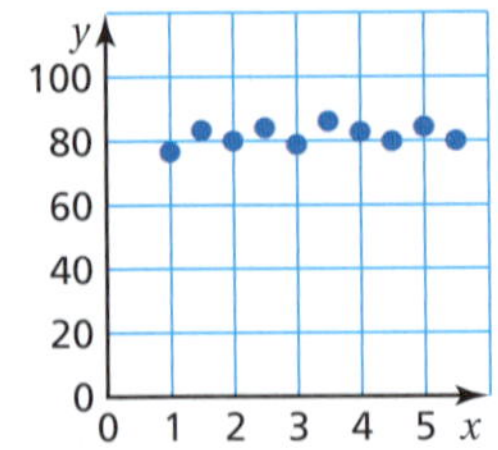

D.

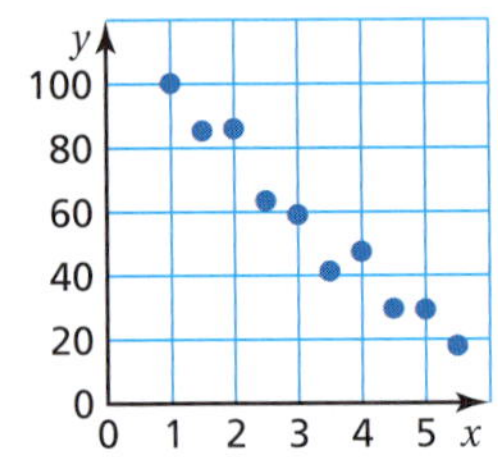

5. The legs of a right triangle have the lengths of 8 centimeters and 15 centimeters. What is the length of the hypotenuse, in centimeters? *(MACC.8.G.2.7)*

6. What is the solution of the equation? *(MACC.8.EE.3.7b)*

$$0.22(x + 6) = 0.2x + 1.8$$

F. $x = 2.4$

G. $x = 15.6$

H. $x = 24$

I. $x = 156$

7. Which triangle is *not* a right triangle? *(MACC.8.G.2.6)*

A.

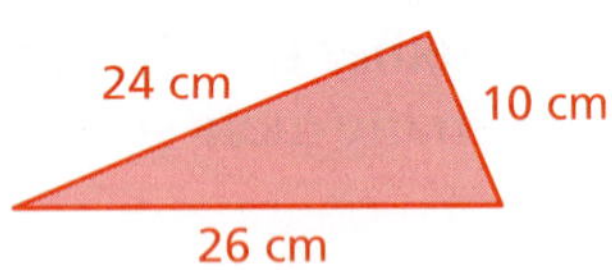

B.

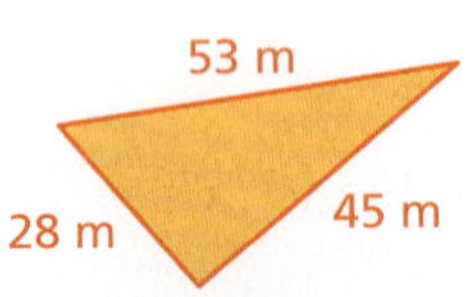

C.

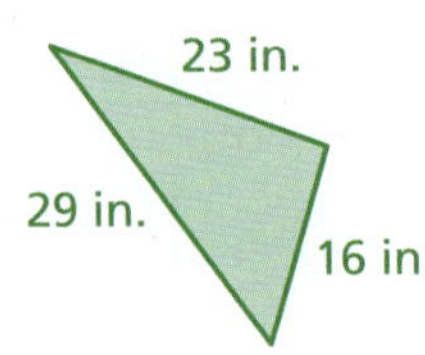

D.

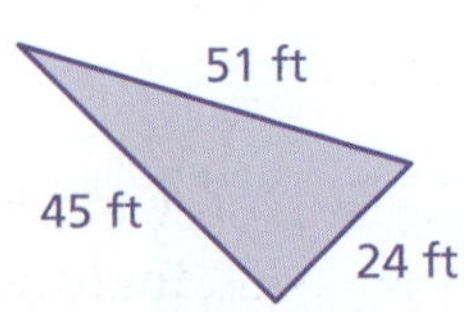

8. A store has recorded total dollar sales each month for the past three years. Which type of graph would best show how sales have increased over this time period? *(MACC.8.SP.1.1)*

F. circle graph

G. line graph

H. histogram

I. stem-and-leaf plot

9. Trapezoid *KLMN* is graphed in the coordinate plane shown.

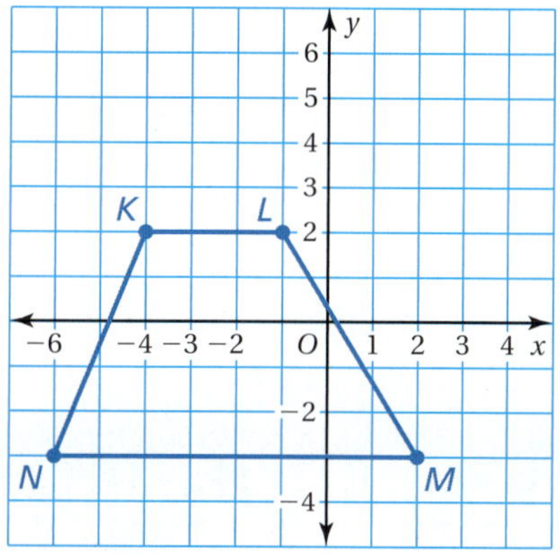

Rotate Trapezoid *KLMN* 90° clockwise about the origin. What are the coordinates of point M', the image of point M after the rotation? *(MACC.8.G.1.3)*

A. $(-3, -2)$

B. $(-2, -3)$

C. $(-2, 3)$

D. $(3, 2)$

10. The table shows the numbers of hours students spent watching television from Monday through Friday for one week and their scores on a test that Friday. *(MACC.8.SP.1.1, MACC.8.SP.1.2)*

Think Solve Explain

Hours of Television, *x*	5	2	10	15	3	4	8	2	12	9
Test Score, *y*	92	98	79	66	97	88	82	95	72	81

Part A Make a scatter plot of the data.

Part B Describe the relationship between hours of television watched and test score.

Part C Explain how to justify your answer in Part B using the linear regression feature of a graphing calculator.

10 Exponents and Scientific Notation

10.1 Exponents
10.2 Product of Powers Property
10.3 Quotient of Powers Property
10.4 Zero and Negative Exponents
10.5 Reading Scientific Notation
10.6 Writing Scientific Notation
10.7 Operations in Scientific Notation

"Here's how it goes, Descartes."
"The friends of my friends are my friends. The friends of my enemies are my enemies."
"The enemies of my friends are my enemies. The enemies of my enemies are my friends."

"If one flea had 100 babies, and each baby grew up and had 100 babies, ..."
"... and each of those babies grew up and had 100 babies, you would have 1,010,101 fleas."

What You Learned Before

"It's called the Power of Negative One, Descartes!"

Using Order of Operations (MACC.6.EE.1.1)

Example 1 Evaluate $6^2 \div 4 - 2(9 - 5)$.

First:	**P**arentheses	$6^2 \div 4 - 2(9-5) = 6^2 \div 4 - 2 \cdot 4$
Second:	**E**xponents	$= 36 \div 4 - 2 \cdot 4$
Third:	**M**ultiplication and **D**ivision (from left to right)	$= 9 - 8$
Fourth:	**A**ddition and **S**ubtraction (from left to right)	$= 1$

Try It Yourself

Evaluate the expression.

1. $15\left(\frac{8}{4}\right) + 2^2 - 3 \cdot 7$ **2.** $5^2 \cdot 2 \div 10 + 3 \cdot 2 - 1$ **3.** $3^2 - 1 + 2(4(3+2))$

Multiplying and Dividing Decimals (MACC.6.NS.2.3)

Example 2 Find $2.1 \cdot 0.35$.

$$\begin{array}{r} 2.1 \\ \times\ 0.35 \\ \hline 105 \\ 63 \\ \hline 0.735 \end{array}$$

2.1 ← 1 decimal place
0.35 ← + 2 decimal places
0.735 ← 3 decimal places

Example 3 Find $1.08 \div 0.9$.

$0.9\overline{)1.08}$ Multiply each number by 10.

$$\begin{array}{r} 1.2 \\ 9\overline{)10.8} \\ -9 \\ \hline 18 \\ -18 \\ \hline 0 \end{array}$$

Place the decimal point above the decimal point in the dividend 10.8.

Try It Yourself

Find the product or quotient.

4. $1.75 \cdot 0.2$ **5.** $1.4 \cdot 0.6$ **6.** $\begin{array}{r} 7.03 \\ \times\ 4.3 \\ \hline \end{array}$ **7.** $\begin{array}{r} 0.894 \\ \times\ 0.2 \\ \hline \end{array}$

8. $5.40 \div 0.09$ **9.** $4.17 \div 0.3$ **10.** $0.15\overline{)3.6}$ **11.** $0.004\overline{)7.2}$

10.1 Exponents

Essential Question How can you use exponents to write numbers?

The expression 3^5 is called a *power*. The *base* is 3. The *exponent* is 5.

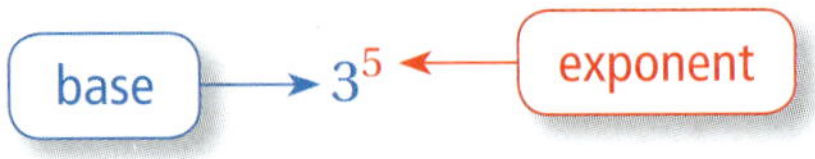

1 ACTIVITY: Using Exponent Notation

Work with a partner.

a. Copy and complete the table.

Power	Repeated Multiplication Form	Value
$(-3)^1$	-3	-3
$(-3)^2$	$(-3) \cdot (-3)$	9
$(-3)^3$		
$(-3)^4$		
$(-3)^5$		
$(-3)^6$		
$(-3)^7$		

b. **REPEATED REASONING** Describe what is meant by the expression $(-3)^n$. How can you find the value of $(-3)^n$?

2 ACTIVITY: Using Exponent Notation

Work with a partner.

a. The cube at the right has \$3 in each of its small cubes. Write a power that represents the total amount of money in the large cube.

b. Evaluate the power to find the total amount of money in the large cube.

COMMON CORE

Exponents

In this lesson, you will

- write expressions using integer exponents.
- evaluate expressions involving integer exponents.

Learning Standard
MACC.8.EE.1.1

3 ACTIVITY: Writing Powers as Whole Numbers

Work with a partner. Write each distance as a whole number. Which numbers do you know how to write in words? For instance, in words, 10^3 is equal to *one thousand.*

a. 10^{26} meters: diameter of observable universe

b. 10^{21} meters: diameter of Milky Way galaxy

c. 10^{16} meters: diameter of solar system

d. 10^{7} meters: diameter of Earth

e. 10^{6} meters: length of Lake Erie shoreline

f. 10^{5} meters: width of Lake Erie

4 ACTIVITY: Writing a Power

Math Practice 1
Analyze Givens
What information is given in the poem? What are you trying to find?

Work with a partner. Write the number of kits, cats, sacks, and wives as a power.

As I was going to St. Ives
I met a man with seven wives
Each wife had seven sacks
Each sack had seven cats
Each cat had seven kits
Kits, cats, sacks, wives
How many were going to St. Ives?

Nursery Rhyme, 1730

What Is Your Answer?

5. **IN YOUR OWN WORDS** How can you use exponents to write numbers? Give some examples of how exponents are used in real life.

Practice Use what you learned about exponents to complete Exercises 3–5 on page 414.

10.1 Lesson

Check It Out
Lesson Tutorials
BigIdeasMath.com

Key Vocabulary
power, *p. 412*
base, *p. 412*
exponent, *p. 412*

A **power** is a product of repeated factors. The **base** of a power is the common factor. The **exponent** of a power indicates the number of times the base is used as a factor.

base → $\frac{1}{2}$; exponent → 5

$$\left(\frac{1}{2}\right)^5 = \frac{1}{2} \cdot \frac{1}{2} \cdot \frac{1}{2} \cdot \frac{1}{2} \cdot \frac{1}{2}$$

power — $\frac{1}{2}$ is used as a factor 5 times.

EXAMPLE 1 Writing Expressions Using Exponents

Study Tip
Use parentheses to write powers with negative bases.

Write each product using exponents.

a. $(-7) \cdot (-7) \cdot (-7)$

Because -7 is used as a factor 3 times, its exponent is 3.

So, $(-7) \cdot (-7) \cdot (-7) = (-7)^3$.

b. $\pi \cdot \pi \cdot r \cdot r \cdot r$

Because π is used as a factor 2 times, its exponent is 2. Because r is used as a factor 3 times, its exponent is 3.

So, $\pi \cdot \pi \cdot r \cdot r \cdot r = \pi^2 r^3$.

On Your Own

Now You're Ready
Exercises 3–10

Write the product using exponents.

1. $\frac{1}{4} \cdot \frac{1}{4} \cdot \frac{1}{4} \cdot \frac{1}{4} \cdot \frac{1}{4}$

2. $0.3 \cdot 0.3 \cdot 0.3 \cdot 0.3 \cdot x \cdot x$

EXAMPLE 2 Evaluating Expressions

Evaluate each expression.

a. $(-2)^4$

The base is -2.

$(-2)^4 = (-2) \cdot (-2) \cdot (-2) \cdot (-2)$ — Write as repeated multiplication.

$= 16$ — Simplify.

b. -2^4

The base is 2.

$-2^4 = -(2 \cdot 2 \cdot 2 \cdot 2)$ — Write as repeated multiplication.

$= -16$ — Simplify.

EXAMPLE 3 Using Order of Operations

Evaluate each expression.

a. $3 + 2 \cdot 3^4 = 3 + 2 \cdot 81$ — Evaluate the power.

$= 3 + 162$ — Multiply.

$= 165$ — Add.

b. $3^3 - 8^2 \div 2 = 27 - 64 \div 2$ — Evaluate the powers.

$= 27 - 32$ — Divide.

$= -5$ — Subtract.

On Your Own

Now You're Ready
Exercises 11–16
and 21–26

Evaluate the expression.

3. -5^4 **4.** $\left(-\frac{1}{6}\right)^3$ **5.** $\left|-3^3 \div 27\right|$ **6.** $9 - 2^5 \cdot 0.5$

EXAMPLE 4 Real-Life Application

In sphering, a person is secured inside a small, hollow sphere that is surrounded by a larger sphere. The space between the spheres is inflated with air. What is the volume of the inflated space?

You can find the radius of each sphere by dividing each diameter given in the diagram by 2.

Outer Sphere		*Inner Sphere*
$V = \frac{4}{3}\pi r^3$	Write formula.	$V = \frac{4}{3}\pi r^3$
$= \frac{4}{3}\pi\left(\frac{3}{2}\right)^3$	Substitute.	$= \frac{4}{3}\pi(1)^3$
$= \frac{4}{3}\pi\left(\frac{27}{8}\right)$	Evaluate the power.	$= \frac{4}{3}\pi(1)$
$= \frac{9}{2}\pi$	Multiply.	$= \frac{4}{3}\pi$

So, the volume of the inflated space is $\frac{9}{2}\pi - \frac{4}{3}\pi = \frac{19}{6}\pi$, or about 10 cubic meters.

On Your Own

7. **WHAT IF?** The diameter of the inner sphere is 1.8 meters. What is the volume of the inflated space?

10.1 Exercises

Check It Out
Help with Homework
BigIdeasMath.com

Vocabulary and Concept Check

1. **NUMBER SENSE** Describe the difference between -3^4 and $(-3)^4$.

2. **WHICH ONE DOESN'T BELONG?** Which one does *not* belong with the other three? Explain your reasoning.

5^3 The exponent is 3.	5^3 The power is 5.	5^3 The base is 5.	5^3 Five is used as a factor 3 times.

Practice and Problem Solving

Write the product using exponents.

(1) 3. $3 \cdot 3 \cdot 3 \cdot 3$

4. $(-6) \cdot (-6)$

5. $\left(-\frac{1}{2}\right) \cdot \left(-\frac{1}{2}\right) \cdot \left(-\frac{1}{2}\right)$

6. $\frac{1}{3} \cdot \frac{1}{3} \cdot \frac{1}{3}$

7. $\pi \cdot \pi \cdot \pi \cdot x \cdot x \cdot x \cdot x$

8. $(-4) \cdot (-4) \cdot (-4) \cdot y \cdot y$

9. $6.4 \cdot 6.4 \cdot 6.4 \cdot 6.4 \cdot b \cdot b \cdot b$

10. $(-t) \cdot (-t) \cdot (-t) \cdot (-t) \cdot (-t)$

Evaluate the expression.

(2) 11. 5^2

12. -11^3

13. $(-1)^6$

14. $\left(\frac{1}{2}\right)^6$

15. $\left(-\frac{1}{12}\right)^2$

16. $-\left(\frac{1}{9}\right)^3$

17. **ERROR ANALYSIS** Describe and correct the error in evaluating the expression.

✗ $-6^2 = (-6) \cdot (-6) = 36$

18. **PRIME FACTORIZATION** Write the prime factorization of 675 using exponents.

19. **STRUCTURE** Write $-\left(\frac{1}{4} \cdot \frac{1}{4} \cdot \frac{1}{4} \cdot \frac{1}{4}\right)$ using exponents.

20. **RUSSIAN DOLLS** The largest doll is 12 inches tall. The height of each of the other dolls is $\frac{7}{10}$ the height of the next larger doll. Write an expression involving a power for the height of the smallest doll. What is the height of the smallest doll?

Evaluate the expression.

3 **21.** $5 + 3 \cdot 2^3$

22. $2 + 7 \cdot (-3)^2$

23. $(13^2 - 12^2) \div 5$

24. $\frac{1}{2}(4^3 - 6 \cdot 3^2)$

25. $\left| \frac{1}{2}(7 + 5^3) \right|$

26. $\left| \left(-\frac{1}{2}\right)^3 \div \left(\frac{1}{4}\right)^2 \right|$

27. **MONEY** You have a part-time job. One day your boss offers to pay you either $2^h - 1$ or 2^{h-1} dollars for each hour h you work that day. Copy and complete the table. Which option should you choose? Explain.

h	1	2	3	4	5
$2^h - 1$					
2^{h-1}					

28. **CARBON-14 DATING** Scientists use carbon-14 dating to determine the age of a sample of organic material.

a. The amount C (in grams) of a 100-gram sample of carbon-14 remaining after t years is represented by the equation $C = 100(0.99988)^t$. Use a calculator to find the amount of carbon-14 remaining after 4 years.

b. What percent of the carbon-14 remains after 4 years?

29. Critical Thinking The frequency (in vibrations per second) of a note on a piano is represented by the equation $F = 440(1.0595)^n$, where n is the number of notes above A-440. Each black or white key represents one note.

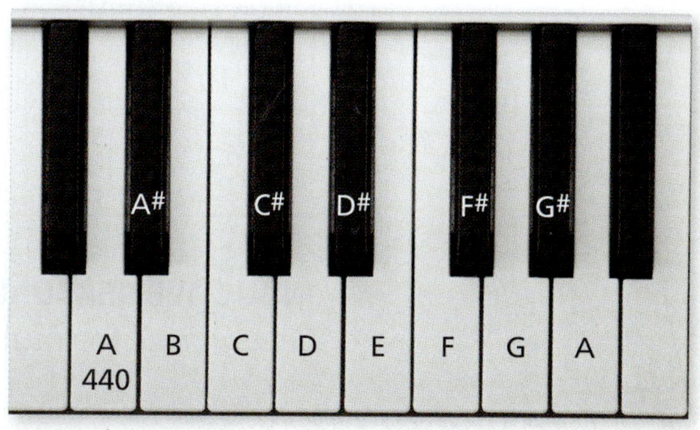

a. How many notes do you take to travel from A-440 to A?

b. What is the frequency of A?

c. Describe the relationship between the number of notes between A-440 and A and the increase in frequency.

Fair Game Review What you learned in previous grades & lessons

Tell which property is illustrated by the statement. *(Skills Review Handbook)*

30. $8 \cdot x = x \cdot 8$

31. $(2 \cdot 10)x = 2(10 \cdot x)$

32. $3(x \cdot 1) = 3x$

33. **MULTIPLE CHOICE** The polygons are similar. What is the value of x? *(Section 2.5)*

Ⓐ 15 Ⓑ 16

Ⓒ 17 Ⓓ 36

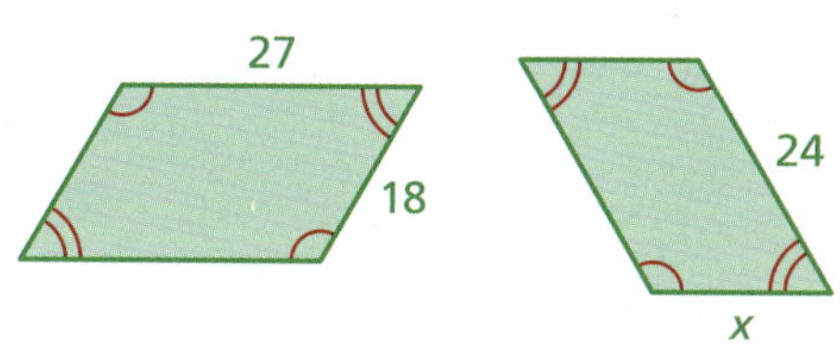

10.2 Product of Powers Property

Essential Question How can you use inductive reasoning to observe patterns and write general rules involving properties of exponents?

1 ACTIVITY: Finding Products of Powers

Work with a partner.

a. Copy and complete the table.

Product	Repeated Multiplication Form	Power
$2^2 \cdot 2^4$		
$(-3)^2 \cdot (-3)^4$		
$7^3 \cdot 7^2$		
$5.1^1 \cdot 5.1^6$		
$(-4)^2 \cdot (-4)^2$		
$10^3 \cdot 10^5$		
$\left(\frac{1}{2}\right)^5 \cdot \left(\frac{1}{2}\right)^5$		

b. **INDUCTIVE REASONING** Describe the pattern in the table. Then write a *general rule* for multiplying two powers that have the same base.

$$a^m \cdot a^n = a^{\square}$$

c. Use your rule to simplify the products in the first column of the table above. Does your rule give the results in the third column?

d. Most calculators have *exponent* keys that you can use to evaluate powers. Use a calculator with an exponent key to evaluate the products in part (a).

COMMON CORE

Exponents

In this lesson, you will

- multiply powers with the same base.
- find a power of a power.
- find a power of a product.

Learning Standard MACC.8.EE.1.1

2 ACTIVITY: Writing a Rule for Powers of Powers

Work with a partner. Write the expression as a single power. Then write a *general rule* for finding a power of a power.

a. $(3^2)^3 = (3 \cdot 3)(3 \cdot 3)(3 \cdot 3) = \square^{\square}$

b. $(2^2)^4 = \square$

c. $(7^3)^2 = \square$

d. $(y^3)^3 = \square$

e. $(x^4)^2 = \square$

3 ACTIVITY: Writing a Rule for Powers of Products

Work with a partner. Write the expression as the product of two powers. Then write a *general rule* for finding a power of a product.

a. $(2 \cdot 3)^3 = (2 \cdot 3)(2 \cdot 3)(2 \cdot 3) = \square^{\square} \cdot \square^{\square}$

b. $(2 \cdot 5)^2 = \square$

c. $(5 \cdot 4)^3 = \square$

d. $(6a)^4 = \square$

e. $(3x)^2 = \square$

4 ACTIVITY: The Penny Puzzle

Work with a partner.

- The rows y and columns x of a chessboard are numbered as shown.
- Each position on the chessboard has a stack of pennies. (Only the first row is shown.)
- The number of pennies in each stack is $2^x \cdot 2^y$.

Math Practice 7

Look for Patterns

What patterns do you notice? How does this help you determine which stack is the tallest?

a. How many pennies are in the stack in location (3, 5)?

b. Which locations have 32 pennies in their stacks?

c. How much money (in dollars) is in the location with the tallest stack?

d. A penny is about 0.06 inch thick. About how tall (in inches) is the tallest stack?

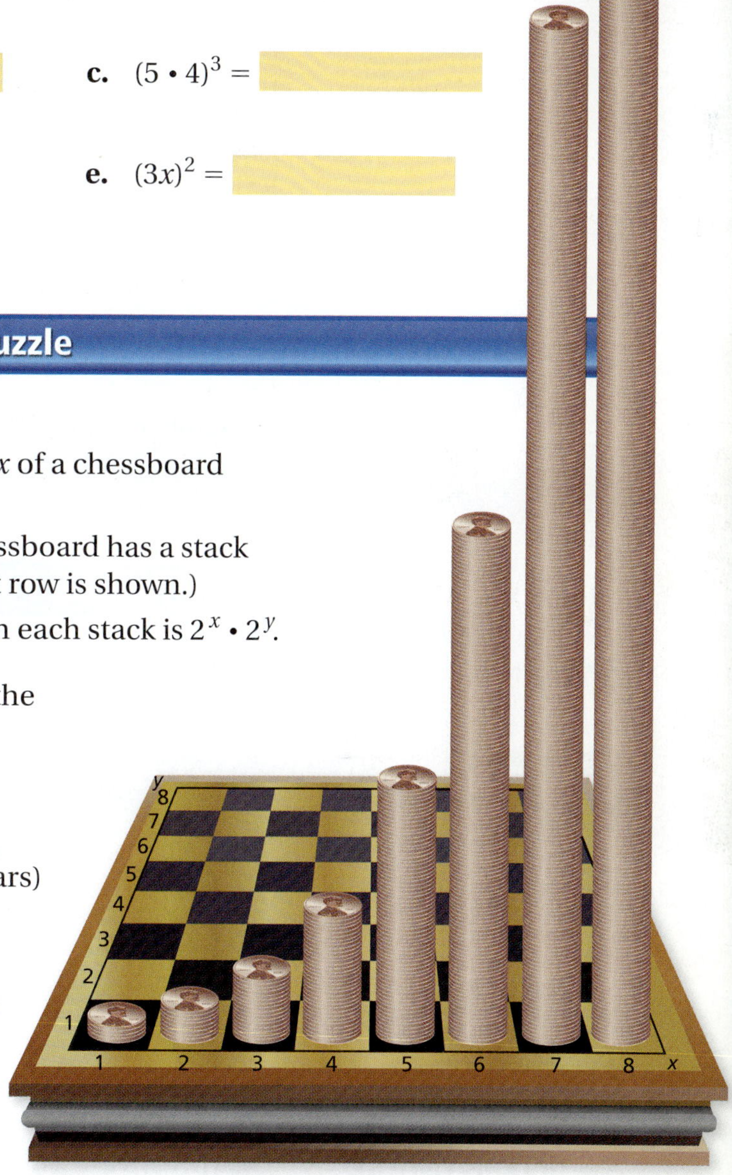

What Is Your Answer?

5. **IN YOUR OWN WORDS** How can you use inductive reasoning to observe patterns and write general rules involving properties of exponents?

Practice Use what you learned about properties of exponents to complete Exercises 3–5 on page 420.

10.2 Lesson

Check It Out
Lesson Tutorials
BigIdeasMath.com

Key Ideas

Product of Powers Property

Words To multiply powers with the same base, add their exponents.

Numbers $4^2 \cdot 4^3 = 4^{2+3} = 4^5$ **Algebra** $a^m \cdot a^n = a^{m+n}$

Power of a Power Property

Words To find a power of a power, multiply the exponents.

Numbers $(4^6)^3 = 4^{6 \cdot 3} = 4^{18}$ **Algebra** $(a^m)^n = a^{mn}$

Power of a Product Property

Words To find a power of a product, find the power of each factor and multiply.

Numbers $(3 \cdot 2)^5 = 3^5 \cdot 2^5$ **Algebra** $(ab)^m = a^m b^m$

EXAMPLE 1 Multiplying Powers with the Same Base

a. $2^4 \cdot 2^5 = 2^{4+5}$ Product of Powers Property

$= 2^9$ Simplify.

b. $-5 \cdot (-5)^6 = (-5)^1 \cdot (-5)^6$ Rewrite -5 as $(-5)^1$.

$= (-5)^{1+6}$ Product of Powers Property

$= (-5)^7$ Simplify.

c. $x^3 \cdot x^7 = x^{3+7}$ Product of Powers Property

$= x^{10}$ Simplify.

Study Tip

When a number is written without an exponent, its exponent is 1.

EXAMPLE 2 Finding a Power of a Power

a. $(3^4)^3 = 3^{4 \cdot 3}$ Power of a Power Property

$= 3^{12}$ Simplify.

b. $(w^5)^4 = w^{5 \cdot 4}$ Power of a Power Property

$= w^{20}$ Simplify.

EXAMPLE 3 Finding a Power of a Product

a. $(2x)^3 = 2^3 \cdot x^3$ — Power of a Product Property

$= 8x^3$ — Simplify.

b. $(3xy)^2 = 3^2 \cdot x^2 \cdot y^2$ — Power of a Product Property

$= 9x^2y^2$ — Simplify.

On Your Own

Now You're Ready
Exercises 3–14 and 17–22

Simplify the expression.

1. $6^2 \cdot 6^4$

2. $\left(-\frac{1}{2}\right)^3 \cdot \left(-\frac{1}{2}\right)^6$

3. $z \cdot z^{12}$

4. $(4^4)^3$

5. $(y^2)^4$

6. $((-4)^3)^2$

7. $(5y)^4$

8. $(ab)^5$

9. $(0.5mn)^2$

EXAMPLE 4 Simplifying an Expression

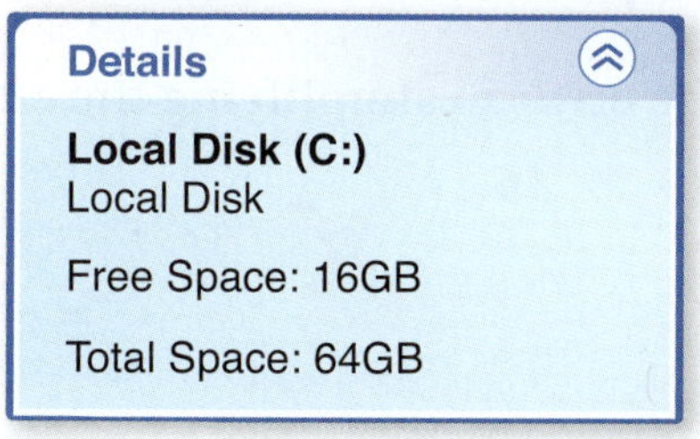

A gigabyte (GB) of computer storage space is 2^{30} bytes. The details of a computer are shown. How many bytes of total storage space does the computer have?

Ⓐ 2^{34} Ⓑ 2^{36} Ⓒ 2^{180} Ⓓ 128^{30}

The computer has 64 gigabytes of total storage space. Notice that you can write 64 as a power, 2^6. Use a model to solve the problem.

Total number of bytes = Number of bytes in a gigabyte • Number of gigabytes

$= 2^{30} \cdot 2^6$ — Substitute.

$= 2^{30+6}$ — Product of Powers Property

$= 2^{36}$ — Simplify.

∴ The computer has 2^{36} bytes of total storage space. The correct answer is Ⓑ.

On Your Own

10. How many bytes of free storage space does the computer have?

10.2 Exercises

Check It Out
Help with Homework
BigIdeasMath.com

Vocabulary and Concept Check

1. **REASONING** When should you use the Product of Powers Property?
2. **CRITICAL THINKING** Can you use the Product of Powers Property to multiply $5^2 \cdot 6^4$? Explain.

Practice and Problem Solving

Simplify the expression. Write your answer as a power.

① ② 3. $3^2 \cdot 3^2$

4. $8^{10} \cdot 8^4$

5. $(-4)^5 \cdot (-4)^7$

6. $a^3 \cdot a^3$

7. $h^6 \cdot h$

8. $\left(\frac{2}{3}\right)^2 \cdot \left(\frac{2}{3}\right)^6$

9. $\left(-\frac{5}{7}\right)^8 \cdot \left(-\frac{5}{7}\right)^9$

10. $(-2.9) \cdot (-2.9)^7$

11. $(5^4)^3$

12. $(b^{12})^3$

13. $(3.8^3)^4$

14. $\left(\left(-\frac{3}{4}\right)^5\right)^2$

ERROR ANALYSIS **Describe and correct the error in simplifying the expression.**

15. ✗ $5^2 \cdot 5^9 = (5 \cdot 5)^{2+9}$
$= 25^{11}$

16. ✗ $(r^6)^4 = r^{6+4}$
$= r^{10}$

Simplify the expression.

③ 17. $(6g)^3$

18. $(-3v)^5$

19. $\left(\frac{1}{5}k\right)^2$

20. $(1.2m)^4$

21. $(rt)^{12}$

22. $\left(-\frac{3}{4}p\right)^3$

23. **PRECISION** Is $3^2 + 3^3$ equal to 3^5? Explain.

24. **ARTIFACT** A display case for the artifact is in the shape of a cube. Each side of the display case is three times longer than the width of the artifact.

    a. Write an expression for the volume of the case. Write your answer as a power.

    b. Simplify the expression.

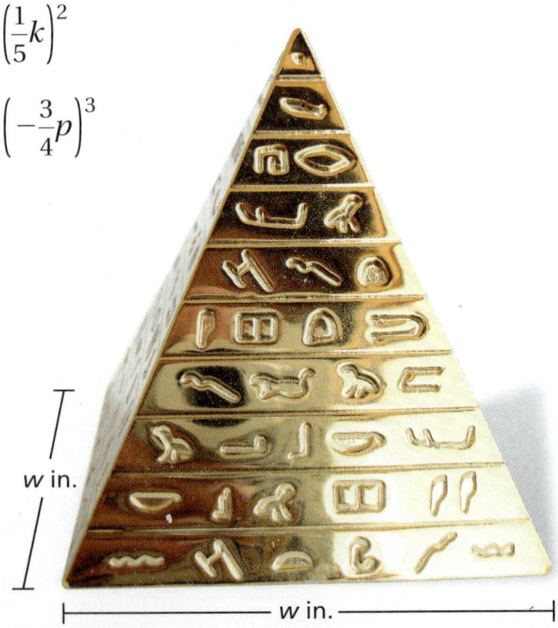

Simplify the expression.

25. $2^4 \cdot 2^5 - (2^2)^2$

26. $16\left(\frac{1}{2}x\right)^4$

27. $5^2(5^3 \cdot 5^2)$

28. CLOUDS The lowest altitude of an altocumulus cloud is about 3^8 feet. The highest altitude of an altocumulus cloud is about 3 times the lowest altitude. What is the highest altitude of an altocumulus cloud? Write your answer as a power.

29. PYTHON EGG The volume V of a python egg is given by the formula $V = \frac{4}{3}\pi abc$. For the python eggs shown, $a = 2$ inches, $b = 2$ inches, and $c = 3$ inches.

a. Find the volume of a python egg.

b. Square the dimensions of the python egg. Then evaluate the formula. How does this volume compare to your answer in part (a)?

30. PYRAMID A square pyramid has a height h and a base with side length b. The side lengths of the base increase by 50%. Write a formula for the volume of the new pyramid in terms of b and h.

31. MAIL The United States Postal Service delivers about $2^8 \cdot 5^2$ pieces of mail each second. There are $2^8 \cdot 3^4 \cdot 5^2$ seconds in 6 days. How many pieces of mail does the United States Postal Service deliver in 6 days? Write your answer as an expression involving powers.

32. Critical Thinking Find the value of x in the equation without evaluating the power.

a. $2^5 \cdot 2^x = 256$

b. $\left(\frac{1}{3}\right)^2 \cdot \left(\frac{1}{3}\right)^x = \frac{1}{729}$

Fair Game Review What you learned in previous grades & lessons

Simplify. *(Skills Review Handbook)*

33. $\frac{4 \cdot 4}{4}$

34. $\frac{5 \cdot 5 \cdot 5}{5}$

35. $\frac{2 \cdot 3}{2}$

36. $\frac{8 \cdot 6 \cdot 6}{6 \cdot 8}$

37. MULTIPLE CHOICE What is the measure of each interior angle of the regular polygon? *(Section 3.3)*

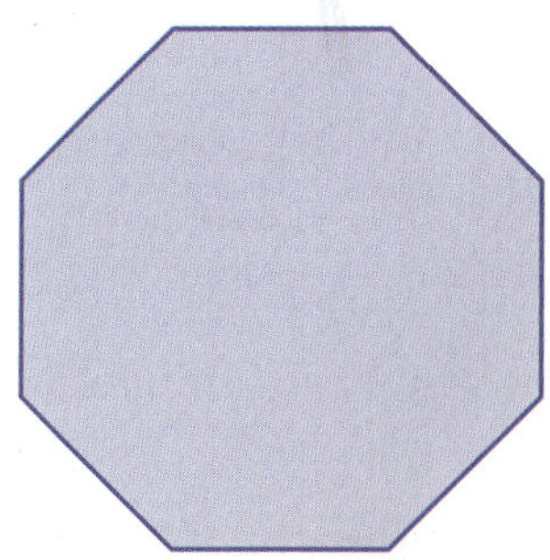

(A) 45°

(B) 135°

(C) 1080°

(D) 1440°

10.3 Quotient of Powers Property

Essential Question How can you divide two powers that have the same base?

1 ACTIVITY: Finding Quotients of Powers

Work with a partner.

a. Copy and complete the table.

Quotient	Repeated Multiplication Form	Power
$\frac{2^4}{2^2}$		
$\frac{(-4)^5}{(-4)^2}$		
$\frac{7^7}{7^3}$		
$\frac{8.5^9}{8.5^6}$		
$\frac{10^8}{10^5}$		
$\frac{3^{12}}{3^4}$		
$\frac{(-5)^7}{(-5)^5}$		
$\frac{11^4}{11^1}$		

b. **INDUCTIVE REASONING** Describe the pattern in the table. Then write a rule for dividing two powers that have the same base.

$$\frac{a^m}{a^n} = a^{\square}$$

c. Use your rule to simplify the quotients in the first column of the table above. Does your rule give the results in the third column?

COMMON CORE

Exponents

In this lesson, you will

- divide powers with the same base.
- simplify expressions involving the quotient of powers.

Learning Standard
MACC.8.EE.1.1

2 ACTIVITY: Comparing Volumes

Math Practice 8

Repeat Calculations

What calculations are repeated in the table?

Work with a partner.

How many of the smaller cubes will fit inside the larger cube? Record your results in the table. Describe the pattern in the table.

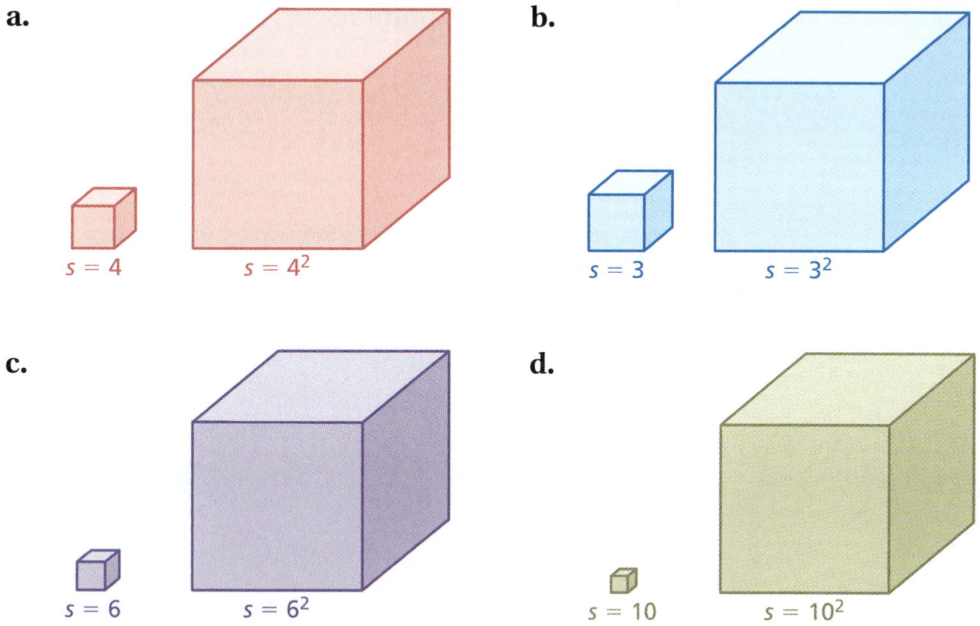

	Volume of Smaller Cube	Volume of Larger Cube	$\frac{\text{Larger Volume}}{\text{Smaller Volume}}$	Answer
a.				
b.				
c.				
d.				

What Is Your Answer?

3. **IN YOUR OWN WORDS** How can you divide two powers that have the same base? Give two examples of your rule.

Practice Use what you learned about dividing powers with the same base to complete Exercises 3–6 on page 426.

10.3 Lesson

Check It Out
Lesson Tutorials
BigIdeasMath.com

Key Idea

Quotient of Powers Property

Words To divide powers with the same base, subtract their exponents.

Numbers $\dfrac{4^5}{4^2} = 4^{5-2} = 4^3$ **Algebra** $\dfrac{a^m}{a^n} = a^{m-n}$, where $a \neq 0$

EXAMPLE 1 Dividing Powers with the Same Base

a. $\dfrac{2^6}{2^4} = 2^{6-4}$ Quotient of Powers Property

$= 2^2$ Simplify.

b. $\dfrac{(-7)^9}{(-7)^3} = (-7)^{9-3}$ Quotient of Powers Property

$= (-7)^6$ Simplify.

c. $\dfrac{h^7}{h^6} = h^{7-6}$ Quotient of Powers Property

$= h^1 = h$ Simplify.

Common Error

When dividing powers, do not divide the bases. $\dfrac{2^6}{2^4} = 2^2$, not 1^2.

On Your Own

Now You're Ready
Exercises 7–14

Simplify the expression. Write your answer as a power.

1. $\dfrac{9^7}{9^4}$
2. $\dfrac{4.2^6}{4.2^5}$
3. $\dfrac{(-8)^8}{(-8)^4}$
4. $\dfrac{x^8}{x^3}$

EXAMPLE 2 Simplifying an Expression

Simplify $\dfrac{3^4 \cdot 3^2}{3^3}$. Write your answer as a power.

The numerator is a product of powers. Add the exponents in the numerator.

$\dfrac{3^4 \cdot 3^2}{3^3} = \dfrac{3^{4+2}}{3^3}$ Product of Powers Property

$= \dfrac{3^6}{3^3}$ Simplify.

$= 3^{6-3}$ Quotient of Powers Property

$= 3^3$ Simplify.

EXAMPLE 3 Simplifying an Expression

Study Tip

You can also simplify the expression in Example 3 as follows.

$$\frac{a^{10}}{a^6} \cdot \frac{a^7}{a^4} = \frac{a^{10} \cdot a^7}{a^6 \cdot a^4}$$
$$= \frac{a^{17}}{a^{10}}$$
$$= a^{17-10}$$
$$= a^7$$

Simplify $\frac{a^{10}}{a^6} \cdot \frac{a^7}{a^4}$. Write your answer as a power.

$\frac{a^{10}}{a^6} \cdot \frac{a^7}{a^4} = a^{10-6} \cdot a^{7-4}$ — Quotient of Powers Property

$= a^4 \cdot a^3$ — Simplify.

$= a^{4+3}$ — Product of Powers Property

$= a^7$ — Simplify.

On Your Own

Now You're Ready
Exercises 16–21

Simplify the expression. Write your answer as a power.

5. $\frac{2^{15}}{2^3 \cdot 2^5}$ **6.** $\frac{d^5}{d} \cdot \frac{d^9}{d^8}$ **7.** $\frac{5^9}{5^4} \cdot \frac{5^5}{5^2}$

EXAMPLE 4 Real-Life Application

The projected population of Tennessee in 2030 is about $5 \cdot 5.9^8$. Predict the average number of people per square mile in 2030.

Use a model to solve the problem.

$$\text{People per square mile} = \frac{\text{Population in 2030}}{\text{Land area}}$$

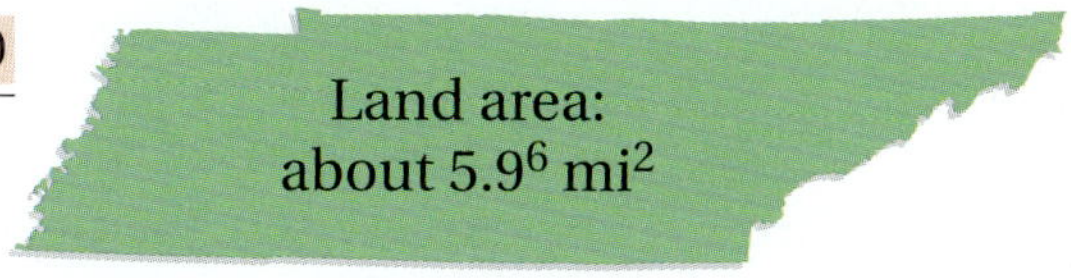

$= \frac{5 \cdot 5.9^8}{5.9^6}$ — Substitute.

$= 5 \cdot \frac{5.9^8}{5.9^6}$ — Rewrite.

$= 5 \cdot 5.9^2$ — Quotient of Powers Property

$= 174.05$ — Evaluate.

So, there will be about 174 people per square mile in Tennessee in 2030.

On Your Own

Now You're Ready
Exercises 23–28

8. The projected population of Alabama in 2030 is about $2.25 \cdot 2^{21}$. The land area of Alabama is about 2^{17} square kilometers. Predict the average number of people per square kilometer in 2030.

10.3 Exercises

Check It Out
Help with Homework
BigIdeasMath.com

Vocabulary and Concept Check

1. **WRITING** Describe in your own words how to divide powers.

2. **WHICH ONE DOESN'T BELONG?** Which quotient does *not* belong with the other three? Explain your reasoning.

$\frac{(-10)^7}{(-10)^2}$ $\frac{6^3}{6^2}$ $\frac{(-4)^8}{(-3)^4}$ $\frac{5^6}{5^3}$

Practice and Problem Solving

Simplify the expression. Write your answer as a power.

3. $\frac{6^{10}}{6^4}$ 4. $\frac{8^9}{8^7}$ 5. $\frac{(-3)^4}{(-3)^1}$ 6. $\frac{4.5^5}{4.5^3}$

1 7. $\frac{5^9}{5^3}$ 8. $\frac{64^4}{64^3}$ 9. $\frac{(-17)^5}{(-17)^2}$ 10. $\frac{(-7.9)^{10}}{(-7.9)^4}$

11. $\frac{(-6.4)^8}{(-6.4)^6}$ 12. $\frac{\pi^{11}}{\pi^7}$ 13. $\frac{b^{24}}{b^{11}}$ 14. $\frac{n^{18}}{n^7}$

15. **ERROR ANALYSIS** Describe and correct the error in simplifying the quotient.

$$\frac{6^{15}}{6^5} = 6^{\frac{15}{5}} = 6^3$$

Simplify the expression. Write your answer as a power.

2 3 16. $\frac{7^5 \cdot 7^3}{7^2}$ 17. $\frac{2^{19} \cdot 2^5}{2^{12} \cdot 2^3}$ 18. $\frac{(-8.3)^8}{(-8.3)^7} \cdot \frac{(-8.3)^4}{(-8.3)^3}$

19. $\frac{\pi^{30}}{\pi^{18} \cdot \pi^4}$ 20. $\frac{c^{22}}{c^8 \cdot c^9}$ 21. $\frac{k^{13}}{k^5} \cdot \frac{k^{17}}{k^{11}}$

22. **SOUND INTENSITY** The sound intensity of a normal conversation is 10^6 times greater than the quietest noise a person can hear. The sound intensity of a jet at takeoff is 10^{14} times greater than the quietest noise a person can hear. How many times more intense is the sound of a jet at takeoff than the sound of a normal conversation?

Simplify the expression.

4 **23.** $\frac{x \cdot 4^8}{4^5}$ **24.** $\frac{6^3 \cdot w}{6^2}$ **25.** $\frac{a^3 \cdot b^4 \cdot 5^4}{b^2 \cdot 5}$

26. $\frac{5^{12} \cdot c^{10} \cdot d^2}{5^9 \cdot c^9}$ **27.** $\frac{x^{15}y^9}{x^8y^3}$ **28.** $\frac{m^{10}n^7}{m^1n^6}$

MP3 Player	Memory (GB)	Price
A	2^1	\$70
B	2^2	\$120
C	2^3	\$170
D	2^4	\$220
E	2^5	\$270

29. MEMORY The memory capacities and prices of five MP3 players are shown in the table.

a. How many times more memory does MP3 Player D have than MP3 Player B?

b. Do memory and price show a linear relationship? Explain.

30. CRITICAL THINKING Consider the equation $\frac{9^m}{9^n} = 9^2$.

a. Find two numbers m and n that satisfy the equation.

b. Describe the number of solutions that satisfy the equation. Explain your reasoning.

Milky Way galaxy
$10 \cdot 10^{10}$ stars

31. STARS There are about 10^{24} stars in the universe. Each galaxy has approximately the same number of stars as the Milky Way galaxy. About how many galaxies are in the universe?

32. Number Sense Find the value of x that makes $\frac{8^{3x}}{8^{2x+1}} = 8^9$ true. Explain how you found your answer.

Fair Game Review What you learned in previous grades & lessons

Subtract. *(Skills Review Handbook)*

33. $-4 - 5$ **34.** $-23 - (-15)$ **35.** $33 - (-28)$ **36.** $18 - 22$

37. MULTIPLE CHOICE What is the value of x? *(Skills Review Handbook)*

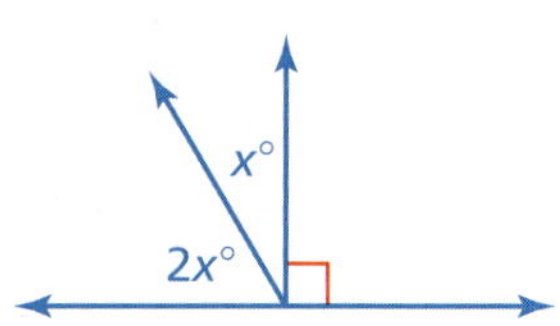

Ⓐ 20 Ⓑ 30

Ⓒ 45 Ⓓ 60

10.4 Zero and Negative Exponents

Essential Question How can you evaluate a nonzero number with an exponent of zero? How can you evaluate a nonzero number with a negative integer exponent?

1 ACTIVITY: Using the Quotient of Powers Property

Work with a partner.

a. Copy and complete the table.

Quotient	Quotient of Powers Property	Power
$\frac{5^3}{5^3}$		
$\frac{6^2}{6^2}$		
$\frac{(-3)^4}{(-3)^4}$		
$\frac{(-4)^5}{(-4)^5}$		

b. **REPEATED REASONING** Evaluate each expression in the first column of the table. What do you notice?

c. How can you use these results to define a^0 where $a \neq 0$?

2 ACTIVITY: Using the Product of Powers Property

Work with a partner.

a. Copy and complete the table.

Product	Product of Powers Property	Power
$3^0 \cdot 3^4$		
$8^2 \cdot 8^0$		
$(-2)^3 \cdot (-2)^0$		
$\left(-\frac{1}{3}\right)^0 \cdot \left(-\frac{1}{3}\right)^5$		

b. Do these results support your definition in Activity 1(c)?

COMMON CORE

Exponents

In this lesson, you will

- evaluate expressions involving numbers with zero as an exponent.
- evaluate expressions involving negative integer exponents.

Learning Standard MACC.8.EE.1.1

3 ACTIVITY: Using the Product of Powers Property

Work with a partner.

a. Copy and complete the table.

Product	Product of Powers Property	Power
$5^{-3} \cdot 5^{3}$		
$6^{2} \cdot 6^{-2}$		
$(-3)^{4} \cdot (-3)^{-4}$		
$(-4)^{-5} \cdot (-4)^{5}$		

b. According to your results from Activities 1 and 2, the products in the first column are equal to what value?

c. **REASONING** How does the Multiplicative Inverse Property help you rewrite the numbers with negative exponents?

d. **STRUCTURE** Use these results to define a^{-n} where $a \neq 0$ and n is an integer.

4 ACTIVITY: Using a Place Value Chart

Math Practice 2

Use Operations

What operations are used when writing the expanded form?

Work with a partner. Use the place value chart that shows the number 3452.867.

Place Value Chart

thousands	hundreds	tens	ones	and	tenths	hundredths	thousandths
10^{3}	10^{2}	10^{1}	$10^{\square}$		$10^{\square}$	$10^{\square}$	$10^{\square}$
3	4	5	2	.	8	6	7

a. **REPEATED REASONING** What pattern do you see in the exponents? Continue the pattern to find the other exponents.

b. **STRUCTURE** Show how to write the expanded form of 3452.867.

What Is Your Answer?

5. **IN YOUR OWN WORDS** How can you evaluate a nonzero number with an exponent of zero? How can you evaluate a nonzero number with a negative integer exponent?

Practice Use what you learned about zero and negative exponents to complete Exercises 5–8 on page 432.

Check It Out
Lesson Tutorials
BigIdeasMath.com

Key Ideas

Zero Exponents

Words For any nonzero number a, $a^0 = 1$. The power 0^0 is *undefined*.

Numbers $4^0 = 1$ **Algebra** $a^0 = 1$, where $a \neq 0$

Negative Exponents

Words For any integer n and any nonzero number a, a^{-n} is the reciprocal of a^n.

Numbers $4^{-2} = \dfrac{1}{4^2}$ **Algebra** $a^{-n} = \dfrac{1}{a^n}$, where $a \neq 0$

EXAMPLE 1 Evaluating Expressions

a. $3^{-4} = \dfrac{1}{3^4}$ Definition of negative exponent

$= \dfrac{1}{81}$ Evaluate power.

b. $(-8.5)^{-4} \cdot (-8.5)^4 = (-8.5)^{-4+4}$ Product of Powers Property

$= (-8.5)^0$ Simplify.

$= 1$ Definition of zero exponent

c. $\dfrac{2^6}{2^8} = 2^{6-8}$ Quotient of Powers Property

$= 2^{-2}$ Simplify.

$= \dfrac{1}{2^2}$ Definition of negative exponent

$= \dfrac{1}{4}$ Evaluate power.

On Your Own

Now You're Ready
Exercises 5–16

Evaluate the expression.

1. 4^{-2}
2. $(-2)^{-5}$
3. $6^{-8} \cdot 6^8$
4. $\dfrac{(-3)^5}{(-3)^6}$
5. $\dfrac{1}{5^7} \cdot \dfrac{1}{5^{-4}}$
6. $\dfrac{4^5 \cdot 4^{-3}}{4^2}$

EXAMPLE 2 Simplifying Expressions

a. $-5x^0 = -5(1)$ — Definition of zero exponent

$= -5$ — Multiply.

b. $\dfrac{9y^{-3}}{y^5} = 9y^{-3-5}$ — Quotient of Powers Property

$= 9y^{-8}$ — Simplify.

$= \dfrac{9}{y^8}$ — Definition of negative exponent

On Your Own

Now You're Ready
Exercises 20–27

Simplify. Write the expression using only positive exponents.

7. $8x^{-2}$ **8.** $b^0 \cdot b^{-10}$ **9.** $\dfrac{z^6}{15z^9}$

EXAMPLE 3 Real-Life Application

A drop of water leaks from a faucet every second. How many liters of water leak from the faucet in 1 hour?

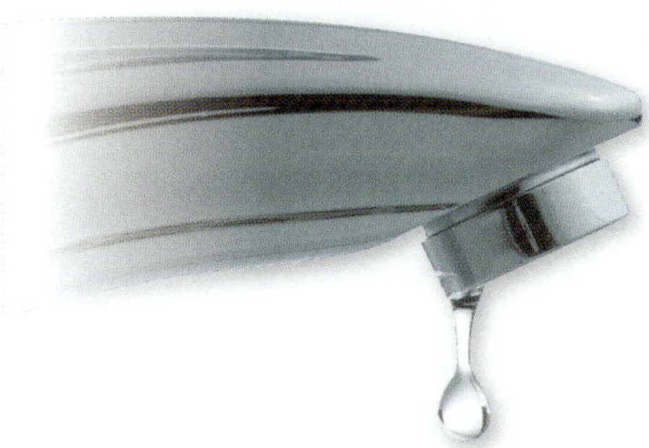

Drop of water: 50^{-2} liter

Convert 1 hour to seconds.

$$1\,\text{h} \times \frac{60\text{ min}}{1\text{ h}} \times \frac{60\text{ sec}}{1\text{ min}} = 3600\text{ sec}$$

Water leaks from the faucet at a rate of 50^{-2} liter per second. Multiply the time by the rate.

$3600\text{ sec} \cdot 50^{-2}\dfrac{\text{L}}{\text{sec}} = 3600 \cdot \dfrac{1}{50^2}$ — Definition of negative exponent

$= 3600 \cdot \dfrac{1}{2500}$ — Evaluate power.

$= \dfrac{3600}{2500}$ — Multiply.

$= 1\dfrac{11}{25} = 1.44\text{ L}$ — Simplify.

So, 1.44 liters of water leak from the faucet in 1 hour.

On Your Own

10. **WHAT IF?** The faucet leaks water at a rate of 5^{-5} liter per second. How many liters of water leak from the faucet in 1 hour?

10.4 Exercises

Check It Out
Help with Homework
BigIdeasMath.com

Vocabulary and Concept Check

1. **VOCABULARY** If a is a nonzero number, does the value of a^0 depend on the value of a? Explain.

2. **WRITING** Explain how to evaluate 10^{-3}.

3. **NUMBER SENSE** Without evaluating, order 5^0, 5^4, and 5^{-5} from least to greatest.

4. **DIFFERENT WORDS, SAME QUESTION** Which is different? Find "both" answers.

Rewrite $\frac{1}{3 \cdot 3 \cdot 3}$ using a negative exponent.	Write 3 to the negative third.
Write $\frac{1}{3}$ cubed as a power.	Write $(-3) \cdot (-3) \cdot (-3)$ as a power.

Practice and Problem Solving

Evaluate the expression.

(1) 5. $\frac{8^7}{8^7}$ 6. $5^0 \cdot 5^3$ 7. $(-2)^{-8} \cdot (-2)^8$ 8. $9^4 \cdot 9^{-4}$

9. 6^{-2} 10. 158^0 11. $\frac{4^3}{4^5}$ 12. $\frac{-3}{(-3)^2}$

13. $4 \cdot 2^{-4} + 5$ 14. $3^{-3} \cdot 3^{-2}$ 15. $\frac{1}{5^{-3}} \cdot \frac{1}{5^6}$ 16. $\frac{(1.5)^2}{(1.5)^{-2} \cdot (1.5)^4}$

17. **ERROR ANALYSIS** Describe and correct the error in evaluating the expression.

$$(4)^{-3} = (-4)(-4)(-4) = -64$$

18. **SAND** The mass of a grain of sand is about 10^{-3} gram. About how many grains of sand are in the bag of sand?

19. **CRITICAL THINKING** How can you write the number 1 as 2 to a power? 10 to a power?

Simplify. Write the expression using only positive exponents.

(2) 20. $6y^{-4}$ 21. $8^{-2} \cdot a^7$ 22. $\frac{9c^3}{c^{-4}}$ 23. $\frac{5b^{-2}}{b^{-3}}$

24. $\frac{8x^3}{2x^9}$ 25. $3d^{-4} \cdot 4d^4$ 26. $m^{-2} \cdot n^3$ 27. $\frac{3^{-2} \cdot k^0 \cdot w^0}{w^{-6}}$

28. **OPEN-ENDED** Write two different powers with negative exponents that have the same value.

METRIC UNITS **In Exercises 29–32, use the table.**

29. How many millimeters are in a decimeter?

30. How many micrometers are in a centimeter?

31. How many nanometers are in a millimeter?

32. How many micrometers are in a meter?

Unit of Length	Length (meter)
Decimeter	10^{-1}
Centimeter	10^{-2}
Millimeter	10^{-3}
Micrometer	10^{-6}
Nanometer	10^{-9}

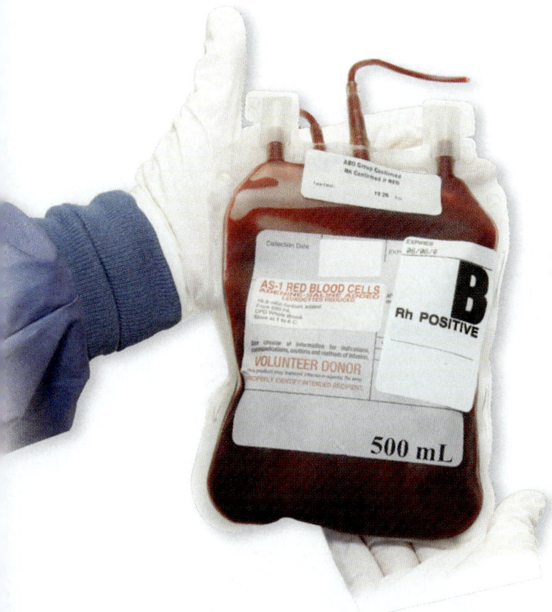

33. **BACTERIA** A species of bacteria is 10 micrometers long. A virus is 10,000 times smaller than the bacteria.

    a. Using the table above, find the length of the virus in meters.

    b. Is the answer to part (a) *less than*, *greater than*, or *equal to* one nanometer?

34. **BLOOD DONATION** Every 2 seconds, someone in the United States needs blood. A sample blood donation is shown. ($1 \text{ mm}^3 = 10^{-3} \text{ mL}$)

    a. One cubic millimeter of blood contains about 10^4 white blood cells. How many white blood cells are in the donation? Write your answer in words.

    b. One cubic millimeter of blood contains about 5×10^6 red blood cells. How many red blood cells are in the donation? Write your answer in words.

    c. Compare your answers for parts (a) and (b).

35. **PRECISION** Describe how to rewrite a power with a positive exponent so that the exponent is in the denominator. Use the definition of negative exponents to justify your reasoning.

36. **Reasoning** The rule for negative exponents states that $a^{-n} = \frac{1}{a^n}$. Explain why this rule does not apply when $a = 0$.

Fair Game Review What you learned in previous grades & lessons

Simplify the expression. Write your answer as a power. *(Section 10.2 and Section 10.3)*

37. $10^3 \cdot 10^6$

38. $10^2 \cdot 10$

39. $\frac{10^8}{10^4}$

40. **MULTIPLE CHOICE** Which data display best orders numerical data and shows how they are distributed? *(Section 9.4)*

    Ⓐ bar graph
    Ⓑ line graph
    Ⓒ scatter plot
    Ⓓ stem-and-leaf plot

Check It Out
Graphic Organizer
BigIdeasMath.com

You can use an **information wheel** to organize information about a topic. Here is an example of an information wheel for exponents.

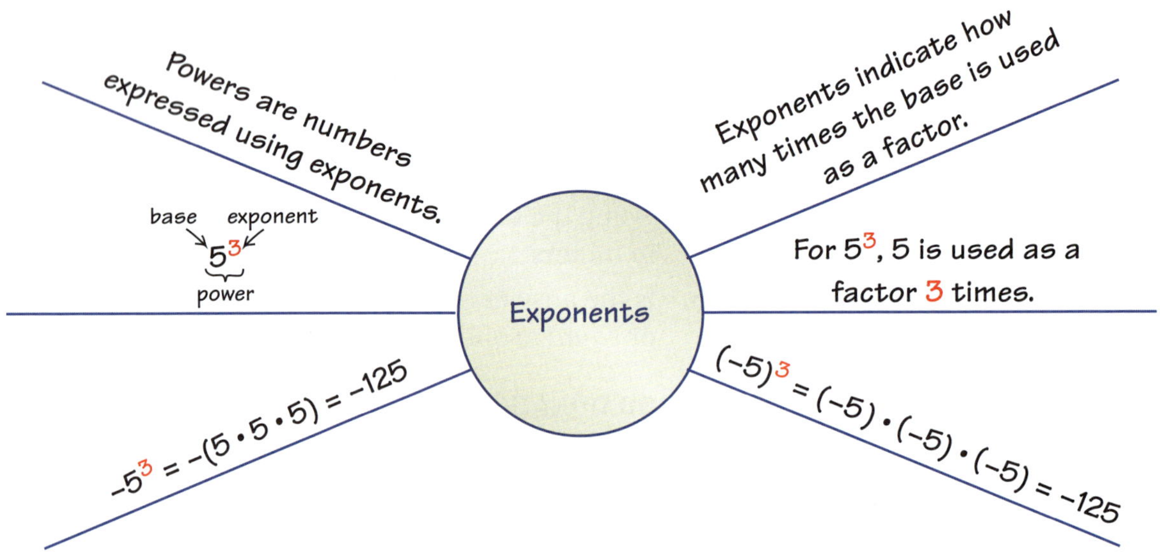

On Your Own

Make information wheels to help you study these topics.

1. Product of Powers Property
2. Quotient of Powers Property
3. zero and negative exponents

After you complete this chapter, make information wheels for the following topics.

4. writing numbers in scientific notation
5. writing numbers in standard form
6. adding and subtracting numbers in scientific notation
7. multiplying and dividing numbers in scientific notation
8. Choose three other topics you studied earlier in this course. Make an information wheel for each topic to summarize what you know about them.

"I decided to color code the different flavors in my information wheel."

10.1–10.4 Quiz

Check It Out
Progress Check
BigIdeasMath.com

Write the product using exponents. *(Section 10.1)*

1. $(-5) \cdot (-5) \cdot (-5) \cdot (-5)$

2. $7 \cdot 7 \cdot m \cdot m \cdot m$

Evaluate the expression. *(Section 10.1 and Section 10.4)*

3. 5^4

4. $(-2)^6$

5. $(-4.8)^{-9} \cdot (-4.8)^9$

6. $\dfrac{5^4}{5^7}$

Simplify the expression. Write your answer as a power. *(Section 10.2)*

7. $3^8 \cdot 3$

8. $(a^5)^3$

Simplify the expression. *(Section 10.2)*

9. $(3c)^4$

10. $\left(-\dfrac{2}{7}p\right)^2$

Simplify the expression. Write your answer as a power. *(Section 10.3)*

11. $\dfrac{8^7}{8^4}$

12. $\dfrac{6^3 \cdot 6^7}{6^2}$

13. $\dfrac{\pi^{15}}{\pi^3 \cdot \pi^9}$

14. $\dfrac{t^{13}}{t^5} \cdot \dfrac{t^8}{t^6}$

Simplify. Write the expression using only positive exponents. *(Section 10.4)*

15. $8d^{-6}$

16. $\dfrac{12x^5}{4x^7}$

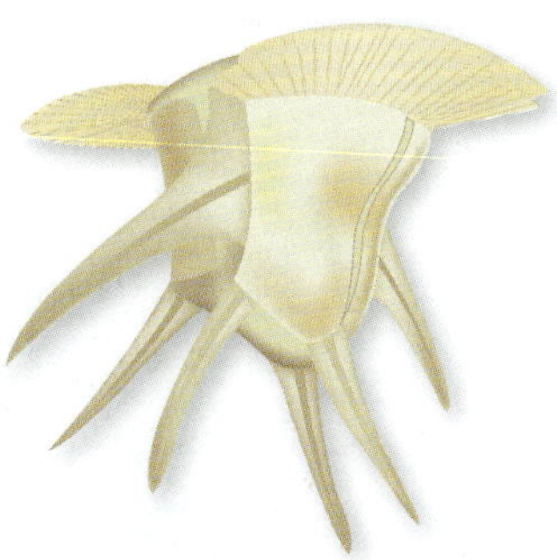

17. ORGANISM A one-celled, aquatic organism called a dinoflagellate is 1000 micrometers long. *(Section 10.4)*

a. One micrometer is 10^{-6} meter. What is the length of the dinoflagellate in meters?

b. Is the length of the dinoflagellate equal to 1 millimeter or 1 kilometer? Explain.

18. EARTHQUAKES An earthquake of magnitude 3.0 is 10^2 times stronger than an earthquake of magnitude 1.0. An earthquake of magnitude 8.0 is 10^7 times stronger than an earthquake of magnitude 1.0. How many times stronger is an earthquake of magnitude 8.0 than an earthquake of magnitude 3.0? *(Section 10.3)*

10.5 Reading Scientific Notation

Essential Question How can you read numbers that are written in scientific notation?

1 ACTIVITY: Very Large Numbers

Work with a partner.

- Use a calculator. Experiment with multiplying large numbers until your calculator displays an answer that is *not* in standard form.
- When the calculator at the right was used to multiply 2 billion by 3 billion, it listed the result as

  6.0E+18.

- Multiply 2 billion by 3 billion by hand. Use the result to explain what 6.0E+18 means.
- Check your explanation by calculating the products of other large numbers.
- Why didn't the calculator show the answer in standard form?
- Experiment to find the maximum number of digits your calculator displays. For instance, if you multiply 1000 by 1000 and your calculator shows 1,000,000, then it can display seven digits.

2 ACTIVITY: Very Small Numbers

Work with a partner.

- Use a calculator. Experiment with multiplying very small numbers until your calculator displays an answer that is *not* in standard form.
- When the calculator at the right was used to multiply 2 billionths by 3 billionths, it listed the result as

  6.0E–18.

- Multiply 2 billionths by 3 billionths by hand. Use the result to explain what 6.0E–18 means.
- Check your explanation by calculating the products of other very small numbers.

COMMON CORE

Scientific Notation

In this lesson, you will

- identify numbers written in scientific notation.
- write numbers in standard form.
- compare numbers in scientific notation.

Learning Standards
MACC.8.EE.1.3
MACC.8.EE.1.4

3 ACTIVITY: Powers of 10 Matching Game

Math Practice 4

Analyze Relationships

How are the pictures related? How can you order the pictures to find the correct power of 10?

Work with a partner. Match each picture with its power of 10. Explain your reasoning.

10^5 m 10^2 m 10^0 m 10^{-1} m 10^{-2} m 10^{-5} m

A.

B.

C.

D.

E.
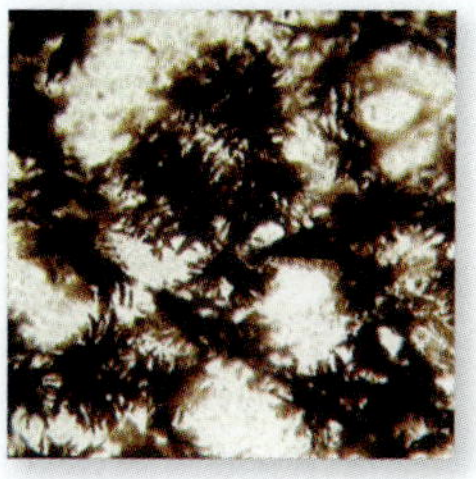

F.

4 ACTIVITY: Choosing Appropriate Units

Work with a partner. Match each unit with its most appropriate measurement.

inches   centimeters   feet   millimeters   meters

A. Height of a door: 2×10^0

B. Height of a volcano: 1.6×10^4

C. Length of a pen: 1.4×10^2

D. Diameter of a steel ball bearing: 6.3×10^{-1}

E. Circumference of a beach ball: 7.5×10^1

What Is Your Answer?

5. **IN YOUR OWN WORDS** How can you read numbers that are written in scientific notation? Why do you think this type of notation is called *scientific notation*? Why is scientific notation important?

Practice Use what you learned about reading scientific notation to complete Exercises 3–5 on page 440.

10.5 Lesson

Check It Out
Lesson Tutorials
BigIdeasMath.com

Key Vocabulary
scientific notation, *p. 438*

Study Tip
Scientific notation is used to write very small and very large numbers.

Key Idea

Scientific Notation

A number is written in **scientific notation** when it is represented as the product of a factor and a power of 10. The factor must be greater than or equal to 1 and less than 10.

The factor is greater than or equal to 1 and less than 10. → 8.3×10^{-7} ← The power of 10 has an integer exponent.

EXAMPLE 1 Identifying Numbers Written in Scientific Notation

Tell whether the number is written in scientific notation. Explain.

a. 5.9×10^{-6}

∴ The factor is greater than or equal to 1 and less than 10. The power of 10 has an integer exponent. So, the number is written in scientific notation.

b. 0.9×10^{8}

∴ The factor is less than 1. So, the number is not written in scientific notation.

Key Idea

Writing Numbers in Standard Form

The absolute value of the exponent indicates how many places to move the decimal point.

- If the exponent is negative, move the decimal point to the left.
- If the exponent is positive, move the decimal point to the right.

EXAMPLE 2 Writing Numbers in Standard Form

a. Write 3.22×10^{-4} in standard form.

$3.22 \times 10^{-4} = 0.000322$ (4) — Move decimal point $|-4| = 4$ places to the left.

b. Write 7.9×10^{5} in standard form.

$7.9 \times 10^{5} = 790,000$ (5) — Move decimal point $|5| = 5$ places to the right.

On Your Own

Now You're Ready
Exercises 6–23

1. Is 12×10^4 written in scientific notation? Explain.

Write the number in standard form.

2. 6×10^7
3. 9.9×10^{-5}
4. 1.285×10^4

EXAMPLE 3 Comparing Numbers in Scientific Notation

An object with a lesser density than water will float. An object with a greater density than water will sink. Use each given density (in kilograms per cubic meter) to explain what happens when you place a brick and an apple in water.

Water: 1.0×10^3 **Brick:** 1.84×10^3 **Apple:** 6.41×10^2

You can compare the densities by writing each in standard form.

Water	Brick	Apple
$1.0 \times 10^3 = 1000$	$1.84 \times 10^3 = 1840$	$6.41 \times 10^2 = 641$

∴ The apple is less dense than water, so it will float. The brick is denser than water, so it will sink.

EXAMPLE 4 Real-Life Application

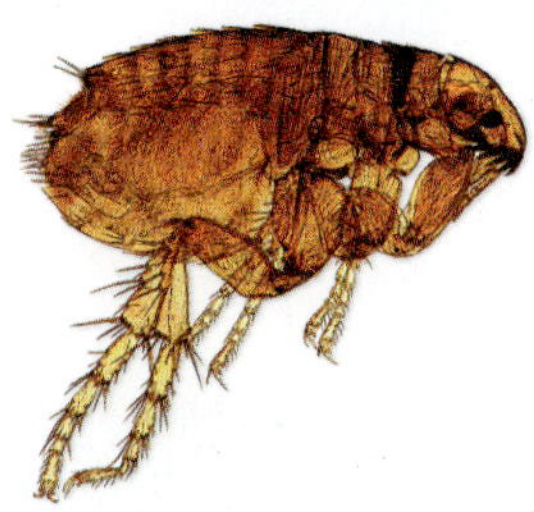

A female flea consumes about 1.4×10^{-5} liter of blood per day.

A dog has 100 female fleas. How much blood do the fleas consume per day?

$$1.4 \times 10^{-5} \cdot 100 = 0.000014 \cdot 100 \quad \text{Write in standard form.}$$
$$= 0.0014 \quad \text{Multiply.}$$

∴ The fleas consume about 0.0014 liter, or 1.4 milliliters of blood per day.

On Your Own

Now You're Ready
Exercise 27

5. **WHAT IF?** In Example 3, the density of lead is 1.14×10^4 kilograms per cubic meter. What happens when you place lead in water?
6. **WHAT IF?** In Example 4, a dog has 75 female fleas. How much blood do the fleas consume per day?

10.5 Exercises

Check It Out
Help with Homework
BigIdeasMath.com

Vocabulary and Concept Check

1. **WRITING** Describe the difference between scientific notation and standard form.

2. **WHICH ONE DOESN'T BELONG?** Which number does *not* belong with the other three? Explain.

2.8×10^{15}	4.3×10^{-30}	1.05×10^{28}	10×9.2^{-13}

Practice and Problem Solving

Write the number shown on the calculator display in standard form.

3. 5.6E12

4. 2.1E-10

5. 8.73E16

Tell whether the number is written in scientific notation. Explain.

1 6. 1.8×10^{9} 7. 3.45×10^{14} 8. 0.26×10^{-25}

9. 10.5×10^{12} 10. 46×10^{-17} 11. 5×10^{-19}

12. 7.814×10^{-36} 13. 0.999×10^{42} 14. 6.022×10^{23}

Write the number in standard form.

2 15. 7×10^{7} 16. 8×10^{-3} 17. 5×10^{2}

18. 2.7×10^{-4} 19. 4.4×10^{-5} 20. 2.1×10^{3}

21. 1.66×10^{9} 22. 3.85×10^{-8} 23. 9.725×10^{6}

24. **ERROR ANALYSIS** Describe and correct the error in writing the number in standard form.

✗ $4.1 \times 10^{-6} = 4{,}100{,}000$

2.7×10^{8} platelets per milliliter

25. **PLATELETS** Platelets are cell-like particles in the blood that help form blood clots.

 a. How many platelets are in 3 milliliters of blood? Write your answer in standard form.

 b. An adult human body contains about 5 liters of blood. How many platelets are in an adult human body?

26. **REASONING** A googol is 1.0×10^{100}. How many zeros are in a googol?

3 27. **STARS** The table shows the surface temperatures of five stars.

a. Which star has the highest surface temperature?

b. Which star has the lowest surface temperature?

Star	Betelgeuse	Bellatrix	Sun	Aldebaran	Rigel
Surface Temperature (°F)	6.2×10^3	3.8×10^4	1.1×10^4	7.2×10^3	2.2×10^4

28. **NUMBER SENSE** Describe how the value of a number written in scientific notation changes when you increase the exponent by 1.

29. **CORAL REEF** The area of the Florida Keys National Marine Sanctuary is about 9.6×10^3 square kilometers. The area of the Florida Reef Tract is about 16.2% of the area of the sanctuary. What is the area of the Florida Reef Tract in square kilometers?

30. **REASONING** A gigameter is 1.0×10^6 kilometers. How many square kilometers are in 5 square gigameters?

31. **WATER** There are about 1.4×10^9 cubic kilometers of water on Earth. About 2.5% of the water is fresh water. How much fresh water is on Earth?

32. **Critical Thinking** The table shows the speed of light through five media.

a. In which medium does light travel the fastest?

b. In which medium does light travel the slowest?

Medium	Speed
Air	6.7×10^8 mi/h
Glass	6.6×10^8 ft/sec
Ice	2.3×10^5 km/sec
Vacuum	3.0×10^8 m/sec
Water	2.3×10^{10} cm/sec

Fair Game Review What you learned in previous grades & lessons

Write the product using exponents. *(Section 10.1)*

33. $4 \cdot 4 \cdot 4 \cdot 4 \cdot 4$

34. $3 \cdot 3 \cdot 3 \cdot y \cdot y \cdot y$

35. $(-2) \cdot (-2) \cdot (-2)$

36. **MULTIPLE CHOICE** What is the length of the hypotenuse of the right triangle? *(Section 7.3)*

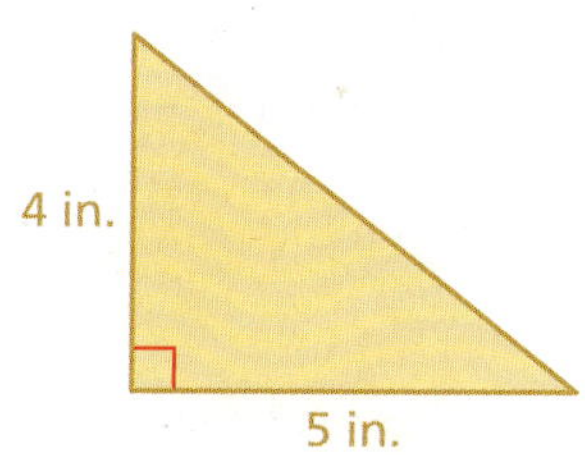

(A) $\sqrt{18}$ in.

(B) $\sqrt{41}$ in.

(C) 18 in.

(D) 41 in.

10.6 Writing Scientific Notation

Essential Question How can you write a number in scientific notation?

1 ACTIVITY: Finding pH Levels

Work with a partner. In chemistry, pH is a measure of the activity of dissolved hydrogen ions (H^+). Liquids with low pH values are called *acids*. Liquids with high pH values are called *bases*.

Find the pH of each liquid. Is the liquid a base, neutral, or an acid?

a. Lime juice:
$[H^+] = 0.01$

b. Egg:
$[H^+] = 0.00000001$

c. Distilled water:
$[H^+] = 0.0000001$

d. Ammonia water:
$[H^+] = 0.00000000001$

e. Tomato juice:
$[H^+] = 0.0001$

f. Hydrochloric acid:
$[H^+] = 1$

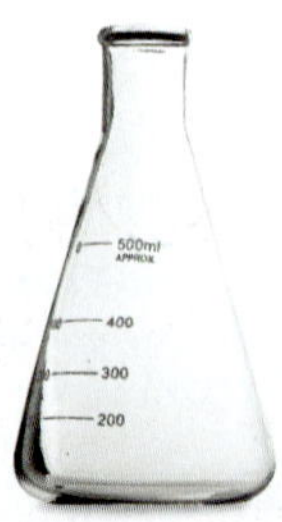

pH	$[H^+]$	
14	1×10^{-14}	Bases
13	1×10^{-13}	
12	1×10^{-12}	
11	1×10^{-11}	
10	1×10^{-10}	
9	1×10^{-9}	
8	1×10^{-8}	
7	1×10^{-7}	Neutral
6	1×10^{-6}	Acids
5	1×10^{-5}	
4	1×10^{-4}	
3	1×10^{-3}	
2	1×10^{-2}	
1	1×10^{-1}	
0	1×10^{0}	

COMMON CORE

Scientific Notation

In this lesson, you will

- write large and small numbers in scientific notation.
- perform operations with numbers written in scientific notation.

Learning Standards
MACC.8.EE.1.3
MACC.8.EE.1.4

2 ACTIVITY: Writing Scientific Notation

Work with a partner. Match each planet with its distance from the Sun. Then write each distance in scientific notation. Do you think it is easier to match the distances when they are written in standard form or in scientific notation? Explain.

a. 1,800,000,000 miles

b. 67,000,000 miles

c. 890,000,000 miles

d. 93,000,000 miles

e. 140,000,000 miles

f. 2,800,000,000 miles

g. 480,000,000 miles

h. 36,000,000 miles

3 ACTIVITY: Making a Scale Drawing

Math Practice 6

Calculate Accurately

How can you verify that you have accurately written each distance in scientific notation?

Work with a partner. The illustration in Activity 2 is not drawn to scale. Use the instructions below to make a scale drawing of the distances in our solar system.

- **Cut a sheet of paper into three strips of equal width. Tape the strips together to make one long piece.**
- **Draw a long number line. Label the number line in hundreds of millions of miles.**
- **Locate each planet's position on the number line.**

What Is Your Answer?

4. **IN YOUR OWN WORDS** How can you write a number in scientific notation?

Practice Use what you learned about writing scientific notation to complete Exercises 3–5 on page 446.

10.6 Lesson

Check It Out
Lesson Tutorials
BigIdeasMath.com

Key Idea

Writing Numbers in Scientific Notation

Step 1: Move the decimal point so it is located to the right of the leading nonzero digit.

Step 2: Count the number of places you moved the decimal point. This indicates the exponent of the power of 10, as shown below.

Number Greater Than or Equal to 10

Use a positive exponent when you move the decimal point to the left.

$8600 = 8.6 \times 10^3$
3

Number Between 0 and 1

Use a negative exponent when you move the decimal point to the right.

$0.0024 = 2.4 \times 10^{-3}$
3

Study Tip

When you write a number greater than or equal to 1 and less than 10 in scientific notation, use zero as the exponent.

$6 = 6 \times 10^0$

EXAMPLE 1 Writing Large Numbers in Scientific Notation

Google purchased YouTube for $1,650,000,000. Write this number in scientific notation.

Move the decimal point 9 places to the left.

$1{,}650{,}000{,}000 = 1.65 \times 10^9$
9

The number is greater than 10. So, the exponent is positive.

EXAMPLE 2 Writing Small Numbers in Scientific Notation

The 2004 Indonesian earthquake slowed the rotation of Earth, making the length of a day 0.00000268 second shorter. Write this number in scientific notation.

Move the decimal point 6 places to the right.

$0.00000268 = 2.68 \times 10^{-6}$
6

The number is between 0 and 1. So, the exponent is negative.

On Your Own

Now You're Ready
Exercises 3–11

Write the number in scientific notation.

1. 50,000 **2.** 25,000,000 **3.** 683

4. 0.005 **5.** 0.00000033 **6.** 0.000506

EXAMPLE 3 Using Scientific Notation

An album receives an award when it sells 10,000,000 copies.

An album has sold 8,780,000 copies. How many more copies does it need to sell to receive the award?

Ⓐ 1.22×10^{-7} Ⓑ 1.22×10^{-6}

Ⓒ 1.22×10^{6} Ⓓ 1.22×10^{7}

Use a model to solve the problem.

Remaining sales needed for award = Sales required for award − Current sales total

$= 10{,}000{,}000 - 8{,}780{,}000$

$= 1{,}220{,}000$

$= 1.22 \times 10^{6}$

∴ The album must sell 1.22×10^{6} more copies to receive the award. So, the correct answer is Ⓒ.

EXAMPLE 4 Real-Life Application

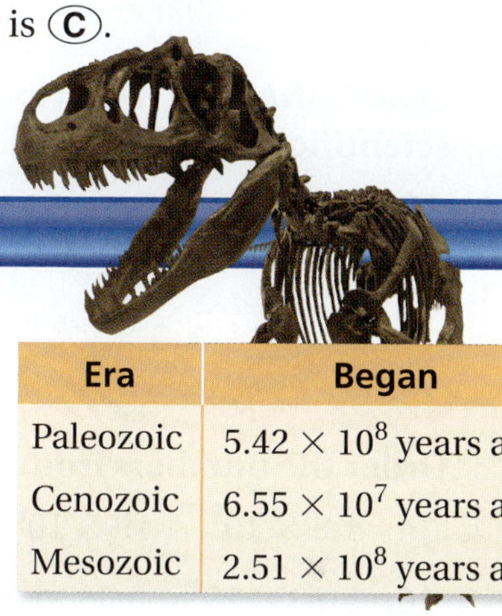

The table shows when the last three geologic eras began. Order the eras from earliest to most recent.

Era	Began
Paleozoic	5.42×10^{8} years ago
Cenozoic	6.55×10^{7} years ago
Mesozoic	2.51×10^{8} years ago

Step 1: Compare the powers of 10.

Because $10^{7} < 10^{8}$,
$6.55 \times 10^{7} < 5.42 \times 10^{8}$ and
$6.55 \times 10^{7} < 2.51 \times 10^{8}$.

Step 2: Compare the factors when the powers of 10 are the same.

Because $2.51 < 5.42$,
$2.51 \times 10^{8} < 5.42 \times 10^{8}$.

From greatest to least, the order is 5.42×10^{8}, 2.51×10^{8}, and 6.55×10^{7}.

∴ So, the eras in order from earliest to most recent are the Paleozoic era, Mesozoic era, and Cenozoic era.

Common Error

To use the method in Example 4, the numbers must be written in scientific notation.

On Your Own

Now You're Ready
Exercises 14–19

7. **WHAT IF?** In Example 3, an album has sold 955,000 copies. How many more copies does it need to sell to receive the award? Write your answer in scientific notation.

8. The *Tyrannosaurus rex* lived 7.0×10^{7} years ago. Consider the eras given in Example 4. During which era did the *Tyrannosaurus rex* live?

10.6 Exercises

Check It Out
Help with Homework
BigIdeasMath.com

Vocabulary and Concept Check

1. **REASONING** How do you know whether a number written in standard form will have a positive or a negative exponent when written in scientific notation?

2. **WRITING** When is it appropriate to use scientific notation instead of standard form?

Practice and Problem Solving

Write the number in scientific notation.

① ② 3. 0.0021

4. 5,430,000

5. 321,000,000

6. 0.00000625

7. 0.00004

8. 10,700,000

9. 45,600,000,000

10. 0.000000000009256

11. 840,000

ERROR ANALYSIS **Describe and correct the error in writing the number in scientific notation.**

12. ✗
0.000036
5
3.6×10^5

13. ✗
72,500,000
6
72.5×10^6

Order the numbers from least to greatest.

④ 14. 1.2×10^8, 1.19×10^8, 1.12×10^8

15. 6.8×10^{-5}, 6.09×10^{-5}, 6.78×10^{-5}

16. 5.76×10^{12}, 9.66×10^{11}, 5.7×10^{10}

17. 4.8×10^{-6}, 4.8×10^{-5}, 4.8×10^{-8}

18. 9.9×10^{-15}, 1.01×10^{-14}, 7.6×10^{-15}

19. 5.78×10^{23}, 6.88×10^{-23}, 5.82×10^{23}

20. **HAIR** What is the diameter of a human hair written in scientific notation?

Diameter: 0.000099 meter

21. **EARTH** What is the circumference of Earth written in scientific notation?

Circumference at the equator: about 40,100,000 meters

22. **CHOOSING UNITS** In Exercise 21, name a unit of measurement that would be more appropriate for the circumference. Explain.

Order the numbers from least to greatest.

23. $\frac{68,500}{10}$, 680, 6.8×10^3

24. $\frac{5}{241}$, 0.02, 2.1×10^{-2}

25. 6.3%, 6.25×10^{-3}, $6\frac{1}{4}$, 0.625

26. 3033.4, 305%, $\frac{10,000}{3}$, 3.3×10^2

27. **SPACE SHUTTLE** The total power of a space shuttle during launch is the sum of the power from its solid rocket boosters and the power from its main engines. The power from the solid rocket boosters is 9,750,000,000 watts. What is the power from the main engines?

Total power = 1.174×10^{10} watts

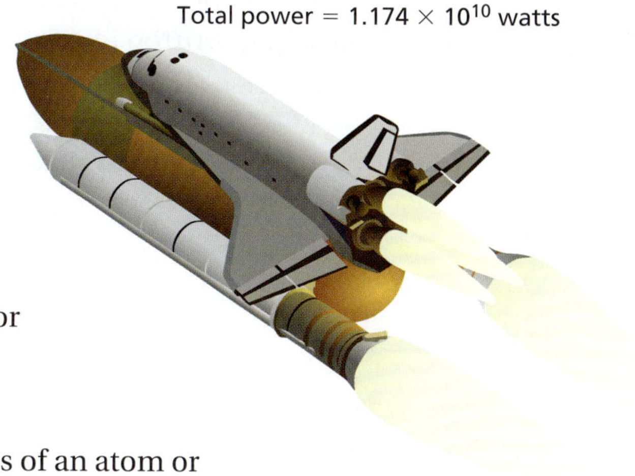

28. **CHOOSE TOOLS** Explain how to use a calculator to verify your answer to Exercise 27.

Equivalent to 1 Atomic Mass Unit
8.3×10^{-24} carat
1.66×10^{-21} milligram

29. **ATOMIC MASS** The mass of an atom or molecule is measured in atomic mass units. Which is greater, a *carat* or a *milligram*? Explain.

30. **Reasoning** In Example 4, the Paleozoic era ended when the Mesozoic era began. The Mesozoic era ended when the Cenozoic era began. The Cenozoic era is the current era.

a. Write the lengths of the three eras in scientific notation. Order the lengths from least to greatest.

b. Make a time line to show when the three eras occurred and how long each era lasted.

c. What do you notice about the lengths of the three eras? Use the Internet to determine whether your observation is true for *all* the geologic eras. Explain your results.

Fair Game Review What you learned in previous grades & lessons

Classify the real number. *(Section 7.4)*

31. 15

32. $\sqrt[3]{-8}$

33. $\sqrt{73}$

34. What is the surface area of the prism? *(Skills Review Handbook)*

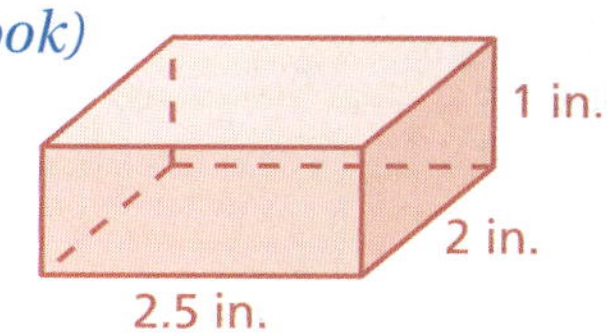

Ⓐ 5 in.^{2}

Ⓑ 5.5 in.^{2}

Ⓒ 10 in.^{2}

Ⓓ 19 in.^{2}

10.7 Operations in Scientific Notation

Essential Question How can you perform operations with numbers written in scientific notation?

1 ACTIVITY: Adding Numbers in Scientific Notation

Work with a partner. Consider the numbers 2.4×10^3 and 7.1×10^3.

a. Explain how to use order of operations to find the sum of these numbers. Then find the sum.

$$2.4 \times 10^3 + 7.1 \times 10^3$$

b. The factor ______ is common to both numbers. How can you use the Distributive Property to rewrite the sum $(2.4 \times 10^3) + (7.1 \times 10^3)$?

$(2.4 \times 10^3) + (7.1 \times 10^3) =$ ______ Distributive Property

c. Use order of operations to evaluate the expression you wrote in part (b). Compare the result with your answer in part (a).

d. **STRUCTURE** Write a rule you can use to add numbers written in scientific notation where the powers of 10 are the same. Then test your rule using the sums below.

- $(4.9 \times 10^5) + (1.8 \times 10^5) =$ ______
- $(3.85 \times 10^4) + (5.72 \times 10^4) =$ ______

2 ACTIVITY: Adding Numbers in Scientific Notation

Work with a partner. Consider the numbers 2.4×10^3 and 7.1×10^4.

a. Explain how to use order of operations to find the sum of these numbers. Then find the sum.

$$2.4 \times 10^3 + 7.1 \times 10^4$$

b. How is this pair of numbers different from the pairs of numbers in Activity 1?

c. Explain why you cannot immediately use the rule you wrote in Activity 1(d) to find this sum.

d. **STRUCTURE** How can you rewrite one of the numbers so that you can use the rule you wrote in Activity 1(d)? Rewrite one of the numbers. Then find the sum using your rule and compare the result with your answer in part (a).

e. **REASONING** Do these procedures work when subtracting numbers written in scientific notation? Justify your answer by evaluating the differences below.

- $(8.2 \times 10^5) - (4.6 \times 10^5) =$ ______
- $(5.88 \times 10^5) - (1.5 \times 10^4) =$ ______

COMMON CORE

Scientific Notation

In this lesson, you will

- add, subtract, multiply, and divide numbers written in scientific notation.

Learning Standards
MACC.8.EE.1.3
MACC.8.EE.1.4

3 ACTIVITY: Multiplying Numbers in Scientific Notation

Math Practice 3

Justify Conclusions

Which step of the procedure would be affected if the powers of 10 were different? Explain.

Work with a partner. Match each step with the correct description.

Step	Description
$(2.4 \times 10^3) \times (7.1 \times 10^3)$	Original expression
1. $= 2.4 \times 7.1 \times 10^3 \times 10^3$	**A.** Write in standard form.
2. $= (2.4 \times 7.1) \times (10^3 \times 10^3)$	**B.** Product of Powers Property
3. $= 17.04 \times 10^6$	**C.** Write in scientific notation.
4. $= 1.704 \times 10^1 \times 10^6$	**D.** Commutative Property of Multiplication
5. $= 1.704 \times 10^7$	**E.** Simplify.
6. $= 17{,}040{,}000$	**F.** Associative Property of Multiplication

Does this procedure work when the numbers have different powers of 10? Justify your answer by using this procedure to evaluate the products below.

- $(1.9 \times 10^2) \times (2.3 \times 10^5) =$ ______
- $(8.4 \times 10^6) \times (5.7 \times 10^{-4}) =$ ______

4 ACTIVITY: Using Scientific Notation to Estimate

Work with a partner. A person normally breathes about 6 liters of air per minute. The life expectancy of a person in the United States at birth is about 80 years. Use scientific notation to estimate the total amount of air a person born in the United States breathes over a lifetime.

What Is Your Answer?

5. **IN YOUR OWN WORDS** How can you perform operations with numbers written in scientific notation?

6. Use a calculator to evaluate the expression. Write your answer in scientific notation and in standard form.

   a. $(1.5 \times 10^4) + (6.3 \times 10^4)$ **b.** $(7.2 \times 10^5) - (2.2 \times 10^3)$

   c. $(4.1 \times 10^{-3}) \times (4.3 \times 10^{-3})$ **d.** $(4.75 \times 10^{-6}) \times (1.34 \times 10^7)$

Practice Use what you learned about evaluating expressions involving scientific notation to complete Exercises 3–6 on page 452.

10.7 Lesson

Check It Out
Lesson Tutorials
BigIdeasMath.com

To add or subtract numbers written in scientific notation with the same power of 10, add or subtract the factors. When the numbers have different powers of 10, first rewrite the numbers so they have the same power of 10.

EXAMPLE 1 Adding and Subtracting Numbers in Scientific Notation

Find the sum or difference. Write your answer in scientific notation.

a. $(4.6 \times 10^3) + (8.72 \times 10^3)$

$= (4.6 + 8.72) \times 10^3$	Distributive Property
$= 13.32 \times 10^3$	Add.
$= (1.332 \times 10^1) \times 10^3$	Write 13.32 in scientific notation.
$= 1.332 \times 10^4$	Product of Powers Property

b. $(3.5 \times 10^{-2}) - (6.6 \times 10^{-3})$

Rewrite 6.6×10^{-3} so that it has the same power of 10 as 3.5×10^{-2}.

$6.6 \times 10^{-3} = 6.6 \times 10^{-1} \times 10^{-2}$	Rewrite 10^{-3} as $10^{-1} \times 10^{-2}$.
$= 0.66 \times 10^{-2}$	Rewrite 6.6×10^{-1} as 0.66.

Subtract the factors.

$(3.5 \times 10^{-2}) - (0.66 \times 10^{-2})$

$= (3.5 - 0.66) \times 10^{-2}$	Distributive Property
$= 2.84 \times 10^{-2}$	Subtract.

Study Tip

In Example 1(b), you will get the same answer when you start by rewriting 3.5×10^{-2} as 35×10^{-3}.

On Your Own

Now You're Ready
Exercises 7–14

Find the sum or difference. Write your answer in scientific notation.

1. $(8.2 \times 10^2) + (3.41 \times 10^{-1})$
2. $(7.8 \times 10^{-5}) - (4.5 \times 10^{-5})$

To multiply or divide numbers written in scientific notation, multiply or divide the factors and powers of 10 separately.

EXAMPLE 2 Multiplying Numbers in Scientific Notation

Find $(3 \times 10^{-5}) \times (5 \times 10^{-2})$. Write your answer in scientific notation.

$(3 \times 10^{-5}) \times (5 \times 10^{-2})$

$= 3 \times 5 \times 10^{-5} \times 10^{-2}$	Commutative Property of Multiplication
$= (3 \times 5) \times (10^{-5} \times 10^{-2})$	Associative Property of Multiplication
$= 15 \times 10^{-7}$	Simplify.
$= 1.5 \times 10^1 \times 10^{-7}$	Write 15 in scientific notation.
$= 1.5 \times 10^{-6}$	Product of Powers Property

Study Tip

You can check your answer using standard form.

$(3 \times 10^{-5}) \times (5 \times 10^{-2})$
$= 0.00003 \times 0.05$
$= 0.0000015$
$= 1.5 \times 10^{-6}$

EXAMPLE 3 Dividing Numbers in Scientific Notation

Find $\frac{1.5 \times 10^{-8}}{6 \times 10^{7}}$. Write your answer in scientific notation.

$\frac{1.5 \times 10^{-8}}{6 \times 10^{7}} = \frac{1.5}{6} \times \frac{10^{-8}}{10^{7}}$	Rewrite as a product of fractions.
$= 0.25 \times \frac{10^{-8}}{10^{7}}$	Divide 1.5 by 6.
$= 0.25 \times 10^{-15}$	Quotient of Powers Property
$= 2.5 \times 10^{-1} \times 10^{-15}$	Write 0.25 in scientific notation.
$= 2.5 \times 10^{-16}$	Product of Powers Property

On Your Own

Now You're Ready
Exercises 16–23

Find the product or quotient. Write your answer in scientific notation.

3. $6 \times (8 \times 10^{-5})$
4. $(7 \times 10^{2}) \times (3 \times 10^{5})$
5. $(9.2 \times 10^{12}) \div 4.6$
6. $(1.5 \times 10^{-3}) \div (7.5 \times 10^{2})$

EXAMPLE 4 Real-Life Application

Diameter = 1,400,000 km

How many times greater is the diameter of the Sun than the diameter of Earth?

Diameter = 1.28×10^4 km

Write the diameter of the Sun in scientific notation.

$$1{,}400{,}000 = 1.4 \times 10^{6}$$

6

Divide the diameter of the Sun by the diameter of Earth.

$\frac{1.4 \times 10^{6}}{1.28 \times 10^{4}} = \frac{1.4}{1.28} \times \frac{10^{6}}{10^{4}}$	Rewrite as a product of fractions.
$= 1.09375 \times \frac{10^{6}}{10^{4}}$	Divide 1.4 by 1.28.
$= 1.09375 \times 10^{2}$	Quotient of Powers Property
$= 109.375$	Write in standard form.

The diameter of the Sun is about 109 times greater than the diameter of Earth.

On Your Own

7. How many more kilometers is the radius of the Sun than the radius of Earth? Write your answer in standard form.

10.7 Exercises

Check It Out
Help with Homework
BigIdeasMath.com

Vocabulary and Concept Check

1. **WRITING** Describe how to subtract two numbers written in scientific notation with the same power of 10.

2. **NUMBER SENSE** You are multiplying two numbers written in scientific notation with different powers of 10. Do you have to rewrite the numbers so they have the same power of 10 before multiplying? Explain.

Practice and Problem Solving

Evaluate the expression using two different methods. Write your answer in scientific notation.

3. $(2.74 \times 10^7) + (5.6 \times 10^7)$
4. $(8.3 \times 10^6) + (3.4 \times 10^5)$
5. $(5.1 \times 10^5) \times (9.7 \times 10^5)$
6. $(4.5 \times 10^4) \times (6.2 \times 10^3)$

Find the sum or difference. Write your answer in scientific notation.

(1) 7. $(2 \times 10^5) + (3.8 \times 10^5)$
8. $(6.33 \times 10^{-9}) - (4.5 \times 10^{-9})$
9. $(9.2 \times 10^8) - (4 \times 10^8)$
10. $(7.2 \times 10^{-6}) + (5.44 \times 10^{-6})$
11. $(7.8 \times 10^7) - (2.45 \times 10^6)$
12. $(5 \times 10^{-5}) + (2.46 \times 10^{-3})$
13. $(9.7 \times 10^6) + (6.7 \times 10^5)$
14. $(2.4 \times 10^{-1}) - (5.5 \times 10^{-2})$

15. **ERROR ANALYSIS** Describe and correct the error in finding the sum of the numbers.

✗
$$(2.5 \times 10^9) + (5.3 \times 10^8) = (2.5 + 5.3) \times (10^9 \times 10^8)$$
$$= 7.8 \times 10^{17}$$

Find the product or quotient. Write your answer in scientific notation.

(2) (3) 16. $5 \times (7 \times 10^7)$
17. $(5.8 \times 10^{-6}) \div (2 \times 10^{-3})$
18. $(1.2 \times 10^{-5}) \div 4$
19. $(5 \times 10^{-7}) \times (3 \times 10^6)$
20. $(3.6 \times 10^7) \div (7.2 \times 10^7)$
21. $(7.2 \times 10^{-1}) \times (4 \times 10^{-7})$
22. $(6.5 \times 10^8) \times (1.4 \times 10^{-5})$
23. $(2.8 \times 10^4) \div (2.5 \times 10^6)$

24. **MONEY** How many times greater is the thickness of a dime than the thickness of a dollar bill?

Thickness = 0.135 cm

Thickness = 1.0922×10^{-2} cm

Evaluate the expression. Write your answer in scientific notation.

25. $5{,}200{,}000 \times (8.3 \times 10^2) - (3.1 \times 10^8)$

26. $(9 \times 10^{-3}) + (2.4 \times 10^{-5}) \div 0.0012$

27. **GEOMETRY** Find the perimeter of the rectangle.

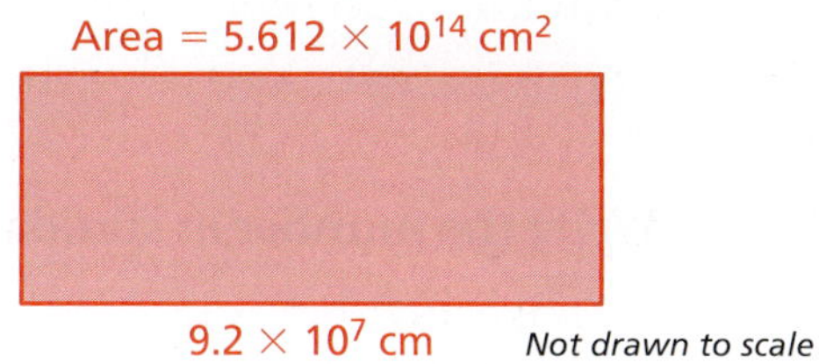

28. **BLOOD SUPPLY** A human heart pumps about 7×10^{-2} liter of blood per heartbeat. The average human heart beats about 72 times per minute. How many liters of blood does a heart pump in 1 year? in 70 years? Write your answers in scientific notation. Then use estimation to justify your answers.

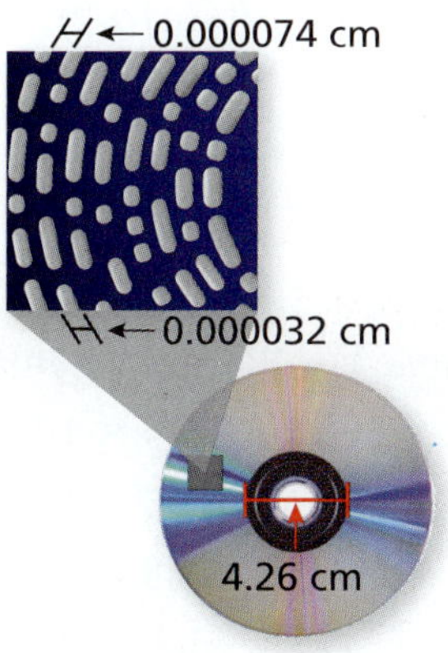

29. **DVDS** On a DVD, information is stored on bumps that spiral around the disk. There are 73,000 ridges (with bumps) and 73,000 valleys (without bumps) across the diameter of the DVD. What is the diameter of the DVD in centimeters?

30. **PROJECT** Use the Internet or some other reference to find the populations and areas (in square miles) of India, China, Argentina, the United States, and Egypt. Round each population to the nearest million and each area to the nearest thousand square miles.

a. Write each population and area in scientific notation.

b. Use your answers to part (a) to find and order the population densities (people per square mile) of each country from least to greatest.

31. Critical Thinking Albert Einstein's most famous equation is $E = mc^2$, where E is the energy of an object (in joules), m is the mass of an object (in kilograms), and c is the speed of light (in meters per second). A hydrogen atom has 15.066×10^{-11} joule of energy and a mass of 1.674×10^{-27} kilogram. What is the speed of light? Write your answer in scientific notation.

Fair Game Review What you learned in previous grades & lessons

Find the cube root. *(Section 7.2)*

32. $\sqrt[3]{-729}$

33. $\sqrt[3]{\frac{1}{512}}$

34. $\sqrt[3]{-\frac{125}{343}}$

35. **MULTIPLE CHOICE** What is the volume of the cone? *(Section 8.2)*

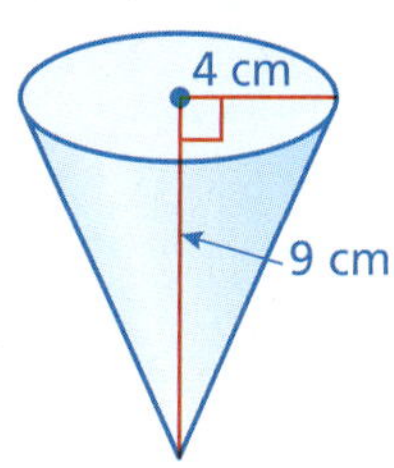

Ⓐ 16π cm^3 Ⓑ 108π cm^3

Ⓒ 48π cm^3 Ⓓ 144π cm^3

10.5 – 10.7 Quiz

Check It Out
Progress Check
BigIdeasMath.com

Tell whether the number is written in scientific notation. Explain. *(Section 10.5)*

1. 23×10^9

2. 0.6×10^{-7}

Write the number in standard form. *(Section 10.5)*

3. 8×10^6

4. 1.6×10^{-2}

Write the number in scientific notation. *(Section 10.6)*

5. 0.00524

6. 892,000,000

Evaluate the expression. Write your answer in scientific notation. *(Section 10.7)*

7. $(7.26 \times 10^4) + (3.4 \times 10^4)$

8. $(2.8 \times 10^{-5}) - (1.6 \times 10^{-6})$

9. $(2.4 \times 10^4) \times (3.8 \times 10^{-6})$

10. $(5.2 \times 10^{-3}) \div (1.3 \times 10^{-12})$

11. **PLANETS** The table shows the equatorial radii of the eight planets in our solar system. *(Section 10.5)*

a. Which planet has the second-smallest equatorial radius?

b. Which planet has the second-largest equatorial radius?

Planet	Equatorial Radius (km)
Mercury	2.44×10^3
Venus	6.05×10^3
Earth	6.38×10^3
Mars	3.4×10^3
Jupiter	7.15×10^4
Saturn	6.03×10^4
Uranus	2.56×10^4
Neptune	2.48×10^4

12. **OORT CLOUD** The Oort cloud is a spherical cloud that surrounds our solar system. It is about 2×10^5 astronomical units from the Sun. An astronomical unit is about 1.5×10^8 kilometers. How far is the Oort cloud from the Sun in kilometers? *(Section 10.6)*

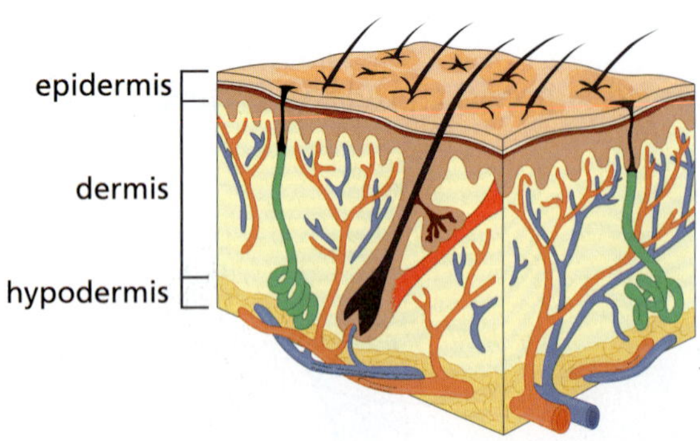

13. **EPIDERMIS** The outer layer of skin is called the *epidermis*. On the palm of your hand, the epidermis is 0.0015 meter thick. Write this number in scientific notation. *(Section 10.6)*

14. **ORBITS** It takes the Sun about 2.3×10^8 years to orbit the center of the Milky Way. It takes Pluto about 2.5×10^2 years to orbit the Sun. How many times does Pluto orbit the Sun while the Sun completes one orbit around the Milky Way? Write your answer in standard form. *(Section 10.7)*

10 Chapter Review

Review Key Vocabulary

power, *p. 412*
base, *p. 412*
exponent, *p. 412*
scientific notation, *p. 438*

Review Examples and Exercises

10.1 Exponents *(pp. 410–415)*

Write $(-4) \cdot (-4) \cdot (-4) \cdot y \cdot y$ using exponents.

Because -4 is used as a factor 3 times, its exponent is 3. Because y is used as a factor 2 times, its exponent is 2.

∴ So, $(-4) \cdot (-4) \cdot (-4) \cdot y \cdot y = (-4)^3y^2$.

Exercises

Write the product using exponents.

1. $(-9) \cdot (-9) \cdot (-9) \cdot (-9) \cdot (-9)$
2. $2 \cdot 2 \cdot 2 \cdot n \cdot n$

Evaluate the expression.

3. 6^3
4. $-\left(\frac{1}{2}\right)^4$
5. $\left|\frac{1}{2}(16 - 6^3)\right|$

10.2 Product of Powers Property *(pp. 416–421)*

a. $\left(-\frac{1}{8}\right)^7 \cdot \left(-\frac{1}{8}\right)^4 = \left(-\frac{1}{8}\right)^{7+4}$ Product of Powers Property

$= \left(-\frac{1}{8}\right)^{11}$ Simplify.

b. $(2.5^7)^2 = 2.5^{7 \cdot 2}$ Power of a Power Property

$= 2.5^{14}$ Simplify.

c. $(3m)^2 = 3^2 \cdot m^2$ Power of a Product Property

$= 9m^2$ Simplify.

Exercises

Simplify the expression.

6. $p^5 \cdot p^2$
7. $(n^{11})^2$
8. $(5y)^3$
9. $(-2k)^4$

10.3 Quotient of Powers Property (pp. 422–427)

a. $\dfrac{(-4)^9}{(-4)^6} = (-4)^{9-6}$ Quotient of Powers Property

$= (-4)^3$ Simplify.

b. $\dfrac{x^4}{x^3} = x^{4-3}$ Quotient of Powers Property

$= x^1$

$= x$ Simplify.

Exercises

Simplify the expression. Write your answer as a power.

10. $\dfrac{8^8}{8^3}$

11. $\dfrac{5^2 \cdot 5^9}{5}$

12. $\dfrac{w^8}{w^7} \cdot \dfrac{w^5}{w^2}$

Simplify the expression.

13. $\dfrac{2^2 \cdot 2^5}{2^3}$

14. $\dfrac{(6c)^3}{c}$

15. $\dfrac{m^8}{m^6} \cdot \dfrac{m^{10}}{m^9}$

10.4 Zero and Negative Exponents (pp. 428–433)

a. $10^{-3} = \dfrac{1}{10^3}$ Definition of negative exponent

$= \dfrac{1}{1000}$ Evaluate power.

b. $(-0.5)^{-5} \cdot (-0.5)^5 = (-0.5)^{-5+5}$ Product of Powers Property

$= (-0.5)^0$ Simplify.

$= 1$ Definition of zero exponent

Exercises

Evaluate the expression.

16. 2^{-4}

17. 95^0

18. $\dfrac{8^2}{8^4}$

19. $(-12)^{-7} \cdot (-12)^7$

20. $\dfrac{1}{7^9} \cdot \dfrac{1}{7^{-6}}$

21. $\dfrac{9^4 \cdot 9^{-2}}{9^2}$

10.5 Reading Scientific Notation (pp. 436–441)

Write (a) 5.9×10^4 and (b) 7.31×10^{-6} in standard notation.

a. $5.9 \times 10^4 = 59{,}000$ (4)

Move decimal point $|4| = 4$ places to the right.

b. $7.31 \times 10^{-6} = 0.00000731$ (6)

Move decimal point $|-6| = 6$ places to the left.

Exercises

Write the number in standard form.

22. 2×10^7

23. 3.4×10^{-2}

24. 1.5×10^{-9}

25. 5.9×10^{10}

26. 4.8×10^{-3}

27. 6.25×10^5

10.6 Writing Scientific Notation (pp. 442–447)

Write (a) 309,000,000 and (b) 0.00056 in scientific notation.

a. $309{,}000{,}000 = 3.09 \times 10^8$ (8)

The number is greater than 10. So, the exponent is positive.

b. $0.00056 = 5.6 \times 10^{-4}$ (4)

The number is between 0 and 1. So, the exponent is negative.

Exercises

Write the number in scientific notation.

28. 0.00036

29. 800,000

30. 79,200,000

10.7 Operations in Scientific Notation (pp. 448–453)

Find $(2.6 \times 10^5) + (3.1 \times 10^5)$.

$(2.6 \times 10^5) + (3.1 \times 10^5) = (2.6 + 3.1) \times 10^5$ — Distributive Property

$= 5.7 \times 10^5$ — Add.

Exercises

Evaluate the expression. Write your answer in scientific notation.

31. $(4.2 \times 10^8) + (5.9 \times 10^9)$

32. $(5.9 \times 10^{-4}) - (1.8 \times 10^{-4})$

33. $(7.7 \times 10^8) \times (4.9 \times 10^{-5})$

34. $(3.6 \times 10^5) \div (1.8 \times 10^9)$

10 Chapter Test

Check It Out
Test Practice
BigIdeasMath.com

Write the product using exponents.

1. $(-15) \cdot (-15) \cdot (-15)$

2. $\left(\frac{1}{12}\right) \cdot \left(\frac{1}{12}\right) \cdot \left(\frac{1}{12}\right) \cdot \left(\frac{1}{12}\right) \cdot \left(\frac{1}{12}\right)$

Evaluate the expression.

3. -2^3

4. $10 + 3^3 \div 9$

Simplify the expression. Write your answer as a power.

5. $9^{10} \cdot 9$

6. $(6^6)^5$

7. $(2 \cdot 10)^7$

8. $\frac{(-3.5)^{13}}{(-3.5)^9}$

Evaluate the expression.

9. $5^{-2} \cdot 5^2$

10. $\frac{-8}{(-8)^3}$

Write the number in standard form.

11. 3×10^7

12. 9.05×10^{-3}

Evaluate the expression. Write your answer in scientific notation.

13. $(7.8 \times 10^7) + (9.9 \times 10^7)$

14. $(6.4 \times 10^5) - (5.4 \times 10^4)$

15. $(3.1 \times 10^6) \times (2.7 \times 10^{-2})$

16. $(9.6 \times 10^7) \div (1.2 \times 10^{-4})$

17. **CRITICAL THINKING** Is $(xy^2)^3$ the same as $(xy^3)^2$? Explain.

18. **RICE** A grain of rice weighs about 3^3 milligrams. About how many grains of rice are in one scoop?

19. **TASTE BUDS** There are about 10,000 taste buds on a human tongue. Write this number in scientific notation.

One scoop of rice weighs about 3^9 milligrams.

20. **LEAD** From 1978 to 2008, the amount of lead allowed in the air in the United States was 1.5×10^{-6} gram per cubic meter. In 2008, the amount allowed was reduced by 90%. What is the new amount of lead allowed in the air?

10 Standards Assessment

Test-Taking Strategy
Use Intelligent Guessing

"It can't be 40 or 5000 because they aren't divisible by 3. So, you can intelligently guess between 3000 and 3072."

1. Mercury's distance from the Sun is approximately 5.79×10^7 kilometers. What is this distance in standard form? *(MACC.8.EE.1.4)*

A. 5,790,000,000 km **C.** 57,900,000 km

B. 579,000,000 km **D.** 5,790,000 km

2. The steps Jim took to answer the question are shown below. What should Jim change to correctly answer the question? *(MACC.8.G.1.5)*

How many degrees are in the largest angle in the triangle below?

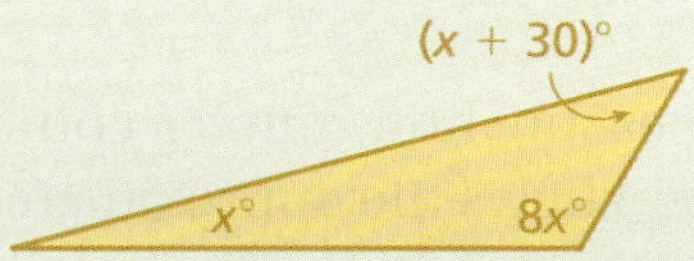

$$x + 8x + x + 30 = 180$$
$$10x = 150$$
$$x = 15$$

F. The left side of the equation should equal 360° instead of 180°.

G. The sum of the acute angles should equal 90°.

H. Evaluate the smallest angle when $x = 15$.

I. Evaluate the largest angle when $x = 15$.

3. Which expression is equivalent to the expression below? *(MACC.8.EE.1.1)*

$$2^4 2^3$$

A. 2^{12} **C.** 48

B. 4^7 **D.** 128

4. In the figure below, $\triangle ABC$ is a dilation of $\triangle DEF$.

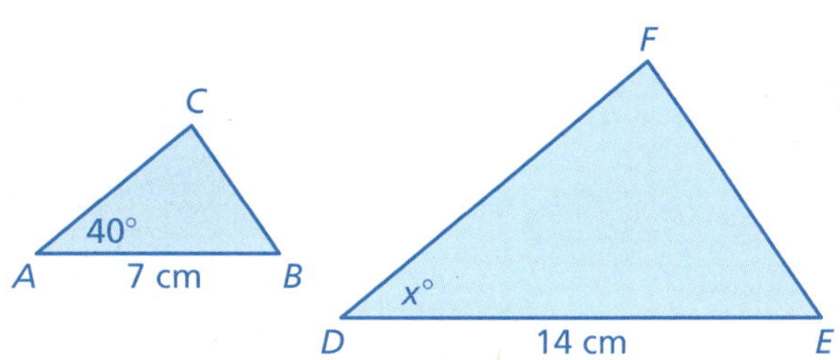

What is the value of x? *(MACC.8.G.1.4)*

5. A bank account pays interest so that the amount in the account doubles every 10 years. The account started with \$5,000 in 1940. Which expression represents the amount (in dollars) in the account n decades later? *(MACC.8.EE.1.1)*

F. $2^n \cdot 5000$

G. $5000(n + 1)$

H. 5000^n

I. $2^n + 5000$

6. The formula for the volume V of a pyramid is $V = \frac{1}{3}Bh$. Solve the formula for the height h. *(MACC.8.EE.3.7b)*

A. $h = \frac{1}{3}VB$

B. $h = \frac{3V}{B}$

C. $h = \frac{V}{3B}$

D. $h = V - \frac{1}{3}B$

7. The gross domestic product (GDP) is a way to measure how much a country produces economically in a year. The table below shows the approximate population and GDP for the United States. *(MACC.8.EE.1.4)*

Think
Solve
Explain

United States 2012	
Population	312 million (312,000,000)
GDP	15.1 trillion dollars (\$15,100,000,000,000)

Part A Find the GDP per person for the United States. Show your work and explain your reasoning.

Part B Write the population and the GDP using scientific notation.

Part C Find the GDP per person for the United States using your answers from Part B. Write your answer in scientific notation. Show your work and explain your reasoning.

8. What is the equation of the line shown in the graph? *(MACC.8.EE.2.6)*

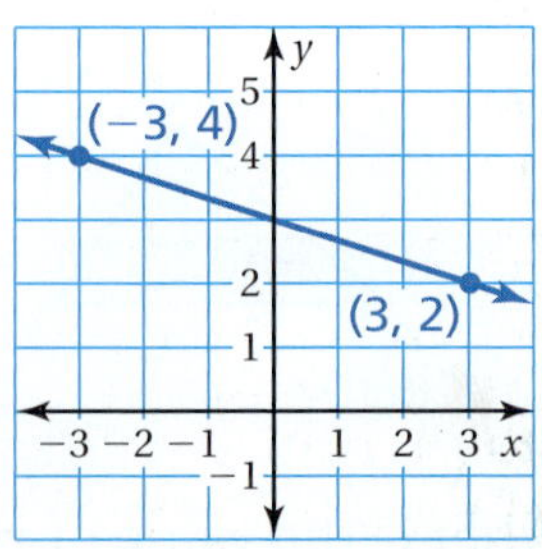

F. $y = -\frac{1}{3}x + 3$

G. $y = \frac{1}{3}x + 1$

H. $y = -3x + 3$

I. $y = 3x - \frac{1}{3}$

9. A cylinder and its dimensions are shown below.

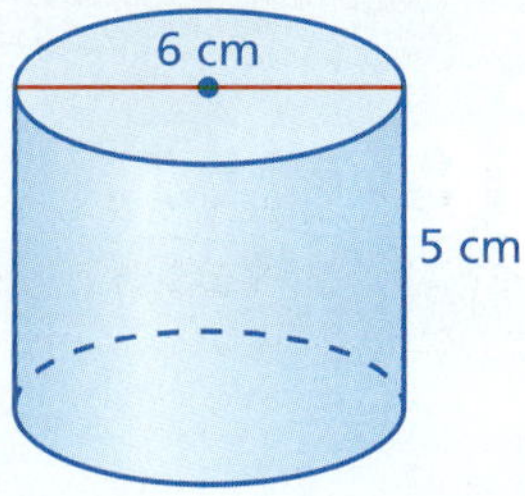

What is the volume of the cylinder? (Use 3.14 for π.) *(MACC.8.G.3.9)*

A. 47.1 cm^3

B. 94.2 cm^3

C. 141.3 cm^3

D. 565.2 cm^3

10. Find $(-2.5)^{-2}$. *(MACC.8.EE.1.1)*

11. Two lines have the same y-intercept. The slope of one line is 1, and the slope of the other line is -1. What can you conclude? *(MACC.8.EE.2.6)*

F. The lines are parallel.

G. The lines meet at exactly one point.

H. The lines meet at more than one point.

I. The situation described is impossible.

12. The director of a research lab wants to present data to donors. The data show how the lab uses a great deal of donated money for research and only a small amount of money for other expenses. Which type of display is best suited for showing these data? *(MACC.8.SP.1.1)*

A. box-and-whisker plot

B. circle graph

C. line graph

D. scatter plot

Appendix A
My Big Ideas Projects

My Big Ideas Projects

Swiss Family Robinson

1 Getting Started

Swiss Family Robinson is a novel about a Swiss family who was shipwrecked in the East Indies. The story was written by Johann David Wyss, and was first published in 1812.

Essential Question How does the knowledge of mathematics provide you and your family with survival tools?

Read *Swiss Family Robinson*. As you read the exciting adventures, think about the mathematics the family knew and used to survive.

Sample: The tree house built by the family was accessed by a long rope ladder. The ladder was about 30 feet long with a rung every 10 inches. To make the ladder, the family had to plan how many rungs were needed. They decided the number was $1 + 12(30) \div 10$. Why?

2 Things to Include

- Suppose you lived in the 18th century. Plan a trip from Switzerland to Australia. Describe your route. Estimate the length of the route and the number of miles you will travel each day. About how many days will the entire trip take?
- Suppose that your family is shipwrecked on an island that has no other people. What do you need to do to survive? What types of tools do you hope to salvage from the ship? Describe how mathematics could help you survive.
- Suppose that you are the oldest of four children in a shipwrecked family. Your parents have made you responsible for the education of your younger siblings. What type of mathematics would you teach them? Explain your reasoning.

3 Things to Remember

- You can download each part of the book at *BigIdeasMath.com.*
- Add your own illustrations to your project.
- Organize your math stories in a folder, and think of a title for your report.

Mathematics in Ancient China

1 Getting Started

Mathematics was developed in China independently of the mathematics that was developed in Europe and the Middle East. For example, the Pythagorean Theorem and the computation of pi were used in China prior to the time when China and Europe began communicating with each other.

Essential Question How have tools and knowledge from the past influenced modern day mathematics?

Sample: Here are the names and symbols that were used in ancient China to represent the digits from 1 through 10.

1	yi	一
2	er	二
3	san	三
4	si	四
5	wu	五
6	liu	六
7	qi	七
8	ba	八
9	jiu	九
10	shi	十

Life-size Terra-cotta Warriors

A Chinese Abacus

2 Things to Include

Ancient Chinese Teapot

- Describe the ancient Chinese book *The Nine Chapters on the Mathematical Art* (c. 100 B.C.). What types of mathematics are contained in this book?
- How did the ancient Chinese use the abacus to add and subtract numbers? How is the abacus related to base 10?
- How did the ancient Chinese use mathematics to build large structures, such as the Great Wall and the Forbidden City?
- How did the ancient Chinese write numbers that are greater than 10?
- Describe how the ancient Chinese used mathematics. How does this compare with the ways in which mathematics is used today?

The Great Wall of China

3 Things to Remember

- Add your own illustrations to your project.
- Organize your math stories in a folder, and think of a title for your report.

Chinese Guardian Fu Lions

A.3 Art Project

Building a Kaleidoscope

1 Getting Started

Mirrors set at 60°

A kaleidoscope is a tube of mirrors containing loose colored beads, pebbles, or other small colored objects. You look in one end and light enters the other end, reflecting off the mirrors.

Essential Question How does the knowledge of mathematics help you create a kaleidoscope?

If the angle between the mirrors is 45°, you see 8 duplicate images. If the angle is 60°, you see 6 duplicate images. If the angle is 90°, you see 4 duplicate images. As the tube is rotated, the colored objects tumble, creating various patterns.

Write a report about kaleidoscopes. Discuss the mathematics you need to know in order to build a kaleidoscope.

Sample: A kaleidoscope whose mirrors meet at 60° angles has reflective symmetry and rotational symmetry.

Reflect

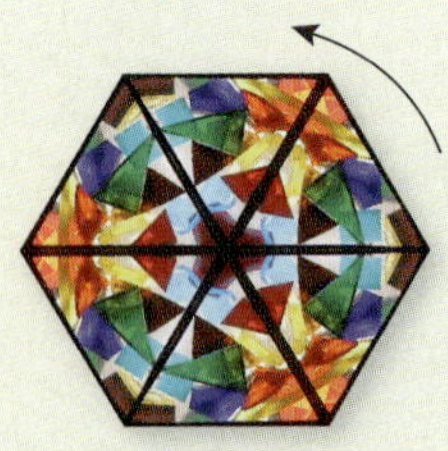

Rotate 120°

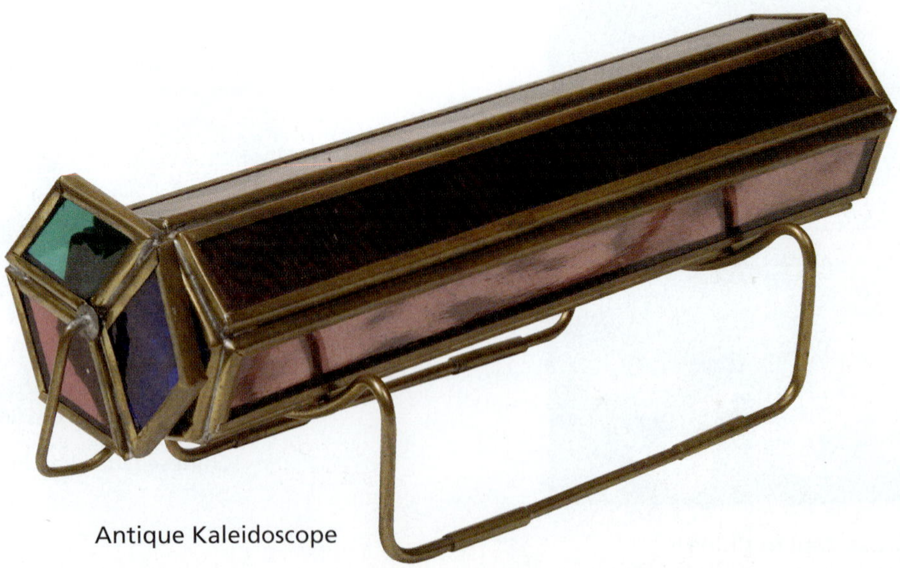

Antique Kaleidoscope

2 Things to Include

- How does the angle at which the mirrors meet affect the number of duplicate images that you see?
- What angles can you use other than 45°, 60°, and 90°? Explain your reasoning.
- Research the history of kaleidoscopes. Can you find examples of kaleidoscopes being used before they were patented by David Brewster in 1816?
- Make your own kaleidoscope.
- Describe the mathematics you used to create your kaleidoscope.

Mirrors set at 90°

Mirrors set at 60°

Giant Kaleidoscope, San Diego harbor

3 Things to Think About

- Add your own drawings and pattern creations to your project.
- Organize your report in a folder, and think of a title for your report.

Mirrors set at 45°

A.4 Science Project

Our Solar System

1 Getting Started

Our solar system consists of four inner planets, four outer planets, dwarf planets such as Pluto, several moons, and many asteroids and comets.

Essential Question How do the characteristics of a planet influence whether or not it can sustain life?

Sample: The average temperatures of the eight planets in our solar system are shown in the graph.

The average temperature tends to drop as the distance between the Sun and the planet increases.

An exception to this rule is Venus. It has a higher average temperature than Mercury, even though Mercury is closer to the Sun.

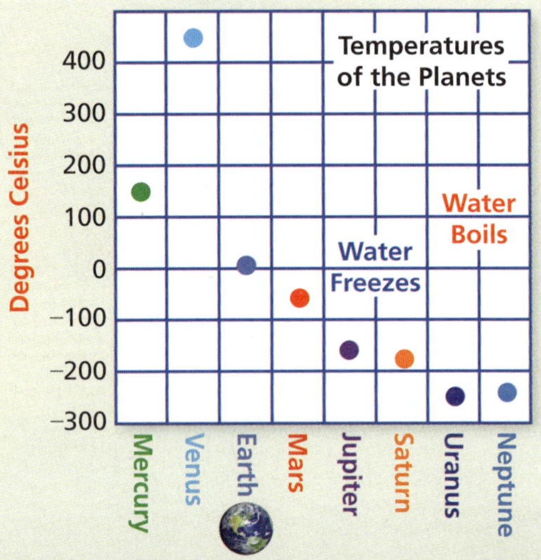

2 Things to Include

- Compare the masses of the planets.
- Compare the gravitational forces of the planets.
- How long is a "day" on each planet? Why?
- How long is a "year" on each planet? Why?
- Which planets or moons have humans explored?
- Which planets or moons could support human life? Explain your reasoning.

Mars Rover

3 Things to Remember

- Add your own drawings or photographs to your report. You can download photographs of the solar system and space travel at *NASA.gov*.
- Organize your report in a folder, and think of a title for your report.

Hubble Image of Space

Hubble Spacecraft

Selected Answers

Section 1.1 Solving Simple Equations
(pages 7–9)

1. + and − are inverses. × and ÷ are inverses.

3. $x - 3 = 6$; It is the only equation that does not have $x = 6$ as a solution.

5. $x = 57$ 7. $x = -5$ 9. $p = 21$ 11. $x = 9\pi$ 13. $d = \frac{1}{2}$ 15. $n = -4.9$

17. a. $105 = x + 14$; $x = 91$

 b. no; Because $82 + 9 = 91$, you did not knock down the last pin with the second ball of the frame.

19. $n = -5$ 21. $m = 7.3\pi$ 23. $k = 1\frac{2}{3}$ 25. $p = -2\frac{1}{3}$

27. They should have added 1.5 to each side.

$$-1.5 + k = 8.2$$
$$k = 8.2 + 1.5$$
$$k = 9.7$$

29. $6.5x = 42.25$; \$6.50 per hour

31. $420 = \frac{7}{6}b$, $b = 360$; \$60

33. $h = -7$ 35. $q = 3.2$ 37. $x = -1\frac{4}{9}$

39. greater than; Because a negative number divided by a negative number is a positive number.

41. 3 mg 43. 12 in. 45. $7x - 4$ 47. $\frac{25}{4}g - \frac{2}{3}$

Section 1.2 Solving Multi-Step Equations
(pages 14 and 15)

1. $2 + 3x = 17$; $x = 5$ 3. $k = 45$; 45°, 45°, 90° 5. $b = 90$; 90°, 135°, 90°, 90°, 135°

7. $c = 0.5$ 9. $h = -9$ 11. $x = -\frac{2}{9}$ 13. 20 watches

15. $4(b + 3) = 24$; 3 in. 17. $\frac{2580 + 2920 + x}{3} = 3000$; 3500 people

19. $<$ 21. $>$

Section 1.3 Solving Equations with Variables on Both Sides
(pages 23–25)

1. no; When 3 is substituted for x, the left side simplifies to 4 and the right side simplifies to 3.

3. $x = 13.2$ in. 5. $x = 7.5$ in. 7. $k = -0.75$

9. $p = -48$ 11. $n = -3.5$ 13. $x = -4$

15. The 4 should have been added to the right side.

$$\begin{aligned} 3x - 4 &= 2x + 1 \\ 3x - 2x - 4 &= 2x + 1 - 2x \\ x - 4 &= 1 \\ x - 4 + 4 &= 1 + 4 \\ x &= 5 \end{aligned}$$

17. $15 + 0.5m = 25 + 0.25m$; 40 mi

19. $x = \frac{1}{3}$

21. no solution

23. infinitely many solutions

25. $x = 2$

27. no solution

29. infinitely many solutions

31. *Sample answer:* $8x + 2 = 8x$; The number $8x$ cannot be equal to 2 more than itself.

33. It's never the same. Your neighbor's total cost will always be \$75 more than your total cost.

35. no; $2x + 5.2$ can never equal $2x + 6.2$.

37. 7.5 units

39. Remember that the box is with priority mail and the envelope is with express mail.

41. 10 mL

43. **a.** 40 ft

b. no;

$$\begin{aligned} 2(\text{white area}) &= \text{black area} \\ 2[5(6x)] &= 4[6(x + 1)] \\ 60x &= 24x + 24 \\ 36x &= 24 \\ x &= \frac{2}{3} \end{aligned}$$

$$5x + 4(x + 1) \stackrel{?}{=} 40$$

Length of hallway is $5\left(\frac{2}{3}\right) + 4\left(\frac{2}{3} + 1\right) \stackrel{?}{=} 40$

$$10 \neq 40$$

45. 15.75 cm^3

47. C

Section 1.4 Rewriting Equations and Formulas
(pages 30 and 31)

1. no; The equation only contains one variable.

3. **a.** $A = \frac{1}{2}bh$ **b.** $b = \frac{2A}{h}$ **c.** $b = 12$ mm

5. $y = 4 - \frac{1}{3}x$

7. $y = \frac{2}{3} - \frac{4}{9}x$

9. $y = 3x - 1.5$

11. The y should have a negative sign in front of it.

$$\begin{aligned} 2x - y &= 5 \\ -y &= -2x + 5 \\ y &= 2x - 5 \end{aligned}$$

13. **a.** $t = \frac{I}{Pr}$

b. $t = 3$ yr

15. $m = \frac{e}{c^2}$

17. $\ell = \dfrac{A - \frac{1}{2}\pi w^2}{2w}$

19. $w = 6g - 40$

21. **a.** $F = 32 + \frac{9}{5}(K - 273.15)$

b. 32°F

c. liquid nitrogen

23. $r^3 = \frac{3V}{4\pi}$; $r = 4.5$ in.

25. $-5\frac{1}{3}$

27. $1\frac{1}{4}$

Section 2.1 Congruent Figures
(pages 46 and 47)

1. **a.** $\angle A$ and $\angle D$, $\angle B$ and $\angle E$, $\angle C$ and $\angle F$

   b. Side AB and Side DE, Side BC and Side EF, Side AC and Side DF

3. $\angle V$ does not belong. The other three angles are congruent to each other, but not to $\angle V$.

5. congruent

7. $\angle P$ and $\angle W$, $\angle Q$ and $\angle V$, $\angle R$ and $\angle Z$, $\angle S$ and $\angle Y$, $\angle T$ and $\angle X$; Side PQ and Side WV, Side QR and Side VZ, Side RS and Side ZY, Side ST and Side YX, Side TP and Side XW

9. not congruent; Corresponding side lengths are not congruent.

11. The corresponding angles are not congruent, so the two figures are not congruent.

13. What figures have you seen in this section that have at least one right angle?

15. **a.** true; Side AB corresponds to Side YZ.

    b. true; $\angle A$ and $\angle X$ have the same measure.

    c. false; $\angle A$ corresponds to $\angle Y$.

    d. true; The measure of $\angle A$ is 90°, the measure of $\angle B$ is 140°, the measure of $\angle C$ is 40°, and the measure of $\angle D$ is 90°. So, the sum of the angle measures of $ABCD$ is $90° + 140° + 40° + 90° = 360°$.

17 and 19.

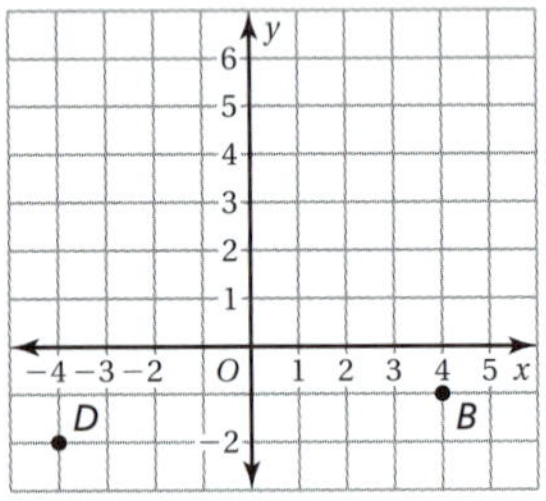

Section 2.2 Translations
(pages 52 and 53)

1. A

3. yes; Translate the letters T and O to the end.

5. no

7. yes

9. no

11. $A'(-3, 0)$, $B'(0, -1)$, $C'(1, -4)$, $D'(-3, -5)$

13.

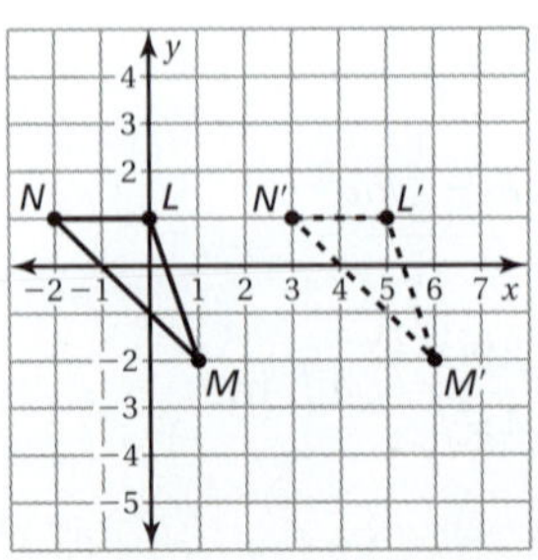

15.

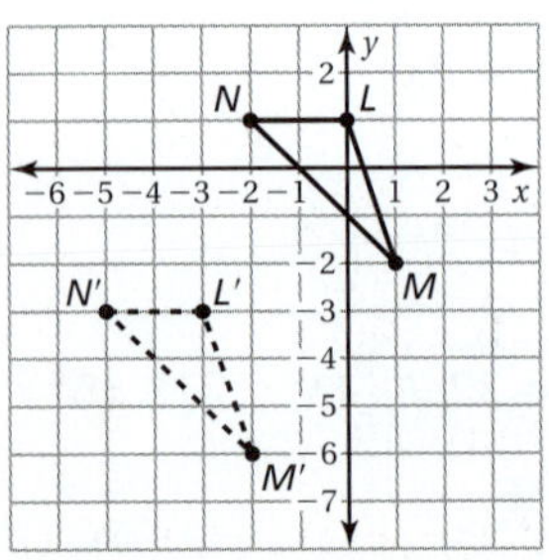

17. 2 units left and 2 units up

19. 6 units right and 3 units down

21. **a.** 5 units right and 1 unit up

b. no; It would hit the island.

c. 4 units up and 4 units right

23. If you are doing more than 10 moves and have not moved the knight to g5, you might want to start over.

25. no

27. yes

Section 2.3 Reflections
(pages 58 and 59)

1. The third one because it is not a reflection.

3. Quadrant IV

5. yes

7. no

9. no

11. $M'(-2, -1), N'(0, -3), P'(2, -2)$

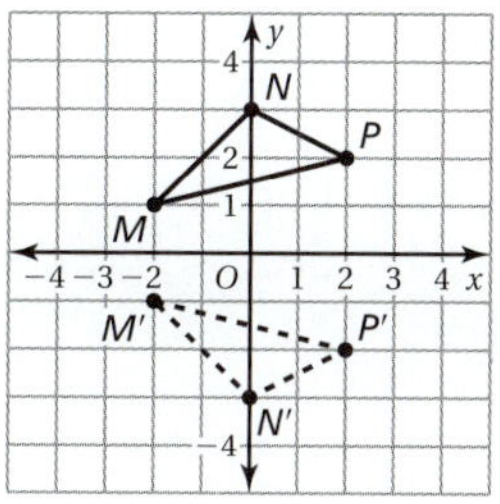

13. $D'(-2, 1), E'(0, 1), F'(0, 5), G'(-2, 5)$

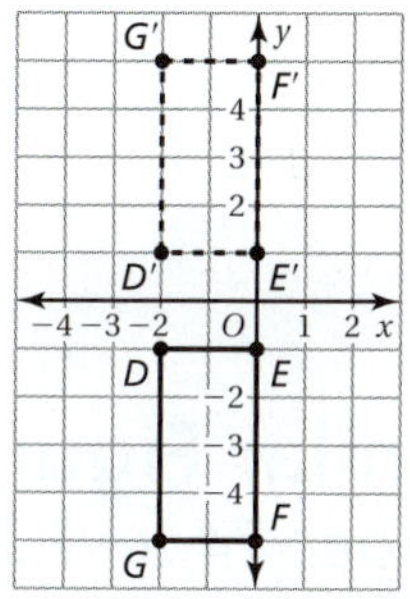

15. $T'(-4, -2), U'(-4, 2), V'(-6, -2)$

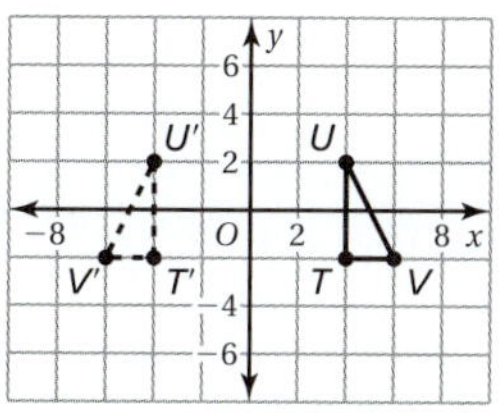

17. $J'(-2, 2), K'(-7, 4), L'(-9, -2), M'(-3, -1)$

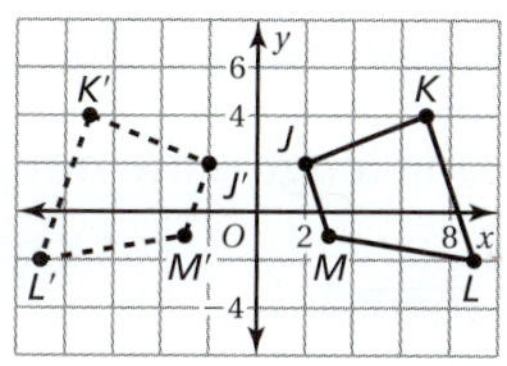

19. x-axis

21. y-axis

23. $R'(3, -4), S'(3, -1), T'(1, -4)$

25. yes; Translations and reflections produce images that are congruent to the original figure.

27. If you are driving a vehicle and want to see who is following you, where would you look?

29. obtuse

31. right

33. B

Section 2.4 Rotations
(pages 65–67)

1. (0, 0); (1, −3)

3. Quadrant IV

5. Quadrant II

7. reflection

9. translation

11. yes; 90° counterclockwise

13. $A'(2, 2), B'(1, 4), C'(3, 4), D'(4, 2)$

15. $J'(0, -3), K'(0, -5), L'(-4, -3)$

17. $W'(-2, 6), X'(-2, 2), Y'(-6, 2), Z'(-6, 5)$

Section 2.4 Rotations *(continued)*
(pages 65–67)

19. It only needs to rotate 120° to produce an identical image.

21. It only needs to rotate 180° to produce an identical image.

23. $J''(4, 4)$, $K''(3, 4)$, $L''(1, 1)$, $M''(4, 1)$

25. *Sample answer:* Rotate 180° about the origin and then rotate 90° clockwise about vertex $(-1, 0)$; Rotate 90° counterclockwise about the origin and then translate 1 unit left and 1 unit down.

27. Use Guess, Check, and Revise to solve this problem.

29. (2, 4), (4, 1), (1, 1)

31. yes

33. no

Section 2.5 Similar Figures
(pages 74 and 75)

1. They are congruent.

3. Yes, because the angles are congruent and the side lengths are proportional.

5. not similar; Corresponding side lengths are not proportional.

7.

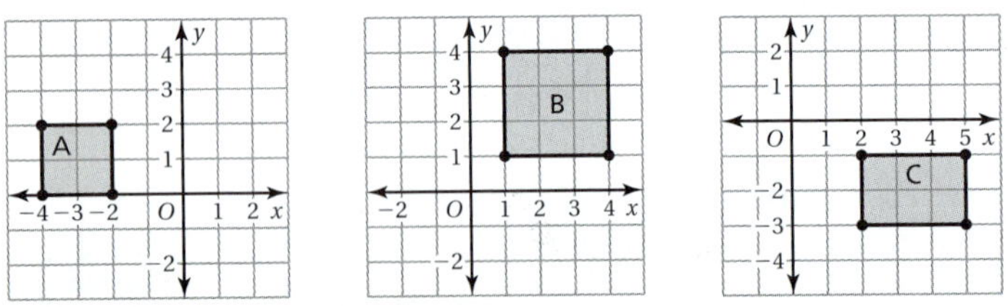

A and B; Corresponding side lengths are proportional and corresponding angles are congruent.

9. $6\frac{2}{3}$

11. 14

13. 30 in.

15. What types of quadrilaterals can have the given angle measures?

17. 3 times

19. **a.** yes

b. yes; It represents the fact that the sides are proportional because you can split the isosceles triangles into smaller right triangles that will be similar.

21. $\frac{16}{81}$

23. $\frac{49}{16}$

25. C

Section 2.6 Perimeters and Areas of Similar Figures
(pages 80 and 81)

1. The ratio of the perimeters is equal to the ratio of the corresponding side lengths.

3. Because the ratio of the corresponding side lengths is $\frac{1}{2}$, the ratio of the areas is equal to $\left(\frac{1}{2}\right)^2$. To find the area, solve the proportion $\frac{30}{x} = \frac{1}{4}$ to get $x = 120$ square inches.

5. $\frac{5}{8}$; $\frac{25}{64}$

7. $\frac{14}{9}$; $\frac{196}{81}$

9. The area is 9 times larger.

11. 25.6

13. 39 in.; 93.5 in.^{2}

15. 108 yd

17. **a.** 400 times greater; The ratio of the corresponding lengths is $\frac{120 \text{ in.}}{6 \text{ in.}} = \frac{20}{1}$. So, the ratio of the areas is $\left(\frac{20}{1}\right)^2 = \frac{400}{1}$.

    b. 1250 ft^2

19. 15 m

21. $x = -2$

23. $n = -4$

Section 2.7 Dilations
(pages 87–89)

1. A dilation changes the size of a figure. The image is similar, not congruent, to the original figure.

3. The middle red figure is not a dilation of the blue figure because the height is half of the blue figure and the base is the same. The left red figure is a reduction of the blue figure and the right red figure is an enlargement of the blue figure.

5. 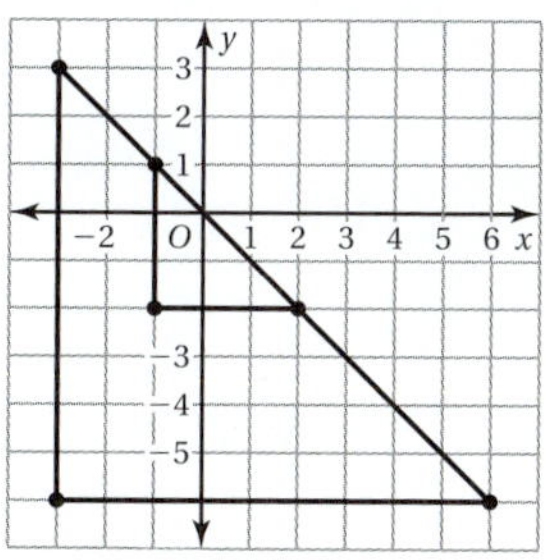

The triangles are similar.

7. yes

9. no

11. yes

13. 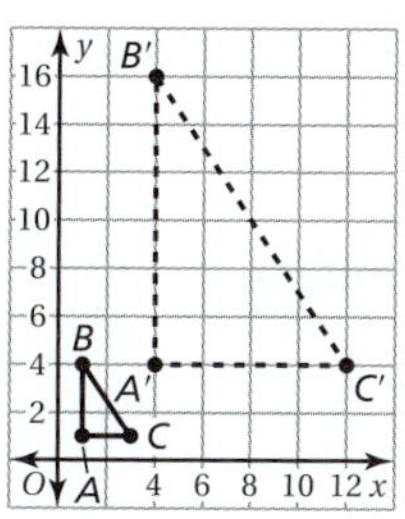

enlargement

15. 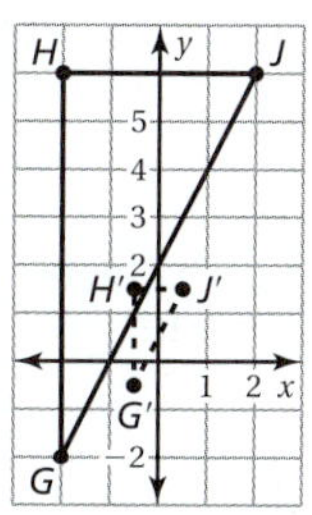

reduction

17. 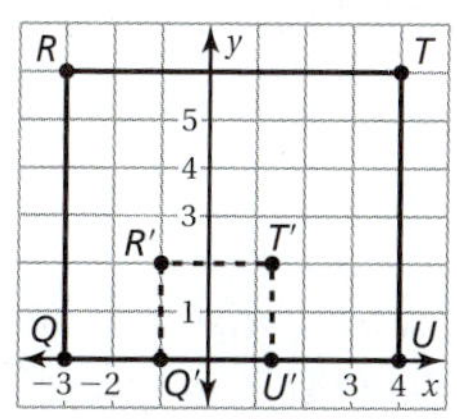

reduction

19. Each coordinate was multiplied by 2 instead of divided by 2. The coordinates should be $A'(1, 2.5)$, $B'(1, 0)$, and $C'(2, 0)$.

Section 2.7 Dilations *(continued)* *(pages 87–89)*

21. reduction; $\frac{1}{4}$

23. $A''(10, 6)$, $B''(4, 6)$, $C''(4, 2)$, $D''(10, 2)$

25. $J''(3, -3)$, $K''(12, -9)$, $L''(3, -15)$

27. *Sample answer:* Rotate 90° counterclockwise about the origin and then dilate with respect to the origin using a scale factor of 2

29. Exercise 27: yes; Exercise 28: no; Explanations will vary based on sequences chosen in Exercises 27 and 28.

31. **a.** enlargement

b. center of dilation

c. $\frac{4}{3}$

d. The shadow on the wall becomes larger. The scale factor will become larger.

33. The transformations are a dilation using a scale factor of 2 and then a translation of 4 units right and 3 units down; similar; A dilation produces a similar figure and a translation produces a congruent figure, so the final image is similar.

35. The transformations are a dilation using a scale factor of $\frac{1}{3}$ and then a reflection in the x-axis; similar; A dilation produces a similar figure and a reflection produces a congruent figure, so the final image is similar.

37. $A'(-2, 3)$, $B'(6, 3)$, $C'(12, -7)$, $D'(-2, -7)$; Methods will vary.

39. supplementary; $x = 16$

41. B

Section 3.1 Parallel Lines and Transversals *(pages 107–109)*

1. *Sample answer:*

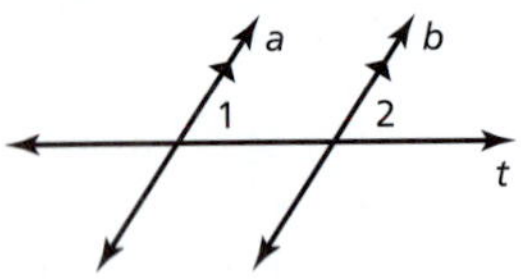

3. m and n

5. 8

7. $\angle 1 = 107°$, $\angle 2 = 73°$

9. $\angle 5 = 49°$, $\angle 6 = 131°$

11. 60°; Corresponding angles are congruent.

13. *Sample answer:* rotate 180° and translate down

15. $\angle 6 = 61°$; $\angle 6$ and the given angle are vertical angles.
$\angle 5 = 119°$ and $\angle 7 = 119°$; $\angle 5$ and $\angle 7$ are supplementary to the given angle.
$\angle 1 = 61°$; $\angle 1$ and the given angle are corresponding angles.
$\angle 3 = 61°$; $\angle 1$ and $\angle 3$ are vertical angles.
$\angle 2 = 119°$ and $\angle 4 = 119°$; $\angle 2$ and $\angle 4$ are supplementary to $\angle 1$.

17. $\angle 2 = 90°$; $\angle 2$ and the given angle are vertical angles.
$\angle 1 = 90°$ and $\angle 3 = 90°$; $\angle 1$ and $\angle 3$ are supplementary to the given angle.
$\angle 4 = 90°$; $\angle 4$ and the given angle are corresponding angles.
$\angle 6 = 90°$; $\angle 4$ and $\angle 6$ are vertical angles.
$\angle 5 = 90°$ and $\angle 7 = 90°$; $\angle 5$ and $\angle 7$ are supplementary to $\angle 4$.

19. 132°; *Sample answer:* $\angle 2$ and $\angle 4$ are alternate interior angles and $\angle 4$ and $\angle 3$ are supplementary.

21. 120°; *Sample answer:* $\angle 6$ and $\angle 8$ are alternate exterior angles.

23. 61.3°; *Sample answer:* $\angle 3$ and $\angle 1$ are alternate interior angles and $\angle 1$ and $\angle 2$ are supplementary.

25. They are all right angles because perpendicular lines form 90° angles.

27. 130

29. **a.** no; They look like they are spreading apart. **b.** Check students' work.

31. 13

33. 51

35. B

Section 3.2 Angles of Triangles
(pages 114 and 115)

1. Subtract the sum of the given measures from 180°.

3. 115°, 120°, 125°

5. 40°, 65°, 75°

7. 25°, 45°, 110°

9. 48°, 59°, 73°

11. 45

13. 140°

15. The measure of the exterior angle is equal to the sum of the measures of the two nonadjacent interior angles. The sum of all three angles is not 180°;

$$(2x - 12) = x + 30$$

$$x = 42$$

The exterior angle is $(2(42) - 12)° = 72°$.

17. 126°

19. sometimes; The sum of the angle measures must equal 180°.

21. never; If a triangle had more than one vertex with an acute exterior angle, then it would have to have more than one obtuse interior angle which is impossible.

23. $x = -4$

25. $n = -3$

Section 3.3 Angles of Polygons
(pages 123–125)

1. *Sample answer:*

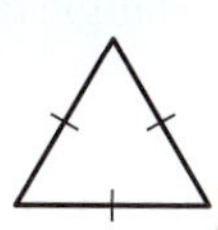

3. What is the measure of an interior angle of a regular pentagon?; 108°; 540°

5. 1260°

7. 360°

9. 1260°

Section 3.3 Angles of Polygons *(continued)*
(pages 123–125)

11. no; The interior angle measures given add up to 535°, but the sum of the interior angle measures of a pentagon is 540°.

13. 90°, 135°, 135°, 135°, 135°, 90°

15. 140°

17. 140°

19. The sum of the interior angle measures should have been divided by the number of angles, 20. $3240° \div 20 = 162°$; The measure of each interior angle is 162°.

21. 24 sides

23. 75°, 93°, 85°, 107°

25. 60°; The sum of the interior angle measures of a hexagon is 720°. Because it is regular, each angle has the same measure. So, each interior angle is $720° \div 6 = 120°$ and each exterior angle is 60°.

27. 120°, 120°, 120°

29. interior: 135°; exterior: 45°

31. 120°

33. a. *Sample answer:*

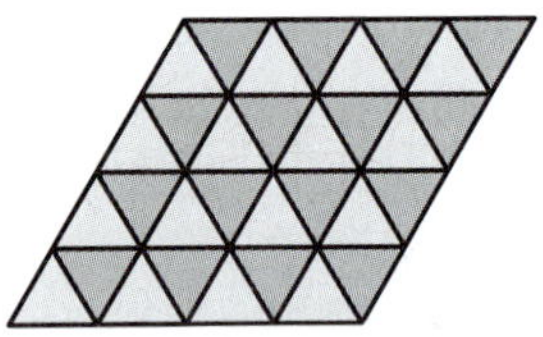

b. *Sample answer:*
square, regular hexagon

c. *Sample answer:*

d. *Answer should include, but is not limited to:* a discussion of the interior and exterior angles of the polygons in the tessellation and how they combine to add up to 360° where the vertices meet.

35. 2

37. 6

Section 3.4 Using Similar Triangles
(pages 130 and 131)

1. Write a proportion that uses the missing measurement because the ratios of corresponding side lengths are equal.

3. *Sample answer:* Two of the angles are congruent, so they have the same sum. When you subtract this from 180°, you will get the same third angle.

5. Student should draw a triangle with the same angle measures as the ones given in the textbook.

If the student's triangle is larger than the one given, then the ratio of the corresponding side lengths, $\frac{\text{student's triangle length}}{\text{book's triangle length}}$, should be greater than 1. If the student's triangle is smaller than the one given, then the ratio of the corresponding side lengths, $\frac{\text{student's triangle length}}{\text{book's triangle length}}$, should be less than 1.

7. no; The triangles do not have two pairs of congruent angles.

9. yes; The triangles have the same angle measures, 81°, 51°, and 48°.

11. yes; The triangles have two pairs of congruent angles.

13. Think of the different ways that you can show that two triangles are similar.

15. 30 ft

17. maybe; They are similar when both have measures of 30°, 60°, 90° or both have measures of 45°, 45°, 90°. They are not similar when one has measures of 30°, 60°, 90° and the other has measures of 45°, 45°, 90°.

19. $y = 5x + 3$

21. $y = 8x - 4$

Section 4.1

Graphing Linear Equations
(pages 146 and 147)

1. a line

3. *Sample answer:*

x	0	1
$y = 3x - 1$	−1	2

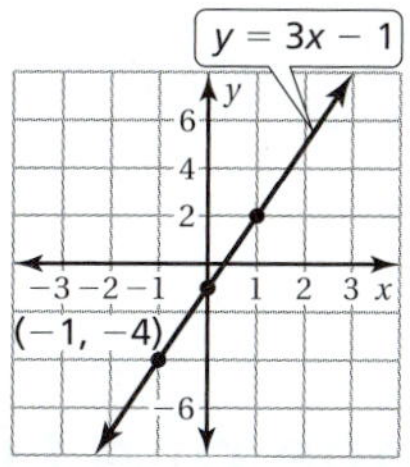

5.

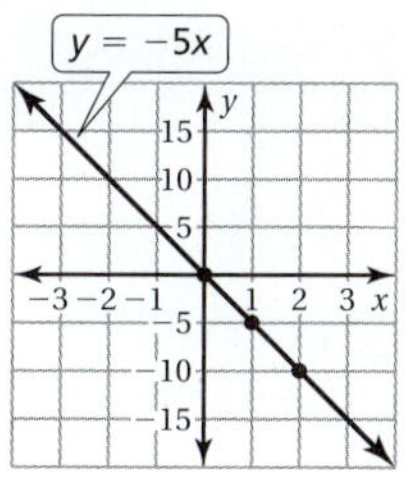

7.

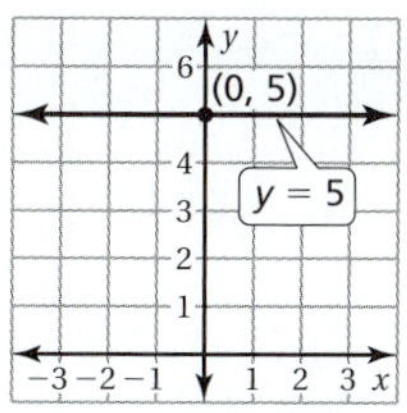

9.

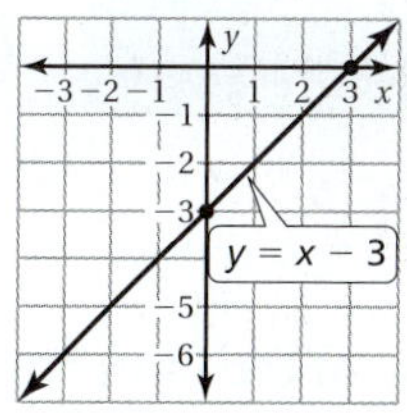

11.

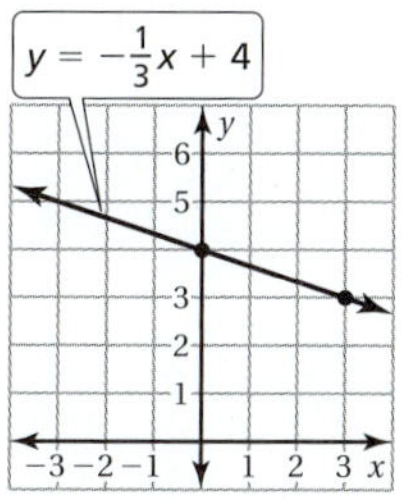

13.

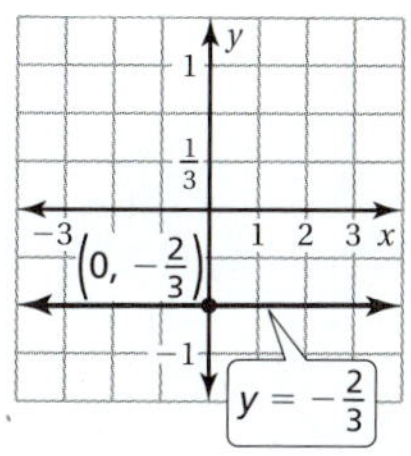

15.

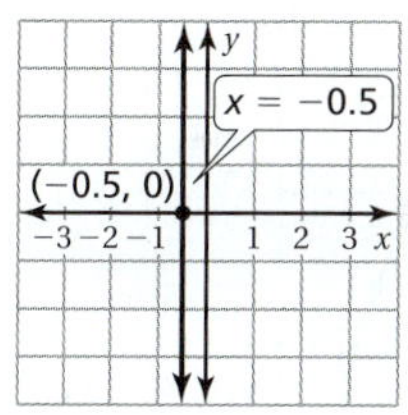

17. The equation $x = 4$ is graphed, not $y = 4$.

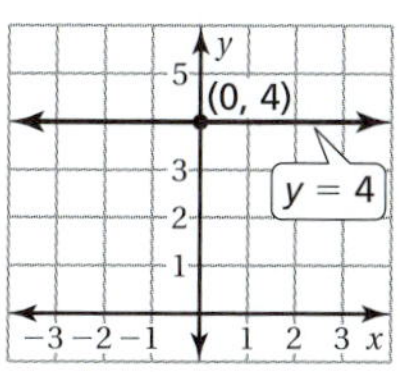

19. a.

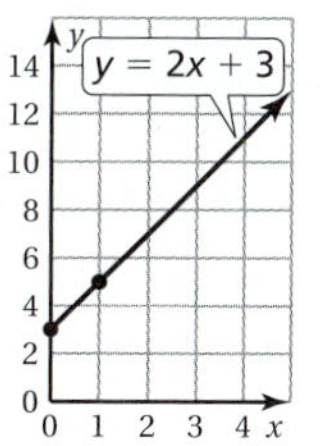

b. about $5

c. $5.25

21. $y = -\frac{5}{2}x + 2$

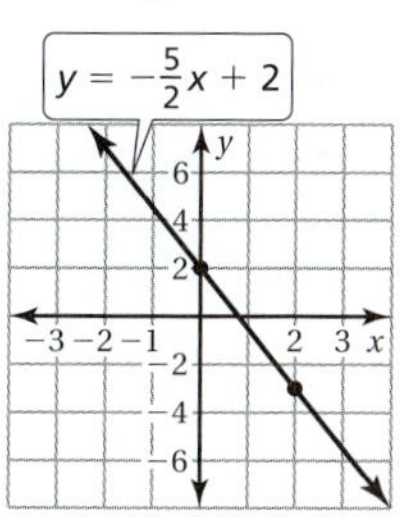

23. $y = -2x + 3$

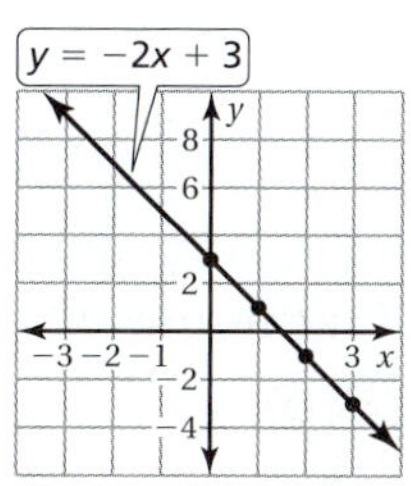

Section 4.1 Graphing Linear Equations *(continued)*

(pages 146 and 147)

25. a. *Sample answer:*

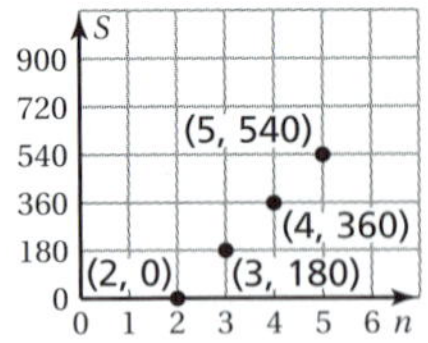

Yes; The graph of the equation is a line.

b. No, $n = 3.5$ does not make sense because a polygon cannot have half a side.

27. Begin this exercise by listing all of the given information.

29. $(-6, 6)$

31. $(-4, -3)$

Section 4.2 Slope of a Line

(pages 153–155)

1. a. B and C

b. A

c. no; None of the lines are vertical.

3. The line is horizontal.

5.

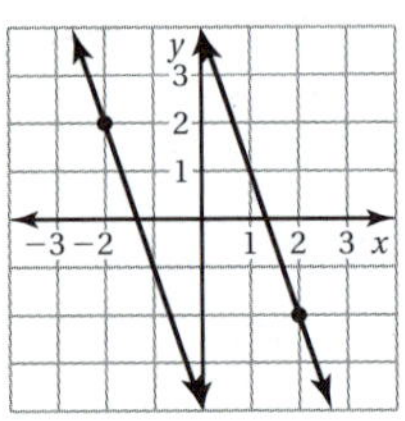

The lines are parallel.

7. $\frac{3}{4}$

9. $-\frac{3}{5}$

11. 0

13. 0

15. undefined

17. $-\frac{11}{6}$

19. The denominator should be $2 - 4$.

$m = -1$

21. 4

23. $-\frac{3}{4}$

25. $\frac{1}{3}$

27. $k = 11$

29. $k = -5$

31. a. $\frac{3}{40}$

b. The cost increases by \$3 for every 40 miles you drive, or the cost increases by \$0.075 for every mile you drive.

33. yes; The slopes are the same between the points.

35. When you switch the coordinates, the differences in the numerator and denominator are the opposite of the numbers when using the slope formula. You still get the same slope.

37. $b = 25$

39. $x = 7.5$

Extension 4.2 Slopes of Parallel and Perpendicular Lines
(pages 156 and 157)

1. blue and red; They both have a slope of -3.
3. yes; Both lines are horizontal and have a slope of 0.
5. yes; Both lines are vertical and have an undefined slope.
7. blue and green; The blue line has a slope of 6. The green line has a slope of $-\frac{1}{6}$. The product of their slopes is $6 \cdot \left(-\frac{1}{6}\right) = -1$.
9. yes; The line $x = -2$ is vertical. The line $y = 8$ is horizontal. A vertical line is perpendicular to a horizontal line.
11. yes; The line $x = 0$ is vertical. The line $y = 0$ is horizontal. A vertical line is perpendicular to a horizontal line.

Section 4.3 Graphing Proportional Relationships
(pages 162 and 163)

1. (0, 0)
3. no; *Sample answer:* The graph of the equation does not pass through the origin.
5. yes; $y = \frac{1}{3}x$; *Sample answer:* The rate of change in the table is constant.
7. 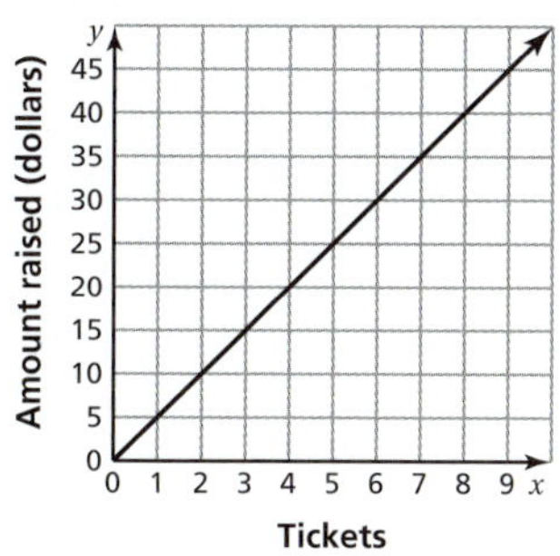

Each ticket costs \$5.

9. **a.** the car; *Sample answer:* The equation for the car is $y = 25x$. Because 25 is greater than 18, the car gets better gas mileage.

   b. 56 miles

11. Consider the direct variation equation and that the graph passes through the origin.

Selected Answers

Section 4.3 Graphing Proportional Relationships *(continued)* *(pages 162 and 163)*

13. **a.** yes; The equation is $d = 6t$, which represents a proportional relationship.

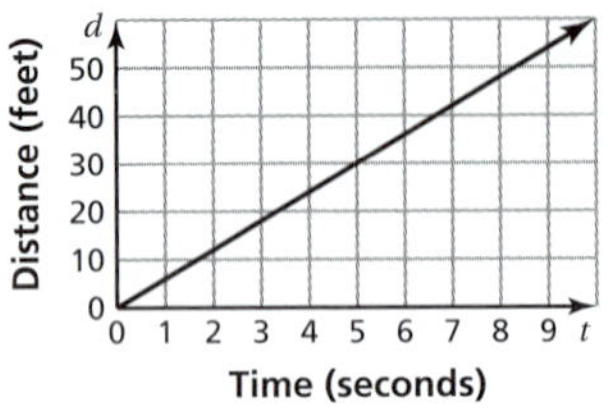

b. yes; The equation is $d = 50r$, which represents a proportional relationship.

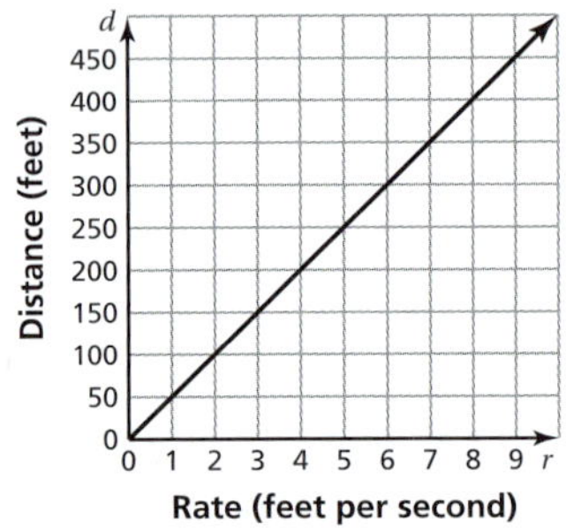

c. no; The equation is $t = \frac{300}{r}$, which does not represent a proportional relationship.

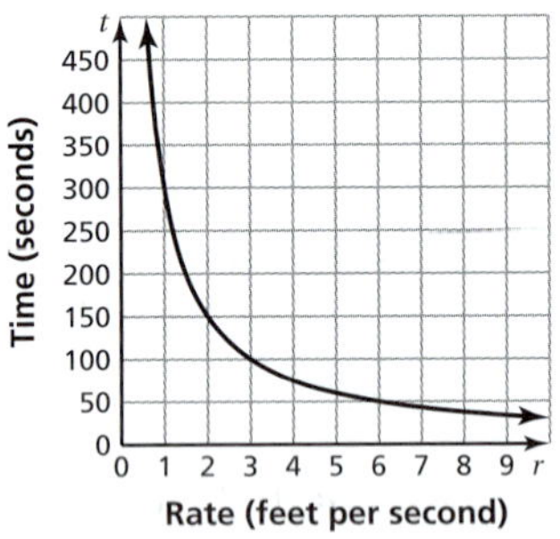

d. part c; It is called inverse variation because when the rate increases, the time decreases, and when the rate decreases, the time increases.

15.

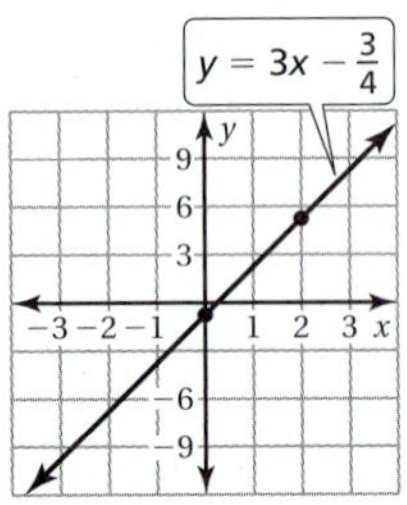

17. B

Section 4.4 Graphing Linear Equations in Slope-Intercept Form *(pages 170 and 171)*

1. Find the x-coordinate of the point where the graph crosses the x-axis.

3. *Sample answer:* The amount of gasoline y (in gallons) left in your tank after you travel x miles is $y = -\frac{1}{20}x + 20$. The slope of $-\frac{1}{20}$ means the car uses 1 gallon of gas for every 20 miles driven. The y-intercept of 20 means there is originally 20 gallons of gas in the tank.

5. A; slope: $\frac{1}{3}$; y-intercept: -2

7. slope: 4; y-intercept: -5

9. slope: $-\frac{4}{5}$; y-intercept: -2

11. slope: $\frac{4}{3}$; y-intercept: -1

13. slope: -2; y-intercept: 3.5

15. slope: 1.5; y-intercept: 11

17. **a.**

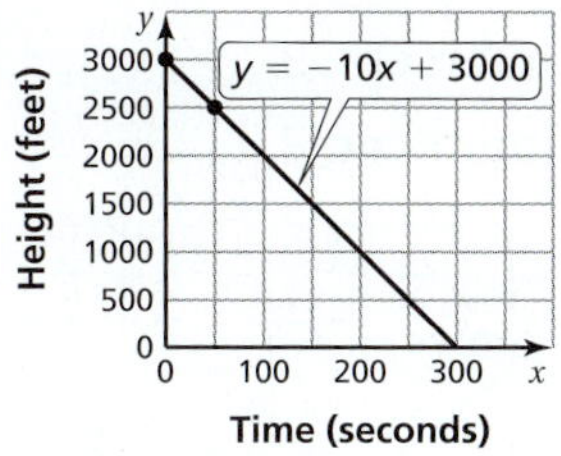

b. The x-intercept of 300 means the skydiver lands on the ground after 300 seconds. The slope of -10 means that the skydiver falls to the ground at a rate of 10 feet per second.

19.

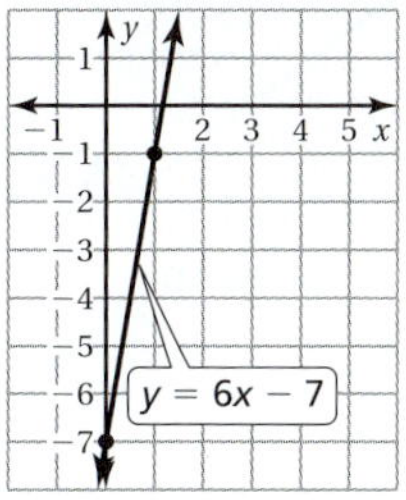

x-intercept: $\frac{7}{6}$

21.

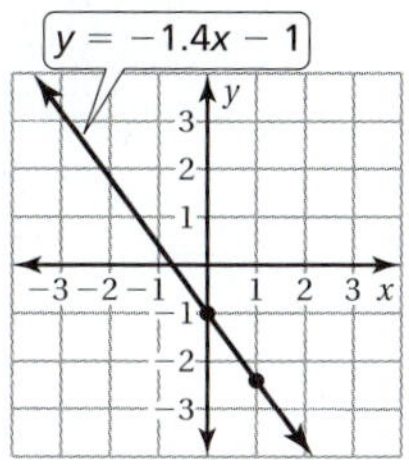

x-intercept: $-\frac{5}{7}$

23.

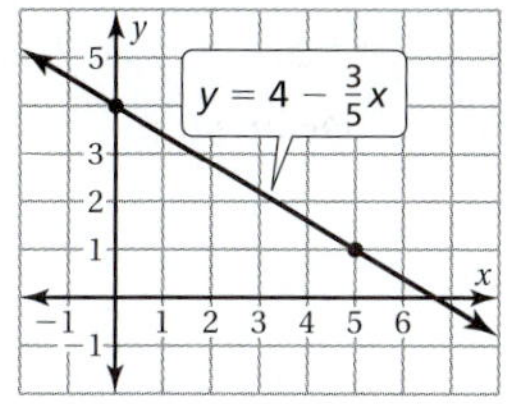

x-intercept: $\frac{20}{3}$

25. **a.** $y = 2x + 4$ and $y = 2x - 3$ are parallel because the slope of each line is 2; $y = -3x - 2$ and $y = -3x + 5$ are parallel because the slope of each line is -3.

b. $y = 2x + 4$ and $y = -\frac{1}{2}x + 2$ are perpendicular because the product of their slopes is -1;

$y = 2x - 3$ and $y = -\frac{1}{2}x + 2$ are perpendicular because the product of their slopes is -1;

$y = -\frac{1}{3}x - 1$ and $y = 3x + 3$ are perpendicular because the product of their slopes is -1.

27. $y = 2x + 3$

29. $y = \frac{2}{3}x - 2$

31. B

Section 4.5 Graphing Linear Equations in Standard Form

(pages 176 and 177)

1. no; The equation is in slope-intercept form.

3. x = pounds of peaches

y = pounds of apples

$y = -\frac{4}{3}x + 10$

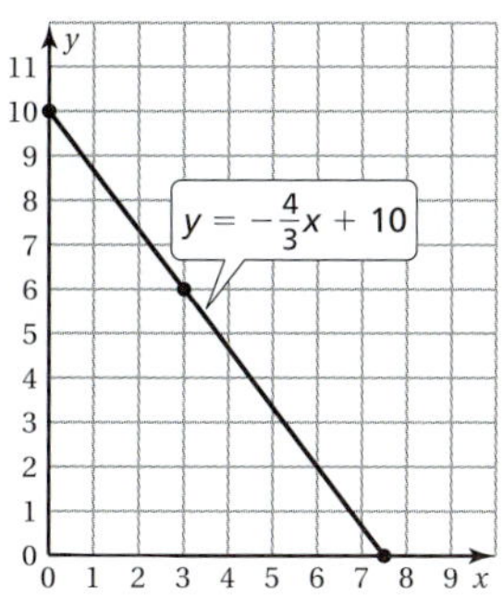

5. $y = -2x + 17$

7. $y = \frac{1}{2}x + 10$

9.

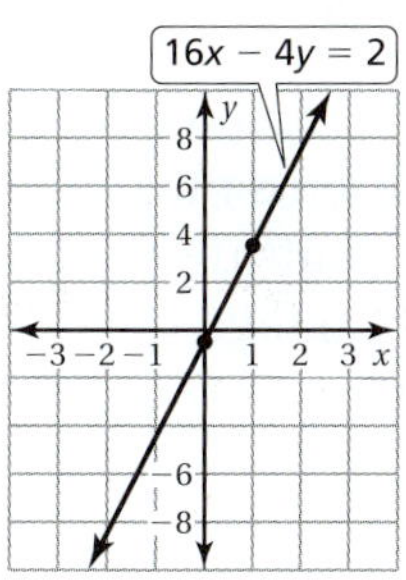

11. B

13. C

Selected Answers

Section 4.5 Graphing Linear Equations in Standard Form

(continued) (pages 176 and 177)

15. a.

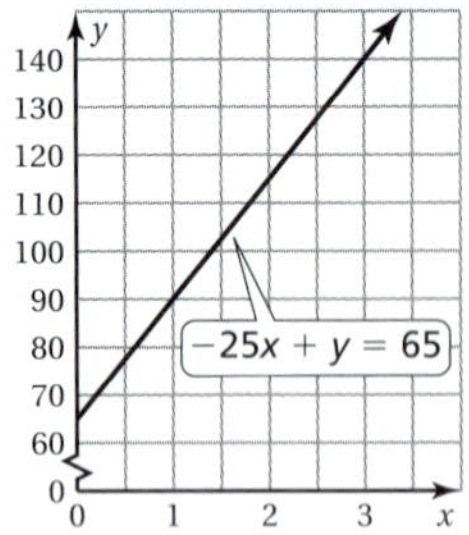

b. \$390

17.

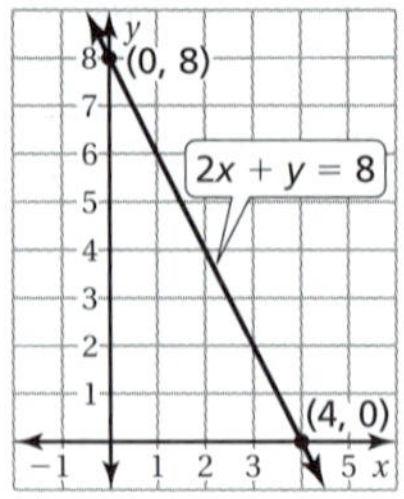

19. x-intercept: 9

y-intercept: 7

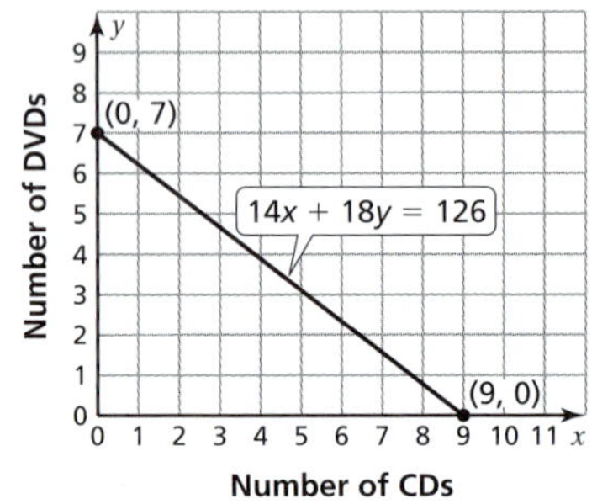

21. a. $9.45x + 7.65y = 160.65$

b.

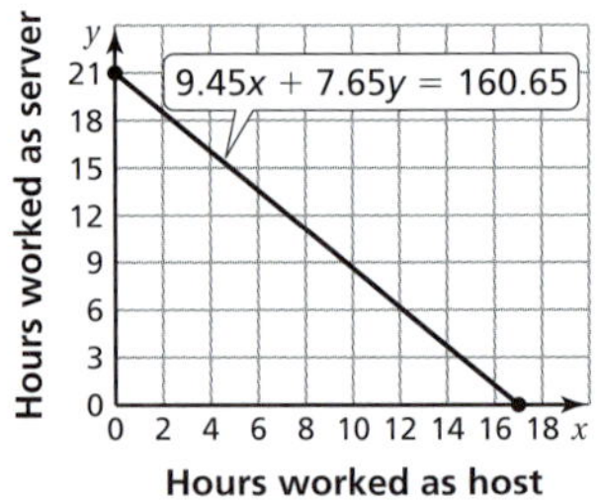

23. a. $y = 40x + 70$

b. x-intercept: $-\frac{7}{4}$; no;

You cannot have a negative time.

c.

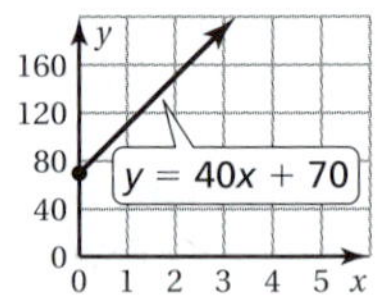

25. $\frac{1}{2}$

Section 4.6 Writing Equations in Slope-Intercept Form

(pages 182 and 183)

1. *Sample answer:* Find the ratio of the rise to the run between the intercepts.

3. $y = 3x + 2$; $y = 3x - 10$; $y = 5$; $y = -1$

5. $y = x + 4$

7. $y = \frac{1}{4}x + 1$

9. $y = \frac{1}{3}x - 3$

11. The x-intercept was used instead of the y-intercept. $y = \frac{1}{2}x - 2$

13. $y = 5$

15. $y = -2$

17. a–b.

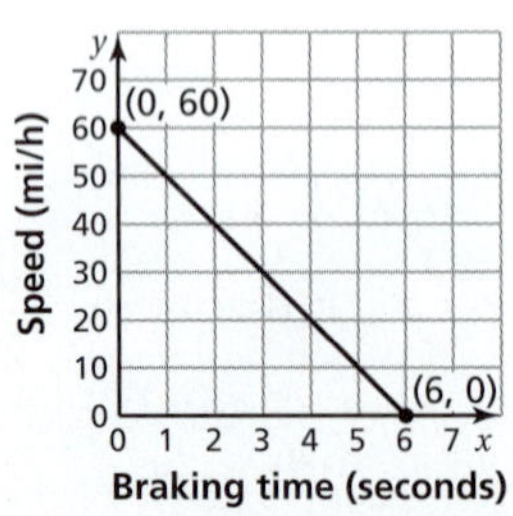

(0, 60) represents the speed of the automobile before braking. (6, 0) represents the amount of time it takes to stop. The line represents the speed y of the automobile after x seconds of braking.

c. $y = -10x + 60$

19. Be sure to check that your rate of growth will not lead to a 0-year-old tree with a negative height.

21 and 23.

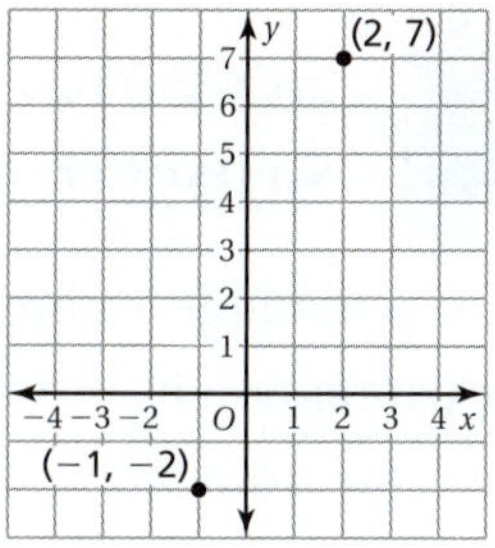

Section 4.7 Writing Equations in Point-Slope Form *(pages 188 and 189)*

1. $m = -2$; $(-1, 3)$

3. $y - 0 = \frac{1}{2}(x + 2)$

5. $y + 1 = -3(x - 3)$

7. $y - 8 = \frac{3}{4}(x - 4)$

9. $y + 5 = -\frac{1}{7}(x - 7)$

11. $y + 4 = -2(x + 1)$

13. $y = 2x$

15. $y = \frac{1}{4}x$

17. $y = x + 1$

19. **a.** $V = -4000x + 30{,}000$

b. $30,000

21. The rate of change is 0.25 degree per chirp.

23. **a.** $y = 14x - 108.5$

b. 4 meters

25.

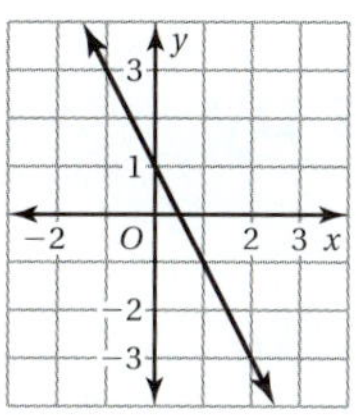

27. D

Section 5.1 Solving Systems of Linear Equations by Graphing *(pages 206 and 207)*

1. yes; The equations are linear and in the same variables.

3. Check whether (3, 4) is a solution of each equation.

5. (4, 176)

7. B; (6, 7)

9. C; (3, −1)

11. (−5, 1)

13. (12, 15)

15. (8, 1)

17. (5, 1.5)

19. (−6, 2)

21. no; Two lines cannot intersect in exactly two points.

23. Make a table to compare your distance to your friend's distance.

25. $c = 8$

27. $x = 11$

Section 5.2 Solving Systems of Linear Equations by Substitution *(pages 212 and 213)*

1. **Step 1:** Solve one of the equations for one of the variables.
   Step 2: Substitute the expression from Step 1 into the other equation and solve.
   Step 3: Substitute the value from Step 2 into one of the original equations and solve.

3. sometimes; A solution obtained by graphing may not be exact.

5. *Sample answer:* $x + 2y = 6$
   $x - y = 3$

7. $4x - y = 3$; The coefficient of y is -1.

9. $2x + 10y = 14$; Dividing by 2 to solve for x yields integers.

11. $(6, 17)$

13. $(4, 1)$

15. $\left(\frac{1}{4}, 6\right)$

17. **a.** $x = 2y$
    $64x + 132y = 1040$
    b. adult tickets: \$8; student tickets: \$4

19. $(-2, 4)$

21. The expression for y was substituted back into the same equation; solution: $(2, 1)$

23. 30 cats, 35 dogs

25. Make a diagram to help visualize the problem.

27. $2x - 5y = -8$

29. B

Section 5.3 Solving Systems of Linear Equations by Elimination *(pages 221–223)*

1. **Step 1:** Multiply, if necessary, one or both equations by a constant so at least one pair of like terms has the same or opposite coefficients.
   Step 2: Add or subtract the equations to eliminate one of the variables.
   Step 3: Solve the resulting equation for the remaining variable.
   Step 4: Substitute the value from Step 3 into one of the original equations and solve.

3. $2x + 3y = 11$
   $3x - 2y = 10$;
   You have to use multiplication to solve the system by elimination.

5. $(6, 2)$

7. $(2, 1)$

9. $(1, -3)$

11. $(3, 2)$

13. The student added y-terms, but subtracted x-terms and constants; solution $(1, 2)$

15. **a.** $2x + y = 10$
    $2x + 3y = 22$
    b. 6 minutes

17. $(5, -1)$

19. $(-2, -1)$

21. $(4, 3)$

23. **a.** ± 4
    b. ± 7

25. yes; The lines are perpendicular.

27. **a.** $23x + 10y = 86$

$28x + 5y = 76$

b. Multiple choice: 2 points each Short response: 4 points each

29. \$95

31. 5 grams of 90% gold alloy, 3 grams of 50% gold alloy

33. $(-1, 2, 1)$

35. yes

37. D

Section 5.4 Solving Special Systems of Linear Equations *(pages 228 and 229)*

1. The graph of a system with no solution is two parallel lines, and the graph of a system with infinitely many solutions is one line.

3. infinitely many solutions; all points on the line $y = 4x + \frac{1}{3}$

5. no solution; The lines have the same slope and different y-intercepts.

7. infinitely many solutions; The lines are identical.

9. $(-1, -2)$

11. infinitely many solutions; all points on the line $y = -\frac{1}{6}x + 5$

13. $(-2.4, -3.5)$

15. no; because they are running at the same speed and your pig had a head start

17. When the slopes are different, there is one solution. When the slopes are the same, there is no solution if the y-intercepts are different and infinitely many solutions if the y-intercepts are the same.

19. $y = 0.99x + 10$

$y = 0.99x$

no; Because you paid \$10 before buying the same number of songs at the same price, you spend \$10 more.

21. Try using the Guess, Test, and Revise method to help you answer this question.

23. $y = 3x$

25. $y = -\frac{1}{2}x + 2$

Extension 5.4 Solving Linear Equations by Graphing *(pages 230 and 231)*

1. $x = \frac{1}{2}$

3. no solution

5. $x = 2$

7. *Sample answer:* $6x - 3 = 6x$; Subtract 3 from the right side.

9. $x = \frac{21}{2}$

11. 6 mo

Section 6.1 Relations and Functions

(pages 246 and 247)

1. the first number; the second number

3. As each input increases by 1, the output increases by 4.

Input Output

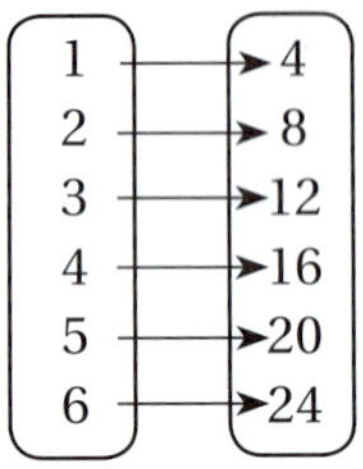

5. As each input increases by 1, the output increases by 5.

Input Output

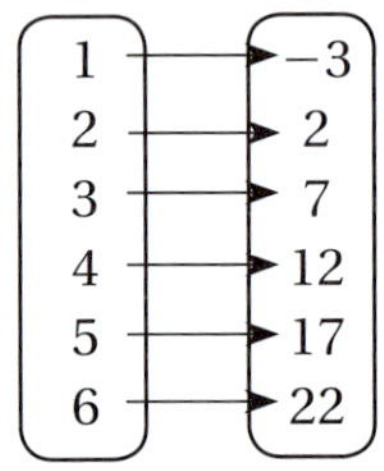

7. (1, 8), (3, 8), (3, 4), (5, 6), (7, 2)

9. no

11. yes

13. Input Output

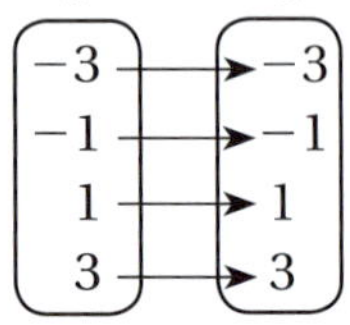

As each input increases by 2, the output increases by 2.

15. Input Output

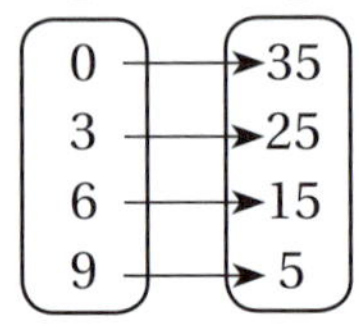

As each input increases by 3, the output decreases by 10.

17. **a.** Input Output

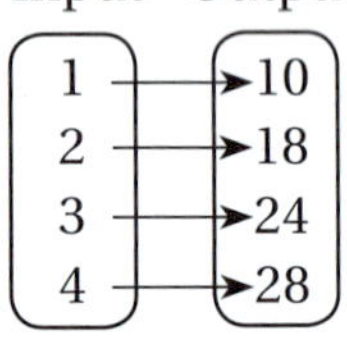

b. yes; Each input has exactly one output.

c. The pattern is that for each input increase of 1, the output increases by \$2 less than the previous increase. For each additional movie you buy, your cost per movie decreases by \$1.

19. y-axis

21. x-axis

Section 6.2 Representations of Functions

(pages 253–255)

1. input variable: x; output variable: y

3. What output is twice the sum of the input 3 and 4?; $2(3 + 4) = 14$; $2(3) + 4 = 10$

5. $y = x + 7$

7. $y = \frac{1}{2}x$

9. $y = x - 3$

11. $y = 6x$

13. 8

15. −17

17. 54

19.

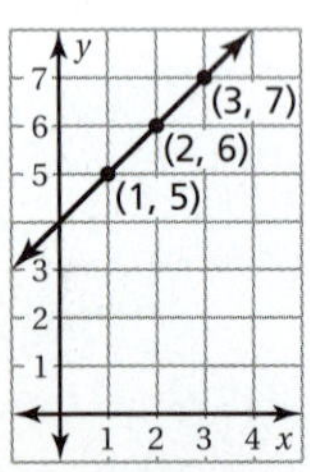

21.

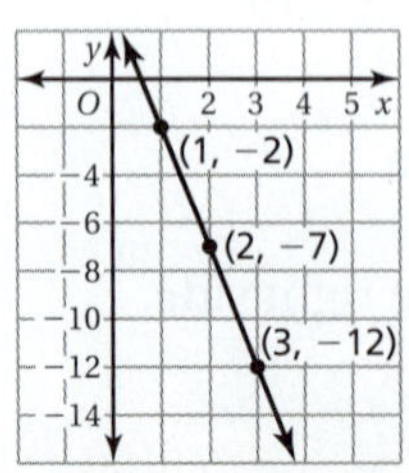

23.

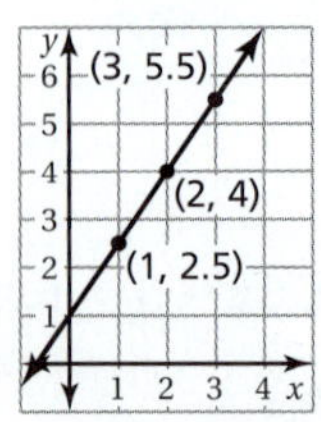

25. The order of the x- and y-coordinates is reversed in each coordinate pair.

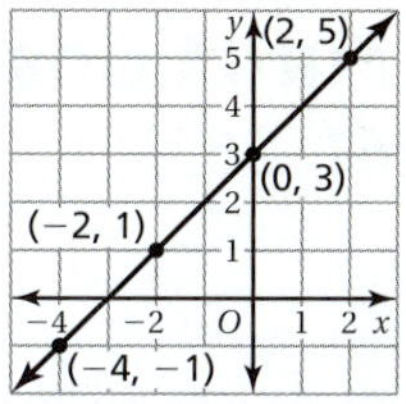

27. B

29. A

31. -4

33. **a.** $P = 3.50b - 84$

b. independent variable: b; dependent variable: P; The profit depends on the number of bracelets sold.

c. 24 bracelets

35. **a.** $G = 35 + 10h$

b. $S = 25h$

c. Snake Tours; For 2 hours, Gator Tours cost \$55 and Snake Tours cost \$50.

37. *Sample answer:*

Side Length	1	2	3	4	5
Perimeter	4	8	12	16	20

Side Length	1	2	3	4	5
Area	1	4	9	16	25

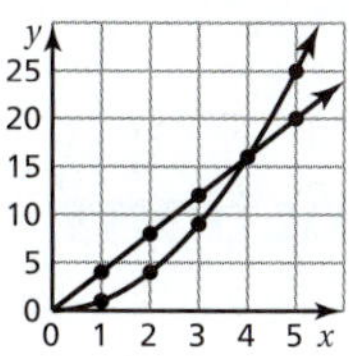

Sample answer: The perimeter function appears to form a line, and the area function appears to form a curve. When the side length is less than 4, the perimeter function is greater. When the side length is greater than 4, the area function is greater. When the side length is 4, the two functions are equal.

39. 1

41. $\frac{1}{3}$

Section 6.3 Linear Functions
(pages 261–263)

1. yes; The graph of $y = mx$ is a nonvertical line, so it is a linear function.

3. $y = \pi x$; x is the diameter; y is the circumference.

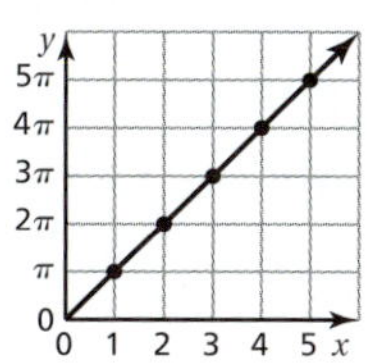

5. $y = \frac{4}{3}x + 2$

7. $y = 3$

9. $y = -\frac{1}{4}x$

11. **a.** independent variable: x; dependent variable: y

b. $y = 3x$; It costs \$3 to rent one movie.

c.

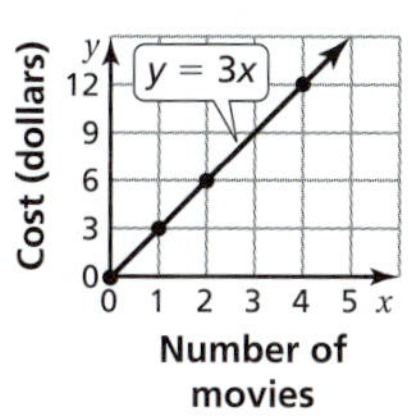

d. \$9

Section 6.3 Linear Functions *(continued)* *(pages 261–263)*

13. **a.** $y = -0.2x + 1$

b. The slope indicates that the power decreases by 20% per hour. The x-intercept indicates that the battery lasts 5 hours. The y-intercept indicates that the battery power is at 100% when you turn on the laptop.

c. 1.25 hours

15. **a.** hiking

b. 67.5 calories

17. yes; A horizontal line is a nonvertical line.

19. **a.**

Temperature (°F), t	94	95	96	97	98
Heat Index (°F), H	122	126	130	134	138

b. independent variable: t; dependent variable: H

c. $H = 4t - 254$

d. 146°F

21. $w = 1.5$

23. C

Section 6.4 Comparing Linear and Nonlinear Functions *(pages 270 and 271)*

1. A linear function has a constant rate of change. A nonlinear function does not have a constant rate of change.

3. 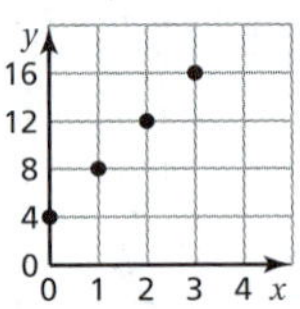

linear

5. 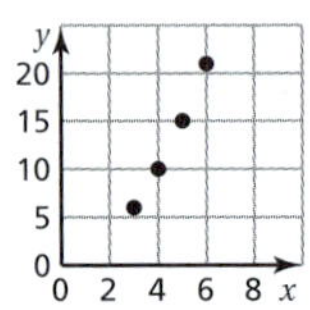

nonlinear

7. linear; The graph is a line.

9. linear; As x increases by 6, y increases by 4.

11. nonlinear; As x increases by 1, V increases by different amounts.

13. linear; You can rewrite the equation in slope-intercept form.

15. nonlinear; As x decreases by 65, y increases by different amounts.

17. **a.** nonlinear; When graphing the points, they do not lie on a line.

b. Tree B; After ten years, the height of Tree A is 20 feet and the height of Tree B is at least 23 feet.

19. **a.** 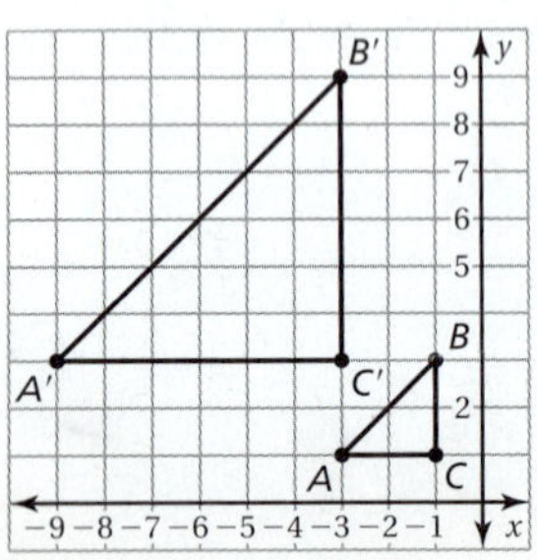

enlargement

21. C

Section 6.5 Analyzing and Sketching Graphs
(pages 276 and 277)

1. F

3. A

5. D

7. The volume of the balloon increases at a constant rate, then stays constant, then increases at a constant rate, then stays constant, and then increases at a constant rate.

9. Horsepower increases at an increasing rate and then increases at a decreasing rate.

11. The hair length increases at a constant rate, then decreases instantly, then increases at a constant rate, then decreases instantly, and then increases at a constant rate.

13. **a.** The usage decreases at an increasing rate.

b. The usage decreases at a decreasing rate.

15.

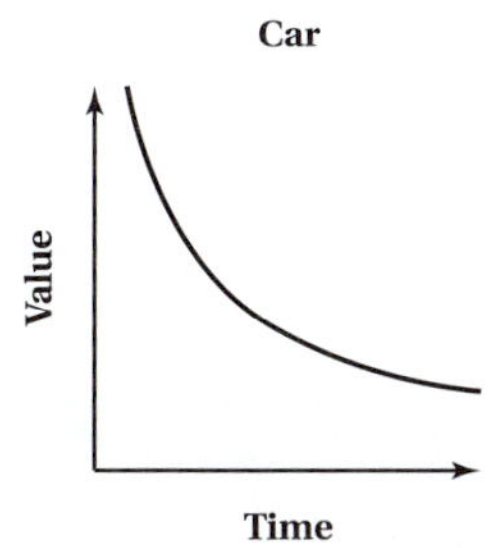

17.

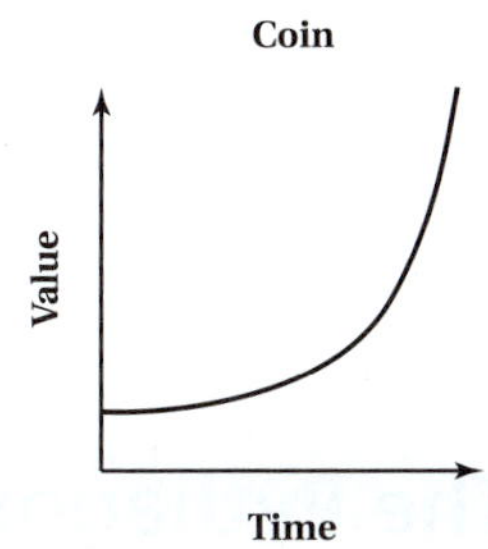

19. Think about the real-life meanings of the words "surplus" and "shortage."

21. $(2, -1)$

23. C

Section 7.1 Finding Square Roots
(pages 292 and 293)

1. no; There is no integer whose square is 26.

3. $\sqrt{256}$ represents the positive square root because there is not a – or a ± in front.

5. $s = 1.3$ km

7. 3 and -3

9. 2 and -2

11. 25

13. $\frac{1}{31}$ and $-\frac{1}{31}$

15. 2.2 and -2.2

17. -19

19. The positive and negative square roots should have been given.

$\pm\sqrt{\frac{1}{4}} = \frac{1}{2}$ and $-\frac{1}{2}$

21. -116

23. 9

25. 25

27. 40

29. because a negative radius does not make sense

31. $=$

33. 9 ft

35. 8 m/sec

37. 2.5 ft

39. $y = 3x - 2$

41. $y = \frac{3}{5}x + 1$

Section 7.2 Finding Cube Roots
(pages 298 and 299)

1. no; There is no integer that equals 25 when cubed.

3. 50 in.

5. 0.4 m

7. -5

9. 12

11. $\frac{7}{4}$

13. $3\frac{5}{8}$

15. $\frac{7}{12}$

17. 74

19. -276

21. 30 cm

23. $>$

25. $<$

27. $-1, 0, 1$

29. The side length of the square base is 18 inches and the height of the pyramid is 9 inches.

31. $x = 3$

33. $x = 4$

35. 289

37. 49

Section 7.3 The Pythagorean Theorem
(pages 304 and 305)

1. The hypotenuse is the longest side and the legs are the other two sides.

3. 29 km

5. 9 in.

7. 24 cm

9. The length of the hypotenuse was substituted for the wrong variable.

$$\begin{aligned} a^2 + b^2 &= c^2 \\ 7^2 + b^2 &= 25^2 \\ 49 + b^2 &= 625 \\ b^2 &= 576 \\ b &= 24 \end{aligned}$$

11. 16 cm

13. Use a right triangle to find the distance.

15. *Sample answer:* length = 20 ft, width = 48 ft, height = 10 ft; $BC = 52$ ft, $AB = \sqrt{2804}$ ft

17. **a.** *Sample answer:*

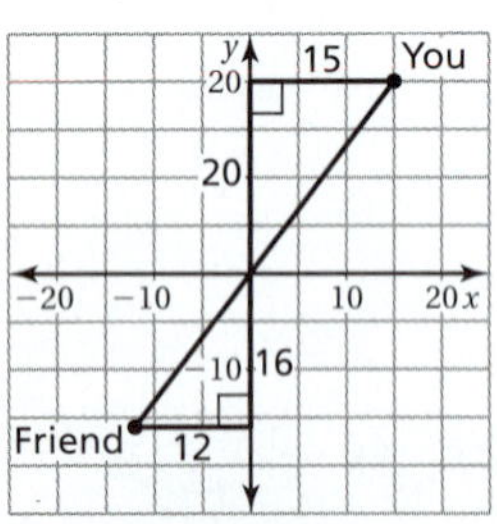

b. 45 ft

19. 6 and -6

21. 13

23. C

Section 7.4 Approximating Square Roots
(pages 313–315)

1. A rational number can be written as the ratio of two integers. An irrational number cannot be written as the ratio of two integers.

3. all rational and irrational numbers; *Sample answer:* $-2, \frac{1}{8}, \sqrt{7}$

5. yes

7. no

9. whole, integer, rational

11. irrational

13. rational

15. irrational

17. 144 is a perfect square. So, $\sqrt{144}$ is rational.

19. **a.** If the last digit is 0, it is a whole number. Otherwise, it is a natural number.
    b. irrational number
    c. irrational number

21. **a.** 26
    b. 26.2

23. **a.** -10
    b. -10.2

25. **a.** -13
    b. -12.9

27. $\sqrt{15}$; $\sqrt{15}$ is positive and -3.5 is negative.

29. $\frac{2}{3}$; $\frac{2}{3}$ is to the right of $\sqrt{\frac{16}{81}}$.

31. $-\sqrt{182}$; $-\sqrt{182}$ is to the right of $-\sqrt{192}$.

33. true

35. 8.1 ft

37. 8.5 ft

39. 20.6 in.

41. Create a table of integers whose cubes are close to the radicand. Determine which two integers the cube root is between. Then create another table of numbers between those two integers whose cubes are close to the radicand. Determine which cube is closest to the radicand; 2.4

43. *Sample answer:* $a = 82$, $b = 97$

45. 1.1

47. 30.1 m/sec

49. Falling objects do not fall at a linear rate. Their speed increases with each second they are falling.

51. 40 m

53. 9 cm

Extension 7.4 Repeating Decimals
(pages 316 and 317)

1. $\frac{1}{9}$

3. $-1\frac{2}{9}$

5. Because the solution does not change when adding/subtracting two equivalent equations; Multiply by 10 so that when you subtract the original equation, the repeating part is removed.

7. $-\frac{13}{30}$

9. $\frac{3}{11}$

11. Pattern: Digits that repeat are in the numerator and 99 is in the denominator; Use 9 as the integer part, 4 as the numerator, and 99 as the denominator of the fractional part.

Selected Answers

Section 7.5 Using the Pythagorean Theorem
(pages 322 and 323)

1. the Pythagorean Theorem and the distance formula
3. If a^2 is odd, then a is an odd number; true when a is an integer; A product of two integers is odd only when each integer is odd.
5. yes
7. no
9. yes
11. $\sqrt{52}$
13. $\sqrt{29}$
15. $\sqrt{85}$
17. The squared quantities under the radical should be added not subtracted; $\sqrt{136}$
19. yes
21. yes
23. no; The measures of the side lengths are $\sqrt{5000}$, $\sqrt{3700}$, and $\sqrt{8500}$ and $(\sqrt{5000})^2 + (\sqrt{3700})^2 \neq (\sqrt{8500})^2$.

25. Notice that the picture is not drawn to scale. Use right triangles.
27. mean: 13; median: 12.5; mode: 12
29. mean: 58; median: 59; mode: 59

Section 8.1 Volumes of Cylinders
(pages 338 and 339)

1. How much does it take to cover the cylinder?; $170\pi \approx 534.1\text{ cm}^2$; $300\pi \approx 942.5\text{ cm}^3$
3. $486\pi \approx 1526.8\text{ ft}^3$
5. $245\pi \approx 769.7\text{ ft}^3$
7. $90\pi \approx 282.7\text{ mm}^3$
9. $252\pi \approx 791.7\text{ in.}^3$
11. $256\pi \approx 804.2\text{ cm}^3$
13. $\frac{125}{8\pi} \approx 5\text{ ft}$
15. $\sqrt{\frac{150{,}000}{19\pi}} \approx 50\text{ cm}$

17. Divide the volume of one round bale by the volume of one square bale.
19. $8325 - 729\pi \approx 6035\text{ m}^3$
21. yes
23. no

Section 8.2 Volumes of Cones
(pages 344 and 345)

1. The height of a cone is the perpendicular distance from the base to the vertex.
3. Divide by 3.
5. $9\pi \approx 28.3\text{ m}^3$
7. $\frac{2\pi}{3} \approx 2.1\text{ ft}^3$
9. $\frac{147\pi}{4} \approx 115.5\text{ yd}^3$
11. $\frac{125\pi}{6} \approx 65.4\text{ in.}^3$
13. The diameter was used instead of the radius;

$$V = \frac{1}{3}(\pi)(1)^2(3) = \pi\text{ m}^3$$

15. 1.5 ft

17. $2\sqrt{\frac{10.8}{4.2\pi}} \approx 1.8$ in.

19. 24.1 min

21. $3y$

23. $A'(-1, 1)$, $B'(-3, 4)$, $C'(-1, 4)$

25. D

Section 8.3 Volumes of Spheres
(pages 352 and 353)

1. A hemisphere is one-half of a sphere.

3. $\frac{500\pi}{3} \approx 523.6$ in.^{3}

5. $972\pi \approx 3053.6$ mm^3

7. $36\pi \approx 113.1$ cm^3

9. 9 mm

11. 4.5 ft

13. 2.5 in.

15. $256\pi + 128\pi = 384\pi \approx 1206.4$ ft^3

17. $r = \frac{3}{4}h$

19. 5400 in.^{2}; 27,000 in.^{3}

21. enlargement; 2

23. A

Section 8.4 Surface Areas and Volumes of Similar Solids
(pages 359–361)

1. Similar solids are solids of the same type that have proportional corresponding linear measures.

3. **a.** $\frac{9}{4}$; because $\left(\frac{3}{2}\right)^2 = \frac{9}{4}$

   b. $\frac{27}{8}$; because $\left(\frac{3}{2}\right)^3 = \frac{27}{8}$

5. no

7. no

9. $b = 18$ m; $c = 19.5$ m; $h = 9$ m

11. 1012.5 in.^{2}

13. 13,564.8 ft^3

15. 673.75 cm^2

17. **a.** 9483 pounds; The ratio of the height of the original statue to the height of the small statue is 8.4 : 1. So, the ratio of the weights, or volumes is $\left(\frac{8.4}{1}\right)^3$.

    b. 221,184 lb

19. **a.** yes; Because all circles are similar, the slant height and the circumference of the base of the cones are proportional.

    b. no; because the ratio of the volumes of similar solids is equal to the cube of the ratio of their corresponding linear measures

21. 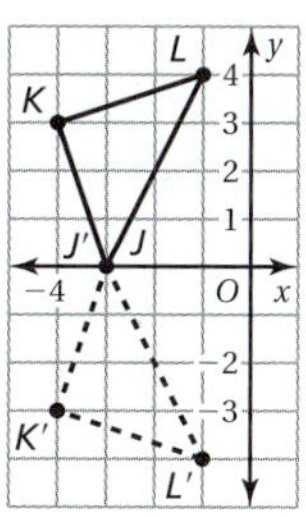

$J'(-3, 0)$, $K'(-4, -3)$, $L'(-1, -4)$

Section 9.1 Scatter Plots
(pages 376 and 377)

1. They must be ordered pairs so there are equal amounts of x- and y-values.

3. no relationship; A student's shoe size is not related to his or her IQ.

5. nonlinear relationship; On each successive bounce, the ball rebounds to a height less than its previous bounce.

7. **a.** (22, 152), (40, 94), (28, 134), (35, 110), (46, 81)

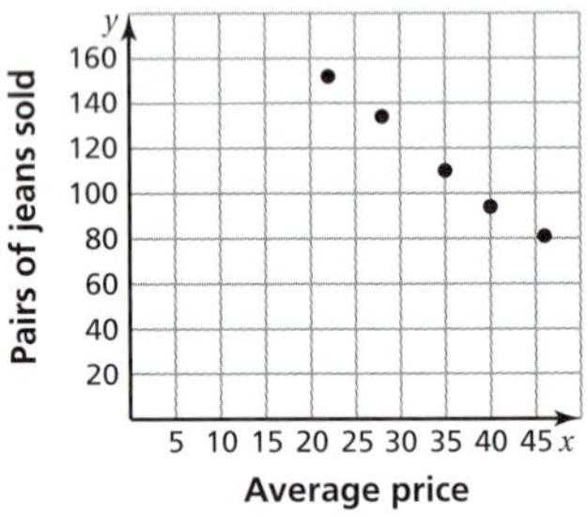

b. As the average price of jeans increases, the number of pairs of jeans sold decreases.

9. **a.** 3.5 h **b.** $85

c. There is a positive linear relationship between hours worked and earnings.

11. nonlinear relationship; no outliers, gaps, or clusters

13. positive linear relationship

15. *Sample answer:* bank account balance during a shopping spree

17. Could there be another event that is causing the sales of both items to increase?

19. 8

21. B

Section 9.2 Lines of Fit
(pages 382 and 383)

1. You can estimate and predict values.

3. -0.98, because it is closer to -1 than 0.91 is to 1. $(|-0.98| > |0.91|)$

5. **a.**

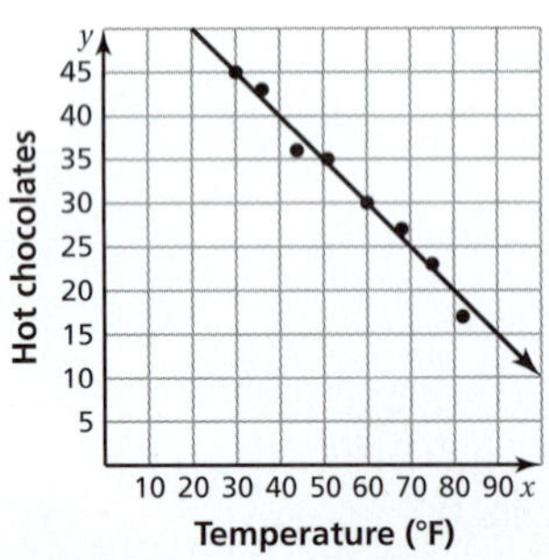

b. *Sample answer:* $y = -0.5x + 60$

c. *Sample answer:* The slope is -0.5 and the y-intercept is 60. So, you could predict that 60 hot chocolates are sold when the temperature is 0°F, and the sales decrease by about 1 hot chocolate for every 2°F increase in temperature.

d. 50 hot chocolates

7. no; There is no line that lies close to most of the points.

9. $y = 0.9x + 4$; $r \approx 0.999$; The relationship between x and y is a strong positive correlation and the equation closely models the data; 4 in.

11. **a.** $y = 48x + 11$; $r \approx 0.98$; The relationship between x and y is a strong positive correlation and the equation closely models the data.

    b. 251 ft

    c. The height of a hit baseball is not linear. The best fit line from part (a) only models a small part of the data.

13. $-2\frac{7}{9}$

15. $\frac{9}{11}$

Section 9.3 Two-Way Tables
(pages 390 and 391)

1. The joint frequencies are the entries in the two-way table that differentiate the two categories of data collected. The marginal frequencies are the sums of the rows and columns of the two-way table.

3. total of females surveyed: 73;
   total of males surveyed: 59

5. 51

7. 71 students are juniors.
   75 students are seniors.
   93 students are attending the school play.
   53 students are not attending the school play.

9. **a.** 19; 42

   b. 72 6th-graders were surveyed.
   74 7th-graders were surveyed.
   65 8th-graders were surveyed.
   112 students chose grades.
   40 students chose popularity.
   59 students chose sports.

   c. about 8.5%

11. **a.**

		Eye Color			
		Green	Blue	Brown	Total
Gender	Male	5	16	27	48
	Female	3	19	18	40
	Total	8	35	45	88

b. 48 males were surveyed.
40 females were surveyed.
8 students have green eyes.
35 students have blue eyes.
45 students have brown eyes.

c.

		Eye Color		
		Green	Blue	Brown
Gender	Male	63%	46%	60%
	Female	38%	54%	40%

Sample answer: About 63% of the students with green eyes are male. 40% of the students with brown eyes are female.

13. Be careful not to count the females with green eyes twice.

15. $y = 5x - 2$

17. B

Section 9.4 Choosing a Data Display
(pages 397–399)

1. yes; Different displays may show different aspects of the data.

3. *Sample answer:*

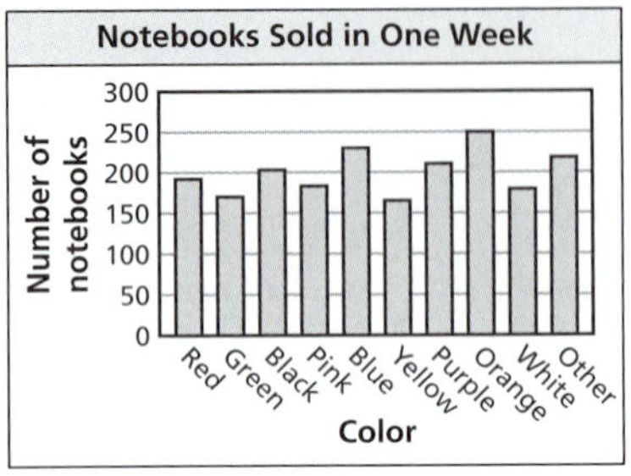

A bar graph shows the data in different color categories.

5. *Sample answer:* line graph; shows changes over time

7. *Sample answer:* line graph; shows changes over time

9. **a.** yes; The circle graph shows the data as parts of the whole.
   b. no; The bar graph shows the number of students, not the portion of students.

11. The pictures of the bikes are the largest on Monday and the smallest on Wednesday, which makes it seem like the distance is the same each day.

13. The intervals are not the same size.

15. *Sample answer:* bar graph; Each bar can represent a different vegetable.

17. *Sample answer:* dot plot

19. Does one display better show the differences in digits?

21. **a.** -9
    b. -8.6

23. A

Section 10.1 Exponents
(pages 414 and 415)

1. -3^4 is the negative of 3^4, so the base is 3, the exponent is 4, and its value is -81. $(-3)^4$ has a base of -3, an exponent of 4, and a value of 81.

3. 3^4

5. $\left(-\frac{1}{2}\right)^3$

7. $\pi^3 x^4$

9. $(6.4)^4 b^3$

11. 25

13. 1

15. $\frac{1}{144}$

17. The negative sign is not part of the base; $-6^2 = -(6 \cdot 6) = -36$.

19. $-\left(\frac{1}{4}\right)^4$

21. 29

23. 5

25. 66

27.

h	1	2	3	4	5
$2^h - 1$	1	3	7	15	31
2^{h-1}	1	2	4	8	16

$2^h - 1$; The option $2^h - 1$ pays you more money when $h > 1$.

29. Remember to add the black keys when finding how many notes you travel.

31. Associative Property of Multiplication

33. B

Section 10.2 Product of Powers Property
(pages 420 and 421)

1. when multiplying powers with the same base

3. 3^4 **5.** $(-4)^{12}$ **7.** h^7

9. $\left(-\frac{5}{7}\right)^{17}$ **11.** 5^{12} **13.** 3.8^{12}

15. The bases should not be multiplied. $5^2 \cdot 5^9 = 5^{2+9} = 5^{11}$

17. $216g^3$ **19.** $\frac{1}{25}k^2$ **21.** $r^{12}t^{12}$

23. no; $3^2 + 3^3 = 9 + 27 = 36$ and $3^5 = 243$

25. 496 **27.** 78,125

29. **a.** $16\pi \approx 50.27$ in.^{3}

b. $192\pi \approx 603.19$ in.^{3} Squaring each of the dimensions causes the volume to be 12 times larger.

31. Use the Commutative and Associative Properties of Multiplication to group the powers.

33. 4 **35.** 3

37. B

Section 10.3 Quotient of Powers Property
(pages 426 and 427)

1. To divide powers means to divide out the common factors of the numerator and denominator. To divide powers with the same base, write the power with the common base and an exponent found by subtracting the exponent in the denominator from the exponent in the numerator.

3. 6^6 **5.** $(-3)^3$ **7.** 5^6

9. $(-17)^3$ **11.** $(-6.4)^2$ **13.** b^{13}

15. You should subtract the exponents instead of dividing them. $\frac{6^{15}}{6^5} = 6^{15-5} = 6^{10}$

17. 2^9 **19.** π^8 **21.** k^{14}

23. $64x$ **25.** $125a^3b^2$ **27.** x^7y^6

29. You are checking to see if there is a linear relationship between memory and price, not if the change in price is constant for consecutive sizes of MP3 players.

31. 10^{13} galaxies **33.** -9

35. 61 **37.** B

Section 10.4 Zero and Negative Exponents
(pages 432 and 433)

1. no; Any nonzero base raised to a zero exponent is always 1.

3. $5^{-5}, 5^0, 5^4$

5. 1

7. 1

9. $\frac{1}{36}$

11. $\frac{1}{16}$

13. $5\frac{1}{4}$

15. $\frac{1}{125}$

17. The negative sign goes with the exponent, not the base. $(4)^{-3} = \frac{1}{4^3} = \frac{1}{64}$

19. 2^0; 10^0

21. $\frac{a^7}{64}$

23. $5b$

25. 12

27. $\frac{w^6}{9}$

29. 100 mm

31. 1,000,000 nanometers

33. **a.** 10^{-9} m **b.** equal to

35. Write the power as 1 divided by the power and use a negative exponent. Justifications will vary.

37. 10^9

39. 10^4

Section 10.5 Reading Scientific Notation
(pages 440 and 441)

1. Scientific notation uses a factor greater than or equal to 1 but less than 10 multiplied by a power of 10. A number in standard form is written out with all the zeros and place values included.

3. 5,600,000,000,000

5. 87,300,000,000,000,000

7. yes; The factor is greater than or equal to 1 and less than 10. The power of 10 has an integer exponent.

9. no; The factor is greater than 10.

11. yes; The factor is greater than or equal to 1 and less than 10. The power of 10 has an integer exponent.

13. no; The factor is less than 1.

15. 70,000,000

17. 500

19. 0.000044

21. 1,660,000,000

23. 9,725,000

25. **a.** 810,000,000 platelets
    b. 1,350,000,000,000 platelets

27. **a.** Bellatrix
    b. Betelgeuse

29. 1555.2 km^2

31. 35,000,000 km^3

33. 4^5

35. $(-2)^3$

Section 10.6 Writing Scientific Notation
(pages 446 and 447)

1. If the number is greater than or equal to 10, the exponent will be positive. If the number is less than 1 and greater than 0, the exponent will be negative.

3. 2.1×10^{-3}

5. 3.21×10^{8}

7. 4×10^{-5}

9. 4.56×10^{10}

11. 8.4×10^{5}

13. 72.5 is not less than 10. The decimal point needs to move one more place to the left. 7.25×10^{7}

15. $6.09 \times 10^{-5}, 6.78 \times 10^{-5}, 6.8 \times 10^{-5}$

17. $4.8 \times 10^{-8}, 4.8 \times 10^{-6}, 4.8 \times 10^{-5}$

19. $6.88 \times 10^{-23}, 5.78 \times 10^{23}, 5.82 \times 10^{23}$

21. 4.01×10^{7} m

23. $680,\ 6.8 \times 10^{3},\ \frac{68{,}500}{10}$

25. $6.25 \times 10^{-3},\ 6.3\%,\ 0.625,\ 6\frac{1}{4}$

27. 1.99×10^{9} watts

29. carat; Because 1 carat $= 1.2 \times 10^{23}$ atomic mass units and 1 milligram $= 6.02 \times 10^{20}$ atomic mass units, and $1.2 \times 10^{23} > 6.02 \times 10^{20}$.

31. natural, whole, integer, rational

33. irrational

Section 10.7 Operations in Scientific Notation
(pages 452 and 453)

1. Use the Distributive Property to group the factors together. Then subtract the factors and write it with the power of 10. The number may need to be rewritten so that it is still in scientific notation.

3. 8.34×10^{7}

5. 4.947×10^{11}

7. 5.8×10^{5}

9. 5.2×10^{8}

11. 7.555×10^{7}

13. 1.037×10^{7}

15. You have to rewrite the numbers so they have the same power of 10 before adding; 3.03×10^{9}

17. 2.9×10^{-3}

19. 1.5×10^{0}

21. 2.88×10^{-7}

23. 1.12×10^{-2}

25. 4.006×10^{9}

27. 1.962×10^{8} cm

29. First find the total length of the ridges and valleys.

31. 3×10^{8} m/sec

33. $\frac{1}{8}$

35. C

Key Vocabulary Index

Mathematical terms are best understood when you see them used and defined *in context*. This index lists where you will find key vocabulary. A full glossary is available in your Record and Practice Journal and at *BigIdeasMath.com*.

Student Index

This student-friendly index will help you find vocabulary, key ideas, and concepts. It is easily accessible and designed to be a reference for you whether you are looking for a definition, real-life application, or help with avoiding common errors.

Student Index

M

N

O

P

Student Index

T

V

W

X

Y

Photo Credits

Cover

Pavelk/Shutterstock.com, Pincasso/Shutterstock.com, valdis torms/Shutterstock.com

Front matter

i Pavelk/Shutterstock.com, Pincasso/Shutterstock.com, valdis torms/Shutterstock.com; **iv** Big Ideas Learning, LLC; **viii** *top* ©iStockphoto.com/Lisa Thornberg, ©iStockphoto.com/Ann Marie Kurtz; *bottom* ©iStockphoto.com/Jane norton; **ix** *top* Kasiap/Shutterstock.com, ©iStockphoto.com/Ann Marie Kurtz; *bottom* wavebreakmedia ltd/Shutterstock.com; **x** *top* ©iStockphoto.com/sumnersgraphicsinc, ©iStockphoto.com/Ann Marie Kurtz; *bottom* Odua Images/Shutterstock.com; **xi** *top* ©iStockhphoto.com/Jonathan Larsen; *bottom* James Flint/Shutterstock.com; **xii** *top* stephan kerkhofs/Shutterstock.com, Cigdem Sean Cooper/Shutterstock.com, ©iStockphoto.com/Andreas Gradin; *bottom* william casey/Shutterstock.com; **xiii** *top* ©iStockphoto.com/ALEAIMAGE, ©iStockphoto.com/Ann Marie Kurtz; *bottom* Edyta Pawlowska/Shutterstock.com; **xiv** *top* ©iStockphoto/Michael Flippo, ©iStockphoto.com/Ann Marie Kurtz; *bottom* PETER CLOSE/Shutterstock.com; **xv** *top* ©iStockphoto.com/ALEAIMAGE, ©iStockphoto.com/Ann Marie Kurtz; *bottom* Kharidehal Abhirama Ashwin/Shutterstock.com; **xvi** *top* ©iStockphoto.com/Alistair Cotton; *bottom* ©iStockphoto.com/Noraznen Azit; **xvii** *top* Varina and Jay Patel/Shutterstock.com, ©iStockphoto.com/Ann Marie Kurtz; *bottom* ©iStockphoto.com/Thomas Perkins; **xviii** Ljupco Smokovski/Shutterstock.com

Chapter 1

1 ©iStockphoto.com/Lisa Thornberg, ©iStockphoto.com/Ann Marie Kurtz; **6** ©iStockphoto.com/David Freund; **7** ©iStockphoto.com/nicolas hansen; **8** amskad/Shutterstock.com; **9** ©iStockphoto.com/Ryan Lane; **12** ©iStockphoto.com/Harley McCabe; **13** ©iStockphoto.com/Jacom Stephens; **14** ©iStockphoto.com/Harry Hu; **15** ©iStockphoto.com/Ralf Hettler, Vibrant Image Studio/Shutterstock.com; **23** ©iStockphoto.com/Andrey Krasnov; **24** Shawn Hempel/Shutterstock.com; **31** *top right* ©iStockphoto.com/Alan Crawford; *center left* ©iStockphoto.com/Julio Yeste; *bottom right* ©iStockphoto.com/Mark Stay; **36** *center right* Ljupco Smokovski/Shutterstock.com; *bottom left* emel82/Shutterstock.com

Chapter 2

40 Kasiap/Shutterstock.com, ©iStockphoto.com/Ann Marie Kurtz; **48** Azat1976/Shutterstock.com; **52** ©iStockphoto.com/Er Ten Hong; **53** *center left* ©iStockphoto.com/Sergey Galushko; *center right* ©iStockphoto.com/Tryfonov levgenii; **54** ©iStockphoto.com/ingmar wesemann; **59** ©iStockphoto.com/Hazlan Abdul Hakim; **67** ©iStockphoto.com/Maksim Shmeljov; **70** *top* ©iStockphoto.com/Viatcheslav Dusaleev; *bottom left* ©iStockphoto.com/Jason Mooy; *bottom right* ©iStockphoto.com/Felix Möckel; **73** gary718/Shutterstock.com; **83** Diego Cervo/Shutterstock.com; **90** *center left* Antonio Jorge Nunes/Shutterstock.com, Tom C Amon/Shutterstock.com; *center right* ©iStockphoto.com/Alex Slobodkin

Chapter 3

100 ©iStockphoto.com/sumnersgraphicsinc, ©iStockphoto.com/Ann Marie Kurtz; **102** PILart/Shutterstock.com, Wildstyle/Shutterstock.com; **103** Estate Craft Homes, Inc.; **114** Marc Dietrich/Shutterstock.com; **120** *bottom left* ©iStockphoto.com/Evgeny Terentev; *bottom right* ©iStockphoto.com/Vadym Volodin; **121** NASA; **124** iStockphoto.com/Evelyn Peyton; **125** *top right* ©iStockphoto.com/Terraxplorer; *top left* ©iStockphoto.com/Lora Clark; *center right* ©iStockphoto.com/Jennifer Morgan

Chapter 4

140 ©iStockphoto.com/Jonathan Larsen; **145** NASA; **146** ©iStockphoto.com/David Morgan; **147** *top right* NASA; *center left* ©iStockphoto.com/jsemeniuk; **154** ©iStockphoto.com/Amanda Rohde; **155** Julian Rovagnati/Shutterstock.com; **159** RyFlip/Shutterstock.com; **162** Luke Wein/Shutterstock.com; **165** AVAVA/Shutterstock.com; **170** ©iStockphoto.com/Dreamframer; **171** *top right* Jerry Horbert/Shutterstock.com; *center left* ©iStockphoto.com/Chris Schmidt; **173** ©iStockphoto.com/biffspandex; **176** ©iStockphoto.com/Stephen Pothier; **177** *top left* Gina Smith/Shutterstock.com; *center left* Dewayne Flowers/Shutterstock.com; **181** Herrenknecht AG; **182** ©iStockphoto.com/Adam Mattel; **183** *top left* ©iStockphoto.com/Gene Chutka; *center right* ©iStockphoto.com/marcellus2070, ©iStockphoto.com/beetle8; **187** ©iStockphoto.com/Connie Maher; **188** ©iStockphoto.com/Jacom Stephens; **189** *top right* ©iStockphoto.com/Petr Podzemny; *bottom left* ©iStockphoto.com/adrian beesley; **190** Richard Goldberg/Shutterstock.com; **196** Thomas M Perkins/Shutterstock.com

Chapter 5

200 stephan kerkhofs/Shutterstock.com, Cigdem Sean Cooper/Shutterstock.com, ©iStockphoto.com/Andreas Gradin; **202** Howard Sandler/Shutterstock.com, ©iStockphoto.com/Dori OConnell; **205** Richard Paul Kane/Shutterstock.com; **206** ©iStockphoto.com/Kathy Hicks; **208** *top right* YuriyZhuravov/Shutterstock.com; *bottom right* Talvi/Shutterstock.com; **211** aguilarphoto/Shutterstock.com; **212** Kiselev Andrey Valerevich/Shutterstock.com; **213** *center left* Susan Schmitz/Shutterstock.com; *center right* akva/Shutterstock.com; **215** Andrey Yurlov/Shutterstock.com; **216** Steve Cukrov/Shutterstock.com; **220** Le Do/Shutterstock.com, Quang Ho/Shutterstock.com, SergeyIT/Shutterstock.com, jon Le-Bon/Shutterstock.com; **221** Ariwasabi/Shutterstock.com; **222** Ewa/Shutterstock.com; **223** *top left* Gordana Sermek/Shutterstock.com; *center right* Rashevskyi Viacheslav/Shutterstock.com; **224** ©iStockphoto.com/walik; **228** ©iStockphoto.com/Corina Estepa; **229** ©iStockphoto.com/Tomislav Forgo; **231** Kateryna Larina/Shutterstock.com; **232** Selena/Shutterstock.com; **236** kostudio/Shutterstock.com

Chapter 6

240 ©iStockphoto.com/ALEAIMAGE, ©iStockphoto.com/Ann Marie Kurtz; **247** ©iStockphoto.com/Kevin Panizza; **249** ©iStockphoto.com/Jacom Stephens; **252** ©iStockphoto.com/DivaNir4a; **254** *top left* ©iStockphoto.com/Manuel Angel Diaz Blanco; *bottom right* ©iStockphoto.com/Sergey Lemeshencko; **255** ©iStockphoto.com/Robert Rushton; **259** General Atomics Aeronautical Systems, Inc.; **262** ©iStockphoto.com/Mlenny Photography; **263** ©iStockphoto.com/medobear; **267** ©iStockphoto.com/PeskyMonkey; **271** ©iStockphoto.com/Tom Buttle; **278** gillmar/Shutterstock.com

Chapter 7

286 ©iStockphoto/Michael Flippo, ©iStockphoto.com/Ann Marie Kurtz; **291** Perfectblue97; **292** ©iStockphoto.com/Benjamin Lazare; **293** *top right* ©iStockphoto.com/iShootPhotos, LLC; *center left* ©iStockphoto.com/Jill Chen, Oleksiy Mark/Shutterstock.com; **298** Gary Whitton/Shutterstock.com; **299** Michael Stokes/Shutterstock.com; **300** ©Oxford Science Archive/Heritage Images/Imagestate; **304** ©iStockphoto.com/Melissa Carroll; **307** *center left* ©iStockphoto.com/Yvan Dubé; *bottom right* Snvv/Shutterstock.com; **308** ©iStockphoto.com/Kais Tolmats; **312** *top left* ©iStockphoto.com/Don Bayley; *center left* ©iStockphoto.com/iLexx; **315** ©iStockphoto.com/Marcio Silva; **319** Monkey Business Images/Shutterstock.com; **327** LoopAll/Shutterstock.com; **328** CD Lanzen/Shutterstock.com

Chapter 8

332 ©iStockphoto.com/ALEAIMAGE, ©iStockphoto.com/Ann Marie Kurtz; **334** ©iStockphoto.com/Jill Chen; **337** ©iStockphoto.com/camilla wisbauer; **339** *Exercises 13 and 14* ©iStockphoto.com/Prill Mediendesigns & Fotografie; *Exercise 15* ©iStockphoto.com/subjug; *center left* ©iStockphoto.com/Matthew Dixon; *center right* ©iStockphoto.com/nilgun bostanci; **345** ©iStockphoto.com/Stefano Tiraboschi; **351** Donald Joski/Shutterstock.com; **352** ©iStockphoto.com/Yury Kosourov; **353** Carlos Caetano/Shutterstock.com; **360** Courtesy of Green Light Collectibles; **361** *top right* ©iStockphoto.com/wrangel; *center left* ©iStockphoto.com/ivanastar; *bottom left* ©iStockphoto.com/Daniel Cardiff; **362** Eric Isselée/Shutterstock.com; **366** ©iStockphoto.com/Daniel Loiselle

Chapter 9

370 ©iStockphoto.com/Alistair Cotton; **372** *baseball* Kittisak/Shutterstock.com; *golf ball* tezzstock/Shutterstock.com; *basketball* vasosh/Shutterstock.com; *tennis ball* UKRID/Shutterstock.com; *water polo ball* John Kasawa/Shutterstock.com; *softball* Ra Studio/Shutterstock.com; *volleyball* vberla/Shutterstock.com; **376** ©iStockphoto.com/Jill Fromer; **377** ©iStockphoto.com/Janis Litavnieks; **378** Gina Brockett; **379** ©iStockphoto.com/Craig Dingle; **381** Sashkin/Shutterstock.com; **382** ©iStockphoto.com/Brian McEntire; **385** Dwight Smith/Shutterstock.com; **386** Aptyp_koK/Shutterstock.com; **391** Alberto Zornetta/Shutterstock.com; **392** *center left* ©iStockphoto.com/Tony Campbell; *bottom right* Eric Isselee/Shutterstock.com; **393** *top right* Larry Korhnak; *bottom right* Photo by Andy Newman; **399** *top left* ©iStockphoto.com/Jane norton; *bottom right* ©iStockphoto.com/Krzysztof Zmij; **400** IrinaK/Shutterstock.com; **404** Lim Yong Hian/Shutterstock.com

Chapter 10

408 Varina and Jay Patel/Shutterstock.com, ©iStockphoto.com/Ann Marie Kurtz; **410** ©iStockphoto.com/Franck Boston; **411** *Activity 3a* ©iStockphoto.com/Manfred Konrad; *Activity 3b* NASA/JPL-Caltech/R.Hurt (SSC); *Activity 3c and d* NASA; *bottom right* Stevyn Colgan; **413** ©iStockphoto.com/Philippa Banks; **414** ©iStockphoto.com/clotilde hulin; **415** ©iStockphoto.com/Boris Yankov; **420** ©iStockphoto.com/VIKTORIIA KULISH; **421** *top right* ©iStockphoto.com/Paul Tessier; *center left* ©iStockphoto.com/subjug, ©iStockphoto.com/Valerie Loiseleux, ©iStockphoto.com/Linda Steward; **426** ©iStockphoto.com/Petrovich9; **427** *top right* Dash/Shutterstock.com; *center left* NASA/JPL-Caltech/L.Cieza (UT Austin); **431** ©iStockphoto.com/Aliaksandr Autayeu; **432** EugeneF/Shutterstock.com; **433** ©iStockphoto.com/Nancy Louie; **435** ©iStockphoto.com/Dan Moore; **436** ©iStockphoto.com/Kais Tolmats; **437** *Activity 3a and d* Tom C Amon/Shutterstock.com; *Activity 3b* Olga Gabay/Shutterstock.com; *Activity 3c* NASA/MODIS Rapid Response/Jeff Schmaltz; *Activity 3f* HuHu/Shutterstock.com; *Activity 4a* PILart/Shutterstock.com; *Activity 4b* Matthew Cole/Shutterstock.com; *Activity 4c* Yanas/Shutterstock.com; *Activity 4e* unkreativ/Shutterstock.com; **439** *top left* ©iStockphoto.com/Mark Stay; *top center* ©iStockphoto.com/Frank Wright; *top right* ©iStockphoto.com/Evgeniy Ivanov; *bottom left* ©iStockphoto.com/Oliver Sun Kim; **440** ©iStockphoto.com/Christian Jasiuk; **441** Microgen/Shutterstock.com; **442** *Activity 1a* ©iStockphoto.com/Susan Trigg; *Activity 1b* ©iStockphoto.com/subjug; *Activity 1c* ©iStockphoto.com/camilla wisbauer; *Activity 1d* ©iStockphoto.com/Joe Belanger; *Activity 1e* ©iStockphoto.com/thumb; *Activity 1f* ©iStockphoto.com/David Freund; **443** NASA; **444** *center* Google and YouTube logos are registered trademarks of Google Inc., used with permission.; **445** *top left* Elaine Barker/Shutterstock.com; *center right* ©iStockphoto.com/breckeni; **446** *bottom left* ©iStockphoto.com/Max Delson Martins Santos; *bottom right* ©iStockphoto.com/Jan Rysavy; **447** *top right* BORTEL Pavel/Shutterstock.com; *center right* ©iStockphoto.com/breckeni; **451** *center left* Sebastian Kaulitzki/Shutterstock.com; *center right* ©iStockphoto.com/Jan Rysavy; **453** ©iStockphoto.com/Boris Yankov; **454** mmutlu/Shutterstock.com; **458** *bottom right* ©iStockphoto.com/Eric Holsinger; *bottom left* TranceDrumer/Shutterstock.com

Appendix A

A0 *background* ©iStockphoto.com/Björn Kindler; *top left* ©iStockphoto.com/mika makkonen; *top right* ©iStockphoto.com/Hsing-Wen Hsu; **A1** *top right* ©iStockphoto.com/toddmedia; *bottom left* ©iStockphoto.com/Loretta Hostettler; *bottom right* NASA; **A4** *top right* ©iStockphoto.com/Hsing-Wen Hsu; *bottom left* ©iStockphoto.com/Thomas Kuest; *bottom right* Lim ChewHow/Shutterstock.com; **A5** *top right* ©iStockphoto.com/Richard Cano; *bottom left* ©iStockphoto.com/best-photo; *bottom right* ©iStockphoto.com/mika makkonen; **A6** *top right* ©iStockphoto.com/Loretta Hostettler; *bottom* ©iStockphoto.com/toddmedia; **A7** *top right* LudmilaM/Shutterstock.com; *center left and bottom right* ©iStockphoto.com/Clayton Hansen; *bottom left* Billwhittaker at en.wikipedia; **A8 and A9** NASA

Cartoon illustrations Tyler Stout

Florida Common Core Standards

Kindergarten

Counting and Cardinality	– Count to 100 by Ones and Tens; Compare Numbers
Operations and Algebraic Thinking	– Understand and Model Addition and Subtraction
Number and Operations in Base Ten	– Work with Numbers 11–19 to Gain Foundations for Place Value
Measurement and Data	– Describe and Compare Measurable Attributes; Classify Objects into Categories
Geometry	– Identify and Describe Shapes

Grade 1

Operations and Algebraic Thinking	– Represent and Solve Addition and Subtraction Problems
Number and Operations in Base Ten	– Understand Place Value for Two-Digit Numbers; Use Place Value and Properties to Add and Subtract
Measurement and Data	– Measure Lengths Indirectly; Write and Tell Time; Represent and Interpret Data
Geometry	– Draw Shapes; Partition Circles and Rectangles into Two and Four Equal Shares

Grade 2

Operations and Algebraic Thinking	– Solve One- and Two-Step Problems Involving Addition and Subtraction; Build a Foundation for Multiplication
Number and Operations in Base Ten	– Understand Place Value for Three-Digit Numbers; Use Place Value and Properties to Add and Subtract
Measurement and Data	– Measure and Estimate Lengths in Standard Units; Work with Time and Money
Geometry	– Draw and Identify Shapes; Partition Circles and Rectangles into Two, Three, and Four Equal Shares

Grade 3

Operations and Algebraic Thinking	– Represent and Solve Problems Involving Multiplication and Division; Solve Two-Step Problems Involving Four Operations
Number and Operations in Base Ten	– Round Whole Numbers; Add, Subtract, and Multiply Multi-Digit Whole Numbers
Number and Operations—Fractions	– Understand Fractions as Numbers
Measurement and Data	– Solve Time, Liquid Volume, and Mass Problems; Understand Perimeter and Area
Geometry	– Reason with Shapes and Their Attributes

Grade 4

Operations and Algebraic Thinking	– Use the Four Operations with Whole Numbers to Solve Problems; Understand Factors and Multiples
Number and Operations in Base Ten	– Generalize Place Value Understanding; Perform Multi-Digit Arithmetic
Number and Operations—Fractions	– Build Fractions from Unit Fractions; Understand Decimal Notation for Fractions
Measurement and Data	– Convert Measurements; Understand and Measure Angles
Geometry	– Draw and Identify Lines and Angles; Classify Shapes

Grade 5

Operations and Algebraic Thinking	– Write and Interpret Numerical Expressions
Number and Operations in Base Ten	– Perform Operations with Multi-Digit Numbers and Decimals to Hundredths
Number and Operations—Fractions	– Add, Subtract, Multiply, and Divide Fractions
Measurement and Data	– Convert Measurements within a Measurement System; Understand Volume
Geometry	– Graph Points in the First Quadrant of the Coordinate Plane; Classify Two-Dimensional Figures

Mathematics Reference Sheet

Conversions

U.S. Customary
1 foot = 12 inches
1 yard = 3 feet
1 mile = 5280 feet
1 acre ≈ 43,560 square feet
1 cup = 8 fluid ounces
1 pint = 2 cups
1 quart = 2 pints
1 gallon = 4 quarts
1 gallon = 231 cubic inches
1 pound = 16 ounces
1 ton = 2000 pounds
1 cubic foot ≈ 7.5 gallons

U.S. Customary to Metric
1 inch = 2.54 centimeters
1 foot ≈ 0.3 meter
1 mile ≈ 1.61 kilometers
1 quart ≈ 0.95 liter
1 gallon ≈ 3.79 liters
1 cup ≈ 237 milliliters
1 pound ≈ 0.45 kilogram
1 ounce ≈ 28.3 grams
1 gallon ≈ 3785 cubic centimeters

Time
1 minute = 60 seconds
1 hour = 60 minutes
1 hour = 3600 seconds
1 year = 52 weeks

Temperature
$C = \frac{5}{9}(F - 32)$

$F = \frac{9}{5}C + 32$

Metric
1 centimeter = 10 millimeters
1 meter = 100 centimeters
1 kilometer = 1000 meters
1 liter = 1000 milliliters
1 kiloliter = 1000 liters
1 milliliter = 1 cubic centimeter
1 liter = 1000 cubic centimeters
1 cubic millimeter = 0.001 milliliter
1 gram = 1000 milligrams
1 kilogram = 1000 grams

Metric to U.S. Customary
1 centimeter ≈ 0.39 inch
1 meter ≈ 3.28 feet
1 kilometer ≈ 0.62 mile
1 liter ≈ 1.06 quarts
1 liter ≈ 0.26 gallon
1 kilogram ≈ 2.2 pounds
1 gram ≈ 0.035 ounce
1 cubic meter ≈ 264 gallons

Number Properties

Commutative Properties of Addition and Multiplication

$a + b = b + a$

$a \cdot b = b \cdot a$

Associative Properties of Addition and Multiplication

$(a + b) + c = a + (b + c)$

$(a \cdot b) \cdot c = a \cdot (b \cdot c)$

Addition Property of Zero

$a + 0 = a$

Multiplication Properties of Zero and One

$a \cdot 0 = 0$

$a \cdot 1 = a$

Distributive Property:

$a(b + c) = ab + ac$

$a(b - c) = ab - ac$

Properties of Equality

Addition Property of Equality

If $a = b$, then $a + c = b + c$.

Subtraction Property of Equality

If $a = b$, then $a - c = b - c$.

Multiplication Property of Equality

If $a = b$, then $a \cdot c = b \cdot c$.

Multiplicative Inverse Property

$n \cdot \frac{1}{n} = \frac{1}{n} \cdot n = 1, n \neq 0$

Division Property of Equality

If $a = b$, then $a \div c = b \div c, c \neq 0$.

Squaring both sides of an equation

If $a = b$, then $a^2 = b^2$.

Cubing both sides of an equation

If $a = b$, then $a^3 = b^3$.

Properties of Exponents

Product of Powers Property: $a^m \cdot a^n = a^{m+n}$

Quotient of Powers Property: $\frac{a^m}{a^n} = a^{m-n}, a \neq 0$

Power of a Power Property: $(a^m)^n = a^{mn}$

Power of a Product Property: $(ab)^m = a^m b^m$

Zero Exponents: $a^0 = 1, a \neq 0$

Negative Exponents: $a^{-n} = \frac{1}{a^n}, a \neq 0$

Slope

$$m = \frac{\text{rise}}{\text{run}} = \frac{\text{change in } y}{\text{change in } x} = \frac{y_2 - y_1}{x_2 - x_1}$$

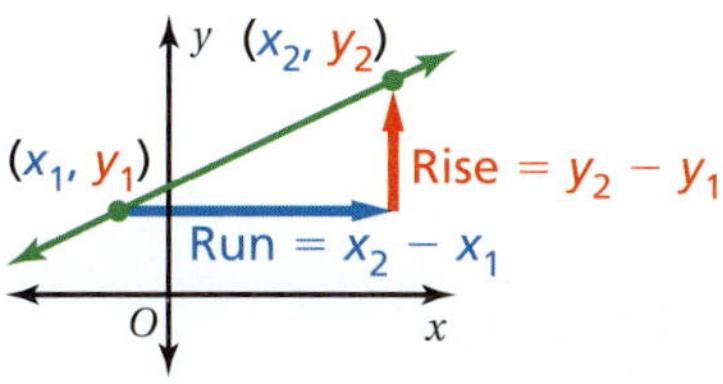

Equations of Lines

Slope-intercept form

$y = mx + b$

Standard form

$ax + by = c, a, b \neq 0$

Point-slope form

$y - y_1 = m(x - x_1)$

Volume

Cylinder

$V = Bh = \pi r^2 h$

Cone

$V = \frac{1}{3}Bh = \frac{1}{3}\pi r^2 h$

Sphere

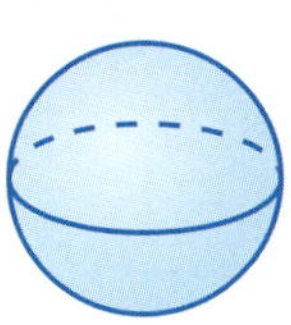

$V = \frac{4}{3}\pi r^3$

Pythagorean Theorem

$a^2 + b^2 = c^2$

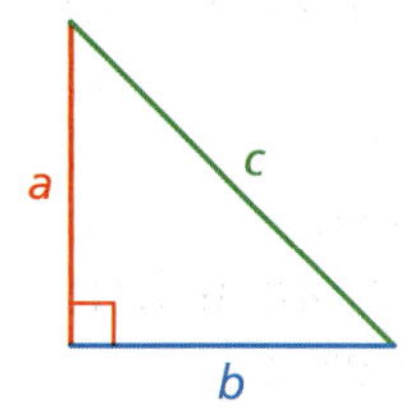

Converse of the Pythagorean Theorem

If the equation $a^2 + b^2 = c^2$ is true for the side lengths of a triangle, then the triangle is a right triangle.

Distance Formula

$$d = \sqrt{(x_2 - x_1)^2 + (y_2 - y_1)^2}$$

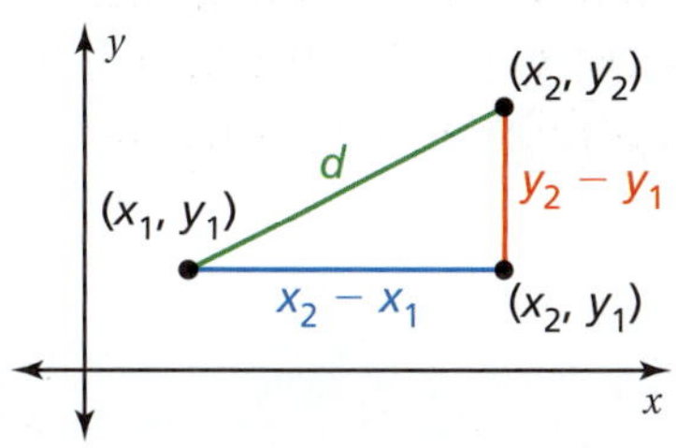

Angles of Polygons

Interior Angle Measures of a Triangle

$x + y + z = 180$

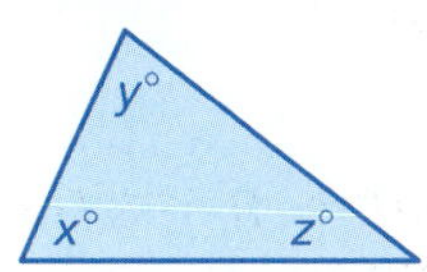

Interior Angle Measures of a Polygon

The sum S of the interior angle measures of a polygon with n sides is $S = (n - 2) \cdot 180°$.

Exterior Angle Measures of a Polygon

$w + x + y + z = 360$

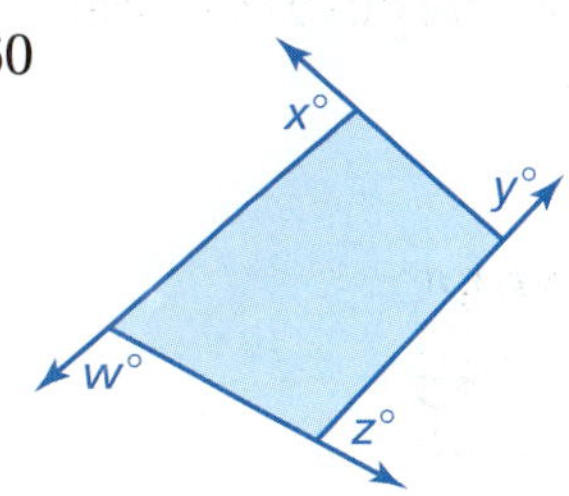